The PMO Theory of Organic Chemistry

The PMO Theory of Organic Chemistry

Michael J. S. Dewar
Department of Chemistry
University of Texas
Austin, Texas

and

Ralph C. Dougherty
Department of Chemistry
Florida State University
Tallahassee, Florida

A PLENUM/ROSETTA EDITION

Library of Congress Cataloging in Publication Data

Dewar, Michael James Steuart.
   The PMO theory of organic chemistry.

   "A Plenum/Rosetta edition."
   Includes bibliographical references.
   1. Molecular orbitals. 2. Perturbation (Mathematics) 3. Chemistry, Organic.
I. Dougherty, Ralph C., 1940-   joint author. II. Title.
[QD461.D49 1974b]   547   74-26609
ISBN 0-306-20010-4 pbk.

A Plenum/Rosetta Edition
Published by Plenum Publishing Corporation
227 West 17th Street, New York, N.Y. 10011

First paperback printing 1975

© 1975 Plenum Press, New York
A Division of Plenum Publishing Corporation

United Kingdom edition published by Plenum Press, London
A Division of Plenum Publishing Company, Ltd.
Davis House (4th Floor), 8 Scrubs Lane, Harlesden, London, NW10 6SE, England

All rights reserved

No part of this book may be reproduced, stored in a retrieval system, or transmitted, in any form or by any means, electronic, mechanical, photocopying, microfilming, recording, or otherwise, without written permission from the Publisher

Printed in the United States of America

Preface

This textbook introduces the perturbation molecular orbital (PMO) theory of organic chemistry.

Organic chemistry encompasses the largest body of factual information of any of the major divisions of science. The sheer bulk of the subject matter makes many demands on any theory that attempts to systematize it. Time has shown that the PMO method meets these demands admirably.

The PMO method can provide practicing chemists with both a pictorial description of bonding and qualitative theoretical results that are well founded in more sophisticated treatments. The only requirements for use of the theory are high school algebra and a pencil and paper.

The treatment described in this book is by no means new. Indeed, it was developed as a complete theory of organic chemistry more than twenty years ago. Although it was demonstrably superior to resonance theory and no more complicated to use, it escaped notice for two very simple reasons. First, the original papers describing it were very condensed, perhaps even obscure, and contained few if any examples. Second, for various reasons, no general account appeared in book form until 1969,* and this was still relatively inaccessible, being in the form of a monograph where molecular orbital (MO) theory was treated mainly at a much more sophisticated level.

The generality of the PMO method is illustrated by the fact that all the new developments over the last two decades can be accommodated in it. Indeed, many of these would have been facilitated had this approach been more generally known. One might add that while the value of perturbation theory has been increasingly realized in recent years, many of its exponents are still unaware of the real potential of the PMO approach.

* M. J. S. DEWAR, *The Molecular Orbital Theory of Organic Chemistry,* McGraw-Hill, New York, 1969.

The lack of publicity concerning the PMO method has of course been reflected in current textbooks of organic chemistry, where it is barely mentioned. An obvious way for us to have remedied this deficiency would have been to write a new textbook based on the PMO theory. However, there are many different views concerning the manner in which the factual background of organic chemistry should be taught; we did not want to commit those who might wish to use the PMO approach to our particular choice. We therefore decided instead to write a general account of the PMO method that could be used in conjunction with any existing textbook of organic chemistry. We see no reason why this should not be done at the undergraduate level and we have therefore included an introduction to quantum mechanics and molecular orbital theory that is based on physical principles and which illustrates the foundations and power of the PMO method (Chapters 1 and 2). Those who wish to delve deeper into the mathematical details of MO theory will find suggestions for further reading at the ends of the chapters. Chapter 1 also discusses bonding in diatomics, the basis for the localized bond approximation, the assumptions involved in the concept of hybridization, and the places where these assumptions break down.

Chapter 3 deals with the electronic structure and relative stabilities of π-electron systems. Chapters 4 and 5, respectively, illustrate the application of the PMO method to the breadth of problems in chemical equilibrium and ground-state reactions in condensed phases.

The last two chapters are at a different level, discussing topics (organic photochemistry, reactions of ions in the gas phase, etc.) that are not adequately treated in current texts. This omission is due partly to the fact that these areas have only recently been explored; however, it is also largely because the lack of a satisfactory theoretical background has made them difficult to teach. As we show, the PMO method can provide such a basis and we hope that this may encourage the inclusion of these topics in organic courses. Since the current textbooks are deficient, we have also included the necessary background information together with numerous examples and references to the literature.

Throughout this text we have attempted to show how the "starring" of alternant hydrocarbon systems (see p. 74), e.g., the benzyl radical,

provides a graphic model for discussions of bonding and also emphasizes the wave mechanical nature of chemical bonds. In the example above the starred atoms are the ones that will carry the bulk of the spin density in the radical. In

Preface

this case, one drawing takes the place of the five lowest energy resonance structures for the benzyl radical. The combination of valence bond-like structures with easily accessible qualitative wave mechanics seems to us to be the major power of the PMO method.

Austin, Texas Michael J. S. Dewar
Tallahassee, Florida Ralph C. Dougherty

Contents

CHAPTER 1. Introduction to MO Theory

1.1.	The Hydrogen Atom; Orbits and Orbitals	1
1.2.	The Orbital Approximation; Helium	7
1.3.	Lithium; the Pauli Principle	10
1.4.	The Atoms Be–Ne; Hund's Rule	12
1.5.	The Hydrogen Molecule; Molecular Orbitals	13
1.6.	The Born–Oppenheimer Approximation	16
1.7.	HHe^+, HHe, and HLi; Perturbation Theory	17
1.8.	Methane; Symmetry Orbitals	21
1.9.	Photoelectron Spectroscopy and Ionization Potentials	24
1.10.	Methane, Continued; Hybridization and Localized Bonds	28
1.11.	Diatomic Molecules	32
1.12.	The Paraffins; Localized Bonds	38
1.13.	Ethylene; π Bonds	41
1.14.	Acetylene	45
1.15.	Breakdown of the Localized Bond Model: Three-Center Bonds, Conjugated Molecules, and Reaction Intermediates	47
1.16.	Relationships between Different Types of Delocalized Systems	50
1.17.	Summary	52
	Problems	53
	Selected Reading	55
	References	55

Chapter 2. Perturbation Theory

- 2.1. The Usefulness of Perturbation Theory 57
- 2.2. Types of Perturbations Involved in the Comparison of Conjugated Systems 59
- 2.3. Monocentric Perturbations 61
- 2.4. Intramolecular Union 64
- 2.5. Intermolecular Union 65
- 2.6. Multiple Union; Additivity of Perturbations 69
- Problems 70
- Selected Reading 71
- Reference 71

Chapter 3. PMO Treatment of Conjugated Systems

- 3.1. Principles of the PMO Method; Alternant and Nonalternant Systems 73
- 3.2. The Pairing Theorem 75
- 3.3. Calculation of NBMO Coefficients 78
- 3.4. Distribution of Formal Charges in AHs 81
- 3.5. Monocentric Perturbations; Correlation of Isoconjugate Systems 83
- 3.6. Intermolecular Union of Even AHs 83
- 3.7. Multiple Union of Even AHs 86
- 3.8. Union of Odd AHs 86
- 3.9. Alternation of Bonds in Polyenes 88
- 3.10. Even Monocyclic Polyenes; Aromaticity and Antiaromaticity; Hückel's Rule 89
- 3.11. Bond Alternation in Annulenes 91
- 3.12. Polycyclic Polyenes 93
- 3.13. Intramolecular Union; Monocyclic Nonalternant Hydrocarbons 96
- 3.14. Essential Single and Double Bonds; General Rules for Aromaticity 99
- 3.15. Significance of Classical Valence Structures 101
- 3.16. Union of an Odd AH with an Even AH 104
- 3.17. Hückel and Anti-Hückel Systems 106
- 3.18. Effect of Heteroatoms 109
- 3.19. Polarization of π Electrons 115
- 3.20. Stereochemistry of Nitrogen 119
- 3.21. Resonance Theory in the Light of the PMO Method 121
- Appendix. π Energy of Union of Even AHs 125
- Problems 127
- Selected Reading 130

Chapter 4. Chemical Equilibrium

4.1.	Basic Principles	131
4.2.	Factors Contributing to the Energy of Reaction	135
4.3.	Reaction of AHs	138
4.4.	Electron Transfer Processes; Redox Potentials	152
4.5.	Nonalternant Systems	157
4.6.	Effect of Heteroatoms	159
4.7.	The π-Inductive Effect	164
4.8.	Classification of Substituents	165
4.9.	Inductive (I) Substituents	165
4.10.	Electromeric Substituents; $\pm E$ Substituents	167
4.11.	$+E$ Substituents	172
4.12.	$-E$ Substituents	175
4.13.	Summary of Substituent Effects	178
4.14.	Cross-Conjugation	179
4.15.	Mutual Conjugation	180
4.16.	The Field Effect	182
4.17.	The Hammett Equation	185
	Problems	191
	Selected Reading	194
	References	195

Chapter 5. Chemical Reactivity

5.1.	Basic Principles	197
5.2.	The Transition State Theory	202
5.3.	Transition States for Aliphatic Substitution	205
5.4.	Reaction Paths and Reaction Coordinates	210
5.5.	The Bell–Evans–Polanyi (BEP) Principle; Relationships between Rates of Reactions and Corresponding Equilibrium Constants	212
5.6.	Reactions Where Intermediates Are Involved	220
5.7.	Solvent Effects; Electrostatic Interactions and the Hellman–Feynman Theorem	222
5.8.	Limitations of the BEP Principle	230
5.9.	Classification of Reactions	235
5.10.	Prototropic Reactions of $ISO_B^{\ddagger}$ Type	240
5.11.	Prototropic Reactions of $EO_B^{\ddagger}$ Type	245
5.12.	Nucleophilic Aliphatic Substitution	253
	A. S_N1 Reactions	253
	B. S_N2 Reactions	256
5.13.	Nucleophilicity and Basicity	265
5.14.	Electrophilic Aliphatic Substitution	270

5.15. Radical Substitution Reactions ($EO_B^{\ddagger}$) 274
5.16. Elimination Reactions 279
5.17. π-Complex Reactions ($E\pi_B^{\ddagger}$) 285
5.18. Electrophilic Addition ($E\pi_B^{\ddagger}$ and $EO_B^{\ddagger}$) 297
5.19. π Complexes vs. Three-Membered Rings 300
5.20. Nucleophilic Addition and Related Reactions ($EO_B^{\ddagger}$) 306
5.21. Radical Addition and Polymerization ($EO_B^{\ddagger}$) 310
5.22. Aromatic Substitution in Even Systems ($EO_B^{\ddagger}$) 316
5.23. Substitution vs. Addition 331
5.24. Neighboring Group Participation 333
5.25. Some $OE^{\ddagger}$ Reactions 336
5.26. Thermal Pericyclic Reactions ($EE_A^{\ddagger}$ and $OO_A^{\ddagger}$) 338
5.27. Examples of Pericyclic Reactions 347
    A. Cycloaddition Reactions ($EE_A^{\ddagger}$ and $OO_A^{\ddagger}$) 347
    B. Some Special Features of the Diels–Alder Reaction 350
    C. Sigmatropic Reactions ($EE_A^{\ddagger}$ and $OO_A^{\ddagger}$) 352
    D. Electrocyclic Reactions 360
    E. Chelotropic Reactions............................. 362
5.28. Alternative Derivations of the Woodward–Hoffman Rules.
      "Allowed" and "Forbidden" Pericyclic Reactions 367
5.29. Catalysis of Pericyclic Reactions by Transition Metals 369
5.30. Reactions Involving Biradical Intermediates ($ER_A^{\ddagger}$) 373
5.31. The $\pm E$ Substituent Technique 378
      Problems 383
      Selected Reading.................................... 388
      References 389

CHAPTER 6. Light Absorption and Photochemistry

6.1. Introduction 391
6.2. The Nature of Electronically Excited States 393
6.3. The Franck–Condon Principle 395
6.4. Singlet and Triplet States 397
6.5. Extinction Coefficients and Transition Moments 398
6.6. Excitation and Deexcitation; Lifetimes of States, Fluorescence,
      and Phosphorescence............................... 401
6.7. Excitation Energies of Even AHs 403
6.8. Excitation Energies of Odd AHs 408
6.9. $\pi \to \pi^*$ and $n \to \pi^*$ in Even, Heteroconjugated Systems 409
6.10. $\pi \to \pi^*$ Transitions in Odd, Heteroconjugated Systems 410
6.11. Effect of Substituents on Light Absorption 413
6.12. Basic Principles of Photochemistry; Types of Photochemical
      Process.. 419
6.13. The Role of the Born–Oppenheimer (BO) Approximation.... 422
6.14. The Role of Antibonding Electrons 426

6.15.	Classification of Photochemical Reactions	428
6.16.	Examples of Photochemical Reactions	438
	A. X-Type Reactions	438
	B. G_R Reactions	452
	C. G_I Reactions	459
	D. G_N Reactions	461
	E. G_A Reactions	462
6.17.	Chemiluminescent Reactions	471
6.18.	Summary	477
	Problems	478
	Selected Reading	481
	References	482

CHAPTER 7. Reactions of Transient Ions

7.1.	Ions in the Gas Phase; the Mass Spectrometer and Ion Cyclotron Spectroscopy	485
7.2.	The Structure of Radical Ions	490
7.3.	Reaction of Cation Radicals	492
	A. Cleavage Reactions	492
	B. Internal Displacement Reactions	496
	C. Pericyclic Reactions	497
7.4.	Radical Anions in the Gas Phase	504
7.5.	Ion–Molecule Reactions in the Gas Phase	509
7.6.	Radical Cations in Solution	517
	A. Oxidation by Electron Transfer	518
	B. Electrochemical (Anodic) Oxidation	521
	C. Photochemical and Radiochemical Oxidation	523
7.7.	Radical Anions in Solution	525
	Problems	538
	Selected Reading	539
	References	539

Answers to Selected Problems 541
Index 555

The PMO Theory of Organic Chemistry

CHAPTER 1

Introduction to MO Theory

1.1 The Hydrogen Atom; Orbits and Orbitals

The hydrogen atom consists of two particles, a proton with a positive charge ($+e$) and an electron with negative charge ($-e$), moving around one another under the influence of their mutual electrostatic attraction. Since this attraction obeys the same inverse square law as gravitational attraction, the system should, according to classical mechanics, be analogous to, e.g., the earth–moon system; the proton and electron should move around their common center of gravity in circles or ellipses. Since the proton is nearly two thousand times heavier than the electron, we can to all intents and purposes regard the electron as moving in the field of a fixed positive charge since the center of gravity of the system will lie so close to the proton. In other words, the electron should move in a circular or elliptic *orbit* with the proton at the center or at one focus (Fig. 1.1).

If this description of the hydrogen atom is correct, there should be no restrictions on the size or ellipticity of the orbit, and if we know the velocity and position of the electron at some instant in time, we should be able to calculate the orbit in which it is moving and so predict its future motion.

The trouble with this picture is that we cannot in fact measure the position and velocity of an electron accurately at the same time. According to Heisenberg's *uncertainty principle*, we are bound to make errors in measuring either the position of a particle or its velocity, or both. If the error in measuring the position is δq and the error in measuring the velocity is δv, then

$$\delta v \, \delta q \geq h/2\pi m \qquad (1.1)$$

where h is Planck's constant and m is the mass of the electron.

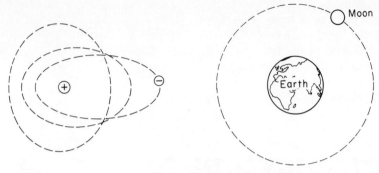

FIGURE 1.1. (a) Circular and elliptic orbits for the electron in a hydrogen atom; electrostatic force $= q_1 q_2/r^2 = -e^2/r^2$. (b) The almost circular orbit of the moon around the earth; gravitational force $= GmM/r^2$.

This difficulty is not peculiar to electrons; the same limitations also hold in the case of the earth and moon. Since, however, the mass of the moon (7.4×10^{28} g) is so very much greater than that of the electron (9.1×10^{-28} g), the product of errors $\delta v \, \delta q$ is completely negligible in the case of the moon but disastrous in the case of the electron, the relative importance of the errors for the electron and for the moon differing by no less than fifty-eight powers of ten. Any attempt to calculate the orbit of the electron in hydrogen is therefore doomed to failure because we cannot get the information needed to do it.

There are, however, certain properties of the hydrogen atom which can be measured accurately, in spite of the uncertainty principle, because they do not depend on the instantaneous position or velocity of the electron. For example, the law of conservation of energy tells us that the total energy of an isolated hydrogen atom must have some fixed value, regardless of the positions and velocities of the proton and electron. The total energy of the atom is the sum of the kinetic energies of the two particles and of the potential energy due to their mutual attraction. The former depend on their velocities and the latter on their positions in space. The uncertainty principle tells us that we cannot find out how the total energy is partitioned between the two, but their sum must be constant. There is therefore no reason why we should not measure this sum. The same argument applies of course to the total energy of any aggregate of particles; it also applies to any other property of such an aggregate that is likewise independent of the instantaneous positions and velocities of the particles. Such quantities obey conservation laws like the law of conservation of energy and for obvious reasons were called *constants of the motion* in classical mechanics. Angular momentum is a good example.

Another very important regularity appears in measurements of the position of the electron in hydrogen. Although the uncertainty principle tells us that we cannot measure this accurately, we can give a definite estimate of the likelihood that the electron will be found in some specified region of space. Suppose we measure the distance r of the electron from the proton a large number of times, making sure that the hydrogen atom is always in the same state, i.e., has the same total energy. Although the value of r will vary, because of the uncertainty principle, the probability of getting some particular value a remains constant. In other words, if we make a large number of measurements of r, the fraction of values lying between $r = a$ and $r = a + \delta a$ is always the same. We can express this by saying that the probability is given by $P(a)\, \delta r$, where $P(r)$ is a function of r that we can determine from our measurements. This suggests that we could specify a given state of the atom by specifying the corresponding function P, a function which one should be able to determine by experiment and which does in fact vary with the energy of the atom.

It is, however, not enough to specify the probability function P, because this fails to take into account another peculiarity of small particles, i.e., the fact that they show *wavelike behavior*. In particular, electrons appear to undergo diffraction and interference just like light.

When we pass monochromatic light through two parallel slits onto a screen, we see on the screen a typical interference pattern of parallel bands of light with dark areas in between (Fig. 1.2). If we close one or the other slit, the interference pattern disappears. This kind of phenomenon is explained by the theory that light is a form of wave motion. The light patches on the screen when both slits are open correspond to areas where the "crests" and "troughs" of the light waves coming through the two slits arrive in step so that they reinforce one another, while in the dark areas the "trough" from one slit arrives at the same time as the "crest" from the other so that they cancel. The pattern is quite different from either of those seen if we close one or the other slit; the "two-slit" pattern is *not* a superposition of the two "one-slit" patterns.

We represent this mathematically by describing a given wave motion by a corresponding wave function ψ which can be positive or negative, the intensity of the wave motion at some point in space, e.g., the brightness of a light or the loudness of a sound, being given by the square (ψ^2) of the corresponding wave function at that point. When two different waves represented

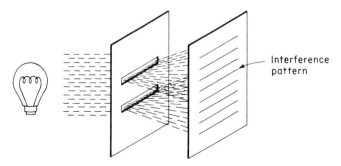

FIGURE 1.2. Interference pattern seen when monochromatic light passes through two parallel slits onto a screen.

by wave functions ψ_1 and ψ_2 meet, the resulting wave function is a sum of the two contributing parts, i.e., $\psi_1 + \psi_2$. The resulting intensity is then $(\psi_1 + \psi_2)^2$, which is *not* the sum $\psi_1{}^2 + \psi_2{}^2$ of the two individual intensities $\psi_1{}^2$ and $\psi_2{}^2$, but differs from this by the term $2\psi_1\psi_2$. Since ψ_1 and ψ_2 can be positive or negative, so can their product $\psi_1\psi_2$. Consequently, the superposition of the two wave motions can lead to a net intensity greater than the sum of the parts (when ψ_1 and ψ_2 have similar signs and so reinforce one another) or to a net intensity less than the sum of the parts (when ψ_1 and ψ_2 have opposite signs and so cancel). Figure 1.3 illustrates this for the experiment indicated in Fig. 1.2.

The fact that light exhibits interference has long been regarded as conclusive evidence that it is a form of wave motion; it is therefore rather startling to find that atomic particles, in particular electrons, show similar interference effects. This was first predicted by de Broglie, who deduced that a particle of mass m traveling at velocity v would have a "wavelength" λ given by

$$\lambda = h/mv \tag{1.2}$$

where h is Planck's constant; crucial experiments confirming this prediction were carried out soon after by Davisson and Germer. One could in principle replace the beam of monochromatic light in the experiment of Fig. 1.3 by a beam of monoenergetic electrons of the same wavelength [equation (1.2)]; a similar interference pattern would appear on the screen, the electrons being most likely to hit it in the places which were illuminated in the experiment with light.

We cannot of course tell where a given electron will hit the screen, because of the uncertainty principle; if, however, we use a beam containing very many electrons, the proportion hitting some particular area will have a definite value given by the corresponding probability function P. The analogy with light indicates, however, that we must specify the behavior of the electrons not by the function P, but by a *wave function* ψ of which P is the square. In the two-slit experiment, the electrons passing through the slits are represented by two distinct wave functions ψ_1 and ψ_2; the probability of their hitting a given area on the screen is then given by $(\psi_1 + \psi_2)^2$.

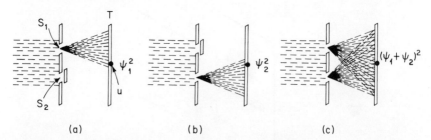

FIGURE 1.3. Effect of passing monochromatic light through two slits S_1 and S_2 onto a screen T. (a) When S_2 is closed, the intensity at some point u on the screen is $\psi_1{}^2$; (b) when S_1 is closed, the intensity at u is $\psi_2{}^2$; (c) when both slits are open, the intensity at u is not $\psi_1{}^2 + \psi_2{}^2$, but $(\psi_1 + \psi_2)^2$.

Introduction to MO Theory

This conclusion does not mean that electrons are composed of waves; it simply means that their motion is governed by mathematical equations similar to those for wave motion. Indeed it now appears that the same is true for light; light can be generated or absorbed only in discrete packets which must be regarded as particles (quanta or photons), the amount of light of frequency v in a single photon being such that its total energy is hv. This result, first discovered by Planck, was the starting point of quantum theory.

Let us now return to the hydrogen atom. The motion of the electron in this will be represented by a wave function ψ. Since space has three dimensions, ψ must be a function of three corresponding coordinates. The probability function P that we introduced earlier is now given by the square (ψ^2) of the wave function. The probability of finding the electron in some small volume element $d\tau$ (which in Cartesian coordinates could be a little cuboid of sides dx, dy, and dz; see Fig. 1.4a) is then $\psi^2 \, d\tau$, or $\psi^2 \, dx \, dy \, dz$.

The function ψ is calculated by solving an equation named after its discoverer, Schrödinger. In the case of the hydrogen atom, it turns out that the equation can be solved only if the atom has one of a series of discrete possible energies. This of course is another respect in which the classical description fails; for, according to the classical picture, the total energy of the atom should be able to vary continuously from zero (when the particles are far apart) to minus infinity (when they coincide). According to wave mechanics, there is a definite state of minimum energy in which the function ψ is spherically symmetric, its variation with distance from the nucleus being indicated in Fig. 1.4(b). We can indicate this state pictorially. If we could photograph the hydrogen atom and take a long time exposure of it, the electron would appear as a blur, the density of which at any point would be proportional to ψ^2. Figure 1.4(c) indicates such a representation of the ground state of hydrogen.

If we supply energy to hydrogen, we can excite it to states of higher energy; these again are discrete states differing by fixed amounts of energy

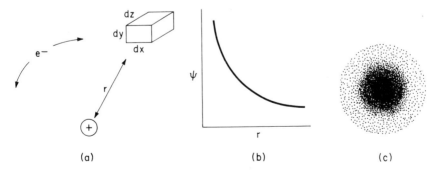

FIGURE 1.4. (a) Diagram illustrating the meaning of the wave function for the electron in a hydrogen atom; the probability of finding the electron in this cube is $\psi^2 \, dx \, dy \, dz = \psi^2 \, d\tau$. (b) Plot of ψ vs. r for the ground state of hydrogen. (c) Pictorial representation of the ground state of hydrogen.

from the ground state, each distinguished by its own characteristic wave function. Because of the obvious analogy between the classical orbits and the wave mechanical wave functions, the latter are referred to as *orbitals*, a term first suggested by Mulliken.

The ground-state orbital of hydrogen has the same sign everywhere (Fig. 1.4b); the other orbitals, however, consist of parts with opposite signs. Where the orbital changes sign, it must of course pass through a *node* where $\psi = 0$. In the case of hydrogen, these nodes are of two kinds: radial nodes, which are spherical in shape, and angular nodes, which pass through the center of the orbital (i.e., the proton). Two states of hydrogen, one with a single radial node and one with a single angular node, are indicated in Fig. 1.5(a, b). The node in the latter is a plane passing through the nucleus. The corresponding orbitals are indicated schematically in Fig. 1.5(c, d). Here the outer line bounds the region in which the electron will be found, say, 95% of the time; as indicated, the wave function changes sign on crossing a node since if ψ changes from positive to negative, it must pass through zero. Note that the absolute signs of wave functions are not significant; only the probability function P can be measured experimentally and of course $\psi^2 = (-\psi)^2 = P$. We could equally well represent the orbital of Fig. 1.4(c) as negative inside and positive outside. What does matter is that ψ has different signs in the two regions, because, as we shall see presently, interference effects occur when orbitals of different atoms interact.

Each atomic orbital (AO) is specified by a *principal quantum number n*, which is simply the total number of nodes plus one, and by an *azimuthal quantum number l*, equal to the number of angular nodes. These quantum numbers come out of the solution of the Schrödinger equation for the individual hydrogen AOs. For the ground-state orbital (Fig. 1.4b), $n = 1$ and $l = 0$; for the orbital of Fig. 1.5(d), $n = 2$ and $l = 1$. Orbitals with $l = 0$ are called *s* orbitals; those with $l = 1$ are *p* orbitals; those with $l = 2$ are *d* orbitals. Each orbital is further specified by the value of n. Thus the orbital of Fig. 1.2(b) is the 1s orbital; that of Fig. 1.4(c) is the 2s orbital; that of Fig. 1.4(d) is a 2p orbital. Note that we said "the" 1s orbital and "the" 2s orbital, but "a" 2p orbital. The *s* orbitals are unique, but orbitals with a given value

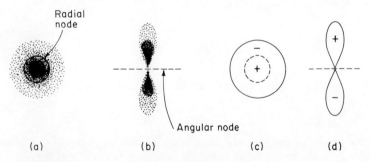

FIGURE 1.5. (a) 2S state of H; (b) a 2P state of H; (c) the 2s orbital; (d) a 2p orbital

Introduction to MO Theory

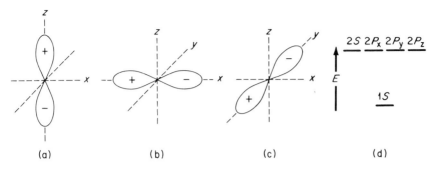

FIGURE 1.6. (a) $2p_z$, (b) $2p_x$, (c) $2p_y$ orbitals of hydrogen; (d) energies E of the five lowest states.

of n and some value of l differing from zero occur in sets of $2l + 1$. There are thus three $2p$ orbitals; each of these is symmetric to rotation about an axis (the z axis in the case of Fig. 1.5d), the three axes being mutually perpendicular. The three $2p$ orbitals can accordingly be designated as $2p_x$, $2p_y$, and $2p_z$, taking their axes as axes of Cartesian coordinates (Fig. 1.6).

In the case of hydrogen, the energy depends only on the principal quantum number. The $2s$ and $2p$ orbitals therefore all correspond to states of the same energy, i.e., *degenerate* states. The states of hydrogen are indicated by a symbol similar to that for the orbital but with a capital letter; thus the state in which the electron occupies the $2s$ orbital is the $2S$ state. The energies of the five lowest states of hydrogen are therefore as indicated in Fig. 1.6d.

1.2 The Orbital Approximation; Helium

In the helium atom, two electrons, each of charge $-e$, circle a nucleus with charge $+2e$ (Fig. 1.7a). The wave function of the system must now allow for the results of simultaneous measurement of the positions of two electrons. Our wave function must tell us the probability of finding electron 1 in some

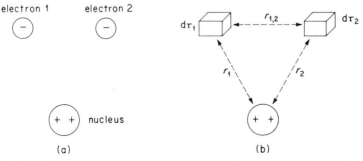

FIGURE 1.7. (a) The helium atom; (b) diagram to illustrate electron correlation in helium.

small volume element $d\tau_1$ at the same time that we find electron 2 in some small volume element $d\tau_2$. We will need three coordinates to specify the motion of each electron. Let us use Cartesian coordinates (x_1, y_1, z_1) for electron 1 and (x_2, y_2, z_2) for electron 2, so that the volume elements $d\tau_1$ and $d\tau_2$ are given by

$$d\tau_1 = dx_1\, dy_1\, dz_1; \qquad d\tau_2 = dx_2\, dy_2\, dz_2 \qquad (1.3)$$

By an obvious extension of our definition of the wave function for the hydrogen atom, the probability of simultaneously finding electron 1 in $d\tau_1$ and electron 2 in $d\tau_2$ must be given by

$$\psi_{He}^2\, d\tau_1\, d\tau_2 \qquad (1.4)$$

where the wave function ψ_{He} of the helium atom is now a function of the six coordinates $(x_1, y_1, z_1; x_2, y_2, z_2)$.

This function can again be found, in principle, by solving the corresponding Schrödinger equation. However, although it is easy to write this equation, it is impossible to solve it by existing mathematical techniques because it contains too many interrelated variables. The difficulties responsible for this failure become rapidly more acute for more complex atoms; therefore although very accurate approximations to the solution of the Schrödinger equation for helium have been obtained by several methods, these cannot be applied to more complicated systems.

The trouble lies in the fact that the Schrödinger equation for helium involves six variables, as opposed to three for hydrogen. If one of the electrons in helium is removed, leaving the ion He^+, the Schrödinger equation can then be solved quite easily; the solutions for He^+ are in fact identical with those for hydrogen except that the orbitals are smaller, and the energies of corresponding states lower, because of the greater attraction between the electron and the helium nucleus (charge $+2e$) than between an electron and a proton (charge $+e$).

If the electrons in helium did not repel each other, so that each moved freely in the field of the nucleus, the motion of each would then be the same as if the other electron were absent; the probabilities of finding electron 1 in $d\tau_1$ and electron 2 in $d\tau_2$ would then be given by $\psi_1^2 d\tau_1$ and $\psi_2^2 d\tau_2$, respectively, where ψ_1 and ψ_2 are "hydrogenlike" orbitals of the ion He^+. Comparison with equation (1.3) shows that in that case

$$\psi_{He}^2 = \psi_1^2 \psi_2^2 \qquad \text{or} \qquad \psi_{He} = \psi_1 \psi_2 \qquad (1.5)$$

The helium atom would then again be representable in terms of an orbital picture, each electron occupying an orbital similar to one of the orbitals of hydrogen.

The trouble with this picture is that electrons *do* repel one another. Therefore as the electrons in helium move around, they try to keep apart. The probability of finding electron 1 in $d\tau_1$ therefore depends on the distance between $d\tau_1$ and electron 2 (r_{12} in Fig. 1.7b); since r_{12} is a function of the

Introduction to MO Theory

coordinates of both electrons, it is impossible to factorize ψ_{He} into a product of two one-electron functions or orbitals as indicated in equation (1.5).

Is there any way in which we can get around this difficulty and yet retain the very simple and appealing orbital picture? We cannot just neglect the interelectronic repulsion, because it is very large (about 30 eV, or 700 kcal/g-atom). Can we then make some allowance for electron repulsion without upsetting the orbital picture?

We can do this by replacing the instantaneous repulsion e^2/r_{12} between the two electrons by an average value. Suppose, for example, that the motion of electron 2 is represented by an orbital function ψ_2. The probability of finding electron 2 in some volume element $d\tau_2$ is then $\psi_2^2 \, d\tau_2$ (Fig. 1.7b). Suppose electron 1 is in some other volume element $d\tau_1$ (Fig. 1.7b) at a distance r_{12} from $d\tau_2$. If electron 2 is in $d\tau_2$, the mutual repulsion energy is e^2/r_{12}. The probability of getting this particular result is $\psi_2^2 \, d\tau_2$. We can now find the average repulsion V between the electrons, averaged over all positions of electron 2, by multiplying the repulsion when electron 2 is in $d\tau_2$ by the probability of its being in $d\tau_2$, and summing the values for all positions of electron 2; i.e.,

$$V = \sum_{d\tau_2} \frac{e^2}{r_{12}} \times \psi_2^2 \, d\tau_2 = \sum_{d\tau_2} \frac{e^2 \psi_2^2 \, d\tau_2}{r_{12}} \tag{1.6}$$

Suppose now we calculate the total repulsion between electron 1 and a cloud of negative charge, the total charge in the small volume element $d\tau_2$ being $e\psi_2^2 \, d\tau_2$. The repulsion δV_R between electron 1 and this small charge is given by

$$\delta V_R = \frac{e \times e\psi_2^2 \, d\tau_2}{r_{12}} = \frac{e^2 \psi_2^2 \, d\tau_2}{r_{12}} \tag{1.7}$$

The total repulsion V_R is given by summing the contributions of all the small volume elements $d\tau_2$, i.e., by

$$V_R = \sum_{d\tau_2} \frac{e^2 \psi_2^2 \, d\tau_2}{r_{12}} \tag{1.8}$$

Comparison of equations (1.6) and (1.8) shows that this is exactly the same as the average repulsion V between the two electrons. The average repulsion V is therefore equal to the repulsion between electron 1 and a fixed cloud of negative charge representing the average motion of electron 2 (Fig. 1.8); V thus depends only on the position of electron 1. Electron 1 is then moving

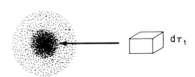

FIGURE 1.8. The repulsion between the two electrons in helium, averaged over all positions of electron 2, is equivalent to the repulsion between electron 1 and a cloud of negative charge representing the average position of electron 2.

in a fixed potential field due to the attraction by the nucleus and to repulsion by a fixed cloud of negative charge; the probability of finding electron 1 in some small volume $d\tau_1$ therefore has a fixed value $P_1 d\tau_1$ which depends only on the position of $d\tau_1$ and so on the coordinates of electron 1. But we know that P_1 is equal to the square of the corresponding wave function ψ_1; it follows that the motion of electron 1 can be represented by an orbital ψ_1 which is a function only of the coordinates of electron 1. Thus if we can represent the motion of electron 2 by an AO ψ_2, we must also be able to represent the motion of electron 1 by an AO ψ_1. Since the converse must also be true, the averaging of interelectronic repulsions thus leads to a self-consistent picture in which the motion of each electron is represented by an individual one-electron orbital function, i.e., an AO. This approach was first introduced by Hartree and Fock and is known as the *Hartree–Fock* or *self-consistent field* (SCF) approximation.

The orbitals ψ_1 and ψ_2 can be found by solving two one-electron Schrödinger equations. Since each of these involves only the three coordinates of one electron, they can be solved. The solutions are a set of orbitals similar in shape to those of hydrogen; we can designate them by the same kind of symbol. The ground state of helium is naturally one in which both electrons occupy the orbital of lowest energy, i.e., the 1s orbital; we accordingly depict this state as $(1s)^2$.

Note that we have not solved the problem of the helium atom exactly; we have neglected the tendency of electrons to synchronize their motions so as to keep apart. This tendency, called *electron correlation*, has the effect of reducing the average interelectronic repulsion; its neglect leads us to underestimate the bonding of the electrons in helium. However, the error is relatively small, about 1 eV, or 1 % of the total energy; this suggests that our orbital picture probably gives a reasonably good representation of the way the electrons in helium behave.

1.3 Lithium; The Pauli Principle

In lithium, three electrons circle a nucleus of charge $+3e$; one might suppose the ground state could be represented by the symbol $(1s)^3$. The energy of such a state is, however, far lower than the observed energy of the helium atom. The reason for this was first stated by Pauli; the capacity of an orbital to hold electrons is limited. Only two electrons can occupy a given orbital. Indeed, there is a further restriction. Electrons do not behave like point charges; they behave like little charged spheres rotating about an axis. Rotation of such a sphere corresponds to a circular flow of charge, i.e., an electric current, around the axis of rotation; electrons therefore behave like tiny magnets, the ends of the axis of rotation being the north and south poles. Furthermore, since *electron spin* is also subject to quantum restrictions, an electron, if it finds itself in a magnetic field, has to orient itself with its spin axis along the field. There are therefore two possibilities. The north pole of the electron will point either toward the north pole of the field or toward the south pole. In our orbital picture of the atom, it is not sufficient to designate the orbitals the electrons occupy; we must also designate their spins. This is usually done by adding a *spin function* α or β to the orbital; the orbital ψ is thus subdivided into two *spin orbitals* $\psi\alpha$ and $\psi\beta$. The Pauli principle states that only one electron can occupy a given spin orbital; consequently, two electrons can occupy a given orbital only if they have opposite spins. The ground state of helium is thus strictly not just $(1s)^2$, but rather $(1s\alpha)(1s\beta)$. If we wish to denote the spin associated with a given spin orbital, it is usual to use an overbar notation instead of αs and βs; an orbital with no overbar (e.g., $1s$) denotes a spin orbital with α spin, one with an overbar ($\overline{1s}$) denotes a spin orbital with β spin. In lithium, only two electrons can occupy the $1s$ orbital; the third electron has to go into an orbital of higher energy, presumably $2s$ or $2p$.

In hydrogen, the $2s$ and $2p$ orbitals are degenerate; however, this is not the case in other atoms. The reason for this can be seen from the nature of the orbital approximation. In this each electron is supposed to move in the field of a nucleus and of a cloud of charge representing the averaged motion of all the other electrons. Now since an atom has spherical symmetry, this cloud of charge will also be spherically symmetric. It can be shown that the potential due to such a cloud at a point P at a distance r from the nucleus (Fig. 1.9) is the same as that due to a point charge at the nucleus, equal in magnitude to the charge of the whole of the cloud within a distance r from the nucleus. If this charge is $-qe$, the potential at P is then the same as it would be if the nucleus and other electrons were replaced by a single point charge at the nucleus of magnitude $Ze - qe$, Z being the atomic number. In the orbital picture, each electron in the atom therefore moves in a central field, i.e., a field equivalent to that of a fixed point charge. This is why the orbitals of other atoms resemble those of hydrogen in shape. In the case of hydrogen, however, the central charge is fixed; the degeneracy of the $2s$ and $2p$ orbitals is an accidental consequence of this. In other atoms the *effective nuclear charge* $Z'e$, where

$$Z' = Z - q \tag{1.9}$$

varies with the distance r from the nucleus because q is the average number of electrons nearer to the nucleus than the one we are considering; the *screening* of the nucleus by the other electrons is less, the smaller r is. Now a $2s$ orbital (Fig. 1.5) has a high density near the nucleus; an electron occupying such an orbital will therefore spend quite a lot of time near the nucleus,

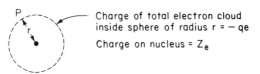

FIGURE 1.9. Shielding of the nucleus by electrons.

where the effective nuclear field is large, q being small. A $2p$ orbital, on the other hand, has a node passing through the nucleus; the chance of finding a $2p$ electron near the nucleus is therefore very small. The average attraction between a $2s$ electron and the nucleus is consequently greater than that for a $2p$ electron; therefore in atoms other than hydrogen the $2s$ orbital lies well below the $2p$ orbitals in energy. The ground state of lithium is therefore $(1s)^2 (2s)$.

1.4 The Atoms Be–Ne; Hund's Rule

The next atom in the periodic table, beryllium, has four electrons and a nuclear charge $+4e$; as one might expect, its ground state can be represented in the orbital approximation as $(1s)^2 (2s)^2$. Likewise the next atom, boron, is represented by $(1s)^2(2s)^2(2p)$. Note that there is no distinction between the three $2p$ orbitals in heavier atoms; since these are identical in shape, differing only in orientation (Fig. 1.6), they remain degenerate.

An ambiguity does, however, appear in attempts to predict the ground state of carbon. Here the first four electrons occupy the $1s$ and $2s$ orbitals, but what about the other two? Do they occupy the same $2p$ orbital, or different $2p$ orbitals? And in the latter case, do they have similar or opposite spins? This ambiguity is resolved by a rule due to Hund; Hund's rule states that, *other things being equal, the lowest energy state is that in which the maximum number of electrons have parallel spins*. The ground state of carbon is thus $(1s)(\overline{1s})(2s)(\overline{2s})(2p_x)(2p_y)$ or $(1s)(\overline{1s})(2s)(\overline{2s})(\overline{2p_x})(\overline{2p_y})$; the two states are indistinguishable in the absence of a magnetic field since one can be derived from the other by turning the atom as a whole through 180° (so that all the α-spin electrons change their spin to β, and conversely). In a magnetic field, the two states differ slightly in energy, due to the interaction between the magnetic moments of the two unpaired p electrons and the applied field; these differences in energy, while too small to be of chemical significance, can be measured, their study being the province of electron spin resonance (esr) spectroscopy. We will see presently that Hund's rule applies with equal force to molecules.

An atom in which all the electrons are paired has of course no net magnetic moment due to electron spin, because the magnetic moments of an α-spin electron and a β-spin electron are equal and opposite. In atoms where there are unpaired electrons, this need not be the case; if the numbers of α- and β-spin electrons differ by n, the atom will have a net spin magnetic moment of n times that of a electron. The magnitude of the unpaired spin in a given state of an atom is denoted by its *multiplicity*, the multiplicity of a state being $n + 1$. States with no unpaired electrons (e.g., He, Be), and so a multiplicity of one, are called *singlet* states; states with a multiplicity of two always have an odd number of electrons (e.g., H, Li) and are called *doublet* states; states with a multiplicity of three (two unpaired electrons with parallel

Introduction to MO Theory

TABLE 1.1. Ground States of Atoms H–Ne in the Orbital Representation

	H	He	Li	Be	B	C	N	O	F	Ne
2p	— — —	— — —	— — —	— — —	↑ — —	↑ ↑ —	↑ ↑ ↑	↑↓ ↑ ↑	↑↓ ↑↓ ↑	↑↓ ↑↓ ↑↓
2s	—	—	↑	↑↓	↑↓	↑↓	↑↓	↑↓	↑↓	↑↓
1s	↑	↑↓	↑↓	↑↓	↑↓	↑↓	↑↓	↑↓	↑↓	↑↓
Atom	H	He	Li	Be	B	C	N	O	F	Ne
Multiplicity	2	1	2	1	2	3	4	3	2	1

spins, C) are called *triplet* states, and states with multiplicities of four, five, or six are called, respectively, *quartet, quintet,* and *sextet* states.

We are now in a position to deduce the ground states of the atoms H–Ne; these are shown in Table 1.1. Arrows pointing up or down indicate electrons of α or β spin, and the horizontal bars represent the AOs.

Hund's rule is due to an additional kind of electron correlation involving pairs of electrons with parallel spins. Such electrons have a peculiar aversion for one another, over and above their mutual electrostatic repulsion, and they put correspondingly more effort into correlating their motions so as to keep apart. Other things being equal, the average repulsion between two electrons of parallel spin is therefore less than that between two electrons of opposite spin since the average interelectronic distance is less for the latter. The states of carbon with spins paired, i.e., $(1s)(\overline{1s})(2s)(\overline{2s})(2p_x)(\overline{2p_x})$ or $(1s)(\overline{1s})(2s)(\overline{2s})(2p_x)(\overline{2p_y})$, are higher in energy than the triplet state of Table 1.1 because in them the mutual repulsion of the $2p$ electrons is greater.

The decrease in energy with parallel electron spins raises another point. One could further reduce the interelectronic repulsion in carbon by *promoting* a $2s$ electron into the empty $2p$ orbital, giving a quintet state $(1s)(\overline{1s})(2s)(2p_x)(2p_y)(2p_z)$. Here there are four electrons of parallel spin. On the other hand, to do this, we will lose the difference in energy between the $2s$ and $2p$ orbitals. Which effect will win? This of course is something that cannot be predicted in qualitative terms since it depends on the relative magnitudes of the two effects. Here the difference in orbital energies in fact dominates, the splitting of the $2s$ and $2p$ levels being large.

1.5 The Hydrogen Molecule; Molecular Orbitals

When two hydrogen atoms come together, they form a molecule, H_2. In the molecule, two electrons move in the field of two protons (Fig. 1.10); here again we can get an approximate representation of the system in terms of individual one-electron functions or orbitals by averaging the interelectronic repulsions, i.e., by neglecting electron correlation. Each orbital represents the motion of an electron in the field of the nuclei and of a cloud

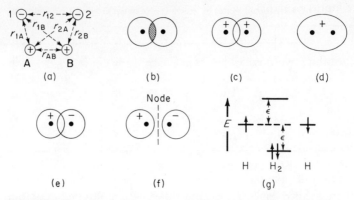

FIGURE 1.10. (a) The hydrogen molecule; (b) overlap of AOs in H_2; (c, d) the bonding MO; (e, f) the antibonding MO; (g) energies of AOs and MOs in the system H + H.

of negative charge representing the averaged motion of the other electron; since these are orbitals of a molecule, we call them *molecular orbitals* (MO), as opposed to the *atomic orbitals* (AO) of electrons in isolated atoms.

Let us consider the form of the MOs of H_2. Since electrons repel one another but are attracted by nuclei, we would expect the two electrons in H_2 to correlate their motions in such a way that each tends to be near a different nucleus. When electron 1 is near nucleus A (Fig. 1a), electron 2 will be near nucleus B. The average distance r_{12} between electron 1 and electron 2 will then be similar to the distance r_{12} between electron 1 and nucleus B; the attraction between electron 1 and nucleus B will then be balanced by the interelectronic repulsion. Electron 1 will then be moving in a field very similar to that for a single hydrogen atom; consequently, the MO it occupies must resemble, in the neighborhood of nucleus A, an AO of atom A. Likewise, in the neighborhood of nucleus B, the MO must resemble an AO of nucleus B. This suggests that the MO ψ may be represented approximately as a linear combination of the two AOs ϕ_A and ϕ_B;

$$\psi = A\phi_A + B\phi_B \tag{1.10}$$

Now in the ground state of H_2, both electrons should occupy the MO of lowest energy. This will presumably be the one that approximates, in the vicinity of each nucleus, to the AO of lowest energy. The AOs ϕ_A and ϕ_B are therefore presumably 1s AOs of the hydrogen atoms. Furthermore, H_2 has a plane of symmetry bisecting the line joining the nuclei; the total time-averaged electron distribution must be symmetric with respect to inversion in this plane. Since both electrons occupy the MO ψ, the total electron distribution is given by $2\psi^2$, or

$$2\psi^2 = 2(A\phi_A + B\phi_B)^2 = 2A^2\phi_A^2 + 4AB\phi_A\phi_B + 2B^2\psi_B^2 \tag{1.11}$$

If we reflect the molecule in the plane of symmetry indicated above, we interchange atoms A and B; this will leave the electron distribution unchanged only if

$$A = \pm B \quad (1.12)$$

There are thus two possible MOs satisfying our conditions*

$$\psi^+ = \phi_A + \phi_B \quad (1.13)$$
$$\psi^- = \phi_A - \phi_B \quad (1.14)$$

The electrons in H_2 must occupy the one of lower energy.

We can arrive at this result in a different way. Consider what happens when two hydrogen atoms approach so that their $1s$ AOs overlap (Fig. 1.10b). From the analogy between waves and wave functions, one might expect an interference effect analogous to that of the experiment indicated in Fig. 1.2. In other words, the orbitals should combine. When they do so, we may get constructive interference, the orbitals adding [cf. equation (1.13)], or we may get destructive interference, the orbitals subtracting [cf. equation (1.14)]. The first situation is indicated in Fig. 1.10(c, d); the electron density is given by

$$(\psi^+)^2 = (\phi_A + \phi_B)^2 = \phi_A^2 + 2\phi_A\phi_B + \phi_B^2 \quad (1.15)$$

This differs from a superposition of the two AOs, i.e., $\phi_A^2 + \phi_B^2$, by the term $2\phi_A\phi_B$; there is therefore an extra increase in electron density in the region between the nuclei due to the "wave behavior" of electrons; this extra cloud of negative charge in the region close to both nuclei should lead to an increase in the nuclear–electron attraction and so lower the energy below that of two isolated hydrogen atoms (corresponding to the electron distribution $\phi_A^2 + \phi_B^2$). Likewise the second possibility (Fig. 1.10e,f) leads to a density distribution

$$(\psi^-)^2 = (\phi_A - \phi_B)^2 = \phi_A^2 - 2\phi_A\phi_B + \phi_B^2 \quad (1.16)$$

which is *less* in the critical region between the nuclei due to destructive interference of the "electron waves." Here the energy should be higher than that of the isolated atoms. When the atoms come together, the two AOs are replaced by the MOs ψ^+ and ψ^-; electrons going into ψ^+ decrease in energy, corresponding to a bonding effect between the atoms, whereas electrons in the latter increase in energy, thus tending to prevent bonding. The MO ψ^+ is therefore called a *bonding* MO, and ψ^- an *antibonding* MO. In H_2, the two electrons can both go into the bonding MO; hence the system is most stable when the atoms are close together.

The extra electron density between the nuclei is clearly determined by the extent to which the AOs ϕ_A and ϕ_B overlap in space, the extra density in some volume element $d\tau$ being given [see equation (1.15)] by $2\phi_A\phi_B\, d\tau$. The total extent of this "overlap cloud" is given by summing the contributions by different volume elements $d\tau$; by definition, this sum is equal to twice

---

* The MOs ψ^+ and ψ^- should strictly be multiplied by constants (normalizing factors; see p. 32).

the *overlap integral* S_{AB}

$$S_{AB} = \int \phi_A \phi_B \, d\tau \tag{1.17}$$

The contribution of the overlap cloud to the total energy is likewise found by summing (i.e., integrating) contributions by individual volume elements $d\tau$; the corresponding integral β_{AB} is called the *resonance integral* between the AOs. An approximation to β_{AB} can be found from the following argument. We can reasonably assume that the average potential energy in the region where the overlap cloud is large, i.e., between the nuclei, is some kind of average of that around each nucleus in the isolated atoms. This in turn can be measured by the energy required to remove an electron from one or the other AO, i.e., their ionization potentials I_A and I_B. Since the total amount of charge on the overlap cloud is equivalent to S_{AB} electrons, the resonance integral β_{AB} should be given approximately by

$$\beta_{AB} = CS_{AB}(I_A + I_B) \tag{1.18}$$

where C is a constant. This is called the *Mulliken* or *Wolfsberg–Helmholtz* approximation. In the case of H_2, I_A and I_B are of course the same; the general expression of equation (1.18) also holds in other molecules where there are bonds between pairs of dissimilar atoms.

1.6 The Born–Oppenheimer Approximation

In the preceding discussion of H_2, we assumed that the two electrons move in the field of two fixed protons. If we calculate the energy E of H_2 for various distances r between the protons, we find that it varies in the way indicated in Fig. 1.11. The energy is a minimum for a certain internuclear distance, corresponding to the equilibrium bond length r_e, and the difference in energy between $r = r_e$ and $r \to \infty$ (corresponding to two isolated hydrogen atoms) is the corresponding bond energy ΔE.

This calculation assumes, however, that the internuclear distance in H_2 is fixed, the molecule behaving like two balls linked by a rigid rod. This in fact is not the case; the system resembles two balls linked by a spring rather than

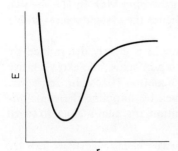

FIGURE 1.11. Energy E of H_2 as a function of the internuclear distance r.

a rigid rod; as a result, the molecule can undergo vibrations in which the internuclear distance changes rhythmically.

Such vibrations can be calculated by wave mechanics if we know the energy of the molecule as a function of the internuclear distance. If therefore the potential function of Fig. 1.11 holds even when the molecule is vibrating, we can carry out our calculation in two steps. First we find the potential function by calculating the energy for various fixed internuclear distances; next we use this function to calculate the course of vibrations. This procedure enables us to treat H_2 as two two-particle systems instead of a four-particle system; the motions of the two electrons are calculated in the first step and the motions of the two nuclei in the second. Since the difficulty of solving Schrödinger equations increases very rapidly with the number of particles, this is an important simplification. It rests on the basic *Born–Oppenheimer approximation*, that the potential energy of a molecule depends only on the positions of the nuclei and not on their velocities, so that the motion of the electrons and of the nuclei can be treated separately.

The basis of the Born–Oppenheimer approximation is that electrons, being much lighter than nuclei, move correspondingly more rapidly. As the heavy nuclei move ponderously to and fro, the nimble electrons can adjust their motions to correspond very closely to the instantaneous positions of the nuclei, only lagging a little behind. This approximation is essential to the structural view of chemistry in which molecules are depicted as collections of atoms held together by static bonds. If we had to allow for the motions of atoms in a molecule in considering its energy and the distribution of electrons in it, no such simple representation would be possible. Equally, the success of such representations suggests that the Born–Oppenheimer approximation is usually valid.

There are, however, cases where it breaks down. The places where this happens are usually ones where the wave function for the electrons changes very rapidly with changes in the position of the nuclei. Here the electrons have to change their motions very drastically during a very small part of the vibration and they may not be able to do this in the time available. A familiar analogy is provided by "bouncing putty." If we apply gentle force to a piece of bouncing putty, it feels soft and plastic; we can mould the putty easily if we do not try to change its shape too rapidly. If, however, we hit it with a hammer or drop it on the floor, it behaves as though it had the rigidity and hardness of a piece of glass; the putty resists attempts to change its shape too rapidly and if we hit it hard enough it will shatter rather than flow.

1.7 HHe^+, HHe, and HLi; Perturbation Theory

The helium ion He^+ has its lone electron in a 1s AO; apart from the greater nuclear charge, it is exactly analogous to a hydrogen atom. When it meets a hydrogen atom, the 1s AOs can interact to form two MOs, one bonding and one antibonding, just as in the case of H_2; the two electrons can go into the low-energy bonding MO (cf. Fig. 1.10) and so H and He$^+$ combine to form HHe$^+$.

There are, however, differences. The ionization potential of He$^+$ (54.14 eV) is far greater than that of H (13.53 eV); the 1s AO of He therefore lies far below that of H in energy (Fig. 1.12a). Consequently, H and He$^+$ can react exothermically by transfer of an electron to give H$^+$ and He (Fig. 1.12b). Since the electron is so much more tightly bound in He$^+$ than in H, one would not expect the sharing of electrons in HHe$^+$ to be even and indeed it is not. The bonding MO in HHe$^+$ is very asymmetric, the electrons in it spending far more time near the helium nucleus than near the proton (Fig. 1.12c). This MO indeed is little different from the 1s AO of He itself and so its energy is also very similar (Fig. 1.12d); the change in energy when H$^+$ and He combine is therefore small. Since most of the He AO is used up in the bonding MO, presumably the antibonding MO will correspondingly resemble the other (i.e., H) AO; this is indeed the case (Fig. 1.12e).

Several points emerge from this discussion.

First, we see that MOs formed from pairs of AOs of unequal energy are asymmetric. The bonding MO resembles the AO of lower energy and the antibonding MO the AO of higher energy. Bonds formed in this way are asymmetric or *polar*.

Second, the formation of HHe$^+$ from H$^+$ + He represents a process radically different from the combination of two hydrogen atoms to H$_2$; in the former, both bond electrons come from the same atom; in the latter, they came from different atoms. Bonds of the latter type are called *covalent*; those of the former type are called *dative*. A dative bond can be regarded as a superposition of an electron transfer and a covalent bond; thus the formation of HHe$^+$ from H$^+$ + He could take place in steps,

$$H^+ + He \longrightarrow H\cdot + He\cdot^+$$

$$H\cdot + \cdot He^+ \longrightarrow (H\text{---}He)^+$$
(1.19)

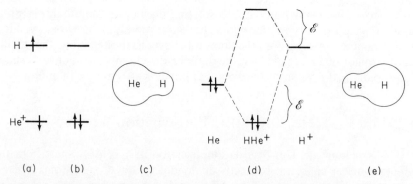

FIGURE 1.12. (a, b) AOs in H + He$^+$ and H$^+$ + He; (c) bonding MO in HHe$^+$; (d) energies of AOs and MOs in system H$^+$ + He; (e) antibonding MO.

Introduction to MO Theory

The first step is exactly analogous to the formation of salts by electron transfer from atoms with weakly bound electrons (e.g., metals) to atoms with a high affinity for electrons (e.g., halogens); a simple example is the formation of lithium fluoride from lithium and fluorine,

$$\text{Li} + \text{F} \longrightarrow \text{Li}^+\text{F}^- \tag{1.20}$$

The second step is the formation of a normal covalent bond.

Third, the combination of atoms to form molecules can be represented as a perturbation of the orbitals in them. The AOs interact and mix to form MOs when the atoms come close enough together for the AOs to overlap in space. This type of process can be studied by a technique known as *perturbation theory*. At this point we need only an intuitive picture and a summary of the results given by it.

When two orbitals interact, two MOs are formed, one of lower energy than either AO, one of higher energy (cf. Figs. 1.10g, 1.11d). It is as though the orbitals pushed each other apart in energy, one going up, and one down, by equal amounts. This analogy suggests that the effect should be greater, the nearer together the AOs are in energy—and indeed it is (cf. Figs. 1.10g and 1.11d). The maximum effect is observed when the AOs are *degenerate*, i.e., have equal energies; the splitting ($\mathscr{E}$ in Fig. 1.10g) is then given by the resonance integral β. If the AOs have very different energies, as was the case in HHe$^+$ (Fig. 1.11e), the splitting ($\mathscr{E}$) is given by

$$\mathscr{E} = \beta^2/\Delta E \tag{1.21}$$

when ΔE is the difference in energy between the AOs. Here $\mathscr{E}$ is called a *second-order perturbation*. One can see that the second-order perturbation decreases as ΔE increases. As ΔE decreases to zero, $\mathscr{E}$ does not tend to infinity but rather to the limit β; there is an intermediate region of small ΔE where neither simple perturbation expression (i.e., the *first-order perturbation* β, or the second-order perturbation $\beta^2/\Delta E$) fits.

When a hydrogen atom with one electron and a neutral helium atom with two electrons approach one another, a molecule HHe is formed. This time there are three electrons to be accounted for. Figure 1.13(a) shows how these electrons occupy the bonding and antibonding MOs of the HeH molecular system. From the Pauli principle only two of the three electrons present can fit into the bonding MO; the extra electron has to go into the antibonding MO. Since, however, the splitting is symmetric, each helium electron gains in energy by the same amount $\mathscr{E}$ that the hydrogen electron loses. There is therefore a net decrease in energy. The resulting *three-electron bond* is, however, weak since the bonding effect of one of the two electrons going into the bonding MO is canceled by the electron forced into the high-energy antibonding MO. An obvious extension of this argument shows (Fig. 1.13b) that two helium atoms should not combine at all since the potential bonding effect of the electrons going into the bonding MO is completely canceled by the two forced into the antibonding MO.

In lithium, two of the electrons occupy the 1s AO, in which they are very

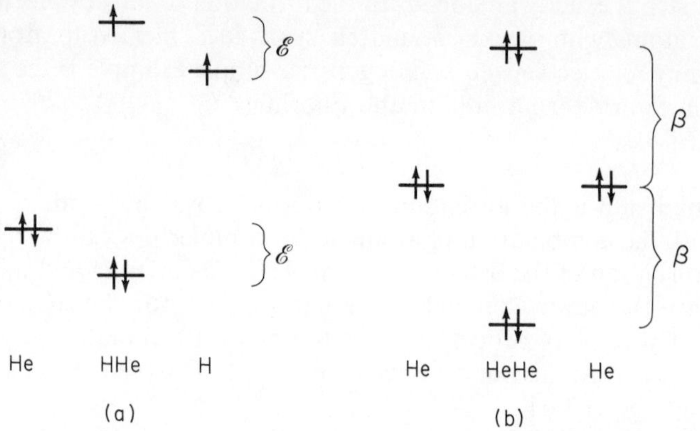

FIGURE 1.13. (a) H + He ⟶ HHe; (b) He + He ⟶ He$_2$.

tightly bound, the nucleus having a charge of $+3e$. The 1s AO in Li is correspondingly smaller than that in H or He since the electrons are held close to the nucleus by its very strong attraction. The third electron occupies the 2s AO, which is much bigger, both because it has a higher principal quantum number and because the two electrons in the small 1s AO shield the third electron very effectively from the nucleus. The ionization potential of Li (5.36 eV) is consequently much less than that (13.53 eV) of H.

When H and Li come together, the 1s AO of H overlaps with both the 1s and 2s AOs of Li. The effect of the 1s–1s overlap is, however, small for three reasons. First, the AOs differ greatly in energy, so it is only a second-order effect. Second, since the difference in energy ΔE is very large, so, too, is the denominator of the expression for the resulting second-order perturbation [equation (1.21)]. Third, since the AOs are so different in size, the overlap between them is small, and consequently so, too, is the resonance integral for their interaction [see equation (1.18)]. The numerator of the expression for the second-order perturbation is therefore correspondingly small.* We can therefore neglect the electrons in the 1s AO of Li; they remain essentially unaffected by the approach of the hydrogen atom. To all intents and purposes, we can regard Li as a one-electron system, the electron moving in the field of a *core* with a single positive charge, composed of the nucleus and two 1s electrons. The bonding MO in LiH must then be formed between the 1s AO of H and the 2s AO of Li. Since these AOs differ in energy, the situation should be similar to that in HHe$^+$ (Fig. 1.12). The resulting bond is a polar one, the hydrogen atom getting more

---

* It is true that the whole of the 1s AO of Li overlaps with the 1s AO of H; however only a very small fraction of the 1s AO of H overlaps with that of Li. The net overlap S is effectively a product of the fractions of each AO in the overlap region; while that for Li is unity, that for H is very small. A good analogy is provided by a fly sitting on an elephant; though the whole of the fly overlaps the elephant, only a tiny fraction of the elephant overlaps the fly.

Introduction to MO Theory

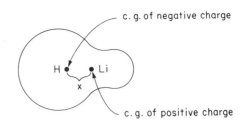

FIGURE 1.14. Polarity of LiH (c.g. = center of gravity).

than its share of the valence electrons. The center of gravity of the electrons therefore lies nearer to H than to Li. Since the center of gravity of positive charge lies halfway between the two nuclei, the molecule should behave like an *electric dipole* (Fig. 1.14); when placed in an electric field, the negative charge (i.e., the H end of the molecule) will be attracted to the positive side of the field, and the positive charge to the negative side. Molecules which have such an internal separation of positive and negative charge are called *polar*. The magnitude of this effect is measured by the *dipole moment*, a product of the charges separated and the distance by which they are separated. In LiH, the dipole moment D is given by

$$D = 2ex \qquad (1.22)$$

x being the distance between the centers of gravity of positive and negative charge and $-e$ the charge on an electron.

1.8 Methane; Symmetry Orbitals

The methane molecule CH_4 is known to have a very symmetric shape. The four hydrogen atoms lie at the corners of a regular tetrahedron with the carbon atom at its center. We can represent this structure by placing the hydrogen atoms (a, b, c, d) at alternate corners of a cube with the carbon at its center (Fig. 1.15). Let us take as coordinate axes the lines joining midpoints of opposite faces of the cube (Fig. 1.15). Our MOs will then be written as linear combinations of the four hydrogen $1s$ AOs (a, b, c, d), and the $1s$ $2s(s)$, $2p_x(x)$, $2p_y(y)$, and $2p_z(z)$ AOs of carbon (Fig. 1.15).

The arguments used for LiH in the previous section show, however, that the $1s$ AOs will stay apart from the rest. We can therefore ignore them, treating the system as one of eight valence electrons moving in the field of a core composed of four protons and a unit C^{4+} consisting of the carbon nucleus with its pair of $1s$ electrons. To accommodate these electrons, we need four MOs constructed from the eight AOs a, b, c, d, s, x, y, z.

Methane has three twofold axes of symmetry; they are the coordinate axes in Fig. 1.15(a). Rotation through 180° about any of these merely interchanges pairs of hydrogen atoms. Since the hydrogen atoms are identical, the molecule remains unchanged. The arguments used for H_2 (Section 1.5) imply that each MO must then be either symmetric or anti-

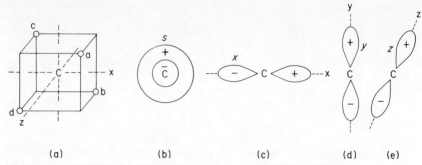

FIGURE 1.15. (a) Demonstration of how CH_4 can be inscribed in a cube; (b–e) $1s$, $2s$ (s), $2p_x(x)$, $2p_y(y)$, and $2p_z(z)$ AOs, respectively, of carbon.

symmetric for each such rotation; this conclusion will assist us in constructing the corresponding MOs. Denoting the three rotations by R_x, R_y, and R_z, respectively, we can divide our MOs into different types according as they are symmetric or antisymmetric with respect to R_x, R_y, and R_z. The original four carbon AOs conform to this symmetry. Thus rotation of s through 180° about any axis, or of x about the x axis, leaves them unchanged. Rotation of x about the y or z axis interchanges its lobes; since these have opposite signs, such rotations replace the AO x by $-x$. Thus s is symmetric with respect to all three rotations while x is symmetric to R_x and antisymmetric to R_y and R_z. These symmetry properties are indicated in Table 1.2 together with those derived in an analogous way for y and z.

If a MO is to have certain symmetry properties, it must obviously be formed from AOs of like symmetry. Thus the combination of AOs $s + x$ is neither symmetric nor antisymmetric with respect to R_y or R_z, both of which convert it into the entirely different combination $s - x$. Each carbon AO can therefore contribute only to MOs of one symmetry type.

Consider the MOs derived from s. These must be of the form

$$\psi_s = Ss + Aa + Bb + Cc + Dd \qquad (1.23)$$

Here S, A, B, C, and D are constants whose relative values can be found from symmetry. R_x interchanges atoms a with b, and c with d. This rotation con-

TABLE 1.2. Symmetry Orbitals for CH_4

Orbitals	Behavior on rotation through 180°		
	R_x	R_y	R_z
$s(a + b + c + d)$	+	+	+
$x(a + b - c - d)$	+	−	−
$y(a - b + c - d)$	−	+	−
$z(a - b - c + d)$	−	−	+

verts ψ_s into ψ_s',

$$\psi_s' = Ss + Ba + Ab + Dc + Cd \tag{1.23'}$$

Since the MO has to be symmetric with respect to R_x because it contains s, we have $\psi_s \equiv \psi_s'$; hence:

$$A = B; \quad C = D \tag{1.24}$$

An obvious extension of this argument, using R_y and R_z, shows that

$$A = B = C = D \tag{1.25}$$

The MO ψ_s can therefore be written as

$$\psi_s = Ss + A(a + b + c + d) \tag{1.26}$$

Likewise the MOs ψ_x, ψ_y, and ψ_z formed, respectively, from x, y, and z must be of the form

$$\psi_x = Px + B(a + b - c - d) \tag{1.27}$$

$$\psi_y = Py + B(a - b + c - d) \tag{1.28}$$

$$\psi_z = Pz + B(a - b - c + d) \tag{1.29}$$

Each of these MOs can be converted into one of the others by suitable rotations. For example, rotation through 90° about the z axis followed by rotation through 90° about the y axis leaves the molecule unchanged but converts ψ_x into ψ_y. The coefficients P and B must therefore be the same for all three. These MOs must, like p AOs, occur in degenerate sets of three.

If we examine equations (1.27)–(1.29), we see that the MOs have exactly the same form as those for HHe$^+$ or HLi except that one AO has been replaced by a combination of hydrogen AOs. We can thus regard the MOs of CH$_4$ as being formed in this way from one of the carbon AOs and a *symmetry orbital* (SO) composed of a combination of hydrogen AOs that conforms to the molecular symmetry. In other words, we start by replacing the four hydrogen AOs a–d, which do not have the right symmetry properties for our problem, with an equivalent set of four linear combinations that do. Since any linear function of the $1s$ AOs can obviously be written as a linear function of the four SOs, the AOs and SOs are clearly equivalent. The SOs and their symmetry properties are indicated in Table 1.2.

Just as in the case of H$_2$ or HHe, the eight MOs that are formed from the four carbon AOs and the four hydrogen AOs will occur in pairs, one bonding and one antibonding, one pair for each symmetry type. The four bonding MOs will hold the eight valence electrons. One of the MOs (ψ_s) must be different from the ψ_x, ψ_y, ψ_z triplet. Presumably it will be of lower energy since the $2s$ AO of carbon lies well below the $2p$ AOs in energy.

1.9 Photoelectron Spectroscopy and Ionization Potentials

If we supply enough energy to an atom or molecule, we may cause it to lose an electron. The minimum energy required to do this is called the *first ionization potential*. Ions, like molecules, can exist in a number of possible energy states; when we ionize a molecule, we may leave the resulting ion not in its ground state (i.e., the state of lowest energy) but in some excited state of higher energy. To do this, we will of course have to supply enough energy both to remove the electron and to excite the resulting ion. A molecule will thus have a series of ionization potentials corresponding to possible states of the resulting ion.

Molecules can be ionized by ultraviolet radiation. Light behaves like a shower of particles (photons) each with energy $h\nu$, h being Planck's constant and ν the frequency of the light. If the energy of a photon is more than sufficient to ionize a molecule, the photon can be absorbed by the molecule and its energy used to ionize it. If the corresponding ionization potential is I_n, there will be an excess of energy $h\nu - I_n$ over and above that needed to ionize the molecule and excite it to its nth excited state. The excess energy must appear as kinetic energy of the ion and the expelled electron. Since the electron is so much lighter than the ion, the law of conservation of momentum requires that nearly all the kinetic energy will go into the electron. To a very good approximation, the kinetic energy T_e of the electron is given by

$$T_e = h\nu - I_n \tag{1.30}$$

If we use light of a known and fixed frequency to ionize a molecule and measure the kinetic energy of the electrons emitted we find the latter confined to certain discrete values, each corresponding to a different I_n. If we measure these kinetic energies, we can then use equation (1.30) to find the various ionization potentials. This is the basis of the recently developed technique known as *photoelectron spectroscopy*. Other methods had been used in the past to measure ionization potentials, but photoelectron spectroscopy is far more efficient. A typical photoelectron spectrometer is indicated schematically in Fig. 1.16. The light source A is generally a discharge through helium at low pressure. Such a source produces mainly a single spectrum line at 486 Å, corresponding to $h\nu = 21.22$ eV. The material, in the form of a gas at low pressure, is ionized in the cylindrical chamber B. The electrons escape through slit C and then pass through the analyzer D. This consists of two cylindrical electrodes with a variable potential applied across them. The electrons are attracted by the inner, positively charged electrode and so travel in a curve, the radius of which depends on how fast they are moving, i.e., on T_e. If their energy is right, they pass through a second slit E to the detector F. If we vary the potential across D and plot it against the current through F, the result is a plot of number of electrons vs. their kinetic energy. Each value of T_e corresponds to a peak in the resulting plot. Figure 1.17 shows such a photoelectron spectrum for H_2. Note that instead of a single peak, corresponding

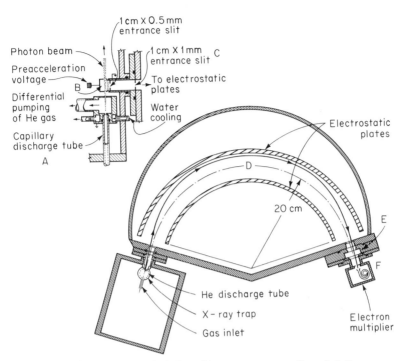

FIGURE 1.16. A typical photoelectron spectrometer (from Ref. 1).

to the first (i.e., lowest) ionization potential, there are a series of peaks. These peaks correspond to changes in vibrational energy. The energy of vibration, like the energy of the electron in a hydrogen atom, is limited to certain definite values. The ion H_2^+ can therefore exist in a number of discrete vibrational states, corresponding to vibrations in which the length of the H—H bond varies periodically. These are therefore a series of different ionization potentials of H_2, corresponding to processes in which the ion is left in its electronic ground state but with varying amounts of vibrational energy.

Note that the peaks on Fig. 1.17 are not all of the same height. When a molecule is ionized, the heavy nuclei do not have time to move while the electron is being expelled (cf. the Born–Oppenheimer approximation, Section 1.6). The ion is therefore most likely to be left in the geometry corresponding to the most favorable state of the molecule. This is an example of the *Franck–Condon principle*. If there is a change in equilibrium geometry in going from the molecule to its ion, the ion will usually be left in a nonequilibrium state, i.e., one of its vibrationally excited states. The corresponding line in the photoelectron spectrum will then be the most intense. Conversely, if there is no change in geometry, there will be little tendency for the molecule to be set vibrating when it is ionized. The photoelectron spectrum will then consist mainly of a single strong line. In the case of H_2, ionization involves removal of an electron from a bonding MO. The bond in H_2^+ is therefore

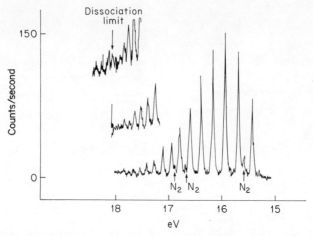

FIGURE 1.17. Photoelectron spectrum of H_2 (from Ref. 2).

only a half bond and is consequently a good deal longer and weaker than that in H_2. Ionization of H_2 therefore leads to strong excitation of stretching vibrations (Fig. 1.17).

In our orbital picture, the electrons in a molecule are supposed to occupy MOs. The energy required to remove an electron from a MO should then be related to the orbital energy. This indeed is the case, a result known as *Koopmans' theorem*. One can get surprisingly good estimates of the ionization potentials of molecules by calculating the orbital energies of the MOs in them and assuming that these are equal to *minus* the corresponding ionization potentials. On this basis, we would expect methane to show three ionization potentials, corresponding to removal of electrons from the carbon 1s AO, the s-type MO, and one or other of the three degenerate p-type MOs, the ionization potentials falling in that order. This is the case. Admittedly, we have to use a different light source to see all three ionizations because the energy needed to remove electrons from two of them is greater than 21.22 eV. Using soft X rays, we can observe ionization of the 1s electrons, the ionization potential being 290.5 eV. The other two ionization potentials are much lower; they can be observed together using a special helium discharge lamp which gives the spectrum line of He^+ at 304 Å (40.7 eV). Figure 1.18 shows the corresponding photoelectron spectrum; as expected, the s-type MO has a much higher ionization potential (27 eV) than the p-type MOs (13 eV); since there are three p-type MOs and only one s-type MO, the probability of ionizing a p-type electron is three times as great and the peak in the photoelectron spectrum is correspondingly bigger.

Photoelectron spectroscopy is thus of particular interest in the present connection since it provides an experimental demonstration of the validity of our orbital picture. The spectrum in Fig. 1.18 also raises another point. Note that the p-type peak is very broad and structureless, unlike the spectrum

Introduction to MO Theory

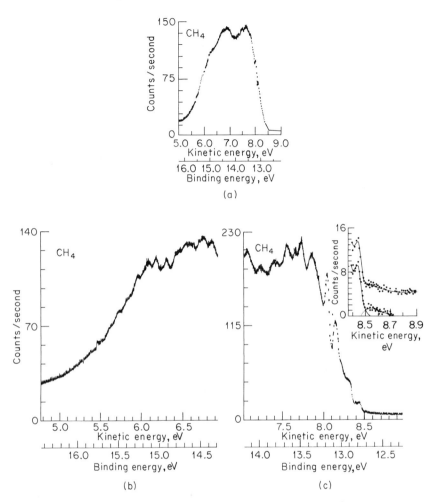

FIGURE 1.18. Photoelectron spectrum of methane (from Ref. 1).

of H_2 (Fig. 1.17). This is because the ionization involves a degenerate level. In the resulting ion, two of the p-type MOs still contain two electrons, but the third contains only one; it can be shown that in such cases there must be distortions of the ion that remove the degeneracy of the MOs, one going up in energy and one down. In the case of the parent molecule, such a distortion has no effect on the total energy since the increase in energy of the two electrons occupying the MO that goes up in energy is balanced by the decrease in energy of the two electrons in the MO that goes down. In the ion, however, if the former MO is singly occupied and the latter is doubly occupied, the distortion will lead to a net decrease in energy, i.e., to stabilization. One cannot therefore have stable structures with partly filled degenerate levels because there will in general* be a distortion that removes the degeneracy and leads to a decrease in energy (the *Jahn–Teller effect*). A further consequence is that the ionization appears as a broad band in the photoelectron spectrum; for since molecules vibrate, and since ionization takes place without any change in geometry, and since small changes in the ground-state geometry lead to large changes in ionization potential, the electron emitted can have any energy in a rather wide range. Conversely, the appearance of a wide band with little structure in the photoelectron spectrum is an indication of a degenerate MO.

1.10 Methane, Continued; Hybridization and Localized Bonds

The picture of methane given in the two previous sections is of course entirely different from the conventional one with four localized equivalent C—H bonds linking the atoms together. Our next task is to examine the relation between these two apparently contradictory pictures and to show that they are compatible.

In the MO description given above, each MO of methane is derived from a single carbon AO and a combination (SO) of hydrogen $1s$ AOs. Let us now try to find a reciprocal description in which each MO is formed from a single hydrogen $1s$ AO and a combination of carbon AOs.

Consider the following four combinations t_{1-4} of the carbon AOs s, x, y, and z:

$$t_1 = s + x + y + z, \quad t_2 = s + x - y - z$$
$$t_3 = s - x + y - z, \quad t_4 = s - x - y + z \quad (1.31)$$

It is easily seen that the functions t_{1-4} are *linearly independent*, i.e., no one of them can be expressed as a linear combination of the others. Equally, the AOs s, x, y, and z can be expressed as linear functions t_{1-4}; e.g.,

$$s = \tfrac{1}{4}(t_1 + t_2 + t_3 + t_4) \quad (1.32)$$

It follows that any linear function of the AOs s, x, y and z can be written as a linear function of t_{1-4}. In constructing our MOs, we can just as well use as our *basis set* the four functions t_{1-4} in place of the AOs s, x, y, and z.

The mathematical form of p AOs is such that any combination of p AOs gives an orbital of the same shape (i.e., another p AO), pointing in some intermediate direction. Each of the orbitals t_{1-4} is therefore identical in shape, being drawn from the s AO and p AOs in similar proportions (i.e., 1:3). The shape of each of these *hybrid* orbitals is indicated in Fig. 1.19(a, b). The combination of the s AO with a p AO leads to a lopsided p-like orbital in which

---

* The only exceptions are linear molecules or ions such as CO_2, NO_2^+, or FCN; here it is possible to have stable structures that are degenerate.

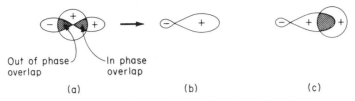

Out of phase overlap In phase overlap

(a)   (b)   (c)

FIGURE 1.19. (a, b) Combination of an s AO and a p AO to form a hybrid AO; (c) formation of a two-center MO from a hybrid AO and an s AO.

one lobe is now bigger than the other. The four hybrid orbitals t_{1-4} point toward the corners of the cube in Fig. 1.15(a), the bigger lobe of each orbital pointing toward the corner where there is a hydrogen atom.

Now the interaction between two orbitals depends on the value of the corresponding resonance integral (Section 1.7) and this in turn is more or less proportional to the extent to which the orbitals in question overlap in space [equations (1.17) and (1.18)]. Since the major lobe of each hybrid orbital points toward just one of the hydrogen 1s AOs, it is evident that the corresponding interactions will be much the greatest. If we ignore the rest, the MOs in methane will be analogous to those in a diatomic molecule such as HHe$^+$ or LiH, being formed from combinations of one of the carbon hybrid AOs and the hydrogen 1s AO with which it overlaps. With the notation of Fig. 1.15(a) and equation (1.31), the resulting MOs are of the following form:

$$\tau_1 = At_1 + Ba = A(s + x + y + z) + Ba$$
$$\tau_2 = At_2 + Bb = A(s + x - y - z) + Bb$$
$$\tau_3 = At_3 + Bc = A(s - x + y - z) + Bc \quad (1.33)$$
$$\tau_4 = At_4 + Bd = A(s - x - y + z) + Bd$$

Since the four hybrid AOs t_{1-4} are identical in shape and point toward the hydrogen atoms, the four MOs τ_{1-4} are also identical in shape and energy. They can indeed be interconverted by the rotations R_x, R_y, and R_z. Each interaction between a hybrid AO t and a hydrogen AO will give rise to one bonding MO and one antibonding MO (cf. p. 14); the four resulting bonding MOs are sufficient to hold the eight valence electrons in CH_4. This picture corresponds to the conventional one, the hydrogen atoms being linked to the central carbon by pairs of electrons occupying two-center bonding MOs.

We now have two apparently dissimilar but reciprocal orbital descriptions of methane. The original MO description constructs each MO from a single AO of carbon and a "hybrid" (SO) of hydrogen 1s AOs, while the description given here constructs each MO from a hybrid of the carbon AOs and a single hydrogen 1s AO. It is also apparent that the description in terms of four identical (and so degenerate) MOs τ_{1-4} cannot account for the photoelectron spectrum of methane, which shows that the valence electrons occupy MOs of two different energies (Fig. 1.18).

The four MOs ψ_s, ψ_x, ψ_y, and ψ_z [equations (1.26)–(1.29)] represent volumes of "orbital space" each able to hold two electrons, so, together, the four MOs represent a volume of "orbital space" able to hold four pairs of electrons. This suggests that we could combine the four MOs and then subdivide their sum into four equal parts in some other way, thus getting an equivalent description of methane.

One such set of combinations τ'_{1-4} can be constructed as follows:

$$\tau'_1 = \frac{B}{A}\psi_s + \psi_x + \psi_y + \psi_z = \frac{BS}{A}s + P(x + y + z) + 4Ba$$

$$\tau'_2 = \frac{B}{A}\psi_s + \psi_x - \psi_y - \psi_z = \frac{BS}{A}s + P(x - y - z) + 4Bb$$

$$\tau'_3 = \frac{B}{A}\psi_s - \psi_x + \psi_y - \psi_z = \frac{BS}{A}s + P(-x + y - z) + 4Bc \quad (1.34)$$

$$\tau'_4 = \frac{B}{A}\psi_x - \psi_x - \psi_y + \psi_z = \frac{BS}{A}s + P(-x - y + z) + 4Bd$$

Each of the orbitals τ' is depicted as a combination of a hybrid carbon orbital (indicated in parentheses) and one of the hydrogen AOs. Indeed, if the s-type MO ψ_s [equation (1.26)] and the p-type MOs [equations (1.27)–(1.29)] had identical coefficients, so that $S/A = P/B$, the orbitals τ'_{1-4} would be identical with τ_{1-4} [cf. equations (1.33) and (1.34)]. The s- and p-type MOs are not in fact identical, because the $2s$ and $2p$ AOs of carbon differ both in shape and energy; however, the differences are not likely to be large. If they can be neglected, our description of methane in terms of two-center bonds formed by the "sp^3" hybrids t_{1-4} [equation (1.31)] and the corresponding hydrogen $1s$ AOs will be valid.

The relation between the two descriptions can be further elucidated by perturbation theory. Let us take as our unperturbed system the picture represented by equation (1.33). The interaction between the four hybrid AOs of equation (1.31) and the four hydrogen $1s$ AOs will give rise to four degenerate bonding MOs that are filled in CH_4 and four degenerate antibonding MOs that are empty (Fig. 1.19a). Let us now introduce the interactions between them by perturbation theory (Section 1.7). The main effect will be first-order perturbations between the four degenerate bonding MOs and the four degenerate antibonding MOs. This has the effect of splitting each level into the (3 + 1) pattern we deduced in our original MO treatment (Fig. 1.20b). Since, however, the interaction between two orbitals has the effect of lowering the energy of one at the expense of an equal increase in the energy of the other, the total energy of the four bonding MOs remains unchanged. Thus if we are concerned only with the *total* energy of methane, the first-order perturbation has no effect. It is important only if we are concerned with the energies of individual orbitals, e.g., in the study of ionization potentials.

If the s- and p-type MOs were in fact equivalent, so that in equations (1.26)–(1.29), $S/A = P/B$, the orbitals given by this first-order perturbation

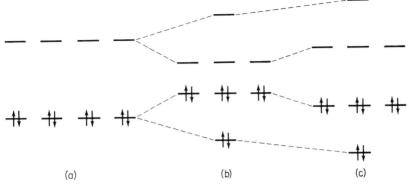

FIGURE 1.20. (a) Unperturbed orbitals [τ_{1-4}; equation (1.32)] for CH_4; (b) effect of first-order interactions; (c) effect of second-order interactions.

would be identical with those from the MO treatment. Since, however, this is not the case, there are also second-order perturbations between the filled and empty MOs (Figure 1.20c) which lead to a depression of the energies of the filled MOs at the expense of an increase in the energy of the empty ones.

If the s-type MO ψ_s [equation (1.26)] and the p-type MOs ψ_x, ψ_y, ψ_z [equations (1.27)–(1.29)] were in fact equivalent, the description given by equations (1.33) would then be the same. Now the density distribution in the hybrid orbital $s \pm x \pm y \pm z$ is given by the corresponding square; i.e.,

$$(s \pm x \pm y \pm z)^2 = s^2 + x^2 + y^2 + z^2 \pm 2sx \pm 2sy$$
$$\pm 2sz \pm 2xy \pm 2xz \pm 2yz \quad (1.35)$$

If we integrate this expression, the contributions of the final terms cancel out because the contributions of regions where the two AOs in question overlap in phase is canceled by regions where they overlap out of phase (so that their product is negative instead of positive; see Fig. 1.19a). The total density distribution is therefore effectively a sum of one part s and three parts p. The s and p orbitals would therefore contribute equally to the final MOs and so to the final electron density. In practice, however, since the $2s$ AO of carbon lies lower in energy than the $2p$ AOs, the coefficient of the $2s$ AO in ψ_s [S in equation (1.26)] is greater than the coefficients of the $2p$ AOs in $\psi_x, \psi_y,$ or ψ_z [P in equations (1.27)–(1.29)]. The utilization of the carbon $2s$ AO in methane is correspondingly more complete than that of the $2p$ AOs and the representation in terms of sp^3 hybrid AOs [equation (1.31)] is correspondingly less accurate.

This argument used, in passing, an important result, i.e., that the overlap integral $\int ab \, d\tau$ between two AOs of a given atom vanishes because the contributions of regions where the product ab is positive are canceled by regions where it is negative. Orbitals which obey this condition, i.e., vanishing of the corresponding overlap integral, are called *orthogonal*.

One final point: If we are to take the square ψ^2 of an orbital as a measure of the probability distribution of an electron occupying ψ, the total probability of finding the electron somewhere or other must of course be equal to unity. This total probability is given by the integral $\int \psi^2 \, d\tau$ integrated over the whole of space. If we set this integral equal to unity by multiplying ψ by a suitable factor, ψ is said to be *normalized*. If the AOs s, x, y, z are normalized, the hybrid $s \pm x \pm y \pm z$ must clearly be multiplied by $\sqrt{\frac{1}{4}}$, or $\frac{1}{2}$, in order that

$$\int (\tfrac{1}{2}s \pm \tfrac{1}{2}x \pm \tfrac{1}{2}y \pm \tfrac{1}{2}z)^2 \, d\tau = 1 \qquad (1.36)$$

1.11 Diatomic Molecules

With a few rare exceptions, the bonding in simple hydrides can be described by analogy with the HeH system. As soon as we look at bonding between atoms other than hydrogen, the situation becomes more complicated because each partner to the bond has more than one AO that can be involved. Molecules of the type XY, when X and Y are second row elements, e.g., N_2 and CO, will illustrate most of the possibilities.

The arguments used for LiH and CH_4 suggest that we can neglect the $1s$ electrons. Our MOs must then be constructed from the $2s$ and $2p$ AOs of the two atoms; i.e., their *valence shell* AOs. The choice of orientation of the $2p$ AOs is arbitrary. For convenience we will assume the axes of the $2p_y$ and $2p_z$ AOs of the two atoms to be parallel and that of the $2p_x$ AOs to lie along the line joining the nuclei. This choice is indicated in Fig. 1.21. For simplicity, we will also first assume X and Y to be the same, forming a symmetric molecule X_2.

The xz and yz planes are planes of symmetry (Fig. 1.21). All MOs must be either symmetric or antisymmetric with respect to reflection in xy or xz. All the component AOs conform to this requirement; see Table 1.3.

The arguments used for methane show that each MO must be constructed entirely from AOs belonging to the same symmetry class. Consequently the

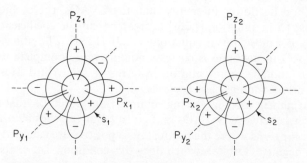

FIGURE 1.21. Valence shell AOs of two second row atoms X, about to form X_2.

TABLE 1.3. Symmetries of AOs in X_2

AOs	Behavior on reflection in	
	xy	xz
s_1, x_1, s_2, x_2	+	+
y_1, y_2	+	−
z_1, z_2	−	+

MOs of p_y-type symmetry must be constructed solely from y_1 and y_2 and the MOs of p_z-type symmetry from z_1 and z_2. The resulting two-center MOs will then be analogous to those in H_2, being given by $(y_1 \pm y_2)$ and $(z_1 \pm z_2)$. In each case one of the resulting MOs is bonding, the other antibonding. The only difference is that here the overlap is sideways (Fig. 1.22) instead of along the axis of the bond. The resulting MOs, like the constituent p AOs, have nodes passing through the nuclei. The MOs formed by such lateral overlap of p AOs are called π MOs and the resulting bonds are called π bonds, in contrast to the σ MOs and σ bonds formed by collinear overlap.

It should be emphasized that there is no essential difference between the two types of bonds other than their geometry. For AOs to fuse to MOs, it is only necessary that they should overlap. The geometry of overlap is not relevant. *Bonding is a matter of the topology of orbital overlap, not geometry; the only important factors are the extent of overlap and whether the phases of the AOs are the same or different in the region where they overlap.*

Next let us consider the symmetric MOs formed from the four AOs s_1, x_1, s_2, x_2. Since the 2s AOs of second row elements lie well below the 2p AOs in energy, we can as a first approximation treat the $(s_1 - s_2)$ and $(x_1 - x_2)$ interactions separately since, according to second-order perturbation theory [equation (1.20)] the interactions between them should be small. On this basis, we will again get two bonding/antibonding pairs of two-center MOs as indicated in Fig. 1.23. We can examine the effect of $(s - x)$ interactions using perturbation theory [see Section 1.7 and equation (1.21)].

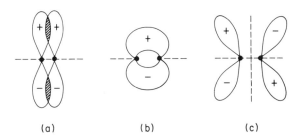

FIGURE 1.22. (a) Interaction of two π AOs to form (b) a bonding π MO, (c) an antibonding π MO. Nodes are indicated by dashed lines.

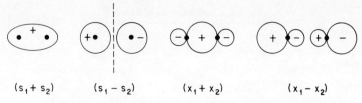

$(s_1 + s_2)$ $\quad$ $(s_1 - s_2)$ $\quad$ $(x_1 + x_2)$ $\quad$ $(x_1 - x_2)$

FIGURE 1.23. MOs of X_2 derived from $2s$ and $2p_x$ AOs.

The molecule X—X has a plane of symmetry bisecting the nuclei. Every MO must be symmetric or antisymmetric with respect to reflection in this plane. All the bonding MOs of X_2, both σ and π, are symmetric and all the antibonding MOs are antisymmetric in this respect. Now when two MOs of different energies interact with each other, the effect is to raise the energy of the upper one and lower the energy of the lower one by equal amounts [see equation (1.21)], and at the same time to mix the MOs together. It is obvious therefore that interactions between MOs of different symmetry must vanish; for, if we mix two MOs of different symmetry, the result will be an orbital in which the symmetry has been lost. We need therefore consider only interactions between $(s_1 + s_2)$ and $(x_1 + x_2)$ and between $(s_1 - s_2)$ and $(x_1 - x_2)$. The former interaction will lower $(s_1 + s_2)$ in energy, making it even more strongly bonding, at the expense of reducing the bondingness of $(x_1 + x_2)$. Likewise, the latter interaction will make $(x_1 - x_2)$ even more antibonding at the expense of $(s_1 - s_2)$. The situation is indicated in Fig. 1.24.

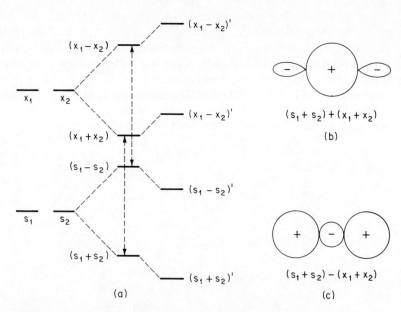

FIGURE 1.24. (a) Energies of MOs arising from s_1, s_2, x_1, and x_2; (b, c) effect of interaction between $(s_1 + s_2)$ and $(x_1 + x_2)$.

Introduction to MO Theory

One can see why this is so, by considering the effect of mixing $(s_1 + s_2)$ with $(x_1 + x_2)$. In the combination $(s_1 + s_2) + (x_1 + x_2)$, there is a reinforcement of the orbital density in the bonding region between the nuclei, with a simultaneous reduction of the nonbonding lobes of the $(x_1 + x_2)$ MO, while in the combination $(s_1 + s_2) - (x_1 + x_2)$ the converse is the case. In practice the effects should be much less extreme than this since the interaction is a second-order perturbation involving two orbitals of very different energies.

Next we must fit the π MOs $(y_1 \pm y_2)$ and $(z_1 \pm z_2)$ into the MO diagram of Fig. 1.21a. Since the $2s$ AOs lie well below $2p$ in energy, one would expect the combinations $(s_1 \pm s_2)$ to lie lowest, since both these MOs are depressed by interaction with $(x_1 \pm x_2)$. One might expect the σ-type interaction of x_1 with x_2 to be more effective than the π-type interactions of y_1 with y_2 or z_1 with z_2 since the lengthwise overlap looks more efficient; this would place $(x_1 + x_2)$ below the bonding π MOs and $(x_1 - x_2)$ above the antibonding ones. Since, however, both the MOs $(x_1 \pm x_2)$ are raised by interaction with $(s_1 \pm s_2)$, the position of $(x_1 + x_2)'$ is uncertain, though $(x_1 - x_2)'$ should certainly remain at the top. The resulting MO energy diagram is shown in Fig. 1.25.

While this diagram was deduced for symmetric molecules X—X, it should also hold qualitatively for asymmetric molecules X—Y. The structure of the second row molecules X—Y can then be deduced by feeding the valence shell electrons into these MOs from the bottom upward, two electrons per MO. Let us consider some examples.

Li—Li. Here there are two valence electrons; both fit into the bonding MO $(s_1 + s_2)'$.

C—C. Carbon contains six electrons, two in the inner shell $1s$ AO; there are consequently eight electrons to be accommodated in the MOs of Fig. 1.25. Four occupy the MOs $(s_1 + s_2)'$ and $(s_1 - s_2)'$. If $(x_1 + x_2)'$ came next, there would be two electrons left over to occupy the degenerate bonding π MOs; from Hund's rule (p. 12), the best arrangement would be for the electrons to go into different π MOs with parallel spin. In that case, C_2 would

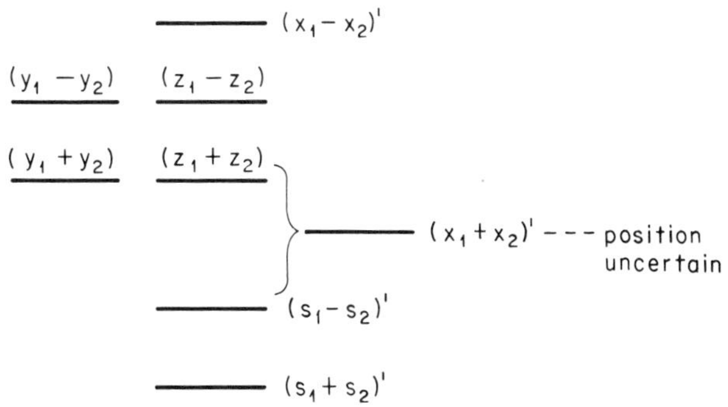

FIGURE 1.25. MO energies in diatomic second row molecules.

have two unpaired electrons and have a triplet ground state. Experiment shows, however, that the ground state of C_2 is a singlet. Thus the MO $(x_1 + x_2)'$ must lie above $(y_1 + y_2)$ and $(z_1 + z_2)$. The net bonding effect of the four electrons in the MOs $(s_1 \pm s_2)'$ should be small (cf. He_2; p. 20); the bond in C_2 is thus effectively double—and rather weak at that, since the four bonding electrons occupy π MOs. The bond length in C_2 (1.48 Å) indicates, as we shall see presently, that this is indeed the case.

N—N. Here there are four electrons more than in C_2; these are enough to fill all three bonding MOs derived from the $2p$ AOs. The bond in N_2 is thus effectively a triple bond (four pairs of bonding electrons *plus* one pair of antibonding electrons). In this case we can deduce the order of the MOs by photoelectron spectroscopy. Figure 1.26 shows the 21.22-eV photoelectron spectrum of N_2; use of higher-energy light shows further ionizations at 25.0 and 27.3 eV, corresponding to ionization from the MOs $(s_1 \pm s_2)'$. The three remaining bands correspond to ionization from $(x_1 + x_2)'$, $(y_1 + y_2)$ and $(z_1 + z_2)$. The latter remain degenerate because of an accident; in a linear molecule, there is no Jahn–Teller splitting because there are no deformations that remove the orbital degeneracy. There is no vibration that can destroy the symmetry of a diatomic molecule. Since the intensity of the band at the highest electron energy (i.e., lowest ionization potential) is about double that of either of the others, it must correspond to the π MOs. Here, therefore, $(x_1 + x_2)'$ lies below $(y_1 + y_2)$ or $(z_1 + z_2)$.

O—O. In O_2 there are two more electrons than in N_2; these must go into the antibonding π level. From Hund's rule, the ground state of O_2 should be a triplet with two unpaired electrons. These, moreover, are in antibonding MOs. The reactivity of O_2, on which life as we now know it depends, is due to the presence of two unpaired and weakly bound electrons with parallel spins. The assigned structure is beautifully confirmed by the

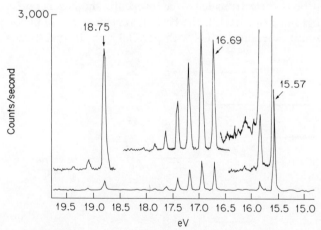

FIGURE 1.26. Photoelectron spectrum of N_2 (from Ref. 2).

Introduction to MO Theory

photoelectron spectrum (Fig. 1.27). The first two ionizations are from the degenerate antibonding and bonding π levels, respectively. Note that the spacing of the vibrational levels is quite different in the first two bands. The spacing corresponds to the frequency of vibration in the ion. The bond in O_2 is effectively double, eight electrons occupying bonding MOs and four occupying antibonding MOs. If we remove one of the antibonding electrons, the bond should then become shorter, stronger, and stiffer. The spacing of lines in the first band of O_2 indicates that the vibration frequency in the ion O_2^+ formed by loss of one of the antibonding electrons is greater than in O_2 itself. This confirms the increase in strength of the bond on ionization. Conversely, removal of a bonding π electron should weaken the O—O bond and lower the vibration frequency. The spacing of lines in the second band of the photoelectron spectrum corresponds to this expectation. There is another important factor in the photoelectron spectrum of oxygen, namely electron spin. Since oxygen's ground state has two parallel unpaired spins, removal of an electron from a doubly occupied oxygen orbital can result in formation of either a doublet state (from removal of an electron with the same spin as the unpaired antibonding π-electrons) or a quartet state (from removal of an electron with the opposite spin of the antibonding π-electrons). Hund's rule clearly predicts that it should take less energy to form the quartet state (maximum multiplicity). The Hund's rule splitting in the case of the degenerate bonding π-electrons is small, so the two bands overlap. Finally, as we have seen, the mixing of $(s_1 - s_2)$ with $(x_1 - x_2)$ should make the latter less bonding (cf. Fig. 1.21b, c); removal of electrons from the MOs $(s_1 - s_2)'$ or $(x_1 + x_2)'$ should therefore have a smaller effect on the vibration frequency. Indeed, the spacing of the lines in the corresponding bands in Fig. 1.27 is clearly intermediate between those in the first and second bands. The assignment of the two highest energy bands in Fig. 1.27 is still in doubt. At this time most of the evidence suggests that both bands arise from the $(x_1 + x_2)'$ orbital. The lower energy band corresponds to ionization to a quartet state, and the high energy band corresponds to ionization to a doublet state. The extreme similarity of the shape of the two bands supports this assignment. If this assignment is correct the Hund's rule splitting for these bands amounts to 2.12 eV.

F—F. All the MOs in F_2 are filled except $(x_1 - x_2)'$. There are still

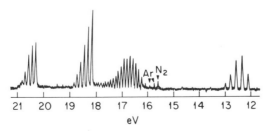

FIGURE 1.27. Photoelectron spectrum of O_2 (from Ref. 2).

only four pairs of bonding electrons, but now there are three pairs of antibonding electrons. The bond in F_2 is therefore effectively single. Since antibonding electrons are in general more antibonding than bonding electrons are bonding, the bond in F_2 is extremely weak. It is indeed the weakest single bond known. The extreme reactivity of fluorine is due to the consequent ease with which it can dissociate into very energetic atoms and the strength of the new bonds it forms with any other atom or radical.

1.12 The Paraffins; Localized Bonds

So far we have managed to deal with all the problems that have confronted us in terms of two-orbital interactions; in more complicated molecules this is no longer possible. The MOs of complex molecules have to be written as combinations of three or more orbitals participating to arbitrary extents. One cannot in general deduce intuitively what the resulting MOs will be like. There is, however, a compensating simplification. In chemistry, we are mainly concerned with the *collective* properties of molecules, i.e., those that depend on all the electrons taken together; the total energy is a good example The analogy of methane suggests that these can be adequately reproduced by an analog model with "localized" bonds. Even when we are concerned with properties involving single electrons, we are usually concerned only with the electrons occupying the MO of highest energy; we may be able to treat these as moving in a "localized" framework of atomic cores linked by two-center bonds. Such a picture would obviously have the further advantage of conforming to the classical representation of molecules in terms of such bonding. Our next object is to see how far this may be justified.

The paraffins (four of which are shown below) are hydrocarbons represented classically by structures in which each carbon atom is linked to four neighbors by "single" bonds and each hydrogen to one neighbor, and in

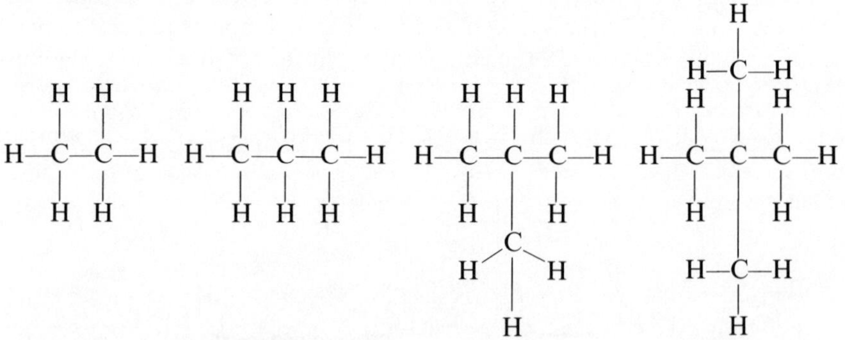

ETHANE PROPANE ISOBUTANE NEOPENTANE

which there are no rings. The MOs of such compounds are formed from combinations of the $2s$ and $2p$ AOs of the carbon atoms and the $1s$ AOs

Introduction to MO Theory

of hydrogen. Now we are perfectly at liberty, as we have seen, to use linear combinations of these as our *basis set* in place of the AOs themselves. If, for example, a given MO ψ_μ can be written as a combination of AOs ϕ_i,

$$\psi_\mu = \sum_i a_{\mu i} \phi_i \qquad (1.37)$$

and if we introduce a set of functions ξ_m that are linear combinations of the ϕ_i,

$$\xi_m = \sum_i b_{mi} \phi_i \qquad (1.38)$$

obviously we can write the ψ_μ as linear combinations of the ξ_m:

$$\psi_\mu = \sum_i c_{\mu i} \xi_m \qquad (1.39)$$

Suppose that we replace the four valence AOs of each carbon atom by sp^3 hybrids, pointing toward its four neighbors. This set of orbitals will overlap in pairs, just like the AOs in the corresponding picture of methane (Fig. 1.28a). The interactions between the orbitals in each such pair will be large, and interactions between orbitals not of the same pair will be small. A good approximation to the actual state of the molecule will then be given by neglecting the latter interactions. Pairs of AOs then interact to form pairs of MOs, one bonding and one antibonding, of two kinds, one corresponding to CC interactions and one to CH interactions. The number of valence electrons is exactly sufficient to just fill all the bonding MOs (Fig. 1.28b). Let us now introduce the smaller interactions we have so far neglected, using perturbation theory.

There will be first-order perturbations between the degenerate sets of CC and CH bonding and antibonding MOs. While these will be relatively large (Fig. 1.28c), they will not affect the total energy of the molecule because the *total* energy of the interacting orbitals remains unchanged (p. 30) and in this case each set of orbitals is either completely filled or completely empty. Next there will be second-order perturbations (Fig. 1.28d) between the filled bonding CC and CH orbitals and between the empty antibonding CC and CH orbitals. These second-order effects will also be relatively large since the interacting orbitals have similar energies, being both bonding or both antibonding [cf. equation (1.21)]. On the other hand, the second-order perturbations involve pair interactions that raise the energy of one interacting orbital at the expense of an equal depression in the energy of the other interacting orbital. If both orbitals are filled with electrons, such an interaction has no net effect on the total energy. Thus these second-order perturbations, like the first-order ones, leave the *total* energy of the molecule unchanged; consequently, they have no effect on the total energy or other collective properties. Finally, we are left with the bonding–antibonding second-order perturbations (Fig. 1.28e). These at last do affect the total energy since in each case they depress the energy of a filled bonding MO at the expense of raising the energy of an empty antibonding MO. Such interactions should, however, be small

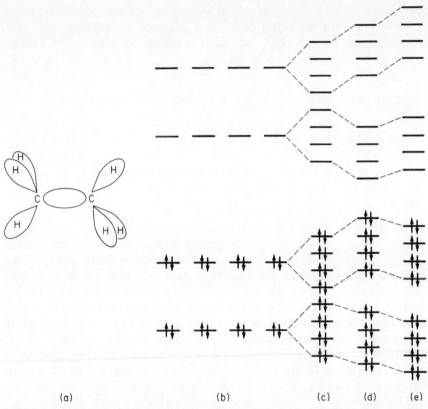

(a)   (b)   (c)   (d)   (e)

FIGURE 1.28. (a) Representation of ethane in terms of hybrid AOs; (b) unperturbed MOs for a paraffin; (c) effect of first-order perturbations; (d) effect of bonding CC–bonding CH interactions; (e) effect of bonding–antibonding interactions.

for two reasons. First, the interacting orbitals differ greatly in energy since one is bonding and the other is antibonding. Second, the pairs of interacting orbitals cannot overlap well since we have chosen our basis set to minimize such overlap. Consequently, we may expect our original simple picture (Figure 1.28b) to give a very good representation of the collective properties of the molecule. Indeed, the situation is even better than this.

The "localized" bond analog represents the bonding in a molecule as a sum of two-center interactions, each depending only on a pair of atoms involved. The heat of atomization should then be expressible as a sum of "bond energies" for the various bonds, these being the energies of the localized bonds in the original model (cf. Fig. 1.28b). In the case of paraffins, the heat of atomization should then be given by $mE_{CH} + nE_{CC}$, where m and n are, respectively, the numbers of CH and CC bonds and E_{CH} and E_{CC} are their respective bond energies. The heat of atomization of paraffins is known to follow such a relationship closely. The "localized bond" model thus provides a very simple scheme for estimating heats of atomization in addition to being much easier to visualize than a model based on delocalized MOs.

Although the localized bond model is only an analog of reality, it is a very useful one. It enables us to think about molecules in terms which are easy to grasp and which at the same time lead to conclusions in agreement with experiment. A good analogy is provided by the use of analog computers in engineering. We can mimic the stresses and strains in the girders of a bridge by analogous electrical circuits in an analog computer and so predict the behavior of the bridge before building it. The value of the analog lies in its simplicity and the only criterion for its use is that it should give the right answer. The localized bond model must be judged the same way. We will describe the bonds in a molecule as "localized" if its heat of atomization and other collective properties can be expressed in terms of bond properties, e.g., bond energies. The bond properties are determined empirically to give the best possible fit to as wide a range of molecules as possible in order that we may be able to use the localized bond model in the largest possible number of cases. These bond properties need not therefore correspond to the properties of the fictitious two-center bonds used in our derivation (cf. Fig. 1.28b). The second-order perturbations of Fig. 1.28(e) are not only small but are also approximately additive functions of the bonds present. If we include an average correction for them in our bond energies, we will get an even better representation of the molecule in terms of our localized bond model. If the bond energies are determined empirically, they will automatically contain such corrections. This is why the localized bond model works so well in practice for compounds of this type.

Note that this picture accounts for *all* collective properties. These include bond lengths, since the length of a bond in a molecule is such as to minimize the total energy. According to the localized bond model, the lengths of bonds should then be independent of the molecule in which they occur. The lengths of CC bonds in paraffins are indeed always the same (1.535 $\pm$ 0.005 Å) and so are those of CH bonds (1.100 $\pm$ 0.003 Å). The properties of paraffins can, in short, be represented very effectively in terms of atoms held together by standard bonds, just like a framework of girders held together by nuts and bolts of standard sizes.

1.13 Ethylene; π Bonds

In ethylene, C_2H_4, the atoms lie in a plane as indicated in Fig. 1.29(a). The MOs must be symmetric or antisymmetric for reflection in this plane. With the choice of coordinates indicated, it is obvious that all the AOs are symmetric except the $2p_z$ AOs of the carbon atoms. These will combine to form a bonding–antibonding pair of π MOs, just as in the diatomic molecules of Section 1.10.

The three remaining AOs of each carbon atom can now be replaced by three equivalent hybrid AOs. With the geometry indicated in Fig. 1.29(b),

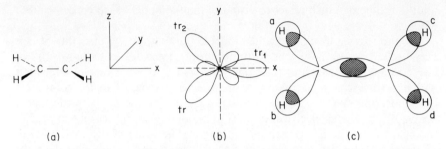

FIGURE 1.29. (a) Geometry of ethylene; (b) sp^2 hybrid AOs formed from AOs s, x, and y; (c) localized bond representation of σ MOs in ethylene

these can be written as

$$tr_1 = (1/\sqrt{3})(s + x\sqrt{2})$$
$$tr_2 = (1/\sqrt{3})[s - (1/\sqrt{2})x + y\sqrt{\tfrac{3}{2}}] \quad (1.40)$$
$$tr_3 = (1/\sqrt{3})[s - (1/\sqrt{2})x - y\sqrt{\tfrac{3}{2}}]$$

The proportion of each AO in a hybrid is given by the square of the corresponding coefficient; it can be seen that each hybrid is one-third s and two-thirds p. All three hybrids are identical in shape, the shapes being similar to those of sp^3 hybrid AOs (Fig. 1.19). The axes of the hybrids in the xy plane are at angles of 120° (Fig. 1.29b) and they can be interconverted by rotation about the z axis through $\pm 120°$. They are therefore called sp^2 hybrid AOs.

As Fig. 1.29(c) shows, we can set up the sp^2 hybrids and $1s$ hydrogen AOs in such a way that they overlap in pairs. The arguments of the previous section show that we can then get a good estimate of the contribution of the σ electrons to the collective properties of ethylene by an analog model in which the σ electrons are "localized" in two-center bonds. There is, however, a complication. We cannot use the CH and CC bond energies appropriate to paraffins because the carbon AOs are not the same. In paraffins, they are sp^3 hybrids; here they are sp^2 hybrids. These differ not only in shape but also in energy. The $2s$ AO of carbon lies well below the $2p$ AOs. A hybrid AO lies between. The nearer it lies to s, the greater is the contribution of s to it. An sp^3 AO, being only one-quarter s, therefore lies higher in energy than an sp^2 AO, which is one-third s. Since the $2s$ AO has a lower energy than a $2p$ AO, electrons in it must be more tightly bound. The $2s$ AO must therefore also be smaller than a $2p$ AO. One would on this basis expect bonds formed by sp^2 hybrid AO's to be shorter and stronger than those formed by sp^3 hybrids.* The CH bonds in ethylene are indeed shorter (1.085 Å) than those in paraffins (1.100 Å) and are also stronger; the energy required to break a

---

* See equation (1.18). The strength of a bond depends on the value of β, which is numerically greater, the greater are the ionization potentials of the component AOs. The electrons in an sp^2 AO are more tightly bound than those in an sp^3 AO, so sp^2 bonds should be stronger.

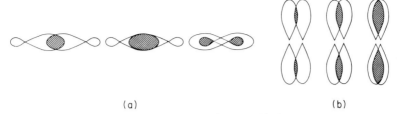

FIGURE 1.30. Effect of internuclear separation on orbital overlap (a) in a σ bond formed by two hybrid AOs; (b) in a π bond.

CH bond in ethylene (105 kcal/mole), i.e., the *bond dissociation energy*, is considerably higher than that for a CH bond in a paraffin (98 kcal/mole).

The CC bond in ethylene differs in another respect from a CC bond in a paraffin. It is a double bond, composed of a σ bond plus a π bond. Now one can easily see that two hybrid AOs on two adjacent carbon atoms will overlap best at some fixed internuclear distance (Fig. 1.30a). On the other hand, the overlap in a π MO increases steadily with decreasing internuclear separation (Fig. 1.30b). In ethylene, a compromise is reached where the σ bond is compressed to less than its optimum length while the π bond fails to satisfy its ambition to join the nuclei into one. The resulting bond is much shorter (1.34 Å) than one would expect from changes in hybridization alone. Hybridization should reduce the length of the CC σ bond only by 0.03 Å from its value (1.54 Å) in paraffins. The rest of the shortening in ethylene is due to the boa constrictor-like action of the π electrons.

While this localized bond or "equivalent orbital" description* of ethylene is the most convenient for practical chemical purposes, it is worthwhile to consider the "correct" MO description. In the notation of Fig. 1.28, both the xy and xz planes are planes of symmetry. Only the AOs z_1 and z_2 are antisymmetric for reflection in xz, while s_1, s_2, x_1, and x_2 are symmetric. The hydrogen AOs a–d are neither symmetric nor antisymmetric; however, the combinations (SOs) $(a + b)$ and $(c + d)$ are symmetric for reflection in xz, while $(a - b)$ and $(c - d)$ are antisymmetric.

The antisymmetric MOs are then combinations of the AOs y_1 and y_2, and the SOs $(a - b)$ and $(c - d)$. Since CH bonds are stronger than single CC bonds, CH interactions are presumably stronger than CC interactions. Let us first ignore the weaker $y_1 - y_2$ interaction in comparison with the $y_1 - (a - b)$ and $y_2 - (c - d)$ interactions and then use perturbation theory to reintroduce the $y_1 - y_2$ interactions.

As Fig. 1.31(a) shows, the AOs y_1 and y_2 do in fact overlap efficiently with the SOs $(a - b)$ and $(c - d)$. This interaction should lead to two strongly bonding MOs and two strongly antibonding MOs (Fig. 1.31b). Introducing the π interaction between y_1 and y_2, each of these pairs of degenerate MOs

---

* The expressive term "equivalent orbital" was introduced by Lennard-Jones to describe the two-center MOs into which the "correct" MOs can often be transformed by procedures analogous to that used for methane in Section 1.10.

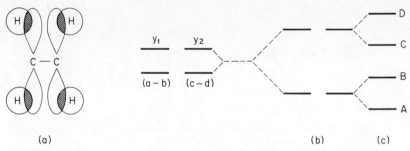

FIGURE 1.31. (a) Overlap of AOs in π_y system of ethylene; (b) effect of CH interactions; (c) further effect of CC interactions.

splits into two. In the lower member of each pair, the CC interaction is bonding; in the upper member, it is antibonding. These perturbed MOs are, respectively, symmetric and antisymmetric for reflection in the third plane of symmetry in ethylene, bisecting the CC bond. Since π interactions are smaller than σ ones, both the MOs (A and B) arising from the bonding CH level should remain bonding (Fig. 1.31c).

Since CH bonds are stronger than CC bonds, and since σ bonds are stronger (as a rule) than π bonds, we could expect *both* the MOs A and B in Fig. 1.31 to lie below the bonding π_z MO in energy.

Next let us consider the symmetric MOs. As in the case of diatomic molecules, we may expect the combinations of $2s$ AOs ($s_1 \pm s_2$) to lie below any of the rest. We are then left with four orbitals, namely, the SOs x_1 and x_2 and the SOs $(a + b)$ and $(c + d)$. In this case the CH interactions are less than in Fig. 1.31 since the carbon $2p_x$ AOs obviously overlap less well with the hydrogen $1s$ AOs (Fig. 1.32a). Equally, the CC interaction should be greater. This suggests (Fig. 1.32b, c) that only one strongly bonding MO should arise from this set of AOs and SOs, two being weakly bonding or antibonding and one very strongly antibonding. Since the first two AOs will

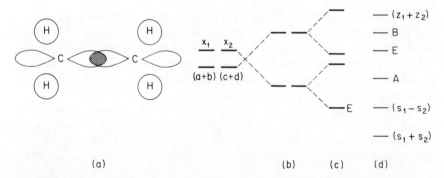

FIGURE 1.32. (a) Overlap of hydrogen AOs with x_1 and x_2; (b) two-center MOs arising from CH σ interactions; (c) effect of $x_1\,x_2$ interaction; (d) other MOs of C_2H_4 included for comparison.

Introduction to MO Theory

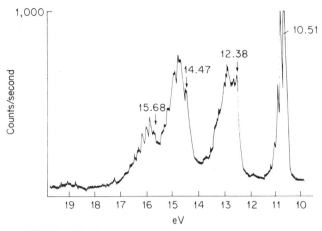

FIGURE 1.33. Photoelectron spectrum of ethylene (from Ref. 2).

be further raised in energy by interaction with $(s_1 \pm s_2)$, only one really bonding MO survives. Since this involves a weaker CH interaction, but a stronger CC interaction, than A or B of Fig. 1.31, it is reasonable to expect it to lie between them. The resulting orbital pattern is indicated in Fig. 1.32(d).

The 12 valence electrons of C_2H_4 should fit into the six MOs indicated in Fig. 1.32(d). In increasing order of energy, these are the 2s bonding MO $(s_1 + s_2)$, the 2s antibonding MO $(s_1 - s_2)$, the CH bonding–CC bonding π_y-type MO A, the CH bonding–CC bonding σ-type MO E, the CH bonding–CC antibonding π_y-type MO B, and the bonding π_z MO.

Figure 1.33 shows the 21.22-eV photoelectron spectra of ethylene. There are four bands only because the MOs $(s_1 \pm s_2)$ have high ionization potentials and can be seen only with higher-energy light sources. The other four bands apparently correspond to the order predicted in Fig. 1.32. The orbital of lowest energy involves removal of an electron from an orbital located solely between the carbon atoms; removal of an electron from this should excite CC vibrations only. This is the case. Ionization from the other MOs leads to weakening of CH bonds, so CH vibrations are also excited. A detailed study of the vibrational structure of the bonds shows that our assignment of ionization potentials is correct.

1.14 Acetylene

Acetylene, C_2H_2, is linear (Fig. 1.34a). The arguments used for diatomic molecules (Section 1.11) show that the $2p_y$ and $2p_z$ as AOs y_1, y_2, z_1, and z_2 must combine to form two degenerate bonding π MOs, $(y_1 + y_2)$ and $(z_1 + z_2)$, and two degenerate antibonding MOs, $(y_1 - y_2)$ and $(z_1 - z_2)$.

As before, the two MOs of lowest energy should be the combinations of 2s AOs, $(s_1 \pm s_2)$. The remaining MOs will be of σ type, formed from

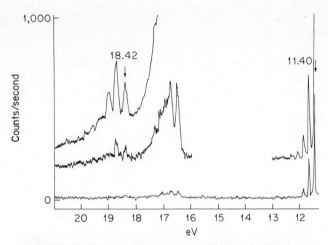

FIGURE 1.34. Photoelectron spectrum of C_2H_2 (from Ref. 2).

the two hydrogen 1s AOs and the $2p_x$ AOs x_1 and x_2. As in the case of ethylene, we would expect to get just one strongly bonding MO, this being both CH bonding and CC bonding and lying between $(s_1 - s_2)$ and the bonding π MOs (cf. A in Fig. 1.32d). The photoelectron spectrum of acetylene should then show four bands with intensities in the ratio 2:1:1:1. The band of lowest ionization potential corresponds to the degenerate π level, followed by the bonding p CH level, followed by $(s_1 - s_2)$, and concluded by $(s_1 + s_2)$. This is exactly what is observed (Fig. 1.34). The spectrum shows only three bands because $(s_1 + s_2)$ can be seen only with a more energetic light source.

The localized bond model for acetylene is precisely analogous to that for ethylene. The π MOs $(y_1 + y_2)$ and $(z_1 + z_2)$ survive as before. Each carbon atom has two AOs left over. We combine them into a pair of sp hybrids, $(s_1 \pm x_1)$ and $(s_2 \pm x_2)$ (Fig. 1.35a), pointing in opposite directions along the x axis. These hybrid AOs are used to form localized CH bonds to the hydrogen atoms, and a localized CC σ bond.

The arguments used for ethylene imply that the CH bonds in acetylene should be even shorter and stronger. This is the case (length 1.057 Å and bond dissociation energy greater than in ethylene). Furthermore, here there are

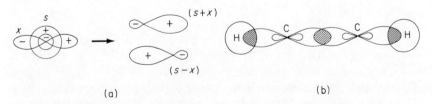

FIGURE 1.35. (a) Formation of sp hybrids; (b) σ bonds in acetylene.

two CC π bonds fighting one CC σ bond. The CC bond length should be correspondingly reduced—and it is (1.20 Å). A final point is that the CC π bonds in acetylene should be stronger than in ethylene since they are shorter and consequently involve better orbital overlap. The first ionization potential of acetylene (11.41 eV), corresponding to removal of a π electron, is indeed much greater than the corresponding value (10.51 eV) for ethylene.

1.15 Breakdown of the Localized Bond Model: Three-Center Bonds, Conjugated Molecules, and Reaction Intermediates

While the majority of bonding situations can be represented in terms of "localized" bonds for reasons indicated in the preceding sections, there are a number of exceptions where bond localization breaks down. The most clear-cut case of this is provided by "electron-deficient" molecules, where there are simply not enough valence electrons to form enough two-center bonds to hold the atoms together. The classic example of this is diborane, B_2H_6. At least seven "classical" bonds would be needed to hold the eight atoms in diborane together. However, boron contains only three valence electrons and hydrogen one, so the total number of valence electrons is 12, enough for only six electron pair bonds. Clearly, some of the electrons must be working overtime.

The structure of diborane and the type of bonding in it are indicated in Fig. 1.36.* The structure (Fig. 1.36a) resembles that of ethylene with the carbon atoms replaced by boron and pulled apart, the two extra hydrogens lying symmetrically between the borons, above and below the B_2H_4 plane. One can write a "localized orbital" picture (Fig. 1.36b) in which each boron has sp^3 hybridization. Two hybrid AOs of each boron are used to form normal two-center bonds to the terminal hydrogen atoms, while the other hybrid AOs overlap simultaneously in pairs with each other and with the 1s AOs of the central hydrogen atoms. The latter overlap gives rise to a set of *three-center* MOs. Only one of these is bonding. There are therefore just six bonding MOs in the molecule, the four bonding MOs of the terminal CH bonds and the two bonding three-center MOs. These are just enough to hold the 12 valence electrons. Diborane therefore contains a new kind of bond, a *three-center bond*, in which a single pair of electrons bonds three atoms.

Another situation where bond localization fails is illustrated by benzene, C_6H_6. In benzene the 12 atoms lie in a plane. Each carbon is attached to a hydrogen atom and the six carbon atoms lie at the vertices of a regular hexagon (Fig. 1.37a). Analogy with ethylene suggests a localized

---

* The structure of diborane was deduced by Longuet-Higgins while an undergraduate in his second year at Oxford University and the type of bonding in it was also deduced by him a few years later.

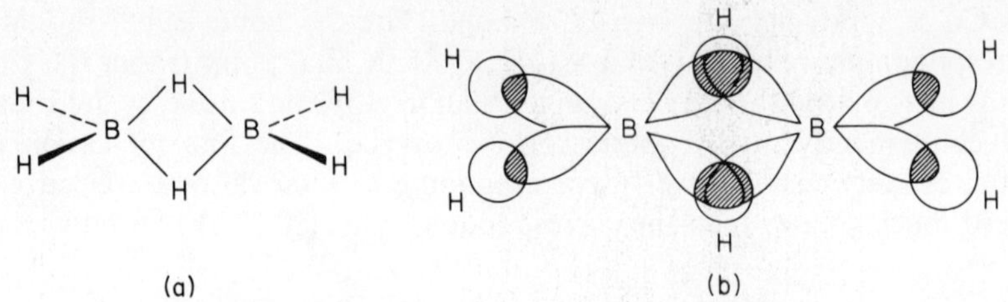

FIGURE 1.36. (a) Structure and (b) MOs in diborane.

bond picture in which each carbon has sp^2 hybridization, being linked to its three neighbors by σ bonds. The fourth $2p$ AO can then be used to form π bonds between the carbon atoms (Fig. 1.37b). When, however, we try to pair up the $2p$ AOs into π bonds, problems arise because we can pair them in two ways (Fig. 1.37c). Each of these would have alternating short (C=C) and long (C—C) bonds, but the bonds that are short in one structure are long in the other and vice versa. In practice benzene solves this dilemma by a compromise, adopting an intermediate structure in which all the C—C bonds have the same length (1.40 Å), intermediate between the lengths of double (1.34 Å) and single (1.48 Å) bonds.* Clearly, one cannot represent benzene by a *classical structure* with localized two-center bonds. The π bonds in it must be formed by electrons occupying six-center π MOs formed from all six $2p$ AOs of the carbon atoms. Similar problems arise in other *conjugated molecules*, i.e., molecules where one has a continuous chain of atoms linked by π bonds. Apart from the special problems that arise in the case of molecules such as benzene where there are two or more possible classical structures, so that a representation in terms of two-center bonds would be ambiguous, one can see that difficulties will in any case arise due to the greater overlap of adjacent π bonds in a conjugated system. In planar butadiene (Fig. 1.38), for example, the overlap between the central pair of $2p$ AOs is clearly comparable with that between each end pair. The arguments for bond localization therefore become much less convincing than in systems where the "localized" bond orbitals overlap only weakly with one another.

A third situation where bond localization must fail is in the intermediate phases of reactions where bonds are simultaneously formed and broken. Consider, for example, the reaction of methyl chloride with iodide ion to give methyl iodide and chloride ion:

$$I^- + CH_3Cl \longrightarrow ICH_3 + Cl^- \tag{1.41}$$

The experimental evidence shows that this is a concerted process in which the new (CI) bond begins to form before the old (CCl) bond has broken. In the intermediate phases of the reaction, carbon is therefore bonded simultaneously to five different atoms. Such a situation cannot be represented in

---

* The lengths of a single C—C bond formed by sp^2 hybrid AOs are in the range of 1.46–1.48 Å; see Chapter 3.

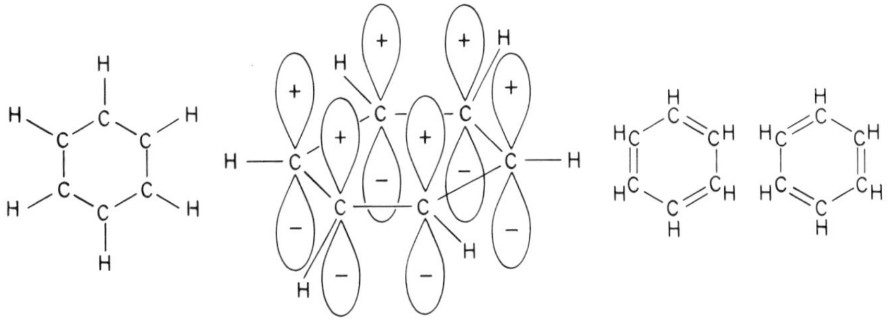

Introduction to MO Theory

FIGURE 1.37. (a) Structure of benzene; (b) orbitals in benzene; (c) classical structures for benzene.

terms of localized two-center bonds because carbon has only four valence shell orbitals for bonding.

In this case, the geometry of the transition state for the reaction has been shown to be that indicated in Fig. 1.39(a). The hydrogen atoms lie approximately in a plane while the carbon, chlorine, and iodine lie in line. An obvious localized orbital interpretation of this structure is that indicated in Fig. 1.39(b). The carbon atom forms σ bonds to the hydrogen atoms by using three sp^2 hybrid AOs. The fourth ($2p$) AO is used to bond the entering and leaving atoms. The bonds to chlorine and bromine are three-center bonds formed by overlapping of the carbon $2p$ AO with AOs of chlorine and iodine. The situation is, however, different from that in diborane in that the chlorine and iodine AOs do not overlap with one another. In diborane, the three-center bonds are formed by simultaneous overlap of three AOs with one another. In diborane all the three-center MOs but one are antibonding. In the transition state of Fig. 1.39, there is still only one bonding MO but there is a second MO which, while not bonding, is not antibonding either. There is therefore room for four electrons in MOs that are not actually offensive. This is fortunate, for a head count shows that all the valence AOs in the I^- ion are filled, so that there are four electrons to be accommodated in the three-center MOs of the transition state, two from the I^- AO and two from the CCl bond orbital.

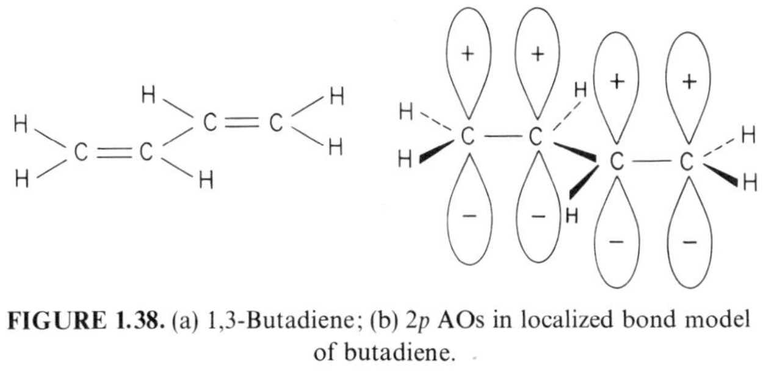

FIGURE 1.38. (a) 1,3-Butadiene; (b) $2p$ AOs in localized bond model of butadiene.

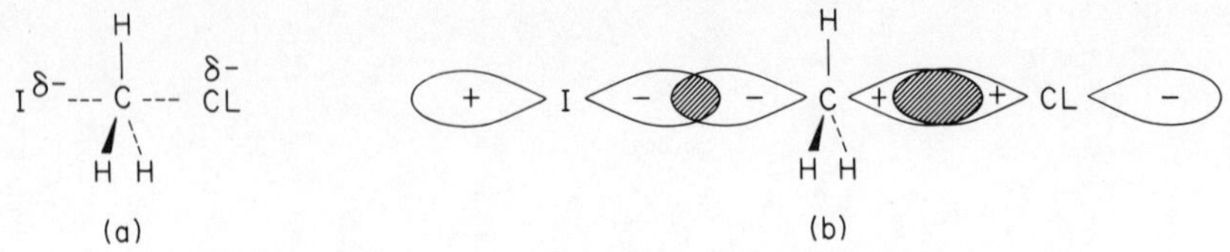

FIGURE 1.39. (a) Geometry of and (b) orbitals in the transition state for the reaction $I^- + CH_3Cl \longrightarrow ICH_3 + Cl^-$.

Mass spectral studies indicate that the three-center bond stabilizes the structure $(I \cdots CH_3 \cdots Cl)^-$ by 9 kcal/mole relative to its components $(CH_3Cl + I^-)$.

1.16 Relationships between Different Types of Delocalized Systems

In the previous section, we met three situations where the localized bond model fails. In each case, we had to represent bonding in terms of MOs covering three or more atoms, formed by overlapping of AOs on them. Now the general rules for formation of such many-center MOs are basically the same as for normal two-center bonds. The AOs involved must be of comparable energies (i.e., all from valence shell AOs of the participating atoms) and they must overlap in space. How they overlap is not important. There is no basic distinction between the π-type overlap of p AOs in benzene (Fig. 1.37), the linear σ-type overlap in the transition state of Fig. 1.39, and the nonlinear σ-type overlap of AOs in diborane (Fig. 1.36). All three situations are equivalent in terms of MO theory. The distinction between them arises solely from quantitative considerations of the efficiency of overlap and the energies of the orbitals involved.

Thus the three-center overlap in diborane is exactly analogous to the overlap of 2p AOs in the cyclopropenium ion $C_3H_3^+$ (Fig. 1.40a). This has a planar triangular structure analogous to that of benzene in which each carbon atom uses three sp^2 AOs to form σ bonds to its neighbors and the fourth 2p AO to form π bonds. Each 2p AO overlaps with both the others. The π MOs in cyclopropenium are therefore *topologically equivalent* to, or *isoconjugate with*, the three-center MOs in diborane (Fig. 1.40b, c). Likewise, the I—C—Cl bonding situation in the transition state of Figure 1.39 is topologically equivalent to, or isoconjugate with, the π MOs of the allyl anion, $(CH_2CHCH_2)^-$, in which four electrons occupy the three-center π MOs formed by overlap of the 2p AOs of the three carbon atoms (Fig. 1.40d). Here the 2p AO of the central carbon atom overlaps with the 2p AOs of the terminal ones, but the latter do not overlap with one another. In the $(I \cdots CH_3 \cdots Cl)^-$ transition state, the carbon 2p AO likewise overlaps with AOs of I and Cl, which do not overlap with one another (Fig. 1.40e, f).

In discussing the topology of orbital interactions, as in the examples

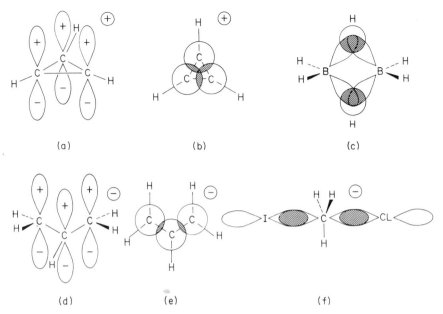

FIGURE 1.40. (a, b) Overlap of carbon $2p$ AOs in cyclopropenium, seen from (a) the side and (b) above; (c) overlap of B and H AOs in three-center bonds of diborane; (d, e) overlap of carbon $2p$ AOs in the allyl anion, seen from (d) the side and (e) above; overlap of carbon $2p$ AO with AOs of I and Cl in the transition state of Fig. 1.39.

above, it is important to remember that the sign of a wave function is arbitrary. The only physical significance of a wave function is its measure of the probability of finding particles at various points in space. This probability is given by the square ψ^2 of the wave function ψ, which of course is the same for its negative $[\psi^2 = (-\psi)^2]$. The choice of signs of AOs in constructing MOs is correspondingly arbitrary. The MOs we get from a given set of AOs is, however, *not* arbitrary. Consider the combination of two hydrogen atoms. If we write their $1s$ AOs ϕ_1 and ϕ_2 as positive, we will find that these are replaced in H_2 by two MOs, a bonding MO ($\phi_1 + \phi_2$) and an antibonding MO ($\phi_1 - \phi_2$). Suppose now that we choose ϕ_1 to be positive and ϕ_2 to be negative. In H_2, the MOs will still be ($\phi_1 \pm \phi_2$), but now ($\phi_1 - \phi_2$) will be the bonding one and ($\phi_1 + \phi_2$) the antibonding one.

Since the choice of signs or *phases* of our *basis set* AOs that we use to construct MOs is arbitrary, it is convenient to use some convention in choosing them to avoid unnecessary confusion. An obvious choice is to pick the phases in such a way that AOs overlap in phase, i.e., positive parts of AOs overlap with positive parts and negative parts with negative parts.

In most cases, we can carry out this program completely so that there are no *phase dislocations* where AOs overlap out of phase with one another; this was true in the cases indicated in Figs. 1.36–1.40. As indicated, we can pick the signs (phases) of the AOs so that positive overlaps with positive and

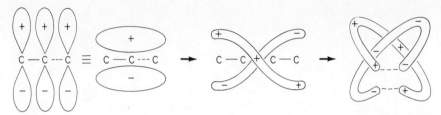

FIGURE 1.41. Construction of a "Möbius strip" system from a chain of conjugated carbon atoms.

negative with negative. In the transition state of Fig. 1.39, this involves opposite signs for the AOs of I and Cl because they overlap with different lobes of the carbon 2p AO and the lobes of a p AO have opposite signs.

There are, however, situations where we cannot avoid phase dislocations, no matter how hard we try. One of the most entertaining of these was suggested first by Heilbronner. Suppose we take a long, conjugated chain of carbon atoms in which the 2p AOs of the individual atoms fuse into long π MOs (Fig. 1.41a), twist this through 180° (Fig. 1.41b), and join the ends up to give a conjugated ring. Obviously there must be a phase dislocation at some point in the resulting cyclic π system (Fig. 1.41c). In a "normal" planar, cyclic, conjugated molecule such as benzene or cyclopropenium (Figs. 1.37 and 1.40), the parallel p AOs fuse into cyclic π MOs composed of two cyclic lobes lying on opposite sides of the nuclei.

No such "Möbius strip" compound has as yet been synthesized, but the concept underlying it is very important. A system in which there is an inevitable phase dislocation is topologically distinct from one in which all phase dislocations can be avoided. Note that this distinction depends on the topology of overlap of the AOs involved in forming a delocalized system; it is quite independent of the MOs that can be constructed from them.

1.17 Summary

In this chapter, we have developed a simple picture of molecules in terms of the orbital approximation. This had to be introduced because the Schrödinger equation cannot be solved exactly for many-electron systems. We first developed a picture of atoms in terms of one-electron functions or atomic orbitals (AO), these being similar to the orbitals of the hydrogen atom but chosen to allow as well as possible for the interelectronic repulsions. In effect, we replace these by average values, the repulsion between two electrons being treated as a repulsion between the two clouds of negative charge that represent the time-averaged motion of the electrons. Atoms are built up by feeding electrons into the AOs from the bottom upward, remembering that no more than two electrons can fit into a given AO and then only if they have opposite signs (Pauli's principle). When there is ambiguity

Introduction to MO Theory

due to two or more AOs having identical energies (i.e., being degenerate), the state of the lowest energy is that of maximum multiplicity (Hund's rule).

The electrons in molecules can be likewise represented as occupying molecular orbitals (MO) that can be derived from combinations of AOs of the participating atoms. The important factors in deriving MOs from AOs are (1) the extent of overlap and (2) the phase topology of the overlap. The AOs used to form MOs must be of similar energies (sizes) if significant overlap is to be obtained. The MOs of a molecule must conform to its symmetry. The AOs used to form MOs of a given symmetry type must belong to the same symmetry type. Wave functions for a molecule can generally be treated as if they were independent of nuclear vibronic motions (Born–Oppenheimer approximation). Transitions between different electronic states caused by the absorption or emission of light occur with virtually no change in the nuclear coordinates or momenta (Franck–Condon principle). Orbital degeneracies in polyatomic systems will be removed by vibronic dislocations (Jahn–Teller effect).

Photoelectron spectroscopy has confirmed the essential features of the MO description of bonding in simple molecules; however, the proper MOs are not always easy to visualize. A localized bond representation, involving hybrid AOs which overlap little with each other, accounts for most of the chemically important properties of most molecules and the localized bonds are easy to visualize. The localized bond picture can be related to the proper (approximate) description through perturbation theory. The localized bond model is generally not applicable to electron-deficient molecules, conjugated systems, or transition states. The application of perturbation theory to the description of these three cases where the localized bond model breaks down is the subject of the following chapters.

Problems

1.1. Give brief definitions for the following terms:
  a. Atomic orbital.
  b. Molecular orbital.
  c. Basis set.
  d. Normalized orbital.
  e. Born–Oppenheimer approximation.
  f. Franck–Condon principle.
  g. Phase dislocation.
  h. Hybrid AO.
  i. Symmetry orbital.
  j. Localized bond.
  k. Degeneracy.
  l. Jahn–Teller effect.

1.2. Calculate the de Broglie wavelength of an electron with an energy of 500 eV.

1.3. Draw MO energy diagrams for BeO, CO, and NO. Indicate the expected ground-state multiplicity for each of these molecules.

1.4. Following the Pauli principle and Hund's rule, give the orbital configuration and multiplicities for the following atoms: S, Ca, Fe, Br.

1.5. $PbBr_2$ is considerably more stable than $PbBr_4$. What does this say about (a) the strength of bonds made with $6s$ orbitals as compared to $6p$ orbitals, and (b) the relative energy gap between $6s$ and $6p$ orbitals as compared to the same gap for the carbon $2s$ and $2p$ orbitals?

1.6. Construct symmetry orbitals for diborane. B_2H_6 has three symmetry planes (see p. 48).

1.7. Construct the MOs for diborane.

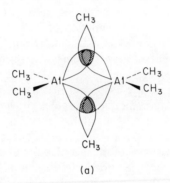

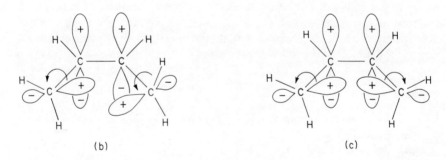

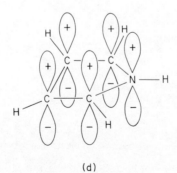

FIGURE 1.42. See Problem 1.9.

1.8. How many bands should there be in the complete photoelectron spectrum of diborane? Give their intensities.

1.9. Draw orbital representations of isoconjugate hydrocarbon π-electron systems for each of the systems in Fig. 1.42.

1.10. Why is it that a Jahn–Teller distortion cannot remove the degeneracy in ground-state O_2 to make the ground state a singlet instead of a triplet? The energy gap between the ground-state O_2 triplet and its first excited singlet is 0.98 eV, which gives some measure of the importance of Hund's rule effects (electron correlation) in molecular systems.

Selected Reading

Two general elementary-level introductory books to MO theory with, respectively, inorganic and organic chemical biases are:

H. B. GRAY, *Electrons and Chemical Bonding*, W. A. Benjamin, New York, 1965.

A. LIBERLES, *Introduction to Molecular Orbital Theory*, Holt, Rinehart and Winston, New York, 1966.

For introductory material at a more advanced level, see:

M. J. S. DEWAR, *The Molecular Orbital Theory of Organic Chemistry*, McGraw-Hill, New York, 1969.

S. R. LAPAGLIA, *Introduction to Quantum Chemistry*, Harper & Row, New York, 1971.

F. L. PILAR, *Elementary Quantum Chemistry*, McGraw-Hill, New York, 1968.

References

1. B. P. PULLEN, T. A. CARLSON, W. E. MODDEMAN, G. K. SCHWETZER, W. E. BULL, AND F. A. GRIMM, *J. Chem. Phys.* **53**, 768 (1970).

2. D. W. TURNER, C. BAKER, A. D. BAKER, AND C. R. BUNDLE, *Molecular Photoelectron Spectroscopy*, John Wiley and Sons, New York, 1970.

CHAPTER 2

Perturbation Theory*

2.1 The Usefulness of Perturbation Theory

Chemical behavior is determined by relatively small differences in energy between pairs of closely related systems. Thus the point of equilibrium in a reversible reaction depends on the difference in energy between the reactants and products, while the rate of an irreversible reaction depends on the difference in energy between the reactants and the transition state.[†] In order to predict the rates and equilibria of chemical reactions, we must be able to estimate these differences in energy with reasonable accuracy. This unfortunately presents a very formidable problem because the differences are so small.

Consider the change in energy involved in the most extreme chemical process an organic molecule can undergo, i.e., complete dissociation into atoms. Using benzene (C_6H_6) as an example, when six carbon nuclei, six protons, and 42 electrons combine to form six carbon atoms and six hydrogen atoms, the total energy liberated is 6230 eV or 144,000 kcal/mole. This is more than 100 times the energy (57 eV, 1330 kcal/mole) required to dissociate benzene into atoms. If we want to estimate the heat of atomization of benzene with reasonable accuracy, say $\pm 2\%$, i.e., ± 25 kcal/mole, we will have to calculate the total energy of benzene and of the atoms from which it is formed to ± 12 kcal/mole or $\pm 0.01\%$. Indeed, the accuracy we need in predicting chemical reactivity is at least ten times greater than this because an error of

---

* In perturbation theory, the effects of small changes, perturbations, are calculated directly by assuming values for the properties of unperturbed systems and directly evaluating the effect of the small change.
[†] The transition state theory of reaction rates is discussed in Chapter 5.

only 1 kcal/mole in the difference in energy between reactants and products, or reactants and transition state, means an error of nearly a factor of ten in the corresponding equilibrium or rate constant. In order to get results accurate enough to serve in the interpretation of chemical phenomena, we would then have to estimate the energies of atoms and molecules to one part in one million.

While progress has been made recently toward the solution of this problem, the methods used are complex and expensive, involving the use of large digital computers. As practicing chemists, we need to develop a simpler procedure that can be used without any special equipment and will at the same time lead to results accurate enough to form the basis of a general theory of chemical structure and reactivity. One can legitimately regard the whole of chemistry as an exercise in perturbation theory on the part of nature and we have indeed already made use of such an approach in several connections in Chapter 1 (Sections 1.7 and 1.12). In order for reaction energies to be significant with reference to the binding energies of electrons to atoms, we must look at hot plasmas, a situation in which all but atomic chemical distinctions are void. One of the conclusions reached in Chapter 1 was that most of the bonding in molecules can be interpreted in terms of localized two-center bonds, the properties of which are essentially independent of their environment. In molecules where all the bonding is of this type (e.g., the paraffins; Section 1.12), the heat of atomization ΔH_a can be written as a sum of bond energies; e.g., for

$$\begin{array}{c} \text{H} \quad \text{H} \\ | \quad \ | \\ \text{H—C—C—H} \longrightarrow 2\text{C} + 6\text{H} \\ | \quad \ | \\ \text{H} \quad \text{H} \end{array}$$

we have

$$\Delta H_a = E_{\text{C-C}} + 6E_{\text{C-H}} = 89 + 6(97) = 671 \text{ kcal/mole} \qquad (2.1)$$

Here $E_{\text{C-C}}$ and $E_{\text{C-H}}$ are the bond energies of the carbon–carbon and carbon–hydrogen single bonds, the bond energies being the same in all molecules where such bonds occur. The bond energies can be found empirically from the measured heats of atomization of a suitable set of molecules; a number of authors have compiled tables of such bond energies.* These can be used to estimate differences in energy between different molecules in which the differences involve only localized bonds. Thus in the hydrogenation of ethylene to ethane, where a C=C double bond and an H—H single bond are

---

* See, for example, MORTIMER.[1]

replaced by two C—H single bonds, we can calculate the heat of reaction ΔH as a perturbation, using the above procedure; thus, for

$$H_2C{=}CH_2 + H_2 \longrightarrow H_3C{-}CH_3$$

we have

$$\Delta H = E_{C=C} + E_{HH} - 2E_{CH} - E_{C-C}$$
$$\Delta H_{calc} = 148.8 + 104.2 - 2(97) - 89 = 33.0 \text{ kcal/mole} \qquad (2.2)$$
$$\Delta H_{direct} = 32.82 \text{ kcal/mole}$$

These simple deductions from perturbation theory then lead to a procedure that solves many of our problems in a very simple manner. We are, however, left with the problem of molecules where the bonding cannot be described completely in terms of localized two-center bonds (Section 1.16). Our next object is to develop a simple way to estimate the differences in energy between molecules due to differences in the energy of the "delocalized" electrons. Again we will use perturbation theory and for convenience we will discuss its application to one particular type of delocalized system, i.e., conjugated molecules where the delocalized electrons occupy π MOs formed by interaction of p AOs of the participating atoms. The principles reached in this way will be equally applicable to delocalized systems of all kinds since, as we have already seen (Section 1.16), there is no *qualitative* difference between different types of delocalized systems.

2.2 Types of Perturbations Involved in the Comparison of Conjugated Systems

Two conjugated systems can differ in any of three respects. The atoms involved may be different; the topology of overlap may be different; and the total number of conjugated atoms may be different. We can then express any overall difference as a sum of individual *perturbations* of three kinds. First, *monocentric perturbations*, in which a given atom is altered or replaced by some other atom; second, *intramolecular* perturbations, in which we alter the connectivity of the conjugated system; and third, *intermolecular* perturbations, in which two smaller conjugated systems unite to form a larger system.

The first type of perturbation, a monocentric perturbation, is seen in the protonation of pyridine (1) to pyridinium ion (2):

Both (1) and (2) have structures similar to that of benzene (Fig. 1.37), with six π electrons occupying six-center π MOs. Indeed, since the ion N^+ and the atom C are isoelectronic,* each having six electrons, (2) is isoelectronic with benzene. In passing from (1) to (2) a new σ bond is formed; the corresponding change in energy will be equal to the NH bond energy E_{NH}. However, there will also be a change in the energy of the π electrons because the nitrogen atom in (2) has a greater affinity for electrons than the carbon atom in (1) (due to the formal positive charge). Since the MOs include a $2p$ AO of nitrogen, their binding energy will be correspondingly greater in (2) than in (1). The change in σ-bonding energy can be estimated from the corresponding bond energies and the change in π-bonding energy can be calculated directly by the methods developed in Section 2.3.

Intramolecular perturbations involve changes in the connectivity of the atoms in a molecule. Chemical bonding depends on the mutual overlapping of AOs of the atoms involved. In the case of a π bond, the atoms concerned cannot get close enough together for their orbitals to overlap effectively unless they are *also* linked by a σ bond. By forming or breaking σ bonds between atoms in a conjugated system, one can consequently alter the connectivity of the π system, giving rise to an intramolecular perturbation. A simple example is provided by 1,3,5-hexatriene (3), in which there is an open-chain, six-atom, conjugated system (4). We can convert this into any of four cyclic conjugated systems (5)–(8) by forming σ bonds between appropriate pairs of carbon atoms. This process, in which two hydrogen atoms are removed and a C—C link formed, is described as *union* and represented by the symbol ←u→ [see (5)–(8)]. When, as here, the union is between two atoms in a single conjugated system, it is described as *intramolecular union*. Procedures for calculating changes in π-electronic energies which result from intramolecular union are developed in Section 2.4.

$$H_2C=CH-CH=CH-CH=CH_2$$

(3)

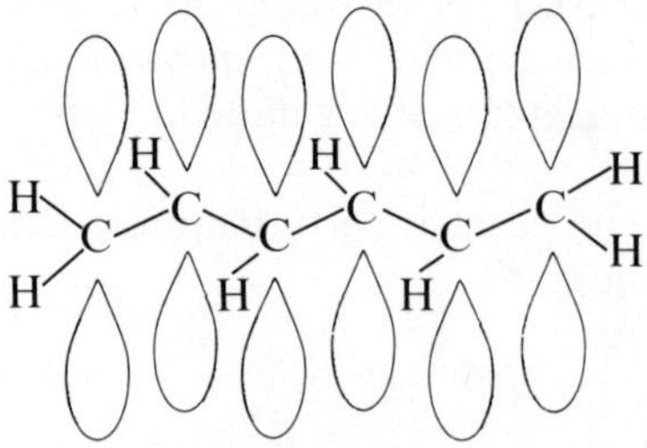

(4) 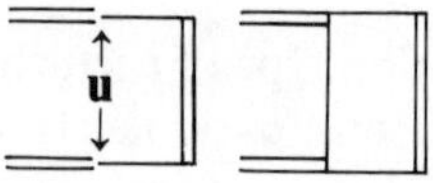   (5)

---

* Isoelectronic refers to atoms or structures that have the same number of valence level electrons; e.g., H_3C-CH_3 is isoelectronic with $H_3B \leftarrow NH_3$.

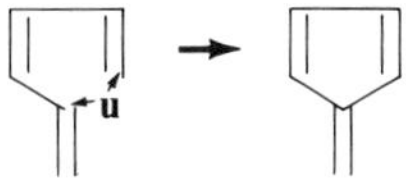

Perturbation Theory

(6)

(7)

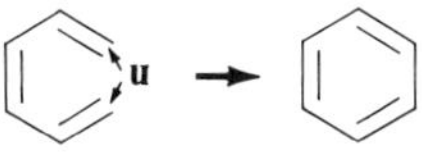

(8)

The third type of perturbation, *intermolecular perturbation*, is illustrated by the coupling of two molecules of benzene (9) to give biphenyl (10); this reaction is used in the manufacture of biphenyl:

$$2 \, C_6H_6 \rightarrow C_6H_5{-}C_6H_5 + H_2 \qquad (2.3)$$

(9)   (10)

During the reaction, two CH bonds are broken and a CC σ bond and an HH bond are formed. The corresponding contribution to the heat of reaction is $2E_{CH} - E_{C-C} - E_{HH}$. However, there is also a change in π energy due to the coalescense of the two six-atom π systems of the molecules of benzene into a single 12-atom π system. In the terminology of the previous paragraph, this change can be represented as an *intermolecular union* between the two molecules of benzene:

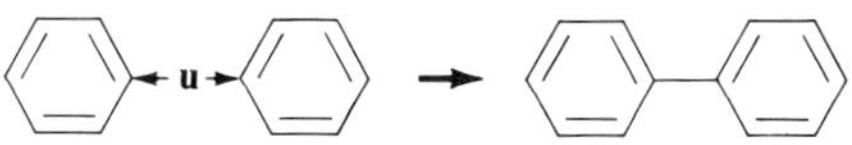

Intermolecular union is treated in Section 2.5.

Once we have developed the means for estimating the changes in π energy due to these three types of perturbation, we will be able to calculate differences in total energy between pairs of molecules or collections of molecules of all kinds, which is the substance of chemical theory.

2.3 Monocentric Perturbations

To develop a quantitative treatment of the perturbations introduced in the previous section, we must examine the effect of these perturbations on the form of the wave functions for the molecules in question. Consider a conjugated system in which a series of atoms are linked by π bonds involving many-center π MOs. In our orbital approximation, each MO ψ_μ will be written as a linear combination of p AOs ϕ_i of the component atoms:

$$\psi_\mu = \sum_i a_{\mu i}\phi_i \qquad (2.4)$$

The corresponding electron density distribution is given by the square of the MO function ψ_μ,

$$\psi_\mu^2 = \left(\sum_i a_{\mu i}\phi_i\right)\left(\sum_j a_{\mu j}\phi_j\right) = \sum_i a_{\mu i}^2\phi_i^2 + 2\sum_{i<j}\sum a_{\mu i}a_{\mu j}\phi_i\phi_j \qquad (2.5)$$

The total π-electron density is given by summing the contributions of the individual π electrons. If there are n_μ electrons in the MO ψ_μ (from the Pauli principle, n_μ must be equal to 0, 1, or 2), then the total density is given by

$$\sum_\mu n_\mu \psi_\mu^2 = \sum_\mu \sum_i n_\mu a_{\mu i}^2 \phi_i^2 + 2\sum_\mu \sum_{i<j}\sum n_\mu a_{\mu i}a_{\mu j}\phi_i\phi_j \qquad (2.6)$$

The total electron distribution is thus dissected into terms $\sum_\mu n_\mu a_{\mu i}^2\phi_i^2$, representing the contribution of $\sum_\mu n_\mu a_{\mu i}^2$ electrons occupying the AO ϕ_i, and terms $\sum_\mu n_\mu a_{\mu i}a_{\mu j}\phi_i\phi_j$, representing the contribution of the "overlap cloud" $\phi_i\phi_j$ to the electron distribution (see Section 1.5). This is illustrated in Fig. 2.1.

The total energy of the electrons should also be dissectable into contributions by the two major terms in equation (2.6). Consider first the monocentric term. The term involving ϕ_i^2 represents the contribution of $\sum_\mu n_\mu a_{\mu i}^2$ electrons occupying a distribution equivalent to the AO ϕ_i. This is the π-electron density q_i of that position; i.e.,

$$q_i = \sum_\mu n_\mu a_{\mu i}^2 \qquad (2.7)$$

The electrons in this region are moving in a field similar to that which they would experience in an analogous isolated atom (see Section 1.5). If we

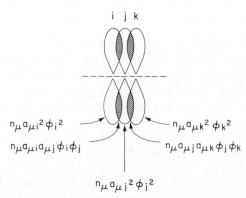

FIGURE 2.1. Dissection of the total π-electron distribution in a conjugated system into contributions by AOs and overlap clouds.

Perturbation Theory

denote the binding energy of an electron in the $2p$ AO of an isolated atom by α_i (the *Coulomb integral*), the total binding energy of the q_i electrons should then be $q_i\alpha_i$.

Consider now the bicentric term in equation (2.6), involving $\phi_i\phi_j$. If atoms i and j are not linked by a σ bond, the overlap of the $2p$ AOs ϕ_i and ϕ_j will be small and so will the product $\phi_i\phi_j$. Such terms may reasonably be neglected. If, on the other hand, atoms i and j are directly linked, the situation will be analogous to that in a compound containing an isolated C_i=C_j double bond. The π-bond energy in such a bond is due to the extra density in the internuclear region due to the "interference" effect represented by the term $\phi_i\phi_j$ [cf. equation (1.11)]. In the case of an isolated double bond, the term in the expression for the electron density between the two atoms will be $4a_{0i}a_{0j}$ [cf. again equation (1.11)] and the bond energy is then given by $4a_{0i}a_{0j}\beta_{ij}$ or simply $2\beta_{ij}$, where β_{ij} is the *resonance integral* between the AOs ϕ_i and ϕ_j (Section 1.5). In the case considered here, the corresponding term [equation (2.6)] is $2\sum_\mu n_\mu a_{\mu i}a_{\mu j}\phi_i\phi_j$. The corresponding contribution E_π^{ij} to to the total π-bond energy should then be

$$E_\pi^{ij} = 2p_{ij}\beta_{ij} \qquad (2.8)$$

where the *bond order* p_{ij} is defined by

$$p_{ij} = \sum_\mu n_\mu a_{\mu i}a_{\mu j} \qquad (2.9)$$

The bond order is a measure of the fraction of a whole π bond that is present between atoms i and j in our conjugated system. Note that there is no reason why p_{ij} should be positive. The wave function of a molecule in its ground state must obviously be such as to minimize the *total* energy. It is entirely possible that this condition may require an antibonding interaction between a given pair of atoms in it, this antibonding interaction being more than compensated by correspondingly greater increases in the bondingness of interactions elsewhere.

Using these results, we can now write a general expression for the total π-electron energy E_π of our conjugated system:

$$E_\pi = \sum_i q_i\alpha_i + 2\sum\sum_{i<j} p_{ij}\beta_{ij} \qquad (2.10)$$

Suppose now that we change one of the atoms k participating in our conjugated system, either by replacing it with a different atom or by attaching some substituent to it. The effect, so far as the π electrons are concerned, will be reflected solely by a change $\delta\alpha_k$ in the corresponding *Coulomb integral* α_k and by changes $\delta\beta_{kl}$ in the *resonance integrals* β_{kl} for pairs of atoms kl that are linked by a σ bond. The other parts of the π-electron cloud are too far away from k for their interactions with the "core" of the molecule to be affected.

There will also be small changes in the charge densities q_i and bond orders p_{ij}; since, however, the original values were such as to minimize the energy of the molecule with respect to changes in them, small changes in the

q's and p's will have no first-order effect on the total energy.* We can therefore, as a first approximation, assume that the electron distribution remains unchanged by the charge at atom k. This is a general result in first-order perturbation theory. The first approximation (i.e., the first-order perturbation) is found by assuming that the electron distribution remains unchanged by the perturbation, the change in energy being due to a change in the potential field in which the electrons move. Here the effect of the change in atom k is to alter the integrals α_k and β_{kl}.

Assuming the electron distribution to remain unchanged, the π-electron energy for the perturbed system, E_π', is then given by

$$E_\pi' = \sum_{i(\neq k)} q_i \alpha_i + q_k(\alpha_k + \delta\alpha_k) + 2\sum_{i(\neq k) < j(\neq k)}\sum p_{ij}\beta_{ij} + 2\sum_l p_{kl}(\beta_{kl} + \delta\beta_{kl}) \quad (2.11)$$

From equations (2.10) and (2.11), the change E in the total energy due to the change in atom k is then given by

$$\delta E_\pi = q_k \, \delta\alpha_k + 2\sum_l p_{kl} \, \delta\beta_{kl} \quad (2.12)$$

This then is the *first-order perturbation* due to the change in atom k. It is possible to calculate higher-order perturbations, but in the simple treatment that concerns us here there will be no need for these. In the case of the protonation of pyridine

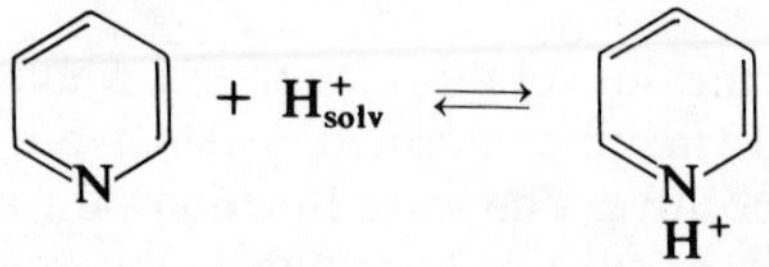

the change in bond energy for the reaction is given by

$$\Delta E_{\text{bond}} = E_{\text{N-H}} + \delta E_\pi = E_{\text{N-H}} + q_N \, \delta\alpha_N + 2P_{\text{C-N}} \, \delta\beta_{\text{C-N}}$$

It is possible to estimate charge densities and bond orders directly in systems like this by the methods of Chapter 3.

2.4 Intramolecular Union

The second type of perturbation, i.e., intramolecular union, can be treated in a manner exactly similar to that for the monocentric perturbations. We are concerned here with a process in which two conjugated atoms (k and l)

---

* The condition that E should be a minimum with respect to some variable x is that $\delta E_\pi/\delta x = 0$. If we change x by a small amount δx, the corresponding change in E_π is given, to a first approximation, by $(\delta E_\pi/\delta x)\,\delta x$ (chain rule for derivatives). Thus if E_π is minimized to x, a small change in x produces no first-order effect in E_π.

Perturbation Theory

which initially are not linked by a σ bond become so linked. The arguments used in Section 2.2 imply that this will result in a change in the corresponding resonance integral from zero in the unperturbed system to the value β_{kl} characteristic of adjacent atoms. If it can again be assumed as a first approximation that there is no corresponding change in the MOs ψ_μ, then it follows from equation (2.12) that the corresponding change δE_π in its energy will be given by

$$\delta E_\pi = 2p_{kl}\beta_{kl} \tag{2.13}$$

The perturbation treatment could easily be extended to higher-order terms, but these are rarely needed; the corresponding formulas are also less easy to derive by the kind of intuitive arguments we are using here. Note that the bond order between two atoms in a conjugated system has a definite value [see equation (2.9)] even if the atoms are not linked by a σ bond; thus the term "bond order" is not confined to pairs of atoms that make contributions to bonding. If the atoms are not linked by a σ bond, there is no contribution to bonding, but this is only because the corresponding resonance integral is then negligible.

In the intramolecular union of 1,3,5-hexatriene to give (5)–(8), the expression for the change in π-electron energy is the same in all cases, namely, (2.13). The 2—5 and 1—4 bond orders in 1,3,5-hexatriene are negative, so δE_π would be positive in the cases (5) and (1). The 1—5 bond order is zero, which accounts for the fact that fulvene (6) behaves like a strained polyene. The 1—6 bond order is positive, which accounts for the aromatic energy of benzene (8). These calculations are discussed in detail in Chapter 3.

2.5 Intermolecular Union

The last type of perturbation involves the intermolecular union of two conjugated molecules to form a single, large, conjugated system, as in the formation of biphenyl from benzene, equation (2.3). Consider first the general case of union of two conjugated systems R and S, by forming a single link between them, through atom r in R to atom s in S (Fig. 2.2). Denote the MOs of R by Φ_μ (of energy E_μ) and those of S by Ψ_ν (of energy F_ν); the MOs are expressed as usual by linear combinations of p AOs ϕ_i of atoms in R and ψ_j of atoms in S, namely

$$\Phi_\mu = \sum_i a_{\mu i}\phi_i \quad \text{energy } E_\mu \tag{2.14}$$

$$\Psi_\nu = \sum_j b_{\nu j}\psi_j \quad \text{energy } F_\nu \tag{2.15}$$

Union involves the formation of a σ bond between atom r in R to atom s in S. The arguments of Section 2.4 indicate that so far as the π electrons are concerned, this corresponds to a change in the rs resonance integral from zero to β_{rs}. Now, since the coefficient of the AO ψ_s is zero in all the MOs

FIGURE 2.2. Union of two conjugated systems R and S to form RS.

Φ_μ, and that of the AO ϕ_r is zero in all the MOs Ψ_v, it is obvious that the rs bond order in the initial (unperturbed) system is zero. There should therefore be no first-order perturbation due to this union. This is because none of the unperturbed MOs has any density in the region between atoms r and s, each MO being confined either to R or to S. This means that the change in π energy on formation of biphenyl from two molecules of benzene is small, but it may not be negligible. In order to determine what happens in this case, we must carry our perturbation treatment to a higher approximation.

In a second approximation, the interacting MOs of R and S begin to mix with one another; as a result, the orbitals now extend across the critical rs region. The second-order perturbations correspond to pair interactions between an MO of R and one of S. The situation is similar to that involved in the combination of H and He$^+$ (Section 1.7). The lower of the two interacting orbitals is reduced in energy and the higher one is raised by equal amounts (cf. Fig. 1.9d). It is as if the interacting orbitals repel one another, with respect to energy. The magnitude of the change in orbital energy is given by the second-order perturbation expression [cf. (1.15)]

$$E = \beta^2/(E_\mu - F_v) \qquad (2.16)$$

For interaction between the MOs Φ_μ and Ψ_v, the numerator is the square of the effective "resonance integral" between them. This is determined by the degree to which the MOs overlap, i.e., by their product $\Phi_\mu \Psi_v$. Using equations (2.14) and (2.15), we have

$$\Phi_\mu \Psi_v = \left(\sum_i a_{\mu i}\phi_i\right)\left(\sum_j b_{vj}\psi_j\right) = \sum_i \sum_j a_{\mu i} b_{vj} \phi_i \psi_j \qquad (2.17)$$

Now *none* of the AOs ϕ_i overlap with any of the AOs ψ_j except for the pair ϕ_r and ψ_s; all the other terms in equation (2.17) therefore vanish, leaving

$$\Phi_\mu \Psi_v = a_{\mu r} b_{vs} \phi_r \psi_s \qquad (2.18)$$

If there were a simple π bond between atoms r and s, the overlap density in it would be given by the product $\phi_r \psi_s$ [cf. equation (1.11) and the discussion following it], and this would correspond to the resonance integral β_{rs}. The overlap of MOs Φ_μ and Ψ_v, being $a_{\mu r} b_{vs}$ times $\phi_r \psi_s$, then corresponds to an effective "resonance integral" $a_{\mu r} b_{vs} \beta_{rs}$. The difference in energy between the MOs Φ_μ and ψ_v is of course just $E_\mu - F_v$. Thus from equation (2.16), the magnitude of the splitting E of the levels due to the second-order perturbation between Φ_μ and Ψ_v is given by

$$E = \frac{a_{\mu r}^2 b_{vs}^2 \beta_{rs}^2}{E_\mu - F_v} \qquad (2.19)$$

Perturbation Theory

The total change δE_μ in the energy of the MO Φ_μ is given by summing the second-order perturbations due to its interaction with all the MOs Ψ_ν, namely,

$$\delta E_\mu = \sum_\nu \frac{a_{\mu r}^2 b_{\nu s}^2 \beta_{rs}^2}{E_\mu - F_\nu} \qquad (2.20)$$

(Note that we have written the denominators so that a given term is negative if $F_\nu > E_\mu$ and positive if $E_\mu > F_\nu$; this ensures that the higher of the interacting MOs increases in energy and the lower one decreases; cf. Fig. 2.3.)

Likewise the total change δF_ν in the energy of the MO Ψ_ν is given by summing the second-order perturbations due to its interactions with the MOs Φ_μ,

$$\delta F_\nu = \sum_\mu \frac{a_{\mu r}^2 b_{\nu s}^2 \beta_{rs}^2}{F_\nu - E_\mu} \qquad (2.21)$$

The total change in π energy δE_{RS} when R and S unite is given by summing the contributions of the electrons occupying these MOs. If the number of electrons in the MO σ is n_σ, then

$$\delta E_{RS} = \sum_\mu n_\mu \delta E_\mu + \sum_\nu n_\nu \delta F_\nu = \sum_\mu \sum_\nu \frac{a_{\mu r}^2 b_{\nu s}^2 \beta_{rs}^2 (n_\mu - n_\nu)}{E_\mu - F_\nu} \qquad (2.22)$$

Note that the interactions between filled MOs ($n_\mu = n_\nu = 2$) vanish; this is because the interaction pushes one MO up in energy by the same amount that it pushes the other down (Fig. 2.3). In ordinary molecules, where all the MOs are either filled ($n = 2$) or empty, the *energy of union* δE_{RS} arises entirely from interactions between filled MOs of R and empty MOs of S, and between filled MOs of S and empty MOs of R. It would be tedious to use expression (2.22) to solve chemical problems like the relative stability of two molecules of benzene and one of biphenyl. In Chapter 3, we will find that the expression (2.22) can be treated as a constant for intermolecular union of hydrocarbons with even-numbered rings or open chains. There is, however, a complication to be considered. What happens to equation (2.22) if one of the MOs of R has the same energy as (i.e., is degenerate with) one of the MOs of S? In that

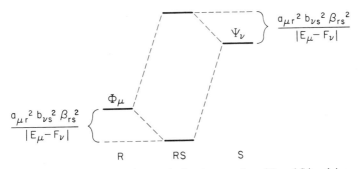

FIGURE 2.3. Second-order perturbation due to union of R and S involving interaction between the MOs Φ_μ and Ψ_ν.

case, the denominator of the corresponding term in equations (2.19)–(2.22) would vanish. This problem was discussed earlier in our treatment of HHe⁺ [see the paragraph following equation (1.16)]. When the MOs Φ_μ and Ψ_ν differ in energy, the second-order perturbation mixes them together, but not symmetrically. Just as in HHe⁺, one of the resulting MOs consists mainly of Φ_μ with a little Ψ_ν, and the other mainly of Ψ_ν with a little Φ_μ added. The resulting perturbed MOs Φ_μ' and Ψ_ν' are given by

$$\Phi_\mu' = \Phi_\mu + \frac{a_{\mu r} b_{\nu s} \beta_{rs}}{E_\mu - F_\nu} \Psi_\nu \qquad (2.23)$$

$$\Psi_\nu' = \Psi_\nu + \frac{a_{\mu r} b_{\nu s} \beta_{rs}}{F_\nu - E_\mu} \Phi_\mu \qquad (2.24)$$

These expressions, however, hold only if the MOs differ considerably in energy. As E_μ and F_ν become more and more nearly equal, the mixing of MOs becomes correspondingly more nearly equal. In the limit when $E_\mu = F_\nu$, the situation is analogous to that in H₂ rather than HHe⁺. The MOs Φ_μ and Ψ_ν must be replaced by the combinations χ^+ and χ^-, where

$$\chi^+ = (1/\sqrt{2})(\Phi_\mu + \Psi_\nu) \qquad (2.25)$$

$$\chi^- = (1/\sqrt{2})(\Phi_\mu - \Psi_\nu) \qquad (2.26)$$

Now these MOs *do* cover the region between atoms r and s. The coefficient of the AO ϕ_r is in each case $a_{\mu r}/\sqrt{2}$ and that of the AO Ψ_s is $\pm b_{\nu s}/\sqrt{2}$. The arguments of Section 2.3 show that a change in the resonance integral between atoms r and s from zero to β_{rs} will then increase the energy of one of the MOs χ^+ or χ^- and decrease that of the other by $2(a_{\mu r}/\sqrt{2})(b_{\nu s}/\sqrt{2})\beta_{rs}$, or $a_{\mu r} b_{\nu s} \beta_{rs}$ (Fig. 2.4). The situation is analogous to that in a diatomic molecule. As the MOs Φ_μ and Ψ_ν approach in energy, the second-order perturbation, given initially by the second-order expression of equation (2.19), increases, but instead of going to infinity as $E_\mu \to F_\nu$, it approaches the first-order perturbation $\pm a_{\mu r} b_{\nu s} \beta_{rs}$ as its limit.

There will be no first-order change in energy on union of R and S if both the degenerate MOs are doubly occupied, for union to RS will decrease the energy of one pair of electrons at the expense of a corresponding increase in that of the other. The energy of union will again depend entirely on second-order interactions. In calculating the second-order perturbations involving the degenerate MOs Φ_μ and Ψ_ν, one should of course use the combinations χ^+ and χ^-. It can, however, be shown that if we are interested only in the total

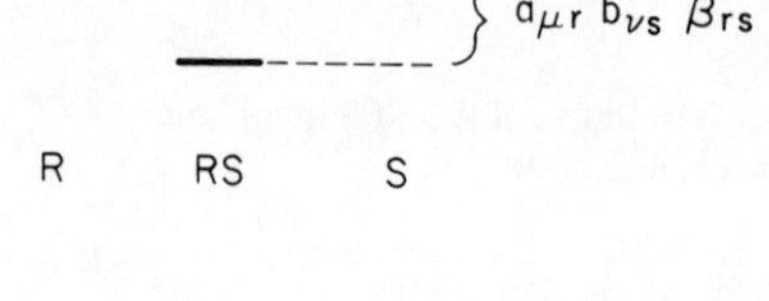

FIGURE 2.4. First-order perturbation on union of R and S due to interaction of a pair of degenerate MOs Φ_μ and Ψ_ν.

change in energy, i.e., the energy of union of R and S to form RS, then the same result is obtained if we use Φ_μ and Ψ_ν instead of χ^+ and χ^- and simply neglect the term involving the interaction between them (which would have a vanishing denominator). Thus the first-order interactions between pairs of filled MOs can be totally ignored. This has been done in equation (2.22).

2.6 Multiple Union; Additivity of Perturbations

We have so far considered only single perturbations, i.e., monocentric perturbations due to a change of one atom and intramolecular and intermolecular union through a single pair of atoms. The results can, however, be extended to multiple perturbations.

First, consider the effect of two simultaneous monocentric perturbations involving atoms k and m. The change in atom k is reflected by changes $\delta\alpha_k$ and $\delta\beta_{kl}$ in the associated Coulomb and resonance integrals (see Section 2.3) and the change in atom m by corresponding changes $\delta\alpha_m$ and $\delta\beta_{mn}$ (where atom n is directly linked to atom m). The argument of Section 2.3 shows that the corresponding change in total π energy δE_π is given by

$$\delta E_\pi = \left(q_k \, \delta\alpha_k + 2 \sum_l p_{kl} \, \delta\beta_{kl} \right) + \left(q_m \, \delta\alpha_m + 2 \sum_n p_{mn} \, \delta\beta_{mn} \right) \quad (2.27)$$

Now the term in the first set of parentheses is simply the change in π energy we calculated originally [equation (2.12)] for the monocentric perturbation due to a change in atom k, atom m remaining the same, while the second set of parentheses represents the effect of changing atom m while atom k remains unchanged. The effect of changing both atoms is therefore just the sum of changing one or other of them; the perturbations are additive. This in fact is a special case of a general result:

First-order perturbations are additive.

It can be seen at once that the same result holds for multiple intramolecular union, the total perturbation energy being given by a sum of terms like that in equation (2.13), one for each new connection.

Consider now the case of multiple intermolecular union. Suppose (cf. Fig. 2.2) that we now unite two conjugate systems R and S at two points, through atom r in R to atom s in S and through atom t in R to atom u in S (Fig. 2.5). The overlap between the MOs Φ_μ and Ψ_ν [cf. equations (2.17) and

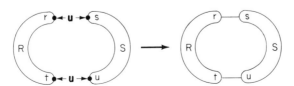

FIGURE 2.5. Double union of two conjugated systems R and S.

(2.18)] will now involve two terms, one for each new bond,

$$\Phi_\mu \Psi_\nu = \left(\sum_i a_{\mu i}\phi_i\right)\left(\sum_j b_{\nu j}\psi_j\right) = a_{\mu r}b_{\nu s}\phi_r\psi_s + a_{\mu t}b_{\nu u}\phi_t\psi_u \tag{2.28}$$

The effective resonance integral β' between the MOs is then given by

$$\beta' = a_{\mu r}b_{\nu s}\beta_{rs} + a_{\mu t}b_{\nu u}\beta_{tu} \tag{2.29}$$

If the MOs Φ_μ and Ψ_ν are degenerate, the first-order interaction between them will then lead to a splitting of $\pm \beta'$ (see Fig. 2.4); this is just the sum of the splittings produced by the two individual unions separately, so that the perturbations are again additive.

If, however, the MOs are not degenerate, the interaction energy E is given by the second-order perturbation expression of equation (2.16),

$$\begin{aligned}E = \frac{(\beta')^2}{E_\mu - F_\nu} &= \frac{(a_{\mu r}b_{\nu s}\beta_{rs} + a_{\mu t}b_{\nu u}\beta_{tu})^2}{E_\mu - F_\nu} \\ &= \frac{a_{\mu r}^2 b_{\nu s}^2 \beta_{rs}^2}{E_\mu - F_\nu} + \frac{a_{\mu t}^2 b_{\nu u}^2 \beta_{tu}'}{E_\mu - F_\nu} + \frac{2a_{\mu r}a_{\mu t}b_{\nu s}b_{\nu u}\beta_{rs}\beta_{tu}}{E_\mu - F_\nu}\end{aligned} \tag{2.30}$$

Here the first term is the second-order perturbation for single union through atoms r and s [equation (2.19)]. Likewise, the second term is the second-order perturbation for single union through atoms t and u. Equation (2.30) shows that the effect of simultaneous union through atom r to atom s and through atom t to atom u is *not* a sum of the effects of the two individual unions separately:

Second-order perturbations are not additive.

As we shall see in the next chapter, this nonadditivity of second-order perturbations has very important consequences.

Problems

2.1. Define the following terms:
   a. Monocentric perturbations.
   b. Intramolecular perturbations.
   c. Intermolecular perturbations.
   d. Charge density.
   e. Bond order.

2.2. What are the values of the charge densities and bond order in the ethylene π-electron system?

2.3. Write an expression for the energy change in the following reaction:

$$CH_2=CH-CH_3 \longrightarrow CH_2=CH-CH_2\cdot + H\cdot$$

2.4. In the union of ethylene and 1,3-butadiene to give 1,3,5-hexatriene, the change in π energy is ~ 3 kcal/mole. Why is it that in the double union of ethylene and 1,3-butadiene to give benzene, the change in π energy is ~ 36 kcal/mole, or roughly ten times that for the first union?

Selected Reading

For further reading on the introduction to perturbation theory, see:

M. J. S. DEWAR, *The Molecular Orbital Theory of Organic Chemistry*, McGraw-Hill Book Co., New York, 1969.

R. L. FLURRY, JR., *Molecular Orbital Theories of Bonding In Organic Chemistry*, Marcel Dekker, New York, 1968.

Reference

1. C. T. MORTIMER, *Reaction Heats and Bond Strengths*, Pergamon Press, Oxford, 1962.

CHAPTER 3

PMO Treatment of Conjugated Systems

3.1 Principles of the PMO Method; Alternant and Nonalternant Systems

In order to apply the formal scheme developed in the previous chapter, we have to have some way of estimating the various quantities that appear in the perturbation expressions. These fall into two categories, the integral terms and the AO coefficients.

The Coulomb integrals α_i and resonance integrals β_{ij} should be characteristic of given atoms and bonds and should carry over from one molecule to another. We may therefore regard them in the same light as bond energies, to be determined empirically from data for a number of molecules containing the atoms or bonds in question. They therefore present no problems in principle.

The situation is quite different in the case of the AO coefficients $a_{\mu i}$, $b_{\nu j}$, etc., and of the charge densities q_i and bond orders p_{ij} calculated from them. These can normally be found only by large-scale calculations using digital computers. Since our object is to develop a simple and general theory of organic chemistry that can be applied without the need for, and cost of, any such computational aids, we have to develop some expedient for bypassing them.

The key to this problem was discovered some time ago by Coulson and Rushbrooke. There is a class of compounds where the π-electron densities and bond orders can be found very simply, by what may be termed "pencil and paper" procedures. The compounds in question are conjugated hydrocarbons in which the conjugated atoms can be divided into two groups, such that each atom is directly linked only to atoms of the opposite group. A conjugated system of this type is termed an *alternant* conjugated system, so the compounds in question are *alternant hydrocarbons* (AH). It is easily seen (see Fig. 3.1) that conjugated systems are alternant if, and only if, they contain no odd-numbered rings. The two groups of atoms are distinguished in Fig. 3.1 by marking one of them with stars. In the first four examples, this *starring* procedure can be carried out successfully, each starred atom being attached only to *unstarred* ones and each unstarred atom only to starred ones. These systems are therefore alternant. Note that the classification includes conjugated systems with both even and odd numbers of atoms (*even* AH, *odd* AH). An even AH is a "normal," neutral hydrocarbon containing an even number of electrons which will normally pair up in the orbitals of lowest energy available to them. Such a system, with all the electrons paired, is called a *closed shell system*. An odd AH must exist as an ion or radical, the latter being neutral but having an odd number of electrons so that one remains unpaired. Allyl provides a simple example, the cation (1) and anion (2) having closed shell structures with

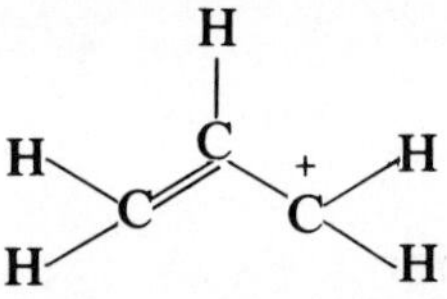

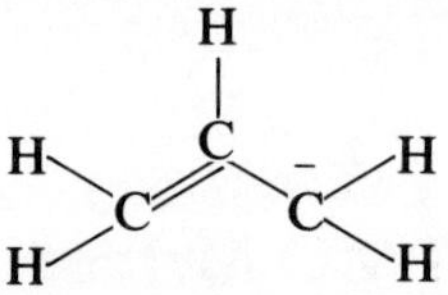

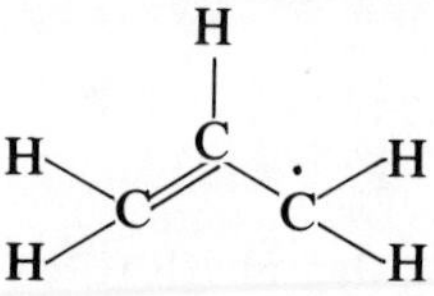

22 and 24 electrons, respectively, while the radical (3) has 23 electrons. Positions of the same type (i.e., both starred or both unstarred) are said to be of *like parity*. The fifth and sixth examples in Fig. 3.1 are nonalternant, the starring procedure failing because they have odd-numbered rings. The last two examples illustrate alternant and nonalternant conjugated systems containing heteroatoms.

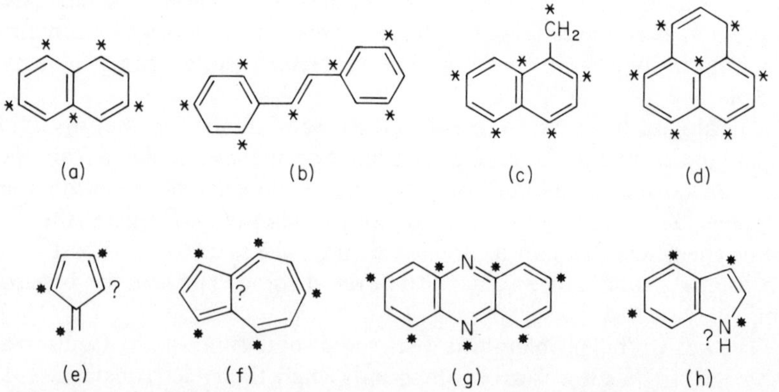

FIGURE 3.1. (a–d) Alternant hydrocarbons; (e, f) nonalternant hydrocarbons; (g) an alternant heteroconjugated system; (h) a nonalternant heteroconjugated system.

PMO Treatment of Conjugated Systems

As we shall see, the special properties of AHs allow us to make very simple predictions concerning their structures and chemical reactivity. This in itself might seem a rather limited achievement, but in fact it is not, for two reasons. First, we can extend the results for hydrocarbons to systems containing heteroatoms by using the monocentric perturbation theory of Section 2.3. We can therefore deal quite generally with problems involving all kinds of π delocalization. Second, in view of the equivalence of delocalized systems of different types (Section 1.16), we can immediately extend the results obtained for conjugated systems to delocalized systems of other kinds. We can therefore deal with problems throughout the whole field of organic chemistry, provided only that they do not involve nonalternant delocalized systems. Even the nonalternant systems can usually be included by using various tricks. This general approach, which has been termed the *PMO method* (PMO = perturbational MO) has therefore proved to be an extremely powerful tool in the study of the behavior of organic molecules.

3.2 The Pairing Theorem

In our simple MO approach, the π MOs Φ_μ of an even AH are written as linear combinations of the $2p$ AOs ϕ_i of the contributing carbon atoms,

$$\Phi_\mu = \sum_i a_{\mu i} \phi_i \tag{3.1}$$

As in the case of methane (Section 1.10), the $2n$ AOs of the $2n$ carbon atoms represent a volume of "orbital space" capable of holding $2n$ pairs of electrons; the MOs derived by scrambling them [equation (3.1)] should therefore between them also be able to hold $2n$ pairs of electrons. Consequently, our $2n$ AOs should give rise to $2n$ MOs.

Coulson and Rushbrooke showed that the $2n$ π MOs of such an even AH show certain striking regularities, comprised in what is now usually termed the *pairing theorem*. We assume first that the conjugated atoms have been divided as above into starred and unstarred sets. The pairing theorem can then be summarized by the following statements:

Pairing Theorem for Even AHs. (1) The π MOs of an even AH appear in pairs Φ_μ^+ and Φ_μ^-, with energies $\alpha + E_\mu$ and $\alpha - E_\mu$, α being the Coulomb integral for carbon.

(2) The coefficients of AOs in the expansions of the paired MOs Φ_μ^+ and Φ_μ^- are numerically the same. Each MO is derived from the other by inverting the signs of one set of coefficients (either those of the starred AOs or those of the unstarred AOs, the choice being arbitrary since the wave functions ψ and $-\psi$ are equivalent).

(3) The π-electron density at each position is unity in neutral AHs.

(4) The bond order between atoms of like parity (both starred or both unstarred) is zero for neutral AHs.

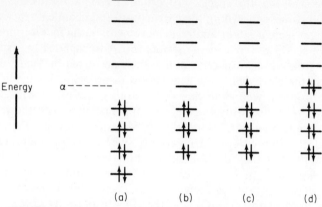

FIGURE 3.2. Orbital energies (a) in an even AH; (b) in an odd AH cation; (c) in an odd AH radical; (d) in an odd AH anion.

The MOs of AHs thus appear in pairs, spaced equally from the central level, α, in energy (Fig. 3.2). If one of the MOs Φ_μ^+ of a pair is expanded in terms of the AOs ϕ_i^* of starred atoms and ϕ_j^0 of unstarred atoms,

$$\Phi_\mu^+ = \sum_i a_{\mu i}^* \phi_i^* + \sum_j a_{\mu j}^0 \phi_j^0 \tag{3.2}$$

then the twin MO Φ_μ^- is found by simply changing the signs of one complete set of coefficients,

$$\Phi_\mu^- = \sum_i a_{\mu i}^* \phi_i^* - \sum_j a_{\mu j}^0 \phi^0 \quad (\text{or} - \sum_i a_{\mu i}^* \phi_i^* + \sum_j a_{\mu j}^0 \phi_j^0) \tag{3.3}$$

The two statements are equivalent since the wave functions ψ^+ and ψ^- are equivalent; one can attach physical meaning only to the square of the wave function, and this is the same for both ψ^+ and ψ^-.* Changing the signs of one set of coefficients has the effect of reversing the bondingness of an orbital because it changes the phases of the overlap between every pair of bonded atoms. This is the unique feature of alternant hydrocarbons. Changing the relative signs of different parts of a wave function is a different matter; cf. the bonding and antibonding MOs of H_2 [equations (1.9) and (1.10)].

In an even AH with $2n$ conjugated atoms, there will be $2n$ electrons. These will occupy in pairs the n MOs of lowest energy. Since there are $2n$ MOs all together, one MO of each pair will be occupied and the other un-

---

* The square of a wave function represents the amplitude of the wave at a given point, since the magnitudes of the coefficients in Φ^+ and Φ^- are the same, $(\Phi^+)^2$ and $(\Phi^-)^2$ will be the same, namely

$$(\Phi^+)^2 = (\Phi^-)^2 = \sum_i (a_{\mu i}^*)^2 (\phi_i^*)^2 + \sum_j (a_{\mu j}^0)^2 (\phi_j^0)^2$$

The cross terms all vanish in this approximation because it is assumed that the basis AOs do not overlap so that the product $\phi_i \phi_j$, $i \neq j$, vanishes.

occupied. Since, moreover, the MOs are disposed symmetrically about the energy α of a carbon $2p$ AO, the occupied MOs are all bonding and the unoccupied ones are antibonding (cf. Fig. 1.8g).

The pairing theorem also applies to odd AHs. In this case, the number of contributing AOs is odd, so the number of MOs must also be odd. If an odd AH obeys the pairing theorem, there must be one unpaired MO left over. It must have energy α and be composed entirely of AOs of one set (starred or unstarred) because only in that way can we get an MO which is identical with its twin. Since an electron in this odd MO has the same energy as that, α, in the $2p$ AO of an isolated carbon atom, such an electron should remain unchanged in energy if the molecule is dissociated into atoms. Electrons in the odd MO are completely neutral insofar as bonding is concerned; they are neither bonding nor antibonding. The odd MO is therefore described as a *nonbonding MO* (NBMO) and we will denote it by the subscript zero (i.e., Φ_0).

In an odd AH, the two sets of atoms (starred and unstarred) must differ in number; we term *the more numerous set the starred set*. The properties of odd AHs can be summarized by the following extension of the pairing theorem:

Pairing Theorem for Odd AHs. (1) The MOs in an odd AH occur in pairs, Φ_μ^+ and Φ_μ^-, showing the same regularities as those in an even AH, except for one nonbonding MO (NBMO) Φ_0 of energy α.

(2) The NBMO is confined entirely to starred atoms, the coefficients of all the unstarred AOs vanishing.

(3) In a neutral, odd AH (i.e., a radical), the π-electron density q_i is unity at each position.

(4) In a neutral, odd AH, the bond order between two starred atoms or between two unstarred atoms, atoms of like parity, is zero.

An odd AH anion differs from the radical in having an extra electron in the NBMO. The π-electron density q_i^- at atom i is therefore greater than that, q_i, in the radical by the term a_{0i}^2 [see equation (2.7)], corresponding to the extra nonbonding electron. But from the pairing theorem, $q_i = 1$. Hence

$$q_i^- = 1 + a_{0i}^2 \qquad (3.4)$$

In the cation, where there is one nonbonding electron less than in the radical, the π-electron density q_i^+ at position i is given by

$$q_i^+ = 1 - a_{0i}^2 \qquad (3.5)$$

Likewise the bond orders between two starred atoms i and j in the anion (p_{i*j*}^-) and cation (p_{i*j*}^+) are given by [cf. equation (2.9)]

$$p_{i*j*}^- = a_{0i}^* a_{0j}^* \qquad (3.6)$$

$$p_{i*j*}^+ = -a_{0i}^* a_{0j}^* \qquad (3.7)$$

since the corresponding bond order in the radical is zero. On the other hand, the bond orders between unstarred atoms are zero in all three species since the NBMO coefficients vanish at unstarred positions,

$$p^+_{i^0j^0} = p^\cdot_{i^0j^0} = p^-_{i^0j^0} = 0 \tag{3.8}$$

Equations (3.4)–(3.8) represent an extension of the pairing theorem to odd AH ions.

3.3 Calculation of NBMO Coefficients

While the pairing theorem provides a wealth of information about the MOs in AHs without the need for detailed calculation, many of the results require a knowledge of the coefficients of AOs in NBMOs. Normally, the coefficients in any one MO can be found only by a full-scale calculation of the whole set of MOs, but the NBMO is fortunately an exception. As Longuet-Higgins first showed, the coefficients in it can be found independently of the rest by a completely trivial "pencil and paper" procedure.

Consider one of the unstarred atoms i in an odd AH. This will be attached by σ bonds to at most three neighboring atoms j, k, l [see (4)]. Since the compound is alternant and atom i is unstarred, any atom attached to atom i must be starred.

$$(a_{0l})C_l \xrightarrow{\beta_{il}} C_i \begin{array}{c} \beta_{ij} \\ \diagup \\ \diagdown \\ \beta_{ik} \end{array} \begin{array}{c} C_j(a_{0j}) \\ \\ C_k(a_{0k}) \end{array}$$

(4)

Longuet-Higgins' rule states that if the NBMO coefficients of atoms j, k, and l are, respectively, a_{0j}, a_{0k}, and a_{0l}, and if the resonance integrals of the ij, ik, and il bonds are, respectively, β_{ij}, β_{ik}, and β_{il}, then

$$a_{0j}\beta_{ij} + a_{0k}\beta_{ik} + a_{0l}\beta_{il} = 0 \tag{3.9}$$

If atom i has only two neighbors or only one neighbor, the corresponding equation has only two terms or only one term, respectively.

The value of the carbon–carbon π resonance integral depends on the extent to which the two contributing $2p$ AOs overlap, and this in turn should be greater, the closer together the atoms are (cf. Fig. 1.26b). The distance between the atoms will in turn depend on how strong the π bond between them is, i.e., on the bond order of the bond (cf. Sections 1.13 and 1.14). The shorter and stronger the bond, the greater is the π resonance integral β.

As a first approximation, however, we may neglect these variations; if so, the β's cancel out of equation (3.9), leaving the simple result

$$a_{0j} + a_{0k} + a_{0l} = 0 \tag{3.10}$$

In other words, to a first approximation, *the sum of the NBMO coefficients at starred atoms adjacent to a given unstarred atom vanishes.*

This rule enables us to find the NBMO coefficients in an odd AH very quickly and easily. The procedure is indicated for α-naphthylmethyl in Fig.

PMO Treatment of Conjugated Systems

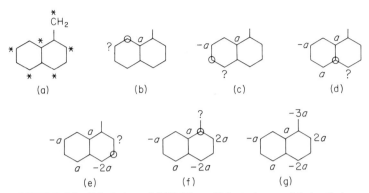

FIGURE 3.3. Calculation of NBMO coefficients in α-naphthylmethyl.

3.3. First one stars the molecule (Fig. 3.3a), making sure that the starred set is the more numerous one; from the pairing theorem, the NBMO is confined to starred atoms. Now assign the value a to the NBMO coefficient of an arbitrary starred atom (Fig. 3.3b) and consider one of the adjacent unstarred atoms (marked with a circle in Fig. 3.3b). If the unstarred atom has only one other neighbor, the coefficient there must be $-a$ because the sum of the coefficients of atoms adjacent to the circled atom must vanish (Fig. 3.3c). Now one takes an unstarred atom adjacent to the starred atom whose coefficient is $-a$ and continues the process. If at any point one encounters an unstarred atom with three neighbors, all is well if two of the NBMO coefficients have already been determined; this is the case in Fig. 3.3 for the steps (d) → (e) and (f) → (g), so the calculation goes through without a hitch.

The last step is to determine a. To do this, we normalize the MO. Since the square of each NBMO coefficient is the fraction of the MO composed of the particular AO, the sum of the squares must add up to unity. In this case,

$$a^2 + (-a)^2 + a^2 + (-2a)^2 + (2a)^2 + (3a)^2 = 1 \qquad (3.11)$$

Hence

$$20a^2 = 1 \quad \text{or} \quad a = 1/\sqrt{20} \qquad (3.12)$$

A complication arises if at some step in the calculation we encounter an unstarred atom with three neighbors, two of whose NBMO coefficients are unknown. This situation is illustrated in Fig. 3.4 by a calculation for α-naphthylmethyl using a different initial atom. In Fig. 3.4(b) the unstarred atom has two neighbors with two undetermined NBMO coefficients. When this happens, we simply call one of the unknowns b (Fig. 3.4c) and continue. There will always be enough equations to determine b in terms of a. Here b is determined in Fig. 3.4(e). Applying Longuet-Higgins' rule to the indicated unstarred atom, we find

$$b + b - a = 0, \quad \text{or} \quad a = 2b \qquad (3.13)$$

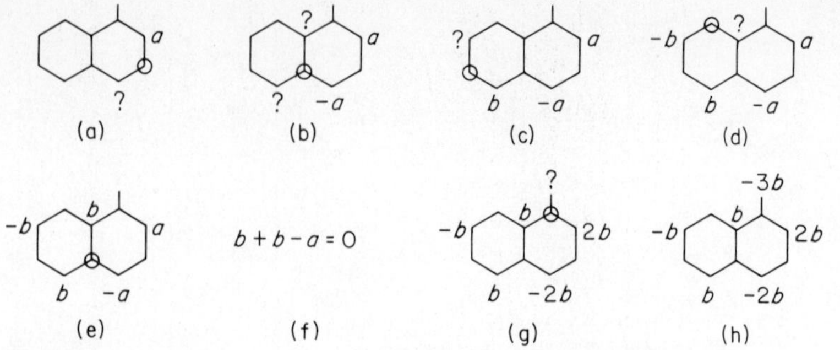

FIGURE 3.4. An alternative calculation of NBMO coefficients for α-naphthylmethyl.

While a good choice of the initial atom may simplify things, the result is always the same (cf. Figs. 3.3g and 3.4h). In very complicated AHs, or with a bad choice of the initial atom, we may encounter a second unstarred atom with two unknown adjacent coefficients. In this case, we follow the same procedure, calling one of the unknowns c and continuing. With a little practice the coefficients can be found very quickly even in quite complicated odd AHs. Figure 3.5 shows a further example.

Note that the absolute signs of the NBMO coefficients are as usual arbitrary. The orbitals ψ and $-\psi$ are equivalent; multiplication of *all* of the NBMO coefficients by -1 leaves the wave function unchanged.

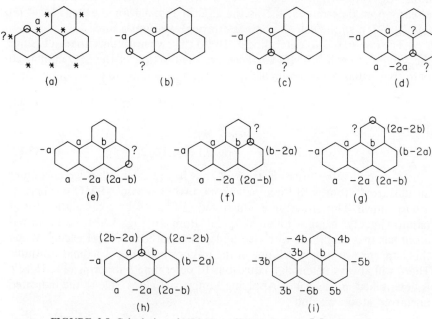

FIGURE 3.5. Calculation of NBMO coefficients in benzo[a]perinaphthenyl.

3.4 Distribution of Formal Charges in AHs

According to the pairing theorem, the π-electron density at each position in an even AH or in an odd AH radical should be unity. The π electrons move in the field of a core where each carbon atom has one unit ($+e$) of charge. Since there is on the average exactly one π electron associated with each carbon atom, the atoms in an even AH or an odd AH radical should be neutral. Such compounds should therefore be nonpolar. This is indeed the case for even AHs, their dipole moments being zero.

One might think this a trivial result on the grounds that the carbon atoms in a neutral hydrocarbon "obviously" must be neutral—but this would be an incorrect intuition. Nonalternant hydrocarbons *are* polar and do have nonvanishing dipole moments. Fulvene (5) and azulene (6) are good examples, each having dipole moments of about 1 D (Debye). In the case of an odd AH ion, the total π-electron density at atom i is $1 + a_{0i}^2$ for an anion and $1 - a_{0i}^2$

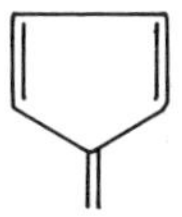

(5) 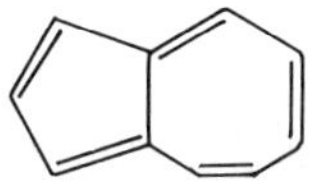   (6)

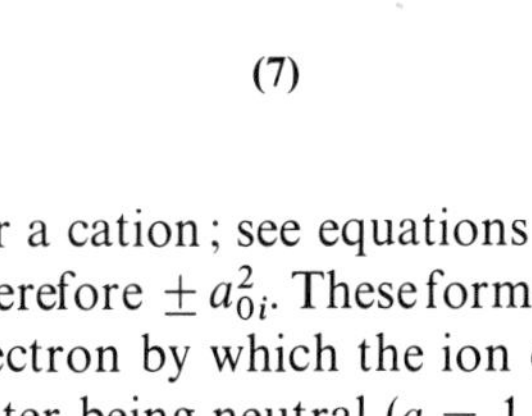

(7) 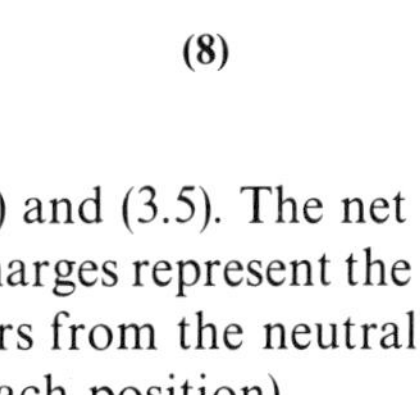   (8) 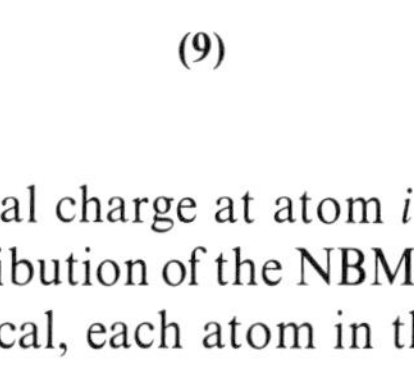   (9)

for a cation; see equations (3.4) and (3.5). The net formal charge at atom i is therefore $\pm a_{0i}^2$. These formal charges represent the distribution of the NBMO electron by which the ion differs from the neutral radical, each atom in the latter being neutral ($q = 1$ at each position).

Consider the benzyl anion and cation, $C_6H_5CH_2^\pm$. The NBMO coefficients in benzyl are easily found [see (7)] by the method of Section 3.3. From these, one arrives at the distributions of formal charge for the anion and cation indicated in (8) and (9), respectively.

It is interesting to compare these with the charge distributions predicted by simple resonance theory, where such an ion is represented by a superposition or hybrid of all the possible classical structures, each structure contributing equally. The formal charge at a given position is then proportional to the number of such structures in which the charge appears there. Figure 3.6 shows the possible classical structures of benzyl anion, the numbers of structures with the formal charge at the various positions being indicated in Fig. 3.6(f). Note that these numbers are *exactly proportional to the corresponding NBMO coefficients in* (7). The distributions of formal charge predicted by resonance theory and the PMO method are therefore qualitatively similar; indeed, if one took the *square* of the number of resonance structures as a measure of charge distribution, both procedures would lead to the same result.

The parallel between numbers of resonance structures and NBMO coefficients is not in fact an accident but holds for the odd AHs, with one exception. It fails at a given position if removal of that atom would leave an even AH with a ring containing $4n$ atoms. Thus in the case of perinaphthenyl, the NBMO coefficients (10) parallel the numbers of resonance structures (11) *except* for the central atom. Removal of the central atom leaves an even AH

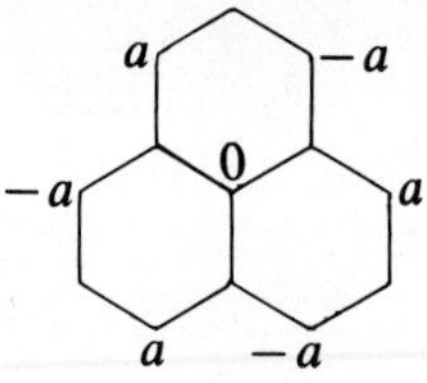

$a = 1/\sqrt{6}$

(10)   (11) 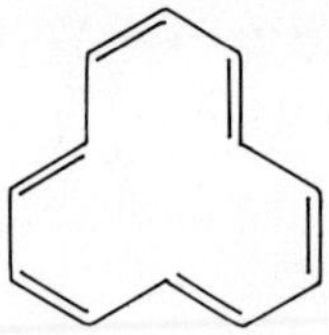   (12)

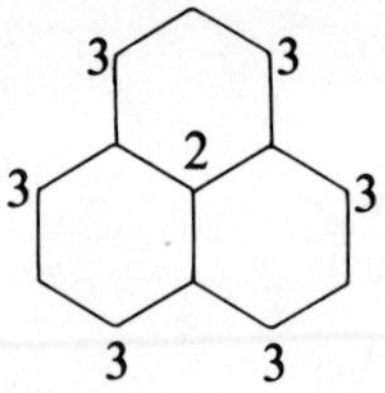

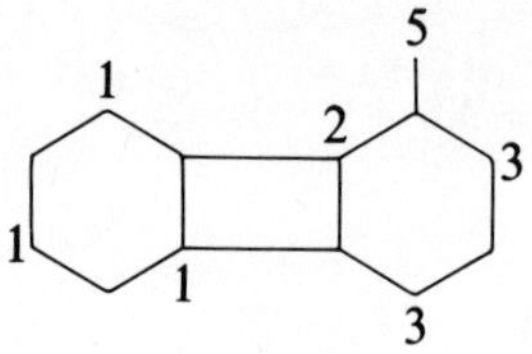

(13) 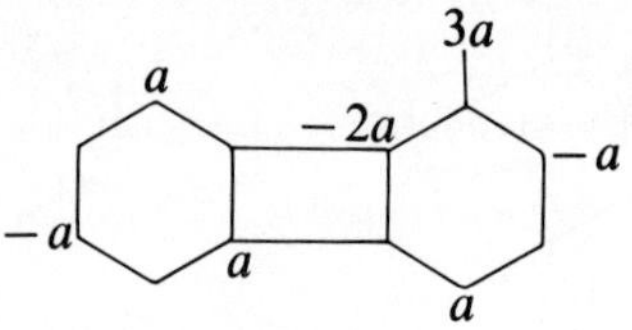   (14)

(12) with a 12-($=4 \times 3$) membered ring. Likewise in the case of odd AHs containing $4n$-membered rings, the parallel fails completely; cf. the number of resonance structures (13) and NBMO coefficients (14) for 1-biphenylenyl-methyl.

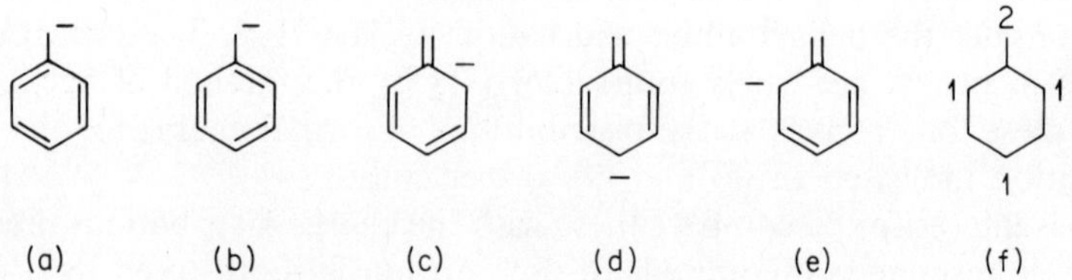

(a)   (b)   (c)   (d)   (e)   (f)

FIGURE 3.6. The benzyl anion in terms of resonance theory.

PMO Treatment of Conjugated Systems

In cases where differences of this kind occur, detailed quantum mechanical calculations show that PMO theory is much closer to the truth. This is not surprising, since PMO theory has a valid basis in quantum mechanics, whereas resonance theory does not. The relationship between them is discussed in detail later in this chapter (Section 3.20).

3.5 Monocentric Perturbations; Correlation of Isoconjugate Systems

In Section 2.3, we derived an expression relating the π energies of two isoconjugate systems, i.e., systems containing similar numbers of conjugated atoms, similarly connected, and with the same total number of π electrons. In the cases that concern us, the most important terms in the corresponding expression [equation (2.12)] are those involving the changes in Coulomb integrals $q_i \delta \alpha_i$, because the strengths of π bonds and so the values of the corresponding resonance integrals β do not vary greatly. Now the basic assumption in this first-order perturbation treatment was that the π-electron densities do not vary with changes in the participating atoms. The π-electron densities will then be the same for a given set of isoconjugate molecules and can be calculated by the methods used in the preceding section for AHs. Using the pairing theorem and Longuet-Higgins' rule, we can calculate the π-electron density at *any* position in *any* AH or any heteroatomic alternant π-electron system. We can then use these π-electron densities to calculate the difference in π energy between any pair of alternant conjugated systems. In particular, we can calculate the π energy of any alternant heteroconjugated system in terms of that of the isoconjugate AH because the energy difference between the two will be simply $\sum q_i \delta \alpha_i$. This result, due to Longuet-Higgins, reduces the problems of chemistry in general to those of hydrocarbon chemistry.

3.6 Intermolecular Union of Even AHs

To complete our program, we need to be able to compare the π energies of two conjugated hydrocarbons which differ either in the number of conjugated atoms or in the way in which the conjugated atoms are linked together. The first of these problems involves estimating the energy of union of two conjugated hydrocarbons to a single larger one (see Section 2.5). Let us then consider the union of two even AHs R and S to a single, larger AH, RS, through atom r in R to atom s in S (see Fig. 2.2).

From the pairing theorem, it follows that all the bonding MOs in R and S are filled with pairs of electrons and all the antibonding MOs are empty (Fig. 3.7a). If an MO of R has the same energy as one of S (i.e., if the two MOs are degenerate), it follows that either both are filled with pairs of electrons or both are empty. The first-order interaction between the MOs therefore leads to no change in the *total* energy of the system (see Section 2.5 and Fig.

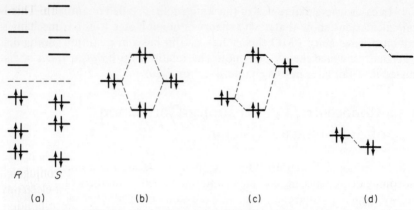

FIGURE 3.7. (a) Filled bonding and empty antibonding MOs of two even AHs R and S; (b) the large first-order interaction between two degenerate bonding MOs has no effect on the total energy since both are filled; (c) the second-order interaction between two bonding AOs, which is moderately large since their energies are comparable, also has no effect; (d) the interaction between a filled bonding MO and an empty antibonding MO has an effect but the interaction is small because the difference in energy is so large.

3.7b). Any change in total energy when R and S unite is therefore due to second-order perturbations.

Since the second-order interaction between two MOs raises one up in energy by the same amount that the other is reduced, the sum of the energies of the two orbitals remains unchanged. The second-order interactions between filled MOs of R and S therefore also fail to alter the total energy (Fig. 3.7c).

We are therefore left with the interactions between filled, bonding MOs of R and empty, antibonding MOs of S, and between filled, bonding MOs of S and empty, antibonding MOs of R. These interactions do lead to a decrease in energy when R and S unite (Fig. 3.7d) but the effect is small because the orbitals in question differ so much in energy and the second-order perturbations depend on the reciprocal of the energy difference [see equations (1.16) and (2.18)]. The change in π energy when two even AHs unite through a single pair of atoms is therefore small.

It can be shown (see the appendix at the end of this chapter) that this overall change in π energy, i.e., the π energy of union, is not only small but also has approximately the same value A_π for union of any two even AHs,

$$\text{RH} \xleftarrow{} \text{u} \xrightarrow{} \text{HS} \longrightarrow \text{RS}, \qquad \delta E_\pi = A_\pi \qquad (3.14)$$

where RH and SH are any two even AHs. We will now show that if this relation holds, then the bonds in an open-chain, conjugated polyene are localized.

Since the bonds in ethylene are localized (Section 1.13), its heat of atomization E_1 can be written as minus a sum of the corresponding bond energies,

$$E_1 = -E_{\text{C=C}} - 4E_{\text{CH}} \qquad (3.15)$$

PMO Treatment of Conjugated Systems

Ethylene can be regarded as the simplest possible even AH, with one starred atom and one unstarred atom ($\overset{*}{C}H_2=CH_2$) and butadiene ($CH_2=CH-CH=CH_2$) can be formed by union of two molecules of ethylene. Now equation (3.14) applies to the union of *any* pair of even AHs; the energy of union δE of ethylene to butadiene is therefore given by

$$\delta E = 2E_{CH} - E_{C-C} - \delta E_\pi = 2E_{CH} - E_{C-C} - A_\pi \quad (3.16)$$

where E_{C-C} is the bond energy of the new C—C σ bond and the π energy of union δE_π has the value A_π common to even AHs. If we write

$$E_{C-C} + A_\pi = E'_{C-C} \quad (3.17)$$

the energy of union becomes

$$\delta E = 2E_{CH} - E'_{C-C} \quad (3.18)$$

We can now find the heat of atomization E_2 of butadiene by combining equations (3.15) and (3.18),

$$E_2 = 2E_1 + \delta E = -2E_{C=C} - 6E_{CH} - E'_{C-C} \quad (3.19)$$

It will be seen that this is the value we would get if we assumed all the bonds in butadiene to be localized, the bond energy of the central "single" bond being E'_{C-C}.

Next let us consider union of butadiene with ethylene to form hexatriene:

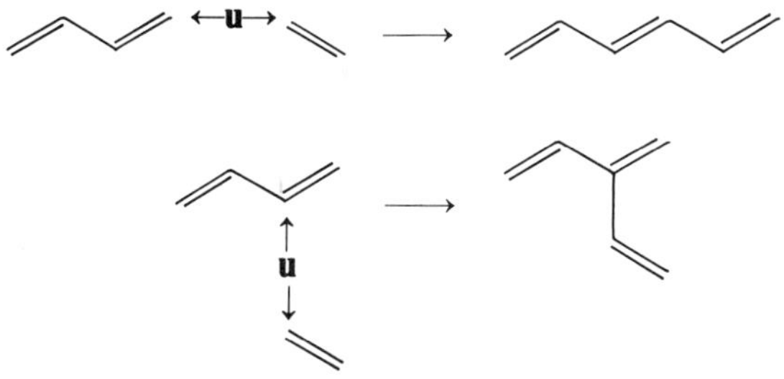

$$(3.20)$$

The energy of union is again given by equation (3.16) because the π energy of union of *any* two even AHs has the same value [$\delta E_\pi = A_\pi$ in equation (2.14)]. The heat of atomization E_3 of hexatriene is then given by adding together the heats of atomization of butadiene and of ethylene and the energy of union; i.e.,

$$E_3 = E_1 + E_2 + A_\pi = -3E_{C=C} - 8E_{CH} - 2E'_{C-C} \quad (3.21)$$

This again is the value we would get if we assumed the bonds in hexatriene to be localized, using for the bond energy of the single bond a "polyene" value E'_{C-C}. Extending this argument, one can see that the heat of atomization of any open-chain polyene, linear or branched, can be represented likewise as a sum of bond energies.

Now as we saw in Chapter 1, the question of bond localization is a purely empirical one. If the heat of atomization of a molecule can be written as a sum of bond energies, the bonds in it are by definition localized. We can therefore regard such polyenes as having localized single (C—C) and double (C=C) carbon–carbon bonds. The bond energy of the "single" bond will contain a π contribution, and so differ from that of a "pure σ" bond, but since the π contribution is constant, it can be included in a fixed constant bond energy for C—C bonds of this type.

It should be emphasized again that this result does *not* imply that the electrons in such polyenes are localized; cf. the discussion of paraffins in Section 1.12. Thus when two molecules of ethylene unite to form butadiene [equation (3.14)], their π MOs do interact very strongly. The two π MOs are of course degenerate, so the interaction involves a large first-order perturbation (Fig. 3.7b). The interaction leads to two π MOs which differ very considerably in energy, as can be seen from the photoelectron spectrum of butadiene. The first two ionization potentials, corresponding to the two π MOs, are at 9.07 and 11.27 eV, showing a splitting of 2.2 eV, or 51 kcal/mole.

3.7 Multiple Union of Even AHs

Since the union of even AHs involves a second-order perturbation, and since second-order perturbations are not additive (Section 2.6), the additivity of bond energies does *not* extend to cyclic polyenes.

Consider, for example, the union of butadiene with ethylene to form hexatriene or benzene;

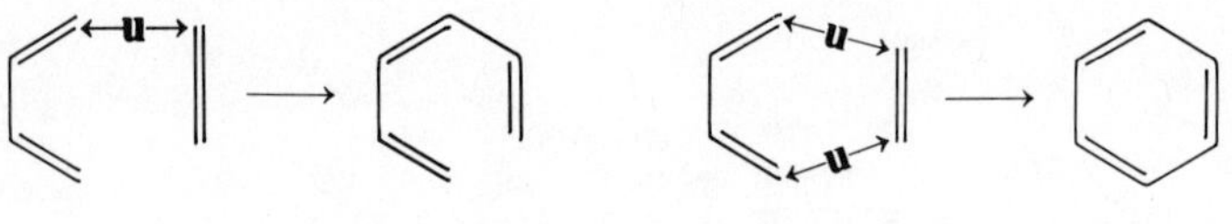

$$\text{(structures)} \tag{3.22}$$

In the first case, the energy of union is as usual A, but in the second it is *not* $2A$. Moreover, the deviation from additivity cannot be estimated in any simple way, because it is a second-order term. We therefore have to find an alternative procedure if we are to estimate the energies of cyclic conjugated systems.

3.8 Union of Odd AHs

When two odd AH radicals unite, their NBMOs are necessarily degenerate and contain between them just two electrons. Here then there is a first-order change in π energy due to union (Fig. 3.8). Since first-order perturbations are much larger than those of second order, the union of odd AHs can be treated in terms of first-order perturbation theory alone. To this approximation, the energy of union arises entirely from interaction of the NBMOs (Fig.3.8).

PMO Treatment of Conjugated Systems

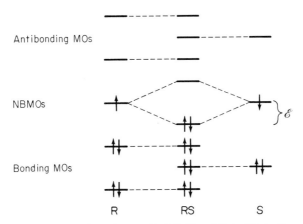

FIGURE 3.8. Union of two odd AHs R and S to RS, in the approximation of first-order perturbation theory.

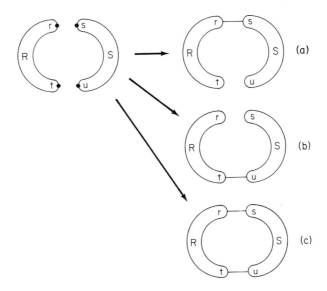

FIGURE 3.9. First-order π energies of union of two odd AHs R and S (a) through atom r in R to atom s in S; $(\delta E_\pi)_1 = 2a_{0r}b_{0s}\beta_{rs}$; (b) through atom t in R to atom u in S; $(\delta E_\pi)_2 = 2a_{0t}b_{0u}\beta_{tu}$; (c) through the atom pairs rs and tu simultaneously; $\delta E_\pi = 2a_{0r}b_{0s}\beta_{rs} + 2a_{0t}b_{0u}\beta_{tu} = (\delta E_\pi)_1 + (\delta E_\pi)_2$.

If R and S are united through atom r in R to atom s in S (see Fig. 2.2), the interaction energy $\mathscr{E}$ (Fig. 3.8) is given (see Fig. 2.4) by

$$\mathscr{E} = a_{0r}b_{0s}\beta_{rs} \tag{3.23}$$

where a_{0r} and b_{0s} are the corresponding NBMO coefficients. The π energy of union δE_π is then given by*

$$\delta E_\pi = 2\mathscr{E} = 2a_{0r}b_{0s}\beta_{rs} \tag{3.24}$$

The NBMO coefficients can be found very simply by using Longuet-Higgins' rule.

The important point is that we are dealing here with a first-order perturbation and first-order perturbations *are* additive (Section 2.6). The π energy of union of two odd AHs at two or more pairs of points *is* therefore a sum of the π energies of union for each pair of points separately. This is illustrated in Fig. 3.9. This approach provides us with the needed entry to cyclic conjugated systems and it can further clarify the properties of open-chain polyenes.

3.9 Alternation of Bonds in Polyenes

Our second-order treatment of polyenes led to a picture in terms of "localized" double and single bonds, the former presumably being stronger; this can be further justified by the following argument.

Rupture of a "single" bond in an open-chain polyene gives rise to two even AHs; e.g.,

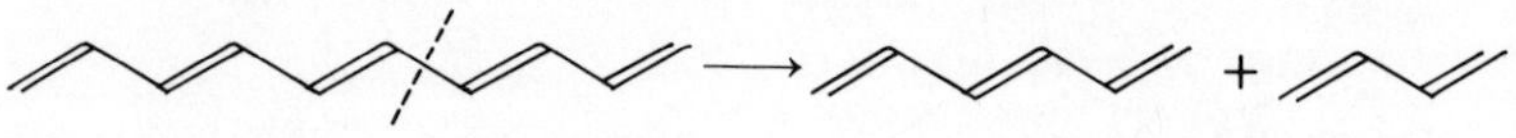

$$\text{(structure)} \longrightarrow \text{(structure)} + \text{(structure)} \tag{3.25a}$$

Union of these to regenerate the original AH involves only a small second-order change in π energy. The resulting bond will therefore be little stronger than its σ component.

Rupture of a double bond in an open-chain polyene leads, on the other hand, to two odd AHs; e.g.,

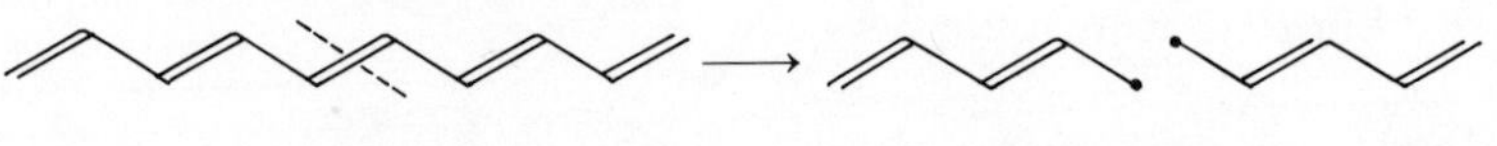

$$\text{(structure)} \longrightarrow \text{(structure)} \cdot \text{(structure)} \tag{3.25b}$$

Union of these to regenerate the original polyene involves a large first-order change in π energy. The resulting bond will therefore be very much stronger than its σ component and so very much stronger than any "single" bond in the polyene.

---

* The signs of the coefficients a_{0r} and b_{0s} are arbitrary; in equation (3.24) it is assumed that we have arranged for their product to be positive. If the product were negative, we would have to set $\delta E_\pi = -2a_{0r}b_{0s}\beta_{rs}$ to ensure that the electrons in RS occupy the lower of the two MOs arising from interaction of the NBMOs; see Fig. 3.8.

This argument accounts for the fact that the bonds in such polyenes do alternate in length and strength, compounds of this type being well represented by "classical" structures with alternating C=C and C—C bonds.

3.10 Even Monocyclic Polyenes; Aromaticity and Antiaromaticity; Hückel's Rule

Let us return to a problem that baffled us earlier (Section 3.7), i.e., the estimation of π energies of cyclic polyenes.

Both 1,3,5-hexatriene and benzene can be formed by union of two allyl radicals,

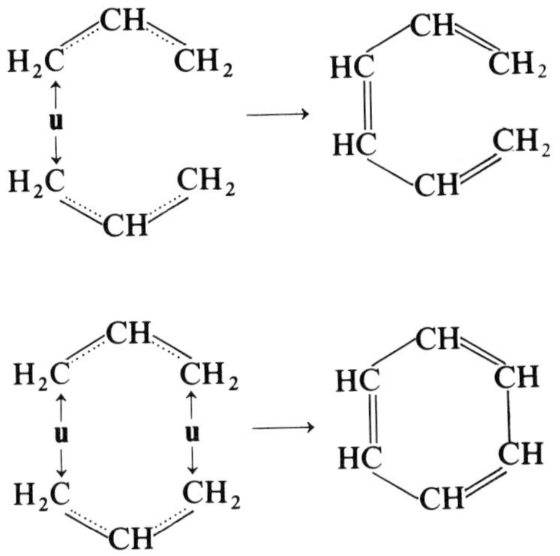

(3.26)

Allyl is an odd AH in which the NBMO coefficients can be found at once by Longuet-Higgins' rule;

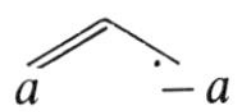

$$a = 1/\sqrt{2} \qquad (3.27)$$

The two π energies of union can now be found from equation (3.24), remembering that the π energy of simultaneous union of two odd AHs at two points is a sum of the two individual unions,

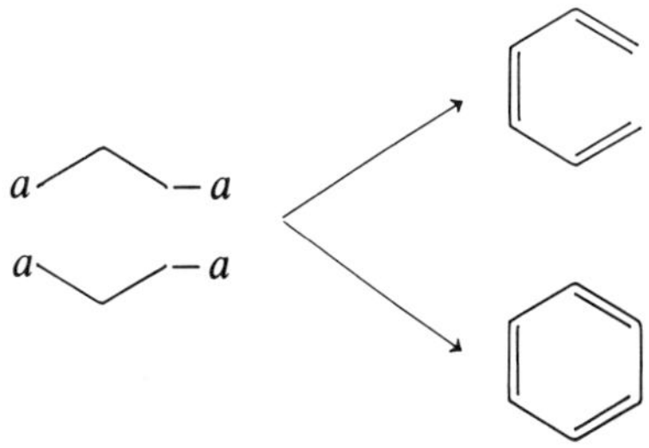

$$\delta E_\pi = 2\beta \cdot a \cdot a = 2a^2\beta = \beta \qquad (3.28a)$$

$$\delta E_\pi = 2a^2\beta + 2a^2\beta = 2\beta \qquad (3.28b)$$

Thus the π energy for double union to benzene is greater than that for single union to hexatriene and the difference is large, being a first-order effect. This of course accounts for the fact that benzene is much more stable than one would expect it to be from analogy with open-chain polyenes, the heat of atomization being greater than that calculated by summing CH, C=C, and "polyene" C—C bond energies. Benzene is an *aromatic* hydrocarbon.

The procedure just described can be used quite generally in the study of π energies of even cyclic systems. By constructing a cyclic system and an open-chain system from the same pair of odd AHs, we can use the easily calculated first-order π energies of union to compare the energies of the two even systems. A particularly convenient procedure is to dissect the two even AHs into a common odd AH radical with one carbon atom less, and methyl. In the methyl radical, the odd electron occupies a carbon $2p$ AO with energy α_c; this is the same as the energy of an NBMO. Methyl can thus be regarded as the limiting case of an odd AH in which there is just one starred atom and of course no unstarred ones; the "NBMO" is now composed entirely of a single carbon $2p$ AO. The corresponding coefficient in equation (3.24) is then unity. We can compare hexatriene and benzene by forming them both by union of the pentadienate radical CH_2=CH—CH=CH—CH_2· and methyl, i.e. (representing methyl by a large dot),

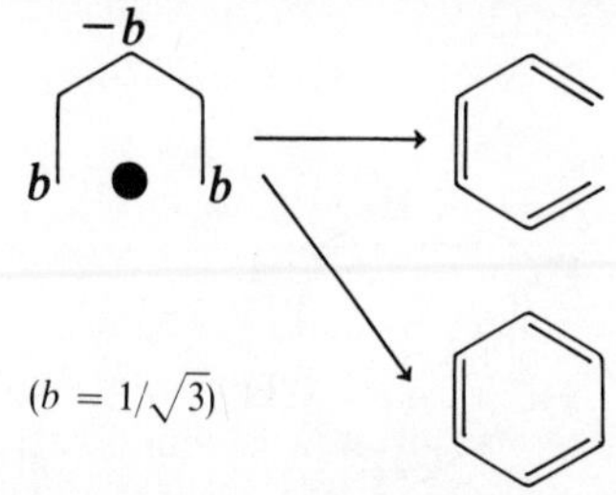

$$\delta E_\pi = 2b \cdot 1 \cdot \beta = 2b\beta$$

$(b = 1/\sqrt{3})$

$$\delta E_\pi = 2b\beta + 2b\beta = 4b\beta$$

(3.29)

Here again the π energy of double union (to benzene) is greater than for single union (to hexatriene).

Next consider union of allyl and methyl to butadiene or cyclobutadiene:

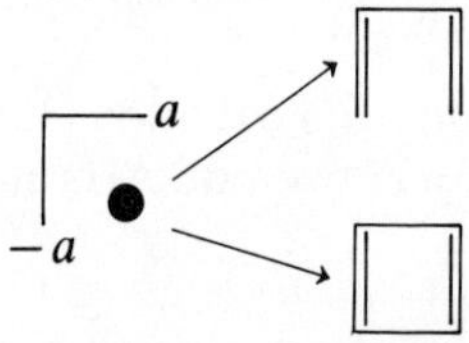

$$\delta E_\pi = 2a\beta$$

$$\delta E_\pi = 2a\beta + 2(-a)\beta = 0$$

(3.30)

This time the π energy of double union is *less* than that for single union so cyclobutadiene should be *less* stable than butadiene. This is because the NBMO coefficients at the ends of the allyl radical have opposite signs. The first-order perturbations are out of phase with one another and so cancel. We therefore predict cyclobutadiene to be as much *less* stable than a corresponding open-chain polyene as benzene is *more* stable. Cyclobutadiene should be *antiaromatic*. Cyclobutadiene is indeed extraordinarily reactive; it dimerizes immediately when it is formed, even in dilute solution at low temperatures.

PMO Treatment of Conjugated Systems

This treatment can be extended at once to the larger even cyclic polyenes. These can be formed by union of a linear, odd AH with methyl,

$$CH_2{=}CH{-}(CH{=}CH)_m{-}CH{=}CH_2 \quad \delta E_\pi = 2c\beta$$

$$CH_2{-}CH{-}(CH{-}CH)_m \cdots CH_2 \leftarrow \mathbf{u} \rightarrow CH_3 \quad (3.31)$$
$$\phantom{CH_2{-}}c \phantom{CH{-}}\pm c \phantom{(CH{-}CH)_m}\mp c$$

$$\begin{array}{c} CH{-}(CH{=}CH)_m{-}CH \\ \| \phantom{(CH{=}CH)_m} \| \\ CH{-\!-\!-\!-\!-\!-\!-\!-}CH \end{array} \quad \delta E_\pi = 2c\beta \pm 2c\beta$$

Whether the cyclic polyene is more or less stable than the open-chain one depends on the relative signs of the NBMO coefficients at the ends of the odd AH. It is easily seen that the signs of the terminal coefficients are the same, and the resulting even AH consequently aromatic, if the odd AH contains $4n + 1$ atoms (n being an integer) and the resulting polyene consequently contains $4n + 2$ atoms [in this case, m in equation (3.31) is an odd integer], and that the signs are opposite if the odd AH contains $4n - 1$ atoms so that the polyene contains $4n$ [m in equation (3.31) is an even integer]. Since the number of π electrons is equal to the number of atoms, we conclude that monocyclic polyenes with $(4n + 2)$ π electrons are aromatic, while those with $4n$ π electrons are antiaromatic. As we shall see, this rule also applies to odd monocyclic systems; it is known after its discoverer as *Hückel's rule*. The PMO method provided the first satisfactory derivation of Hückel's rule and also the first clear prediction that cyclic polyenes with $4n$ atoms should be not merely not aromatic, but *antiaromatic*, i.e., actually destabilized by their cyclic structure. This prediction has been confirmed for the whole series of cyclic polyenes, now termed *annulenes*, up to [30]annulene, with the exception of [28]annulene, which has not yet been studied.

It is worth noting that the magnitude of the NBMO coefficients in equation (3.31) decreases as m increases, suggesting that the aromaticity of a $[4n + 2]$ annulene should decrease as the size of the ring increases. This appears to be the case.

3.11 Bond Alternation in Annulenes

We have so far calculated NBMO coefficients by using the simplified form of Longuet-Higgins' rule in which all carbon–carbon β's are assumed equal; this in effect assumes that the CC bonds all have similar lengths. Since the CC π bond becomes stronger the shorter it is (Section 1.13), and since the π-bond energy of such a bond is -2β, β must be numerically greater the shorter is the CC bond.

We have seen (Section 3.9) that the bond lengths in open-chain polyenes alternate. The CC resonance integral must then have a larger value (β'') for pairs of atoms forming "double" bonds than (β') for pairs of atoms forming single bonds. Will this situation also hold in a cyclic polyene, i.e., an annulene? In other words, will an alternation of bond lengths raise or lower its energy?

In order to compare the energies of linear and cyclic polyenes in which the bond lengths alternate, we must study the union of methyl with an odd AH radical in which the bond lengths alternate. Consider, for example, allyl:

$$\overset{b}{C}H_2 = CH - \overset{a}{C}H_2 \tag{3.32}$$

The values of the NBMO coefficients are given by the rigorous form of Longuet-Higgins' rule [equation (3.9)]

$$a\beta' + b\beta'' = 0 \tag{3.33}$$

or

$$b = -a\zeta \tag{3.34}$$

where

$$\zeta = \beta'/\beta'' \tag{3.35}$$

The alternation of bond lengths in the allyl radical means that β'' is greater than β'. Thus ζ is less than unity.

Extending this argument and assuming equal alternation down the chain, one can see that in a large, odd AH, the NBMO coefficients are as follows:

$$\cdots -\overset{a\zeta^4}{C}H=CH-\overset{-a\zeta^3}{C}H=CH-\overset{a\zeta^2}{C}H=CH-\overset{-a\zeta}{C}H=CH-\overset{a}{C}H_2 \tag{3.36}$$

The π energies of single and double union with methyl to form a polyene or an annulene are then

$$\text{polyene:} \quad \delta E_\pi = 2a\beta \tag{3.37}$$

$$\text{annulene:} \quad \delta E_\pi = 2\beta[a + (-\zeta)^n] \tag{3.38}$$

where the number of conjugated atoms in the odd AH is $2n + 1$. The "aromatic energy" of the annulene E_{an} is then given by

$$E_{an} = -2\beta(-\zeta)^n \tag{3.39}$$

Since $\zeta < 1$, this is numerically less than the corresponding value for equal bond lengths (when $\zeta = 1$). Alternation of bond lengths makes aromatic compounds less aromatic and antiaromatic compounds less antiaromatic. Thus we arrive at the general rule:

Bonds in aromatic compounds tend not to alternate but bonds in antiaromatic compounds alternate.

The available evidence is entirely in accord with this prediction. The bonds in aromatic compounds such as benzene or [18]annulene all have similar lengths, close to 1.40 Å, whereas the bonds in antiaromatic compounds seem to alternate. A good example of the latter statement is provided by cyclooctatetraene, which has the "tub" geometry (15). Cyclooctatetraene can undergo a configurational inversion to the mirror image tub structure (16), presumably via the planar intermediate (17). If the bond lengths in (17) were equal, there would be a scrambling of single and double bonds during the inversion. It has, however, been shown by Anet that during the inversion the single and double bonds retain their identity, showing that the most stable form of planar cyclooctatetraene is one in which the bond lengths alternate.

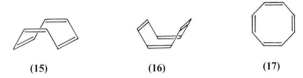

(15)   (16)   (17)

3.12 Polycyclic Polyenes

Bicyclic polyenes, even nonalternant ones, can be treated in a similar manner since they can be derived by union of an open-chain, odd AH and methyl. Figure 3.10 illustrates the treatment of the C_{10} series by union of methyl and nonatetraenyl.

It will be seen that the aromatic energy of naphthalene $(4d\beta)$ is double that of the monocyclic systems $(2d\beta)$; naphthalene should therefore be a bicyclic aromatic system, and its chemistry is consistent with this. Phenylbutadiene and [10]annulene are predicted to be similar in stability; in practice [10]annulene is unstable because it cannot exist in a planar geometry, due to mutual interference by the central hydrogen atoms (18). The internal methylene derivative (19) of (18) is, however, stable, being almost planar.

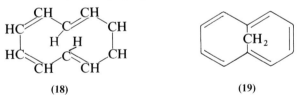

(18)   (19)

Another interesting way in which (18) can be obtained in a stable form is by triple union of nonatetraenyl and methyl, as indicated in Fig. 3.10(d). The third bond involves an *inactive* atom in the odd AH, i.e., one where the NBMO coefficient vanishes; this union therefore makes no contribution to the first-order π energy [equation (3.23)]. The resulting bicyclic hydrocarbon, azulene, should therefore be in effect a monocyclic aromatic compound, the central bond playing no part in the aromatic system. Azulene is indeed much less stable than the isomeric naphthalene and the length of the central bond (1.48 Å) is similar to that of "single" bonds in open-chain polyenes, e.g., the central bond in butadiene.

This argument can be extended to other even, nonalternant cyclic polyenes in which there are two adjacent odd-numbered rings. Thus Fig. 3.11

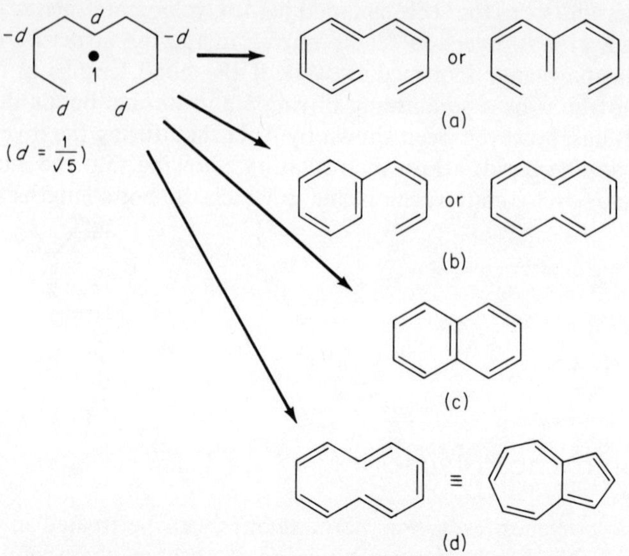

FIGURE 3.10. Union of methyl and nonatetraenyl to form (a) two isomeric decapentaenes; (b) 1-phenylbutadiene or [10] annulene; (c) naphthalene; (d) azulene. Values of $\delta E_\pi (d = 1/\sqrt{5})$: (a) $2d\beta$, (b) $4d\beta$, (c) $6d\beta$, (d) $4d\beta$.

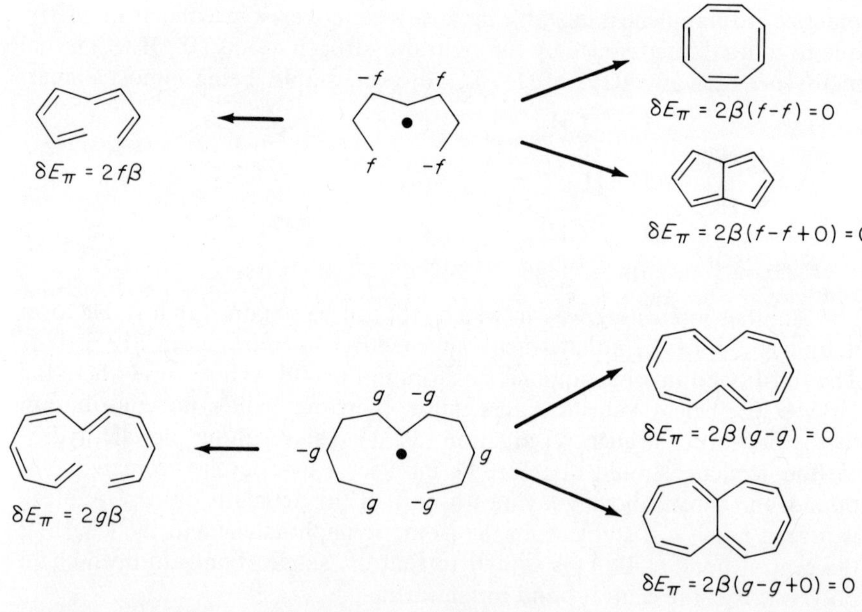

FIGURE 3.11. PMO analysis leading to the prediction that both pentalene and heptalene should be antiaromatic.

PMO Treatment of Conjugated Systems

shows that pentalene is antiaromatic, being an analog of cyclooctatetraene, while heptalene is likewise an antiaromatic analog of [12]annulene. These conclusions are of some interest since it had originally been assumed on the basis of Hückel MO calculations that pentalene and heptalene would prove to be aromatic. Indeed, they were given these names by the theoreticians responsible for the calculations in the confident expectation that they would both prove of importance in chemistry. At that time neither compound was known. Recently heptalene and derivatives of pentalene have been synthesized; both are extremely reactive, confirming their antiaromatic nature.

The PMO treatment can be extended to individual rings in polycyclic systems. Consider, for example, one of the terminal rings in anthracene (20). Anthracene can be constructed by union of methyl with the odd AH radical below in which the NBMO coefficients are as indicated. The π energies of union to the two bicyclic systems (21) and (22) and to (20) are as indicated. It will be seen that anthracene is more stable than (21) or (22), implying that that the terminal ring is aromatic.

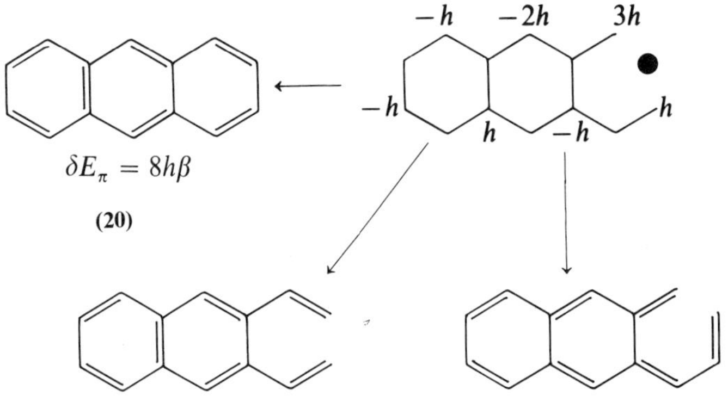

$\delta E_\pi = 8h\beta$

(20)

$\delta E_\pi = 6h\beta$

(21)

$\delta E_\pi = 2h\beta$

(22)

This result can be generalized. In order to study the aromaticity of a given ring in an even AH, one considers the odd AH R obtained from it by loss of one carbon (Fig. 3.12a). Union of this odd AH with methyl can give rise to the parent cyclic even AH (Fig. 3.12b) or to one or the other (Fig. 3.12c,d) of two even AHs in which the ring in question has been opened. Figure 3.12 shows that the cyclic compound will be more stable than either open-chain analog if, and only if, the NBMO coefficients a_{0r} and a_{0s} (Fig. 3.12a) at the relevant atoms in the odd AH have similar signs. If they have opposite signs, the ring is antiaromatic.

It can be shown that the NBMO coefficients have similar signs if the ring in question has $4n + 2$ atoms and opposite signs if it has $4n$ atoms. This leads to an extension of Hückel's rule:

In a polycyclic AH, a ring with $(4n + 2)$ atoms is aromatic and one with $4n$ atoms is antiaromatic.

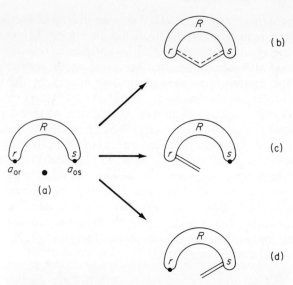

FIGURE 3.12. Analysis of aromaticity in a given even AH by considering union of (a) methyl and an odd AH with one atom removed from the ring to form (b) the parent cyclic hydrocarbon; (c, d) two open-chain analogs. Values of δE_π: (b) $2\beta(a_{0r} + a_{0s})$, (c) $2\beta a_{0r}$, (d) $2\beta a_{0s}$. The ring is aromatic if $(a_{0r} + a_{0s}) > a_{0r}$ or a_{0s}, which is so only if a_{0r} and a_{0s} have similar signs.

3.13 Intramolecular Union; Monocyclic Nonalternant Hydrocarbons

The procedure developed above can be used quite generally to study the stability of cyclic even AHs, and also of nonalternant even hydrocarbons in which there are two adjacent odd-numbered rings. It cannot, however, be extended to other odd systems since an odd conjugated system cannot be derived by union of two odd AHs. Here we need the third type of perturbation considered in Section 2.4, i.e., intramolecular union between two positions in an AH.

Intramolecular union between positions r and s in an AH corresponds to a change in the corresponding resonance integral from zero to some final value β_{rs}. The corresponding change in π energy [equation (2.13)] is $2p_{rs}\beta_{rs}$, where p_{rs} is the rs bond order in the original AH. We can represent the intermolecular union in AH as follows:

We are concerned only with the case where union gives rise to an odd-numbered ring, because even conjugated rings can be treated very simply by the methods developed in the preceding sections. If the new ring is odd-numbered, then atoms r and s must be of similar type (both starred or both unstarred). The rs bond order is then given by the pairing theorem, namely

$$\left.\begin{array}{l}\text{even AH:}\\ \text{odd AH radical:}\\ \text{unstarred positions in an odd AH:}\end{array}\right\} \quad p_{rs} = 0 \hspace{2cm} (3.40)$$

starred positions in an odd AH anion: $\quad p_{rs} = a_{0r}b_{0s}$
starred positions in an odd AH cation: $\quad p_{rs} = -a_{0r}b_{0s}$

We call a ring *aromatic* if opening it leads to an increase in π energy, and *antiaromatic* if opening it leads to a decrease in π energy. If opening the ring has no effect, we call it *nonaromatic*. On this basis, we can see that the following systems are nonaromatic:

1. Even systems containing an odd-numbered ring; *example*: fulvene:

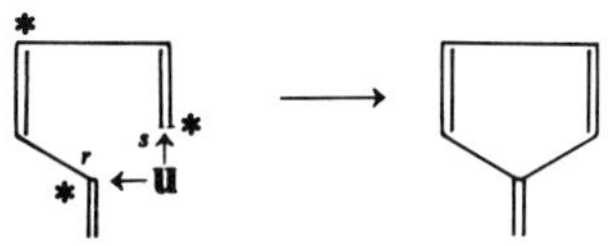

   $p_{rs} = 0; \quad \delta E_\pi = 0$

2. Odd, nonalternant, monocyclic radicals; *example*: the cyclopentadienate radical:

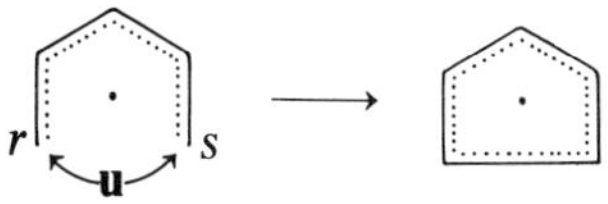

   $p_{rs} = 0; \quad \delta E_\pi = 0$

3. Odd, nonalternant ions formed by union through unstarred positions; *example*: the 1,5-bismethylenecyclopentadienate anion:

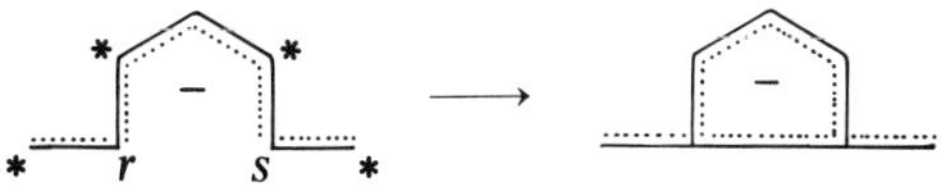

   $p_{rs} = 0; \quad \delta E_\pi = 0$

Aromatic and antiaromatic systems arise only by union between starred positions in an odd AH ion. As Fig. 3.13 shows, the cyclopentadienate anion is aromatic and the cation is antiaromatic, while the tropylium cation is aromatic and the corresponding anion is antiaromatic.

This argument can be generalized.

A cyclic, nonalternant ion can be derived by union of a corresponding linear, odd AH ion through the terminal atoms. If the corresponding NBMO

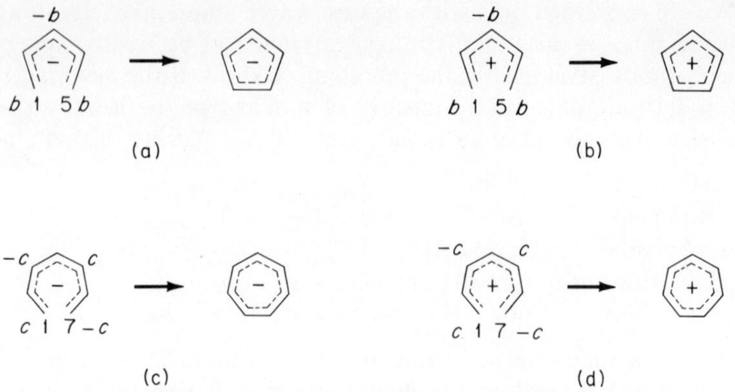

FIGURE 3.13. PMO analysis of aromaticity in the [5] and [7] annulenium ions. (a) $p_{15} = a_{01}a_{05} = b^2$, $\delta E_\pi = 2p_{15}\beta = 2b^2\beta$; anion aromatic. (b) $p_{15} = -a_{01}a_{05} = -b^2$, $\delta E_\pi = -2p_{15}\beta = -2b^2\beta$; cation antiaromatic. (c) $p_{17} = a_{01}a_{07} = c(-c) = -c^2$, $\delta E_\pi = 2p_{15}\beta = -2c^2\beta$; anion antiaromatic. (d) $p_{17} = -a_{01}a_{07} = -c(-c) = c^2$, $\delta E_\pi = 2p_{17}\beta = 2c^2\beta$; cation aromatic.

coefficients are a_{0r} and a_{0s}, the energy of union is

$$\text{anion} \quad \delta E_\pi = 2p_{rs}\beta = 2a_{0r}a_{0s}\beta$$
$$\text{cation} \quad \delta E_\pi = 2p_{rs}\beta = -2a_{0r}a_{0s}\beta \quad (3.41)$$

Thus if a_{0r} and a_{0s} have similar signs, the anion is aromatic and the cation antiaromatic, while if a_{0r} and a_{0s} have opposite signs, the cation is aromatic and the anion is antiaromatic. The signs of the coefficients can be seen from equation (3.31). The signs are the same if the odd AH has $4n + 1$ atoms and opposite if it has $4n + 3$ atoms. Thus anions with $4n + 1$ atoms and cations with $4n + 3$ atoms are aromatic, while cations with $4n + 1$ atoms and anions with $4n + 3$ atoms are antiaromatic. Since the anion has one π electron more than the radical and the cation one π electron less, and since the radical has as many π electrons as conjugated carbon atoms, one can see that the ions with $(4n + 2)$ π electrons are aromatic, while those with $4n$ π electrons are antiaromatic. Thus Hückel's rule applies equally well to odd, cyclic conjugated ions.

This analysis can also be extended to hydrocarbons containing a single odd-numbered ring in an otherwise alternant system. Opening the odd-numbered ring then gives rise to an AH from which the parent compound can be derived by intramolecular union, i.e.,

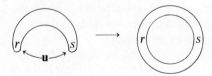

The π energy of union is given as before by $2p_{rs}\beta$. Everything depends on the value of the rs bond order. If the AH is even or is an odd radical, or if r and s are unstarred positions in an odd AH ion, p_{rs} vanishes and the ring is non-aromatic. If, however, r and s are starred positions in an odd AH ion, the bond order is given as before by

$$p_{rs} = a_{0r}a_{0s} \quad \text{for an anion} \tag{3.42}$$
$$p_{rs} = -a_{0r}a_{0s} \quad \text{for a cation}$$

If the coefficients have the same sign, the anion is aromatic and the cation is antiaromatic. If they have opposite signs, the cation is antiaromatic and the anion is aromatic. Now the relative signs of such coefficients have already been established at the end of Section 3.12, where exactly the same odd AH was considered in the analysis of the aromaticity of an even ring in an arbitrary even AH. The coefficients a_{0r} and a_{0s} have similar signs if the odd ring in the hydrocarbon has $4n + 1$ atoms (so that the corresponding even ring formed by union with methyl has $4n + 2$; cf. Section 3.12) and they have opposite signs if it has $4n - 1$ atoms. A $(4n + 1)$-membered ring in an otherwise alternant compound will therefore be aromatic as the anion and antiaromatic as the cation, while the converse will be true for a $(4n - 1)$-membered ring. The available experimental evidence supports these predictions; thus the fluorenyl anion (23) is aromatic and the cation (24) is antiaromatic, while benzotropylium (25) is aromatic and the corresponding anion (26) is antiaromatic.

185702

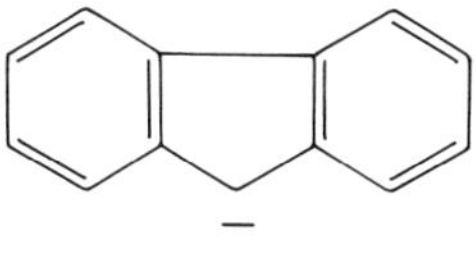

(23) 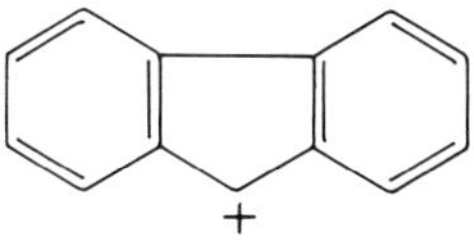   (24)

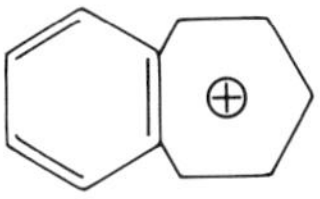

(25) 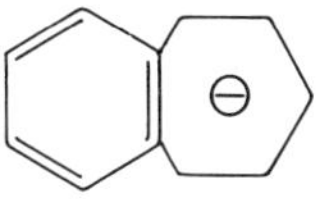   (26)

3.14 Essential Single and Double Bonds; General Rules for Aromaticity

One can usually write two or more classical structures for a given cyclic conjugated molecule; the Kekulé structures for benzene are of course the classic example. However, one often finds that even in the case of molecules containing conjugated rings, certain bonds are single or double in all the possible classical structures that one can write. Such a bond is described as an

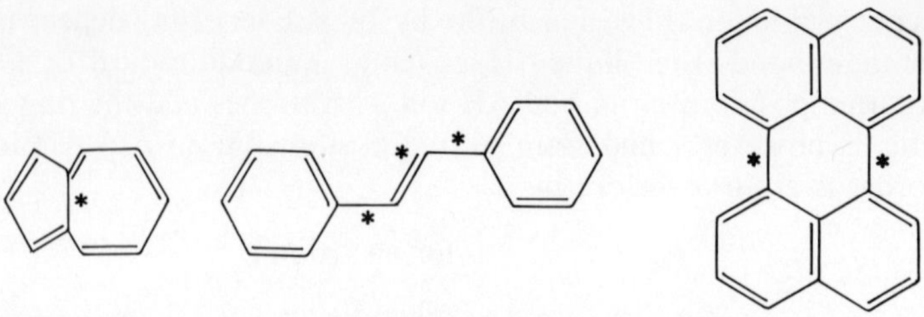

FIGURE 3.14. The bonds marked with asterisks are single or double in all the possible classical structures that one can write.

essential single or *essential double* bond. Figure 3.14 shows some typical examples.

It can be shown by an extension of the argument in the appendix (see Section 3.6) that the contribution of such bonds to the heat of atomization of a molecule can be represented in terms of the corresponding C=C and C—C bond energies, the latter of course having the "polyene" value $[E'_{C-C}$; equation (3.17)]. The total π energy of delocalization therefore remains unchanged if such bonds are removed. Using this result and the results obtained in Sections 3.12 and 3.13, we arrive at the following general rules for aromaticity in a conjugated hydrocarbon.

1. Delete all essential single and essential double bonds.
2. If charges or unpaired electrons are present, write a structure in which the charges or odd electrons are localized in odd-numbered rings.
3. Even rings with $4n + 2$ atoms are then aromatic; even rings with $4n$ atoms are antiaromatic.
4. Odd rings with $4n + 1$ atoms are aromatic if they contain a negative charge and antiaromatic if they contain a positive charge.
5. Odd rings with $4n + 3$ atoms are aromatic if they contain a positive charge and antiaromatic if they contain a negative charge.
6. Odd radical rings are nonaromatic.

Some examples follow:

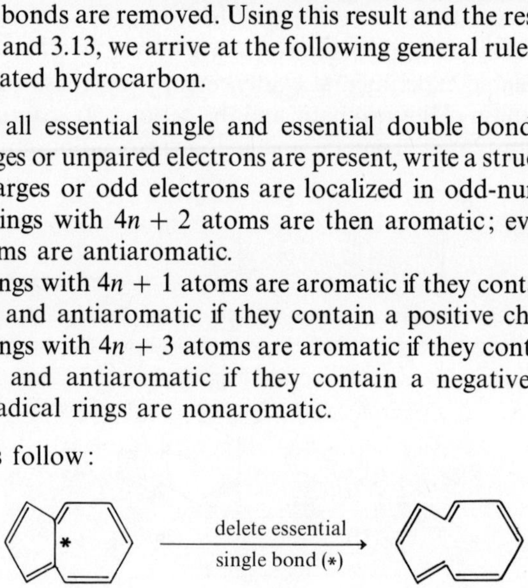

delete essential single bond (∗)

$4 \times 2 + 2 = 10$ atoms; aromatic

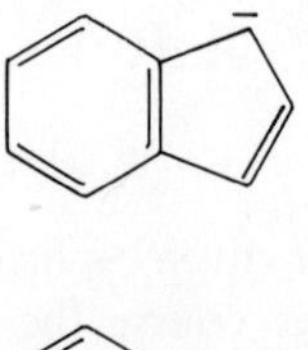

both rings aromatic; one with $4 + 2 = 6$ atoms, the other an anion with $4 + 1 = 5$ atoms

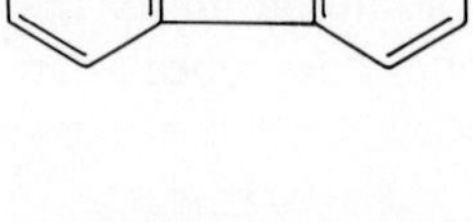

terminal rings aromatic $4 + 2 = 6$ atoms, central ring antiaromatic (4 atoms)

Earlier attempts to generalize Hückel's rule were unsuccessful since they relied on the magic "$(4n + 2)$-electron" formula. This fails. Thus pyrene (27) is aromatic, though it contains $4 \times 4 = 16$ electrons. According to the analysis given here, all four rings are aromatic since each contains $4 + 2 = 6$ atoms.

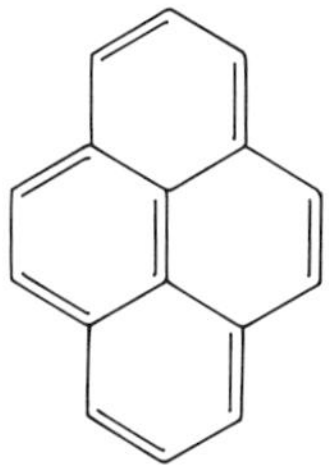

(27)

3.15 Significance of Classical Valence Structures

The rules for aromaticity in the previous section do need qualification in one respect. We have throughout this chapter assumed implicitly that it is possible to write at least one classical structure, i.e., a structure obeying the rules of valence and stereochemistry, for each molecule. Now this is not necessarily the case. Consider, for example, triangulene (28). However one struggles, one cannot write a structure in which all the carbon atoms are linked in pairs by double bonds. There are always two nonadjacent atoms left over. The reason for this can be seen at once if we star the molecule (29). There are two more starred atoms (total 12) than unstarred ones (total 10). Since each double bond in a classical structure must by definition link a starred atom to an unstarred one, it is clearly impossible to pair up the atoms into doubly bonded pairs unless the numbers of starred and unstarred atoms are the same.

It is not immediately apparent that this should make any difference, since we have shown that aromatic compounds are not in any case represented by classical structures. The bonds in them do not alternate in length in the way that the bonds in a polyene do and the π electrons in them cannot at all be regarded as localized in specific bonds. Yet in practice no stable hydro-

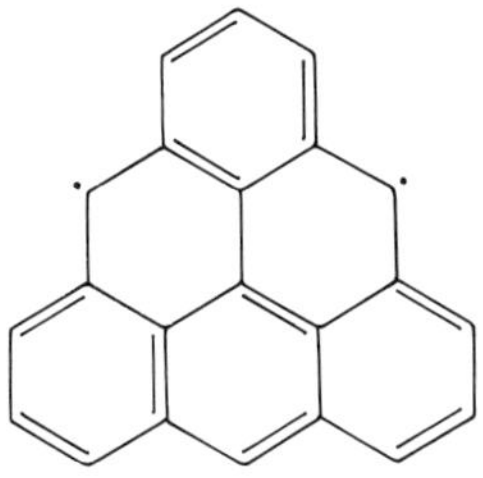

(28) 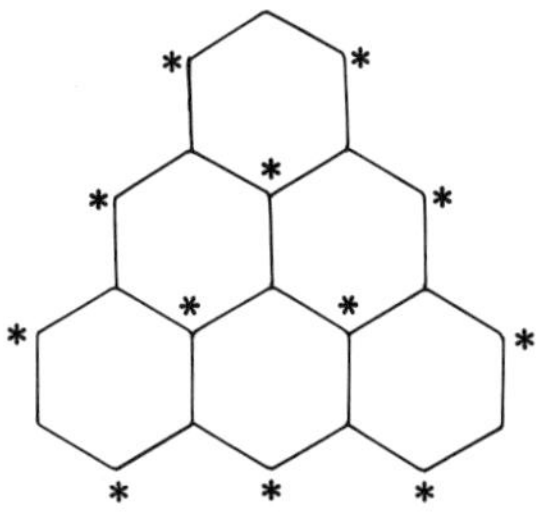   (29)

carbon is known for which it is impossible to write at least one classical structure. Thus while Clar probably succeeded in making triangulene, it polymerized immediately, as one would expect if it had the biradical structure implied in (28). Why should classical structures be so important even in the case of molecules which would not in any case be expected to conform to them?

The answer is given in the following theorem, which we will now prove:

An even AH for which no classical structure can be written is much less stable than a "classical" isomer and also will exist as a biradical with two unpaired electrons.

First let us consider the case of an even AH with two more starred atoms than unstarred ones, e.g., (28), or *m*-quinodimethane (Fig. 3.15a). Removal of one of the starred atoms from such an even AH will leave an odd AH (e.g., Fig. 3.15b). Union of the odd AH with methyl to regenerate the original even AH will involve no first-order change in π energy, because the NBMO of the odd AH vanishes at unstarred atoms and the union is taking place through an unstarred atom. On the other hand, union through a starred atom, giving an isomer of the original even AH, will involve a first-order change in π energy; moreover, the new carbon atom, being adjacent to a starred atom in the odd AH, will be unstarred, so that the new even AH contains equal numbers of starred and unstarred atoms and can be represented by a classical structure (Fig. 3.15c,d). Any such classical isomer will therefore be more stable than the original "nonclassical" AH and the difference in energy, being a first-order perturbation, will be large.

Since an NBMO has zero density at unstarred atoms, one would expect it to be unaffected by union at such points. This is in fact the case. In a perturbation treatment of union, all the terms for perturbations involving an

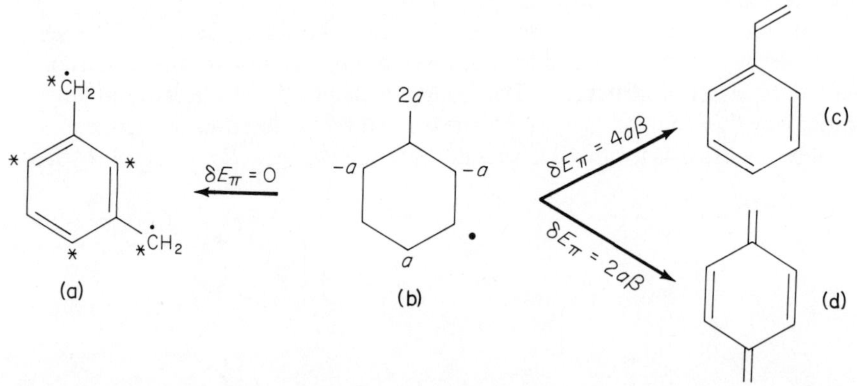

FIGURE 3.15. Union of an odd AH (b) with methyl to form a "nonclassical," even AH (a) involves no first-order change in π energy. The nonclassical AH is therefore less stable than classical isomers (c, d).

MO vanish if the coefficient at the point of union vanishes. The nonclassical, even AH therefore also possesses an NBMO. But from the pairing theorem, *all* the MOs in an even AH are paired. NBMOs in such a system must occur in pairs. The nonclassical, even AH therefore possesses two NBMOs. These, however, have to accommodate only two electrons between them. From Hund's rule, the stable state of the system will then be a triplet in which the electrons occupy different MOs with parallel spins. Such a structure, with two unpaired electrons, should behave as a biradical.

This argument shows that an NBMO in an odd AH will be unaffected by attachment of any group to an unstarred atom. Figure 3.16 shows a simple example. The NBMO of benzyl (a) remains unchanged in its *m*-phenyl derivative (*b–i*). The phenyl group is therefore an irrelevant excrescence so far as the NBMO is concerned. All the NBMO coefficients in it vanish even at starred positions. This result corresponds to the phenomenon described by organic chemists as *cross-conjugation*. In view of the importance of the distinction between positions in NBMOs where the coefficients do or do not vanish, and in view of this demonstration that coefficients of starred atoms vanish in certain cases, it is useful to have a specific term. We call positions in an odd AH with vanishing NBMO coefficients *inactive*, and positions with nonvanishing NBMO coefficients *active*. All unstarred positions are of course necessarily inactive. A portion of an odd AH where all the NBMO coefficients vanish (e.g., the phenyl group in Fig. 3.16i) is termed an *inactive segment*.

There is one type of even AH in which the numbers of starred and unstarred atoms are the same but for which one can nevertheless write no classical structure. These are AHs derived by union of two odd AHs through an inactive position in each; 3,3'-bibenzyl (Fig. 3.17a) is a simple example, containing equal numbers of starred and unstarred positions (Fig. 3.17b). The reason for the lack of a classical structure for these molecules is easily seen. Any classical structure for an odd AH radical necessarily has the odd electron at a starred position. In order to get a classical, even AH from it, we must unite it with another odd AH through one of its starred atoms. The

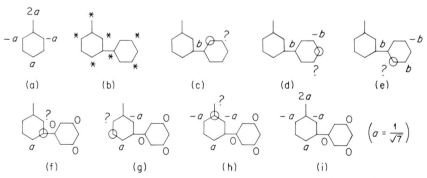

FIGURE 3.16. Demonstration that the phenyl group in *m*-phenylbenzyl is an inactive segment $(a = 1/\sqrt{7})$.

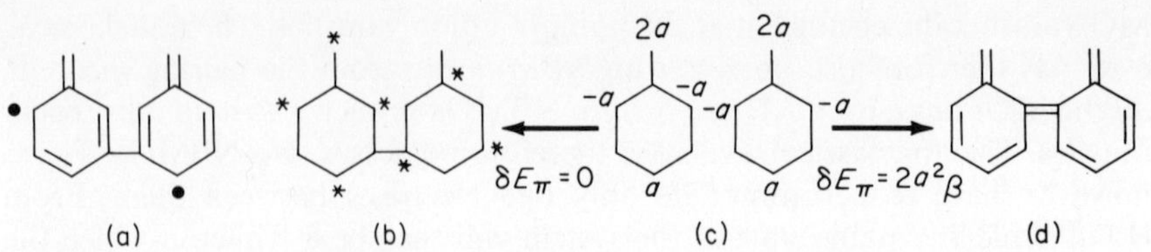

FIGURE 3.17. (a, b) Example of a nonclassical, even AH with equal numbers of starred and unstarred atoms; (c, d) demonstration that such an even AH is less stable than a classical isomer.

argument of Fig. 3.17 shows that nonclassical, even AHs of this type also contain a pair of NBMOs, the NBMOs of the component odd AHs remaining unaffected by union. The nonclassical, even AH can be converted to more stable classical isomers by splitting up the odd AHs and reuniting them through active atoms (Fig. 3.17c,d). Thus our theorem also holds for nonclassical, even AHs of this type.

3.16 Union of an Odd AH with an Even AH

One type of intermolecular union which we have not yet considered is that of an even AH to an odd AH. The discussion will be given only for the simplest case, that of union of an even AH with methyl, since the extension to other odd AHs will be self-evident.

The situation we are considering is indicated in Fig. 3.18(a). Since the MOs of the even AH are all either bonding or antibonding while the AO of methyl is nonbonding, there is no degeneracy and the interaction is a second-order perturbation. It is, however, larger than the perturbation involved in the union of even AHs because the interacting orbitals are closer together in energy. In an even AH, the π energy of union arises from interactions between bonding MOs and antibonding MOs. Here the interaction is between bonding MOs and a nonbonding MO, so the difference in energy is about half that in the union of even AHs. The bond linking the new carbon

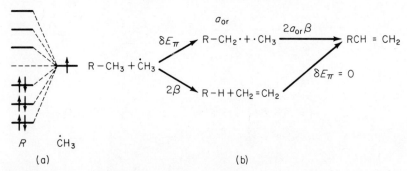

FIGURE 3.18. (a) Interaction of orbitals in union of an even AH R with methyl; (b) calculation of the π energy of union.

to the even AH will therefore be stronger than a normal "polyene" C—C bond, and the π energy of union will be greater than the usual value [A in equation (3.17)].

As usual, these second-order perturbations are difficult to handle and it is fortunate that we can estimate the energy of union by the following trick, which enables us to use first-order perturbation theory. Let us consider the successive unions of an even AH R with two methyl groups to form RCH=CH$_2$ (Fig. 3.18b). If we unite R with one methyl to form RCH$_2\cdot$, the energy of union δE_π will be the quantity we are trying to calculate and the product will be an odd AH. If the NBMO coefficient of the methylene group in this is a_{0r}, the π energy of union of RCH$_2$ with methyl to form RCH=CH$_2$ is $2a_{0r}\beta$. The total energy of union of R + 2CH$_3$ to RCH=CH$_2$ is then $\delta E_\pi + 2a_{0r}\beta$. Alternatively, we could first unite the two methyl radicals to form ethylene, the change in π energy being then the π-bond energy in ethylene, (2β). Union of R with ethylene to form RCH=CH$_2$ involves no further change in π energy since both are even AHs and the change has already been taken into account in the bond energy of the new C—C bond. The total energy of union of R + 2CH$_3$ to RCH=CH$_3$ by this route is thus 2β. Equating these two estimates of the overall energy of union, we have

$$\delta E_\pi + 2a_{0r}\beta = 2\beta \quad (3.43)$$

or

$$\delta E_\pi = 2\beta(1 - a_{0r}) \quad (3.44)$$

The π energy of union is therefore greater, the smaller a_{0r} is. This result is what one might have expected intuitively. In the radical RCH$_2$, the odd electron occupies the NBMO. The density of the NBMO at the methylene group is given by a_{0r}^2. The greater the interaction between R and CH$_2$ in RCH$_2$, the more the unpaired electron will tend to be delocalized into R. Such delocalization should lead to stabilization of RCH$_2$ (i.e., to an increase in the energy of union of R with CH$_3$) and it will of course be reflected by a decrease in a_{0r}.

Some examples follow.

$$\delta E_\pi = 2(1 - 1/\sqrt{2}) = 0.586\beta$$

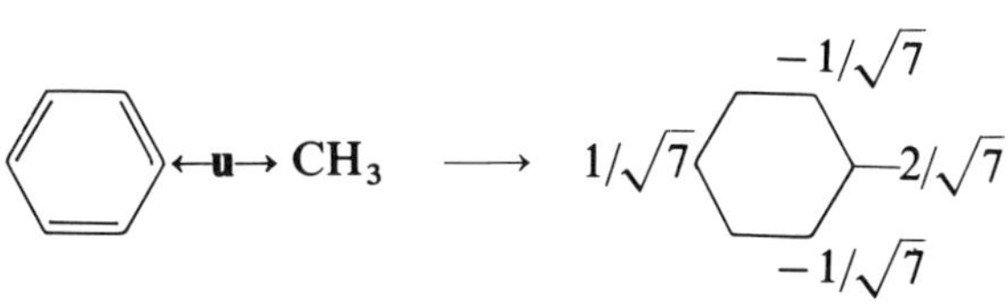

$$\delta E_\pi = 2(1 - 2/\sqrt{7}) = 0.488\beta$$

$$\delta E_\pi = 2(1 - 3/\sqrt{20}) = 0.658\beta$$

Thus the stabilization of the corresponding radicals by delocalization of the π electrons increases in the order benzyl < allyl < α-naphthylmethyl. As we shall see later, the rates of bond-breaking reactions for the aryl methyl systems should follow the same order.

3.17 Hückel and Anti-Hückel Systems

In our approach, the MOs Φ_μ in a molecule are written as linear combinations of AOs ϕ_i,

$$\Phi_\mu = \sum_i a_{\mu i} \phi_i \tag{3.45}$$

Now the choice of signs or *phases* of our *basis set* of AOs is quite arbitrary; in a p AO, we are free to choose which lobe we will label positive and which negative. The choice can make no difference to our treatment because the coefficients $a_{\mu i}$ in equation (3.45) can adapt to them. If we replace the AO ϕ_k in our basis set by $-\phi_k$, we will find when we calculate the MOs Φ_μ that the coefficient $a_{\mu k}$ is now replaced by $-a_{\mu k}$. Thus in H_2, if we took one AO (ϕ_1) to be positive and the other (ϕ_2) to be negative, the MOs would still be $\phi_1 \pm \phi_2$, but now $\phi_1 - \phi_2$ would be the bonding combination. The change appears in our MO treatment as a change in the signs of the integrals involving that AO. Thus with this choice of signs of the AOs in H_2, the overlap integral would be negative instead of (as usual) positive, and the resonance integral β would be positive instead of (as usual) negative.

Since the choice of phases of our AOs is arbitrary and since it would cause great confusion if our β's kept varying in sign, it is convenient to choose the phases in such a way as to make the overlap between pairs of AOs positive so that the β's are negative. In other words, we choose the AOs to overlap in phase. This of course can always be done in a normal π system by taking all the lobes of p AOs on one side of the plane passing through the nuclei as positive and all those on the other side as negative (Fig. 3.19a). That our arguments are unaffected by choice of phases of AOs is illustrated in Fig. 3.19(b–d) by a calculation of the energy of union of pentadienate radical and methyl to form benzene when one of the $2p$ AOs in the pentadienate is out of phase with the rest. Note that the NBMO of pentadienate (Fig. 3.19b) is unchanged, the coefficient of the aberrant AO changing sign, and that the energy of union remains the same as before.

PMO Treatment of Conjugated Systems

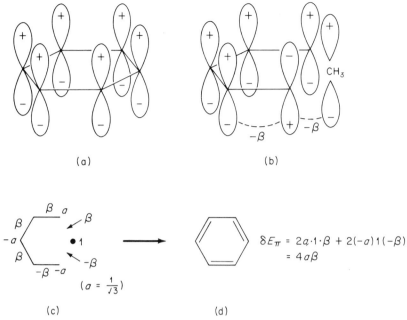

FIGURE 3.19. (a) The AOs in a typical π system can be chosen to overlap in phase; (b) inverting the phase of one AO in the pentadienate radical changes resonance integrals for adjacent π interactions from β to $-\beta$; (c) calculation of NBMO coefficients with the AOs of (b); (d) calculation of the π energy of union of (b) with methyl to form benzene. The result is the same as before [equation (3.29)].

One can, however, envisage two types of π system where this simple scheme breaks down. Suppose, for example, we replace one p AO in a cyclic conjugated system by a d AO (Fig. 3.20a). Since the lobes on one side of the d AO have opposite phases to those on the other side, it is obviously impossible to make the d AO overlap in phase with the AOs on either side of it. There is nothing we can do about this because changing the overall sign of an AO in a ring changes the signs of the overlap integrals on both sides of it. We can increase or decrease the number of *phase dislocations*, i.e., places where AOs overlap out of phase, two at a time in this way but we can do nothing about a single phase dislocation. A system of this kind, as Craig first pointed out, is topologically different from a normal conjugated system containing only $p\pi:p\pi$ bonds.

Another very ingenious system of this type has been proposed by Heilbronner. If we take a conjugated chain of atoms in which the AOs all overlap in phase (Fig. 3.20b), twist it through 180°, and then unite the ends to form a ring (Fig. 3.20c), the AOs at the point of union will overlap out of phase. Here again, we have a ring with one phase dislocation, topologically equivalent to the Craig system of Fig. 3.20(a) and topologically distinct from normal π systems. The π MOs in the twisted ring of Fig. 3.20(c), instead of

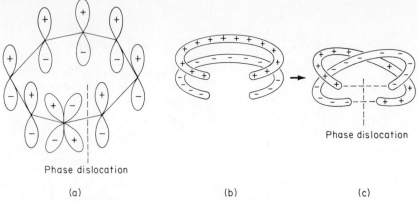

(a)   (b)   (c)

FIGURE 3.20. Anti-Hückel π systems; (a) with a d AO; (b, c) Möbius-strip type.

separating into two distinct lobes, form a single, continuous ring with the topology of a Möbius strip.

While neither of these theoretical possibilities has been realized for stable π-electron systems, it must be remembered that delocalized systems in chemistry are not all of π type. There is no reason why delocalized systems should be two dimensional or involve exclusively π-type overlap of p or d AOs. In a three-dimensional delocalized system involving σ- as well as π-type overlap, we have much more flexibility and, as we shall see presently, examples of this alternative type of topology become important. Since the MO treatment of π systems was first introduced by Hückel, the normal class of delocalized systems with no required phase dislocations may be described as *Hückel* systems. The second topological type with one inevitable phase dislocation may then be called *anti-Hückel* systems.*

The PMO treatment of anti-Hückel systems presents no problems. The distinction between Hückel and anti-Hückel types applies only to cyclic conjugated systems, since in an open-chain system it is always possible to choose the phases of AOs so that there are no phase dislocations. The only problem then is that of aromaticity; under what conditions are the cyclic anti-Hückel systems more or less stable than the open-chain analogs?

It can be seen at once from the argument indicated in Fig. 3.19(c,d), and from equations (3.31) and (3.36) that if we introduce a phase dislocation into an odd AH, we reverse the relative signs of the NBMO coefficients of the terminal AOs. Thus,

---

* Certain authors have referred to these as Möbius systems. This term seems unsatisfactory, for two reasons. First, it is only in the case of the specific type of π system considered by Heilbronner that the MOs have a topology even remotely resembling that of a Möbius strip. Second, the first example of such a system was the one of Fig. 3.20a, which had been pointed out some time previously by Craig. It would be historically more correct, though equally unsatisfactory, to describe systems of this type as $p\pi : d\pi$ systems.

PMO Treatment of Conjugated Systems

Hückel case:
$$\cdots \overset{a}{C}-\overset{-a}{C}-\overset{a^*}{C}-\overset{-a^*}{C}-\overset{a}{C}-C-C-C-C$$

anti-Hückel case:
$$\cdots \overset{-a}{C}-\overset{a}{C}-\overset{-a^*}{C}-\overset{-a^*}{C}-\overset{a}{C}-C-C-C-C$$
$$-\beta$$

Note the inversion of sign between the asterisked coefficients when an intervenening resonance integral changes sign through a phase dislocation. It follows immediately that *the rules for aromaticity in anti-Hückel systems are exactly the opposite of those in Hückel systems*. Every cyclic conjugated system that is aromatic in the Hückel series is antiaromatic in the anti-Hückel series and every cyclic conjugated system that is antiaromatic in the Hückel series is aromatic in the anti-Hückel series. Thus in the anti-Hückel series, neutral rings with $4n$ atoms, anionic rings with $4n + 3$ atoms, and cationic rings with $4n + 1$ atoms are aromatic, while neutral rings with $4n + 2$ atoms, anionic rings with $4n + 1$ atoms, and cationic rings with $4n + 3$ atoms are antiaromatic.

This simple extension of the rules given above (p. 100) extends the PMO treatment to hydrocarbons of all types, both Hückel and anti-Hückel. The distinction between the aromaticity of Hückel and anti-Hückel systems is crucial to the understanding of the stereochemistry of pericyclic reactions (Sections 5.6 and 6.16) which prefer to proceed through σ–π delocalized aromatic transition states.

The treatment of heteroconjugated systems which follows can be applied to both Hückel and anti-Hückel systems in precisely the same way.

3.18 Effect of Heteroatoms

As we have now developed a general theory of π-electron structure in AHs, we can extend this to heteroconjugated molecules using the results for monocentric perturbations of Section 3.5.

First consider the effect of replacing a carbon atom in an even AH by a heteroatom that contributes just one π electron. Thus in pyridine (30), the nitrogen atom, like the carbon atoms in benzene, has sp^2 hybridization; two of the hybrid AOs and two of its five valence electrons are used to form σ bonds. Two more electrons occupy the third hybrid AO, forming a *lone pair* in the plane of the ring. The fifth electron occupies the remaining p AO which forms part of the π system. In pyridine, there are therefore six π electrons, moving in a core of six positive charges (31); the resulting structure is clearly isoconjugate with benzene (cf. Section 1.15, p. 49).

(30)   (31)

We assume as usual that the π bonds formed by heteroatoms do not differ significantly in strength from those formed by carbon atoms (cf. Section 3.5). The change δE_π in π energy on passing from an even AH to an isoconjugate system containing heteroatoms is then attributed entirely to the corresponding changes $\delta\alpha_i$ in the Coulomb integrals α_i, being given [see equation (2.12)] by

$$\delta E_\pi = \sum_i q_i \, \delta\alpha_i \qquad (3.46)$$

In the case of an even AH, the pairing theorem tells us that $q_i = 1$ at each position; hence for even, alternant heteroatomic systems

$$\delta E_\pi = \sum_i \delta\alpha_i \qquad (3.47)$$

But $\delta\alpha_i$ is the energy of an electron in the p AO of atom i relative to the energy of an electron in a carbon $2p$ AO, which we take to be zero. If the molecule were dissociated into atoms, the π electrons would now occupy the p AOs of the atoms, one electron in each [since each atom contributes one π electron; see e.g., (31)]. The energy of these relative to that of a corresponding number of electrons in carbon $2p$ AOs would be just the sum of the corresponding differences $\delta\alpha_i$, i.e., $\sum_i \delta\alpha_i$. Thus, replacing carbon atoms in an even AH by heteroatoms changes both the π energy of the molecule and the total p energy of the corresponding p electrons in the atoms of which it is composed, both by equal amounts. The total π-bond energy of the heteroconjugated system is therefore the same as that of the isoconjugate, even AH. This argument is illustrated in Fig. 3.21.

It follows at once that all the rules derived above for even AHs apply equally well to isoconjugate systems containing heteroatoms. Essential single and double bonds are localized, the rules for aromaticity are unchanged, bond lengths in aromatic systems tend not to alternate while those in antiaromatic systems alternate, and compounds for which one can write

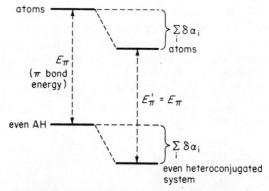

FIGURE 3.21. The total π-bond energies of an even AH and of an isoconjugate system containing heteroatoms are the same (E_π).

no classical structure should be unstable. All these conclusions are supported by experiment. Thus pyridine is aromatic, like benzene, and the extra stabilization ("aromatic energy") relative to a corresponding open-chain system is the same for both. The lengths of bonds in aza analogs of polyenes show the same alternation of "single" and "double" types, whereas in pyridine all the bonds have intermediate lengths, those of the carbon–carbon bonds being similar to the length (1.40 Å) of the bonds in benzene. And finally, whereas o-benzoquinone (32) and p-benzoquinone (33) are stable compounds which have long been known, the *meta* isomer (34), isoconjugate with the *m*-quinodimethane of Fig. 3.15(a), has never been obtained as a stable species:

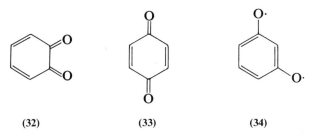

(32)　　　　　　(33)　　　　　　(34)

Similar arguments apply to odd AH radicals, where again $q_i = 1$ at each position. The corresponding heteroconjugated radicals should be similar to the isoconjugate AHs in stability. A reaction producing such a radical from an even, conjugated system should occur equally easily in the case of an isoconjugate system containing heteroatoms. As we shall see (Chapter 5), the experimental evidence suggests that this is certainly the case.

The argument can be extended to odd AH ions. Here there is a complication because in such an ion, the number of π electrons is not the same as the number of atoms. In a cation, there is one less π electron, and in an anion, one more, than there are atoms. Thus in an odd AH anion, one carbon atom contributes two π electrons, whereas the rest contribute one only. In the allyl anion (35), the π MOs are formed by interaction of two carbon atoms with singly occupied $2p$ AOs and one with a doubly occupied $2p$ AO [see (36)].

The two former carbon atoms are neutral, whereas the latter has a negative charge. If we replace a neutral carbon atom by nitrogen (37), the change in energy of the atoms will be equal to the difference $\delta\alpha_N$ between the Coulomb integrals in carbon and nitrogen. If we replace the negatively charged atom (38), the change in energy will be $2\,\delta\alpha_N$. We must allow for this difference in calculating the effect of heteroatoms on π-bonding energies.

$$\underset{(35)}{\overset{\ominus}{\mathrm{H_2C\!\cdots\!CH\!\cdots\!CH_2}}} \qquad \underset{(36)}{\mathrm{H_2C\!-\!CH\!-\!CH_2^{\ominus}}} \qquad \underset{(37)}{\mathrm{H_2C\!-\!N\!-\!CH_2^{\ominus}}}$$

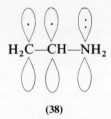

(38)

The difference in π energy between the AH and the hetero analog is given by equation (3.46). For an odd AH anion, using Equation (3.4),

$$\delta E_\pi = \sum_i (1 + a_{0i}^2)\, \delta\alpha_i \qquad (3.48)$$

If each heteroatom contributes one π electron, the change δE_p in the total energy of the p electrons is given as before

$$\delta E_\pi = \sum_i \delta\alpha_i \qquad (3.49)$$

This time there is then a change ΔE_π in the total π-bond energy given by

$$\Delta E_\pi = -\delta E_\pi + \delta E_p = -\sum_i a_{0i}^2\, \delta\alpha_i \qquad (3.50)$$

Since most heteroatoms (e.g., N, O, F, Cl) are more electronegative than carbon and their $\delta\alpha$ values consequently negative, there will be an increase in the total π-bond energy of the anion on passing from the AH to the hetero-conjugated analog. This result, which is illustrated in Fig. 3.22, implies that a reaction converting an even AH to an odd AH anion should be accelerated if we introduce heteroatoms at active positions, assuming that the heteroatoms contribute only one π electron each.

This is certainly the case, as will appear in the next chapter. A simple example is the fact that α-picoline (39) is much more acidic than toluene (41), implying that the process

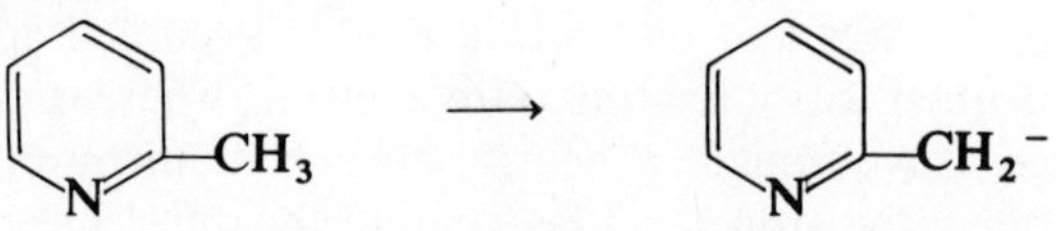

$$\text{(39)} \longrightarrow \text{(40)} \qquad (3.51)$$

occurs much more readily than

$$\text{(41)} \longrightarrow \text{(42)} \qquad (3.52)$$

The anion (40) is derived from the odd AH anion (42) by replacing a starred atom by a heteroatom which contributes just one π electron.

PMO Treatment of Conjugated Systems

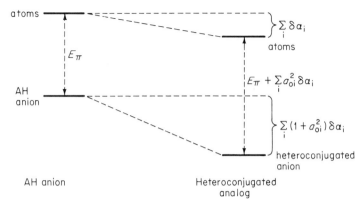

FIGURE 3.22. Effect of replacing atoms in an odd AH anion by heteroatoms, each of which contributes just one π electron.

Now consider the case of a heteroconjugated system isoconjugate with an odd AH anion and containing just one heteroatom which, however, contributes *two* π electrons. A good example is provided by aniline (43), which is isoconjugate with the benzyl anion (44). In aniline, the nitrogen atom uses its three sp^2 hybrid AOs and three of its valence electrons to form σ bonds. The remaining two electrons are therefore contributed to the π system. The core in which the π electrons move is therefore as indicated in (45), nitrogen having two charges. The reason for this is that (43) and (44) are in fact isoelectronic and so have the same total number of electrons. The only difference is that in aniline one of the nuclei has an additional unit of charge.

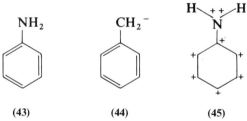

(43)   (44)   (45)

In this case, then, the difference in energy between the contributing p electrons in the heteroconjugated system and that of those in the AH anion is not $\delta\alpha_i$ but $2\delta\alpha_i$. The difference in total π energy is given as before by

$$\delta E_\pi = (1 + a_{0i}^2)\, \delta\alpha_i \tag{3.53}$$

The difference ΔE_π in π-bond energy between the AH anion and the heteroconjugated system is then

$$\Delta E_\pi = -(1 + a_{0i}^2)\, \delta\alpha_i + 2\, \delta\alpha_i = (1 - a_{0i}^2)\, \delta\alpha_i \tag{3.54}$$

Since a_{0i}^2 is necessarily less than unity, ΔE_π is negative; the bond energy of the odd AH anion is therefore reduced by introducing a two-electron heteroatom.

The reason for this is easily seen. The total π-electron density in the heteroatom in the molecule is $1 + a_{0i}^2$. If the heteroatom contributes one π electron, it then gains a_{0i}^2 electrons on forming the molecule; these come from carbon atoms where the binding energy of an electron is less. The extra bond energy ΔE_π in equation (3.50) represents the gain in energy due to transfer of electrons from carbon to the more electronegative heteroatoms. If, however, the heteroatom contributes *two* π electrons, the electron density at it *decreases* on passing from the atoms to the molecule. Since electrons are now transferred from the heteroatom to carbon, i.e., uphill in energy, the effect is to reduce the net bond energy in the molecule [equation (3.54)]. As a result, the energy of union of benzene and ammonia to form (46),

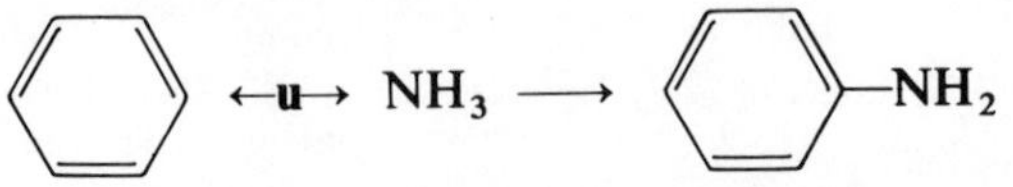

$$\text{C}_6\text{H}_6 \xleftarrow{u} \text{NH}_3 \longrightarrow \text{C}_6\text{H}_5\text{-NH}_2 \qquad (3.55)$$

(46)

is much less than the energy of union of benzene and methyl anion to form (47),

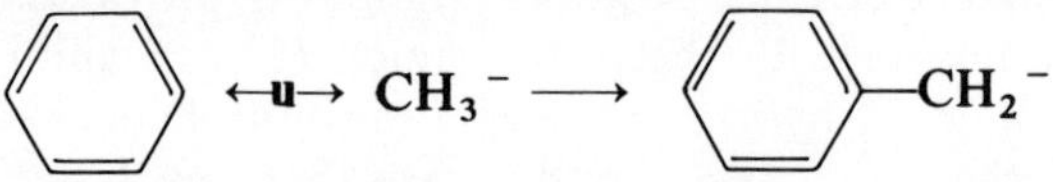

$$\text{C}_6\text{H}_6 \xleftarrow{u} \text{CH}_3^- \longrightarrow \text{C}_6\text{H}_5\text{-CH}_2^- \qquad (3.56)$$

(47)

The π-bond order of the CN bond in aniline is therefore low. This means that its resonance integral will be small and this in turn implies an increase in the associated NBMO coefficient [cf. equation (3.36)] and so in the π-electron density at nitrogen. In other words, aniline is much more like a classical structure with a localized CN bond than is the benzyl anion. The same effect is seen in other cases, too. Thus the π energy of union of ammonia and butadiene to form pyrrole is much less than that for the corresponding carbanion,

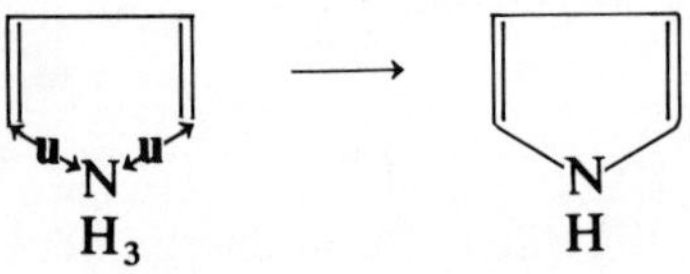

$$\text{butadiene-NH}_3 \longrightarrow \text{pyrrole} \qquad (3.57)$$

$$\text{butadiene-CH}_3^- \longrightarrow \text{C}_5\text{H}_5^- \qquad (3.58)$$

Pyrrole approximates quite closely to a classical structure in which the carbon–carbon bonds alternate strongly in length, whereas the $C_5H_5^-$ ion is completely symmetric.

PMO Treatment of Conjugated Systems

One can of course combine the two types of substitution, introducing into an odd AH anion heteroatoms contributing single π electrons as well as one heteroatom contributing two. Since the effects on the π-bond energy are now opposite [compare equations (3.50) and (3.54)], the net result will depend on their relative magnitudes. It would be asking too much of such a simple treatment as this to make such quantitative assessments; qualitatively, however, the effect of added heteroatoms is clear. Thus the strength of the C—NH$_2$ bond in aniline is greatly increased by introducing additional atoms into the ring, increasing in the series

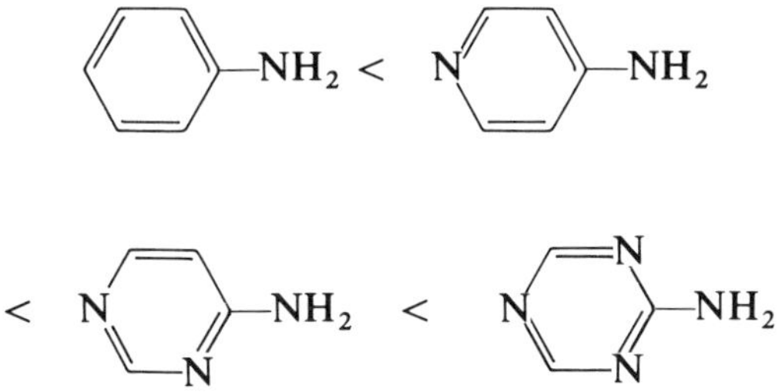

$$\text{C}_6\text{H}_5\text{—NH}_2 < \text{N-C}_5\text{H}_4\text{—NH}_2 \qquad (3.59)$$

$$< \text{N}_2\text{C}_4\text{H}_3\text{—NH}_2 < \text{N}_3\text{C}_3\text{H}_2\text{—NH}_2$$

We have so far considered only AH anions; the same arguments obviously apply in reverse to AH cations. In an odd AH cation,

$$q_i = 1 - a_{0i}^2 \qquad (3.60)$$

Equations (3.48)–(3.54) hold for cations if we replace a_{0i}^2 by $-a_{0i}^2$. Now ΔE_π is *negative*, so heteroatoms reduce the π-bond energy. Such substitution should reduce the ease of a reaction in which an even AH is converted to an odd AH cation. This effect is very well known experimentally and numerous examples will be found in the following chapter. A simple one is provided by the solvolysis of 2-methylallyl chloride, which takes place readily by an S_N1 process involving conversion to allyl cation;

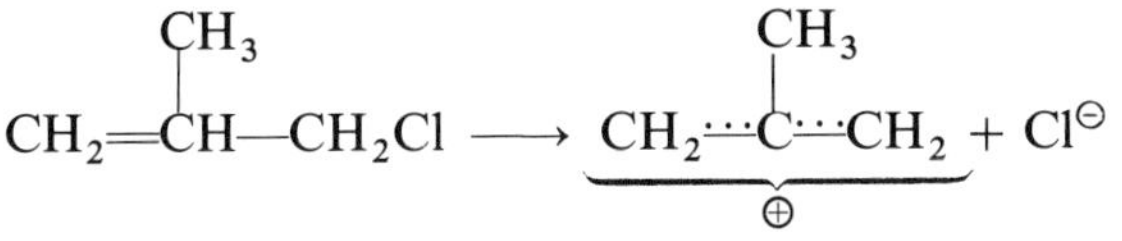

$$\text{CH}_2=\overset{\underset{|}{\text{CH}_3}}{\text{C}}\text{—CH}_2\text{Cl} \longrightarrow \underbrace{\text{CH}_2\cdots\overset{\underset{|}{\text{CH}_3}}{\text{C}}\cdots\text{CH}_2}_{\oplus} + \text{Cl}^\ominus \qquad (3.61)$$

This reaction does not take place at all if we replace the terminal methylene by oxygen:

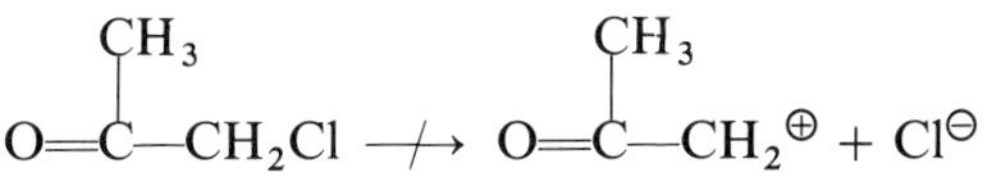

$$\text{O}=\overset{\underset{|}{\text{CH}_3}}{\text{C}}\text{—CH}_2\text{Cl} \;\not\longrightarrow\; \text{O}=\overset{\underset{|}{\text{CH}_3}}{\text{C}}\text{—CH}_2^\oplus + \text{Cl}^\ominus \qquad (3.62)$$

3.19 Polarization of π Electrons

In our discussion of heteroconjugated systems, we have so far assumed that the distribution of the electrons is the same as it is in the corresponding isoconjugate hydrocarbons. This is not in fact the case. In an even AH, the π-electron density at each position is unity. If we now change the electronegativity of one atom, k, for example, by replacing it with a heteroatom, the π-electron density will change throughout. If we make atom k more electronegative, corresponding to a change in the Coulomb integral from zero to a negative value α_k, atom k will attract π electrons to itself at the expense of the remaining less electronegative carbon atoms. Our problem is to estimate how large the resulting charges will be and how they will be distributed.

The even AH R can be constructed by union of methyl with the odd AH R' derived from R by removal of atom k (Fig. 3.23a). To a first approximation, this union involves only an interaction between the NBMO Φ_0 of R and the $2p$ AO ψ of methyl (Fig. 32.3b). Since these orbitals are degenerate, the interaction between them is analogous to that between two hydrogen atoms when they combine to H_2 (Section 1.5; p. 13). The resulting bonding MO Ξ in Fig. 3.23(b) will then be composed equally of the two interacting orbitals,

$$\Xi = (1/\sqrt{2})(\Phi_0 + \psi) = (1/\sqrt{2})(\sum_i a_{0i}\phi_i + \psi) \qquad (3.63)$$

where $1/\sqrt{2}$ is a normalizing factor and a_{0i} is the NBMO coefficient of atom i in R'.

In this approximation, the only MO in R that includes the AO ψ is Ξ. The charge density at atom k is given by $\sum_\mu n_\mu a_{\mu k}^2$, or

$$q_k = 2(1/\sqrt{2})^2 = 1 \qquad (3.64)$$

It is also easily seen that the charge density at each other position in R is also unity,

$$q_i = 2\left[\sum_\mu^{occ} a_{\mu i}^2 + (a_{0i}/\sqrt{2})^2\right] = 2\sum_\mu^{occ} a_{\mu i}^2 + a_{0i}^2 = 1 \qquad (3.65)$$

The last result follows from the pairing theorem since R' is an odd AH [see equation (2.7)]. Thus our approximation reproduces correctly the π-electron

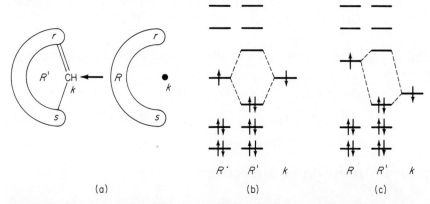

FIGURE 3.23. (a) Union of methyl with odd AH R to regenerate the even AH R'; (b) first-order perturbation of orbitals during union; (c) corresponding perturbation when atom k is replaced by a heteroatom electronegative relative to carbon.

PMO Treatment of Conjugated Systems

distribution in R. The higher-order perturbations scramble the MOs together but do not alter the q's.

Next let us replace the methyl carbon in Fig. 3.23(a) by a more electronegative atom of Coulomb integral α_k relative to carbon. Union will now lead to an even heteroconjugated molecule in which atom k in R has been replaced by a heteroatom. The main interaction will still be between the NBMO of R' and the $2p$ AO ψ of the extra atom, but this time the two orbitals are not degenerate (Fig. 3.23c). The interaction between them is now analogous to that between H and He$^+$ in forming HHe$^+$ (Section 1.7). The resulting bonding MO Ξ' will be of the form

$$\Xi' = \frac{1}{(A^2 + B^2)^{1/2}} (A\Phi_0 + B\psi) = \frac{1}{(A^2 + B^2)^{1/2}} \left(A \sum_i a_{0i}\phi_i + B\psi\right) \quad (3.66)$$

where $B > A$ because ψ is lower in energy than Φ. Thus the MO Ξ' is composed mainly of the AO ψ. The π-electron density at atom k of the resulting even heteroconjugated system is given by $\sum_\mu n_\mu a_{\mu k}^2$; since atom k participates in only one MO, this is

$$q_k' = \frac{2B^2}{A^2 + B^2} = 1 + \frac{B^2 - A^2}{B^2 + A^2} > 1 \quad (3.67)$$

There will be a net negative charge δq_k at atom k given by

$$\delta q_k = \frac{B^2 - A^2}{B^2 + A^2} \quad (3.68)$$

The π-electron density at atom i in R is given in the same form as equation (3.65) as

$$q_i' = 2\left(\sum_\mu^{occ} a_{\mu i}^2 + \frac{A^2 a_{0i}^2}{A^2 + B^2}\right) = 1 - a_{0i}^2 \frac{B^2 - A^2}{A^2 + B^2} \quad (3.69)$$

The positive charges on the carbon atoms are proportional to the squares a_{0i}^2 of the corresponding NBMO coefficients in R' and so are proportional to the corresponding charges in (R')$^+$. The sum of the positive charges must of course equal the negative charge on the heteroatom. This provides an extremely simple way of estimating the charge distribution in even heteroconjugated molecules (Fig. 3.24). Note that this result follows from an extension of first-order perturbation theory, so the perturbations due to two or more heteroatoms should be additive.

This treatment tells us how the charges will be distributed, but it does not tell us how large the charges will be. They must be proportional, at any rate in a first approximation, to the difference in electronegativity between carbon and the heteroatom, i.e., to α_i, and also to the polarizability of the π electrons. As a first approximation, we can then write the total formal charge q_i in the form

$$q_i = \pi_{i,i}\alpha_i \quad (3.70)$$

where $\pi_{i,i}$ is the *self-polarizability* of atom i in R'. The rigorous expression for $\pi_{i,i}$ is too complicated to be useful but a simple approximation can be

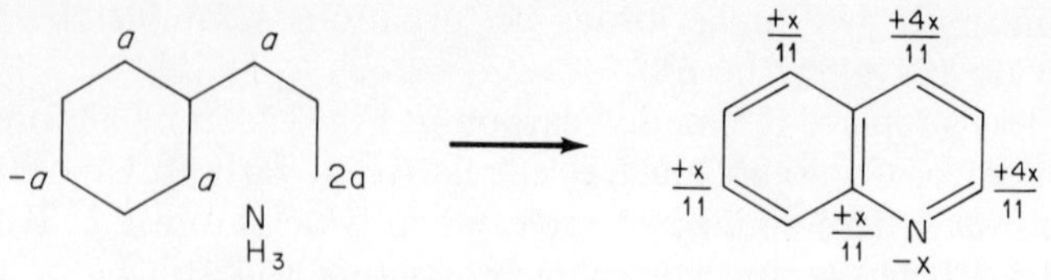

FIGURE 3.24. Distribution of formal charge in quinoline.

given in terms of the model indicated in Fig. 3.23(c), where the interaction between the heteroatom and R is treated as a two-orbital interaction (for a complete discussion, see the sources cited at the end of this chapter). It can be shown that

$$\pi_{i,i} \simeq 1/(2\beta \sum a_{or}) \tag{3.71}$$

where β is the resonance integral (assumed the same for all π bonds) and the sum of NBMO coefficients a_{or} in R is over all atoms adjacent to atom i in R'. Figure 3.25 shows some examples.

The excess negative charge in nitrogens in quinoline and isoquinoline should thus be $0.553\alpha_N/\beta$ and $0.471\alpha_N/\beta$, respectively. The expectation that isoquinoline should be less polar than quinoline is supported by both the physical and chemical properties of these molecules.

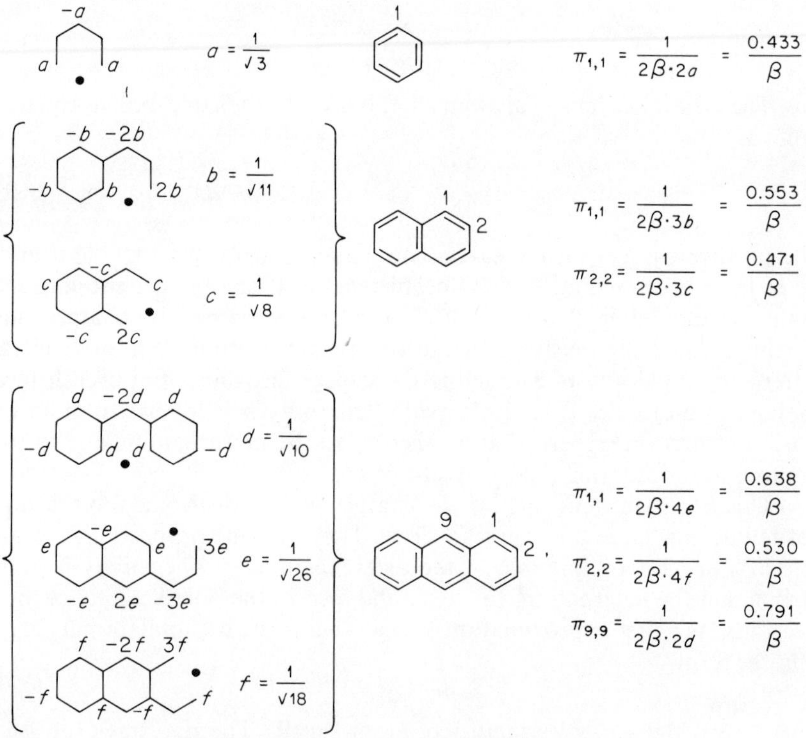

FIGURE 3.25. Calculation of self-polarizabilities.

PMO Treatment of Conjugated Systems

The approach above applies only to even systems. In odd systems, isoconjugate with odd AH anions and in which a heteroatom contributes two π electrons, the π-electron distribution will resemble qualitatively that in the isoconjugate AH. However, for reasons indicated in Section 3.16, the heteroatom will tend to get more than its share of the π electrons, so the charges in the rest of the system will be correspondingly smaller. Thus in the α-naphthylmethyl anion, the formal charge is distributed as indicated in Fig. 3.26a. In the isoconjugate amine, the charges in the ring are proportional to those in the anion but are smaller (Fig. 3.26b; $x < 1$).

In this case, there is unfortunately no simple way of estimating how large the decrease in charge will be.

3.20 Stereochemistry of Nitrogen

Conjugated nitrogen, in compounds where it is bound to three other atoms, presents an interesting stereochemical problem which raises a number of points which were not covered in Chapter 1.

The ground state of nitrogen is $(1s)^2(2s)^2(2p)^3$; from Hund's rule, the three p electrons occupy different $2p$ AOs with parallel spins. In this state, nitrogen has three singly occupied AOs that can be used to form covalent bonds (Fig. 3.27a). Since the $2p$ AOs lie along mutually perpendicular axes, the bonds formed to the p AOs will be strongest when the bonded atoms lie along those axes. Compounds of this type, e.g., ammonia, should then have bond angles of 90°. This, however, is not the case. The hydrogen atoms in ammonia, not being directly bonded, repel one another. The HNH bond angles (106.8°) are consequently close to the tetrahedral angle (109.47°).

In order to form bonds effectively in these directions, the nitrogen atom needs to use tetrahedral hybrid AOs (cf. methane, Section 1.10). We therefore need to *promote* the nitrogen atom into a *valence state* where its five valence electrons occupy tetrahedral hybrid AHs, each one-fourth s and three-fourths p (Fig. 3.27b). The effective occupation of the $2s$ and $2p$ AOs in such a state is then five-fourths $2s$ and fifteen-fourths $2p$. To get there from the ground state involves in effect the promotion of three-fourths of an electron from the $2s$ AO to the $2p$. The energy required to do this is supplied by the

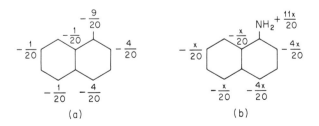

FIGURE 3.26. Distribution of formal charge (a) in the α-naphthylmethyl anion; (b) in α-naphthylamine.

relief of mutual repulsion between the hydrogen atoms when the HNH bond angles increase from 90° to 106.8°.

The repulsion between the hydrogen atoms could be still further reduced if the molecule became planar. The nitrogen atom would then have sp^2 hybridization (Fig. 3.27c) with three electrons in orbitals which are ($\frac{1}{3}s + \frac{2}{3}p$) and two in the remaining $2p$ AO. The effective valence state would then be sp^2. To get here from the tetrahedral configuration would require further promotion of one-fourth of an electron from the $2s$ to the $2p$ AO; apparently the hydrogen atoms are sufficiently content with their situation when nitrogen is tetrahedral for this further promotion not to be worthwhile. There is therefore a barrier to inversion in ammonia and in its alkyl derivatives of about 6 kcal/mole. This is the energy required to convert the pyramidal molecule into the planar intermediate.

If one of the groups attached to nitrogen is conjugated, the nitrogen atom can join in the conjugation, using its lone pair of electrons. The conjugation will of course be more efficient if the nitrogen atom is planar, so that conjugation involves a $2p$ AO (Fig. 3.28a), than if it is pyramidal, so that conjugation involves a hybrid AO (Fig. 3.28b), since the overlap with the adjacent $2p$ AO is better in the former case. This raises an interesting question: Will the extra conjugation pay for the 6 kcal/mole required to flatten out the nitrogen?

The hybrid AO T of Fig. 3.28b can be expressed (Fig. 3.28c) as a sum of contributions by the $2s$ AO, the $2p_x$ AO (along the bond), and the $2p_y$ AO (alone able to π-bond to the adjacent carbon); i.e.,

$$T = 0.5(2s) + 0.15(2p_x) + 0.82(2p_y) \tag{3.72}$$

Since only the $2p_y$ AO can π-bond to the adjacent carbon atom, the overlap of T with the carbon $2p$ AO is thus 0.82 times the overlap of $2p$ AOs in Fig. 3.28a. Since the CN resonance integral should be more or less proportional to the overlap of the AOs in question, the π interaction energy for tetrahedral

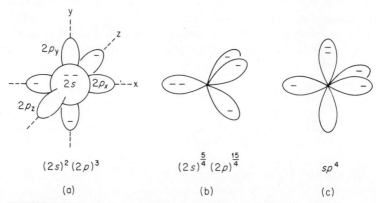

FIGURE 3.27. Promotion involved in converting the $(2s)^2 (2p)^3$ ground state (a) of nitrogen into tetrahedral (b) or trigonal (c) valence states.

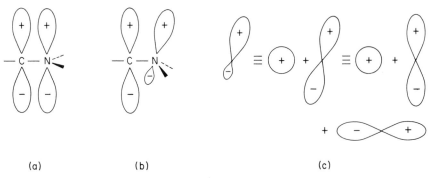

(a)　　　　　　　　(b)　　　　　　　　　　　　　(c)

FIGURE 3.28. (a) Conjugation involving planar N; (b) conjugation involving tetrahedral N; (c) dissection of an sp^3 hybrid AO into components.

nitrogen (Fig. 3.28b) should then be about four-fifths that for planar nitrogen (Fig. 3.28a).

Now unless the *difference* between the two interaction energies is at least 6 kcal/mole, it will not be enough to make the planar geometry on balance worthwhile. The total interaction energy must therefore be at least ~30 kcal/mole if the planar geometry is to be favored. It follows that such nitrogen atoms will be planar only in situations where the π interaction is very large. This is not the case in compounds such as aniline, $PhNH_2$, where nitrogen is attached to an even AH, and it may well not be the case in compounds such as pyrrole, where nitrogen is the only heteroatom in a five-membered aromatic ring. If nitrogen is not planar, the resonance integral between it and the adjacent carbon will be correspondingly reduced and so too will the π interaction between them. This will further increase the tendency of nitrogen to hold onto its pair of unshared electrons and so reduce the transfer of negative charge into the adjacent conjugated system.

3.21 Resonance Theory in the Light of the PMO Method

Prior to the development of the PMO method, the only qualitative treatment available to organic chemists was that based on resonance theory. Resonance theory was originally developed by Pauling as an intuitive generalization of an earlier theoretical treatment of molecules, the so-called valence bond (VB) method. Later developments showed that the intuitive arguments used in deriving resonance theory were wrong and its relative success therefore became something of a mystery. Studies by Dewar and Longuet-Higgins indicated that this success was due to an almost fortuitous correspondence between resonance theory and the PMO method. The success of resonance theory therefore in no way justifies the premises on which it is based. The correspondence between resonance theory and the PMO method does not always hold and this can lead to numerous mistakes. By comparison, the

PMO method has a firm foundation in theory, and there are no phenomena that cannot be explained as well (usually better) in terms of the PMO method as in terms of resonance theory. One must thus regard resonance theory as obsolete.

Since the potentialities of the PMO method have not as yet been sufficiently recognized, most textbooks of organic chemistry are still based on resonance theory. Thus a brief summary of the connection between resonance theory and the PMO method may still be of interest. The arguments will be given only for conjugated hydrocarbons; it is easy to extend them to systems containing heteroatoms.

In the original version of resonance theory, a conjugated hydrocarbon was represented as a hybrid or superposition of the possible classical structures that could be written for it. Later the theory was modified by introducing contributions by excited structures, but the arguments we are about to give indicate that this was in fact an error. The original version of resonance theory can be summarized by the following rules:

1. If only a single classical structure can be written for a given conjugated hydrocarbon, that structure will represent it satisfactorily.
2. If two or more such structures can be written, the molecule can be represented as a superposition or hybrid of the structures in equal amounts.
3. As a corollary of rule 2, the strength of a given bond in a conjugated hydrocarbon is proportional to the percentage of classical structures in which it is double.
4. The stabilities of isomeric conjugated hydrocarbons run parallel to the numbers of classical structures that can be written for them.
5. In an odd, conjugated hydrocarbon ion, the formal charge at each position is equal to the fraction of classical structures in which the charge is located at that position.

The success of resonance theory depends, as we shall now show, on the parallel between NBMO coefficients and numbers of resonance structures in odd AHs which was noted in Section 3.4 (cf. Fig. 3.6). This parallel leads immediately to a qualitative correspondence between the distributions of formal charge in odd AH ions calculated by resonance theory and the PMO method (see rule 5). The correspondence fails if there are $4n$-membered rings present (see Section 3.4, p. 81). In such cases, resonance theory should fail— and it does. We will assume that no $4n$-membered rings are present, so that the number N_i of classical structures for an odd AH with the odd atom at position i is proportional to the corresponding NBMO coefficient a_{0i}; i.e.,

$$N_i = A|a_{0i}| \tag{3.73}$$

where the proportionality factor A is the same for different positions in a given molecule but may differ for different molecules. Figure 3.29 illustrates this relationship for a typical odd AH.

The results of the preceding discussion are summarized below in a

PMO Treatment of Conjugated Systems

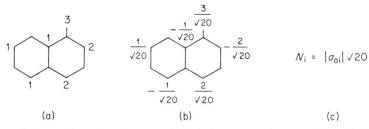

FIGURE 3.29. (a) Numbers N_i of classical structures for α-naphthylmethyl with the odd atom located in various positions; (b) NBMO coefficients a_{0i} for α-naphthylmethyl; (c) relation between N and a_{0i}.

series of rules regarding the relationship between the PMO and resonance methods:

Rule 1. If only one classical structure can be written for a molecule, all the bonds in it are essential single or essential double bonds. Such bonds are localized. This has already been shown to hold (see Sections 3.6 and 3.12).

Rule 2. Bonds that are not essential single bonds or essential double bonds have intermediate strengths and so lengths. This was established in Section 3.10, p. 89.

Rule 3. The total π-bond energy of a molecule can be written as a sum of terms $2p_{ij}\beta_{ij}$ [equation (2.8)]. The bond energy $(E_\pi)_{ij}$ of the π bond between two atoms i and j is thus

$$(E_\pi)_{ij} = 2p_{ij}\beta_{ij} = 2p_{ij}\beta \tag{3.74}$$

if we make the simplifying assumption that all β's are equal. Thus if we can find $(E_\pi)_{ij}$, we can at once find p_{ij}.

Consider one atom r in an even AH and the atoms, at most three (s, t, u), to which it is directly linked (Fig. 3.30a). Removal of atom r leaves an odd AH in which atoms s, t, and u are necessarily starred; denote their NBMO coefficients by a_{0s}, a_{0t}, and a_{0u} (see Fig. 3.30b). The π energies of the rs, rt, and ru bonds in the original AH can be estimated at once by considering the change in π energy when the odd AH unites with methyl to regenerate the original AH (Fig. 3.30),

$$(E_\pi)_{rs} = 2\beta|a_{0s}|, \qquad (E_\pi)_{rt} = 2\beta|a_{0t}|, \qquad (E_\pi)_{ru} = 2\beta|a_{0u}| \tag{3.75}$$

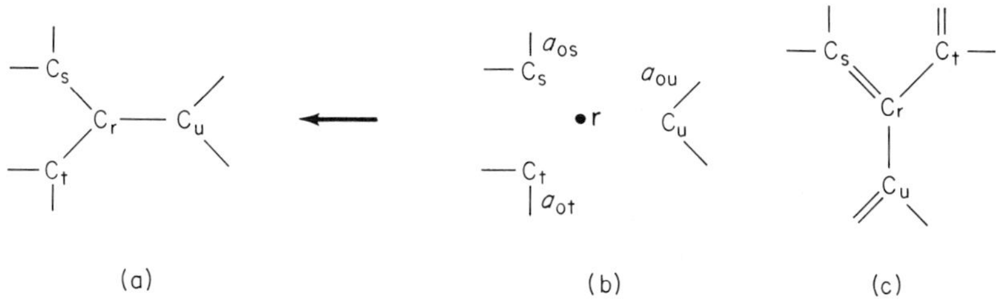

FIGURE 3.30. (a, b) Calculation of bond orders of bonds to an atom r in an even AH; $\delta E_\pi = 2\beta(a_{0s} + a_{0t} + a_{0u})$; (c) one classical structure for such an AH.

From equation (3.74), we can then find the corresponding bond orders and we can express them in terms of the numbers of resonance structures by using equation (3.73),

$$p_{rs} = \frac{(E_\pi)_{rs}}{2\beta} = |a_{0s}| = \frac{1}{A} N_s$$

$$p_{rt} = \frac{(E_\pi)_{rt}}{2\beta} = |a_{0t}| = \frac{1}{A} N_t \qquad (3.76)$$

$$p_{ru} = \frac{(E_\pi)_{ru}}{2\beta} = |a_{0u}| = \frac{1}{A} N_u$$

Now in any classical structure for the original even AH, atom r must be linked to one of its neighbors by a double bond and to the others by single bonds; cf. Fig. 3.30c. The number of such structures with the double bond between atoms r and s (Fig. 3.30c) must obviously be N_s, the number of structures for the odd AH with the odd atom at position s, because in both cases we have the number of permutations of bonds in a system derived from the original even AH by removing both atoms r and s. Likewise the numbers of structures with rt double or ru double are respectively N_t and N_u. It follows from equation (3.76) that the bond orders, and so the strengths and lengths, of the bonds should run parallel to the proportion of classical structures in which they are double.

Rule 4. We will show that if we have two isomeric AHs $(RS)_1$ and $(RS)_2$ derived by union of two odd AHs R and S at one point, the π energies of union will be proportional to the numbers of classical structures that can be written for them.

Suppose that the odd AHs R and S are linked in $(RS)_1$ through atom r in R to atom s in S, and in $(RS)_2$ through atom t in R to atom u in S (Fig. 3.31). The π energies of union are then

$$E_\pi(RS)_1 = 2\beta |a_{0r} b_{0s}| = \frac{2\beta}{A^2} N_r N_s \qquad (3.77)$$

$$E_\pi(RS)_2 = 2\beta |a_{0t} b_{0u}| = \frac{2\beta}{A^2} N_t N_u \qquad (3.78)$$

the last results following from equation (3.73). But since R and S are both odd, the bond between r and s in $(RS)_1$, or between t and u in $(RS)_2$, must be double (see Fig. 3.31). The number of classical structures that can be written for the R moiety in $(RS)_1$ is therefore N_r, and the number for the S moiety is

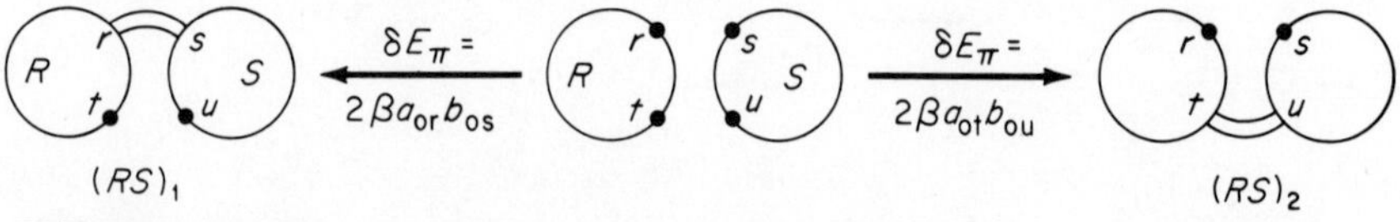

FIGURE 3.31. Union of two odd AHs R and S to form either of two isomeric even AHs RS_1 or RS_2.

N_s. The total number of classical structures for $(RS)_1(N(RS)_1)$ is therefore $N_r N_s$. Likewise the total number of classical structures for $(RS)_2(N(RS)_2)$ is $N_t N_u$. It follows from equations (3.77) and 3.78) that

$$\frac{E_\pi(RS)_1}{E_\pi(RS)_2} = \frac{N(RS)_1}{N(RS)_2} \tag{3.79}$$

This result can be generalized to cases where the odd AHs R and S unite at more than one point. Since any pair of even AHs can be compared by dissection into odd AHs followed by reunion, it can be seen that their stabilities must in all cases run parallel to the numbers of resonance structures that can be written for them.

Note that this conclusion is based on the assumption that no $4n$-membered rings are present. If they are, resonance theory fails because the parallel between numbers of resonance structures and NBMO coefficients no longer holds. Consider, for example, styrene (48) and cyclooctatetraene (49). There are in each case two possible classical structures; yet styrene is aromatic, while cyclooctatetraene is antiaromatic. This situation is quite general. Thus one can write four classical structures for biphenyl (50), corresponding to the Kekulé structures for each benzene ring, but *five* for biphenylene (51), where there is an additional quinonoid structure (52). This should imply that the central ring in biphenylene is aromatic; in fact, it is antiaromatic. Thus resonance theory should not be allowed to survive even as a poor substitute for the PMO method since there are cases where it leads to qualitatively incorrect results. The reason for the failure of resonance theory in these cases stems from the fact that resonance theory has no firm foundation in the wave properties of matter.

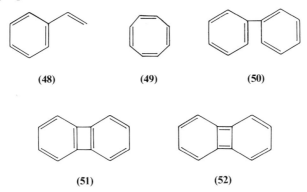

Appendix. π Energy of Union of Even AHs

In this appendix, we will complete the proof that the union of two even AHs results in a small and constant change in π energy. Consider the union of two even AHs R and S through atom r in R to atom s in S (see Fig. 2.2).

We use the notation of the previous chapter (Section 2.5). Since the bonding MOs in an even AH are all filled and the antibonding ones are empty, any degeneracy between R and S involves pairs of filled MOs or pairs of empty MOs. The π energy of union ΔE_{RS} is then a sum of second-order perturbations [equation (2.19)], involving interactions between filled MOs of R and empty MOs of S, and between filled MOs of S and empty MOs of R; i.e.,

$$\Delta E_{RS} = 2 \sum_\mu^{occ} \sum_\nu^{unocc} \frac{a_{\mu r}^2 b_{\nu s}^2 \beta_{rs}^2}{E_\mu - F_\nu} + 2 \sum_\mu^{unocc} \sum_\nu^{occ} \frac{a_{\mu r}^2 b_{\nu s}^2 \beta_{rs}^2}{F_\nu - E_\mu} \qquad (3.80)$$

From the pairing theorem the two paired MOs $\Phi_\mu^\pm$ of R have energies $\pm E_\mu$ and the two paired MOs $\Psi_\nu^\pm$ of S have energies $\pm F_\nu$, both with reference to the energy α of a carbon $2p$ AO. The coefficients $a_{\mu r}^\pm$ for the MOs $\Phi_\mu^\pm$, moreover, differ at most in sign, so their squares are equal. Likewise, the squares of the coefficients $b_{\nu s}^\pm$ for the MOs $\Psi_\nu^\pm$ are identical. We can therefore replace the sum over unoccupied MOs Φ_μ in equation (3.80) by a sum over occupied MOs if we replace E_μ by $-E_\mu$, and the sum over unoccupied MOs Ψ_ν by a sum over occupied MOs if we replace F_ν by $-F_\nu$, because the numerators of the expressions, containing only squares of coefficients, will remain unchanged. Both double sums then become identical, so that

$$\Delta E_{RS} = 4 \sum_\mu^{occ} \sum_\mu^{occ} \frac{a_{\mu r}^2 b_{\nu s}^2 \beta_{rs}^2}{E_\mu + F_\nu} \qquad (3.81)$$

The energies of bonding MOs in an even AH do not vary greatly; it is therefore reasonable to replace the energies E_μ and F_ν in equation (3.81) by mean values $\bar{E}$ and $\bar{F}$. Since these values hardly vary at all between different AHs, we may use a common mean value ε. In that case,

$$\begin{aligned}
\Delta E_{RS} &= 4 \sum_\mu^{occ} \sum_\nu^{occ} \frac{a_{\mu r}^2 b_{\nu s}^2 \beta_{rs}^2}{2\varepsilon} \\
&= \frac{2\beta_{rs}^2}{\varepsilon} \sum_\mu \sum_\nu a_{\mu r}^2 b_{\nu s}^2 \\
&= \frac{\beta_{rs}^2}{2\varepsilon} \sum_\mu 2 a_{\mu r}^2 \sum_\nu 2 b_{\nu s}^2 \\
&= \frac{\beta_{rs}^2}{2\varepsilon} q_r q_s \quad \text{[equation (2.7)]} \\
&= \frac{\beta_{rs}^2}{2\varepsilon} \quad \text{(from the pairing theorem, } q_r = q_s = 1\text{)}
\end{aligned} \qquad (3.82)$$

Thus to this approximation, the π energy of union has a constant value, independent of the even AHs.

PMO Treatment of Conjugated Systems

Problems

3.1. Calculate NBMO coefficients for the following odd AHs:

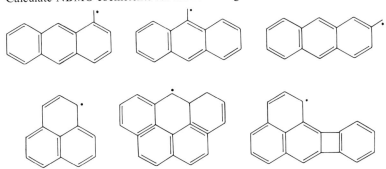

3.2. Deduce the order of stability in each of the two following series of isomeric hydrocarbons:

a.

b.

3.3. Discuss the aromaticity of the following hydrocarbons, using PMO theory rather than merely the rules for aromaticity:

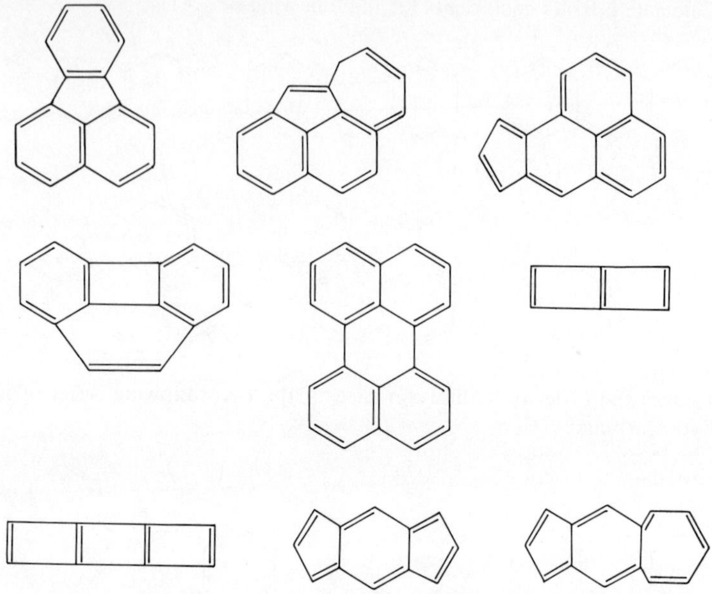

3.4. Calculate the changes in π energy in the following processes (assume $\alpha_N = \beta$):

(i) [naphthalene]–CH$_3$ $\longrightarrow$ [naphthalene]–CH$_2^-$ + H$^+$

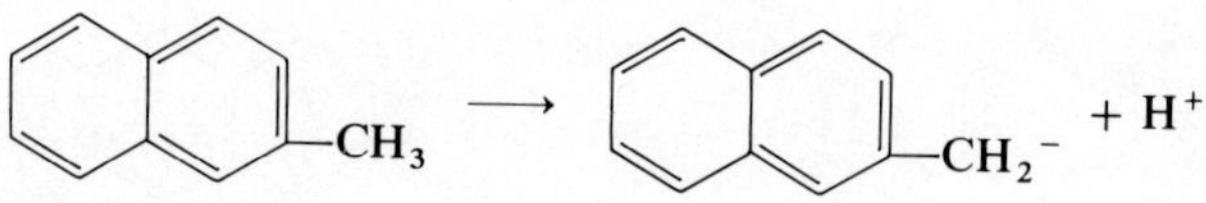

(ii) 1-methylnaphthalene $\longrightarrow$ 1-(CH$_2^-$)naphthalene + H$^+$

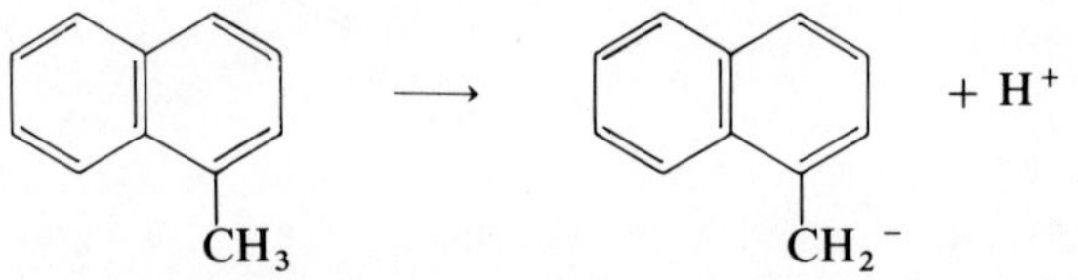

(iii) 4-methylquinoline (CH$_3$) $\longrightarrow$ 4-(CH$_2^-$)quinoline + H$^+$

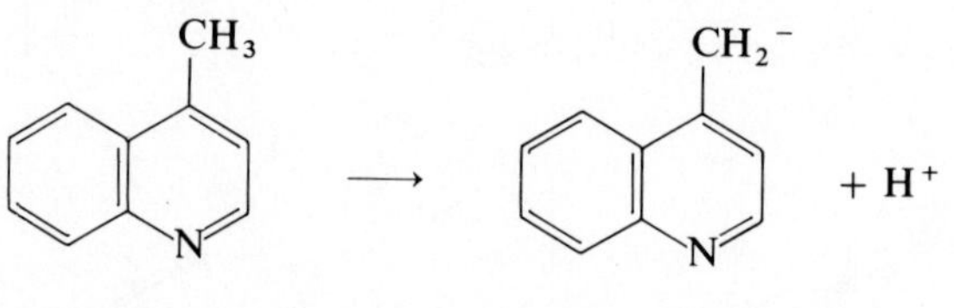

(iv) 2-methylquinoline–CH$_3$ $\longrightarrow$ 2-(CH$_2^-$)quinoline + H$^+$

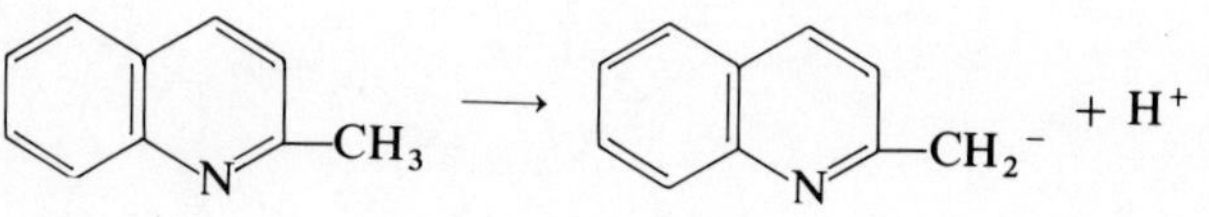

PMO Treatment of Conjugated Systems

(v) 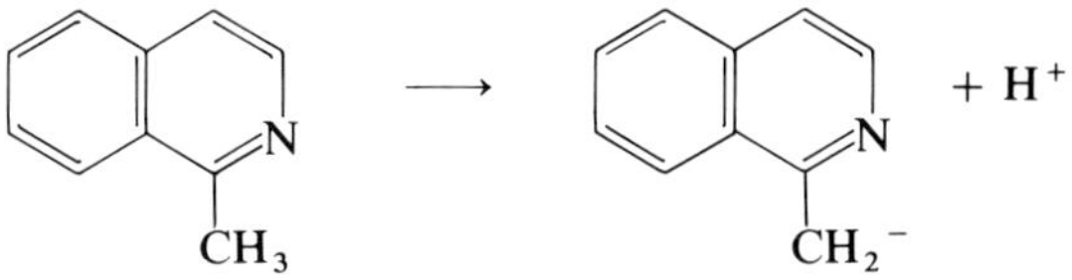 [isoquinoline-CH₃] → [isoquinoline-CH₂⁻] + H⁺

(vi) 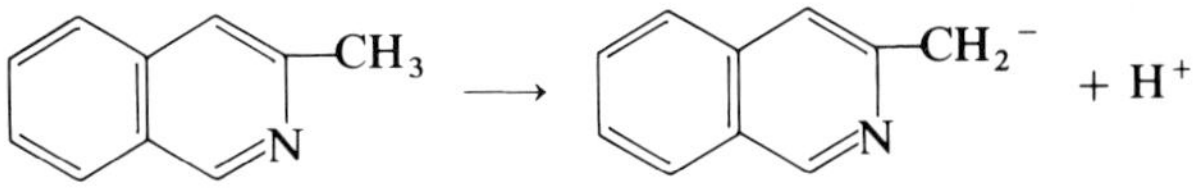 [isoquinoline with -CH₃] → [isoquinoline with -CH₂⁻] + H⁺

3.5. In the following pairs of isomers, deduce which should be the more stable.

a. 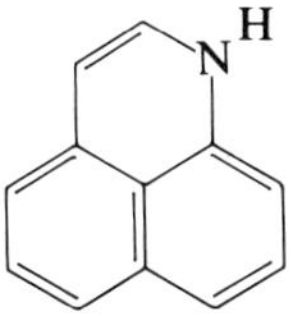 [structure with NH] and [structure with N]

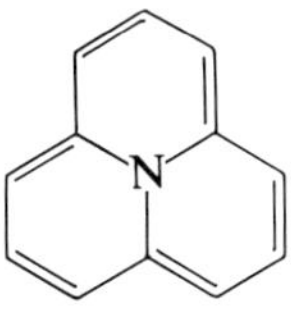

b. 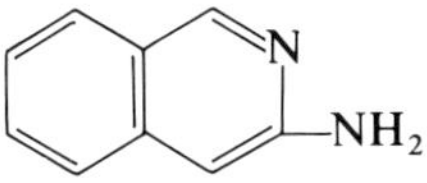 [isoquinoline-NH₂] and [quinoline-NH₂]

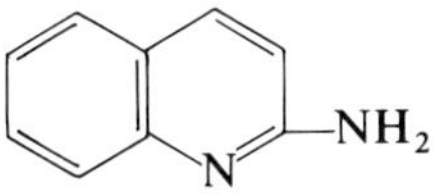

c. 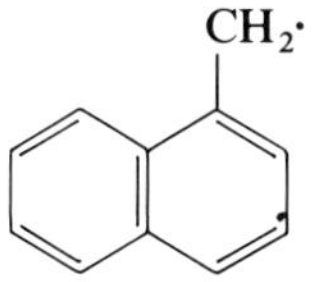 [naphthalene-CH₂·] and [naphthalene-CH₂·]

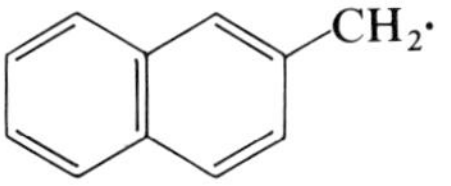

d. 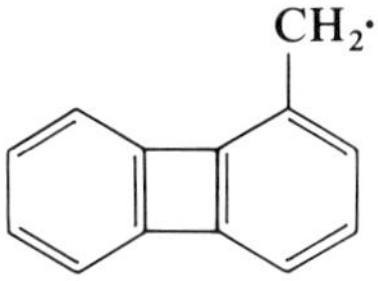 [biphenylene-CH₂·] and [biphenylene-CH₂·]

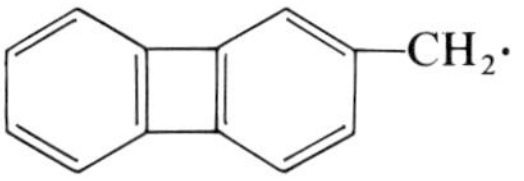

3.6. Predict the distribution of π electrons in the following azines (assume $\alpha_N = \beta$):

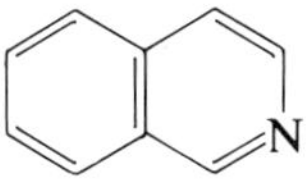

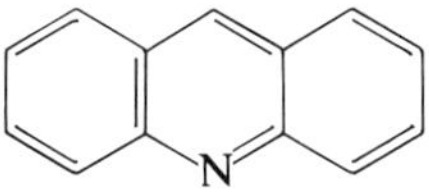

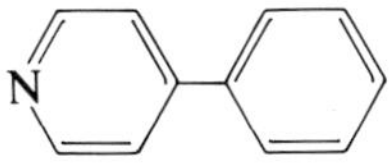

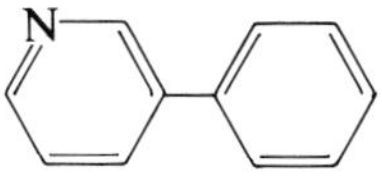

[isoquinoline] [acridine]

[pyridyl-phenyl] [pyridyl-phenyl]

3.7. Compare NBMO coefficients with the numbers of classical resonance structures that can be written with the odd atom in various places for the compounds listed in Problem 3.1.

3.8. Compare the relative stabilities of the following anti-Hückel hydrocarbons.

a.

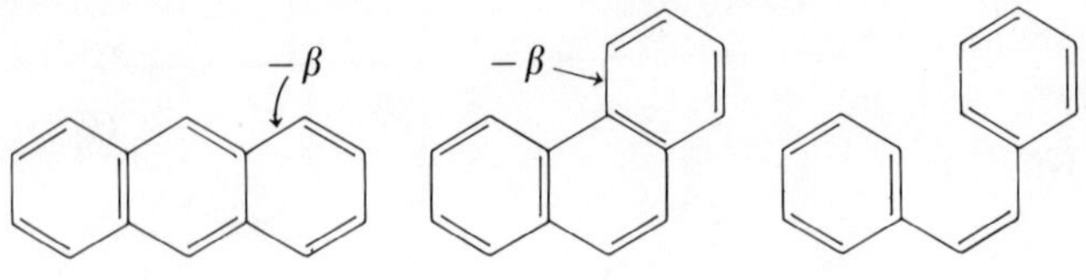

b.

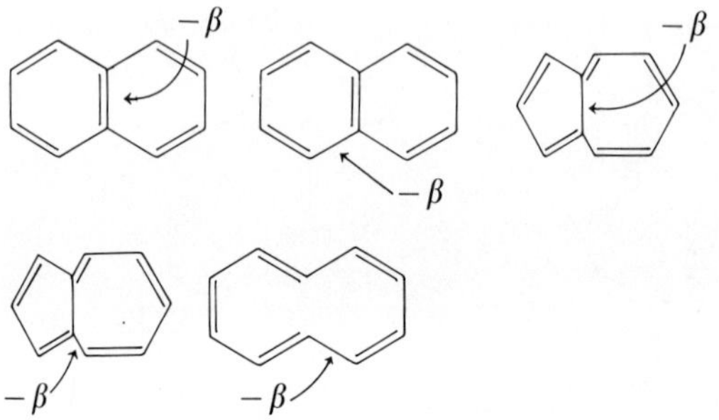

c.

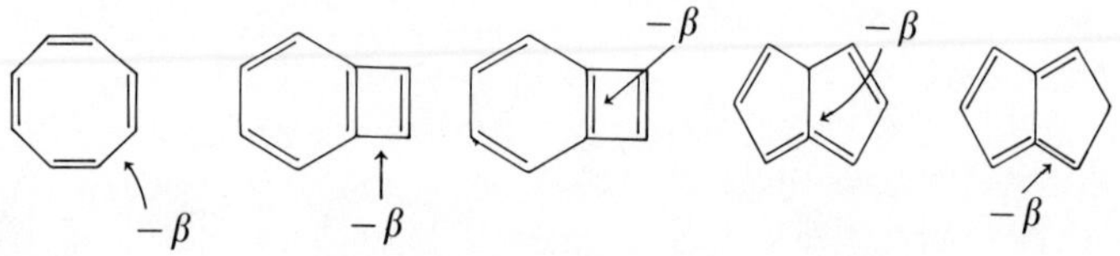

Selected Reading

For a more complete discussion of these topics, including the derivation of the equations for self-polarizability, see:

M. J. S. DEWAR, *The Molecular Orbital Theory of Organic Chemistry*, McGraw-Hill Book Co., New York, 1969.

For alternate discussions of aromaticity and the properties of conjugated hydrocarbons, see:

A. LIBERLES, *Introduction to Theoretical Organic Chemistry*, Macmillan, New York, 1968.

R. L. FLURRY, JR., *Molecular Orbital Theories of Bonding in Organic Molecules*, Marcel Dekker, New York, 1968.

CHAPTER 4

Chemical Equilibrium

4.1 Basic Principles

The most important function of theory in organic chemistry is to provide a guide to the relationship between structure and chemical reactivity, a guide which will help us in planning syntheses and predicting the course of chemical reactions. This chapter and the following three are concerned with the use of PMO theory in this connection.

The law of microscopic reversibility* tells us that all chemical reactions must be reversible. A basic requirement that must be satisfied if a reaction is to be of practical value is that it should be faster than the corresponding reverse reaction, because otherwise the products will be converted back into the reactants faster than they are formed. It is also necessary that the rate of the forward reaction should be reasonable. For the moment, we will be concerned only with the effect of reversibility. The rates of reactions will be considered in the next chapter.

Consider a reversible reaction between reactants $A_1, A_2, ..., A_m$ to form products $B_1, B_2, ..., B_n$,

$$A_1 + A_2 + \cdots + A_m \rightleftharpoons B_1 + B_2 + \cdots + B_n \tag{4.1}$$

If the reactions proceed until equilibrium is reached, the rates of the forward and backward reactions will then be equal. The concentrations [X] of reac-

---

* The law of microscopic reversibility is a consequence of the special theory of relativity which states that the equations of motion of a mechanical system must be invariant for a change in the sign of the time variable. This means that the mechanisms of the forward and reverse reactions in a closed chemical system must be identical.

tants and products will then obey the *mass law*,

$$\frac{[B_1][B_2]\cdots[B_n]}{[A_1][A_2]\cdots[A_m]} = K \tag{4.2}$$

where K is the *equilibrium constant*. The extent to which the conversion of the A's to B's can take place is thus determined by the value of K.

The equilibrium constant is in turn determined by the changes in free energy ΔG, heat content ΔH, and entropy ΔS when one mole of each reactant is converted to one mole of each product and when the concentration of each species is unity. These quantities are called the *free energy of reaction*, *heat of reaction*, and *entropy of reaction*, respectively. The relation between them and K is as follows:

$$-RT \ln K = \Delta G = \Delta H - T\Delta S \tag{4.3}$$

where R is the gas constant and T is the absolute temperature. In order to predict K, we need to be able to estimate heat contents and entropies of molecules in their standard states.

The entropy of a molecule depends in a complicated way on its modes of internal vibration and rotation and these cannot be predicted theoretically for any but the simplest molecules. Thus we cannot hope to predict equilibrium constants absolutely.

This is not in itself too serious from a practical point of view, because of the framework of empirical theory that has been established by the work of generations of organic chemists. Organic chemists have learned that reactions can be classified into types, the reactions of a given type all involving the same functional groups. The reactants $A_1, \ldots, A_m$ in a reaction of a given type can then be written in the form $(R_1X_1), (R_2X_2),\ldots, (R_mX_m)$, where the X_i are the same for different reactions of the group and where the reaction involves the X's only. Thus we can write equation (4.1) in the form

$$R_1X_1 + R_2X_2 + \cdots \rightleftharpoons R_1Y_1 + R_2Y_2 + \cdots \tag{4.4}$$

the Y's being the groups into which the X's are converted during the reaction. A simple example is the reaction of a carboxylic acid with an amine to form a salt,

$$\underset{X_1}{R_1\text{—COOH}} + \underset{X_2}{H_2N\text{—}R_2} \rightleftharpoons \underset{Y_1}{R_1\text{—COO}^-} + \underset{Y_2}{H_3N\text{—}R_2} \tag{4.5}$$

Here the groups R_1 and R_2 play no direct part in the reaction, although they may, and in fact do, affect the value of the equilibrium constant.

In certain cases, such as that of equation (4.5), the reactive centers X_i are fixed. In others, however, we may have a family of groups of reactions following the same general course where each group involves a fixed set of X_i but where the sets of X_i are different for different groups. A good example is the protonation of aromatic hydrocarbons to form arenonium ions; e.g.,

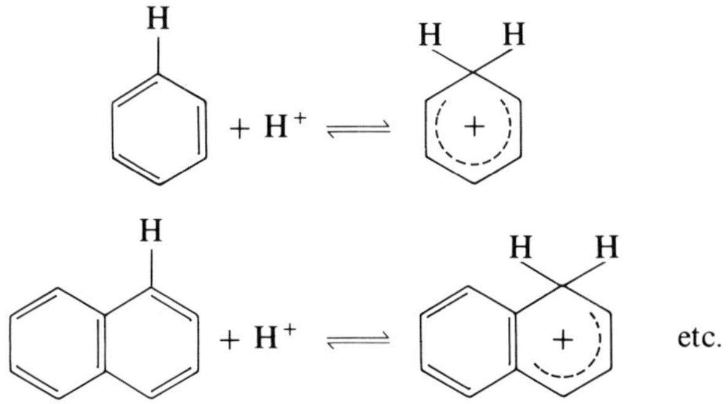

$$\text{C}_6\text{H}_6 + \text{H}^+ \rightleftharpoons \text{C}_6\text{H}_7^+ \quad (4.6)$$

$$\text{C}_{10}\text{H}_8 + \text{H}^+ \rightleftharpoons \text{C}_{10}\text{H}_9^+ \quad \text{etc.}$$

In each case, the reaction involves addition of a proton to one of the carbon atoms in the hydrocarbon, converting that carbon atom from a planar sp^2-hybridized state to a tetrahedral sp^3-hybridized state and so removing it from the conjugated system. The net effect is to convert the original hydrocarbon, an even AH, into an odd AH cation. The reactions are all similar, but of course the equilibrium constant will be different for different hydrocarbons—and also for different points of attachment of the proton in a given hydrocarbon in which there are two or more nonequivalent positions. We can next replace one or more hydrogen atoms in a given hydrocarbon by *substituents* R_i and so generate a series of *derivatives* of the hydrocarbon, each of which can in turn undergo protonation. The reactions will then fall into groups, each group involving protonation of derivatives of a given hydrocarbon at a given position in it.

Organic chemists have established the types of reaction that take place between organic molecules and have classified them in the way indicated above. We do not therefore need to predict whether or not a given type of reaction will take place. What we do need is some way of predicting how the course of a given reaction will vary with changes in the structure of the reactants. In the case of equilibrium, we need to predict the relative equilibrium constants for the parent members of the reactions forming a given family, and then the effect on these equilibrium constants when various types of substituents are introduced at different positions in the reactants.

The ratio of equilibrium constants K_1/K_2 for a pair of such processes is given [see equation (4.3)] by

$$-RT\ln(K_1/K_2) = -RT(\ln K_1 - \ln K_2)$$
$$= \Delta G_1 - \Delta G_2 = (\Delta H_1 - \Delta H_2) - T(\Delta S_1 - \Delta S_2) \quad (4.7)$$

This expression still depends on the changes in entropy involved in the two reactions, changes which we cannot estimate. Here, however, we are dealing with two reactions of very similar type, the main differences between the reactants being confined to groups which play no direct part in the reaction. It is not unreasonable to suppose that in such cases the entropy of reaction may be the same for both reactions, i.e.,

$$\Delta S_1 = \Delta S_2 \quad (4.8)$$

If this is in fact the case, it can be shown that the changes in ΔH are due solely to corresponding changes ΔE in the internal energy,

$$\Delta H_1 - \Delta H_2 = \Delta E_1 - \Delta E_2 \tag{4.9}$$

Equation (4.7) then becomes

$$-RT \ln(K_1/K_2) = \Delta E_1 - \Delta E_2 \tag{4.10}$$

Since the energies of molecules are quantities that we can estimate by the methods developed in the preceding chapters, equation (4.10) will then enable us to attain our objective of calculating relative equilibrium constants for related reactions.

There is, however, a complication. If we measure the changes in entropy that occur during a series of related reactions, the values we get are *not* in general the same. Indeed, they often vary greatly from one member of the series to another. How then can we possibly justify the use of equation (4.10)?

The reactions in which such variations in ΔS occur are generally ones in which ionic species are involved. The variations in cases where the reactants and products are all neutral are usually quite small. This suggests that the variations arise from changes in the degree of solvation. The energies of solvation of ions are far greater than those of neutral molecules, so changes in solvation have a much greater effect in the former case. Solvation of an ion leads to a decrease in heat content. It also leads to a decrease in entropy because the solvent molecules involved are now tied to the ion and are no longer free to move about. The more tightly the solvent molecules are bound, the lower is the heat content H of the system but the lower also is the entropy S because the more restricted will be the motions of the participating solvent molecules. An increase in the number of solvent molecules involved in solvation, or in the tightness of binding to the ion, will therefore reduce both H and S. Now the free energy of the system is given by

$$G = H - TS \tag{4.11}$$

The effect of the changes in H and S will therefore tend to cancel. Moreover, the equilibrium state of a chemical system must be such as to minimize the free energy G. Since, then, G must be a minimum, and since changes in solvation affect H and S more than they affect G, a plot of G vs. S for various degrees of solvation is likely to be fairly flat (Fig. 4.1a), the minimum corresponding to the equilibrium state of the system.

Similar considerations should apply to the changes (ΔG and ΔS) in G and S that occur during a reaction. Changes in the degree of solvation of the products should lead to large changes in ΔS but only small changes in ΔG, a plot of ΔG vs. ΔS being again a rather flat curve (Fig. 4.1b). The minimum in this plot corresponds to the product actually formed.

Next let us consider the effects of solvation on two similar reactions. Plots of ΔG vs. ΔS for each should be flat and the minima may correspond to very different values (ΔS_1 and ΔS_2) of ΔS (Fig. 4.1b). The ratio K_1/K_2 of the two equilibrium constants will be determined [equation (4.7)] by the

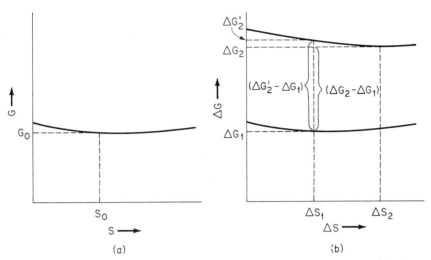

FIGURE 4.1. (a) Plot of G vs. S for various degrees of solvation of a system containing ions; (b) analogous plot for changes in G and S during two related reactions.

difference $\Delta G_2 - \Delta G_1$ between the values of ΔG at the minima. Suppose now that we make the simplifying assumption that ΔS for the second reaction has the same value (ΔS_1) as for the first. The corresponding difference in ΔG will now be $\Delta G_2' - \Delta G_1$ instead of $\Delta G_2 - \Delta G_1$ (see Fig. 4.1b). The resulting error may, however, be quite small because the plots of ΔG vs. ΔS are so flat. If so, we will get quite a good estimate of the ratio of equilibrium constants if we neglect the changes in ΔS even when these changes are in fact large. The fact that experimentally measured entropies of reaction vary greatly may not therefore seriously affect the use of equation (4.10) for predicting relative equilibrium constants.

This argument does not of course tell us that the errors so introduced must be small, only that they may be small. In practice they do seem to be small since theories of reactivity based on equation (4.10) have proved successful and useful. One point should be emphasized in this connection. While our arguments justify the use of equation (4.10), they do not justify the assumption (of constant ΔS) on which it was based. We must not therefore try to correlate calculated energies of reaction ΔE with experimentally measured heats of reaction ΔH, because the latter, like entropies of reaction, contain large and variable contributions from solvation. We can correlate ΔE only with ΔG, in which the effects of solvation more or less cancel.

4.2 Factors Contributing to the Energy of Reaction

The change in internal energy during a reaction, i.e., the energy of reaction ΔE, can be divided into contributions from (1) changes in the energy of localized bonds ΔE_{loc}; (2) changes in the total energy of delocalized bonds

ΔE_{deloc}; (3) changes in internal electrostatic interactions ΔE_{elec}; and (4) changes in the energy of solvation ΔE_{solv}; i.e.,

$$\Delta E = \Delta E_{\text{loc}} + \Delta E_{\text{deloc}} + \Delta E_{\text{elec}} + \Delta E_{\text{solv}} \qquad (4.12)$$

Of these, the first and last terms are quickly dealt with.

In pairs of similar reactions, the bonds involved in the two reactions are those in the corresponding reaction centers, which are by nature similar. The term ΔE_{loc} will therefore be the same or nearly the same for both processes and will cancel in the expression for the relative equilibrium constant [equation (4.10)].

This point can be easily illustrated. Take, for example, the protonation of a pair of even AHs [cf. equation (4.6)]. The only localized bonds undergoing any change in the reaction are those formed by the carbon atom to which the proton is to be attached; i.e.,

$$\text{(structural equation)} \qquad (4.13)$$

Here we form a new localized CH bond (bond energy E_{CH}) and we also alter the preexisting σ bonds by changing the hybridization of carbon (changes in bond energy δE_{CC} and δE_{CH}). Thus the change in total energy of localized bonds during the reaction is given by

$$\Delta E_{\text{loc}} = -E_{\text{CH}} + 2\delta E_{\text{CC}} + \delta E_{\text{CH}} \qquad (4.14)$$

Now E_{CH} and δE_{CH} will be the same for substitution at different positions in different hydrocarbons. Since the CC bond lengths in aromatic hydrocarbons vary very little, the terms δE_{CC} will also be much the same since they represent the difference in bond energy between a CC sp^2–sp^2 σ bond compressed to the "aromatic" length of ~ 1.4 Å and a normal sp^3–sp^2 σ bond. All the terms in equation (4.14) should therefore be much the same for attack at different points in different hydrocarbons, so their sum, ΔE_{loc}, should be likewise constant.

The last term in equation (4.12) refers to changes in solvation energies.

Absolute energies of solvation cannot be calculated; however, as we shall see in Chapter 5, relative energies of solvation can be estimated. In an equilibrium, the solvation terms refer to the reactant and product states. For a series of related equilibria, the charge size and orbital energies of the reactants and products should be similar, so we can assume that the changes in solvation energy for a pair of reactions will be the same. This is a very reasonable assumption since it is inherently consistent with our neglect of changes in ΔS due to changes in the degree of solvation. With it, the term ΔE_{solv} also cancels out of equation (4.10). Hence, we have

$$-RT\ln(K_1/K_2) = [(\Delta E_{\text{deloc}})_1 - (\Delta E_{\text{deloc}})_2] + [(\Delta E_{\text{elec}})_1 - (\Delta E_{\text{elec}})_2] \tag{4.15}$$

The term E_{elec} requires some explanation. It refers in essence to electrostatic interactions between charges developed or destroyed in the reaction centers during the course of the reaction and charges in other parts of the reactants, in particular in substituent groups. A simple example is provided by the two dissociation constants of a dicarboxylic acid; e.g.,

$$\text{HOOC—CH}_2\text{—COOH} \rightleftharpoons \text{HOOC—CH}_2\text{—COO}^- + \text{H}^+;$$

$$pK_A = 2.83 \tag{4.16}$$

$$^-\text{OOC—CH}_2\text{—COOH} \rightleftharpoons {}^-\text{OOC—CH}_2\text{—COO}^- + \text{H}^+;$$

$$pK_A = 5.69$$

The second dissociation constant is much smaller, and pK_A correspondingly greater, because removal of the second proton leads to an ion with two negative charges which repel one another. The energies of ionization for the first and second protons differ by the corresponding electrostatic potential energy due to the repulsion between the negative charges in the dianion.

The terms ΔE_{elec} in equation (4.12) refer primarily to the effects of substituents and will not be present in the case of the parent members of two groups of reactions in a given family, e.g., two different aromatic hydrocarbons in the case of the reaction indicated in equation (4.6). Here the relative equilibrium constants are consequently determined solely by the terms ΔE_{deloc}, namely

$$-RT\ln(K_1/K_2) = (\Delta E_{\text{deloc}})_1 - (\Delta E_{\text{deloc}})_2 \tag{4.17}$$

Since this relation holds for any pair of compounds in our set, it can be written more succinctly as

$$-RT\log K = \Delta E_{\text{deloc}} + C \tag{4.18}$$

where C is a constant for each series of reactions. A plot of $\log K$ vs. ΔE_{deloc} should therefore be linear.

In the case of an equilibrium between "normal" molecules or ions, the delocalized systems will be of π type. Here the terms ΔE_{deloc} refer to the π energies of conjugated systems. The right-hand side of equation (4.17) therefore represents the difference in π energy between two conjugated systems, a quantity that we can estimate using the procedures developed in Chapter 3.

In the PMO approximation, the π energy of a conjugated system is referred to that of the analogous AH from which it is derived by replacement of carbon atoms by heteroatoms or by internal union to form a nonalternant system. We follow the same procedure here. Our treatment of a family of reactions will therefore involve the following steps:

1. Prediction of the relative equilibrium constants for the reactions of a series of AHs.

2. Analysis of the effect of replacement of atoms in those AHs by heteroatoms.
3. Analysis of the effect of internal union to form analogous nonalternant systems.
4. Analysis of the effect of added groups (i.e., substituents).

4.3 Reactions of AHs

The ratio of equilibrium constants for related reactions of a pair of AHs is given by equation (4.17), where the terms on the right represent the corresponding difference in π energies of reaction, i.e., difference in π energy between the reactants and products. We can immediately distinguish two basic types of equilibria: *isoconjugate* equilibria, in which the conjugated systems in the reactants and products are the same size, and *heteroconjugate* equilibria, in which the π system changes size.

Isoconjugate Equilibria. Isoconjugate reactions of hydrocarbons are rare and the literature is devoid of data on isoconjugate equilibria. There is, however, a substantial amount of literature on isoconjugate equilibria of heterocycles. Consider, for example, salt formation by an even heteroatomic AH, e.g., pyridine or quinoline,

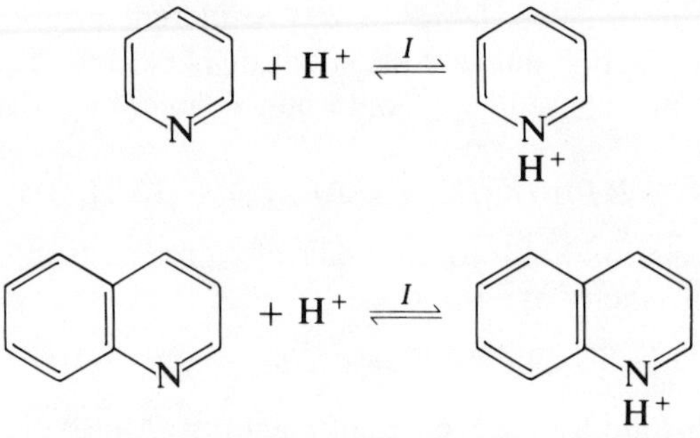

$$\text{pyridine} + H^+ \xrightleftharpoons{1} \text{pyridinium}^+ \tag{4.19}$$

$$\text{quinoline} + H^+ \xrightleftharpoons{1} \text{quinolinium}^+$$

In this case, the reactant and product are isoconjugate, differing only by the change in electronegativity of nitrogen when it acquires a positive charge. If we denote the corresponding change in Coulomb integral by $\delta\alpha_N$, the π energy of reaction δE_π is given (cf. Section 3.18) in terms of that for the isoconjugate* hydrocarbon $(\delta E_\pi)_0$ by

$$\delta E_\pi \simeq (\delta E_\pi)_0 + q_N \delta\alpha_N = q_N \delta\alpha_N \tag{4.20}$$

---

* The hydrocarbons isoconjugate with the azines are carbanions such as those in which one carbon atom has a pair of electrons occupying a σ-type AO:

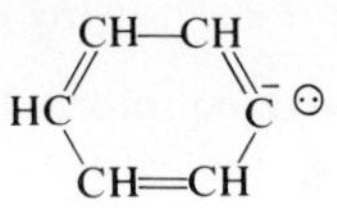

$$\text{cyclic carbanion}^-$$

where q_N is the π-electron density at the atom in the isoconjugate AH that is replaced by nitrogen in the azine. Now since the AH is even, the π-electron density at each position is unity. Hence

$$\delta E_\pi \simeq \delta \alpha_N \qquad (4.21)$$

Thus the terms ΔE_{deloc} in equation (4.12) are the same, so the equilibrium constants should be identical. The pK_A of bases of this type in fact show little variation, all lying within the range of one pK_A unit.

In general the positions of isoconjugate equilibria are determined by the values of ΔE_{deloc}, ΔE_{elec}, and ΔE_{solv}, all of which should be similar for a related series of reactions. The term symbol for these equilibria is I, which indicates that there is no change in the size of the π-electron system in the equilibrium. We will indicate the term symbols for reactions and equilibria over the reaction arrows for the reactions which follow in this book.

Heteroconjugate Equilibria. In the case of a heteroconjugate equilibrium, the conjugated systems of the reactants and products differ in size. One can therefore be derived from the other by union with an appropriate extra bit. Furthermore, for reasons that will become clear in the following chapter, chemical reactions rarely involve the formation and/or breaking of more than two bonds at a time and the majority of reactions involve the breaking of just one bond. The conjugated systems of the reactant and product will therefore nearly always differ by either one or two atoms. If, moreover, the two systems differ by only one atom, one must be even and the other odd. We can therefore divide heteroconjugated equilibria into four categories: (1) those where the two conjugated systems differ by one atom, the odd one being the smaller; (2) analogous equilibria in which the odd system is the larger; (3) equilibria where both systems are even, differing by two atoms; (4) equilibria where both systems are odd, again differing by two atoms. The term symbols for these four types of equilibria are EO_s, EO_l, EE, and OO, respectively. Examples of each type, with illustrations of the application of PMO theory to each, follow.

EO_s Equilibria Reactions. These involve the interconversion of an even AH and an odd AH with one atom less. A good example of an EO_s equilibrium is the protonation of an even AH [equation (4.6)]. Here the change in π energy can be found at once by the first-order perturbation treatment of Section 3.8 [see equation (3.24)], i.e.,

$$\begin{array}{c} \diagdown C_r \quad C_s \diagup \\ CH_3 \\ \diagup C_t \diagdown \\ \text{odd AH} + CH_3 \end{array} \quad \xrightleftharpoons[\delta E_\pi' = -2\beta(a_{0r} + a_{0s} + a_{0t})]{\delta E_\pi = 2\beta(a_{0r} + a_{0s} + a_{0t})} \quad \begin{array}{c} \text{even AH} \end{array} \qquad (4.22)$$

The energy required for the converse process, i.e., removal of one atom from a conjugated system (E'), is called the *localization energy* LE. Our argument

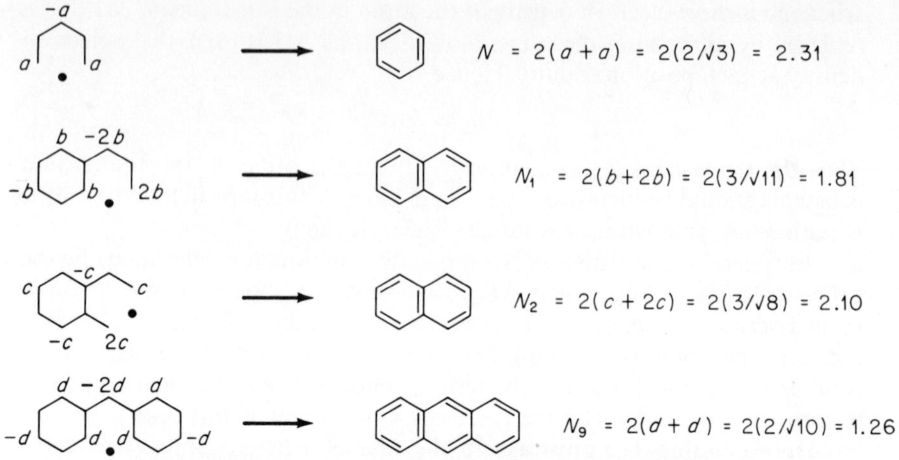

FIGURE 4.2. Calculation of reactivity numbers in even AHs.

shows that the localization energy of atom u in an even AH is given by

$$LE_u = -2\beta \sum_r^{(u)} a_{0r} \qquad (4.23)$$

where the sum is over atoms r directly linked to atom u in the even AH. This can be written in the form

$$LE_u = -\beta N_u \qquad (4.24)$$

where the *reactivity number* N_u is defined by

$$N_u = 2\sum_r^{(u)} a_{0r} \qquad (4.25)$$

Figure 4.2 illustrates the calculation of reactivity numbers for several AHs.

Consider the protonation of a series of even AHs,

$$ArH + H^+ \rightleftharpoons ArH_2^+ \qquad (4.26)$$

The corresponding equilibrium constant K is given by

$$K = \frac{[ArH_2^+]}{[ArH][H^+]} \qquad (4.27)$$

In discussions of acid–base reactions, it is more usual to use the equilibrium constant K_A for the reverse reaction,

$$ArH_2^+ \rightleftharpoons ArH + H^+; \quad K_A = \frac{[ArH][H^+]}{[ArH_2^+]} = \frac{1}{K} \qquad (4.28)$$

The strength of a given acid is determined by the corresponding dissociation constant K_A. This is usually quoted on a logarithmic scale as pK_A, the pK_A of an acid being defined by

$$pK_A = -\log_{10} K_A \qquad (4.29)$$

Consider addition of a proton to some position in an even AH with reactivity number N_u. From equations (4.17) and (4.23)–(4.25), we have

$$-RT \log K = -\beta N_u + C \qquad (4.30)$$

where C will be the same for protonation reactions of even AHs. Now in equation (4.29), K refers to protonation of one specific position in the hydrocarbon. If we measure pK_A for dissociation of the conjugate acid ArH_2^+ experimentally, the value we get will refer to a composite of all possible protonation processes. It is reasonable as a first approximation to consider only protonation at the most reactive positions, i.e. those for which K is greatest and N_u consequently smallest, e.g., the α positions in naphthalene (Fig. 4.2). There is, however, a minor complication. Our calculation [equations (4.24) and (4.30)] refers to protonation at *one* such position. If there are two or more equivalent positions in the molecule, the measured value for K will be correspondingly too great. Thus in benzene, K for a single position is one-sixth the value for overall protonation and K for one α position in naphthalene is one-quarter the overall value. The experimentally measured dissociation constants must be corrected for these *statistical factors*. Making this correction and using equations (4.28)–(4.30), we find

$$-2.303 RT(pK_A + \log_{10} n) = -\beta N_u + C \qquad (4.31)$$

where 2.303 is the factor for conversion from natural logarithms to logarithms to the base ten and n is the statistical factor. Since β is in any case a parameter, this can be written in the simpler form

$$pK_A + \log_{10} n = AN_u + B \qquad (4.32)$$

where A and B are constants for measurements of constant temperature. Thus a plot of $(pK_A + \log_{10} n)$ vs. N_u should be a straight line.

Values of pK_A for a number of hydrocarbons have been measured by Mackor et al.[3] Figure 4.3a shows a corresponding plot of equation (4.32). The points lie reasonably close to a straight line. Some scatter would in any case be expected not only because our calculation of ΔE_{deloc} involves first-order perturbation theory, but also because the assumption that the solvation energies of the ions ArH_2^+ is the same cannot be entirely correct. Figure 4.3(b) shows for comparison a plot of the left-hand side of equation (4.32) against localization energies calculated by a more accurate and more sophisticated but much more complicated MO method. The points certainly lie closer to a straight line than do those in Fig. 4.3(a). On the other hand, the calculation of the reactivity numbers in equation (4.32) was a "pencil and paper" operation needing only a few minutes, whereas the calculation of the localization energies plotted in Fig. 4.3(b) involved the use of a large digital computer and cost several thousand dollars.

EO_l Equilibria Reactions. These involve the interconversion of an even AH and an odd AH with one atom more. In this case, the odd AH is derived

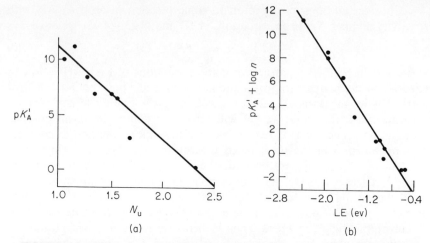

FIGURE 4.3. (a) Plot of equation (4.32) for even AHs; pK_A' is the pK_A value relative to one position in benzene; (b) plot of $(pK_A' + \log n)$ vs. localization energies calculated by an SCF MO procedure (experimental values are from Ref. 3).

by the union of an even AH with methyl,

$$\boxed{\text{even AH}} \leftarrow \mathbf{u} \rightarrow CH_3 \longrightarrow \boxed{\text{even AH}}\text{---}CH_2 \qquad (4.33)$$

the π energy of union in such a process is given [see equation (3.44)] by

$$\delta E_\pi = 2\beta(1 - a_{0r}) \qquad (4.34)$$

where a_{0r} is the NBMO coefficient of the extra atom in the resulting odd AH. We distinguish between EO_s and EO_t because of the difference between the two expressions for δE_π.

A simple example of an EO_t reaction is the deprotonation of arylmethanes (ArCH$_3$, ArH being an even AH) by base,

$$ArCH_3 + B \xrightleftharpoons{EO_t} ArCH_2^- + \overset{+}{B}H \qquad (4.35)$$

The equilibrium constants for such reactions, involving a common base B, should be related to the acid dissociation constants K_A of the compounds ArCH$_3$ by

$$K = CK_A \qquad (4.36a)$$

or

$$pK_A = -\log_{10} K_A + \text{constant} \qquad (4.36b)$$

The corresponding π energy of reaction [i.e., ΔE_{deloc} in equation (4.17)] is given by equation (4.34). An argument similar to that used in the preceding section then shows that the following relation should hold between pK_A and a_{0r}, this being the NBMO coefficient of the methylene carbon in ArCH$_2$:

$$pK_A = Aa_{0r} + B \qquad (4.37)$$

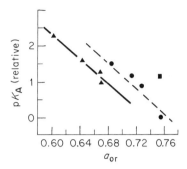

FIGURE 4.4. Plot of pK_A (relative to toluene) for compounds $ArCH_3$ (ArH being an even AH) vs. the NBMO coefficient a_{0r} at methylene in $ArCH_3$ (experimental values are from Ref. 4). (▲) "α-naphthyl-like"; (●) "β-naphthyl-like; (■) 2-pyrenyl.

The pK_A of a number of methyl derivatives of aromatic hydrocarbons have been estimated by Streitwieser and Longworthy[4]; Figure 4.4 shows a plot of their values against the corresponding NBMO coefficients a_{0r}. While the points show much more scatter than the corresponding plot of Fig. 4.3(a), they do reproduce the general trend observed.

Another example of such a process is the dissociation of arylethanes into radicals, e.g.,

$$PhCH_2CH_2Ph \xrightleftharpoons{EO_t} PhCH_2\cdot + \cdot CH_2Ph \qquad (4.38)$$

Here two even AHs are converted simultaneously into odd AHs each with one additional carbon atom; the π energy of reaction δE_π is given by

$$\delta E_\pi = 2\beta(1 - a_{0r}) + 2\beta(1 - b_{0s}) = 2\beta(2 - a_{0r} - b_{0s}) \qquad (4.39)$$

where a_{0r} and b_{0s} are the NBMO coefficients of the added carbon atom in the two resulting odd AH radicals. The equilibrium constant K for the dissociation of arylethanes [equation (4.38)] should then follow the relation

$$-RT \log K = 2\beta(2 - a_{0r} - b_{0s}) + c \qquad (4.40)$$

This can be abbreviated to

$$\log K = (2\beta/RT)(a_{0r} + b_{0s}) + A \qquad (4.41)$$

where A is a constant. Since β is negative, K should be larger, and dissociation of the ethane into radicals consequently more complete, the smaller is the sum of NBMO coefficients $a_{0r} + b_{0s}$.

Figure 4.5 shows NBMO coefficients for several arylmethyl, diarylmethyl, and triarylmethyl radicals. It will be seen that the NBMO coefficients at the exocyclic carbon atoms decrease rapidly with the number of attached groups. Dissociation should therefore occur most readily in the case of hexaarylethanes. This is the case. Solutions of hexaphenylethane at room temperature contain free triphenylmethyl radicals, whereas tetraarylmethanes dissociate to an appreciable extent only at much higher temperatures and diarylethanes are still more stable.

$$Ph_3C-CPh_3 \xrightleftharpoons{EO_t} 2Ph_3C\cdot$$
$$Ph_2CH-CHPh_2 \xrightleftharpoons{EO_t} 2Ph_2CH\cdot \qquad (4.42)$$

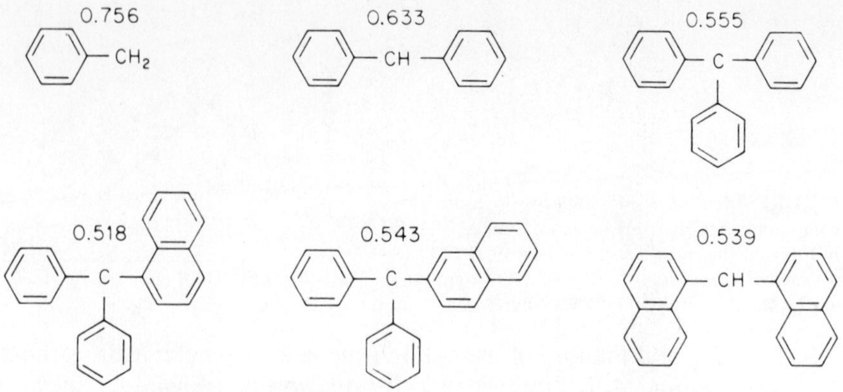

FIGURE 4.5. NBMO coefficients in some arylmethyl radicals.

As predicted from Fig. 4.5, naphthyl is even more effective than phenyl at promoting dissociation and α-naphthyl is, as predicted, more effective than β-naphthyl.

It should be added, however, that an important factor in these reactions is the overcrowding that takes place when five or six aryl groups are introduced into ethane. It is impossible to construct space-filling models of hexaarylethanes and indeed it now appears that such compounds do not in fact exist. The dimer of triphenylmethyl, which was formerly thought to be hexaphenylethane (Ph_3C—CPh_3) in fact has the quinonoid structure (1). Formation of this from triphenylmethyl is much less favorable than formation of Ph_3CCPh_3 (see the corresponding NBMO coefficients in Fig. 4.5) so far as the π electrons are concerned, but the steric effects outweigh this difference.

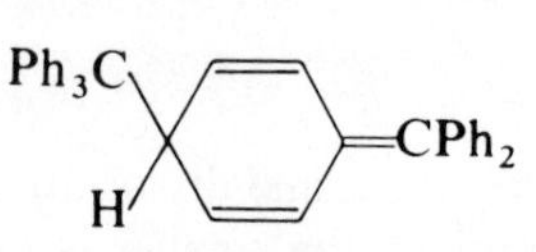

(1) 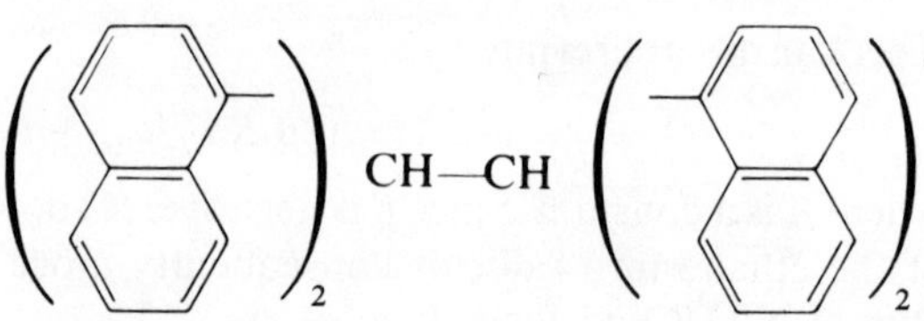   (2)

This result also accounts for the fact that tetra-α-naphthylethane (2) does not dissociate into radicals as easily as does "hexaphenylethane," even though the NBMO coefficient at the exocyclic carbon is less in di-α-naphthylmethyl than in triphenylmethyl (Fig. 4.5). The smaller steric interactions allow (2) to exist as such.

EE Equilibria Reactions. These involve the interconversions of two even systems. The classic example of a reversible reaction involving the interconversion of two even AHs of different sizes is the Diels–Alder reaction, a reaction in which a 1,3-diene and an ethylene derivative combine to form a

Chemical Equilibrium

cyclohexene,

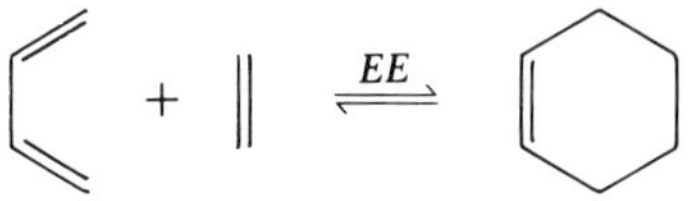

$$\text{cyclohexadiene} + \text{ethylene} \underset{EE}{\rightleftarrows} \text{cyclohexene} \quad (4.43)$$

The diene system can form part of an aromatic ring; e.g.,

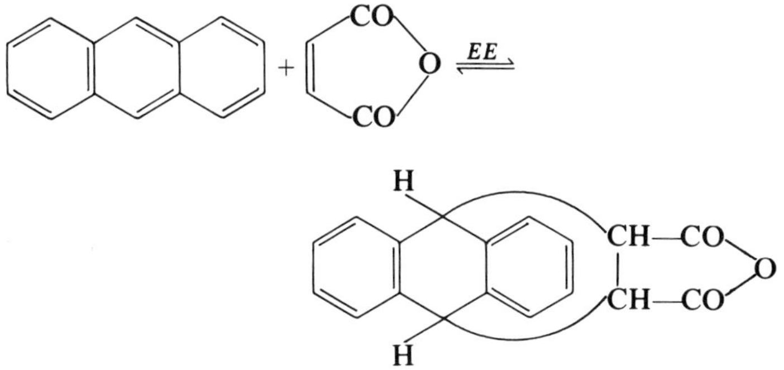

$$\text{anthracene} + \text{maleic anhydride} \underset{EE}{\rightleftarrows} \text{adduct} \quad (4.44)$$

Reactions of this type involve the interconversion of two even AHs differing by two carbon atoms.

The change in π energy during the reaction can be estimated by the first-order perturbation theory used in Section 3.12 (p. 93). To a first approximation, the π-bond energy $E_{\pi b}$ of a given bond is given by

$$E_{\pi b} = 2p\beta \quad (4.45)$$

where p is the corresponding bond order. In the case of a bond between atom t and atom u in an even AH,

$$P_{tu} = a_{0u} \cdot (1) \quad (4.46)$$

where a_{0u} is the NBMO coefficient of atom u in the odd AH formed by removing atom t. Thus the loss in π-bond energy on removing atom t is given by

$$(\delta E_\pi)_t = 2\beta \sum_r^{(t)} a_{0r} = \beta N_t \quad (4.47)$$

where the sum is over NBMO coefficients of atoms r in the odd AH formed by removing atom t, the sum being over atoms r directly linked to atom t in the original even AH. The final equality follows from equation (4.25). The π energy lost on removing two atoms t and u is given by

$$\delta E_\pi = \beta(N_t + N_u) \quad (4.48)$$

This is equal to the sum of the corresponding localization energies [equation (4.24)].

In the case of aromatic systems, this change in π energy corresponds to the removal of two carbon atoms *para* to one another in a benzene ring [cf. equation (4.44)]. It has therefore been termed the *paralocalization energy*. Since the same principles should apply to open chain systems or to Diels–

TABLE 4.1. 1,4-Bislocalization Energies for a Series of Aromatic Hydrocarbons

Compound	1,4-Bislocalization energy ($\times \beta^{-1}$)
 benzene	4.62
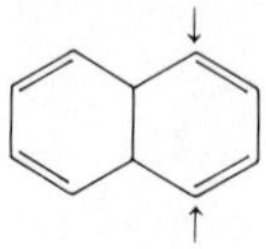 naphthalene	3.62
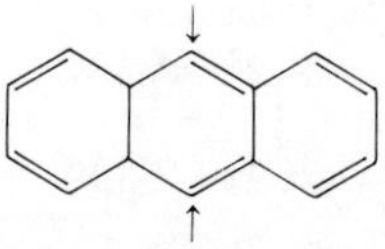 anthracene	2.53
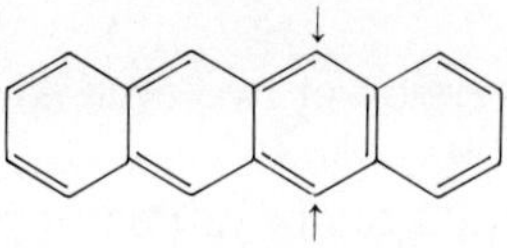 tetracene (central ring)	2.05
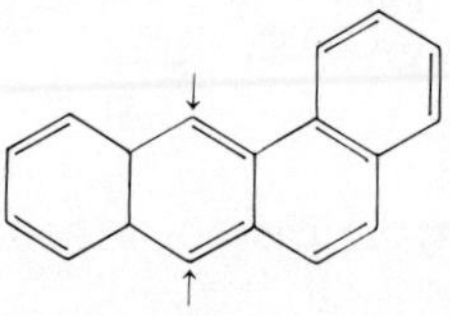 benz[a]anthracene	2.79
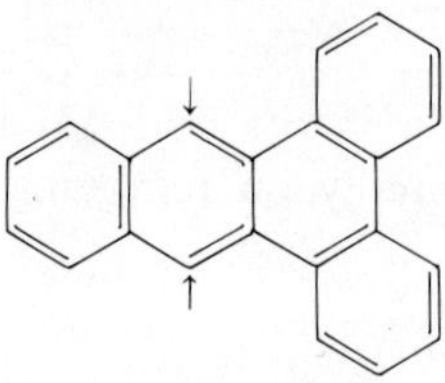 dibenz[a,c]anthracene	3.01
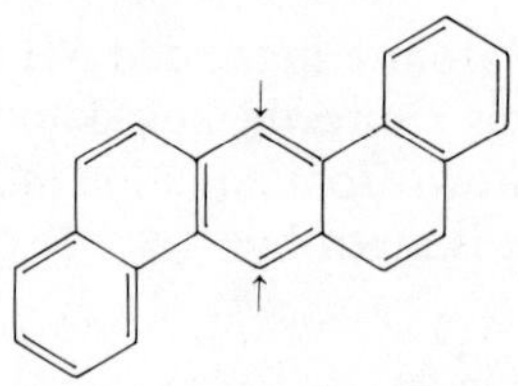 dibenz[a,h]anthracene	3.02
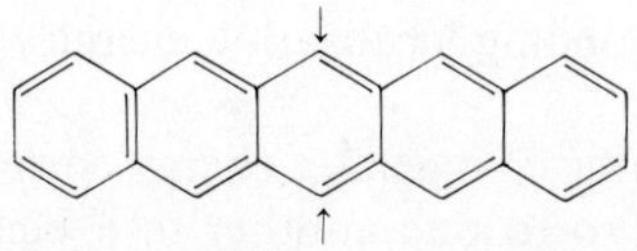 pentacene	1.60

Alder reactions involving vinyl groups attached to an aromatic ring, e.g.

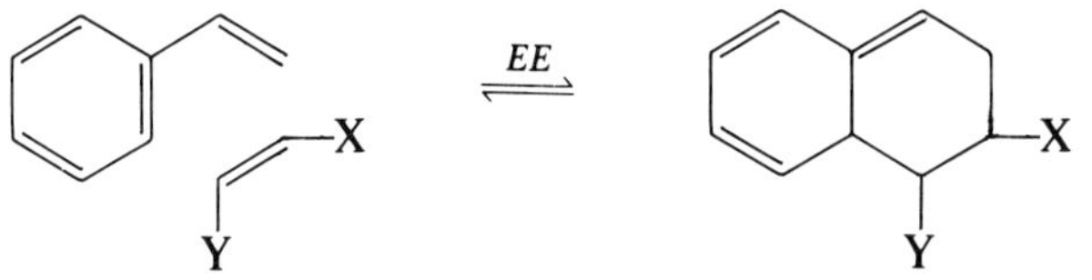

$$\begin{array}{c}\text{[styrene with vinyl-X,Y]}\end{array} \underset{}{\overset{EE}{\rightleftarrows}} \begin{array}{c}\text{[dihydronaphthalene with -X, Y]}\end{array} \quad (4.49)$$

it seems better to call it the *1,4-bislocalization energy*.

Values for 1,4-bislocalization energies of a number of conjugated and aromatic systems are shown in Table 4.1. The possibility of Diels–Alder reactions rests on a favorable equilibrium, i.e., a low enough value for the corresponding 1,4-bislocalization energy. In the reactions of dienes and aromatic compounds with maleic anhydride, [maleic anhydride structure] a typical *dienophile*, reaction occurs if, and only if, the 1,4-bislocalization energy found in this way is less than 3.7β. Thus naphthalene, which lies slightly below the limit, does seem to react reversibly with maleic anhydride to form the adduct (3), but the equilibrium is very unfavorable. On the other hand 1,3-dibenzanthracene reacts very readily to give the adduct (4).

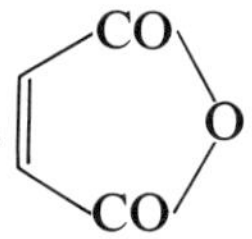

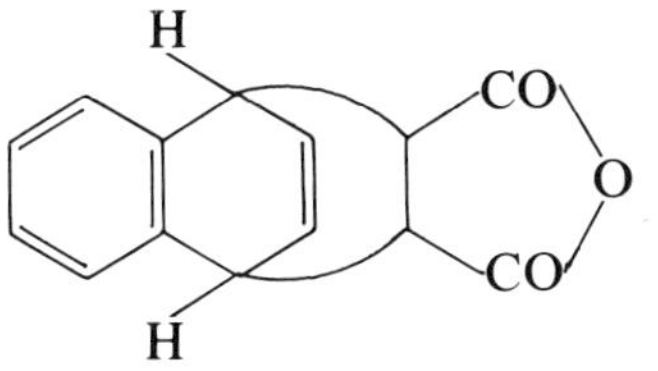

(3) 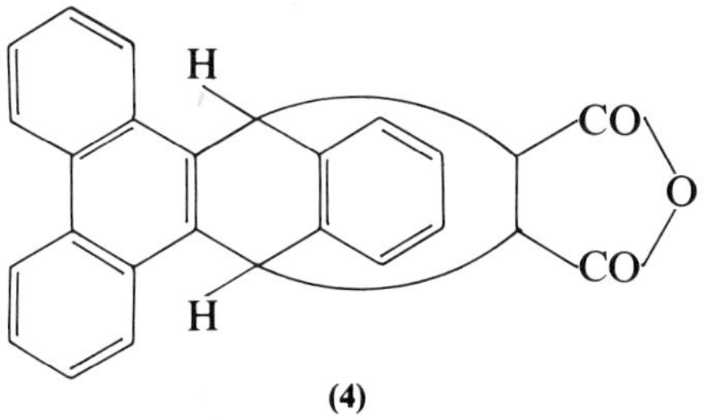   (4)

Hardly any quantitative data are available to check the predictions of equilibrium constants that can be made from the results in Table 4.1. However, the 1,4-bislocalization energies can be used to predict the preferred points of attack in an even AH and these are invariably the ones at which reaction is observed to take place. The values in Table 4.1 refer to the positions giving the lowest 1,4-bislocalization energy. In cases where this is low enough for reaction to occur, it occurs at those positions.

The ease of addition to an aromatic compound should be a measure of the aromaticity of the corresponding ring whose cyclic conjugation is destroyed in the reaction. It is interesting to compare the 1,4-bislocalization energies in Table 4.1 with aromatic energies estimated by the PMO method. Figure 4.6 illustrates the procedure for doing this, the parent compound being compared with an analog in which the bonds originally forming part

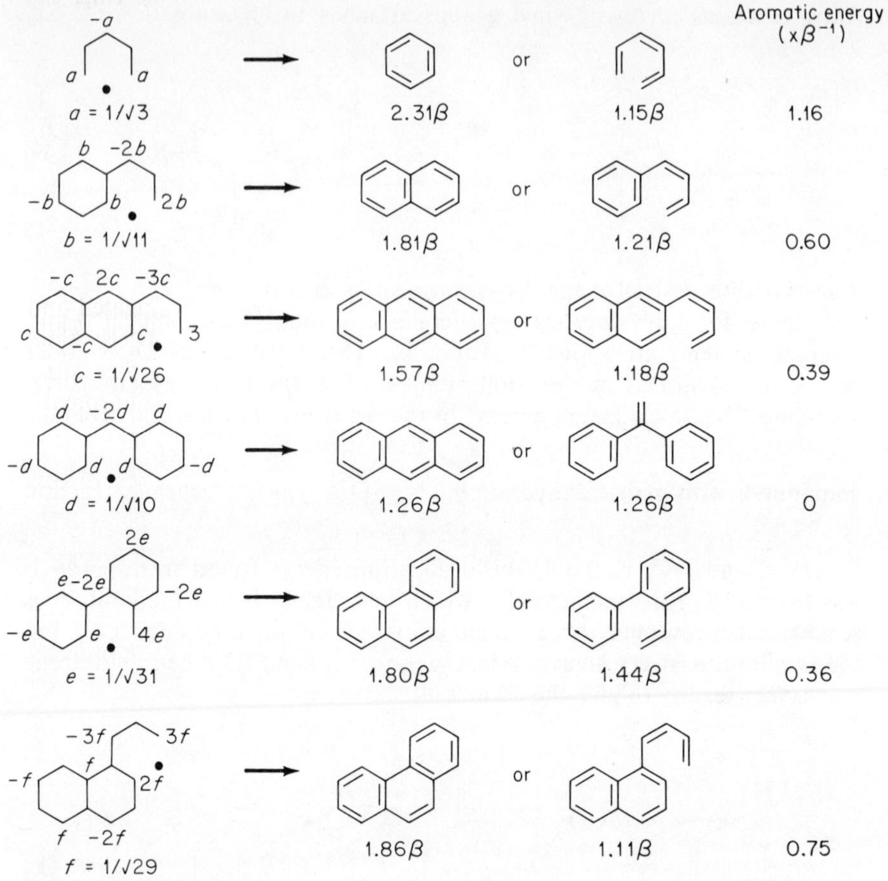

FIGURE 4.6. Calculation of aromatic energies of individual rings in aromatic hydrocarbons.

of the aromatic ring are now localized. The difference in π energy between the two species should be a measure of the aromaticity of the ring. We may term it the *aromatic energy* of the ring. Note that the central ring in anthracene has zero aromatic energy; one would expect addition across the 9,10 positions of anthracene to be as easy as 1,4 addition to a diene. This is indeed the case. In phenanthrene, on the other hand, all the rings are strongly aromatic, so Diels–Alder reactions are hardly ever observed.

Another analogous reaction is the concerted addition across a single π bond in a conjugated system. The Diels–Alder reaction also exemplifies this kind of process, the addition of the diene to the *dienophile* involving 1,2 addition to a π bond in the latter. Another similar reaction is the 1,2 addition of osmium tetroxide to double bonds,

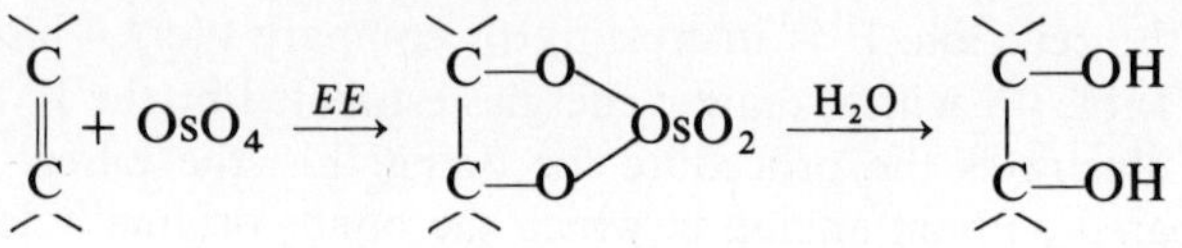

$$\underset{\text{C}}{\overset{\text{C}}{\|}} + \text{OsO}_4 \xrightarrow{EE} \underset{\text{C—O}}{\overset{\text{C—O}}{|}}\text{OsO}_2 \xrightarrow{H_2O} \underset{\text{C—OH}}{\overset{\text{C—OH}}{|}} \qquad (4.50)$$

Chemical Equilibrium

Such reactions can in principle take place at any two adjacent positions in a conjugated or aromatic system and osmium tetroxide can indeed oxidize many aromatic hydrocarbons in this way; e.g.,

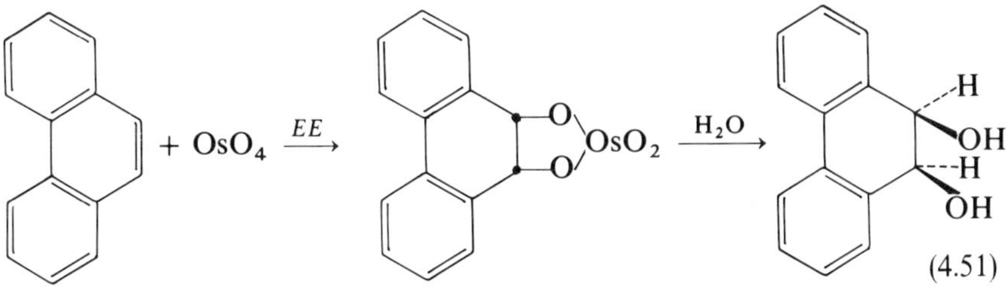

$$\text{(4.51)}$$

Since the addition is a one-step concerted process, it must for steric reasons take place *cis*, so the product is a *cis* diol.

Here again the energy of reaction ΔE can be written in the form

$$\Delta E = \text{constant} + \Delta E_{\text{deloc}} \quad (4.52)$$

for reactions involving concerted 1,2 addition. The delocalization energy of reaction ΔE_{deloc} is in turn the difference in π energy between two conjugated systems differing by two atoms adjacent to one another. In the case of aromatic systems [cf. equation (4.51)] these are *ortho* to one another in a benzene ring and the corresponding difference in π energy has consequently been termed the *ortholocalization energy*. Since the same principles apply likewise to open-chain systems, a better term seems the *1,2-bislocalization energy*.

Consider two atoms s and t in an even AH, linked to atoms r and u respectively; i.e.,

$$C_r \cdots C_s \cdots C_t \cdots C_u$$

The arguments used above show that the 1,2-bislocalization energy $(\delta E_\pi)_{st}$ of atoms s and t is given by

$$(\delta E_\pi)_{st} = 2\beta(p_{rs} + p_{st} + p_{tu}) \quad (4.53)$$

Now the localization energies of atoms s and t, $(LE)_s$ and $(LE)_t$, are given by

$$LE_s = 2\beta(p_{rs} + p_{st}) = \beta N_s \quad (4.54)$$

$$LE_t = 2\beta(p_{st} + p_{tu}) = \beta N_t \quad (4.55)$$

where N_s and N_t are corresponding reactivity numbers. Hence

$$(\delta E_\pi)_{st} = \beta N_s + \beta N_t - 2\beta p_{st} \quad (4.56)$$

In this case, the bislocalization energy is therefore *not* a sum of the corresponding localization energies $(\beta N_s$ and $\beta N_t)$ since removing either atom s or atom t destroys the st π bond, so that a sum of the corresponding localization energies contains the st π-bond energy twice over. The last term in equation (4.52) is a correction for this. Now the π bond order of bond st in an even AH is given approximately by

$$p_{st} \simeq \tfrac{1}{2}(a_{0s} + b_{0t}) \quad (4.57)$$

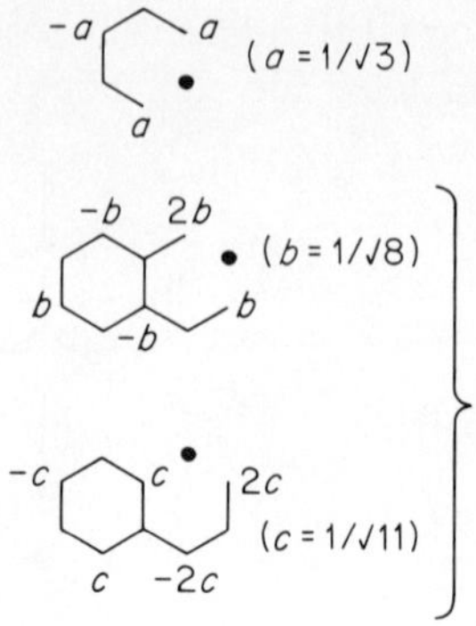

$$(\delta E_\pi)_{12} = \beta(2a + a + a + 2a)$$
$$= 6\beta/\sqrt{3} = 3.46\beta$$

$(a = 1/\sqrt{3})$

$(b = 1/\sqrt{8})$

$$(\delta E_\pi)_{12} = \beta(2b + 2b + 2c + 2c)$$
$$= 4\beta\left(\frac{1}{\sqrt{8}} + \frac{1}{\sqrt{11}}\right) = 2.62\beta$$

$(c = 1/\sqrt{11})$

$$(\delta E_\pi)_{23} = \beta(4b + b + b + 4b)$$
$$= 10\beta/\sqrt{8} = 3.54\beta$$

FIGURE 4.7. Calculation of 1,2-bislocalization energies ("ortholocalization energies").

where a_{0s} is the NBMO coefficient at atom s in the odd AH obtained by removing atom t, and b_{0t} is the NBMO coefficient at atom t in the odd AH obtained by removing atom s. Moreover, from the definition of reactivity number [equation (4.25)], we have

$$N_s = 2(a_{0r} + a_{0t}) \qquad (4.58)$$

$$N_t = 2(b_{0s} + b_{0u}) \qquad (4.59)$$

Combining equations (4.56)–(4.59), we have

$$(\delta E_\pi)_{st} = \beta(2a_{0r} + a_{0t} + b_{0s} + 2b_{0u}) \qquad (4.60)$$

Figure 4.7 illustrates the calculation of 1,2-bislocalization energies, using equation (4.60), and Table 4.2 shows values for various bonds in several conjugated systems. Since hardly any of these reactions are reversible, no experimental values of equilibrium constants are available; however, the additions take place most easily at bonds where the 1,2-bislocalization energy is low. This admittedly is more a matter of the rates of addition rather than equilibria, a problem which will be discussed in the next chapter.

OO Equilibria Reactions. These involve the interconversion of two odd systems. Reactions involving the interconversion of two odd systems differing by two atoms are quite rare. Most of those known involve opening of a saturated bond in an otherwise unsaturated odd-numbered ring to give an open, odd, conjugated system. For example, various substituted epoxides and amines can undergo reversible ring opening in this way,

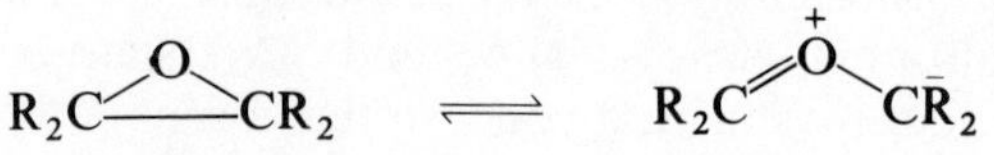

$$R_2C\overset{O}{-\!\!\!-\!\!\!-}CR_2 \rightleftharpoons R_2C\overset{\overset{+}{O}}{}\bar{C}R_2 \qquad (4.61)$$

$$R_2C\overset{\overset{R}{N}}{-\!\!\!-\!\!\!-}CR_2 \rightleftharpoons R_2C\overset{\overset{R}{\overset{+}{N}}}{}\bar{C}R_2 \qquad (4.62)$$

Chemical Equilibrium

TABLE 4.2. 1,2-Bislocalization Energies in Hydrocarbons[a]

Hydrocarbon[b]	1,2-Bislocalization energy[c]
 benzene	3.46
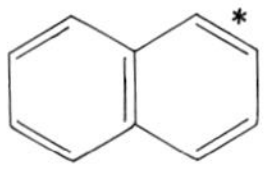 naphthalene	2.68
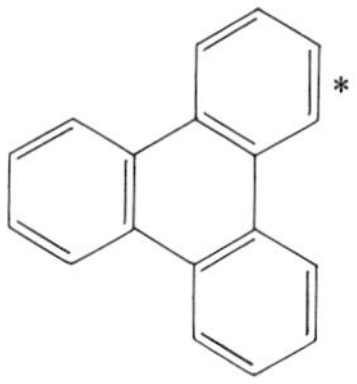 triphenylene	2.99
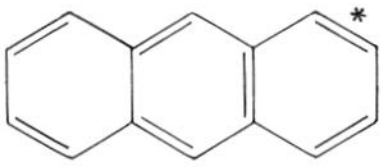 anthracene	2.16
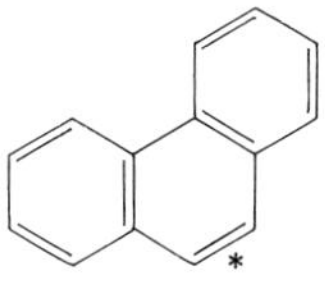 phenanthrene	2.15
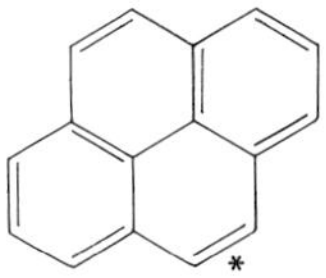 pyrene	2.35
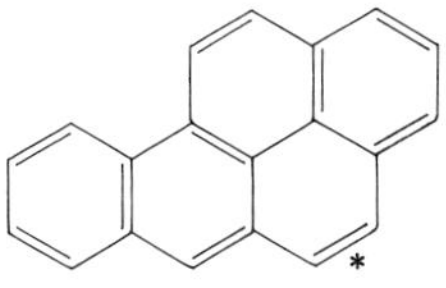 benzo[a]pyrene	2.02
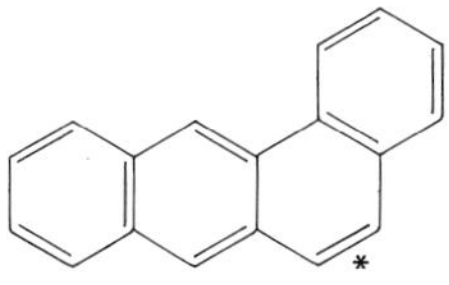 benz[a]anthracene	1.90

[a] Values from M. J. S. Dewar, *J. Am. Chem. Soc.* **74**, 3357 (1952).
[b] Bond with lowest 1,2-bislocalization energy marked with asterisk.
[c] For asterisked bond, in units of β.

The equilibria have not been studied as such but the ring opening has been shown to occur reversibly by trapping of the unstable open chain species by suitable reagents. In these simple cases the cyclic compound is isoconjugate with the cyclopropyl anion, ▷—, a derivative of the simplest odd AH anion (CH_3^-), whereas the open-chain isomer is isoconjugate with the allyl anion [CH_2=CH—CH_2^- or (CH_2⋯CH⋯$CH_2)^-$], an odd AH with three conjugated atoms. In the case of the isoconjugate AHs, the reaction involves conversion of an odd AH to a larger odd AH by union with ethylene,

$$CH_3^- \xleftarrow{u} CH_2=CH_2 \rightarrow (CH_2\cdots CH\cdots CH_2)^- \tag{4.63}$$

The corresponding change in π energy can be found by the PMO method indicated in Section 3.16 (p. 104). Thus the π energy of the union indicated in equation (4.63) is given by

$$\delta E_\pi = 2\beta[1 - (1/\sqrt{2})] = 0.586\beta \tag{4.64}$$

(since the NBMO coefficient of a terminal atom in allyl is $1/\sqrt{2}$). We can get a rough estimate of the difference in energy between the two species from equation (4.64), assuming that $\beta \sim 20$ kcal/mole, so that $\delta E_\pi \sim 12$ kcal/mole. The bond energy of a CC bond is ~ 81 kcal/mole, while opening of the three-membered ring leads to a decrease in strain energy of 27 kcal/mole. The difference in energy δE between the cyclic and open-chain species should then be

$$\delta E = 81 - 27 - 12 = 42 \text{ kcal/mole} \tag{4.65}$$

This is rather high, so the ring-opening reaction is not observed in the case of cyclopropyl anion. There is also no advantage to be had in replacing the carbon at the carbanion center by a heteroatom [cf. equations (4.61) and (4.62)], because the central atom in allyl is unstarred; the heteroatom actually loses p electrons during the ring opening and so opening of ethylene oxide or azirine should be even harder than that of cyclopropyl anion. Reactions of this type are observed only in cases where substituents [R in equations (4.61)–(4.62)] are present that favor the open-chain isomer. Further discussion may therefore be deferred until we have considered the effects of substituents of various types.

4.4 Electron Transfer Processes; Redox Potentials

Before we continue with the main theme of this chapter, we may consider briefly an entirely different type of process that may lead to equilibrium, namely electron transfer. Processes of this kind are very familiar in inorganic chemistry in the form of the typical oxidation–reductions undergone by metal ions; e.g.,

$$Cr^{2+} + Co^{3+} \rightleftharpoons Cr^{3+} + Co^{2+} \tag{4.66}$$

$$Sn^{2+} + 2Ag^+ \rightleftharpoons Sn^{4+} + 2Ag \tag{4.67}$$

$$2Fe^{2+} + Cl_2 \rightleftharpoons 2Fe^{3+} + 2Cl^- \tag{4.68}$$

Chemical Equilibrium

In organic chemistry, similar *electron transfer* reactions can be observed, interconverting species that differ only by one or two electrons; e.g.,

$$O=\!\!\langle\rangle\!\!=O + 2e^- \rightleftharpoons {}^-O-\!\!\langle\rangle\!\!-O^- \tag{4.69}$$

$$Me_2N-\!\!\langle\rangle\!\!-N=\!\!\langle\rangle\!\!=\overset{+}{N}Me_2 + H^+ + 2e^-$$

$$\rightleftharpoons Me_2N-\!\!\langle\rangle\!\!-NH-\!\!\langle\rangle\!\!-NMe_2 \tag{4.70}$$

When a metal *electrode* is placed in a solution of one of its salts, an equilibrium is set up between the metal and its ion; e.g.,

$$Zn \rightleftharpoons Zn^{2+} + 2e^- \tag{4.71}$$

The rates of the forward and backward reactions in equation (4.71) will depend on the potential difference between the metal and the solution. The more positive the metal, the more easily will its atoms be oxidized to ions, the electron remaining behind in the electrode. The more negative the metal, the more easily will it give up electrons to ions, reducing them to atoms. Since each of these processes leads to production on the electrode of charges which oppose them, equilibrium will be reached when the difference in potential between the metal and the solution has some definite value (the *electrode potential*). This potential will of course depend on the concentration of the ion since the reverse reaction involves the ion.

In order to measure the potential difference, we have to connect the electrode and the solution to a voltmeter or potentiometer. We cannot of course use a metal connection to the solution since the metal would itself behave as an electrode. What we can do is to take a second metal in a solution of one of its salts and connect the two solutions together electrically. The potential difference between the two metal electrodes will then be equal to the difference between the corresponding electrode potentials. Such measurements do not of course tell us the absolute values of electrode potentials, but these in fact are of little importance. The relative values are sufficient. Since the electrode potentials vary with concentration, it is also necessary to specify this. The potential when the electrolyte is present in molar concentration is called the *standard electrode potential*.

In connecting together two such "half-cells," one cannot just put the solutions in contact, because they would mix. Nor of course can they be connected by a metallic conductor, since this would itself act as an electrode. One can, however, use as a bridge between the two half-cells a solution of a salt since such solutions conduct by ionic migration and not by transfer of electrons. Diffusion of solutions into or out of the bridge can be restrained

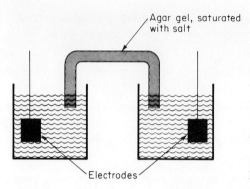

FIGURE 4.8. Simple potentiometric cell.

by various expedients, e.g., immobilizing the salt solution in the bridge by converting it to a gel by adding agar to it, as indicated in Fig. 4.8.

Since only relative values of electrode potentials can be measured, it is usual to quote them relative to that of some standard electrode. The two most commonly used are the hydrogen electrode, which depends on the oxidation of hydrogen absorbed on a platinum electrode, i.e.,

$$H_2 \rightleftharpoons 2H^+ + 2e^- \qquad (4.72)$$

and the calomel electrode, which depends on the oxidation of mercury to calomel (Hg_2Cl_2),

$$2Hg + 2Cl^- \rightleftharpoons Hg_2Cl_2 + 2e^- \qquad (4.73)$$

In the former, the standard electrode corresponds to hydrogen gas at 1 atm (1 bar) in normal acid. In the latter, the standard electrode corresponds to a saturated solution of calomel in sodium chloride solution. The concentration of calomel in such a solution is extremely small because it is almost insoluble in water.

In reactions of this kind, the metal of the electrode forms part of the reacting system [see equations (4.66)–(4.68)]. A similar effect can be observed in the case of oxidation–reduction (*redox*) systems in which both components are in solution [equations (4.69) and (4.70)]. If an inert electrode (platinum, gold, graphite, etc.) is immersed in such a solution, it will acquire a potential due to a balance between the two competing electrode processes,

$$A \longrightarrow A^+ + e^-, \quad A^+ + e^- \longrightarrow A \qquad (4.74)$$

Here the electrode acts solely as a source of, or sink for, electrons. The corresponding potential when the reagents are at unit concentration is termed the *redox potential*. Redox potentials are usually given relative to that of the standard calomel electrode.

Suppose now that we connect together two such redox half-cells, corresponding to the reactions

$$A \rightleftharpoons A^+ + e^- \qquad (4.75)$$

$$B \rightleftharpoons B^+ + e^- \qquad (4.76)$$

Chemical Equilibrium

If everything is at unit concentration, the potential difference v across the complete *cell* will be given by

$$v = E_A - E_B \tag{4.77}$$

where E_A and E_B are the redox potentials of the systems A–A$^+$ and B–B$^+$. Suppose now that $E_A > E_B$, the electrode in solution A being more positive. If we connect the electrodes together through an external electric circuit so that current can flow, electrons will flow from the electrode in B to that in A. The effect will be to reduce and oxidize equivalent amounts of A and B, respectively, the overall reaction being

$$A^+ + B \longrightarrow A + B^+ \tag{4.78}$$

If we allow one gram equivalent of A$^+$ and B to react, the total current flowing will be Ne, where N is the Avogadro number and e the charge on an electron. This quantity (the *Faraday equivalent*, F) amounts to 96,487 C. If we connect the electrodes through an electric motor and if we keep all the species at unit concentration, the work done w will be given by

$$w = vF = 96,487(E_A - E_B) \quad \text{joules} \tag{4.79}$$

(assuming of course that none is lost through friction, electrical resistance, etc.). Now the maximum amount of work that can be carried out by a system when it undergoes some change is equal to the corresponding decrease in free energy $-\Delta G$. By definition, the free energy of a reaction is the change in free energy when one mole of each reactant reacts, all being at unit concentration. Hence,

$$G = -w = -96,487(E_A - E_B) \quad \text{joules} \tag{4.80}$$

$$= -23,061(E_A - E_B) \quad \text{cal} \tag{4.81}$$

Measurements of redox potentials thus enable one to deduce the changes in free energy during redox reactions.

If we can assume that there is no change in the energy of solvation or entropy during the redox reaction of equation (4.78), the change in free energy ΔG should be equal to the corresponding change in energy ΔE. Now the ionization potentials of A (I_A) and B (I_B) are defined as the changes in energy during the processes in equations (4.75) and (4.76). It follows that for equation (4.78),

$$\Delta E = I_B - I_A \tag{4.82}$$

Ionization potentials are usually expressed in electron volts (eV), 1 eV being the energy acquired by an electron falling through a potential difference of 1 V and the ionization potential being expressed in terms of the energy required to ionize one molecule. The corresponding difference in energy ΔE, in joules/mole, is then given by

$$\Delta E = F(I_B - I_A) = 96,487(I_B - I_A) \quad \text{joules/mole} \tag{4.83}$$

where I_A and I_B are expressed in eV. If then our assumptions are correct, so

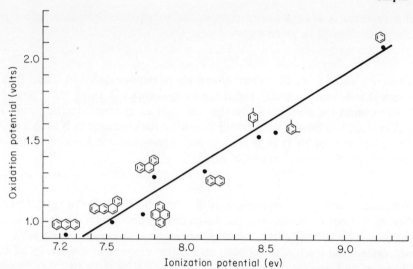

FIGURE 4.9. Plot of half-wave oxidation potentials vs. ionization potentials for aromatic hydrocarbons. The electrochemical potentials are from Ref. 5 and the ionization potentials from Ref. 6.

that $\Delta G = \Delta E$, comparison of equations (4.80) and (4.83) shows that

$$E_A - E_B = I_A - I_B \qquad (4.84)$$

Thus for a series of redox processes of the type indicated in equation (4.75), the variations in redox potential should be equal to the corresponding variations in ionization potential.

Reversing the argument, one can use a comparison of ionization potentials with redox potentials to check the validity of the assumptions made in deriving equation (4.84), i.e., that the energies of solvation of the different ions A^+ are the same and that the entropies of ionization of different molecules A are the same. Such a check can be carried out in the case of aromatic hydrocarbons, the ionization potentials of a number of these having been measured accurately by photoelectron spectroscopy. Figure 4.9 shows a plot of E_A vs. I_A for several of these compounds. If equation (4.84) holds, the points should lie on a line of unit slope. As can be seen, this relation does hold reasonably well; the line in Fig. 4.9 has a slope of 0.7. The deviation from unit slope is virtually certain to be a solvent effect.

While redox equilibria are not of much significance in organic chemistry, reactions involving electron transfer are important and will be considered in Chapter 7 (p. 517). The facility of electron transfer depends on the ionization potentials of the molecules involved and these in turn depend on the energies of the highest occupied MOs (Koopmans' theorem; see Section 1.9, p. 26). The ionization potential is therefore a one-electron property and its calculation involves problems distinct from those that arise in the treatment of collective properties such as heats of formation, total electron distributions,

Chemical Equilibrium

and molecular geometries. For example, although the bonds in paraffins are "localized," their collective properties being additive functions of bond properties, the ionization potentials of paraffins vary over a wide range in an apparently erratic manner. Fortunately, normal chemical equilibria are controlled primarily by the collective properties of molecules, so we will dispense with a detailed discussion of ionization potentials at this time.

4.5 Nonalternant Systems

Returning now to "normal" chemical equilibria, our next problem is to extend the treatment of Section 4.3 to nonalternant hydrocarbons. This can be done (see Sections 3.12 and 3.13, pp. 93 and 96), if the nonalternant systems in question can be derived from alternant ones by intramolecular or intermolecular union.

Consider, for example, the deprotonation of cyclopentadiene (5) to the anion (6) by base,

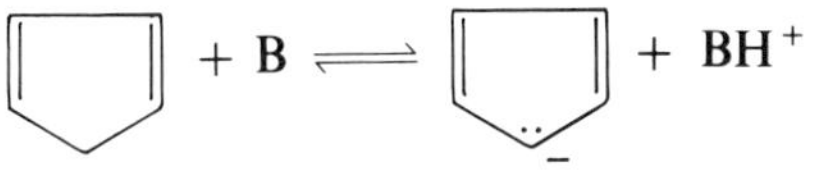

$$\text{(5)} + B \rightleftharpoons \text{(6)} + BH^+ \qquad (4.85)$$

We can derive (5) and (6) by intramolecular union from 1,3-pentadiene (7) and the corresponding anion (8),

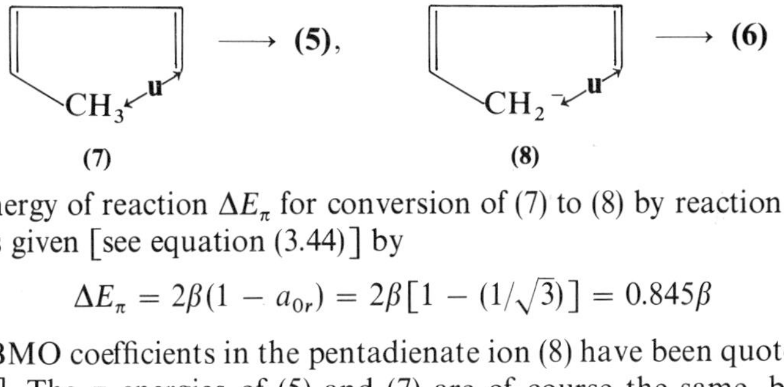

$$(7) \longrightarrow (5), \qquad (8) \longrightarrow (6) \qquad (4.86)$$

The π energy of reaction ΔE_π for conversion of (7) to (8) by reaction with the base B is given [see equation (3.44)] by

$$\Delta E_\pi = 2\beta(1 - a_{0r}) = 2\beta[1 - (1/\sqrt{3})] = 0.845\beta \qquad (4.87)$$

[The NBMO coefficients in the pentadienate ion (8) have been quoted often: $\pm 1/\sqrt{3}$]. The π energies of (5) and (7) are of course the same, both containing 1,3-butadiene systems. The π energies of (6) and (8) are, however, different, the difference ΔE_π being given [see equation (2.13)] by

$$\delta E_\pi = 2p_{1,5}\beta = 2\left(\frac{1}{\sqrt{3}}\frac{1}{\sqrt{3}}\right)\beta = 0.667\beta \qquad (4.88)$$

Thus the π energy of reaction $\Delta E_\pi'$ for the deprotonation of (5) to (6) is given by

$$\Delta E_\pi' = \Delta E_\pi + \delta E_\pi = 1.512\beta \qquad (4.89)$$

The gain in π energy on conversion of (5) to (6) is consequently much greater than that in the conversion of (7) to (8). This is because (6), unlike (8), is aromatic and its formation from the nonaromatic (5) is correspondingly favored. Cyclopentadiene is in fact a very much stronger acid than 1,3-pentadiene and is converted to salts by bases of moderate strength (e.g., potassium hydroxide), whereas 1,3-pentadiene shows no acidic properties under normal conditions.

Similar considerations apply to the open-chain chloride (9) and its cyclic analog (11). Both (9) and (11) can undergo reversible ionization to carbonium ion chlorides (10) and (12). In the case of (9) and (10), the equilibrium, even in polar solvents, is right over on the side of the covalent chloride (9). In the case of (11) and (12), however, it is right over on the side of the ion pair (12). Indeed this compound exists as a crystalline salt, benzotropylium chloride, no detectable amount of the covalent isomer (11) being present in the solid or its solutions.

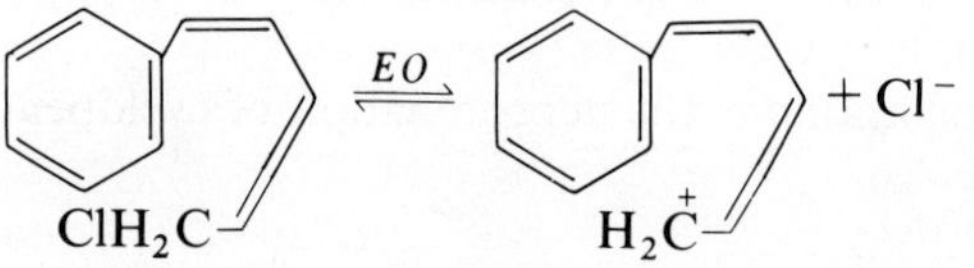

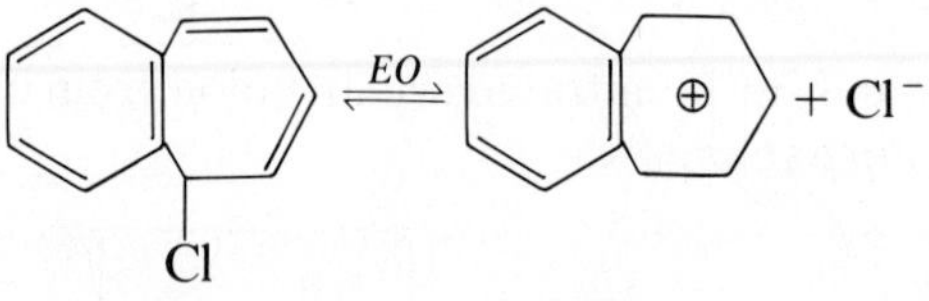

As we have seen (Section 3.13, p. 96), tropylium (13) is aromatic, i.e., more stable than the corresponding open-chain carbonium ion (14). The same is true of its benzo derivative (12) (see the rules for aromaticity; Section 3.14, p. 99). The formation of (12) from (11) is therefore correspondingly more favorable than the formation of (10) from (9).

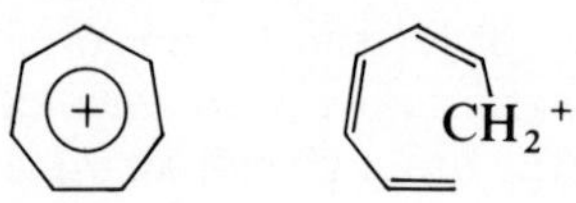

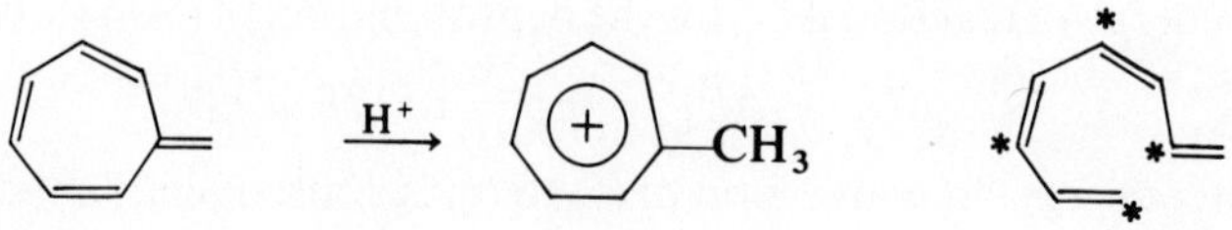

The ready conversion of heptafulvene (15) to methyltropylium (16) by acid is another example of an equilibrium involving nonalternant HCs. Heptafulvene is easily seen to be nonaromatic, being derived by intramolecular union between positions of like parity in octatetraene (17). The conversion of (15) to (16) is therefore assisted by the aromatic energy of the product (16).

4.6 Effect of Heteroatoms

In the preceding sections, we have developed a theory of equilibrium for reversible reactions involving hydrocarbons. Now we will extend this to systems containing heteroatoms. We can do so by the procedures developed in Sections 3.18 and 3.19 (pp. 109 and 115), the π energies of the heteroconjugated species being deduced from those of the isoconjugate hydrocarbons by using the first-order perturbation expression of equation (3.46) [see equation (4.92)]. As usual, we will assume that the only significant terms in this are the terms involving changes in the Coulomb integrals; the resonance integrals of π bonds involving heteroatoms do not differ greatly from those for carbon–carbon bonds.

Suppose then that we have a reversible reaction between molecules containing heteroatoms,

$$A_1 + A_2 + \cdots \rightleftharpoons B_1 + B_2 + \cdots \tag{4.90}$$

the corresponding reaction of the isoconjugate hydrocarbons being

$$A_1' + A_2' + \cdots \rightleftharpoons B_1' + B_2' + \cdots \tag{4.91}$$

Suppose that we replace carbon atom i in one of these species by a heteroatom with Coulomb integral α_i relative to carbon. The change in π energy of the reactants $\delta E_\pi(\text{reac})$ is given by

$$\delta E_\pi(\text{reac}) = q_i \alpha_i \tag{4.92}$$

where q_i is the π-electron density at atom i in the reactants. Likewise the change in energy of the products $\delta E_\pi(\text{prod})$ is given by

$$\delta E_\pi(\text{prod}) = q_i' \alpha_i \tag{4.93}$$

where q_i' is the π-electron density at the same atom i in the product. If the π energy of reaction for the hydrocarbons [equation (4.89)] is $\Delta E_\pi'$, then that (ΔE_π) for the system containing heteroatoms is given by

$$\Delta E_\pi = \Delta E_\pi' + (q_i' - q_i)\alpha_i \tag{4.94}$$

This result is shown diagrammatically in Fig. 4.10. The heteroatom lowers the energies of reactants and products by amounts proportional to the π-electron densities at the position in question.

In the case of AHs, the π-electron densities are easily found by using the pairing theorem. One can see from equation (4.94) or Fig. 4.10 that

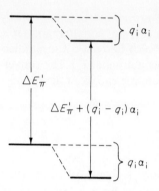

FIGURE 4.10. Effect of introducing a heteroatom into one of the reactants for a reversible reaction involving hydrocarbons.

introduction of a heteroatom will favor the species in which the π-electron density is greater. This of course is what one would expect intuitively, the more electronegative heteroatom trying to acquire electrons at the expense of the less rapacious carbon atoms.

From the pairing theorem, the π-electron density at any position in any neutral AH (even AH or odd AH radical) is unity. If atom i is at such a position in both reactants and products, replacing it by a heteroatom will have no first-order effect on the equilibrium. If, however, atom i is active in a reaction involving odd AH ions, the final term in equation (4.92) will *not* vanish and so the equilibrium will be disturbed by introduction of the heteroatom.

There are very few quantitative data to test this prediction but the general chemical evidence certainly seems to suggest that it is essentially correct. For example, polycyclic heteroaromatic compounds undergo Diels–Alder reactions if, and only if, the isoconjugate hydrocarbons do. A very nice illustration is provided by 10,9-borazaronaphthalene (18), a compound isoelectronic with naphthalene (19), since the ions N^+ and B^- are isoelectronic with neutral carbon. Since, however, nitrogen (and still more, N^+) is much more electronegative than carbon, and since boron (and still more, B^-) is less electronegative than carbon, one might expect the distribution of π electrons in (18) to be significantly different from that in naphthalene, (18) approximating to the uncharged "classical" structure (20) with two localized butadiene moieties. Yet (18) does *not* undergo a Diels–Alder reaction with maleic anhydride, even on heating.

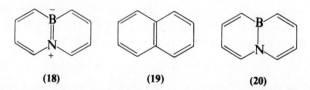

(18)   (19)   (20)

The effect of heteroatoms on equilibria involving radicals is still less well documented. One example is the dimerization of triphenylmethyl radicals ($Ph_3C \cdot$, see p. 144), the tendency to dimerize being little affected if the phenyl groups are replaced by pyridyl. The main evidence for the small

Chemical Equilibrium

effect of heteroatoms in this connection comes from studies of the rates of such reactions, a topic which will be discussed in the next chapter.

The effect of heteroatoms on reactions involving odd AH ions has, on the other hand, been exhaustively studied. A simple example is provided by a comparison of the reactions of toluene and its aza derivatives with base,

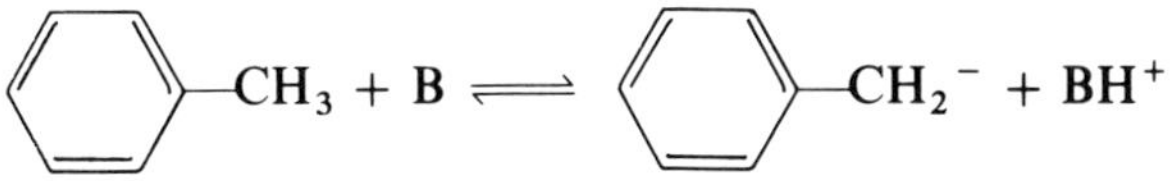

$$\text{Ph-CH}_3 + \text{B} \rightleftharpoons \text{Ph-CH}_2^- + \text{BH}^+ \qquad (4.95)$$

$$\text{(m-Py)-CH}_3 + \text{B} \rightleftharpoons \text{(m-Py)-CH}_2^- + \text{BH}^+ \qquad (4.96)$$

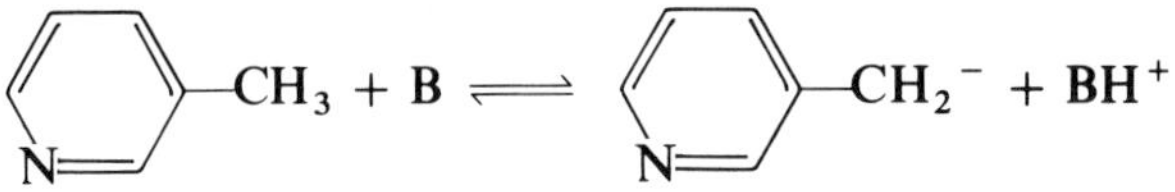

$$\text{(p-Py)-CH}_3 + \text{B} \rightleftharpoons \text{(p-Py)-CH}_2^- + \text{BH}^+ \qquad (4.97)$$

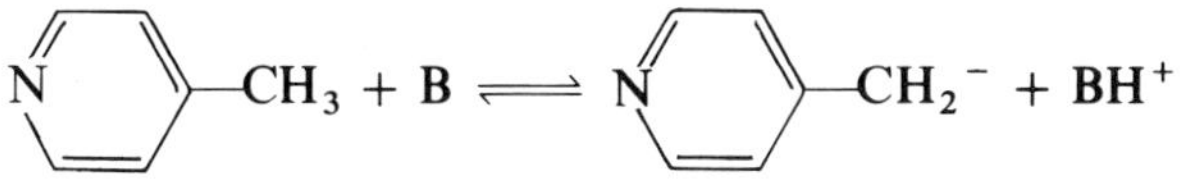

The π-electron densities in the benzyl anion have already been calculated (see p. 81). The position *meta* to methylene is inactive and the π-electron density is accordingly unity, whereas at the *para* position the additional electron density q is 1/7. One would therefore expect a heteroatom at the *meta* position to have little effect on the equilibrium of equation (4.95), while a heteroatom at the *para* position should favor formation of the conjugate anion. γ-Picoline is in fact a much stronger acid than toluene or β-picoline, the equilibrium of equation (4.97) being pushed toward the right by the gain in π-electron density at the nitrogen atom.

Equation (4.94) can be used to predict quantitatively the effect of heteroatoms in different positions in a given system. One can, for example, predict from the distribution of formal charge in the α-naphthylmethyl anion (21) that heteroatoms should have a greater effect in the ring carrying the exocyclic group. Thus the methyl groups in 1-methylisoquinoline (22) and

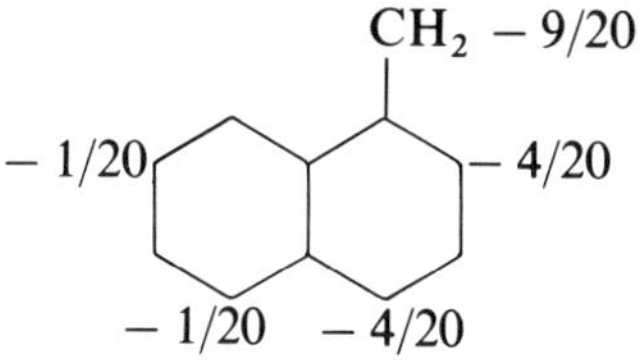

(21) naphthalene with CH$_2$ −9/20, −4/20, −1/20, −4/20, −1/20

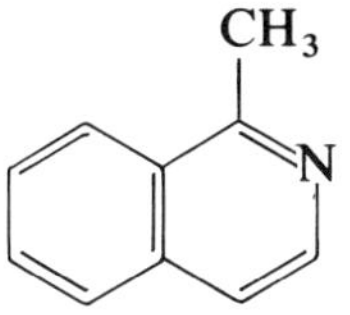

(22) 1-methylisoquinoline

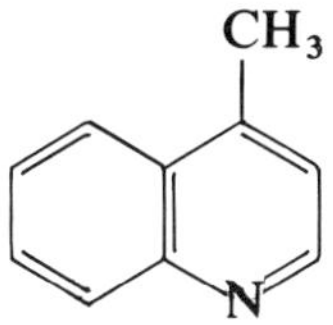

(23) 1-methylquinoline

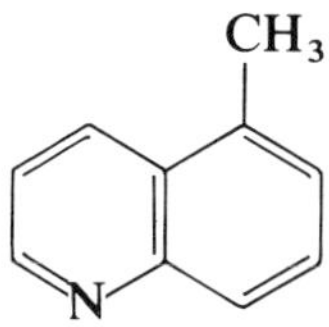

(24) 5-methylquinoline

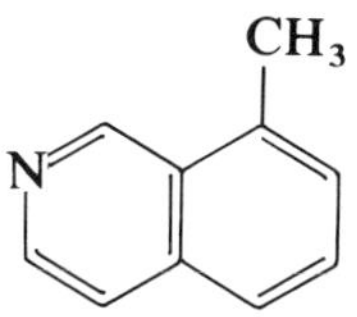

(25) 5-methylisoquinoline

4-methylquinoline (23) should be much more acidic than those in 5-methylquinoline (24) or 8-methylisoquinoline (25). While no quantitative equilibrium data are available, the chemical evidence indicates that (22) and (23) are much more easily deprotonated to the corresponding arylmethyl anions than are (24) and (25).

The same arguments can of course be applied to reactions involving carbonium ions where heteroatoms at an active position will have a destabilizing effect ($q < 1$). A simple example is the failure of acridinium ion (26) to protonate at the position para to nitrogen to form an analog of the anthracenonium ion (28). The ion (26) is isoconjugate with anthracene (27), which readily protonates to (28).

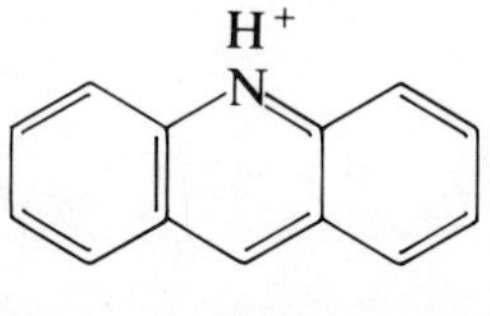

(26)

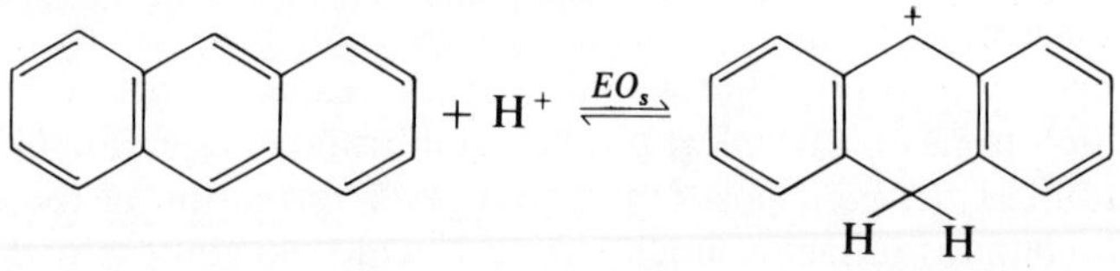

(27)   (28)

Slightly different considerations apply to situations where a reaction involves interconversion of a smaller, even system and a larger, odd one, the extra atom being the heteroatom. A simple example is the protonation of aromatic amines,

$$\text{Ar}\ddot{\text{N}}\text{H}_2 + \text{H}^+ \xrightleftharpoons{OE_s} \text{ArNH}_3^+ \qquad (4.98)$$

In the product, one of the two electrons that formally constitute the nitrogen lone pair in the amine is effectively lost; the nitrogen atom now shares those electrons with the proton (see the discussion of dative bonds in Section 1.7, p. 18; the combination of the amine and proton involves the formation of a dative bond from N to H^+). There is therefore a decrease in electron density at the nitrogen atom on passing from the amine to its salt.

In the isoconjugate AH reaction, i.e.,

$$\text{ArCH}_2^- + \text{H}^+ \xrightleftharpoons{OE_s} \text{ArCH}_3 \qquad (4.99)$$

the π-electron density at the methylene carbon in the anion is $1 + a_{0r}^2$, with a_{0r} the corresponding NBMO coefficient. In the arylmethane, the methyl carbon is neutral. The change in electron density at the exocyclic carbon atom during the reaction is therefore $-a_{0r}^2$. The π energy of reaction ΔE_π for the equilibrium of equation (4.98) is therefore given in terms of that

($\Delta E_\pi'$) from the isoconjugate AH equilibrium [equation (4.98)] by

$$\Delta E_\pi = \Delta E_\pi' - a_{0r}^2 \alpha_X \qquad (4.100)$$

where α_X is the difference between the Coulomb integrals of the heteroatom X and carbon. Combining this with the expression already derived for $\Delta E_\pi'$ [equation (4.94)], we find

$$\Delta E_\pi' = 2\beta(1 - a_{0r}) - a_{0r}^2 \alpha_X \qquad (4.101)$$

Since the pK_A of the ion $ArNH_3^+$ is defined as the negative logarithm (base ten) of the equilibrium constant for equation (4.98), it follows that

$$pK_A = A + B a_{0r} + C a_{0r}^2 \qquad (4.102)$$

where A, B, and C are constants. This parabolic relation between pK_A and a_{0r} has been shown to hold for a number of amines of this type.

The arguments and illustrations given above show how quantitative relationships between structure and equilibrium constant can be derived for reactions involving heteroconjugated species, using the results derived earlier for hydrocarbons. In many connections, however, a qualitative knowledge of the effect of heteroatoms is sufficient. We want to know the direction in which a given equilibrium will be displaced if we replace a carbon atom by a heteroatom, the magnitude of the effect being less important. In order to make such a prediction, we need to be able to deduce the effect of the heteroatom on the relative π energies of the reactants and products, i.e., on the π energy of reaction.

Since we are concerned only with the relative effects of the heteroatom on the energies of two different species, we do not need to know the absolute magnitudes of either. We can specify the effect of the heteroatom by reference to its effect in some standard system. If the heteroatom lowers the energy of the system in question more than it does that of the standard system, we will say that the heteroatom has a stabilizing effect. If the decrease in energy is less than that in the standard system, the heteroatom will be said to have a destabilizing effect. The "standard system" will of course be a neutral AH since the electron density at any position in any such AH is unity and the effect of replacing the corresponding carbon atom by a heteroatom is consequently the same, i.e., α_X, where α_X is the Coulomb integral of the heteroatom X relative to carbon (see Fig. 4.10). From equations (4.92)–(4.94), it is evident that the heteroatom will have a stabilizing effect if it occupies a position where the π-electron density is greater than unity, i.e., an active position in an odd AH anion, and a destabilizing effect if it occupies a position where the π-electron density is less than unity, i.e., an active position in an odd AH cation. Hence we arrive at the following rule:

Rule 1. (a) Introduction of a heteroatom stabilizes an odd AH anion and destabilizes an odd AH cation if the heteroatom occupies an active position. The effect is proportional to the formal charge at that position.

(b) Heteroatoms in even AHs or at inactive positions in odd AH ions or any positions in odd AH radicals have no first-order effect.

Thus the introduction of a heteroatom at an active position tends to favor the formation of a carbanion and to disfavor production of a cation. For example, such substitution in an odd AH radical increases both its electron affinity and its ionization potential.

4.7 The π-Inductive Effect

The last five words of Rule 1 involve a necessary proviso because heteroatoms at inactive positions in odd AH ions *do* exert stabilizing and destabilizing effects. While these are smaller than the first-order effects due to heteroatoms at active positions, they are nevertheless very significant. Thus while β-picoline is a weaker acid than γ-picoline, it is considerably stronger than toluene [see equations (4.95)–(4.97)]. Evidently heteroatoms can exert some additional influence on conjugated systems over and above that implied in equation (4.94).

The explanation follows from the discussion of polar covalent bonds in Section 1.7 (p. 17). Since heteroatoms are more electronegative than carbon,* the σ bond between such an atom X and carbon will be polar in the sense $C^{\delta+}\text{---}X^{\delta-}$. The resulting positive charge on carbon will make it more electronegative. Such a carbon atom will in turn behave qualitatively like a heteroatom. If the carbon atom forms part of a conjugated system, there will be a corresponding change in the π energy given by equation (4.94). If now we examine β-picoline and the anion derived from it [equation (4.96)], we can see that the carbon atoms adjacent to nitrogen occupy active positions in the anion. Their effect on the π energy will therefore be qualitatively similar to, though of course smaller than, the effect of a heteroatom in such a position. The change in electronegativity of atoms due to polarity of bonds formed by them is called the *inductive effect* and the resulting changes in the energies of π electrons due to inductive changes in the electronegativity of atoms comprised in a conjugated system is called the *π-inductive* effect.

The π-inductive effect can be included in equation (4.94). Since this represents the effect of a first-order perturbation, and since first-order perturbations are additive, we can write a generalized form of the equation as follows:

$$\Delta E_\pi = \Delta E_\pi' + \sum_i (q_i' - q_i)\alpha_i \quad (4.103)$$

Here the individual terms represent the effect of simultaneous changes in electronegativity at any number of positions in a conjugated system, the changes being due either to replacement of carbon atoms by heteroatoms or to the π-inductive effect.

---

* This is true of the "usual" heteroatoms, i.e., nitrogen, oxygen, sulfur, and the halogens. Boron and silicon are less electronegative than carbon and for them the conclusions of Rule 1 (Section 4.6) would be inverted. However, these elements play a relatively small role in conjugated systems.

4.8 Classification of Substituents

Organic chemists have found it convenient to divide molecules into a central core and attached *substituents*. Since our purpose is to provide a general qualitative theory of reactivity that will be of practical value to organic chemists, we must follow the same classification. Thus a very large number of molecules are known which are derived from benzene by replacing hydrogen atoms by other atoms or groups of atoms. Our next object is to see how such substituents alter the chemical properties of the central core (here benzene) to which they are attached.

Substituents can be divided into two main types, depending on whether or not they can interact directly with the π electrons of an adjacent conjugated system to give one in which the π orbitals differ in size from those in the substrate. Now a substituent R is a group derived from a molecule RH by removal of the hydrogen atom. If we introduce such a substituent into a molecule SH, we will get a corresponding *derivative* RS. The conjugated system in this will differ from that in SH only if three conditions are satisfied:

(a) The atom in RH through which R is linked to S in RS must form part of the conjugated system in RH.
(b) The substituent S must contain a conjugated system or an atom with unshared electrons.
(c) The atom in S through which it is linked to R in RS must form part of the conjugated system in S or be the atom with the unshared electrons.

Substituents which conform to these conditions are called *electromeric* (*E*) substituents. This term carries over from earlier theories of organic chemistry in which the π interactions of these substituents were described as the electromeric effect, ("resonance active" has also been used to describe these substituents). Vinyl (29), phenyl (30), formyl (31), and amino (32) are typical electromeric substituents. Substituents which do not conform to the conditions indicated above, and which do not therefore alter the topology of an adjacent π system, are called *inductive* (*I*) substituents. Such substituents (e.g., —CH_3, —CF_3, —$CH_2CH_2CH_3$) can influence the reactivity of an adjacent conjugated system only by the π-inductive effect or by direct electrostatic interactions across space, a phenomenon discussed later in this chapter.

—CH=CH₂   ⟨phenyl⟩   —CH=O   —NH₂
 (29)   (30)   (31)   (32)

4.9 Inductive (*I*) Substituents

An inductive substituent can affect an adjacent conjugated system only through the π-inductive effect. The character of this effect will depend on the

polarity of the bond linking the substituent X to an adjacent carbon atom. If the bond is polarized in the sense $C^{\delta+}$—$X^{\delta-}$, the substituent effectively withdraws electrons from the carbon atom and so makes it more electronegative. Since an entity deficient in electrons is described as *positive* in the accepted terminology of physics, such a substituent is described as a $+I$ substituent.*
The $+I$ substituents present a positive charge to the ring or chain to which they are attached. Likewise a substituent Y, forming CY bonds polarized in the sense C^-—Y^+, is called a $-I$ substituent. The $-I$ substituents present a negative charge to the ring or chain to which they are attached.

The effect of inductive substituents on the π energy of an adjacent conjugated system can be immediately deduced from Rule 1 (Section 4.6). Introduction of a $+I$ substituent at position i in a conjugated system will make that carbon atom more electronegative and so will qualitatively mimic the effect of replacing that carbon atom by a heteroatom. Equally, a $-I$ substituent will have a diametrically opposite effect. One can sum up these conclusions in the following rule:

Rule 2. (a) Inductive substituents in even AHs, or at inactive positions in odd AHs, have no first-order effect on the π energy.

(b) $+I$ substituents at active positions stabilize odd AH anions and destabilize odd AH cations.

(c) $-I$ substituents at active positions destabilize odd AH anions and stabilize odd AH cations.

The following examples show the effects of inductive substituents on equilibria.

(a) *Basicity of amines.* Consider the protonation of aniline to anilinium ion (33):

$$\text{C}_6\text{H}_5\text{—NH}_2 + \text{H}^+ \rightleftharpoons \text{C}_6\text{H}_5\text{—NH}_3^+ \qquad (4.104)$$

(33)

Aniline is isoconjugate with the benzyl anion, the π-electron density in which is greater than unity at the *para* position (see Section 3.4, p. 81). In the anilinium ion, the π-electron density in the *para* position is unity. A $-I$ substituent

---

* The sign convention used here was the one proposed by Lapworth and Robinson in their original "electronic theory" of organic chemistry. Subsequently, for reasons that need not be elaborated, Ingold introduced the opposite sign convention and this has been adopted by many chemists. The resulting confusion has been a source of much inconvenience to theoretical organic chemistry and little can be done about it at this stage. One can only hope that the original terminology of Lapworth and Robinson will eventually be generally adopted both because it has historical priority and is in accordance with the conventions of physics concerning the signs used to designate electric charge, and also because it conforms to the sign convention of the Hammett equation. Substituents that withdraw electrons have a plus sign and also a plus substituent constant (see p. 185).

(e.g., CH$_3$) in the *para* position should therefore destabilize aniline relative to the anilinium ion, while a $+I$ substituent (e.g., CF$_3$) should have the opposite effect. The $-I$ substituent should increase the basicity of aniline, while the $+I$ substituent should decrease it. The pK_A values of the anilinium ion and its *p*-methyl derivative are 4.65 and 5.08, respectively, in agreement with the predicted trend. The π-inductive effect operates in these cases, so that the pK_A of the *m*-trifluoromethyl anilinium ion is 3.49.

(b) *Acidity of phenols*. Both phenol (PhOH) and the phenoxide ion (PhO$^-$) are isoconjugate with the benzyl anion and so might be expected to have analogous π-electron distributions; however, the argument of Section 4.3 shows that the delocalization of the lone pair of electrons into the ring will be less for phenol or the phenoxide ion than for the benzyl anion because oxygen is more electronegative than carbon (see p. 112). Furthermore, the delocalization of electrons should be much less for phenol than for the phenoxide ion because neutral oxygen is much more electronegative than the ion O$^-$. Indeed, the CO bonds in phenols are effectively localized single bonds in the sense of Sections 1.12–1.14, the amount of π bonding being so small that variations in it from one phenol to another are essentially negligible. In the ion PhO$^-$, on the other hand, the delocalization of electrons from oxygen into the ring, while less than in the benzyl anion, is still appreciable. Consequently, the π-electron density at the position *para* to oxygen is greater in phenoxide ion than in phenol. The effect of inductive substituents on the equilibrium should therefore be qualitatively similar to their effect on the corresponding reaction of aniline [equation (4.104)]. The pK_A values of phenol and methylphenol, 9.98 and 10.17, respectively, show that the substituent effects are indeed similar in both cases.

$$\text{Ph-O}^- + \text{H}^+ \rightleftharpoons \text{Ph-OH} \qquad (4.105)$$

(c) *Stability of carbonium ions*. Triphenylcarbinol can act as a base, reacting reversibly with protons to form the triphenylmethyl cation,

$$\text{Ph}_3\text{COH} + \text{H}^+ \rightleftharpoons \text{Ph}_3\text{C}^+ + \text{H}_2\text{O} \qquad (4.106)$$

The cation is an odd AH cation in which the *ortho* and *para* positions in the benzene rings are active. The equilibrium of equation (4.106) should therefore be displaced to the right by $-I$ substituents in the *para* positions and to the left by $+I$ substituents. Experimental examination of these equilibria in sulfuric acid has shown this to be the case.

4.10 Electromeric Substituents; $\pm E$ Substituents

A $\pm E$ substituent is a group derived from an even AH by removal of a hydrogen atom; vinyl (34), phenyl (35), and α-naphthyl (36) are typical examples, being derived in this way from ethylene, benzene, and naphthalene, respectively:

$$CH_2=CH-$$

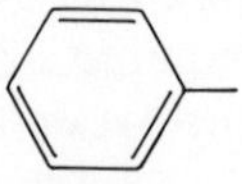

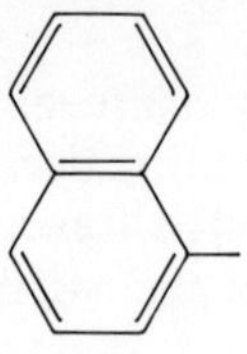

(34)   (35)   (36)

Introduction of such a substituent into an AH corresponds to the union of two AHs, so the corresponding change in π energy can be deduced by the procedures developed in Chapter 3. In the case of an even AH or an inactive position in an odd AH, there is (by our definition) no change in π energy because any such change is automatically absorbed into the bond energy of the sp^2-sp^2 C—C bond linking the substituent to the substrate, this bond being an essential single bond. In the case of an active position in an odd AH, however, there will be a decrease in π energy because although the π energy of union is a second-order perturbation, it will be greater than that involved in the union of two even AHs (see Section 3.16, p. 104).

Since single union of an even AH and an odd AH necessarily produces an odd AH, and since a given odd AH anion, radical, and cation differ only in the number of electrons occupying the NBMO, the total π-bond energy must be the same in all three forms. The stabilizing effect of the $\pm E$ substituent should therefore be the same for a given odd AH, regardless of whether it exists as an anion, radical, or cation. This is why such substituents are termed $\pm E$ substituents. Hence, we have the following rule.

Rule 3. A $\pm E$ substituent at an active position stabilizes an odd AH equally in all three forms (anion, radical, and cation). A $\pm E$ substituent in an even AH or at an inactive position in an odd AH has no effect.

The following examples illustrate the operation of this rule.

(a) *Dissociation of ethane into radicals.* Phenyl substituents enormously facilitate the dissociation of ethane into methyl radicals (methyl being the simplest odd AH) by stabilizing the latter. The heat of reaction for a series of reactions like

$$Ph-CH_2-CH_2-Ph \longrightarrow 2PhCH_2 \cdot \qquad (4.107)$$

decreases monotonically until one reaches hexaphenylethane, which is less stable than the quinoid structure:

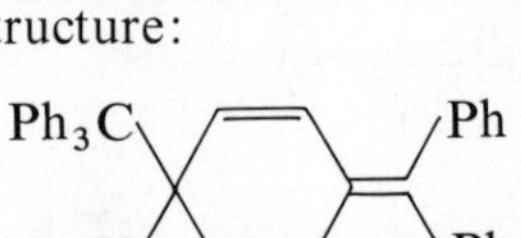

(b) *Deprotonation of methane.* Phenyl substituents greatly increase the acidity of methane by stabilizing the resulting methyl anion (the simplest odd

Chemical Equilibrium

AH anion), as the following pK_A values show:

$$\begin{aligned} CH_4 &\longrightarrow CH_3^- + H^+ & pK_A &\sim 80 \\ PhCH_3 &\longrightarrow PhCH_2^- + H^+ & pK_A &\sim 37 \\ Ph_2CH_2 &\longrightarrow Ph_2CH^- + H^+ & pK_A &\sim 34.1 \\ Ph_3CH &\longrightarrow Ph_3C^- + H^+ & pK_A &\sim 32.5 \end{aligned}$$

(4.108)

(c) *Formation of carbonium ions.* Phenyl substituents greatly facilitate the conversion of methanol to methyl cation [cf. equation (4.106)], as the following observations show:

$CH_3OH + H^+ \longrightarrow CH_3^+ + H_2O$ ionization not observed

$PhCH_2OH + H^+ \longrightarrow PhCH_2^+ + H_2O$ ionization observed in extremely strong acids* (e.g., FSO_3H/SbF_5)

$Ph_2CHOH + H^+ \longrightarrow Ph_2CH^+ + H_2O$ ionization complete in strong nonaqueous acids

$Ph_3COH + H^+ \longrightarrow Ph_3C^+ + H_2O$ ionization observed even in strong aqueous acids

(4.109)

(d) *Basicity of amines.* Phenyl substituents on nitrogen reduce the basicity of amines. Amines are isoconjugate with the anions considered in (b); the classic example of this effect is the relative acidities of the ammonium and anilinium ions:

$$\begin{array}{ccc} & NH_4^+ & PhNH_3^+ \\ pK_A: & 9.25 & 4.63 \end{array}$$

(4.110)

The relative effects of different $\pm E$ substituents on equilibria involving the interconversion of odd and even AHs can be estimated by the following extension of the procedures used in Section 3.16.

Consider an odd AH RH and a $\pm E$ substituent S which is to be introduced at position r in RH (Fig. 4.11a). Let the NBMO coefficient at position r in RH be a_{0r}. Consider the odd AH SCH_2. This will have a NBMO. Calculate the NBMO coefficients in this, setting the coefficient at CH_2 equal to unity (so that the NBMO is not normalized; the procedure is illustrated for benzyl in Fig. 4.11b). Let the resulting coefficient at position j in S be b_{0j} (Fig. 4.11c). In RS, S is attached to R at a position where the NBMO coefficient is a_{0r}. The NBMO of R must spill over into S so that the coefficient at atom j in S will now be $a_{0r}b_{0j}$ (Fig. 4.11d).

The NBMO coefficients found in this way for RS are not of course

---

* The mixture of fluorosulfonic acid and antimony pentafluoride is the strongest liquid acid presently known. This mixture will protonate methane to give hydrogen gas and ultimately the *t*-butyl cation, $C_4H_9^+$.

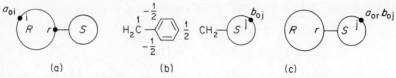

FIGURE 4.11. (a) Odd AH RH substituted by a $\pm E$ substituent S at position r; (b) NBMO coefficients in benzyl, chosen to make the coefficient at methylene unity; (c, d) corresponding NBMO coefficients in S–CH$_2$ and in R–S.

normalized. The next step is to normalize them by multiplying them by a normalizing factor A given by

$$A^2\left(\sum_i a_{0i}^2 + \sum_j a_{0r}^2 b_{0j}^2\right) = 1 \qquad (4.111)$$

Now the first sum in equation (4.111) is equal to unity because we assume the NBMO of RH to be normalized. Hence,

$$A = \frac{1}{(1 + a_{0r}^2 \sum_j b_{0j}^2)^{1/2}} = \frac{1}{(1 + a_{0r}^2 A^{\pm})^{1/2}} \qquad (4.112)$$

where

$$A^{\pm} = \sum_j b_{0j}^2 \qquad (4.113)$$

Thus $A^{\pm}$ is a quantity characteristic of the substituent and independent of the AH to which it is attached. Figure 4.12 illustrates the calculation of $A^{\pm}$ for several $\pm E$ substituents. The NBMO coefficient at atom i in R is there-

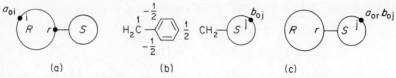

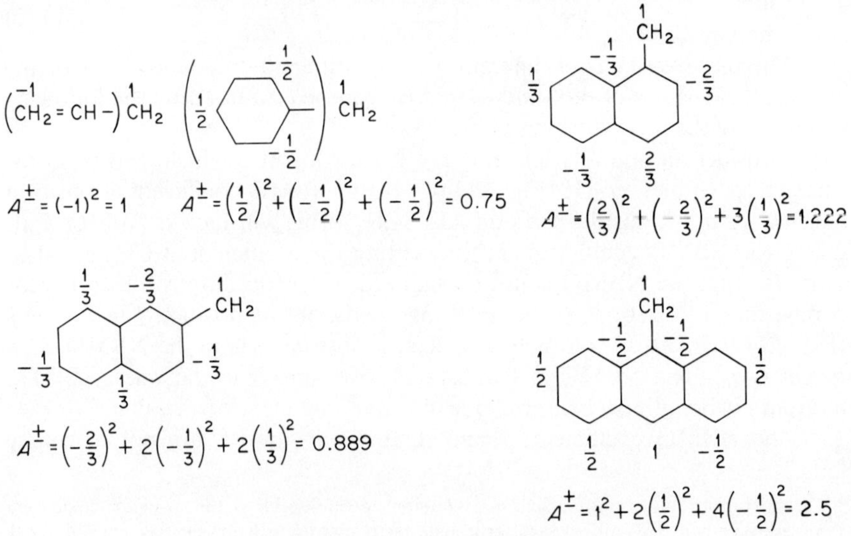

FIGURE 4.12. Calculation of $\pm E$ activities.

fore decreased by introduction of the substituent, the new value a'_{0i} being given by

$$a'_{0i} = Aa_{0i} = \frac{a_{0i}}{(1 + a_{0r}^2 A^{\pm})^{1/2}} \qquad (4.114)$$

Consider now a reaction in which the odd AH RH is produced from an even AH with one atom more. The change in π energy is then equal to the localization energy of the extra atom t in the even AH (see Section 4.3, p. 138); this is given [see equation (4.25)] by

$$\text{LE} = \beta N_t = 2\beta \sum_i^{(t)} a_{0i} \qquad (4.115)$$

where the sum is over atoms adjacent to atom t. Comparison of equations (4.114) and (4.115) shows that the substituent reduces the localization energy, the new values LE' (N_t') being given by

$$\text{LE}' = -\beta N_t' = -\frac{\beta N_t}{(1 + a_{0r}^2 A^{\pm})^{1/2}} \qquad (4.116)$$

The substituent will therefore facilitate conversion of the even AH to the odd one, effectively stabilizing the latter relative to the former (cf. Rule 3). Moreover, the magnitude of the stabilization is greater, the greater is the density a_{0r}^2 of the NBMO at the point of attachment of the substituent and the greater is the value of $A^{\pm}$ which is characteristic of the substituent. We term $A^{\pm}$ the $\pm E$ *activity* of the substituent since it is evidently a measure of the substituent's ability to stabilize an odd AH. The effect of the substituent is to dilute the NBMO of R and so reduce the individual coefficients a_{0i}. The magnitude of the dilution depends (very understandably) on the efficacy $A^{\pm}$ of the substituent and on the density a_{0r}^2 of the NBMO at the point where it is attached.

Similar considerations apply to a reaction in which RH is converted to an even AH by loss of one atom s. The change in energy [δE_π; see equation (4.34)] is given by

$$\delta E_\pi = 2\beta(1 - a_{0s}) \qquad (4.117)$$

The effect of the substituent is to replace a_{0s} by a'_{0s}, given by equation (4.114). The new value for the π-energy difference $\delta E_\pi'$ is then

$$\delta E_\pi' = 2\beta \left[1 - \frac{a_{0s}}{(1 + a_{0r}^2 A^{\pm})^{1/2}} \right] \qquad (4.118)$$

Here $\delta E_\pi'$ is numerically greater than δE_π since the substituent again stabilizes the odd AH RH relative to the even AH derived by loss of atom s. The magnitude of the effect again depends on the product $a_{0r}^2 A^{\pm}$.

There are few quantitative data to check these predictions, though examples of their application to rates of reaction will be found in the next chapter. Using the results in Fig. 4.12, we can compare the basicities of aromatic amines $ArNH_2$, ArH being an even AH. This comparison shows that the pK_A values of the conjugate acids do run parallel to the $\pm E$ activity of the group Ar; e.g.,

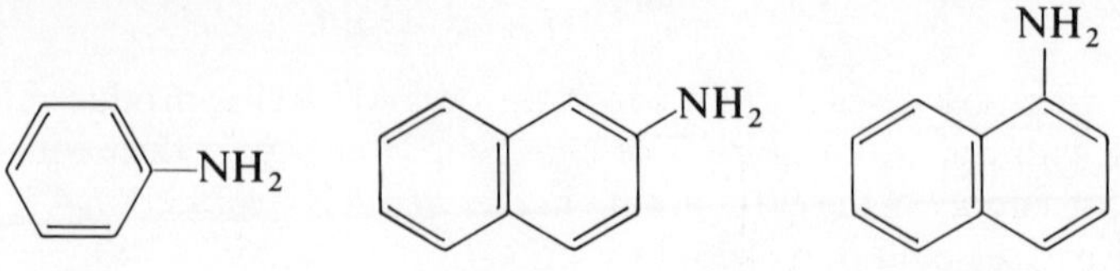

(4.119)

$A^{\pm}$:	0.75	0.89	1.22
pK_A:	4.63	4.16	3.92

Likewise [cf. equation (4.39)] replacement of a phenyl group in triphenylmethyl by α-naphthyl greatly reduces the tendency of the radical to dimerize. Naphthyldiphenylmethyl is monomeric in solution at concentrations where triphenylmethyl exists mainly as the dimer. The numerical values of the dissociation constants are, however, not too significant in view of the steric effects present in these very crowded systems (see p. 144).

4.11 $+E$ Substituents

A $+E$ substituent is one derived from a $\pm E$ substituent by replacing one or more carbon atoms by heteroatoms. Thus the formyl group (37) is isoconjugate with vinyl (34), the γ-pyridyl group (38) with phenyl (35), and the β-acetylvinyl group (39) with pentadienyl (40).

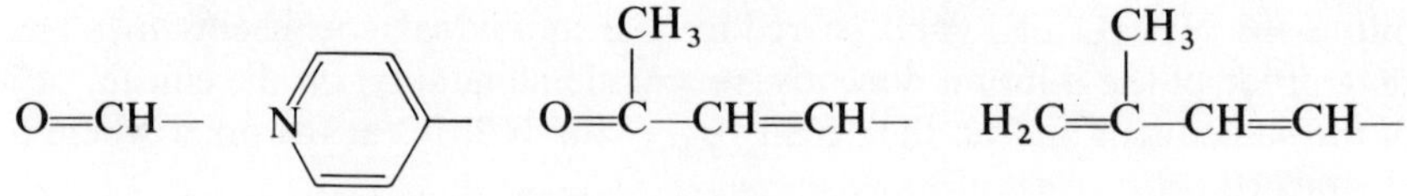

O=CH— (37)   (38) $O=\overset{CH_3}{\underset{|}{C}}-CH=CH-$ (39) $H_2C=\overset{CH_3}{\underset{|}{C}}-CH=CH-$ (40)

The effect of such a substituent can be deduced from that of the isoconjugate $\pm E$ one by the procedures used to study heteroconjugated systems (Sections 3.18 and 3.19).

If the substituent is attached to an even AH or is at an inactive position in an odd AH, the π-electron density at all positions in the $\pm E$ substituent will remain equal to unity, so replacement of the corresponding carbon atoms by heteroatoms will have no first-order effect (cf. Rule 1, Section 4.6).

If the substituent S is attached to an active position in an odd AH anion R^-, the NBMO will flow out into the substituent (cf. Fig. 4.10) and atoms of like parity to the atom r in R at which S is attached (see Fig. 4.11) will therefore acquire negative formal charges equal to the square of the corresponding NBMO coefficients. The value Q_j of the excess charge at atom j is given by [see Fig. 4.10(d) and equation (4.112)]

$$Q_j = A^2 a_{0r}^2 b_{0j}^2 = -\frac{a_{0r}^2 b_{0j}^2}{1 + a_{0r}^2 A^{\pm}} \qquad (4.120)$$

The stabilizing effect of the $+E$ substituent on the odd AH anion is therefore greater than that of the corresponding $\pm E$ substituent by an amount δE^+ given by

$$\delta E^+_{\text{anions}} = \sum_j Q_j \alpha_j = -\sum_j \frac{a_{0r}^2 b_{0j}^2 \alpha_j}{1 + a_{0r}^2 A^\pm} \tag{4.121}$$

where the sum is over atoms j in S that are replaced by heteroatoms in converting S to the $+E$ substituent S'. The extra stabilizing effect is again greater, the greater is the NBMO coefficient a_{0r} at the point of attachment of the substituent. The rest of the expression does not, however, depend solely on quantities characteristic of the substituent since the denominator contains the product $A^\pm a_{0r}^2$. Thus the greater is $A^\pm$, the smaller is the extra stabilizing effect of the substituent in comparison with the isoconjugate $\pm E$ substituent.

If the substituent is attached to an odd AH radical, the resulting combination RS' will of course be isoconjugate with an odd AH radical RS in which the electron density is unity at each point. Here the $+E$ substituent will have exactly the same effect as the isoconjugate $\pm E$ substituent.

Finally, if the substituent is attached at an active position in an odd AH cation, there will now be positive formal charges Q_j [equation (4.120)] at the atoms j in S or S'. This time, the $+E$ substituent will be *less* effective than the $\pm E$ substituent by the same amount δE^+ [equation (4.121)] that it was *more* effective in the case of an odd AH anion; for the cation, the sign of δE^+ [equation (4.121)] is reversed because the sign of the excess charge is reversed.

The $+E$ substituent therefore shows a decreasing stabilizing effect for odd AHs in the order anion > radical > cation, just as does a $+I$ substituent (see Rule 2, Section 4.9). The effects of $+E$ substituents on AH cations are complex, being a combination of the stabilizing $+E$ effect and the destabilizing $+E$ effect.

In the case of an odd AH cation, the destabilizing effect of a $+E$ substituent is inescapable even if it outweighs the stabilizing effect of the isoconjugate $\pm E$ substituent. The destabilizing effect is due to a decrease in electron density at an atom of high electronegativity and this effect is inexorably associated with the stabilizing effect of a $+E$ substituent, which is also due to dilution of the NBMO of R onto the substituent S'. It is entirely possible for a $+E$ substituent to have a net stabilizing effect on an odd AH cation.

The net effect can be predicted only by a quantitative assessment of the two contributing factors and this is somewhat outside the scope of the present first-order treatment. It is clear from equations (4.116), (4.118), and (4.121) that the stabilizing contribution will be greater, and the destabilizing contribution less, the greater is $A^\pm$. In general $A^\pm$ will tend to be the greater, the larger is the substituent. Moreover, the larger the substituent S, the more spread out will be the NBMO in SCH_2 and consequently the smaller the individual coefficients b_{0j} and the smaller the quantity δE^+ [equation (4.121)]. Other things being equal, one would therefore expect the smaller $+E$ substituents to be the ones that tend to destabilize odd AH cations, the larger ones having a net stabilizing effect. While equilibrium data are scanty, the evidence presented in the next chapter will show that in fact only the smallest $+E$ substituents (e.g., RCO and $-NO_2$) have a net destabilizing effect on odd AH cations.

These conclusions lead to our next rule:

Rule 4. $+E$ substituents at active positions have the same stabilizing effect on odd AH radicals as do the isoconjugate $\pm E$ substituents, a greater stabilizing effect on odd AH anions, and no stabilizing effect on even AHs or at inactive positions in odd AHs. At an active position in an odd AH cation, small $+E$ substituents are destabilizing, while larger ones are stabilizing but less so than the isoconjugate $\pm E$ substituents.

Some examples will illustrate the operation of this rule.

(a) *Deprotonation of methane.* The following pK_A values show the dramatic effect of $+E$ substituents:

$$CH_4 \quad CH_2=CH-CH_3 \quad O=\underset{\underset{CH_3}{|}}{C}-CH_3 \quad O_2N-CH_3 \quad (4.122)$$

pK_A: ~80 ~37 20 11

A particularly striking comparison is provided by 1,4-pentadiene (41) and acetylacetone (42). The former shows virtually no acid properties, being converted to the salt only by exceptionally strong bases, whereas the latter is a stronger acid than phenol (pK_A 9.3; pK_A of phenol, 9.9).

$$CH_2=CH-CH_2-CH=CH_2 \qquad O=\underset{\underset{CH_3}{|}}{C}-CH_2-\underset{\underset{CH_3}{|}}{C}=O$$

(41) (42)

(b) *Bond strengths.* The energy required to cleave a single bond in a molecule is termed the *bond strength* of the bond. This is quite distinct from the corresponding bond energy because the fragments resulting may reorganize themselves and so acquire modes of stabilization not present in the original molecule. Thus the CH bonds in the methyl group of toluene (PhCH$_3$) and the CC bond linking methyl to the ring are normal localized bonds. Their contribution to the heat of atomization is just a sum of the corresponding bond energies. If, however, we break one of the CH bonds, we get a radical PhCH$_2$ in which the CH$_2$—C bond is no longer localized but has a large π contribution. The resulting gain in π energy reduces the energy required to break the CH bond, so the bond strengths of the CH bonds in the methyl group of toluene are much less than their bond energies.

We can compare the stabilizing effects of substituents on radicals by comparing their effect on the strengths of, e.g., the CH bonds in methane. Addition of a vinyl or a carbonyl substituent to methane decreases the bond strength of the CH bond by approximately 13 kcal/mole. The effect is virtually independent of whether the substituent is of the $\pm E$ or $+E$ type.

Chemical Equilibrium

This is in marked contrast to the immensely greater stabilizing effect of $+E$ substituents on odd AH anions [equation (4.122)].

(c) *Effect on carbonium ions.* The destabilization of odd AH cations by small $+E$ substituents is shown rather clearly by the effect of such substituents on the last equilibrium in equation (4.109); i.e., the equilibrium lies very far to the left under conditions that would produce essentially complete disassociation of the unsubstituted carbinol.

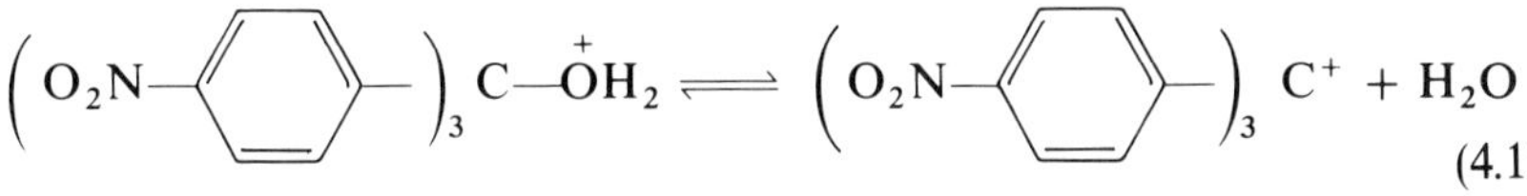

$$\left(O_2N-\!\!\left\langle\!\!\bigcirc\!\!\right\rangle\!\!-\right)_3 C-\overset{+}{O}H_2 \rightleftharpoons \left(O_2N-\!\!\left\langle\!\!\bigcirc\!\!\right\rangle\!\!-\right)_3 C^+ + H_2O \tag{4.123}$$

4.12 $-E$ Substituents

A $-E$ substituent is a group isoconjugate with an odd AH anion and attached through an active atom to the parent molecule. In principle, carbanionoid substituents such as $-CH_2^-$ (derived from the simplest odd AH anion, CH_3^-), $-\bar{C}H-Ph$, and $-\!\!\left\langle\!\!\bigcirc\!\!\right\rangle\!\!-CH_2^-$ are $-E$ substituents, but the only important $-E$ substituents are the neutral substituents derived from carbanionoids by introduction of heteroatoms. Typical examples are amino (43) and methoxy (44) (isoconjugate with $-CH_2^-$), acetamido (45) and β-methoxyvinyl (46) (isoconjugate with $-CH-CH=CH_2$), and *p*-aminophenyl (47) (isoconjugate with $-\!\!\left\langle\!\!\bigcirc\!\!\right\rangle\!\!-CH_2^-$):

$$-NH_2 \qquad -OCH_3 \qquad -NH-\overset{\overset{\displaystyle CH_3}{|}}{C}=O$$

(43)   (44)   (45)

$$-CH=CH-OCH_3 \qquad -\!\!\left\langle\!\!\bigcirc\!\!\right\rangle\!\!-NH_2$$

(46)   (47)

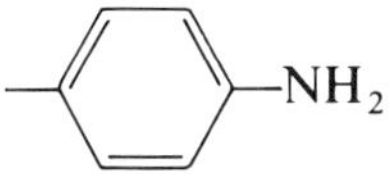

The effects of these substituents can be deduced in the usual way by considering first the isoconjugate carbanionoid ones and then examining the effect of replacement of carbon atoms in these by heteroatoms. Let us then first consider the effects of introducing into an AH, SH, a substituent R^-, where HR^- is an odd AH anion. The change in π energy will be the π energy of union of HR^- and SH and this can be deduced immediately from the arguments of Chapter 3.

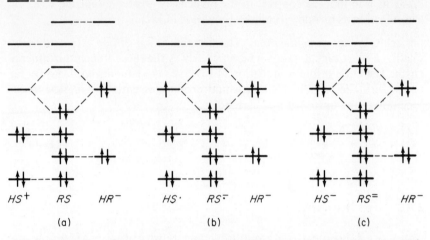

FIGURE 4.13. Effect of introducing a $-E$ substituent R^- into (a) an odd AH cation HS^+; (b) an odd AH radical $HS\cdot$; (c) an odd AH anion HS^-.

Consider first the case where the substrate is an odd AH cation HS^+ and R is attached to S in RS through an active atom in S. In this case, union will lead to an interaction of the NBMOs of R and S (Fig. 4.13a) to give a pair of MOs, one bonding and one antibonding. Union of HR^- and HS^+ therefore leads to a large decrease in π energy, the nonbonding electrons of HR^- being transferred to a bonding MO of RS. Indeed, the π energy of union will be the same as that for the corresponding radicals HR and HS.

Next consider union of HR^- with an odd radical $HS\cdot$. In this case the NBMOs of R and S again interact in RS but now there are *three* electrons to be accommodated in the resulting pair of MOs (Fig. 4.13b). While two of these can be accommodated in the lower, bonding MO of the pair, the third has to go into the upper, antibonding MO. While there is a decrease in π energy on union of $HS\cdot$ and HR^-, it is much less than in the case of the cation HS^+.

Finally, in the case of the anion HS^-, both the NBMOs are filled. Consequently (Fig. 4.13c), there is no first-order change in π energy on union. The situation is analogous to the union of even AHs in which two of the filled bonding MOs are degenerate. In the case of two even AHs, the second-order π energy of union is more or less constant (p. 83) and can be absorbed into the bond energy of the CC σ bond linking them. In the union of odd AH anions, the energy of union can be shown to be somewhat less than in the case of even AHs. A carbanionoid substituent R^- therefore destabilizes an odd AH anion.

Simple (carbanionoid) $-E$ substituents at an active position therefore have a very large stabilizing effect on odd AH cations and a smaller one on odd AH radicals, while odd AH anions are somewhat destabilized. This order is exactly the opposite shown by $+E$ substituents.

Let us next consider the effect of introducing heteroatoms. To get a neutral substituent rather than an ionic one, it is necessary that one of the active atoms in the carbanion HR⁻ should be replaced by a heteroatom. The π-electron density at that position will then be $1 + a_{0r}^2$, with a_{0r} the corresponding NBMO coefficient. Consider now introduction of the substituent into an odd AH cation HS⁺. The resulting derivative RS will be an even AH (identical with that produced by union of the odd AH radicals HR· and HS·; Fig. 4.15a); the π-electron density of the heteroatom will now be unity. The energy of union $\delta E_\pi'$ of the heteroatomic substituent (HR')⁻ with HS⁺ will therefore be less than that (δE_π) of the carbanion HR⁻ with HS⁺, the difference between the two energies being given by

$$\delta E_\pi' - \delta E_\pi = a_{0r}^2 \, \delta \alpha_X \qquad (4.124)$$

where α_X is the Coulomb integral (relative to carbon) of the heteroatom. The heteroatomic $-E$ substituent R' will therefore be less effective than the corresponding carbanionoid substituent, the difference being greater, the more electronegative the heteroatom (or heteroatoms) in R' and the more active the positions they occupy. Similar considerations apply to the case where the substituent is introduced into an AH radical. There is again a stabilizing effect but this is less than that produced by the isoconjugate carbanionoid substituent. In the case of carbanions, the effect of $-E$ substituents is destabilizing.

So far we have considered only cases where the $-E$ substituent is attached to an active position in an odd AH. If the substituent is attached to an even AH, HS, the change in π energy will be the π energy of union of HS with an odd AH anion. This will be the same as the change in π energy when the $\pm E$ substituent S is introduced into the odd AH, HR⁻. Now this change in π energy depends on the $\pm E$ activity of the $\pm E$ substituent and so is *not* the same for different even AHs. This raises difficulties because our definition of the activity of a substituent is based on its effects in a given system relative to its effect in an even AH, the latter being assumed to be constant. In practice, the first-order effects that occur when a $-E$ substituent is attached to an odd AH cation or radical are much greater than the second-order effects that occur when it is attached to an even AH, so the conclusions reached above hold in a qualitative sense, i.e.,

Rule 5. A $-E$ substituent in an active position has a very large stabilizing effect on odd AH cations and a smaller stabilizing effect on odd AH radicals. It has no effect in even AHs, odd AH anions, or inactive positions in other odd AHs.

Examples of the operation of Rule 5 follow.

(a) *Formation of odd AH cations.* The reaction of triphenylmethyl carbinol with acid to form triphenylmethyl cation [equation (4.109)] is enormously facilitated by $-E$ substituents. Thus whereas triphenylmethyl chloride (Ph₃CCl) is a normal covalent alkyl chloride which does not dissociate appreciably into ions even in polar solvents, the bis-*p*-dimethylamino (47) and tris-*p*-dimethylamino (49) derivatives exist as salts, no detectable amount of the covalent chlorides being present either in the solid or in solution. These salts are highly colored and were used as dyes until more stable materials were discovered; (48) is Malachite Green and (49) is Crystal Violet.

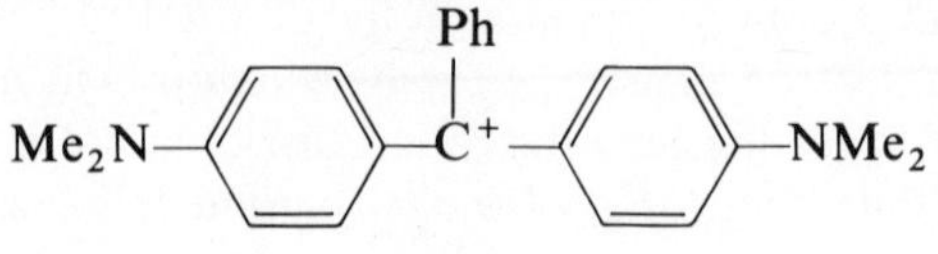

$$Me_2N-\!\!\left\langle\;\right\rangle\!\!-\underset{\underset{Ph}{|}}{C^+}\!\!-\!\!\left\langle\;\right\rangle\!\!-NMe_2$$

(48)

$$\left(Me_2N-\!\!\left\langle\;\right\rangle\!-\right)_3 C^+$$

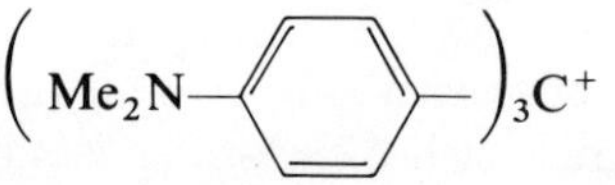

(49)

$$Ph_2N\!-\!O$$

(50)

(b) *Formation of odd AH radicals.* The dissociation of the dimer of triphenylmethyl (see p. 144) into triphenylmethyl radicals is assisted by $-E$ substituents; replacement of the phenyl groups by *p*-anisyl (*p*-methoxyphenyl) greatly increases the amount of monomeric radical at equilibrium in solution. Another good example of the stabilization of radicals by $-E$ substituents is provided by the nitroxide radicals, e.g., diphenylnitroxide (50). Radicals of the type RO·, where R is an innocuous group, dimerize to peroxides ROOR. Here the free radical is stabilized by the $-E$ substituent Ph_2N.

(c) *Basicity of aniline.* The protonation of aniline to anilinium ion [equation (4.110)] involves the conversion of a molecule isoconjugate with an odd AH anion (i.e., $PhCH_2^-$) to an even AH. This process should be facilitated by $-E$ substituents in active positions in aniline, as indeed it is. Thus *p*-anisidine (MeO$-\!\!\left\langle\;\right\rangle\!\!-NH_2$) is a stronger base than aniline, the pK_A of the conjugate acids being 5.34 and 4.63, respectively.

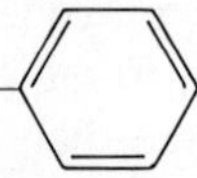

4.13 Summary of Substituent Effects

The conclusions reached in the preceding discussion of substituent effects can be summarized in a simple way. Table 4.3 indicates the stabilizing effect of a substituent at an active position in various types of odd AHs relative to the effect of the substituent in an even AH or at an inactive position in an odd AH. Here + implies stabilization, − destabilization, the relative magnitude of the effect being indicated by the number of plusses or minuses.

One point should perhaps be cleared up. The only types of substituents

TABLE 4.3. Relative Stabilizing Effects of Substituents on Odd AHs

Substituent	Odd AH cation	Odd AH radical	Odd AH anion
$+I$	−	0	+
$-I$	+	0	−
$+E$	?[a]	+ +	+ + + +
$\pm E$	+ +	+ +	+ +
$-E$	+ + + +	+ +	(−)[b]

[a] Small $+E$ substituents destabilize odd AH cations while larger ones stabilize them (see p. 173).
[b] Small second-order destabilizing effect.

considered thus far are those isoconjugate with even AHs or odd AH anions. What about odd AH radicals and cations? The answer is that such groups could be treated in the same manner but they are of no importance in organic chemistry. Substituents derived from odd AH anions are not in themselves important because such systems are too reactive; however, they can give rise to isoconjugate systems containing heteroatoms that are neutral and stable. This is not the case for analogous radicals or cations. Radicals are inherently too reactive, containing as they do unpaired electrons, while the only neutral analogs of carbonium ions contain boron.* Organoboron compounds in general are highly reactive, so boron-containing substituents are also of little interest.

4.14 Cross-Conjugation

In the preceding discussion, the effects of E-type substituents on the properties of odd AHs have been deduced by examining the substituent effect on the properties of the NBMOs of odd AHs. If an E-type substituent R is attached to an inactive position in an odd AH, HS, it will not affect the NBMO of the latter. In the resulting derivative RS, R will form an inactive segment. Since the specific effects of E-type substituents arise entirely from their influence on the NBMOs of adjacent odd AHs, such a substituent at an inactive position will show no E-type activity. It will to all intents and purposes behave as an I-type substituent. In this sense, there will be no conjugation between R and S in RS. Such a molecule is called *cross-conjugated*.

Another type of cross-conjugation is observed in the case of odd conjugated substituents attached to an adjacent substrate through an inactive atom. Consider for example the substituent

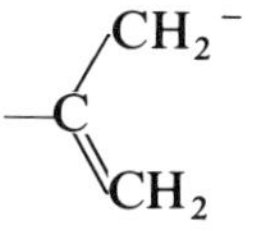

$$-\overset{\displaystyle CH_2^-}{\underset{\displaystyle CH_2}{C}}\!\!\!\diagup\!\!\!\diagdown$$

* Just as $-NH_2$ is isoconjugate with $-CH_2^-$, so $-BH_2$ is isoconjugate with $-CH_2^+$.

This is derived from an odd AH anion by loss of a hydrogen atom, but the hydrogen atom is lost from an inactive position. The NBMO of such a substituent cannot therefore interact with the NBMO of an adjacent odd AH, so none of the first-order effects indicated in Fig. 4.13 can operate. Such a substituent is therefore not of $-E$ type.

If we introduce such a substituent into an even AH, HS, to form an odd AH anion, RS$^-$, the group S will form an inactive segment. The π-electron densities of the atoms in S will therefore still be unity. The substituent R therefore behaves just like a $\pm E$ substituent, and indeed it can be shown that the second-order π energy of union is the same as that between two even AHs. A substituent derived from an odd AH anion by loss of a hydrogen atom from an inactive position therefore behaves as a $\pm E$ substituent.

It follows that substituents derived from such carbanionoid systems by the introduction of heteroatoms must be of $+E$ type. This is the case. Thus the nitro (51), carbethoxy (52), and amide (53) groups are of $+E$ type, all being isoconjugate with (54):

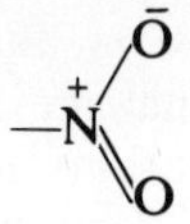

(51) 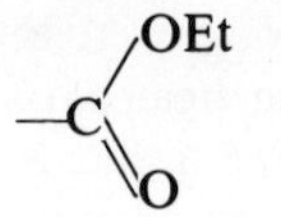   (52) 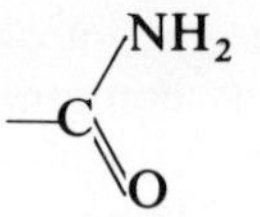   (53) 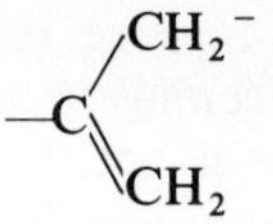   (54)

The $+E$ effect of such substituents is shown very clearly by their influence on the stability of Ph$_3$C$^+$ and Ph$_3$C$^-$. p-Nitro groups greatly increase the acidity of triphenylmethane and greatly decrease the tendency of trityl chloride (Ph$_3$CCl) to ionize, i.e., they substantially stabilize anions and destabilize cations.

4.15 Mutual Conjugation

The preceding sections have considered the stabilizing and destabilizing effects of single substituents on conjugated systems. Let us now consider the consequences of attaching two such substituents simultaneously to the same system. Will their effects be additive or not? In other words, will the net stabilizing or destabilizing effect be the sum of those for the two substituents acting separately?

We need consider only the case where the conjugated system is an AH, because it can be shown that the conclusions we reach will also hold qualitatively for isoconjugate systems containing heteroatoms. Let us then consider the effect of attaching two substituents R and S to atoms i and k, respectively, in an even AH, T;

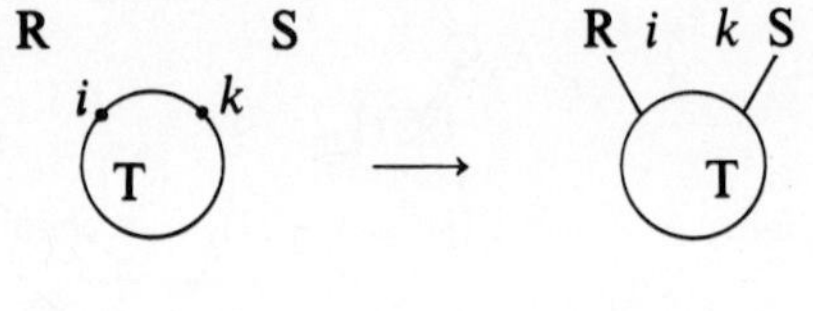

(4.125)

If R and S are $\pm E$ substituents, or $+I$ or $-I$ substituents, introducing one or the other into T to form RT or ST leads to no stabilization or destabilization (Table 4.3). The product in each case is an even AH. Consequently, the same argument applies for further substitution by S in RT or R in ST to form RTS. The effects of substituents in these cases are additive. The same arguments hold in other cases provided neither R nor S is of $-E$ type. Here again RT, ST, and RTS are all even alternant systems, so the effect of S on RT or of R on ST is the same as its effect on T itself. In all these cases, the effects of R and S are additive.

Additivity also holds even if R and/or S are of $-E$ type, provided that positions i and j are of like parity (e.g., a pair of *meta* positions in benzene). If R is a $-E$ substituent, RT is now isoconjugate with an odd AH anion in which, however, atom k is inactive; e.g.,

$$H_2N-\bigcirc \qquad H_2\overset{*}{N}-\bigcirc-* \qquad H_2N-\bigcirc^{S} \qquad (4.126)$$

R T RT RTS

Since atom k is inactive, RT behaves toward S as an even conjugated system (cf. Section 4.14; RTS is cross-conjugated). Here again the substituents act independently of one another.

The one place where additivity breaks down is where one at least of the substituents (say R) is of $-E$ type and positions i and j are of opposite parity (e.g., a pair of *para* positions in benzene). In this case, RT is isoconjugate with an odd AH anion and S is introduced into this at an active position (Fig. 4.14; the position *para* to NH_2 in aniline is active). If S is a $\pm E$ or $-I$ substituent, introduction of S into RT to give RTS will then lead to a large stabilization (Table 4.3), while if S is of $+E$ type, the extra stabilization will be very large indeed. Since there is no change in stabilization when these substituents are introduced into the even AH, T, to form ST or RT, it is evident that introduction of both substituents simultaneously to form RTS leads to a net stabilization which is greater than the sum of those produced by R or S individually. This phenomenon, which is illustrated in Fig. 4.14, is described as *mutual conjugation*. Mutual conjugation occurs when a $-E$ substituent and a $\pm E$, $+I$, or $+E$ substituent are attached to positions of opposite parity in an even alternant system.

The effect of mutual conjugation on the structure of *p*-nitroaniline is easily seen by examining dipole moments. The dipole moment of *p*-nitroaniline (Fig. 4.14) is significantly larger than that calculated from the individual dipole moments of aniline and nitrobenzene (5.5D). This comparison is further supported by the fact that the dipole moment of nitroaminodurene (55) is only 4.98 D. Mutual conjugation is prohibited in (55) because the nitro group is twisted out of planarity with the ring by the adjacent methyl groups.

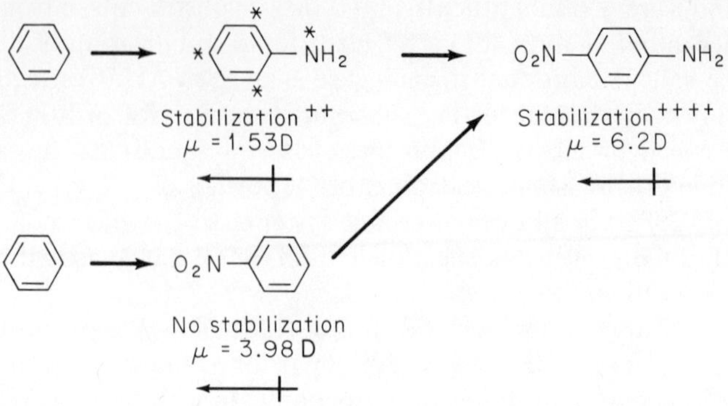

[Diagram showing:
- Benzene → *-C6H4-NH2 (with * marks); Stabilization ++, μ = 1.53D
- Benzene → O2N-C6H5; No stabilization, μ = 3.98D
- Both → O2N-C6H4-NH2; Stabilization ++++, μ = 6.2D]

FIGURE 4.14. Mutual conjugation between the nitro and amino groups in *p*-nitroaniline. The effect of $NH_2 + NO_2$ is greater than the sum of the effects of NH_2 and of NO_2.

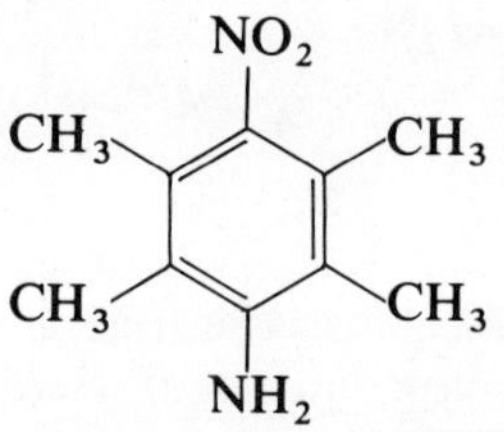

[Structure: benzene ring with NO2 (top), NH2 (bottom), and four CH3 groups at the remaining positions]

(55)

It is not necessary for two substituents to constructively reinforce each other. If the first and second substituents are of $-I$ or $-E$ type, then the combined effect of R + S will be *less* than the sum of the effects of R and of S. Here there is *negative mutual conjugation*. A simple example is provided by resorcinol and hydroquinone. Reaction of these with bases gives the corresponding phenate ions. Since $-O^-$ is a more powerfully $-E$ group than $-OH$, the negative mutual conjugation is greater in HO—C6H4—O$^-$ than in the corresponding phenol. Since there is no mutual conjugation in the *meta* isomer, the net effect should be to make hydroquinone a weaker acid than resorcinol. This is the case (pK_A of *para* isomer, 10.0; of *meta* isomer 9.4).

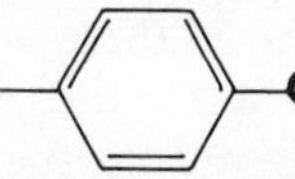

4.16 The Field Effect

Sections 4.9–4.13 were concerned with the effects of substituents on chemical equilibria in cases where the reactant and product contained conjugated systems that differed in some respect. Substituents can, however, also affect equilibria in cases where neither reactant nor product is conjugated or in which both contain identical conjugated systems.

Suppose, for example, that the substituent has an electronic charge (e.g., the substituent —COO⁻) and that the reaction in question leads to a change in charge (e.g., ionization of a neutral carboxylic acid RCOOH to the ion RCOO⁻). There will be a direct electrostatic interaction between the substituent and the substrate to which it is attached and this interaction will be different in the reactant and product. The resulting change in electrostatic energy will alter the heat of reaction correspondingly and hence the equilibrium constant.

Consider, for example, the successive ionization of the two carboxyl groups in malonic acid,

$$\text{HOOC—CH}_2\text{—COOH} \rightleftharpoons \text{HOOC—CH}_2\text{—COO}^- + \text{H}^+ \quad (4.127)$$

$$\text{HOOC—CH}_2\text{—COO}^- \rightleftharpoons \bar{\text{O}}\text{OC—CH}_2\text{—COO}^- + \text{H}^+ \quad (4.128)$$

The second ionization leads to a dianion in which the two negatively charged groups repel one another strongly. In order to bring about this ionization, we must supply enough energy to overcome this repulsion. Consequently, the equilibrium constant for equation (4.128) is much smaller than that for equation (4.127), malonic acid being a much stronger acid (pK_A 2.77) than is the monomalonate ion (pK_A 5.66).

We can regard both these compounds as derivatives XCH_2COOH of acetic acid, X in one case being COOH and in the other COO⁻. The reaction involves conversion of the neutral carboxyl group of XCH_2COOH to carboxylate (XCH_2COO^-). If X has a negative charge (as in $\bar{\text{O}}\text{OCCH}_2\text{COOH}$), this will lead to an increase in the total energy, due to the electrostatic repulsion between X and COO⁻.

Even if X does not have a net electric charge, it may still affect the course of the reaction if it has a dipole moment. This is true for the carboxyl group itself, in which the carbon atom carries a positive charge and the oxygen atoms carry negative charges. This time ionization of XCH_2COOH leads to a *decrease* in electrostatic energy since the charge on COO⁻ and the dipole on X will attract each other in XCH_2COO^-,

$$\begin{array}{c} \overset{\delta^-}{\text{O}} \\ \diagdown \\ \text{C—CH}_2\text{—COO}^- \\ \diagup{\scriptstyle\delta^+} \\ \underset{\delta^-}{\text{HO}} \end{array}$$

Therefore malonic acid (pK_A 2.77) is not only a stronger acid than the semimalonate ion, but is also a stronger acid than acetic acid (pK_A 4.76).

Charged substituents or dipolar substituents (i.e., those with nonvanishing dipole moments) can therefore affect the equilibrium in a chemical reaction in which there is a change in net charge at the *reaction center*, i.e., the group of atoms directly concerned in the reaction. Effects of this kind, due to direct electrostatic interactions across space, are described as the *field effect*. A substituent will stabilize a distant negative charge in this way if it is itself positively charged or if it is dipolar with the positive end of the dipole nearer the reaction center. For obvious reasons (cf. Section 4.9, p. 165), we term this a positive field effect $+F$. Likewise a substituent with a net negative charge, or a dipolar substituent with the negative end of the dipole nearer the reaction center, is said to exert a negative field effect $-F$.

In a vacuum, the electrostatic energy of interaction between charges q_1 and q_2 at a distance r is given by $q_1 q_2/r$. In the cases we are considering, however, the charges are not in a vacuum. They are separated by the molecule in which they reside and they are generally surrounded by a polar solvent. The electrostatic energy E has then to be reduced by the effective (i.e., average) dielectric constant D of the medium separating them, and is given by

$$E = q_1 q_2/Dr \qquad (4.129)$$

In the case of a dipolar substituent, the part away from the substrate will be immersed in the solvent. If the solvent is polar, it will have a relatively high dielectric constant. The value of D will then be correspondingly greater, and the effect of the charge consequently less, than for the part of the dipole nearer the substrate. Indeed, in most cases the bond (CX) linking the substituent X to the adjacent carbon atom in the substrate is itself polar and the main contribution to the field effect then arises from the charge on that carbon atom. The effect of a dipolar substituent on a reaction in a polar solvent should vary roughly as the inverse first power of the distance r_{iR} separating the carbon atom i to which the substituent is attached from the reaction center, the corresponding electrostatic energy E_F being given by

$$E_F = q_S q_R/Dr_{iR} \qquad (4.130)$$

where q_S is the charge at the carbon atom i adjacent to the substituent and q_R is that at the reaction center. In this case, the medium separating the charges is essentially the interior of the molecule. Now molecules are not very polarizable, so the average dielectric constants inside them are small (2–3) and are, moreover, much the same for different molecules. The high dipole moments of polar solvents (e.g., water, $D = 80$) are due to orientation of the molecules in an electric field, not to polarization of individual molecules. Thus D in equation (4.130) will be much the same for different positions of a given substituent and reaction centers, either in the same molecule or in different molecules. Since q_S is a property of the substituent only, we can therefore write equation (4.130) in the form

$$E_F = f_S q_R/r_{iR} \qquad (4.131)$$

where f_S ($= q_S/D$) is a quantity characteristic of the substituent and is a measure of its field effect.

Suppose now that the charge q_R at the reaction center changes during the reaction by an amount δq_R. The electrostatic energy of interaction [E_F in equation (4.131)] will then change by an amount δE_F given by

$$\delta E_F = f_S \, \delta q_R/r_{iR} \qquad (4.132)$$

The corresponding quantity for the unsubstituted compound (i.e., where substituent S = H) is given by equation (4.132) with $f_S = f_H$. If the substituent alters the equilibrium only through this change in the electrostatic energy, the equilibrium constants for the unsubstituted (K_0) and substituted (K_s) compounds will be related by

$$-RT \log \frac{K_s}{K_0} = \Delta G_s - \Delta G_0 = \Delta E_s - \Delta E_0 = \frac{(f_S - f_H)\delta q_R}{r_{iR}} \qquad (4.133)$$

where ΔG_0 and ΔG_s are the free energies of reaction and ΔE_s and ΔE_0 the corresponding energies of reaction, and where $\Delta G_s - \Delta G_0$ is equal to $\Delta E_s - \Delta E_0$, since the entropies of reaction are assumed to be equal (see p. 133). This in turn can be written in the form

$$\log_{10} \frac{K_s}{K_0} = \frac{F_S \rho_R{}^F}{r_{iR}} \qquad (4.134)$$

where

$$F_S = \frac{f_S - f_H}{2.303RT}; \qquad \rho_R{}^F = \delta q_R \qquad (4.135)$$

Here F_S is a quantity characteristic of the substituent and is a measure of its field effect.

4.17 The Hammett Equation

Suppose we have a series of compounds $S_i TR_k$, where R_k is a reaction center (i.e., a group of atoms that undergoes some reaction independently of the backbone T to which it is attached) and S_i is a substituent. If the substituent and reaction center are attached to fixed points in T, r in equation (4.133) will be constant. We can then write equation (4.133) in the form

$$\log(K_s/K_0) = \sigma_{S_i} \rho_{R_k} \qquad (4.136)$$

where σ_{S_i} is a constant characteristic of the substituent S_i and ρ_{R_k} is a constant characteristic of the substituent R_k. Here the constant σ_{S_i} will have the same value for the different reactions comprised in the reaction centers R_k, while the constant ρ_{R_k} will equally have a fixed value for the given reaction corresponding to R_k for different substituents S_i. This equation was discovered by Hammett, who showed it to hold for a wide variety of substituents and reactions where T is benzene, R and S being *meta* or *para* to one another. Equation (4.136) is therefore called the *Hammett equation*. It has since been shown to hold for a number of other backbone structures.

Experimental values for equilibrium constants give only values for the product $\sigma\rho$. Each is therefore uncertain by a numerical factor. We can multiply all the ρ's by a common factor A if we divide all the σ's by the same factor. In order to resolve this ambiguity, we have to assume the value of some specific ρ or σ. Hammett did this in the case of his benzene derivatives by assuming that $\rho = 1$ for ionization of carboxylic acids in water. If then the pK_A values for benzoic acid and for a substituted benzoic acid are $(pK_A)_0$

and $(pK_A)_s$, respectively, the σ constant for the substituent is given by

$$\sigma_S = (pK_A)_0 - (pK_A)_s \tag{4.137}$$

The Hammett equation needs different sets of constants not only for different substrates T, but also for different pairs of positions in a given substrate occupied by the substituent and the reaction center. This did not matter in the context in which Hammett originally formulated his equation (i.e., *meta* and *para* disubstituted benzenes), but generalization to other systems would involve an unacceptable pullulation of parameters. Indeed the situation is even worse than this because the use of a single substituent constant σ for a given substituent, even for a given substitution pattern in a given substrate, proved to be unsatisfactory.

Our derivation of the Hammett equation rested on the assumption that the sole mode of interaction between substituent S and reaction center R is by direct electrostatic interaction across space. This is true if the substrate T is saturated, but not if it is conjugated. If T is conjugated, there may be mutual conjugation between R and S in RTS. Even if there is not, π interactions between S and T may lead to changes in the latter that in turn can influence R by the field effect. If these alternative processes are important, the Hammett equation will fail if we use in it σ constants based on the field effect alone.

Consider, for example, the protonation of *p*-nitraniline,

$$O_2N-\langle\text{ring}\rangle-NH_2 + H^+ \longrightarrow O_2N-\langle\text{ring}\rangle-NH_3^+ \tag{4.138}$$

The nitro group exerts both a $+E$ effect and a $+I$ effect, the latter through the polarity of the CN bond linking it to the ring. Consequently, *p*-nitraniline is stabilized by mutual conjugation between the $(+E, +I)$ nitro group and the $-E$ amino group. The group $-NH_3^+$, on the other hand, is of $+I$ type and cannot conjugate mutually with NO_2. Protonation therefore involves a loss of the stabilization due to mutual conjugation in *p*-nitraniline and so occurs with more difficulty than would be expected on the basis of the field effect alone. Consequently, *p*-nitraniline is a weaker base than would be expected from the Hammett equation, using for NO_2 a σ constant derived from reactions where mutual conjugation is unimportant. The pK_A of the conjugate acid of *p*-nitraniline is 1.2 units higher than that predicted on the basis of the substituent constant σ for *p*-NO_2.

Substituent effects due to mutual conjugation between $+E$ and $-E$ groups are described as the *mesomeric effect* and those due to mutual conjugation between a $+I$ substituent and $-E$ reaction center as the *π-inductive effect* or *inductoelectromeric effect*. The stabilizing effects of $+E$ and $+I$ substituents on an odd AH anion are both proportional to the formal negative charge at the position to which these are attached. If a substituent exerts both $+E$ and $+I$ effects, the stabilizations due to them will then be in fixed proportion to one another for all odd AH anions. It follows that mutual conjugation between such a substituent and a $-E$ reaction center will then also be due to the mesomeric effect and to the π-inductive effect in the same fixed proportions. The two effects are therefore indistinguishable in most practical connections and can be referred to collectively as the mesomeric (M) effect.

The other complication is illustrated by the ionization of *m*-aminobenzoic acid,

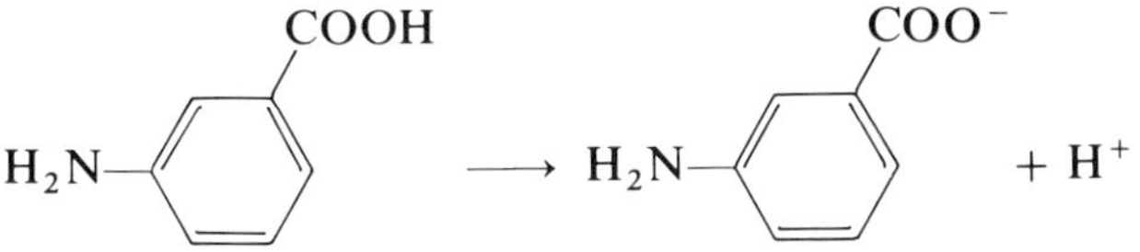

$$H_2N-C_6H_4-COOH \longrightarrow H_2N-C_6H_4-COO^- + H^+ \quad (4.139)$$

Since nitrogen is more electronegative than carbon, the CN bond is polar with C positive, $C^{\delta+}-N^{\delta-}$. The corresponding $\overset{\delta-}{H_2N}-\overset{\delta+}{C_6H_5}$ dipole in *m*-aminobenzoic acid is oriented with its positive end toward the carboxyl. Therefore the amino group should exert a $+F$ effect and so increase the ease of ionization, i.e., lower pK_A. Analogous effects are seen in similar connections in aliphatic compounds. Here, however, the observed effect is exactly the opposite of that predicted, the pK_A of *m*-aminobenzoic acid (4.35) being *greater* than that (4.19) of benzoic acid. In this case, the substituent and reaction center are attached to positions *meta* to one another, i.e., of like parity, so there can be no mutual conjugation. Since, however, NH_2 is a $-E$ group, introduction of it into benzene gives a molecule (i.e., aniline) which is isoconjugate with an odd AH anion (i.e., $PhCH_2^-$). Here there are large negative charges in the ring, due to the interactions with CH_2^-. While these interactions should be reduced by replacement of CH_2^- by $\dot{N}H_2$, they should still persist in aniline and so also in *p*-aminobenzoic acid. The direct electrostatic interactions between these charges and the COOH or COO$^-$ group are opposite to, and greater than, the direct field effect interaction with the CN dipole; see Fig. 4.15.

This indirect interaction, through secondary electrostatic interaction with charges produced by polarization of a π system, is described as the *mesomeric field* (*MF*) *effect*.

All these problems can be avoided in the following general treatment of substituents (FMMF method), based on the PMO approach. In this we take account of all three substituent effects (F, M, MF) and calculate them explicitly

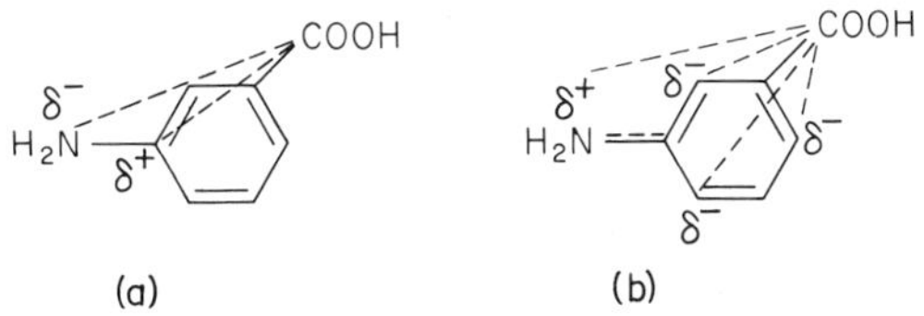

FIGURE 4.15. Interactions of *m*-aminobenzoic acid due to (a) the field effect and (b) the mesomeric field effect.

for each situation. The calculation of the field effect has of course been considered already; we use the approximate result embodied in equation (4.134), the field effect constant F_S being found from data for reactions involving saturated substrates where no other effect operates. The only problem here is to ensure that the distance r_{iR} between substituent and substrate has a fixed value. This means that we must use data for some system that is rigid. Appropriate examples are the acid dissociation constants of rigid bicyclic acids such as (56) or (57):

(56)   (57)

The presence of benzene rings in the latter is not important since they cannot interact with either the substituent S or the reaction center.

The mesomeric effect can be estimated in the following way. Attachment of CH_2^- to an even AH, T, at position i gives an odd AH anion, TCH_2^-, for which the formal charge q_{ik} at position k can be found by Longuet-Higgins' method (see Section 3.3, p. 78). It is easily shown (see Section 4.10, p. 170) that the corresponding excess charge produced at position k by a $-E$ substituent S at position i is given by $A_S q_{ik}$, where A_S is a constant characteristic of S. Now the stabilization brought about by introduction into ST of a $+E$ or $+I$ substituent at position k is proportional to the formal negative charge there (see the discussion of mutual conjugation, Section 4.15, p. 180), and is also proportional to the activity of the substituent. The net stabilization is thus given by

$$A_S q_{ik} B_R \tag{4.140}$$

where A_S and B_R are constants, independent of the substrate T or the positions i and j in it. It follows that the effect of such mutual conjugation on an equilibrium involving the group R must be likewise given by a similar triple product. Since the mesomeric effect arises from a change in this quantity during the reaction, the contribution of the mesomeric effect must be of the form

$$M_S q_{ik} \rho_R^M \tag{4.141}$$

where M_S is a constant characteristic of the substituent S and ρ_R^M is a constant characteristic of the reaction involving R.

The contribution of the mesomeric field effect can be estimated by combining equations (4.134) and (4.140). The charge produced at position j by our substituent S at position i is, as we have seen, given by $A_S q_{ij}$. This charge exerts a field effect given by equation (4.134). The total mesomeric field effect

Chemical Equilibrium

is then given by summing these contributions by the various atoms in T, i.e., by

$$\sum_{j \neq i,k} \frac{MF_S q_{ij} \, \delta q_R}{r_{jR}} \qquad (4.142)$$

where MF_S is a constant characteristic of substituent S. The contribution of atom k is omitted since it has already been effectively included in the expression for the mesomeric effect [(4.141)].

Combining (4.134), (4.141), and (4.142), we arrive at the following general expression for the effect of substituents on equilibria:

$$\log_{10} \frac{K_s}{K_0} = \left(\frac{F_S}{r_{iR}} + \sum_{j \neq i,k} \frac{MF_S q_{ij}}{r_{jR}} \right) \rho_R^F + M_S q_{ik} \rho_R^M \qquad (4.143)$$

Since the r_{jR} and q_{ij} can be calculated for any system, there are just three parameters to be determined for each substituent (F_S, M_S, MF_S) and two for each reaction center (ρ_R^F, ρ_R^M). These are found by fitting suitable experimental data, e.g., dissociation constants of various substituted carboxylic acids. In the original Hammett treatment, there were just two parameters for each substituent (σ_m, σ_p) and one for each reaction center (ρ); however, this approach was limited to benzene derivatives only. Equation (4.143) should apply to molecules of all kinds, saturated as well as conjugated, and it does indeed seem to do so in quite an effective manner. This approach has been termed the FMMF method.

Table 4.4 lists the substituent constants F, M, and MF for 13 common substituents. This set of substituent constants replaces the individual substituent constants that must be obtained for every substituent site in a series of molecules and it also replaces the series of different substituent constants $(\sigma, \sigma^n, \sigma^0, \sigma^+, \text{ and } \sigma^-)$ that have been established to account for differences in

TABLE 4.4. Substituent Constants in the FMMF Method[a]

Substituent	F	M	MF
CH_3	−0.87	−0.910	−0.355
F	4.85	−2.10	−0.577
CH	4.95	−1.11	−0.419
Br	4.92	−1.14	−0.303
I	4.57	−1.26	−0.347
OH	2.48	−3.70	−0.593
OCH_3	3.16	−3.14	−0.982
CH	5.57	1.06	0.342
NO_2	7.09	0.927	0.421
CO_2H	3.13	0.944	0.512
$CO_2C_2H_5$	3.18	0.932	0.489
NH_2	0.317	−4.11	−1.16
NHAc	3.22	−1.70	−0.464

[a] Data from Ref. 2.

electromeric and field effects in reactions of different species, e.g., anions, cations, radicals, etc.

With the exception of methyl, all of the substituents have group electronegativities greater than carbon, and the sign of the field effect constant is positive, in agreement with the convention that the substituent constant should have the same sign as the charge which the substituent presents to the reaction center. The change of sign of M and MF in going from $-E$ to $+E$ groups is also obvious in Table 4.4.

Although the FMMF method dramatically reduces the number of substituent constants that one must consider, it suffers from the disadvantage that it is not possible to visually compare two reactions with a three-parameter equation, i.e., one cannot simply plot logs of rates or equilibrium constants against a single parameter as one does in the normal Hammett treatment. It is possible, however, to calculate Hammett substituent constants for any position using the FMMF parameters and the following relationship:

$$\sigma_{Sim} = \frac{F_S}{r_{im}} + M_S q_{im} + MF_S \sum_{k \neq m} \frac{q_{ik}}{r_{km}} \quad (4.144)$$

where σ_{Sim} is the σ constant for substituent S at position i when the reaction occurs at position m. The substituent constants obtained in this way can be compared with the substituent constants obtained from the pK_A values of substituted carboxylic acids, as in Table 4.5, or the values can be used to correlate equilibrium properties or reaction rates for a wide variety of reactions. Figure 4.16 illustrates the correlation of the logarithms of the rates of solvolysis of m- and p-cumyl chlorides in 90% aqueous acetone by the FMMF substituent constants.

In calculating the substituent constants for the plot in Fig. 4.16, it was assumed that the field effect operates in the developing carbonium ion, which is assumed to be one standard bond length from the ring. The deviations from linearity in Fig. 4.16 are all due to mutual conjugation between $-E$ and $-I$ substituents and the positive reaction center. The deviations in Fig. 4.16 are roughly proportional to the M_S values for the substituents, and

TABLE 4.5. Calculated σ Constants for α-Naphthoic Acids Compared with Experimental Values[a]

Substituent	1,3	1,4	1,5	1,6	1,7
CH_3	−0.06 (−0.05)	−0.23 (−0.14)	−0.11 (0.01)	−0.06 (−0.05)	−0.11 (−0.07)
Cl	0.38 (0.30)	0.15 (0.26)	0.21 (0.29)	0.28 (0.17)	—
Br	0.40 (0.34)	0.16 (0.30)	0.23 (0.30)	0.29 (0.18)	0.21 (0.07)
OH	0.13 (0.06)	−0.59 (−0.52)	−0.11 (−0.06)	−0.08 (−0.08)	−0.15 (−0.10)
OCH_3	—	−0.47 (−0.36)	−0.10 (−0.01)	0.07 (−0.06)	−0.13 (−0.08)
CN	0.55 (−0.59)	0.73 (0.79)	0.50 (0.46)	0.44 (0.34)	0.48 (0.31)
NO_2	0.70 (0.61)	0.84 (0.86)	0.61 (0.54)	0.55 (0.41)	0.59 (0.36)

[a] Data from Ref. 2.
Experimental values in parentheses.

Chemical Equilibrium

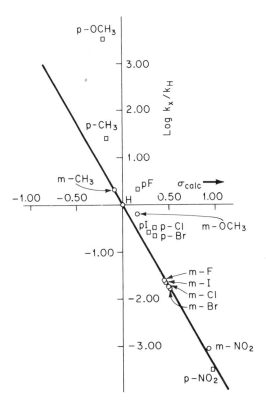

FIGURE 4.16. Logarithms of relative rates of solvolysis of (○) *meta-* and (□) *para*-substituted cumyl chlorides in 90% aqueous acetone at 25°C (relative to the unsubstituted compound) plotted against σ constants calculated by the FMMF method (rate data from Ref. 7).

are not substantially larger than the deviations that are encountered whenever two different reactions are compared by a Hammett treatment. The difference between the reaction in Fig. 4.16 and the ionization of arylcarboxylic acids is large enough that the reaction in Fig. 4.16 was used as the basis for a new set of substituent constants σ^+. The need for extra constants for different reaction types has been vitiated by the FMMF method. The application of the Hammett equation and the FMMF method to reaction rates in Fig. 4.16 provides a direct introduction to the subject of the next chapter.

Problems

4.1. Indicate the signs of the substituent constants F, M, and MF for specific $+I$, $-I$, $\pm E$, $+E$, and $-E$ substituents.

4.2. Explain the stability of the following radicals:

a. 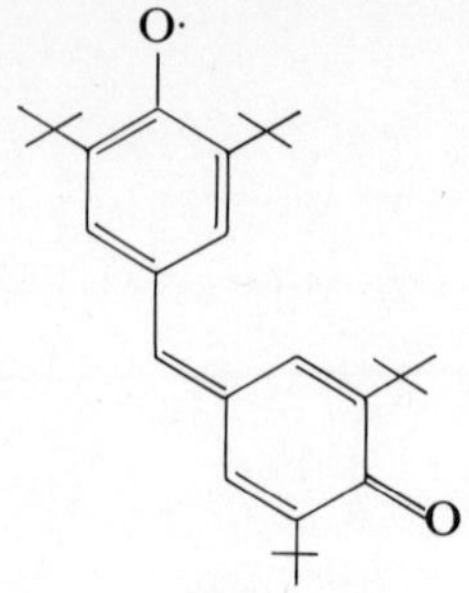 [2,6-di-tert-butyl-4-(3,5-di-tert-butyl-4-oxocyclohexa-2,5-dien-1-ylidenemethyl)phenoxyl radical structure]

b. $(CF_3)_2N-O\cdot$

c. 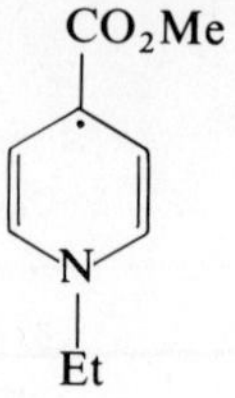 [1-ethyl-4-(methoxycarbonyl)-1,4-dihydropyridin-4-yl radical structure]

4.3. Give three examples each of the following types of equilibrium: EO_l, EO_s, EE, and I.

4.4. a. Estimate the orders of the equilibrium constants for the following EO_l equilibria in strong acids:

$$ArCH_2OH + H^+ \rightleftharpoons ArCH_2^+ + H_2O$$

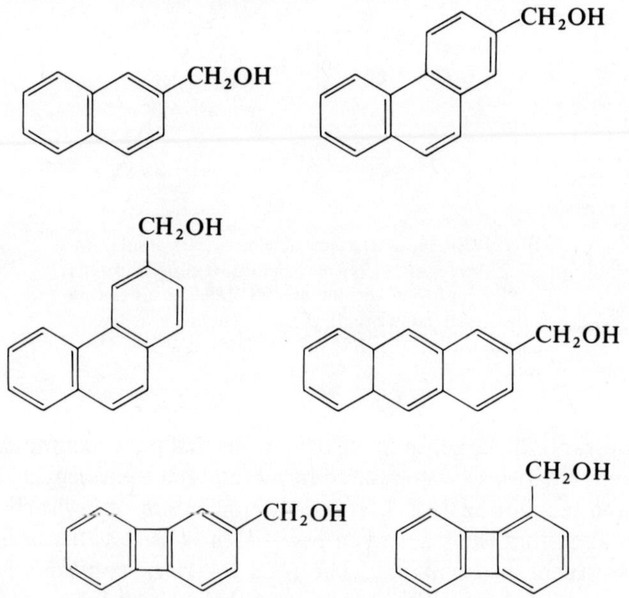

[Structures: 2-naphthylmethanol, 2-phenanthrylmethanol, 3-phenanthrylmethanol, 2-anthrylmethanol, biphenylenylmethanol, biphenylenylmethanol isomer]

b. Would you expect the equilibrium constants for the reaction

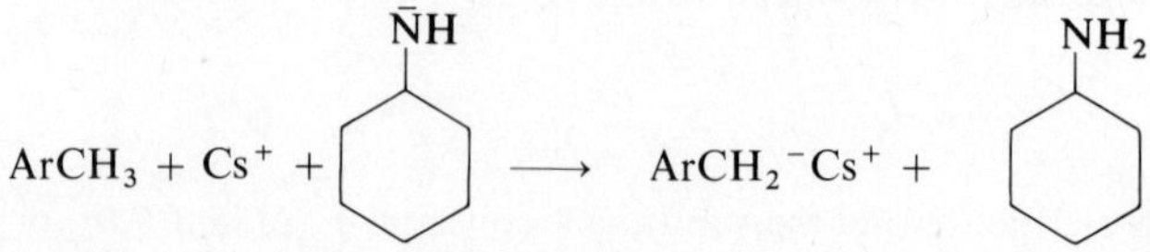

$$ArCH_3 + Cs^+ + \underset{\text{(cyclohexylamide)}}{C_6H_{11}\bar{N}H} \longrightarrow ArCH_2^- Cs^+ + \underset{\text{(cyclohexylamine)}}{C_6H_{11}NH_2}$$

to follow the same order? Explain.

4.5. **a.** Using the data in Table 4.4 and equation (4.136), calculate the substituent constants σ_{Sim} for the ionization of phenols with the following substituents (neglect delocalization of charge in the phenoxide ion and assume all bond lengths to be 1.40 Å):

para	F, Cl, Br, OCH_3, CH_3
meta	F, Cl, Br, OCH_3, CH_3

b. Given the pK_A of phenol as 12.80 and a ρ of $+1.74$, estimate the pK_A's of the substituted phenols. Actual values of the pK_A's can be found in Ref. 1.

4.6. Estimate the order of the equilibrium constants for the following EO_s equilibria in strong acid media (statistical corrections for the number of equivalent positions should be made):

$$ArH + H^+ \rightleftharpoons ArH_2^-$$

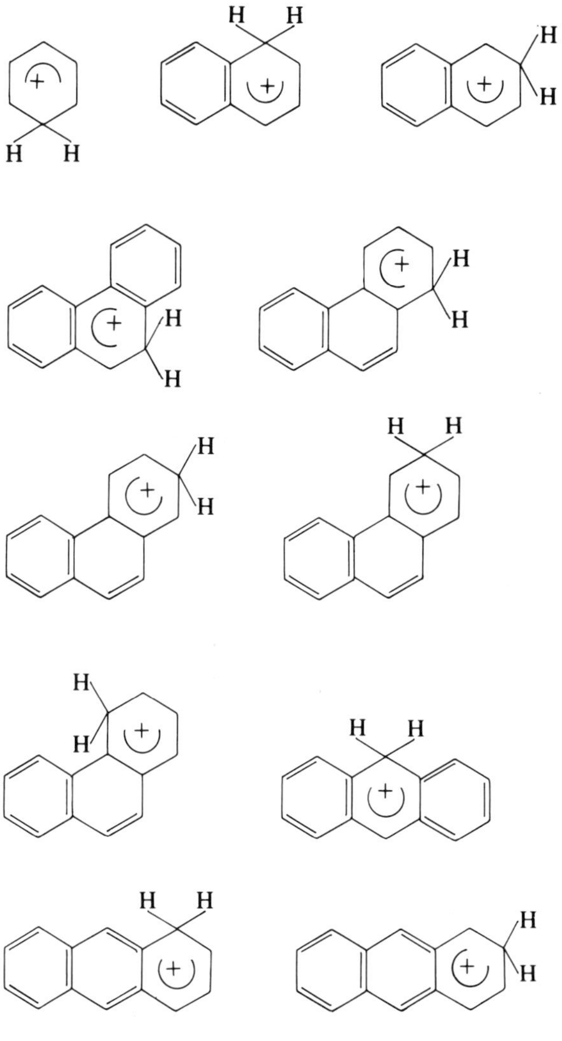

4.7. In a closed vial at high temperature, hexahelicene can be equilibrated with its Diels–Alder dimer,

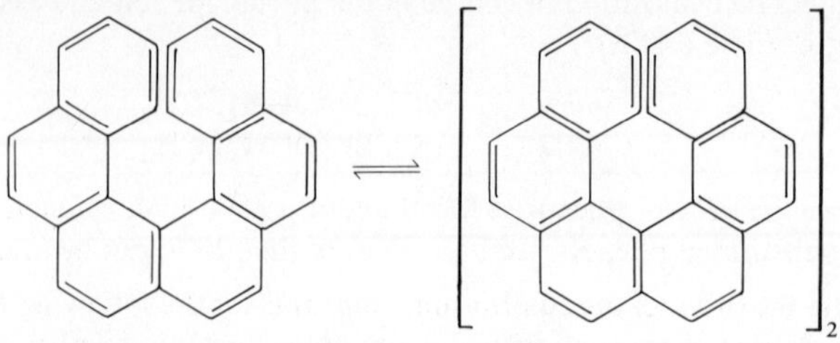

Predict the structure of the dimer, using the 1,2 and 1,4 bislocalization energies for hexahelicene.

4.8. The ionization potentials of substituted benzenes are well correlated by a Hammett relationship. What would be the effect of the substituents in Problem 4.5 on the half-wave electrode potentials of a series of substituted benzenes?

4.9. **a.** Estimate the order of the equilibrium constants for the following equilibria:

$$ArBF_2 + Me_3N \rightleftharpoons Ar\bar{B}F_2 - \overset{+}{N}Me_3$$

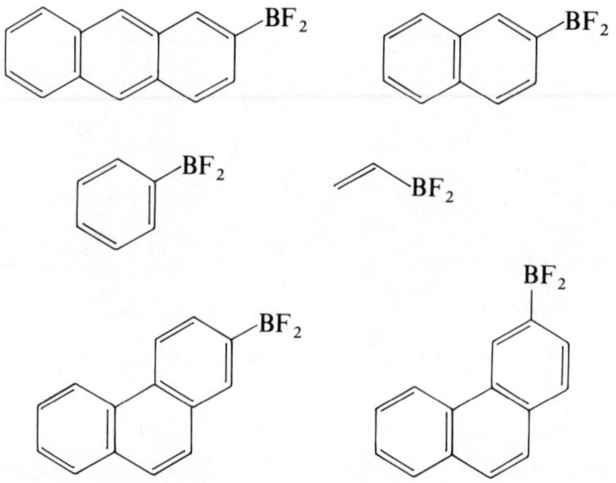

b. How would substituents $+I$, $-I$, $\pm E$, $+E$, and $-E$ affect the position of the equilibrium?

4.10. **a.** Pyridine N-oxide undergoes both electrophilic and nucleophilic substitution more readily than benzene. Explain this.
b. Explain why nitrosobenzene also behaves in this way.

Selected Reading

General discussions of organic equilibria:

L. P. HAMMETT, *Physical Organic Chemistry,* 2nd ed., McGraw-Hill Book Co., New York, 1970.

E. M. KOSOWER, *An Introduction to Physical Organic Chemistry*, John Wiley and Sons, New York, 1968.

Advanced summaries of specific areas can be obtained from the two series:

Advances in Physical Organic Chemistry, Academic Press, New York.

Progress in Physical Organic Chemistry, Wiley–Interscience, New York.

Application of PMO theory to equilibria, substituent effects, and stabilities of reaction intermediates:

M. J. S. DEWAR, *The Molecular Orbital Theory of Organic Chemistry*, McGraw-Hill Book Co., New York, 1969.

References

1. H. C. BROWN, D. H. MCDANIEL, AND E. A. HÀFLINGER, *Determination of Organic Structures by Physical Methods*, E. A. Broude and F. C. Nachod, eds. Academic Press, New York, 1955, p. 589.
2. M. J. S. DEWAR, R. GOLDEN, AND J. M. HARRIS, *J. Am. Chem. Soc.* **93**, 4187 (1971).
3. E. L. MACKOR, A. HOFSTRA, AND J. H. VAN DER WAALS, *Trans. Faraday Soc.* **54**, 66 (1958).
4. A. STREITWIESER, JR., AND W. C. LONGWORTHY, *J. Am. Chem. Soc.* **85**, 1757 (1963).
5. C. K. MANN AND K. K. BARNES, *Electrochemical Reactions in Nonaqueous Systems*, Marcel Dekker, New York, 1970.
6. R. W. KISER, *Introduction to Mass Spectrometry and Its Applications*, Prentice-Hall, New York, 1965.
7. H. K. JAFFÉ, *Chem. Rev.* **53**, 191 (1953).

CHAPTER 5

Chemical Reactivity

5.1 Basic Principles

Chemical reactions take place by individual processes, each of which involves the almost instantaneous conversion of one or more molecules of reactant into one or more molecules of product. In order to predict the course of a chemical reaction, we need to be able to predict the way in which these basic conversions take place and the frequency with which they take place under any given set of conditions.

When we write an equation representing an observed chemical process, there is no guarantee that this corresponds to the individual processes occurring on the molecular level. The overall reaction may very well involve two or more successive molecular processes. A good example is the addition of bromine to ethylene in water,

$$CH_2\!\!=\!\!CH_2 + Br_2 \longrightarrow CH_2Br\!\!-\!\!CH_2Br \qquad (5.1)$$

The reaction does not take place in a single step by direct addition of individual molecules of bromine to individual molecules of ethylene. It takes place by two successive steps in the following way:

$$C_2H_4 + Br_2 \longrightarrow C_2H_4Br^+ + Br^- \qquad (5.2)$$

$$C_2H_4Br^+ + Br^- \longrightarrow C_2H_4Br_2 \qquad (5.3)$$

Our object in interpreting the *mechanism* of an overall reaction such as that of equation (5.1) is first to establish the individual steps involved on the molecular level; second, to establish the way in which each of these individual molecular conversions takes place; and third, to predict the rates at which they take place.

Two or more molecules can react only if they are in contact with one another, i.e., during a collision between them. If a reaction involves three or more molecules, it can therefore take place only if all the molecules of reactant collide simultaneously. Now the duration of a collision between two molecules is extremely small. If the molecules fail to react, they immediately fly apart again. The probability of a third molecule hitting the first two while they are still in contact is therefore very small. A reaction involving such a three-body collision will therefore be slow since the number of suitable collisions will be small. Indeed, no reactions are known which involve the simultaneous conversion of three or more separate molecules into products in a single collision. The unit processes with which we are concerned are therefore of two kinds; *unimolecular* reactions, in which a single molecule undergoes fission or rearrangement, and *bimolecular* reactions, in which two molecules of reactant interact with each other.

Consider a bimolecular reaction

$$X + Y \longrightarrow \text{products} \tag{5.4}$$

The rate of the reaction will be proportional to the rate at which molecules of X and Y collide, i.e., to the *collision frequency* v, which we define as the number of X–Y collisions in unit volume per second. If the concentration of Y is kept constant, v will be proportional to the concentration [X] of X because each molecule of X will undergo collisions with molecules of Y at the same constant rate. A similar argument shows that v must also be proportional to the concentration [Y] of Y. Hence

$$v = v_0[X][Y] \tag{5.5}$$

where v_0 is the collision frequency when X and Y are present at unit concentration. The overall rate of reaction is then given by

$$\text{rate} = Pv = Pv_0[X][Y] \tag{5.6}$$

where P is the fraction of X–Y collisions that lead to reaction.

A given collision may fail to lead to reaction, for two reasons. First, it may be necessary for a specific part of X to hit a specific part of Y. For example, the reaction

$$CH_3COOH + N(CH_3)_3 \longrightarrow CH_3COO^- + HN^+(CH_3)_3 \tag{5.7}$$

can occur only if the molecules of acetic acid and trimethylamine collide in such a way that the hydroxylic hydrogen of the acid hits the nitrogen atom of the base. Only a certain fraction σ of the collisions will satisfy this condition.

Second, since a reaction involves a rearrangement of bonds, energy may be needed to weaken the relevant bond or bonds in the reactants before reaction can occur. Unless the reactants contain enough extra energy, either as kinetic, vibrational, or rotational energy, a collision between them will not then lead to reaction. The energy in a molecule of course increases with increasing temperature. The fraction of reactant molecules with energy greater than some critical value E must therefore likewise increase. Indeed, it can be shown that the fraction of reactant molecules with energy greater than $\mathscr{E}$ is given by $\exp(-\mathscr{E}/kT)$, where T is the absolute temperature and k the Boltzmann constant.

Thus the quantity P in equation (5.6) is given by

$$P = \sigma \exp(-\mathscr{E}/kT) \qquad (5.8)$$

Here $\mathscr{E}$ is the energy per pair of molecules reacting. It is more usual to quote such energies as energies per mole rather than per molecules. Since one mole contains N molecules, N being the Avogadro number,

$$P = \sigma \exp(-N\mathscr{E}/NkT) = \sigma \exp(-E/RT) \qquad (5.9)$$

where E is the critical energy in joules per mole and R is the gas constant ($R = Nk$). The overall rate is then given by

$$\text{rate} = v_0 \sigma \exp(-E/RT)[X][Y] \qquad (5.10)$$

If the temperature is kept constant, this becomes

$$\text{rate} = k[X][Y] \qquad (5.11)$$

where k is a constant. This is the familiar rate law that is satisfied in practice by bimolecular reactions. The *rate constant* k, which can be found by measuring the rate at known concentrations of X and Y, is given by

$$k = v_0 \sigma \exp(-E/RT) \qquad (5.12)$$

Since the collision frequency does not vary much with temperature, and since σ is clearly independent of temperature, this can be written in the form

$$k = A \exp(-E/RT) \qquad (5.13)$$

or

$$\log k = (\log A) - (E/RT) \qquad (5.14)$$

A plot of $\log k$ against $1/T$ should therefore be a straight line of slope $-E/R$. The measured rate constants of bimolecular reactions follow this relation closely. Equation (5.13) is called the *Arrhenius equation*, Arrhenius having been the first to establish it by experiment. The quantity E is called the *activation energy* of the reaction.

The situation in the case of unimolecular reactions is at first sight rather different since here reaction does not involve a collision between two molecules of reactant. A unimolecular reaction can occur slowly only if appreciable energy is required to bring about the corresponding molecular fission or rearrangement. If the amount of energy required (in joules per mole) is E, the fraction of reactant molecules possessing this energy at absolute temperature T is again $\exp(-E/RT)$. However, even if a reactant molecule possesses this amount of energy, it may not in fact react. The reason for this is that the total energy of a molecule is distributed between the various possible modes of vibration and rotation. For a unimolecular reaction to occur, the energy must be concentrated in a particular mode. Thus the reaction

$$CH_3\text{---}CH_3 \longrightarrow CH_3\cdot + \cdot CH_3$$

will occur only if the energy of vibration of the CC bond is greater than the corresponding bond energy. Even if a molecule of ethane contains more energy as a whole than this, it will not dissociate unless enough of the energy is concentrated at some instant in vibration of the CC bond. The total vibrational energy of a molecule is shuffled around among the various degrees of freedom; sooner or later all or a large part of it will be concentrated in any given mode. If therefore a molecule of ethane contains enough vibrational energy to dissociate, it will do so eventually, but only after a certain length of time. In the meantime, it may lose the excess energy by collision with other molecules before it can react. The overall rate of reaction is therefore a product of the number of molecules containing the necessary energy E and the rate at which such an *activated* molecule reacts. This rate is the inverse of the average time τ taken for the energy to concentrate itself in the right place. The concentration of reactant molecules X that are activated is $[X]\exp(-E/RT)$. The rate of reaction, i.e., the rate of conversion of X to products in moles per liter, is then given by

$$\text{rate} = (1/\tau)[X]\exp(-E/RT) \tag{5.15}$$

At constant temperature

$$\text{rate} = k[X] \tag{5.16}$$

where k is the corresponding rate constant. Since the rate of redistribution of energy among the various degrees of freedom in a molecule is independent of temperature, τ must also be independent of temperature. It follows that the rate constant k again obeys the Arrhenius equation (5.12)–(5.13). The variations in the rates of unimolecular reactions with concentration and temperature do follow equation (5.15).

By studying the rate of a given chemical reaction as a function of the concentration of the reactants, we can determine empirically the dependence of the rate on those concentrations. In many cases, the resulting expression is of the form

$$\text{rate} = k[X_1]^k[X_2]^l \cdots [X_n]^m \tag{5.17}$$

where $[X_1],[X_2],\ldots,[X_n]$ are the concentrations of the reactants. In this case, the reaction is said to be of the $(k + l + \cdots + m)$th order. Thus a unimolecular reaction is a first-order reaction and a bimolecular reaction is a second-order reaction. The arguments given above indicate that reactions of higher order are not of higher molecularity but rather take place in steps, each of which is unimolecular or bimolecular.

In the case of complex reactions of this kind, intermediates are necessarily formed since the reactants are not converted to the products in a single step. One of the commonest reaction schemes in organic chemistry is such a process in which a single intermediate is formed reversibly from the reactants, usually by a bimolecular reaction, and then reacts with a third reagent to form the product. Thus

$$X + Y \longrightarrow Q, \quad \text{rate constant } k_1$$
$$Q \longrightarrow X + Y, \quad \text{rate constant } k_{-1} \quad (5.18)$$
$$Q + Z \longrightarrow \text{product}, \quad \text{rate constant } k_2$$

Here Q is the intermediate, the overall reaction being

$$X + Y + Z \longrightarrow \text{products} \quad (5.19)$$

Initially the concentration of Q is zero. As the reaction proceeds, Q is formed at a rate given by

$$\text{rate of formation of Q} = k_1[X][Y] \quad (5.20)$$

As the concentration of Q begins to build up, it begins to be removed by reversion to X + Y and by reaction with Z to form the product,

$$Q \longrightarrow X + Y, \quad \text{rate} = k_{-1}[Q] \quad (5.21)$$
$$Q + Z \longrightarrow \text{product}, \quad \text{rate} = k_2[Q][Z] \quad (5.22)$$

Presently the concentration of Q reaches a stationary value when the rate of formation is equal to the rate of disappearance. If this equilibrium concentration $[Q]_0$ is small, as is usually the case, we can find it by equating the rates of formation of Q [equation (5.20)] and the rates of removal of Q [equations (5.21) and (5.22)], i.e.,

$$k_1[X][Y] = k_{-1}[Q]_0 + k_2[Q]_0[Z] \quad (5.23)$$

Hence,

$$[Q]_0 = \frac{k_1[X][Y]}{k_{-1} + k_2[Z]} \quad (5.24)$$

The overall rate of the reaction of equation (5.19), i.e., the rate of formation of products, is then given [see equation (5.22)] by

$$\text{rate of formation of product} = k_2[Q][Z]$$
$$= \frac{k_1 k_2[X][Y][Z]}{k_{-1} + k_2[Z]} \quad (5.25)$$

The rate of the overall reaction will not therefore in general follow the simple expression (5.17). There are, however, two limiting cases where it does. In the first, Q reverts to X + Y much faster than it reacts with Z. In this case [see equations (5.21) and (5.22)],

$$k_{-1}[Q]_0 \gg k_2[Q]_0[Z] \quad (5.26)$$

If then we neglect the smaller term in the denominator of equation (5.25), we have

$$\text{rate of formation of products} \simeq (k_1 k_2 / k_{-1})[X][Y][Z] \quad (5.27)$$

The reaction is then of third order, as it would be if it were a simple thermomolecular process, which in fact it is not. Since the rate of reversion of Q to X + Y is fast compared with the rate of reaction of Q with Z, the concentration of Q is determined essentially by the equilibrium

$$X + Y \rightleftharpoons Q \tag{5.28}$$

The concentration of Q is then given by

$$[Q]_0 = K[X][Y] \tag{5.29}$$

when the corresponding equilibrium constant K is equal to the ratio of the rate constants for the forward and backward reactions,

$$K = k_1/k_{-1} \tag{5.30}$$

The rate of the overall reaction is then determined by that of the slow or *rate-determining* step of equation (5.22),

$$\text{rate} = k_1[Q]_0[Z] = k_1 K[X][Y][Z] \tag{5.31}$$

The other limiting case occurs when Q reacts with Z much faster than it reverts to X + Y:

$$k_2[Q]_0[Z] \gg k_{-1}[Q]_0 \tag{5.32}$$

In this case, the first term in the numerator of equation (5.25) becomes negligible, so the overall rate is now given by

$$\text{rate} \simeq \frac{k_1 k_2 [X][Y][Z]}{k_2[Z]} = k_1[X][Y] \tag{5.33}$$

The overall rate is now equal to that of the first step, equation (5.20), which is now *rate determining*. Each time a molecule of Q is formed, it reacts with Z to form the product. The overall rate is therefore equal to the rate of formation of Q from X + Y.

The large majority of reactions in organic chemistry are either simple unimolecular or bimolecular reactions or follow the scheme indicated above. Organic reactions are therefore usually either of first, second, or third order or follow a rate law corresponding to equation (5.25).

5.2 The Transition State Theory

While this *collision theory* of reaction provides a good physical explanation of the observed variations in the rates of organic reactions with temperature and the concentrations of the reactants, it is difficult on this basis to estimate the vital parameters A and E in the Arrhenius equation. We will now consider an alternative picture of the processes leading to chemical reaction which will serve us better in this respect.

A chemical reaction involves a reorganization of a given collection of n atoms from the grouping corresponding to the molecule or molecules of reactant to that corresponding to the product. We can in principle calculate the energy of the collection of atoms as a function of their geometric arrangement in space. Three coordinates are needed to describe the position of a single atom, so $3n$ coordinates will be needed to represent the individual positions of our collection of n atoms.* Any possible arrangement of the atoms can therefore be represented by a point in a $3n$-dimensional space. The variation of energy with geometry can in turn be represented by a contour map in this $3n$-dimensional space, the contours joining points that correspond to similar total energies. The resulting map will be exactly analogous to ordinary topographical contour maps. In the latter, the contours link points where the gravitational potential energy is the same. In order to pass from a state of lower energy to one of higher energy, we will in each case have to supply the necessary excess energy; we have to work hard to climb a mountain against the pull of gravity and equally we will have to supply extra energy to our collection of atoms if they are to rearrange themselves in a manner corresponding to motion "uphill" in our contour map.

The minima in our $3n$-dimensional contour map (or *potential surface*) will correspond to possible stable arrangements of the atoms, i.e., to a stable molecule or collection of molecules that can be formed from them. In particular, one such minimum (R) will correspond to the reactants in our reaction and another (P) to the products. The situation is indicated schematically in Fig. 5.1 for a simplified case with just two geometric variables.

Since R is a minimum, we have to go uphill to get to P. The reaction R → P can therefore occur only if the reactant molecules R contain enough extra energy to carry them over the intervening high ground. The minimum amount of excess energy needed should correspond to the activation energy in the Arrhenius equation. This in turn should be equal to the difference in energy between R and the highest point in the easiest path from R to P. In Fig. 5.1 this path is marked with a horizontal dashed line. The highest point in the path is the lowest point (T) in the ridge (dashed vertical line) separating R from P, i.e., the col at the heads of valleys leading from R and P up to the ridge.

If then we can calculate the potential surface for the reaction and so identify the transition state, we can estimate the parameter E (i.e., the activation energy) in the Arrhenius equation (5.13); what about the other parameter, A?

It is easily seen that the transition state for a reaction also represents an equilibrium configuration of the reacting system. The energy is a minimum for any geometric displacement other than motion to and fro along the reaction path, while for the latter it is a maximum. The equilibrium is stable except for displacements along this one *reaction coordinate*. We can therefore treat the formation of the transition state from the reactants as an

---

* In the case of a chemical reaction, only the relative positions of the atoms are important. Translation or rotation of the collection of n atoms as a whole is irrelevant. We can therefore ignore six of the $3n$ coordinates, only $3n - 6$ being needed to describe the positions of the atoms relative to one another.

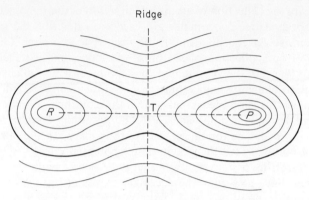

FIGURE 5.1. Two-dimensional contour map representing schematically the reaction R → P.

equilibrium

$$R \rightleftharpoons T \tag{5.34}$$

The concentration [T] of the transition state will then be given by

$$[T] = K[R] \tag{5.35}$$

where K is the corresponding equilibrium constant and R is either the concentration of the single reactant in a unimolecular reaction or the product of the concentrations of both reactants in a bimolecular reaction. The equilibrium constant can in turn be expressed in terms of the differences in free energy $\Delta G^\ddagger$, heat content $\Delta H^\ddagger$, and entropy $\Delta S^\ddagger$ between R and T:

$$K = \exp(-\Delta G^\ddagger/RT) = \exp(-\Delta H^\ddagger/RT)\exp(\Delta S^\ddagger/R) \tag{5.36}$$

Here $\Delta G^\ddagger$, $\Delta H^\ddagger$, and $\Delta S^\ddagger$ are called the free energy of activation, heat of activation, and entropy of activation, respectively. The overall rate of reaction is then given by

$$\text{rate} = k_r[T] = k_r K[R] = k_r\{\exp(\Delta S^\ddagger/R)\exp(-\Delta H^\ddagger/RT)\}[R] \tag{5.37}$$

where k_r is the unimolecular rate constant for the conversion of T to the product. It can be shown that k_r has the same value in all cases, being given by

$$k_r = kT\tau/h \tag{5.38}$$

where k is the Boltzmann constant, h is Planck's constant, and τ is the so-called transmission coefficient, which is usually assumed to be equal to unity. In that case, combining equations (5.37) and (5.38), we find

$$\text{rate} = \left(\frac{kT}{h}\exp\frac{\Delta S^\ddagger}{R}\right)\left(\exp\frac{-\Delta H^\ddagger}{RT}\right)[R] \tag{5.39}$$

This equation is clearly very similar to the Arrhenius equation (5.13), the heat of activation corresponding to E and the term in the first set of parentheses to A. It is true that the latter term contains T and so is not independent of temperature; however, in deriving equation (5.13), we assumed that the collision frequency v_0 was independent of temperature, and this is not the case. In fact, $v_0 \sim T$, so A in equation (5.13) should really be written as BT, B being a temperature-independent constant. The advantage of equation (5.39) is that it allows us, in principle at least, to calculate B. If we knew the potential surface for the reacting system in detail, we could calculate the vibration frequencies of the reactants and transition state and so find their entropies. Then we could estimate the first term in equation (5.39). While we cannot hope to do this by the crude procedures we are using, this approach is useful because it shows that the problem of estimating reaction rates is essentially the same as that of estimating equilibrium constants. The rate of a reaction depends on an equilibrium between the reactants and the transition state. We can therefore use the same general procedures that we used in the last chapter in our discussion of chemical equilibrium.

For reasons also indicated in Chapter 4, our main problem is to predict relative rates of reaction, not absolute rates of reaction. Given the knowledge that a certain reaction can occur in practice, our main concern is to predict how its rate will vary with changes in the reactants. As in the case of equilibrium, these changes in rate are likely to be due mainly to changes in the activation energy. Changes in the entropy of activation on passing from one reaction to another closely related reaction are likely to be due to changes in solvation—and these will be compensated more or less by corresponding changes in the heat of activation. Likewise, the changes in activation energy on passing from one reaction to another are likely to be due mainly to changes in the *delocalization energy of activation*, i.e., the difference in the energy of delocalized electrons on passing from the reactant to the transition state. These can be estimated as before by the PMO method.

Admittedly, the problem is more difficult in the case of reaction rates than chemical equilibria because we do not know the structures of transition states. There is no experimental technique by which such structures can be determined. For the present purpose, where we are concerned primarily with the qualitative effects of structural changes rather than with attempts to estimate rates quantitatively, two procedures can be used. In the first, we try to estimate the structure of the transition state directly, using the knowledge that it must be intermediate between those of the reactants and products and must lie on the easiest path from one to the other. This approach is illustrated by an example in the following section. The other procedure is based on a relation between activation energies of reactions and heats of reaction, discussed in Section 5.4.

5.3 Transition States for Aliphatic Substitution

The kind of arguments used to deduce the structures of transition states are best indicated by an example. We will consider one of the basic reactions of organic chemistry, aliphatic substitution. Aliphatic substitution is a process in which a group X in a compound RX is replaced by another group Y, R being an alkyl or substituted alkyl group,

$$Y + RX \longrightarrow YR + X \qquad (5.40)$$

Processes of this kind are of two main types.

In *nucleophilic substitution*, Y is a *nucleophile*, i.e., a group with a pair of electrons in an orbital available for forming a dative bond. The RY bond in the product is formed in this way, using a vacant AO of R. To provide this vacant AO, the RX bond must undergo heterolytic cleavage to give a cation R^+ and an anion X^-;

$$R\frown\!\!X \longrightarrow R^+ + X^- \qquad (5.41)$$

The curved arrow indicates in a self-evident way the fate of the two electrons that formed the R—X bond. Both are acquired by X, the corresponding AO of R being left empty. The presence of this empty AO allows R^+ to combine directly with the reagent Y^-,

$$Y^-: + R^+ \longrightarrow Y\text{---}R \qquad (5.42)$$

In *electrophilic substitution*, the reagent Y^+ has an empty AO. The R—Y bond in the product is formed by interaction of this with a filled AO at R. To provide this filled AO, the R—X bond in RX must undergo heterolysis in the opposite sense to that of equation (5.41),

$$\overset{\frown}{R}\text{---}X \longrightarrow R:^- + X^+ \qquad (5.43)$$

$$Y^+ + :R^- \longrightarrow Y\text{---}R \qquad (5.44)$$

The pair of electrons used to form the R—Y bond in the product can be formally regarded as the same pair that formed the R—X bond in RX.

Nucleophilic and electrophilic substitution reactions can indeed take place in two steps as indicated in equations (5.41)–(5.42) and (5.43)–(5.44). In the Ingold terminology, the former mechanism is designated $S_N 1$ and the latter $S_E 1$. Such substitution reactions can, however, also occur by single bimolecular processes involving concerted formation of the new RY bond and breaking of the old RX one. These reactions, designated $S_N 2$ and $S_E 2$, respectively, can be represented as follows in the curved arrow symbolism:

$$Y:^{\frown} \quad R\overset{\frown}{\text{---}}X \longrightarrow Y\text{---}R : X^- \quad (S_N 2) \qquad (5.45)$$

$$Y^+ \,{}^{\frown} R\overset{\frown}{\text{---}}X \longrightarrow Y\text{---}R \, X^+ \quad (S_E 2) \qquad (5.46)$$

The alkyl group R can be written in the form C*abc* where *a*, *b*, and *c* are hydrogen atoms or univalent groups. Since *a*, *b*, and *c* remain bonded throughout to the central carbon atom and play no part in the reaction, nothing can be gained energetically by tampering with the localized σ bonds linking them to the central carbon. Since our object is to find the path of minimum energy leading from the reactants Y + RX to the products YR + X, we must then ensure that the bonds linking a, b, and c to the central carbon remain intact throughout. Three of the valence shell AOs of the latter must consequently be reserved to form localized Ca, Cb, and Cc bonds. The fourth valence shell AO ϕ alone is directly involved in the reaction.

If substitution is concerted, both Y and X must be linked to the central carbon in the transition state. Since the latter has only one AO available (i.e., ϕ), the bonding must involve three-center MOs formed from ϕ and the relevant AOs of X and Y. This can be achieved most easily if the carbon atom adopts sp^2 hybridization, the group $Cabc$ becoming planar and ϕ being the remaining $2p$ AO. The two lobes of the latter can then be used to interact with the AOs of X and Y (Fig. 5.2a).

Consider now the π MOs in allyl. These are formed by π-type interactions of three $2p$ AOs, one from each carbon atom. The $2p$ AO of the central carbon atoms overlaps with the $2p$ AOs of the terminal atoms but the latter do not overlap significantly with one another (Fig. 5.2b). There is clearly an exact parallel with the three-center MOs of the transition state of Fig. 5.2(a). As can be seen from Fig. 5.2, both systems are moreover of Hückel type. The topology of the orbital overlap involved in forming the three-center σ-type MOs in the S_N2 transition state of Fig. 5.2(a) is therefore the same as that of the π-type orbital overlap of p AOs in allyl.

Now the interaction of AOs to form MOs depends on the *extent* to which the orbitals overlap in space but it does *not* depend on the *geometry* of overlap. From this point of view, σ- and π-type MOs are equivalent and indeed our description of two-center σ and π bonds indicated a complete equivalence between them. The three-center MOs of the transition state in Fig. 5.2(a) are therefore topologically equivalent to, or isoconjugate with, the π MOs of allyl. This extension of the concept of isoconjugation is extremely important since it allows us to apply the PMO treatment developed for π systems to the isoconjugate σ-type delocalized structures present in transition states.

This analogy enables us to see immediately that the S_N2 and S_E2 mechanisms are reasonable. Union of ethylene with CH_3^- or CH_3^+ to form allyl anion or cation leads of course to a decrease in π energy (see Section 3.16). The isoconjugate interactions of the filled AO of a nucleophile Y^- or of the

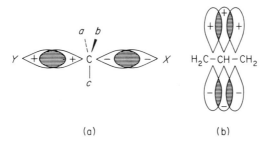

(a) (b)

FIGURE 5.2. (a) σ-type overlap of AOs in an S_N2 transition state; (b) π-type overlap of AOs in allyl.

empty AO of an electrophile Y^+ with the R—X σ bond of RX must likewise lead to a decrease in energy on forming the corresponding S_N2 or S_E2 transition state. It is therefore entirely reasonable that the substitutions should occur as concerted one-step processes.

If the interaction of Y^- with RX leads to a decrease in energy, it might seem at first sight that the ion $(YRX)^-$ formed by their association should be a stable species rather than a transition state. This argument, however, overlooks the difference in geometry between RX and $(YRX)^-$. In RX, the central carbon atom is tetrahedral and the R—X bond is formed via a sp^3 hybrid AO of the central carbon. In the transition state, a $2p$ AO is used. The bond in the latter should be weaker than that in RX both because the $2p$ AO has a lower binding energy than an sp^3 hybrid AO (see p. 28) and because the overlap between it and the AO of X must be less, only half the $2p$ AO being able to take part (see Fig. 5.2a). A third factor operating in the same direction is the steric repulsion between X and the groups a, b, c, the distances between them being much less in the transition state than in RX. Thus while the ion $(YRX)^-$ should be formed exothermically from Y^- and RX if the latter is deformed to a geometry in which the Cabc moiety is planar, as in Fig. 5.2(a), the same may not be true for RX in its equilibrium state.

While this argument shows that the ions $(YRX)^-$ need not be stable, recent work has shown that certain ions of this type do in fact occur as stable species in the gas phase; e.g.,

$$Cl^- + CH_3Cl \longrightarrow (ClCH_3Cl)^- \quad \text{(stable)} \quad (5.47)$$

$$Br^- + CH_3Br \longrightarrow (BrCH_3Br)^- \quad \text{(stable)} \quad (5.48)$$

Yet the same species occur as transition states for S_N2 halide exchange reactions in solution, e.g.,†

$$(Cl^*)^- + CH_3Cl \longrightarrow (Cl^* \cdots CH_3 \cdots Cl)^- \longrightarrow Cl^*\text{—}CH_3 + Cl^- \quad (5.49)$$

The reason for this apparent discrepancy is that our treatment neglects the the effects of solvation. Evidently the energy of solvation of halide ions X^- is much greater than that of the intermediate ions $CH_3X_2^-$. The relative energies of $X^- + CH_3X$ and of $CH_3X_2^-$ are therefore reversed on passing from the gas phase to solution.

For similar reasons, it is impossible to make any predictions concerning the relative rates of the S_N1 and S_N2 processes in solution for any given reaction because the transition states are entirely different and will have entirely different energies of solvation. We have no way of estimating these differences since they involve a knowledge of the absolute energies of solvation of different species. The most we can hope to do is to estimate the changes in the rates of the S_N2 and the S_N1 reactions with changes in the structures of the reactants or the solvent.

---

† The asterisk in equation (5.49) denotes a radioactive isotope (^{36}Cl) of chlorine, since isotopic labeling is needed to enable the replacement of one chlorine atom by another to be observed.

Chemical Reactivity

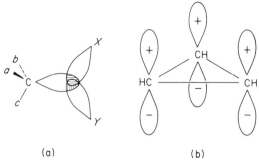

(a)   (b)

FIGURE 5.3. Overlap of AOs to form three-center MOs (a) in a possible transition state for bimolecular aliphatic substitution; (b) in cyclopropenium.

This problem will be considered presently. There is, however, a further aspect of aliphatic substitution that we may examine at this stage. The one-step bimolecular reactions involve transition states where Y, R, and X are linked by bonds formed by three-center overlap of AOs. One possible geometry for achieving this is that of Fig. 5.2(a), but there is an alternative. Instead of having the AOs of Y and X overlap with different lobes of a $2p$ AO of R, they may overlap with the same larger lobe of a hybrid AO, as indicated in Fig. 5.3(a). The situation here is topologically different from that in Fig. 5.2(a) in that the AOs of Y and X now overlap with one another. All three AOs overlap. The resulting structure is therefore topologically equivalent to, or isoconjugate with, the π MOs of cyclopropenium (Fig. 5.3b), where again each of the three relevant $2p$ AOs overlap with both the other two.

The π energies of union of ethylene with CH_3^+ and CH_3^- to form allyl cation or anion are the same because the allyl ions differ only by two non-bonding electrons. The π energies of double union to form cyclopropenium are, however, different. The π energy of double union of ethylene with CH_3^+ to form cyclopropenium cation is much greater than that of single union to form allyl cation, because the cyclopropenium cation is aromatic. Likewise, the π energy of double union of ethylene with CH_3^- to form the cyclopropenium anion is much less than that of single union to form the allyl anion because the cyclopropenium anion is antiaromatic. Similar relationships should hold between the corresponding transition states. Thus nucleophilic substitution, where the transition state is isoconjugate with $(CH_2CHCH_2)^-$ or $(CH)_3^-$, should take place preferentially via the transition state isoconjugate with allyl anion, i.e., that of Fig. 5.2(a), while electrophilic substitution, when the transition state is isoconjugate with $(CH_2CHCH_2)^+$ or $(CH)_3^+$, should take place preferentially via the transition state isoconjugate with cyclopropenium, i.e., that of Fig. 5.3(a).

If the groups a, b, and c are different, $CabcX$ and $CabcY$ are optically active. Substitution via the transition state of Fig. 5.2(a) must then lead to a product whose configuration is the mirror image of that in the reactant:

$$Y\quad\overset{b}{\underset{c}{\overset{a}{{\diagdown}}}}C-X \longrightarrow Y\cdots\overset{b}{\underset{c}{\overset{a}{|}}}C\cdots X \longrightarrow Y-\overset{b}{\underset{c}{\overset{a}{C}}} \underset{\text{images}}{\overset{\text{mirror}}{\longleftrightarrow}} \overset{b}{\underset{c}{\overset{a}{{\diagdown}}}}C-Y \quad (5.50)$$

Substitution via the transition state of Fig. 5.3(a) should, on the other hand, lead to retention of configuration:

$$\overset{b}{\underset{c}{\overset{a}{{\diagdown}}}}C-X \longrightarrow \overset{b\quad Y}{\underset{c\quad X}{\overset{a}{{\diagdown}}}}C \longrightarrow \overset{b}{\underset{c\quad X}{\overset{a}{{\diagdown}}}}C-Y \quad (5.51)$$

The $S_N 2$ reaction should therefore lead to inversion of configuration at the central carbon, while the $S_E 2$ reaction should lead to retention of configuration.

Many years ago, Walden discovered that what we now call nucleophilic substitution reactions could lead to inversion of configuration. This phenomenon, which seemed remarkable at the time, was called the *Walden inversion*. We now know that such inversion always occurs during $S_N 2$ reactions. The situation in the case of $S_E 2$ reactions is even more interesting because it had been predicted on the basis of the arguments indicated above that they should lead to retention of configuration before any experimental evidence was available. We now know that this prediction is also correct.

5.4 Reaction Paths and Reaction Coordinates

The potential surfaces for many reactions seem to be of the general type indicated in Fig 5.1, a single valley leading from the minimum corresponding to the reactants to that corresponding to the products. In this case, one can define a unique *minimum gradient reaction path* (MGRP) along the bottom of that valley (indicated by the dotted line in Fig. 5.1). The MGRP corresponds to the path of minimum gradient at each point, like that taken by a lazy mountaineer who always moves uphill in the direction where the gradient is least and downhill where it is greatest.

It is tempting, but quite incorrect, to assume that the reaction will, like our lazy mountaineer, tend to follow the MGRP. The mountaineer has to plod up the valley step by step. At each step, he has to summon enough energy for the next one. Our reaction, however, takes place in a period so short that there is no time for the reacting system to gain additional energy through molecular collisions. Moreover, the reacting system can pass freely through any geometry that is energetically accessible to it. In our analogy, the reaction behaves like a bird that can soar to and fro across the valley at a a given level. If the reactants in Fig. 5.1 have enough energy for reaction to occur, they can pass freely through any geometry within the contour that passes through the transition state (indicated in Fig. 5.1 by a heavier line).

The overall form of the potential surface is in fact irrelevant. The rate of the reaction is determined only by two points on it, corresponding, respectively, to the reactants and the transition state. The difference in energy between these is equal to the activation energy, while the curvatures of the surface at the two points enable us to find the corresponding entropies and hence the entropy of activation. In our simplified treatment, where entropy is neglected, even the curvatures are irrelevant. The only purpose served by the calculation of such a potential surface, so far as our present study is concerned, would therefore be to provide a mechanism for identifying the transition state and so finding its energy.

The essential features of the surface can then be represented more simply and more clearly in the following way. Suppose that the potential surface for a reaction is of the type indicated in Fig. 5.1, so that we can define a MGRP. In most cases, there will be one or more geometric variables x_i (e.g., bond lengths and bond angles) that either increase steadily or decrease steadily throughout the reaction, in particular along the MGRP. If now for each value of x_i we find the geometry that minimizes the energy, the resulting energy and geometry will correspond to a point on the MGRP; for the MGRP corresponds to a line along the bottom of the valley linking the reactants and products and x_i increases (or decreases) steadily along it. A plot of energy vs. x_i will then correspond to a section of the potential surface along the MGRP and the maximum will correspond to the transition state. This enables us to represent the essential features of the reaction in terms of an easily reproduced and easily visualized two-dimensional plot.

One difficulty here is that of establishing that a given coordinate x_i has the required properties. Even if x_i increases during the reaction, it may not increase uniformly along the MGRP. It may, for example, first increase, then decrease, and then increase again. Over a certain range of x_i, there will then be three different geometries that minimize the energy. This difficulty can be avoided only by calculating the complete potential surface so that the behavior of each coordinate can be seen. In this case we might as well take as the reaction coordinate the distance along the MGRP measured from the minimum corresponding to the reactants. Indeed, in many recent papers, this quantity is referred to as "the" reaction coordinate for the reaction.

A further and less defensible extension is to describe plots of energy against distance along the MGRP as plots of the reaction coordinate. This of course is a misnomer since such a plot has no geometric connotations. Difficulties also arise in certain cases because the shapes of the very complex multidimensional potential surfaces for reactions involving complex molecules often differ greatly from the ideal simplicity of Fig. 5.1. Indeed, it is sometimes impossible to define a unique MGRP, the minimum gradient path upward from the reactants either not leading to the products or leading to them via intermediates higher in energy than the true transition state.

Another problem that can occur in complex multidimensional potential

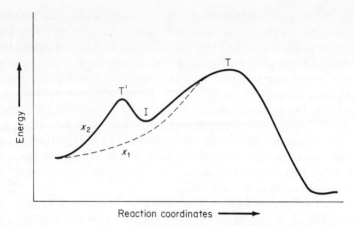

FIGURE 5.4. Reaction coordinate for a reaction that can proceed directly from starting materials or through an intermediate I to the transition state T.

surfaces is the existence of more than one coordinate that approaches the transition state in the MGRP energy range. Figure 5.4 illustrates such a case.

For this case, it makes little difference if the reaction proceeds through the intermediate I (Fig. 5.4) or around it. In any event, the equilibrium-determined concentration of the intermediate should be detectable during the reaction, and reactant molecules with energies in excess of T′ should be able to adopt the structure of the intermediate.

Since the plots of energy against different "suitable" reaction coordinates are necessarily similar in shape, the choice of reaction coordinate is really a matter of convenience. In the rest of this book, we will frequently cite plots of energy against (unspecified) reaction coordinates; the term "reaction coordinate" implies "any coordinate that changes monotonically along the MGRP." We will also often describe such plots as representations of "reaction paths," the term reaction path implying in this connection the MGRP.

5.5 The Bell–Evans–Polanyi (BEP) Principle; Relationships between Rates of Reactions and Corresponding Equilibrium Constants

According to the transition state theory, the problem of calculating reaction rates is in principle no different from that of calculating equilibria. Both depend on an estimation of the difference in energy between the reactants and another equilibrium species, i.e., either the transition state or the reaction product. The only difficulty, and that unfortunately is a severe one, is that we cannot determine the structures of transition states by experiment. These have to be deduced theoretically and we have no experimental data to check the validity of such predictions.

Chemical Reactivity

It has long been known that there is a general correspondence between rates of reaction and heats of reaction. Very exothermic reactions usually tend to take place faster than less exothermic ones. This implies that the activation energy $\Delta E^\ddagger$ of a reaction tends to be the smaller, the more negative is the heat of reaction ΔH. If it could be shown that a quantitative relation held between them, this could be used to estimate activation energies, and hence rates of reaction, from heats of reaction. Rates of reaction could then be estimated without having first to determine structures of transition states.

Polanyi and his collaborators found that a relation of this kind did indeed hold for the reactions of sodium vapor with alkyl halides in the gas phase, leading to sodium halide and an alkyl radical, e.g.,

$$\text{Na} + \text{CH}_3\text{Cl} \longrightarrow \text{NaCl} + \text{CH}_3\cdot \qquad (5.52)$$

The rates of reaction varied over a wide range and the variations were shown to be due essentially to changes in the activation energy $\Delta E^\ddagger$. The changes in $\Delta E^\ddagger$ were, moreover, proportional to the corresponding changes in the heat of reaction ΔH; i.e.,

$$\Delta E^\ddagger = A + B\,\Delta H \qquad (5.53)$$

where A and B are constants, the proportionality factor B being approximately one-half. In an attempt to explain this relation, Evans and Polanyi developed the following approach.

The sodium halides formed in the reaction are of course salts, $\text{Na}^+ \text{X}^-$. At some point in the reaction, an electron must therefore be transferred from a sodium atom to a molecule of RX. Up to that point, the electronic structure is similar to that in the reactants, Na + RX, the R—X bond being stretched with little interaction between it and the neutral sodium atom. After electron transfer has taken place, however, the structure resembles that in the product, $\text{Na}^+\text{X}^-\text{R}\cdot$. The energy now decreases as the ions Na^+ and X^- approach one another, neither interacting much with the neutral radical R $\cdot$. If therefore we plot the energy of the system against a suitable reaction coordinate, the resulting plot will fall into two parts. The first (A in Fig. 5.5) corresponds essentially to stretching of the R—X bond, the sodium atom being more or less irrelevant. The second (B in Fig. 5.5) corresponds to the change in energy as the ions Na^+ and X^- approach one another, the radical R $\cdot$ being in turn superfluous. At the point (C) where the curves cross, both electronic structures have the same energy. One can therefore cross from A to B at this point by an electron transfer from Na to RX. The energy of the system therefore increases along the curve A up to the point C and then decreases along the curve B; C therefore corresponds to the transition state.

The terminus (D) of curve A in Fig. 5.5 corresponds to complete dissociation of RX into R $\cdot$ + $\cdot$X without reaction with Na. The energy of D is therefore $E_0 + E_{\text{RX}}$, E_0 being the energy of the reactants and E_{RX} the R—X bond energy. The starting point of curve B corresponds to a situation where an electron has been transferred to RX to give R $\cdot$ + X$^-$ and Na^+ but the two ions are far apart. This point therefore lies above D in energy by an amount

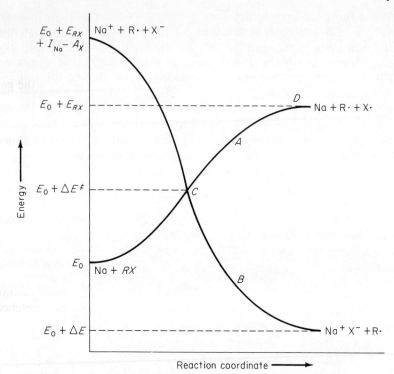

FIGURE 5.5. Evans–Polanyi plot for the reaction of a sodium atom with an alkyl halide RX in the gas phase.

equal to the energy of transfer of an electron from an isolated atom of sodium to an isolated atom of X. This is $I_{Na} - A_X$, where I_{Na} is the ionization potential of sodium and A_X is the electron affinity of X. Curve B terminates at a point with energy $E_0 + \Delta E$, where ΔE is the difference in energy between the reactants and products, i.e., the energy of reaction.

Consider now plots of this kind for the reactions of sodium with a number of different alkyl halides R_1X, R_2X, R_3X, etc. These are shown in Fig. 5.6. Here the bond-breaking curves A_i are different since the strength of the R_iX bond varies with the alkyl group R_i. The bond-forming curves B_i are all of the same shape, representing the formation of Na^+X^- from isolated ions, but are displaced up or down since each must start at a point $(I_{Na} - A_X)$ above the terminus of the corresponding bond-breaking curve.

If the central parts of the curves are reasonably straight, it is fairly obvious that the height of the crossing point C above the reactants must increase or decrease in proportion to the corresponding changes in the energy of the products. Since the height of C above the reactants is equal to the activation energy $\Delta E_i^{\ddagger}$, while that of the products is equal to the corresponding energy of reaction ΔE_i, it follows that

$$\Delta E_i^{\ddagger} = A + B \, \Delta E_i \tag{5.54}$$

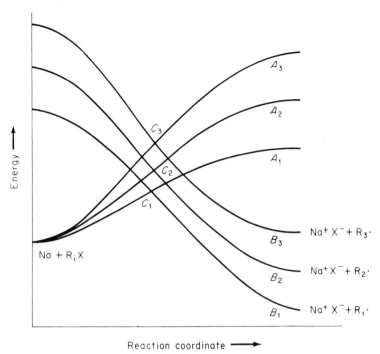

FIGURE 5.6. Evans–Polanyi plots for reactions of sodium vapor with three different alkyl halides R_1X, R_2X, and R_3X.

where A and B are constants. If we neglect changes in the entropy of activation or of reaction (cf. Section 4.1), this implies an analogous linear relation between $\Delta E_i^\ddagger$ and ΔH_i, i.e., equation (5.53).

Equation (5.53) therefore appears as a consequence of the bond-breaking and bond-forming processes being independent of one another, so that we can represent the overall reaction path as a composite of two segments, one involving bond breaking, the other bond forming. Evans and Polanyi suggested that an analysis of this kind could be used generally in reactions where bonds are broken and formed in a single step.

This argument was developed in some detail by Bell as an explanation of an empirical relationship that had been discovered by Brønsted between the rates of acid–base-catalyzed reactions and the strengths of the catalyzing acids or bases.

A number of reactions are catalyzed by undissociated acids or bases through transfer of a proton from or to the catalyst. A very good example is the decomposition of nitramide into water and nitrous oxide,

$$H_2NNO_2 \longrightarrow H_2O + N_2O \qquad (5.55)$$

This is catalyzed by a base (B) in the following way:

$$B + H_2NNO_2 \longrightarrow BH^+ + H\bar{N}NO_2 \qquad (5.56)$$

$$H\bar{N}NO_2 \longrightarrow HO^- + N_2O \quad (5.57)$$

$$BH^+ + HO^- \longrightarrow B + H_2O \quad (5.58)$$

The first step is a typical acid–base reaction, nitramide functioning as an acid, and would normally lead to an equilibrium. Here, however, the conjugate base of nitramide decomposes so rapidly [equation (5.57)] that it has no time to undergo the back reaction. This indeed is a typical example of the general reaction scheme considered in equations (5.18)–(5.33) in which the back reaction is slow, k_{-1} being small. The overall rate is then equal to that of the first rate-determining step [cf. equation (5.33)], i.e.,

$$\text{rate} = k[B][H_2NNO_2] \quad (5.59)$$

The rate constant k is thus a measure of the efficiency of the base as a catalyst and is therefore called the *catalytic constant*. One might expect k to be bigger, the stronger B is as a base. Not only is this the case, but there is also a quantitative relation between catalytic activity and basicity, i.e., the Brønsted relation,

$$\log k = C + \alpha \log K_B \quad (5.60)$$

where C and α are constants and K_B is the basic dissociation constant of the base, i.e., the equilibrium constant for its reversible reaction with water:

$$B + H_2O \rightleftharpoons BH^+ + HO^- \quad (5.61)$$

The linear relation between $\log k$ and $\log K_B$ holds generally for base-catalyzed reactions, each reaction having its characteristic values for C and α. Moreover, similar relations hold in the case of acid-catalyzed reactions between the catalytic constants and the acid dissociation constants of the catalysts.

Suppose the reaction of equation (5.56) were reversible and led to an equilibrium. The equilibrium constant K would be given by

$$K = \frac{[BH^+][H\bar{N}NO_2]}{[B][H_2NNO_2]} \quad (5.62)$$

We can expand this as follows

$$K = \frac{[BH^+][HO^-]}{[B][H_2O]} \frac{[H\bar{N}NO_2][H^+]}{[H_2NNO_2]} \frac{[H_2O]}{[H^+][HO^-]}$$

$$= K_B K_A \frac{1}{K_w} \quad (5.63)$$

where K_A is the acid dissociation constant of nitramide and K_w is the self-dissociation constant of water. Hence,

$$K_B = KK_w/K_A \quad (5.64)$$

Substituting this in equation (5.60), we find

$$\log k = C + \alpha \log K + \alpha \log(K_w/K_A)$$

$$= C' + \alpha \log K \quad (5.65)$$

where C' is a constant,

$$C' = C + \log(K_w/K_A) \tag{5.66}$$

We can express k and K in terms of the free energy of activation $\Delta G^\ddagger$ and free energy of reaction ΔG of the corresponding reaction [e.g., equation (5.56)]; cf. equations (5.36) and (5.39). Thus,

$$\log k = \log \frac{kT}{h} - \frac{\Delta G^\ddagger}{RT} \tag{5.67}$$

$$\log K = -\Delta G/RT \tag{5.68}$$

If we substitute these expressions in equation (5.65), and if we make our usual assumption that entropies of activation and reaction are the same for similar reactions, we find the following relation between the activation energy ΔE and the energy of reaction ΔE:

$$\Delta E^\ddagger = C'' + \alpha \Delta E \tag{5.69}$$

This of course is exactly analogous to the corresponding relation [equation (5.54)] found by Polanyi, and Bell explained it in a similar way. We dissect the overall proton transfer [cf. equation (5.56)],

$$B_i + HX \longrightarrow B_iH^+ + X^- \tag{5.70}$$

into two steps, a *bond-breaking* step

$$HX \longrightarrow H^+ + X^- \tag{5.71}$$

and a *bond-forming* step,

$$B_i + H^+ \longrightarrow BH^+ \tag{5.72}$$

During the reaction, the HX bond breaks and the BH bond forms. A reaction coordinate appropriate to the overall reaction will therefore also serve as a reaction coordinate for equations (5.71) and (5.72). In Fig. 5.7, the line A shows a plot of energy against such a reaction coordinate for the bond-breaking reaction of equation (5.71), while the lines B_i are similar plots for the bond-forming reactions of equation (5.72). Thus A corresponds to the change in energy with bond length that would occur if no reaction took place, the system remaining in the state $B_i + HX$ throughout, while the B_i likewise correspond to the products $B_iH^+ + X^-$. At the points C_i where the lines cross, both systems have the same energy, so we can cross from A to B_i. Here proton transfer plays the same role that electron transfer does in the reactions corresponding to Figs. 5.5 and 5.6 and the C_i likewise correspond to the transition state for the overall reaction of equations (5.70). Again, if the lines are reasonably straight in the regions where they cross, there should be a linear relation between the variations in the energy of the points C_i and the corresponding charges in the overall heat of reaction.

The resulting relation is of course equation (5.69). Reversing the argument that led to it, we immediately recover the Brønsted relation of equation (5.65).

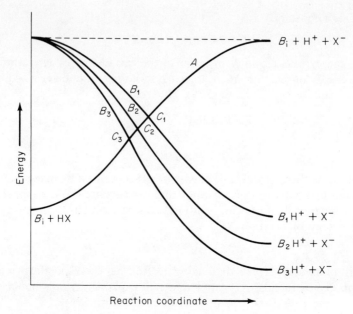

FIGURE 5.7. Bell–Evans–Polanyi plot of energy vs. a reaction coordinate for the reaction of equation (5.70).

This type of analysis can be applied quite generally to reactions where bonds are simultaneously formed and broken. We plot first a *bond-breaking* curve showing the change in energy as a function of reaction coordinates for a situation when the bonds break but no new bonds form. Then, starting with the fragments found by the bond-breaking process, we plot a corresponding *bond-forming* curve that shows the progressive decrease in energy with bond formation. The point where the lines cross then corresponds to the transition state.

If this representation is satisfactory, several corollaries follow immediately.

First, there will be an approximately linear relation between activation energy and heat of reaction. This enables us to estimate relative rates of reactions from the corresponding heats of reaction, which are of course much easier to estimate.

Second, the proportionality constant will tend to decrease, the more exothermic is the reaction. This is illustrated in Fig. 5.8 for four reactions where the bond-breaking process is the same in each case. The difference in heat of reaction between B_1 and B_2 is the same as that between B_3 and B_4; however, the difference in energy between the transition states C_1 and C_2 is much greater than that between C_3 and C_4.

Third, as is again obvious from Fig. 5.8, the crossing point C moves to progressively smaller values of the reaction coordinate, the more exothermic is the reaction. Very energetic reactions should not only be fast, having low

Chemical Reactivity

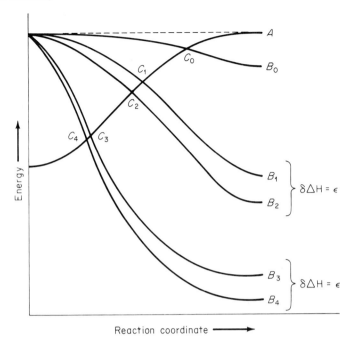

FIGURE 5.8. Bell–Evans–Polanyi plots showing the decrease in selectivity with increasing exothermicity and the corresponding change in structure of the transition state.

activation energies, but should also have transition states that resemble the reactants, corresponding as they do to values of the reaction coordinate close to that for the reactants.

As the reactions become less exothermic, the structure of the transition state changes progressively. In the case of a very endothermic reaction (cf. B_0 in Fig. 5.8), the transition state (C_0) will be very similar in structure to the product.

While these relations have been rediscovered from time to time, they are immediate consequences of the Bell–Evans–Polanyi treatment and have been recognized and used as such in the literature for many years. It is therefore convenient to refer to them collectively as the Bell–Evans–Polanyi (BEP) principle, which can be stated as follows.

BEP Principle. A number of reactions can be represented satisfactorily in terms of independent bond-breaking and bond-forming processes. The course of such a reaction can be represented by corresponding plots of energy for bond breaking and bond forming against a common reaction coordinate. The crossing point corresponds to the transition state. Several corollaries follow for such BEP reactions:

1. For a group of related BEP reactions, there is an approximately linear relation between activation energy and energy of reaction.

2. The proportionality factor in this linear relation is smaller, the more exothermic is the reaction.
3. The more exothermic a reaction, the nearer is the transition state to the reactants in structure. The transition states of highly endothermic reactions correspondingly resemble the products in structure.

Corollary 1 is the basis of a number of so-called linear free energy relationships, the Brønsted relation being a typical example.

Corollary 2 is the basis of Brown's selectivity rule in aromatic substitution (see Section 5.7).

Corollary 3 is commonly referred to as the *Hammond postulate*.

5.6 Reactions Where Intermediates Are Involved

The argument on which the BEP principle is based rests on the assumption that we are dealing with a one-step reaction. If a reaction takes place in two or more steps, via one or more stable intermediates, the arguments fail and the BEP principle does not apply.

A classic example of this is provided by the reactions of carbonyl compounds with semicarbazide to form semicarbazones,

$$\begin{array}{c} R_1 \\ \diagdown \\ C=O + H_2NNHCONH_2 \longrightarrow \\ \diagup \\ R_2 \end{array} \begin{array}{c} R_1 \\ \diagdown \\ C=NNHCONH_2 + H_2O \\ \diagup \\ R_2 \end{array}$$

(5.73)

If the overall reaction followed the BEP principle, the rate of formation of a given semicarbazone should be greater, the more stable it is. In other words, the rate constants k_1 for the forward reaction should run parallel to the corresponding equilibrium constants K. This is not the case, as the results in Table 5.1 show. Thus acetaldehyde reacts nearly 200 times faster with semicarbazide than does benzaldehyde, but the equilibrium constant favors benzaldehyde semicarbazone by a factor of ten.

If we react equimolar amounts of acetaldehyde, benzaldehyde, and semicarbazide, the product initially formed will contain 99.5% of acetaldehyde semicarbazone and only 0.5% of benzaldehyde semicarbazone. Under these conditions, the reaction is subject to *kinetic control*, the products being determined by the rates at which they are formed since the backward reactions are much slower. If, however, the reaction mixture is allowed to stand, so that the backward reactions can take place and lead to equilibrium, the product now contains 90% of benzaldehyde semicarbazone and only 10% of acetaldehyde semicarbazone. The reaction is then said to be subject to *thermodynamic control*, the proportion of products being determined by their relative stabilities rather than their rates of formation.

The explanation of this discrepancy is that the reaction takes place in several steps, one of the intermediates being the product formed by addition

TABLE 5.1. Semicarbazone Reactions

$$R_1R_2C=O + H_2N-NH-CONH_2 \underset{K_{-1}}{\overset{K_1}{\rightleftharpoons}}$$

$$R_1R_2C=N-NHCONH + H_2O$$

Carbonyl compound	Formation K_1	Hydrolysis K_{-1}	K_{eq} for formation
CH_3CHO	6.02	1.73×10^{-4}	3.4×10^4
$PhCHO$	0.0342	1.03×10^{-7}	3.3×10^5
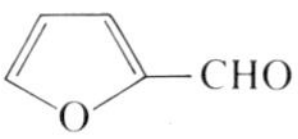 furyl-CHO	0.0122	9.2×10^{-8}	1.32×10^5
$(CH_3)_3C-CHO$	0.333	6.2×10^{-6}	5.4×10^4
CH_3-COCO_2H	0.123	6.3×10^{-7}	1.96×10^5
CH_3COCH_3	0.1004	3.0×10^{-4}	3.09×10^3
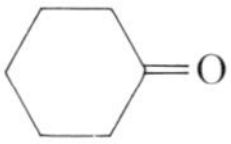 cyclohexanone (C=O)	0.600	1.27×10^{-3}	4.67×10^3
$(CH_3)_3-COCH_3$	0.00113	1.43×10^{-5}	79.0

of semicarbazide to the C=O bond, e.g.,

$$CH_3-\underset{O}{\overset{\|}{C}}-H + H_2N-NH\underset{O}{\overset{\|}{C}}-NH_2 \rightleftharpoons CH_3-\underset{\underset{(1)}{}}{\overset{OH}{C}H}-NH-NH\underset{O}{\overset{\|}{C}}-NH_2$$

$$\underset{-H_2O}{\rightleftharpoons} CH_3CH=N-NH\underset{O}{\overset{\|}{C}}-NH_2 \quad (5.74)$$

$$Ph\underset{O}{\overset{\|}{C}}-H + H_2N-NH\underset{O}{\overset{\|}{C}}-NH_2 \rightleftharpoons Ph\underset{\underset{(2)}{}}{\overset{OH}{C}H}-NH-NH\underset{O}{\overset{\|}{C}}-NH_2$$

$$\underset{-H_2O}{\rightleftharpoons} PhCH=N-NH\underset{O}{\overset{\|}{C}}-NH_2 \quad (5.75)$$

In cases of this kind, the BEP principle cannot be expected to apply to the overall reaction, although it may apply to the individual steps. Thus the formation of (2) is less favorable than that of (1) because in forming (2) from benzaldehyde, one has to convert the sp^2–sp^2 Ph—CHO bond into a much weaker sp^2–sp^3 bond. In forming (1) from acetaldehyde, one loses only the smaller difference between the sp^2–sp^3 C—C bond in acetaldehyde and the sp^3–sp^3 C—C bond in the product. Thus the BEP principle does apply to the formation of (1) and (2), the more exothermic reaction taking place more rapidly. Under conditions of kinetic control, the formation of (1) or (2) is rate determining since the dehydration of the intermediate to a semicarbazone is much faster than its dissociation to the carbonyl compound and semicarbazide.

5.7 Solvent Effects; Electrostatic Interactions and the Hellmann–Feynman Theorem

If the free energy of the reactants in a reaction is different from that in the transition state, the rate of the reaction in solution will be different from that in the gas phase. Such *solvent effects* on reactivity are indeed extremely important, as we have already seen in the discussion of the $S_N 2$ reaction in Section 5.3. The $S_N 2$ halide exchange reactions of methyl halides occur slowly in solution, via a transition state of the form $(X \cdots CH_3 \cdots Y)^-$. In the gas phase, however, the transition state becomes a stable species, formed exothermically from methyl halide and halide ion. Here solvent effects completely alter the course of the reaction.

In order to discuss solvation theoretically, we should solve the Schrödinger equation for the substrate and solvent together. This of course is not practicable. However, chemical evidence suggests that the molecules of solvent and solute in a solution are not usually greatly affected by the interactions between them; this therefore should be a favorable case for the use of PMO theory. We therefore regard the interactions as a perturbation, the electrons of each molecule now moving in a field due to the nuclei and electrons of solute as well as solvent, instead of those of solvent or solute alone.

The first-order perturbation is found by assuming that the MOs occupied by the electrons remain unchanged. The resulting change in energy can then be found by using a very important result which was discovered by Hellmann and Feynman and is therefore known as the *Hellmann–Feynman theorem*. This states that if we know the distribution of electrons in a molecule, i.e., the average number of electrons that will be found at any instant in any given volume element $d\tau$ so that we can treat the electrons as a cloud of negative charge of known density at each point, then we can calculate the total energy by using classical electrostatics, summing the electrostatic repulsions between the nuclei, the attractions between the nuclei and the cloud of electrons, and the self-repulsion between different parts of the cloud. The result we get is *exactly* that given by solving the Schrödinger equation.

In the present case, we assume that we know the total energy of the solvent and solute molecules when they are isolated. The interaction energy can then be found by a straightforward classical electrostatic calculation. In the MO approximation, this is particularly simple because we regard

Chemical Reactivity

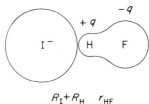

FIGURE 5.9. Calculation of the charge–dipole interaction between I^- and HF.

the electron distribution as being built up from contributions of AOs around each atom. The total electric field due to the molecule can therefore be regarded as a sum of contributions by individual atoms, the charge on an atom being $(Z - Q)e$, where Z is the atomic number, Q is the number of electrons in the vicinity of that nucleus in the molecule, and e is the electronic charge. Thus $(Z - Q)e$ is the formal charge on the atom in the molecule. A simple example will illustrate this. Consider the interaction between an iodide ion and a molecule of hydrogen fluoride (Fig. 5.9). The HF molecule is polar in the sense $H^{\delta-}-F^{\delta-}$. The molecule thus forms a dipole with charges $+q$ and $-q$ on hydrogen and fluorine, respectively. In our MO approximation, we can treat it as a pair of point charges $+q$ and $-q$ at a distance r_{HF} (the HF bond length) apart. The negative I^- ion (charge $-e$) will be attracted to the positive end of the dipole. The attraction will be strongest when all three atoms are in line (so that the negative charges keep as far apart as possible) and when H and I^- approach as closely as possible, the separation between them being the sum of their radii $R_I + R_H$. The total energy of interaction E_I is then given by

$$E_I = -\frac{eq}{R_I + R_H} + \frac{eq}{R_I + R_H + r_{HF}}$$

$$= -\frac{\mu e}{(R_I + R_H)^2 + (R_I + R_H)r_{HF}} \qquad (5.76)$$

where μ is the dipole moment of HF,

$$\mu = q \times r_{HF} \qquad (5.77)$$

The denominator of equation (5.76) can be written in the form

$$(R_I + R_H)^2 + (R_I + R_H)r_{HF} = (R_I + R_H + \tfrac{1}{2}r_{HF})^2 - \tfrac{1}{4}r_{HF}^2 \qquad (5.78)$$

Since the last term is small, we can ignore it. Equation (5.76) then becomes

$$E_I \simeq -\frac{\mu e}{(R_H + R_I + \tfrac{1}{2}r_{HF})^2} \equiv -\frac{\mu e}{R^2} \qquad (5.79)$$

where R is the distance from the iodine nucleus to the midpoint of the HF bond. This is the expression we would get if we assumed that the moment of the HF molecule arose from a point dipole at the center of the bond. An extension of this argument shows that we should be able to approximate the *charge dipole* interaction between an ion and a polar molecule by similar expression, R being the distance from the electrical center of gravity of the ion to that of the polar molecule.

Charge–dipole interactions between ions and polar molecules are very large and are responsible for the ability of polar solvents to dissolve salts. In a salt, the ions are held together by extremely strong electrostatic interactions. These can be overcome, and the ions thus separated, only if the energies of solvation of the separable ions are correspondingly large.

Ion–dipole interactions vary rapidly with distance [equation (5.79)]. The most effective solvents are therefore those with dipoles of small radius.

In the case of anions, the best solvents are those of the type $H^{\delta+}-X^{\delta-}$, since hydrogen is the smallest of all atoms. In order to make the HX bond polar in the required sense, the atom in X adjacent to H must be halogen, oxygen, or a positively charged nitrogen or carbon atom. Other atoms, even neutral nitrogen, are insufficiently electronegative to give rise to suitably polar HX bonds. Nitrogen can conveniently be given a positive charge by attaching $+E$ groups, as in formamide

$$\overset{\delta-}{O}\text{---}CH\text{---}\overset{\delta+}{N}H_2$$

Formamide is isoconjugate with the allyl anion and the resulting allyl-type π delocalization gives nitrogen a positive charge. Carbon can be made positive by $+I$ substituents as in chloroform

$$H\text{---}\overset{\delta+}{C}(Cl^{\delta-})_3$$

Both formamide and chloroform can therefore solvate anions effectively, by *hydrogen bonding*.

Cations are best solvated by a special type of dipole, that arising from pairs of electrons in hybrid AOs (see p. 119). The negative end of such a dipole

$$\overset{\delta+}{X}\!\!\overset{\delta-}{\Longleftarrow\!\!\uparrow\downarrow}$$

can naturally approach the cation more closely than could a negatively charged atom. Dipoles of this type are found exclusively in compounds of nitrogen, oxygen, and fluorine since lone-pair electrons of elements later in the periodic table do not occupy hybrid AOs (p. 119). The effectiveness of oxygen in this connection can of course be augmented by giving it a negative charge. Formamide

$$\overset{\delta-}{O}\text{---}CH\text{---}\overset{\delta+}{N}H_2$$

is therefore effective at stabilizing both anions and cations.

Two polar molecules can likewise attract one another by *dipole–dipole* interaction. The best geometry for this is a linear arrangement of the dipoles, e.g.,

$$\overset{\delta^-}{F}\!\!-\!\!\overset{\delta^+}{H} \quad (\updownarrow) \quad \overset{\delta^-}{O}\!\!\!\underset{\diagdown H}{\overset{\diagup H}{\overset{\delta^+}{}}}$$

The attraction between two point dipoles varies as the inverse cube of the distance between them and so falls off even more rapidly than do charge–dipole interactions [equation (5.79)]. The strongest interactions are therefore those involving a hydrogen bond between a good anion solvent and a molecule with a strong hybrid AO dipole. Dipole–dipole interactions are smaller than charge–dipole interactions since the charges involved in a polar molecule are only fractions of the electronic charge.

An intermediate case is seen in neutral molecules where there is a large separation of charge, a zwitterion (such as $Me_3\overset{+}{N}$—CH_2—COO^-) being an extreme example. Here the molecule is regarded as having two localized charges, each of which undergoes charge–dipole interactions with the solvent.

The free energy of solvation by any of these electrostatic processes decreases as the size of the charges increases. This can be seen from a simple example, solvation at a spherical ion Y^- by a polar molecule HX (Fig. 5.10).

As X^- increases in size, the number of solvent molecules around it increases. The solvent molecules lie on a sphere of radius R [equation (5.79)] about the ion. Since the surface area of this sphere varies as R^2, so will the number of solvent molecules. However, the potential energy of attraction of each solvent molecule varies as R^{-2} [equation (5.70)]. The total attractive energy is therefore independent of R and so is independent of the size of the anion. On the other hand, the molecules of HX attached to Y^- are lined up with their dipoles parallel, so they will repel one another. The repulsion will increase with the number of molecules attached to Y^-, i.e., with the size of the anion. The energy of solvation is therefore less, the bigger is the anion. Furthermore, attachment of a solvent molecule to Y^- reduces its freedom of movement and so lowers the entropy. The decrease in entropy will again be greater, the greater is the number of solvent molecules bound to Y^- and so the larger is Y^-. The free energy of solvation of a large ion will therefore be less than that of a small one, both because the energy of solvation is less

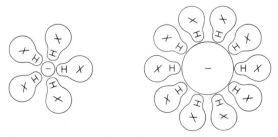

FIGURE 5.10. Solvation of anions of different size by a polar molecule HX.

and because the entropy of solvation is more negative. This argument also shows how compensating changes can occur in the heat and entropy of solvation (see p. 135). If we remove one or more solvent molecules from Y^-, the net decrease in solvation energy may be quite small since the remaining molecules of HX can spread themselves out so that there may be a large reduction in the HX repulsions. The resulting change in binding energy may then be compensated by the increase in entropy when one or two molecules of HX are set free from Y^-.

The arguments given above provide a further insight into the nature of substituent effects. If we treat the interaction between a substituent S and a substrate R in terms of perturbation theory, the first-order perturbation will correspond to a situation where the MOs of R and of S remain intact. The interaction between them will then be exactly analogous to the first-order interactions between molecules of solvent and solute and can likewise be treated by a classical electrostatic model. This is the theoretical basis of the field effect (p. 182). Other substituent effects correspond to higher-order perturbations in which the wave functions of R and S mix.

The second-order perturbation between two molecules is a sum of pair interactions of the type [see equation (1.21)]

$$P_{\mu\nu}^\pi = P_{\mu\nu}^2 / \Delta E_{\mu\nu} \tag{5.80}$$

where $P_{\mu\nu}$ is the interaction integral between MOs μ and ν due to the perturbation, and $\Delta E_{\mu\nu}$ is the difference in energy between the MOs. Since the interaction between two MOs raises one in energy by the same amount that it depresses the other, the net effect is zero if both MOs are filled with pairs of electrons. The only effective terms are therefore those that correspond to interactions between filled bonding MOs and empty antibonding MOs (p. 66).

The terms of equation (5.80) are of two types, those involving pairs of MOs in the same molecule and those involving pairs of MOs in different molecules. Interactions of the former type alter the electron distribution in the molecules of solvent or molecules of solute by mixing bonding and antibonding MOs, but they do not mix the MOs of the solvent with those of the solute. The latter type of interaction leads to solvent–solute orbital mixing. In other words, the first type of interaction leads to a polarization of the molecule in question, the latter to a definite bonding interaction between two different molecules.

If the net interaction between solvent and solute is small, their molecules will remain far apart since the van der Waals radii of molecules are relatively large. Their MOs will not therefore overlap significantly, so the corresponding terms $P_{\mu\nu}$ in equation (5.80) will be small. The main second-order effect therefore corresponds to a polarization of solvent and/or solute by the electric fields due to the other. Thus if the solvent is nonpolar and the solute is polar, the solute will remain unperturbed (since the solvent molecules have no net fields), but the solvent will be polarized by the fields of the polar solute molecules. The polarization is such as to lead to an attraction between solute

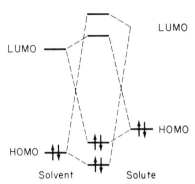

FIGURE 5.11. HOMO–LUMO interactions between molecules of solvent and solute.

and solvent. However, while *dipole–induced dipole* interactions of this kind contribute to the van der Waals forces between neutral molecules, they are relatively unimportant in the present connection since they are relatively small so variations in them during a reaction can be neglected. The same is true of the still smaller *dispersion forces** between neutral nonpolar molecules.

If the interaction between solute and solvent is large, we must include the terms in equation (5.80) that refer to interaction between MOs of the solvent and MOs of the solute. As we have already seen (p. 38), interactions between filled MOs have no effect on the total energy. The only relevant terms are those involving a filled, bonding MO of solvent and an empty, antibonding MO of solute or a filled, bonding MO of solute and an empty, antibonding MO of solvent. Furthermore, since the denominator of the perturbation term is equal to the difference in energy between the MOs, the interaction is small unless the MOs are close together in energy. The main interactions are therefore between the highest occupied MO (HOMO) of the solvent and lowest unoccupied MO (LUMO) of solute, and between the HOMO of the solute and LUMO of the solvent (Fig. 5.11). Interactions of this kind will be large if the HOMO of the solvent is close to the LUMO of the solute or if the HOMO of the solute is close to the LUMO of the solvent. Now from Koopmans' theorem, the energies of the HOMO and LUMO of a molecule are equal to *minus* its ionization potential and electron affinity, respectively. Thus the interactions will be large if the solvent has a low ionization potential and the solute a high electron affinity, or conversely. A cation with an empty valence shell AO must have a very high electron affinity be-

---

* A nonpolar molecule is nonpolar only in a statistical sense. At any instant, the random motion of the electrons will lead to a dipole moment, but this fluctuates as the electrons move around and its average value is zero. Thus a hydrogen atom at any given instant has a dipole moment since the electron and proton are at different points in space; as the electron moves around the proton, the direction and size of the dipole fluctuate, the average value being zero. When two neutral molecules are close together, the electrostatic interactions between them cause the electrons to correlate these motions so that the instantaneous dipole moments of the molecules lead to a net attraction. These attractive forces, due to electron correlation between adjacent molecules, are called *Heitler–London dispersion forces*, usually abbreviated to *dispersion forces*.

cause the binding energy of an electron in a valence shell AO of a neutral atom is high and that in a corresponding positive ion must be much greater. Cations should therefore interact very strongly with solvents with low ionization potentials, in particular molecules with unshared valence electrons, since such nonbonding electrons are necessarily less strongly bound than those in bonding MOs. Solvent–solute interactions of this kind are therefore very strong and lead to definite chemical bonding, i.e., coordination of the solvent to the cation. The only cations that may survive unbonded to solvent molecules are those where all the valence AOs are used in bonding, e.g., ammonium ions, R_4N^+. Here the LUMO is an antibonding MO and therefore of relatively high energy in spite of the positive charge.

The converse effect is seen in the case of anions dissolved in solvents of high electron affinity, anions having very low ionization potentials. None of the usual solvents have empty valence shell AOs, so here an antibonding MO is involved. For the energy of this to be low and the electron affinity of the solvent consequently to be high, it is necessary that the MO should be weakly antibonding and formed from AOs with large binding energies. These requirements are best met by π bonds formed from electronegative atoms. Thus nitromethane, CH_3NO_2, is a good solvating agent for anions.

These strong solvent–solute interactions are really of covalent type, involving interactions between specific orbitals. The number of solvent molecules that can be attached in this way to a given solute is therefore limited. Thus in the case of a carbonium ion dissolved in an alcohol ROH, two solvent molecules can be bound to the lobes of the empty $2p$ AO of the ion; namely

$$\begin{array}{c} a \quad b \\ R \diagdown \quad \diagdown \; \diagup \quad \diagup R \\ O \; \text{⬭} \; \overset{+}{C} \; \text{⬭} \; O \\ H \diagup \quad | \quad \diagdown H \\ c \end{array}$$

Caution should be exercised in using a mixture of MO and electrostatic descriptions of solvation. The Hellmann–Feynman theorem guarantees that the electrostatic and quantum mechanical desciptions will be equivalent; however, approximations using mixtures of the two descriptions may be worse than either the approximate electrostatic or the approximate molecular orbital description above.

We are now in a position to draw some general conclusions concerning the effects of solvents on rates of reaction.

The simplest case is that where the reactants and products are neutral and nonpolar. Here the reactants and products are solvated only by weak dipole–induced dipole interactions and dispersion forces. The effect of the solvent should therefore be small and the reaction should consequently take place at much the same rate in solution as in the gas phase. This is true for a number of reactions of hydrocarbons, e.g., the Diels–Alder dimerization of cyclopentadiene,

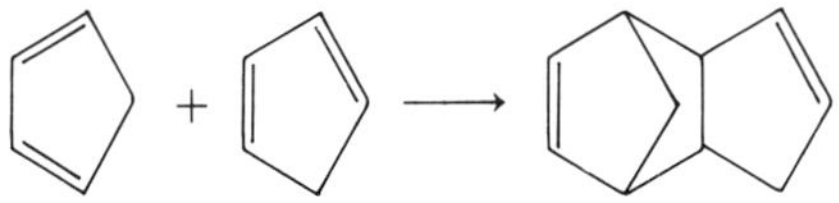

$$\text{cyclopentadiene} + \text{cyclopentadiene} \longrightarrow \text{dicyclopentadiene} \qquad (5.81)$$

and the Cope rearrangement of 3-phenyl-1,5-hexadiene,

$$\text{Ph-CH(CH=CH}_2\text{)CH}_2\text{CH=CH}_2 \longrightarrow \text{Ph-CH=CHCH}_2\text{CH}_2\text{CH=CH}_2 \qquad (5.82)$$

Next we have reactions where the reactants and products are neutral but polar. Here again solvents have no large effect but the reactions are faster in solution than in the gas phase and are faster, the more polar is the solvent. The reason for this is that the transition states of such reactions usually contain delocalized systems that are both more polar and more polarizable than the less delocalized systems in the reactants. A simple example is the Diels–Alder reaction of isoprene with maleic anhydride,

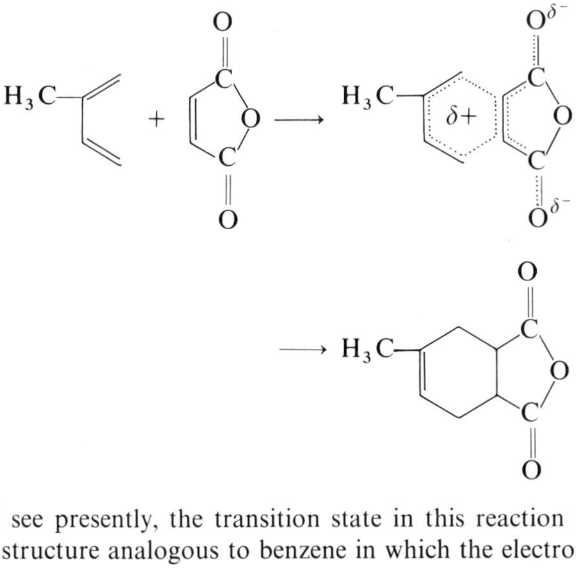

$$\text{isoprene} + \text{maleic anhydride} \longrightarrow [\text{TS with } \delta+ \text{ ring and } O^{\delta-}] \longrightarrow \text{adduct} \qquad (5.83)$$

As we shall see presently, the transition state in this reaction has a cyclic delocalized structure analogous to benzene in which the electrons are more easily polarized than in the reactants and in which there is a consequent increase in the charges on the carbonyl oxygens, due to a mesomeric interaction with the ring. The reaction is therefore faster, the more polar is the solvent.

In a third class of reaction, neutral reactants give rise to ionic products. A typical example is the S_N1 reaction of an alkyl halide, where the rate-determining step is its ionization; e.g.,

$$\text{Me}_3\text{CCl} \longrightarrow \text{Me}_3\text{C}^+ + \text{Cl}^- \qquad (5.84)$$

TABLE 5.2. Effects of Solvents on Rates of Reaction

Reactants	Products	Effects of polar solvent on rate
Neutral, nonpolar	Neutral, nonpolar	0
Neutral, polar	Neutral, polar	(+)
Neutral	Ionic	+ +
Ionic	Neutral	− −
Ionic	Ionic	−

Here the solvation energy of the reactant is very small, while that of the product is very large, particularly if the solvent is polar and so is capable of stabilizing ions efficiently. The solvation energy of the transition state will also be large since it is intermediate in structure between reactants and products. Reactions of this type therefore take place very much faster in solution than in the gas phase and their rate increases with the ability of this solvent to stabilize ions.

Conversely, reactions in which ionic reagents lead to neutral products are retarded by solvents and take place more slowly, the more polar is the solvent; for the energy of solvation of the transition state will be less than that of the reactants since it is part way to the products and so the charges in it are smaller. A good example is the reaction of trimethylsulfonium ion with ethoxide ion,

$$Me_3S^+ + EtO^- \longrightarrow Me_2S + MeOEt \tag{5.85}$$

Finally, we have reactions where both reactants and products carry similar charges. Here the transition state is also charged, but since it has a structure intermediate between those of the reactants and products, the charge in it is more dispersed. A good example is the Finkelstein reaction of iodide ion with an alkyl chloride RCl,

$$I^- + RCl \longrightarrow \overset{\delta^-}{I} \cdots R \cdots \overset{\delta^-}{Cl} \longrightarrow IR + Cl^- \tag{5.86}$$

Since the charge in the transition state is spread over a larger volume than that in the reactants, its energy of solvation is correspondingly less. Reactions of this type are therefore also retarded by an increase in polarity of solvent, though to a lesser extent than those in the previous category [cf. equation (5.85)]. This retardation is indeed responsible for the fact that S_N2 halide replacements occur slowly in solution via transition states that are stable intermediates in the gas phase [cf. equation (5.47)].

These conclusions are summed up in Table 5.2.

5.8 Limitations of the BEP Principle

Our derivation of the BEP principle was based on the assumption that the bond-breaking and bond-forming processes in a reaction occur inde-

Chemical Reactivity

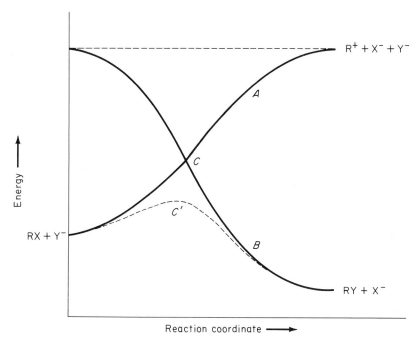

FIGURE 5.12. Interaction between bond breaking and bond forming in an S_N2 reaction.

pendently of one another, so that at the transition state the reaction suddenly switches from a bond structure corresponding to the reactants to one corresponding to the products. This, however, is not usually true, so we must now consider the consequences of an interaction between the bond-forming and bond-breaking processes.

We have indeed already considered a reaction where interactions of this kind occur, i.e., bimolecular nucleophilic (S_N2) substitution (Section 5.3). Consider, for example, the S_N2 reaction

$$Y^- + RX \longrightarrow YR + X^- \tag{5.87}$$

In a BEP plot for this (Fig. 5.12), the bond-breaking line A corresponds to the reaction of equation (5.41) and the bond-forming line B to equation (5.42). According to the BEP analysis, the transition state should correspond to the crossing point C of the two lines, the reaction at that point suddenly switching from a structure $Y^- + RX$ corresponding to the reactants to the structure $YR + X^-$ corresponding to the products. This, however, is not the case. In the transition state (see Fig. 5.2), *both* Y and X are attached to R. In other words, formation of the new bond YR begins while the old one is weakening and the transition state is correspondingly lowered in energy. The reaction therefore follows the broken line in Fig. 5.12, the true transition state C' lying well below the point C. Indeed, as we have seen [equation (5.47)], in certain cases at least, the interaction is so strong that the "transition state" becomes a stable intermediate, formed exothermically from $Y^- + RX$ in the gas phase.

Nevertheless, the S_N2 reactions of simple alkyl derivatives do obey the BEP principle and a number of authors have in the past discussed such reactions successfully in such terms. The reason for this is that the transition state $(Y \cdots R \cdots X)^-$ for such a reaction is genuinely intermediate in structure between the reactants and products, containing no features that are not present in one or the other. Any change in the reactants that alters the heat of reaction therefore tends to alter the energy of the transition state in proportion and such changes will of course conform to the BEP principle.

The way in which this compensation occurs can be seen clearly from a more detailed examination of the S_N2 reaction of equation (5.87). We can estimate the heat of reaction by dissecting it into steps in the following way:

$$R\text{---}X \longrightarrow R\cdot + \cdot X; \qquad \Delta E = E_{RX} \tag{5.88}$$

$$Y^- + \cdot X \longrightarrow Y\cdot + X^-; \qquad \Delta E = A_Y - A_X \tag{5.89}$$

$$Y\cdot + \cdot R \longrightarrow Y\text{---}R; \qquad \Delta E = -E_{RY} \tag{5.90}$$

Here E_{RX} and E_{RY} are the RX and RY bond energies and A_X and A_Y are the electron affinities of the radicals $X\cdot$ and $Y\cdot$. Adding the three equations, we find

$$Y^- + R\text{---}X \longrightarrow YR + X^-; \qquad \Delta E = (E_{RX} - E_{RY}) + (A_Y - A_X) \tag{5.91}$$

Now the bond energies are equal to *minus* twice the corresponding resonance integrals (β_{RX}, β_{RY}), and the electron affinities of X and Y are equal to minus the corresponding Coulomb integrals (α_X, α_Y). The energy of reaction ΔH can therefore be written as

$$\Delta E = 2(\beta_{RY} - \beta_{RX}) + (\alpha_X - \alpha_Y) \tag{5.92}$$

Next let us consider the transition state $(Y \cdots R \cdots X)^-$. Let us denote the RX and RY bond orders in this by $p_{RX}^\ddagger$ and $p_{RY}^\ddagger$ and the formal charges on X and Y by $q_{X^\ddagger}$ and $q_{Y^\ddagger}$. Since the transition state is isoconjugate with the allyl anion (Section 5.4), the negative charge in it will, to a first approximation, be confined to the terminal atoms. Hence

$$q_{X^\ddagger} + q_{Y^\ddagger} = -1 \tag{5.93}$$

Let us now calculate the activation energy ΔE, i.e., the difference in energy between the reactants and the transition state. The relevant charges and bond orders are shown in Fig. 5.13. Using first-order perturbation theory (p. 64), we find

$$\begin{aligned}\Delta E &= 2\beta_{RY}\,\delta p_{RY} + 2\beta_{RX}\,\delta p_{RX} + \alpha_Y\,\delta q_Y + \alpha_X\,\delta q_X \\ &= 2\beta_{RY}p_{RY} + 2\beta_{RX}(p_{RX}^\ddagger - 1) + \alpha_Y(1 - q_{Y^\ddagger}) + \alpha_X q_{X^\ddagger}\end{aligned} \tag{5.94}$$

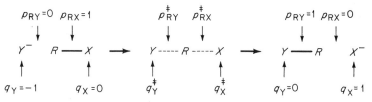

FIGURE 5.13. Formal charges and bond orders in reactants, transition state, and products of an S_N2 reaction.

Suppose now that we change one of the reactants somewhat and suppose that we calculate the corresponding changes in ΔE and $\Delta E^{\ddagger}$ by using first-order perturbation theory. In this approximation, the wave function is supposed to remain unchanged, so the bond orders ($p_{RX}^{\ddagger}, p_{RY}^{\ddagger}$) and charges ($q_X^{\ddagger}, q_Y^{\ddagger}$) in the transition state are assumed to be the same as before. Any change in the reactants corresponds to a change in the four parameters ($\beta_{RX}, \beta_{RY}, \alpha_X, \alpha_Y$) specifying them. Suppose first that we change the parameter β_{RX} by $\delta\beta_{RX}$, corresponding to a change $-2(\delta\beta_{RX})$ in the RX bond energy. Using equations (5.92) and (5.94), we can at once find the corresponding changes in ΔE and $\Delta E^{\ddagger}$.

$$\delta(\Delta E) = -2\,\delta\beta_{RX}; \qquad \delta(\Delta E^{\ddagger}) = 2\,\delta\beta_{RX}(p_{RX}^{\ddagger} - 1) \qquad (5.95)$$

Hence,

$$\delta(\Delta E^{\ddagger})/\delta(\Delta E) = 1 - p_{RX}^{\ddagger} \qquad (5.96)$$

The variations in activation energy for variations in the RX bond energy are therefore proportional to the corresponding variations in the energy (heat) of reaction, as the BEP principle requires. It is easily seen that the same conclusion follows for variation of each of the other parameters ($\beta_{RY}, \alpha_X, \alpha_Y$).

A change in one of the reactants, say RX, will correspond to a simultaneous change in two of the parameters (here β_{RX} and α_X). The corresponding changes in ΔE and $\Delta E^{\ddagger}$ are then

$$\delta(\Delta E) = -2\,\delta\beta_{RX} + \delta\alpha_X; \qquad \delta(\Delta E^{\ddagger}) = 2\,\delta\beta_{RX}(p_{RX}^{\ddagger} - 1) + q_X^{\ddagger}\,\delta\alpha_X \qquad (5.97)$$

Consider now the charges $q_X^{\ddagger}$ and $q_Y^{\ddagger}$ in the transition state. The closer the transition state is to the reactants, the stronger will be the RX bond and the less charge will have been transferred from Y^- to X. It is reasonable to suppose* that the fractional change in the bond strength, i.e., in p_{RX}, is the same as the fractional charge transfer; i.e.,

$$1 - p_{RX}^{\ddagger} = q_X^{\ddagger} \qquad (5.98)$$

If so,

$$\delta(\Delta E^{\ddagger})/\delta(\Delta E) = 1 - p_{RX}^{\ddagger} = q_X^{\ddagger} \qquad (5.99)$$

* This assumption can in fact be justified by a more detailed MO analysis.

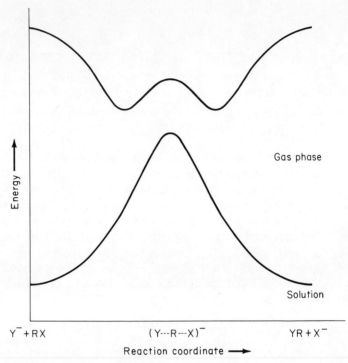

FIGURE 5.14. Reaction paths for an S_N2 reaction in solution and in the gas phase.

Thus the BEP principle still holds and a similar argument shows that the same is true for variations in the nucleophile Y^-.

This kind of argument can be extended to other reactions where the transition state is again genuinely intermediate between reactants and products. It can also be extended to reactions in solution because if the transition state is intermediate in structure between reactants and products, its solvation energy must be proportional to some average of their solvation energies. Thus in the S_N2 reaction of equation (5.87), the formal charge is shared between X and Y. The solvation energy due to attachment of solute molecules to X in the transition state will then be proportional to the solvation energy of the ion X^-, and the solvation energy due to attachment of solvent molecules to Y in the transition state likewise proportional to the solvation energy of Y^-. An interesting incidental further effect of solvation in this reaction is seen from Fig. 5.14. The solvation energies of X^- and Y^- are greater than that of the transition state $(Y \cdots R \cdots X)^-$ in which the charge is more dispersed; the solvation energy will therefore decrease during the first part of the reaction until the transition state is reached, and increase again thereafter. The effect is to neutralize the depression of the transition state due to interaction between bond formation and bond breaking (compare points C and C' in Fig. 5.11) and so to make the reaction path approxi-

Chemical Reactivity

mate more closely to the ideal BEP picture (compare the BEP path in Fig. 5.11 with the solution path in Fig. 5.14).

These arguments are based on the assumption that the transition state is truly intermediate in structure between the reactants and products. If this is not true, the arguments fail and so does the BEP principle. The main case when this happens is when electromeric substituents are attached to a carbon atom which is tetrahedral and saturated in both reactants and products but has sp^2 hybridization and forms part of a delocalized system in the transition state. Compare, for example, the S_N2 reactions of iodide ion with benzyl chloride and with a saturated alkyl chloride RCH_2Cl, where R is an alkyl group,

$$I^- + PhCH_2Cl \longrightarrow (I \cdots \underset{Ph}{\overset{H\diagdown \diagup H}{C}} \cdots Cl)^- \longrightarrow ICH_2Ph + Cl^- \quad (5.100)$$

$$I^- + RCH_2Cl \longrightarrow (I \cdots \underset{R}{\overset{H\diagdown \diagup H}{C}} \cdots Cl)^- \longrightarrow ICH_2R + Cl^- \quad (5.101)$$

Since the phenyl group is linked, in both reactants and products, to a saturated carbon atom, the intervening bonds are localized and should have similar bond energies. The heats of reaction should therefore be the same in both cases. However, the central carbon atom in the transition state has sp^2 hybridization and forms part of a delocalized allyl-like structure. The transition state in equation (5.100) is therefore stabilized both by the change in phenyl–carbon bond type (to sp^2–sp^2) and by the conjugative interaction between phenyl and the delocalized system in the transition state. Therefore benzyl chloride reacts faster with iodide ion than does a simple primary alkyl chloride, although the heats of reaction are similar.

Since the transition states in these Finkelstein reactions have negative charges on the entering and leaving halogens, they should be especially well stabilized by substituents of $+E$ type. This is indeed the case. Thus the reaction of chloroacetaphenone ($PhCOCH_2Cl$) with iodide ion is three orders of magnitude faster than that of the analogous but saturated β-phenylethyl chloride ($PhCH_2CH_2Cl$), the $+E$ carbonyl group in the former selectively stabilizing the transition state.

5.9 Classification of Reactions

Reactions were classified in Chapter 4 in terms of the structures of the reactants and products. Here we will introduce a corresponding classification based on relationships between the reactants and the transition state.

As we have seen (p. 205), the relative rates of a series of related reactions are determined by the corresponding changes in the delocalization energy of activation $\Delta E_{\text{deloc}}^{\ddagger}$, i.e., the difference in delocalization energy between the reactants and the transition state. The calculation of these differences by the PMO method is no different in principle from the corresponding calculation of ΔE_{deloc}, the delocalization energy of reaction. The essential distinction, so far as the PMO method is concerned, is once more the oddness or evenness of the conjugated systems being compared. We can therefore classify reactions by an O—E symbolism precisely analogous to that used in Chapter 4, the first symbol referring to the reactants and the second to the transition state. A superscript dagger is added at the end to show that the classification is based on reaction rates rather than equilibria. Thus in the $S_N 2$ reaction (Section 5.4), all the bonding electrons in the reactants and products are in two-center bonds. Both are therefore even systems (two is an even number), so the reaction is of EE type. In the transition state, however, the entering and leaving groups and the central carbon atom are linked by electrons in MOs delocalized over all three atoms. The reaction is therefore of $EO^{\ddagger}$ type.

Transition states of reactions can be further divided into two main classes.

In the first, the delocalized system in the transition state is isoconjugate either with that in the reactants or with that in the products. Such transition states involve no bonding that is not found either in the reactants or products and as a result they necessarily obey the BEP principle. This is apparent from the PMO approach, in which isoconjugate structures are essentially equivalent. The PMO calculations of ΔE_{deloc} and $\Delta E_{\text{deloc}}^{\ddagger}$ are therefore similar and the BEP principle appears as a necessary consequence of this correlation.

The $S_N 1$ reaction of an arylmethyl chloride $ArCH_2Cl$, where Ar is a $\pm E$ group, provides a good example. The rate-determining step in such a reaction is the ionization of the chloride,

$$ArCH_2Cl \longrightarrow ArCH_2^+ Cl^- \tag{5.102}$$

In the gas phase, the reaction would be endothermic throughout, the resulting ions recombining without activation. In a polar solvent, however, the ions are very strongly stabilized, whereas the solvation energy of the neutral chloride is small. During the reaction, as the C—Cl bond stretches and charges consequently build up ($ArCH_2^{\delta+} \cdots Cl^{\delta-}$), the solvation energy increases. However, the main contribution in the cation comes from bonding interactions with the solvent (see p. 228), which become appreciable only when the CCl bond has been greatly weakened. The reaction therefore passes through an energy maximum before reaching the ionic product, the maximum corresponding of course to the $S_N 1$ transition state. These relationships are indicated in Fig. 5.15.

The overall change in delocalization energy ΔE_{deloc} for the reaction of

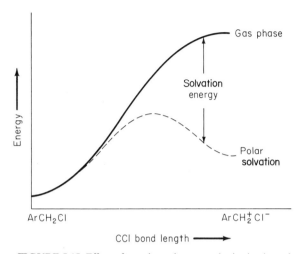

FIGURE 5.15. Effect of a polar solvent on the ionization of an arylmethyl chloride, ArCH$_2$Cl.

equation (5.102) was estimated in Chapter 3 (p. 104) as

$$\Delta E_{\text{deloc}} = 2\beta_{\text{CC}}(1 - a_{0r}) = 2\beta_{\text{CC}} - 2a_{0r}\beta_{\text{CC}} \quad (5.103)$$

where a_{0r} is the NBMO coefficient of the methylene carbon in the odd AH ArCH$_2$ and β_{CC} is the normal carbon–carbon π resonance integral. In the transition state for the reaction, Cl is still partly bonded to this carbon atom, leading to a structure isoconjugate with the arylethylene. This can be seen from Fig. 5.16. The overlap of the chlorine AO with the $2p$ AO of the methylene group is equivalent to the overlap of the two terminal $2p$ AOs in the arylethylene. The difference in delocalization energy between ArCH$_2{}^+$ and the transition state can therefore be found in the same way as the corresponding difference between ArCH$_2{}^+$ and ArCH=CH$_2$,

$$\text{ArCH}_2{}^+ + \text{Cl}^- \longrightarrow \text{Ar}\overset{\delta+}{\text{CH}}_2 \cdots \overset{\delta+}{\text{Cl}}; \quad \Delta E_{\text{deloc}} = 2a_{0r}\beta_{\text{CCl}} \quad (5.104)$$

where β_{CCl} is the C—Cl resonance integral in the transition state. We can now estimate the difference in delocalization energy between the reactant (ArCH$_2$Cl) and the transition state by combining equations (5.103) and (5.104),

$$\text{ArCH}_2\text{Cl} \longrightarrow \text{ArCH}_2^{\delta+} \cdots \text{Cl}^{\delta+}; \quad \Delta E_{\text{deloc}} = (2\beta_{\text{CC}} - 2a_{0r}\beta_{\text{CC}}) + 2a_{0r}\beta_{\text{CCl}}$$

$$= 2\beta_{\text{CC}} - 2a_{0r}(\beta_{\text{CC}} - \beta_{\text{CCl}}) \quad (5.105)$$

The transition states for a series of reactions of this type should have similar structures, so β_{CCl} should have the same value along the series. Thus equation (5.105) becomes

$$\text{ArCH}_2\text{Cl} \longrightarrow \text{ArCH}_2^{\delta+} \cdots \text{Cl}^{\delta-}; \quad \Delta E^{\ddagger}_{\text{deloc}} = 2\beta_{\text{CC}} - 2a_{0r}\beta \quad (5.106)$$

FIGURE 5.16. Topological equivalence of the transition state for $S_N 1$ reaction of $ArCH_2Cl$ and the corresponding arylethylene, $ArCH\!=\!CH_2$.

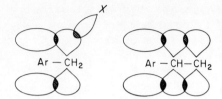

where β has a constant value along the series. This expression is clearly of exactly the same form as that for ΔE_{deloc} [equation (5.103)]. Moreover, if we change to another arylmethyl chloride where the NBMO coefficient at methylene is now a'_{0r}, the corresponding changes in ΔE_{deloc} and $\Delta E^{\ddagger}_{deloc}$ are given [see equations (5.103) and (5.106)] by

$$\delta(\Delta E_{deloc}) = (2\beta_{CC} + 2a_{0r}\beta_{CC}) - (2\beta_{CC} - 2a'_{0r}\beta_{CC})$$
$$= 2(a_{0r} - a'_{0r})\beta_{CC} \tag{5.107}$$

$$\delta(\Delta E^{\ddagger}_{deloc}) = (2\beta_{CC} + 2a_{0r}\beta) - (2\beta_{CC} - 2a'_{0r}\beta)$$
$$= 2(a_{0r} - a'_{0r})\beta \tag{5.108}$$

Thus

$$\frac{\delta(\Delta E^{\ddagger}_{deloc})}{\delta(\Delta E_{deloc})} = \frac{2(a_{0r} - a'_{0r})\beta}{2(a_{0r} - a'_{0r})\beta} = \frac{\beta}{\beta_{CC}} = \text{constant} \tag{5.109}$$

Therefore in this series of reactions, the change in energies of activation on passing from one to another will be a constant fraction of the corresponding change in the energy of reaction. The reactions therefore follow the BEP principle and their rates can be safely interpreted in terms of it.

Reactions of the second class have transition states in which the delocalized system is different both from that in the reactants and that in the products. Charges in the reactants are therefore liable to change the energy of the transition state in a way that bears no relation to the corresponding changes in the energies of the reactants or products. Such reactions do not therefore in general obey the BEP principle.

The $S_N 2$ reaction is a good example. As our discussion in Section 5.8 showed, substituents at the reaction center can greatly alter the rate of such a reaction without significantly changing the overall heat of reaction. This is because the transition state has an allyl-like delocalized system which is present neither in the reactants nor in the products. The central carbon atom has sp^2 hybridization in the transition state and can therefore conjugate with E-type substituents. No such interactions are possible in the reactants or products where that carbon atom is quadricovalent and has sp^3 hybridization.

BEP reactions, which follow the BEP principle, and for which the PMO calculations of ΔE_{deloc} and $\Delta E^{\ddagger}_{deloc}$ are consequently equivalent, are logically designated by the same symbol as in Chapter 4. If we wish to stress that our analysis is concerned with reaction rates rather than equilibria, we may add a superscript dagger (to show that we are interested in the transition state) and a subscript B (for BEP). Thus the reaction of equation (5.102), where the even arylmethyl chloride $ArCH_2Cl$ is converted to the odd cation $ArCH_2^+$, would be designated $EO_B^\ddagger$.

In the case of reactions which do not follow the BEP principle, the reaction symbol must refer explicitly to the transition state. We call such reactions anti-BEP reactions* and designate them by a subscript A (for anti). Thus the S_N2 reaction is described as $EO_A^\ddagger$.

In an anti-BEP reaction, the transition state contains some unique feature that is present neither in the reactants nor in the products. There are four main ways in which this can come about.

In the first, the transition state contains a conjugative interaction which is present neither in the reactants nor in the products. This usually arises from a change of hybridization of one or more atoms on passing from the reactants to the transition state, the change being reversed on passing from the transition state to the products. The S_N2 reaction is a typical example, the central carbon atom having sp^3 hybridization in the reactants and products but sp^2 hybridization in the trigonal bipyramidal transition state. Consequently, E-type substituents at that point can interact conjugatively with the delocalized system in the transition state but not with the localized bonds in the reactants or products.

In the second, the deviations are due to steric effects. Here the transition state contains some steric interaction which is absent in the reactants and products and which consequently raises the activation energy while leaving the heat of reaction more or less unchanged. A good example is the reluctance of t-butyl halides to undergo S_N2 reactions. In the transition state for such a reaction, e.g.,

$$I^- + \begin{array}{c} H_3C \\ H_3C \\ H_3C \end{array}\!\!\!>\!\!C\!-\!Cl \longrightarrow I\cdots\overset{\overset{\displaystyle CH_3}{|}}{\underset{\underset{\displaystyle H_3C\ CH_3}{}}{C}}\cdots Cl \longrightarrow I\!-\!C\!\!<\!\!\!\begin{array}{c} CH_3 \\ CH_3 \\ CH_3 \end{array} + Cl^- \quad (5.110)$$

the halogen—C—methyl bond angles are 90°, much less than in the tetrahedral reactant and product. The resulting increase in steric repulsion raises the activation energy, while the steric repulsions in the reactant and product cancel.

The third type of deviation is due to solvent effects specific to the transition state. Either the transition state is more polar than the reactants or products and so has a higher solvation energy, or steric effects are present which selectively hinder solvation in the transition state. S_N2 halide exchange

---

* Note that the prefix "anti" simply means "against." Anti-BEP reactions refuse to accept the restrictions of the BEP principle. They need not follow a course opposite to its predictions.

reactions are a good example, the transition state having a lower solvation energy than the reactants or products (p. 225).

The fourth type of anti-BEP reaction involves a transition state containing a cyclic delocalized system that is present neither in the reactants nor in the products. The classic example is the Diels–Alder reaction in which an 1,3-diene adds to ethylene or an ethylene derivation to form a cyclohexene,

$$\text{diene} + \text{ethylene} \longrightarrow [\text{cyclic TS}] \longrightarrow \text{cyclohexene} \tag{5.111}$$

Each carbon atom remains bound throughout to three neighbors by σ bonds. Three of its valence shell AOs are used in this way. The dotted lines in the cyclic transition state must therefore represent MOs formed by interaction of the remaining six carbon AOs, one from each atom. As we shall see in more detail, the cyclic interaction between them is topologically equivalent to the cyclic interaction of six $2p$ AOs in forming the π MOs of benzene. The transition state for the Diels–Alder reaction is therefore isoconjugate with benzene and should be stabilized accordingly. The reaction therefore takes place readily since the activation energy is correspondingly low. On the other hand, the analogous cyclodimerization of olefins, e.g.,

$$\begin{array}{c}\text{CH}_2 \\ \| \\ \text{CH}_2\end{array} + \begin{array}{c}\text{CH}_2 \\ \| \\ \text{CH}_2\end{array} \longrightarrow \begin{array}{c}\text{CH}_2\cdots\text{CH}_2 \\ \| \quad \| \\ \text{CH}_2\cdots\text{CH}_2\end{array} \quad \begin{array}{c}\text{CH}_2-\text{CH}_2 \\ | \quad | \\ \text{CH}_2-\text{CH}_2\end{array} \tag{5.112}$$

does not take place thermally because the corresponding transition state would be isoconjugate with cyclobutadiene and so antiaromatic. The resulting destabilization makes the activation energy so high that alternative reactions take place instead.

A large number of *pericyclic reactions* are now known in which the transition states have analogous cyclic delocalized structures and the ease with which they occur is determined in all cases by the aromaticity or antiaromaticity of the transition states. Since the aromatic energies of these are unrelated to the heats of reaction, the BEP principle fails.

Pericyclic reactions will be discussed in detail later (Section 5.25).

5.10 Prototropic Reactions of $ISO_B^\ddagger$ Type

Prototropic reactions involve the transfer of a proton from an acid HX to a base B,

$$\text{B} + \text{HX} \longrightarrow \text{BH}^+ + \text{X}^- \tag{5.113}$$

Reactions of this kind, involving the common acids and bases, are exceedingly fast and their rates have been measured only quite recently with the development of relaxation methods.* Here we will be concerned with prototropic reactions in which the rate of proton transfer is slow enough to be measured by more conventional techniques.

The simplest reactions of this kind are ones where the electrons of B, used to bind H in BH^+, initially occupy a localized AO and where the electrons forming the HX bond in HX end up in a similar localized AO in X^-; e.g.,

$$H_3\overset{+}{N}\!\!-\!\!H + :\overset{-}{O}H \longrightarrow H_3N + H\!-\!OH \qquad (5.114)$$

Reactions of this kind are clearly analogous to simple S_N2 reactions, H replacing the alkyl group and the 1s AO of hydrogen playing the role of the central carbon atom. Here there is no question of attaching substituents to the central atom (here hydrogen), so the arguments on p. 231 show that the rates should follow the BEP relation. Indeed, reactions of this type in polar solvents involve no specific activation. If the reaction is exothermic, its rate is limited only by the rate at which the reactants can diffuse together (i.e., it is *diffusion controlled*), while if it is endothermic, the activation energy is equal to the (positive) heat of reaction. Such reactions therefore obey the BEP relation of equation (5.69) with the proportionality factor α equal to unity. They take place slowly only if they are extremely endothermic, i.e., if the reaction involves an extremely weak acid or an extremely weak base.

This behavior is of course different from that of the S_N2 reaction, where solvation strongly destabilizes the intermediate phases so that specific activation is needed. The reason for this is that protic acids and bases can by their nature form strong hydrogen bonds with one another, $XH \cdots B$, analogous to those by which the polar solvent solvates one or both of them. Nothing is then lost when XH and B decide to solvate one another and in the resulting hydrogen-bonded complex $XH \cdots B$ only a small displacement of the proton is needed to give the correspondingly hydrogen-bonded product $(X^- \cdots HB^+)$. Since, moreover, there is no other change in bonding, the arrangements of solvent molecules around XH and B will be similar to those around X^- and BH^+. The reaction can therefore take place in the hydrogen-bonded HX—B pair without disturbing the surrounding solvent shell. The situation is entirely different in the case of an S_N2 reaction, where there is no significant attraction between a nucleophile Y^- and an alkyl derivative RX. Here energy is lost in desolvating both components, particularly the nucleophile, in order that the reactants can approach within bonding distance. Also the solvent shell around the reactants $Y^- + RX$ is very different from that around the

---

* The main problem in measuring the rates of very fast reactions is that the reactants cannot be mixed quickly enough, the reaction reaching equilibrium in the time it takes to mix them. Relaxation methods avoid this difficulty by using a premixed solution of the reactants at equilibrium. At a given instant, the conditions are suddenly changed; for example, one can discharge a capacitor through the solution so that it is suddenly heated by the pulse of current. In the new situation, the equilibrium constant has a different value from that in the original one, so the reactants are now no longer present at equilibrium concentrations. Reaction therefore takes place to restore equilibrium and one can follow the course of this by monitoring some property (e.g., light absorption) that changes during the reaction.

products YR + X⁻, so that considerable reorganization must take place during the reaction. The S_N2 reaction therefore involves intermediate phases of higher energy than either the reactants on the products, whereas simple prototropic reactions do not.

The energy of reaction of an acid HX with a given base B in the gas phase can be estimated by the following dissection of equation (5.113):

$$\begin{aligned} HX &\longrightarrow H\cdot + \cdot X; & \Delta H &= E_{HX} \\ B: &\longrightarrow B\cdot^+ + e; & \Delta H &= I_B \\ H\cdot + \cdot B^+ &\longrightarrow BH^+; & \Delta H &= -E_{BH^+} \\ X\cdot + e &\longrightarrow :X^-; & \Delta H &= -A_X \end{aligned} \quad (5.115)$$

Adding the four equations together, we find

$$B + HX \longrightarrow BH^+ + X^-; \quad \Delta H = (E_{HX} - A_X) - (E_{HB} - I_B) \quad (5.116)$$

The energies of reaction ΔE_i of a series of acids HX_i with a given base B are therefore of the form

$$\Delta E_i = \text{constant} + E_{HX_i} - A_{X_i} \quad (5.117)$$

The acid will therefore be stronger, the smaller is the HX_i bond energy E_{HX_i} or the lower is the electron affinity A_{X_i} of X_i. Thus acidity increases rapidly along the series $CH_4 < NH_3 < OH_2 < FH$ since the bond energies are not too different, while electronegativity rises very rapidly in the order $C < N < O < F$. Likewise, acidity increases in the series $HF < HCl < HBr < HI$ since the electronegativities of the halogens are similar while the HX bond energies decrease very rapidly.

In the case of reactions in solution, we will need to add to equation (5.117) the difference in solvation energy between X⁻ and HX. Thus a solvent that solvates X⁻ very efficiently, e.g., water or alcohol, will make the energy of reaction more negative, i.e., increase the acidity of HX, whereas a solvent that solvates HX but not X⁻ will have the opposite effect. This is seen very clearly in the case of water and alcohols, which alone are quite acidic, so that the conjugate bases are fairly weak. Water or an alcohol is an excellent solvent for anions such as HO⁻ or CH_3O^-. However, in dimethylsulfoxide, $(CH_3)_2SO$, commonly called DMSO, the reverse is true. DMSO can solvate acids very efficiently by hydrogen bonding, $Me_2SO \cdots HX$, whereas it is a poor solvent for anions. Bases such as HO⁻ and CH_3O^- are far stronger in DMSO solution than they are in water or methanol.

Prototropic reactions of simple acids with "normal" bases are slow only if they are extremely endothermic, i.e., if the acids are extremely weak. Since the reactions follow the BEP principle, one can then use the rates of proton transfer to a given base to estimate relative acidities in the case of acids so weak that normal acid–base equilibria cannot be observed. Streitwieser et al. have used this procedure to estimate the pK_A of a number of hydrocarbons; some of their results are shown in Table 5.3. Note the enormous

TABLE 5.3. Estimated Equilibrium pK_A for Hydrocarbons

Compound	Estimated pK_A
CH_4	47
CH_3—CH_3	48
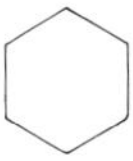 ⬡ (cyclohexane)	49
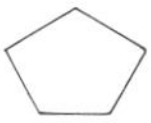 ⬠ (cyclopentane)	48
 ▢ (cyclobutane)	47
△ (cyclopropane)	44
CH_2=CH_2	42
 benzene	41
HC≡CH	25
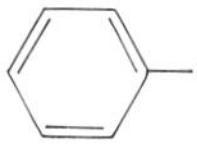 Ph—C≡CH	20

increase in acidity along the series paraffin < ethylene < acetylene. This is due to the increase in the binding energy of a hybrid AO with percentage of s character (see p. 41). The electron affinity of a carbon sp^3 hybrid AO (25% s) is less than that of an sp^2 hybrid (33% s) and this in turn is less than that of an sp (50% s). Note also the increase in acidity along the series cyclohexane < cyclopentane < cyclobutane < cyclopropane. Since the angle between two hybrid AOs is less, the less s character there is in them, there is a tendency for the CC bonds in small rings to be formed by hybrid AOs with less than the usual amount of s character. This means that the exocyclic bonds have correspondingly more s character, the effect being greater, the smaller is the ring. The CH bonds in cyclopropane are formed from carbon AOs which are estimated to be 29.4% s.

Prototropic reactions are also very endothermic, and consequently slow, if the proton acceptor is a very weak base. Such reactions involve no new principles since a very weak base is the conjugate base of a very strong acid and we have already considered the relation between structure and acid strength. Since reactions of this kind all follow the BEP principle, their activation energies should be linearly related to their energies of reaction. If now we have a very weak base whose conjugate acid reacts immediately to form some product, the rate of formation of this will be equal to the rate at which the base becomes protonated. The reaction should therefore show general acid catalysis (p. 216) and the activation energy of the catalyzed reaction should be linearly related to the energy of reaction, i.e., to the equilibrium constant for the reversible protonation. This in turn means that there should be a linear relation between the logarithm of the catalytic constant and the pK_A of the catalyzing acid. As we have seen (p. 216), such a relation (i.e., the Brønsted relation) does exist.

In the acids and bases so far considered, the electrons involved in prototropy occupy localized H—X σ MOs or localized AOs. This continues to be the case if conjugated systems are present, provided that these are insulated from the acidic center by intervening saturated atoms or that the HX bond lies in a nodal plane of the π system. Two examples of the latter type are included in Table 5.3, namely benzene and phenylacetylene. In each case, the lone-pair electrons in the conjugate base,

occupy an AO in the nodal plane of the π MOs and are therefore unable to interact with the π electrons. In reactions of this type, the conjugated system remains intact throughout. They are therefore classified as $ISO_B^\ddagger$. Other examples are acetic acid (CH$_3$C(=O)OH) and acetate ion (CH$_3$C(=O)O$^-$), both isoconjugate with the allyl anion (HC(=CH$_2$)CH$_2^-$), and phenol (PhOH) and phenate ion (PhO$^-$), both isoconjugate with the benzyl anion (PhCH$_2^-$).

In acids of this kind, the conjugated system can influence acid strength, and so the rate of proton transfer, only through changes in π energy due to changes in electronegativity of the conjugated atoms. Since ionization of HX leaves the HX bond electron in a lone-pair AO of the atom (Z) originally attached to hydrogen, the negative charge in X$^-$ is localized on atom Z. This will lead to a decrease in its electronegativity, reflected by a change $\delta\alpha_Z$ in its Coulomb integral. If, then, Z forms part of the conjugated system in HX or X$^-$, there will be a change in π energy ΔE_π on a conversion of HX to X$^-$ given [see equation (2.12)] approximately by

$$\delta E_\pi = q_Z \delta\alpha_Z \tag{5.118}$$

where q_Z is the π-electron density at atom Z in HX. This can as usual be set equal to the π-electron density q_Z' of the corresponding atom in the hydrocarbon isoconjugate with HX. If this hydrocarbon is alternant, q_Z' can of course be found very simply from the pairing theorem.

Two simple examples will indicate this procedure.

Consider first the heteroaromatic bases derived from even AH's by replacing carbon atoms by nitrogen; pyridine and quinoline are typical examples. In compounds of this type, $q_Z' = 1$ at each position. The change in π energy on ionization [equation (5.118)] is therefore the same for all, so all such bases should be similar in strength. This is the case, their pK_A varying by no more than one unit.

Next let us consider the three hydroxyl derivatives, methanol, phenol, and acetic acid. These are isoconjugate with CH_3^-, $PhCH_2^-$, and $CH_2\!=\!CH\!-\!CH_2^-$, respectively. The corresponding π-electron densities are 2, 11/7, and 3/2, respectively. Since $\delta\alpha$ in equation (5.118) is positive, atom 2 becoming less electronegative when it acquires a negative charge, the term $q_Z\,\delta\alpha_Z$ is positive and so acts to destabilize the anion. Consequently, the destabilization should be less, and the anion formed more easily, the smaller is q_Z. We would therefore expect acidity to increase in the order $CH_3OH < PhOH < CH_3COOH$, as indeed is the case.*

5.11 Prototropic Reactions of $EO_B^{\ddagger}$ Type

As we have seen (p. 174), the acidity of methane can be enormously enhanced by $\pm E$ or $+E$ substituents. Thus acetylacetone ($CH_3COCH_2COCH_3$) is a stronger acid than phenol. This increase in acidity is due to an increase in π binding energy on passing from the acid to its conjugate base. In the acid, the relevant hydrogen is attached to a saturated carbon atom that can play no part in conjugation. In the anion, this atom can adopt sp^2 hybridization, so that its doubly occupied AO, now the unhybridized $2p$ AO, can interact with an adjacent π system. The conversion of the acid to its conjugate base is thus an EO reaction, involving an extension of the even system in the acid by one extra atom.

Acids of this kind differ in one striking respect from those we have so far considered in that their deprotonations are always activated and so are relatively slow. Thus although acetylacetone is a stronger acid than phenol, its rate of reaction with hydroxide ion in water at room temperature is about 10^{16} times slower than that of phenol under the same conditions. Clearly, some quite new principle is involved.

---

* It would be unwise to try to use equation (5.118) in a quantitative way in cases such as this because the assumption that q_Z has the same value in HX, X$^-$, and the isoconjugate AH is a poor approximation. Thus, since oxygen is far more electronegative than carbon, the π-electron density on oxygen in phenol is very much greater than on the methylene group in $PhCH_2^-$. Likewise, q_Z will have significantly different values in HX and X$^-$.

The geometry of acetylacetone (3) differs significantly from that of the acetylacetonate ion (4). In (3), the central carbon atom is tetrahedral and the adjacent CC bonds are single. In (4), the central carbon atom is planar, while the adjacent CC bonds have a large π component and are shortened accordingly. When we start to remove a proton from (3) with base, the geometry initially corresponds to (5), not (4). The change in energy as the reaction proceeds is initially the same as it would be if we were forming an anion (5) with the anionic center pyramidal, not planar, and with the adjacent CC bond lengths having the value (1.50 Å) typical of localized sp^3–sp^2 C—C bonds. The structure can relax to that corresponding to the stable form (4) of the resulting anion only when the CH bond has been weakened. We can indeed represent the situation rather clearly by a BEP-type plot. In Fig. 5.17, line A represents the change in energy that would occur during the reaction if the geometry remained unchanged. This is analogous to the bond-breaking curve in a normal BEP plot. Curve B represents the change in energy of the resulting ion as its geometry relaxes. The actual course of the reaction will correspond to a composite of A and B. One can see that even though the overall reaction may be exothermic, as represented here, the reaction may nevertheless initially have to proceed uphill in energy since the stabilizing effect of mesomerism can only come into full play after the reaction has proceeded some way.

Many reactions are known where an analogous situation holds, i.e., where an atom changes its hybridization during the course of the reaction with a consequent change in delocalization energy. It is clear that an analysis similar to that of Fig. 5.17 will apply in all such cases and that specific activation should therefore always be required. Even if the reaction is exothermic, it should still require activation. This effect is further accentuated in solution by solvent effects. Thus the conjugative stabilization that makes acetylacetone a relatively strong acid involves a redistribution of charge. In passing from the reactants [i.e., base + (3)] to the product [i.e., (4)], extensive reorganisation of the solvent is required to stabilize the latter. This stabilization can come into play only when the charge is redistributed, i.e., only after the reorganization indicated by curve B in Fig. 5.17.

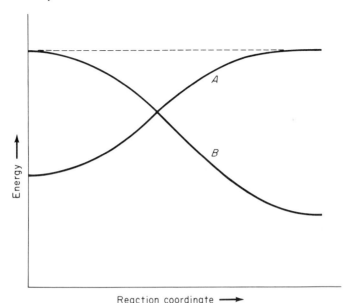

FIGURE 5.17. Plot of energy vs. a reaction coordinate; A, conversion of (3) to (5); B, conversion of (5) to (4).

The PMO treatment of these reactions can be seen best from a specific example. We will consider the deprotonation of arylmethanes, $ArCH_3$, where ArH is an even AH. The reaction of $ArCH_3$ with a standard base B gives an odd AH anion $ArCH_2^-$,

$$ArCH_3 + B \longrightarrow ArCH_2^- + BH^+ \qquad (5.119)$$

For a series of such reactions, the energies of reaction ΔE will be given (see p. 104) by

$$\Delta E = \Delta E_\pi + \text{constant} \qquad (5.120)$$

where ΔE_π is the π energy of reaction, i.e., the difference in π energy between $ArCH_3$ and $ArCH_2^-$. From equation (3.44), we have

$$\Delta E_\pi = 2\beta(1 - a_{0r}) \qquad (5.121)$$

where a_{0r} is the NBMO coefficient at the methylene carbon in $ArCH_2^-$. From equations (5.120) and (5.121), we have

$$\Delta E = \text{constant} - 2a_{0r}\beta \qquad (5.122)$$

Next let us consider the transition state (6) for the deprotonation. This is clearly isoconjugate with the arylallyl anion (7). The difference in delocalization energy between the reactant and the transition state (i.e., the delocalization energy of activation $\Delta E_{\text{deloc}}^\ddagger$) should be analogous to the difference in π energy between $ArCH_3$ and (7) and it can consequently be found by the following modification of the argument used on p. 104 (to find the difference in π energy between an even AH and an odd AH with one additional conjugated atom).

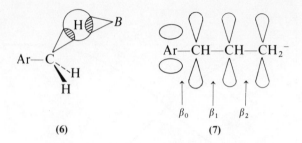

(6)　　　　　　　(7)

We start with three fragments, ArH, allyl, and methyl. We can unite these in two successive steps, to form ArCH=CH—CH=CH$_2$, in two different ways:

$$\text{ArH} \leftarrow\!\mathbf{u}\!\rightarrow \text{CH}_2\text{CHCH}_2 \xrightarrow{\delta E_1} \text{ArCHCHCH}_2 \leftarrow\!\mathbf{u}\!\rightarrow \text{CH}_3$$
$$\xrightarrow{\delta E_2} \text{ArCH=CH—CH=CH}_2 \quad (5.123\text{a})$$

$$\text{CH}_2\text{CHCH}_2 \leftarrow\!\mathbf{u}\!\rightarrow \text{CH}_3 \xrightarrow{\delta E_3} \text{CH}_2\text{=CH—CH=CH}_2 \leftarrow\!\mathbf{u}\!\rightarrow \text{Ar}$$
$$\xrightarrow{\delta E_4} \text{ArCH=CH—CH=CH}_2 \quad (5.123\text{b})$$

The total π-energy change must be the same in each case, so we have

$$\delta E_1 + \delta E_2 = \delta E_3 \quad (5.124)$$

since δE_4 vanishes, being the energy of union of two even AHs.

As a first approximation, we assume that all the resonance integrals in the delocalized system are equal ($= \beta$). The NBMO coefficients in allyl are then

$$\begin{array}{ccc} -1/\sqrt{2} & & 1/\sqrt{2} \\ \text{CH}_2\!\!-\!\!\text{CH}\!\!-\!\!\text{CH}_2 & & \end{array} \quad (5.125)$$

The energy of union with methyl [δE_3 in equation (5.124)] is then

$$\delta E_3 = 2\beta(1/\sqrt{2}) = \beta\sqrt{2} \quad (5.126)$$

The NBMO coefficients in (7) are clearly of the form

$$\begin{array}{ccc} \text{Ar}\!\!-\!\!\text{CH}\!\!-\!\!\text{CH}\!\!-\!\!\text{CH}_2 \\ Aa_{0i} \quad Aa_{0r} \quad -Aa_{0r} \end{array} \quad (5.127)$$

where a_{0i} is the corresponding coefficient in ArCH$_2$ and A is a normalizing factor,

$$\sum (Aa_{0i})^2 + (Aa_{0r}) = 1 \quad (5.128\text{a})$$

Since the coefficients a_{0i} are normalized, the sum of their squares is unity. Hence,

$$\sum (Aa_{0i})^2 = A^2 \sum a_{0i}^2 = A^2 \quad (5.128\text{b})$$

Consequently,

$$A^2 + A^2 a_{0r}^2 = 1$$

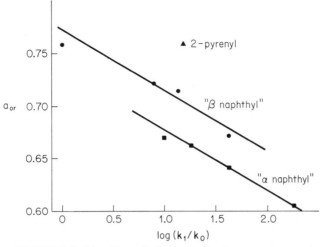

FIGURE 5.18. Plot of $\log_{10}(k_1/k_0)$ vs. a_{0r} for deprotonation reactions of an ArCH$_3$ series (data from Ref. 1).

or

$$A = 1/(1 + a_{0r}^2)^{1/2} \tag{5.129}$$

The energy of union of (7) with methyl to form ArCH=CH—CH=CH$_2$ [δE_2 in equation (5.124)] is then

$$\delta E_2 = 2(Aa_{0r})\beta = \frac{2a_{0r}\beta}{(1 + a_{0r}^2)^{1/2}} \tag{5.130}$$

Combining equations (5.124), (5.126), and (5.130), we find

$$\delta E_1 = \delta E_3 - \delta E_2 = \beta\sqrt{2} - \frac{2a_{0r}\beta}{(1 + a_{0r}^2)^{1/2}} \tag{5.131}$$

Since a_{0r}^2 is usually less than one-half, δE_1 is approximately a linear function of a_{0r}. This conclusion is not changed if one allows for the fact that the resonance integrals of the various bonds in (6) will not be equal, provided that the transition states are similar so that the CC, CH, and HB resonance integrals have constant values. Since δE_1 is a measure of the difference in π energy between ArH and the transition state (6), i.e., of ΔE_{deloc}, it follows that the activation energy $\Delta E^{\ddagger}$ should also be approximately a linear function of a_{0r}. Since we have already shown that the energy of reaction ΔE is a linear function of a_{0r} [equation (5.122)], there must be an approximately linear relation between the activation energy $\Delta E^{\ddagger}$ and the energy of reaction ΔE. The reaction therefore follows the BEP principle.* Since the reactant is even and the transition state odd, it is classed as $EO_B^{\ddagger}$.

---

* This is still true if the $\pm E$ aryl group is replaced by other groups of E type. The proof is straightforward, the effects of groups containing heteroatoms being deduced from those of isoconjugate AH substituents by the procedures indicated in Chapter 3; see, e.g., p. 109.

The activation energies, and so the logarithms of the rate constants, of deprotonation reactions of arylmethanes (ArCH$_3$) should [equation (5.131)] be linear functions of the NBMO coefficients a_{0r} of the methylene carbon in the corresponding odd AH anion (ArCH$_2^-$); Fig. 5.18 shows such a plot. It will be seen that, with one exception, the points lie on two parallel straight lines, one characteristic of compounds in which both the positions *ortho* to the methyl group in ArCH$_3$ are unsubstituted, the other to compounds in which one of the positions is taken up by a second fused ring [as in 1-methylnaphthalene, (8)]. This could be explained in terms of steric hindrance between the methylene group and the *peri* hydrogen in the anion ArCH$_2^-$; see (9). Part of this strain energy would also appear in the transition state which is intermediate between ArCH$_3$ and ArCH$_2^-$ in structure. The spacing of the lines in Fig. 5.18 corresponds to such a strain energy of 0.9 kcal/mole in the "α-naphthyl" compounds.

(8) (9)

Since these reactions follow the BEP principle, substituents that stabilize carbanions also accelerate their formation by deprotonation of corresponding substituted methanes. Substituents of $+E$ type are therefore especially efficient (see Table 4.2), acetone (CH$_3$COCH$_3$) and nitromethane (CH$_3$NO$_2$) undergoing deprotonation quite rapidly with bases even of moderate strength. Reactions of this kind are, however, very much slower than corresponding $ISO_B^\ddagger$ reactions of acids of comparable strength, so acids such as acetylacetone do not obey the Brønsted relation.

A further interesting consequence follows in the case of carbon acids where acidity is induced by a $+E$ substituent. In the conjugate base of such an acid, the carbanion interacts with the $+E$ substituent to form an odd conjugated system isoconjugate with an odd AH anion. Such an *ambident* anion can react with a proton at more than one point, so protonation may lead to an isomer of the original methane derivative.

Consider, for example, ethyl acetoacetate (10). All the bonds in the CH$_3$COCH$_2$ moiety of (10) are localized. However, when (10) reacts with base, it forms the mesomeric or ambident anion (11), which is isoconjugate with the pentadienate anion (13). Protonation of (11) may regenerate (10). On the other hand, it may lead instead to the isomeric enol (12). Thus (10) and (12) can be interconverted by deprotonation to the common anion (11) followed by reprotonation. The interconversion of isomers in this way, via deprotonation to a common ambident anion, is called *prototropic tautomerism*.

$$\underset{(10)}{CH_3-\overset{O}{\overset{\|}{C}}-CH_2-\overset{O}{\overset{\|}{C}}-OEt} \underset{H^+;\text{slow}}{\overset{B;\text{slow}}{\rightleftharpoons}} \underset{(11)}{CH_3-\overset{O}{\overset{\|}{C}}\cdots\overset{\ominus}{CH}\cdots\overset{O}{\overset{\|}{C}}-OEt} \underset{B;\text{fast}}{\overset{H^+;\text{fast}}{\rightleftharpoons}}$$

$$\underset{(12)}{CH_3-\overset{OH}{\overset{|}{C}}=CH-\overset{O}{\overset{\|}{C}}-OEt} \qquad \underset{(13)}{(CH_2\cdots CH\cdots CH\cdots CH\cdots CH_2)^-}$$

When acid is added to an alcoholic solution of a salt of the anion (11), the product first formed will be determined by the relative rates of proton attack at the various possible positions. Since attack on carbon to regenerate (10) involves a change in hybridization of the carbon atom in question, this reaction should require specific activation. Attack on oxygen, however, involves no specific activation since this is an $ISO_B^\ddagger$ reaction, the enol (12) still being isoconjugate with (13) and so with (11). Therefore protonation of (11) under conditions of kinetic control gives (12), not (10).

If (10) and (12) are allowed to equilibrate under conditions where deprotonation and reprotonation can both occur, the product obtained will then be the more stable one (thermodynamic control). The sum of the bond energies of the bonds in (10) is greater than that for (12); while the enol (12) is stabilized by mesomerism, it is still less stable than the ketone (10) in ethanolic solution. The equilibrium between (10) and (12) in ethanol therefore favors the ketone, only 5% of enol being present.

A similar situation arises in all cases of this kind. Protonation of such ambident anions on nitrogen or oxygen is always much faster than protonation on carbon. Under kinetic control, the product is therefore invariably an enol or analog of an enol, e.g., the *aci* isomer $CH_2=NOOH$ in the case of the conjugate base $(CH_2\cdots NO_2)^-$ from nitromethane. At equilibrium, i.e., under thermodynamic control, the more stable isomer is favored. In the case of simple ketones, this is the ketonic isomer. However, in the case of dimethyl oxaloacetate (14), the enol is the more stable form in ethanolic solution and it is therefore almost the sole product under conditions either of kinetic control or thermodynamic control.*

$$\underset{(14)}{CH_3OOC-CO-CH_2-COOCH_3} \qquad \underset{(15)}{CH_3OOC-\overset{OH}{\overset{|}{C}}=CH-COOCH_3}$$

---

* At one time, it was thought that the protonation of ambident ions always leads initially to the less stable isomer. This misconception (which can still be found in some texts) arose from the case of keto–enol tautomerism, where the enol is formed initially and where the ketone is usually the more stable isomer. However, the enol is still the initial product even in cases where it is more stable than ketone. Reaction on carbon is slow because it involves a change in hybridization, not because it leads to a more stable product.

The slowness of prototropic reactions involving a change in hybridization of carbon is a very general phenomenon. An interesting example is provided by (17), which is an aza derivative of the fluorenyl anion (19). If a solution of (17) in ether is shaken with cold dilute hydrochloric acid, it remains in the ether layer. However, (17) will dissolve in dilute hydrochloric acid on warming and it can then no longer be extracted with ether, even after neutralizing the solution. This apparently curious behavior is due to the slowness of protonation on carbon. Protonation of (17) on nitrogen, to form (16), would involve a large decrease in resonance energy (since it would destroy the aromaticity of two rings) and it would also incidentally involve a change in hybridization of the nitrogen atom. Consequently (17) cannot react with dilute acid in this way. Protonation to (18), on the other hand, is quite favorable since the resonance energy of (18) is greater than that of (17). This reaction is, however, so slow that it does not take place on shaking an etheral solution of (17) with cold acid. At equilibrium, however, (17) is completely protonated to (18); being a salt, (18) is much more soluble in water than in ether.

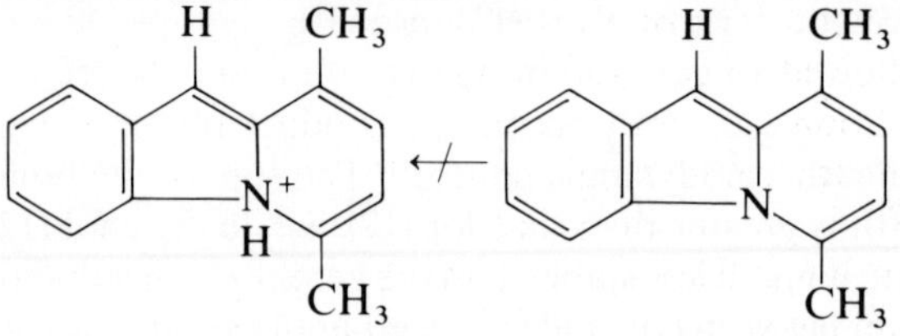

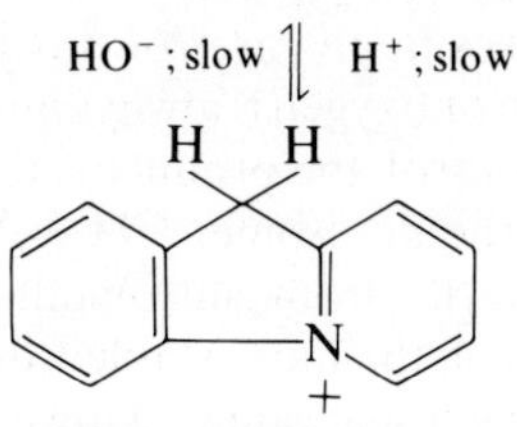

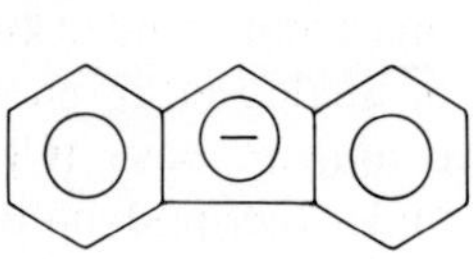

A number of substitution reactions can be brought about by reaction of these stabilized ambident anions with electrophiles. Thus acetone undergoes deuteration and bromination on treatment with base in presence of deuterium oxide or bromine,

$$CH_3COCH_3 + B \longrightarrow CH_3COCH_2^- \quad (5.132)$$

$$CH_3COCH_2^- + D_2O \longrightarrow CH_3COCH_2D + DO^- \quad (5.133)$$

$$CH_3COCH_2^- + Br_2 \longrightarrow CH_3COCH_2Br + Br^- \quad (5.134)$$

In these cases, the reaction of the electrophile with the intermediate anion is fast, so the rate of reaction is equal to the rate of formation of the anion, following equation (5.132). The rate of bromination is consequently independent of the concentration of bromine, i.e., is of *zeroth order* with respect to it.

Chemical Reactivity

The $+E$ activity of carbonyl can be greatly enhanced by protonation to $>\!\!C\!\!=\!\!\overset{+}{O}H$, O^+ being much more electronegative than neutral O. The acidifying effect of such a group on an adjacent CH bond is so great that it can be deprotonated quite rapidly by quite weak bases such as water. The enolization of ketones can therefore be brought about by acids as well as bases, the reaction being both base-catalyzed and acid-catalyzed. Since the enol can recombine with a proton to regenerate the original protonated ketone, acids can catalyze keto–enol tautomerism. The enol can likewise react with other electrophiles on carbon, so reactions such as bromination are also subject to acid catalysis; e.g.,

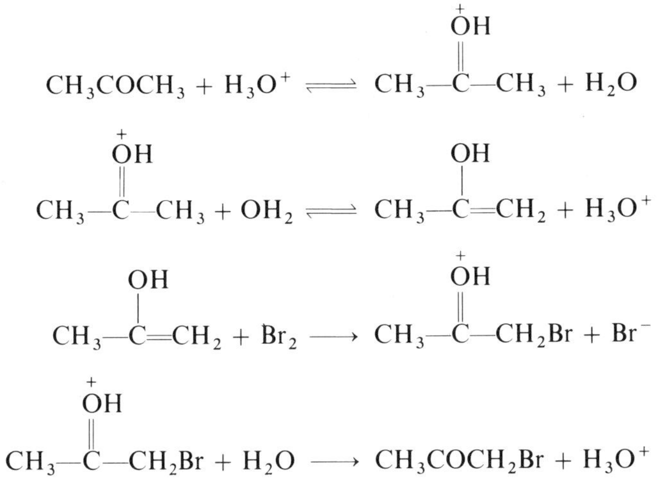

$$CH_3COCH_3 + H_3O^+ \rightleftharpoons CH_3-\overset{\overset{+}{O}H}{\underset{\|}{C}}-CH_3 + H_2O$$

$$CH_3-\overset{\overset{+}{O}H}{\underset{\|}{C}}-CH_3 + OH_2 \rightleftharpoons CH_3-\overset{OH}{\underset{|}{C}}=CH_2 + H_3O^+ \qquad (5.135)$$

$$CH_3-\overset{OH}{\underset{|}{C}}=CH_2 + Br_2 \longrightarrow CH_3-\overset{\overset{+}{O}H}{\underset{\|}{C}}-CH_2Br + Br^-$$

$$CH_3-\overset{\overset{+}{O}H}{\underset{\|}{C}}-CH_2Br + H_2O \longrightarrow CH_3COCH_2Br + H_3O^+$$

5.12 Nucleophilic Aliphatic Substitution

A. $S_N 1$ Reactions

Nucleophilic aliphatic substitution is a process in which a nucleophile Y^- displaces a group X from an alkyl derivative RX to form RY and X^-,

$$Y^- + RX \longrightarrow YR + X^- \qquad (5.136)$$

Many years ago, Hughes and Ingold showed that most reactions of this kind fall into two distinct categories. The rates of reactions of the first type (designated $S_N 1$) are independent of the concentration of the nucleophile Y^-. The rate-determining step must therefore involve RX alone and is in fact an ionization,

$$RX \longrightarrow R^+ + X^- \qquad (5.137)$$

The resulting cation R^+ combines with the nucleophile to form RY. Reactions of the second type ($S_N 2$) are bimolecular, their rates being proportional to the concentrations both of Y^- and of RX. These take place by a concerted one-step process which we have already considered (Section 5.3).

Ionization of, e.g., alkyl halides in the gas phase would be very endothermic. The ionization is greatly facilitated in solution by solvation of the resulting ions (p. 237). Since the ions are solvated much more efficiently than the intermediate phases of the reaction, the reaction passes through an intermediate energy maximum or transition state (see Fig. 5.15). In a series of analogous S_N1 reactions in which X and the solvent are kept constant while R varies, the transition states should have similar structures. The reaction should then be of BEP type. Since it leads to the conversion of a neutral, even compound into an odd cation, it is classed as $EO_B{}^\ddagger$.

We can regard RX as a derivative of CH_3X in which hydrogen atoms are replaced by substituents. The reaction should then be accelerated by substituents that stabilize $CH_3{}^+$ relative to CH_3X and retarded by substituents that have the opposite effect. These effects of substituents can be deduced qualitatively from Table 4.2. Thus $-I$ and $\pm E$ substituents should accelerate the reaction strongly and $-E$ substituents should accelerate it very strongly indeed. This is the case. Thus the rates of S_N1 reactions of alkyl halides increase in the order $CH_3X < CH_3CH_2X < (CH_3)_2CHX < (CH_3)_3CX$, methyl being a $-I$ group. Likewise, the rates of S_N1 reactions increase in the series $CH_3X < PhCH_2X < Ph_2CHX < Ph_3CX$, phenyl being a $\pm E$ group. Finally, α-chloroethers undergo S_N1 reactions with extreme ease, alkoxy being a $-E$ group. Indeed, powerful $-E$ groups may stabilize the cation to such an extent that it becomes a stable species. Thus α-chloroamines such as 2-chloro-N-methylpiperidine (20) do not exist as such, undergoing irreversible ionization to salts [e.g., (21)].

The effect of E-type substituents on the rates of S_N1 reactions can be estimated by the PMO method. Consider, for example, the S_N1 reactions of a series of compounds of the type $ArCH_2X$, where ArH is an even AH. The energy of ionization ΔE to $ArCH_2{}^+ + X^-$ is given (see p. 104) by

$$\Delta E = \text{constant} + \Delta E_{\text{deloc}} \quad (5.138)$$

where ΔE_{deloc} is the difference in delocalization energy between the even AH ArH and the odd AH cation $ArCH_2{}^+$. From equation (3.44), we have

$$\Delta E_{\text{deloc}} = 2\beta(1 - a_{0r}) \quad (5.139)$$

where a_{0r} is the NBMO coefficient of the methylene carbon of $ArCH_2$.

Chemical Reactivity

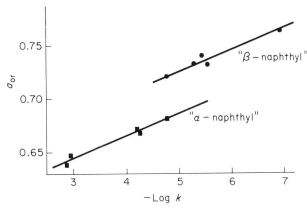

FIGURE 5.19. Plot of $-\log_{10}k$ against a_{0r} for the solvolysis of arylmethyl chlorides in moist formic acid at 25°. Data from Ref. 2.

From equations (5.138) and (5.139), we have

$$\Delta E = \text{constant} - 2a_{0r}\beta \tag{5.140}$$

Since the reaction obeys the BEP principle, the activation energy $\Delta E^{\ddagger}$ is given by

$$\Delta E^{\ddagger} = \text{constant} + \alpha\,\Delta E = \text{constant} - 2a_{0r}\alpha\beta \tag{5.141}$$

where α is the BEP proportionality factor [equation (5.69)]. Since the activation energy is proportional to the logarithm of the rate constant k, a plot of log k against a_{0r} should be linear.

Figure 5.19 shows such a plot for the S_N1 solvolyses of a series of arylmethyl chlorides in moist formic acid. It will be seen that the points lie on two parallel straight lines. As in the corresponding plot for deprotonation of arylmethanes (Fig. 5.18), one of the lines corresponds to "α-naphthyl-type" compounds and the other to "β-naphthyl-type" compounds. Here again the difference can be attributed to steric strain in the cation $ArCH_2^+$ due to interactions between the methylene group and the *peri* hydrogen [see formula (9)]. The spacing of the lines would then correspond to a strain energy in the "α-naphthyl-type" transition states of 2.3 kcal/mole.*

The effects of other substituents can be deduced in the usual way, as the following examples show:

$$\text{ClCH}_2-\underset{\underset{\text{CH}_3}{|}}{\overset{\overset{\text{CH}_3}{|}}{\text{C}}}-\text{Cl} \quad \text{reacts more slowly than} \quad \text{CH}_3-\underset{\underset{\text{CH}_3}{|}}{\overset{\overset{\text{CH}_3}{|}}{\text{C}}}-\text{Cl}$$

(retardation by a $+I$ substituent, i.e., Cl).

---

* It should be added that this explanation is by no means certain since the values of ΔE_{deloc} calculated by a more elaborate MO procedure led to plots analogous to Figs. 5.18 and 5.19 in which all the points lie on a single line. See Ref. 3.

PhCOCH$_2$Cl reacts more slowly than CH$_3$Cl
(retardation by a small $+E$ substituent; see p. 172).

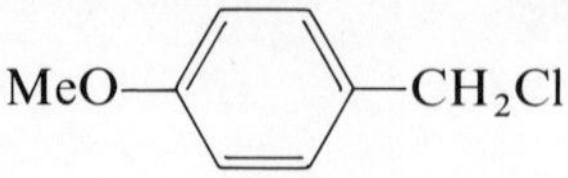

MeO—C$_6$H$_4$—CH$_2$Cl reacts faster than PhCH$_2$Cl

(PhCH$_2^+$ is stabilized by a *para* $-E$ substituent).

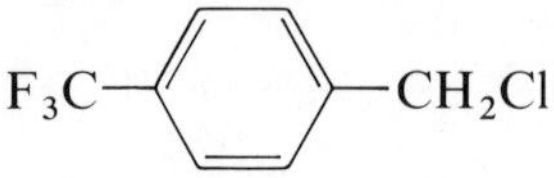

F$_3$C—C$_6$H$_4$—CH$_2$Cl reacts more slowly than PhCH$_2$Cl

(PhCH$_2^+$ is destabilized by a $+I$ substituent).

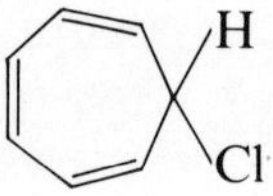

[cycloheptatriene–HCl] exists only as the salt [tropylium$^+$] Cl$^-$

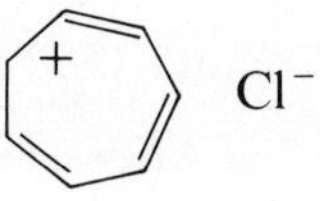

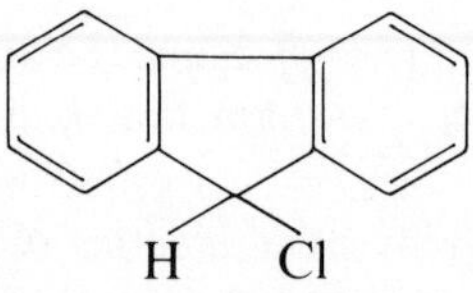

(tropylium is aromatic).

[9-chlorofluorene with H, Cl] reacts more slowly than [Ph$_2$CHCl with H, Cl]

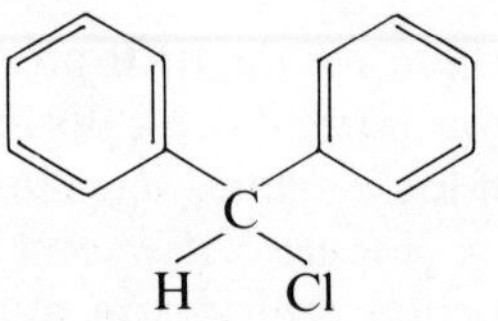

(the central ring in [fluorenyl cation] is antiaromatic).

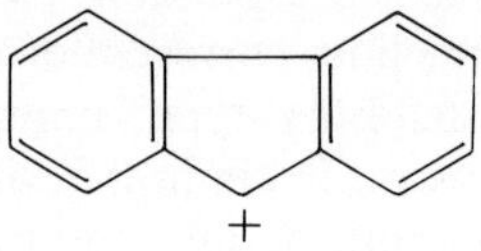

B. $S_N 2$ Reactions

The stereochemistry of the $S_N 2$ reaction has already been discussed (Section 5.3). The $S_N 2$ reaction of a nucleophile Y$^-$ with an alkyl derivative RX takes place in a one-step process via a trigonal bipyramidal transition state (Fig. 5.2). We have also seen that reactions of this type are anti-BEP reactions of class $EO_A^\ddagger$. The activation energy must therefore be found by comparing the energies of the reactants and the transition state.

Let us first examine the $S_N 2$ reactions of the compounds considered in the previous section (p. 255), i.e., arylmethyl derivatives ArCH$_2$X where ArH is an even AH. Let us consider their $S_N 2$ reactions with the nucleophile X$^-$; this could, for example, correspond to chloride exchange between an

Chemical Reactivity

arylmethyl chloride and radioactive chloride ion. The transition state should then have the symmetric structure (22).

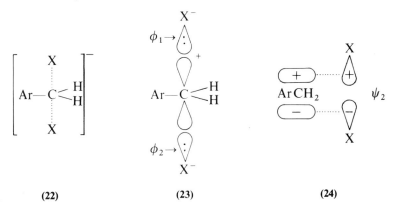

(22)   (23)   (24)

We can form (22) from the cation $ArCH_2^+$ and the ions X^- (23), the relevant AOs of X^- (ϕ_1 and ϕ_2) being doubly occupied. Let us assume as a first approximation that X has the same Coulomb integral as carbon. In that case, there will be first-order perturbations between the AOs of X and the NBMO of $ArCH_2$ since all three will be degenerate.

We can treat this threefold interaction by a trick analogous to that used in the discussion of substituents such as nitro and carbethoxy (p. 179). The transition state (22) has a plane of symmetry corresponding to the π nodal plane of $ArCH_2$ in which the CH_2 group lies. All the MOs of the transition state must be either symmetric or antisymmetric for reflection in this plane. Since the AOs ϕ_1 and ϕ_2 do not have this property, we replace them by an equivalent pair of symmetry orbitals (p. 21) ψ_1 and ψ_2 given by

$$\psi_1 = (\phi_1 + \phi_2); \quad \psi_2 = (\phi_1 - \phi_2) \quad (5.142)$$

Of these, ψ_1 is symmetric with reflection in the plane of symmetry, while ψ_2 is antisymmetric. Now the π MOs of $ArCH_2$ are antisymmetric; they can therefore combine only with ψ_2, not with ψ_1. Consequently, ψ_1 remains as a MO of the transition state. Since the two X atoms are too far apart for their AOs to interact, ψ_1 is a nonbonding MO, having the same energy as one of the AOs (ϕ_1 or ϕ_2) of X. The other symmetry orbital (ψ_2) can interact in π fashion (23) with the NBMO of $ArCH_2$. The interaction is exactly analogous to that between the NBMO of $ArCH_2$ and the $2p$ AO of methyl in union of $ArCH_2\cdot$ with $\cdot CH_3$ to form $ArCH=CH_2$. The resulting first-order interaction therefore leads to a pair of MOs, one bonding and one antibonding, spaced from the central level by $2a_{0r}\beta^{\ddagger}$, where $\beta^{\ddagger}$ is the resonance integral for the $C-X_2$ interaction (Fig. 5.20).

The difference in energy between $ArCH_2X$ and $ArCH_2^+X^-$ is simply the energy of reaction ΔE for the corresponding S_N1 reaction; this is given [see equation (5.139)] by

$$\Delta E = \text{constant} - 2a_{0r}\beta \quad (5.143)$$

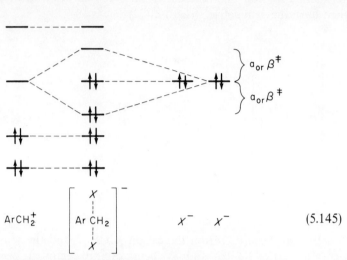

(5.145)

FIGURE 5.20. Combination (not union!) of $ArCH_2^+$ with $2X^-$ to form the transition state $(ArCH_2X_2)^-$.

where a_{0r} is the NBMO coefficient of the methylene carbon in $ArCH_2$. The difference in energy δE between $ArCH_2^+$ and the transition state is given (see Fig. 5.20) by

$$\delta E = 2a_{0r}\beta^{\ddagger} \tag{5.144}$$

We can then find the activation energy ΔE for the $S_N 2$ reaction from the following cycle:

$$ArCH_2 + X^- \xrightarrow{\Delta E} ArCH_2^+ + 2X^-$$

$$\Delta E^{\ddagger} \searrow \begin{bmatrix} X \\ | \\ ArCH_2 \\ | \\ X \end{bmatrix}^- \nearrow \delta E \tag{5.145}$$

Clearly,

$$\Delta E^{\ddagger} = \Delta E + \delta E = \text{constant} - 2a_{0r}(\beta - \beta^{\ddagger}) \tag{5.146}$$

For a series of such reactions, the transition states should have similar structures. Hence $\beta^{\ddagger}$ should have the same value throughout, so equation (5.146) can be written

$$\Delta E^{\ddagger} = \text{constant} - 2a_{0r}\beta_{\text{eff}} \tag{5.147}$$

where

$$\beta_{\text{eff}} = \beta - \beta^{\ddagger} \tag{5.148}$$

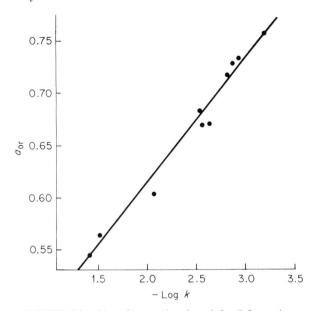

FIGURE 5.21. Plot of a_{0r} against $\log_{10} k$ for $S_N 2$ reactions of arylmethyl chlorides with iodide in dry acetone. The rate constants are from Ref. 4.

Equation (5.148) is of exactly the same form as the expression for the corresponding $S_N 1$ reaction [equation (5.141)]. A plot of $\log k$ against a_{0r} should again be linear. This indeed is the case, as the example in Fig. 5.21 shows.

Thus both the $S_N 1$ and the $S_N 2$ reactions show similar variations in rate with changes in the aryl group. The plots of Figs. 5.19 and 5.21 do, however, differ in two respects. First, the slope of the line in Fig. 5.21 is less than that of either line in Fig. 5.19. Second, all the points in Fig. 5.21 lie on a single line regardless of whether Ar is of "α-naphthyl" or "β-naphthyl" type.

The ionization of $CH_3 X$ to CH_3^+ and X^- will be facilitated by any factor that stabilizes the trigonal CH_3^+ ion relative to the tetrahedral $CH_3 X$. This can be done either by introducing $\pm E$ substituents or by using the lobes of the empty $2p$ AO of CH_3^+ to bind nucleophiles. The more CH_3^+ is stabilized by one of these factors, the less it will gain from the other.* This can be seen from equations (5.144), (5.147), and (5.148). Thus the more active Ar is

---

* The combination of CH_3 with X_2^- is, as we have seen, analogous to the formation of a π bond when CH_3^+ and CH_3^- unite to form $CH_2=CH_2$. The group $C-X_2^-$ in the transition state is, in effect, an even group and the interaction between it and Ar in $ArCH_2 X_2^-$ will be correspondingly less than in $ArCH_2^+$, the difference being greater, the stronger is the $C-X_2^-$ interaction. Conversely, the energy of combination of X_2^- with $ArCH_2^+$ is less $(2a_{0r}\beta^{\ddagger})$ than with CH_3^+ $(2\beta^{\ddagger})$, the difference being greater, the smaller is a_{0r}, i.e., the greater is the $\pm E$ activity of the aryl group. Thus the stronger the Ar—C interaction, the weaker is the $C-X_2^-$ interaction, and conversely.

as a $\pm E$ substituent, the smaller will be the coefficient a_{0r} at the methylene carbon in $ArCH_2$ and the smaller will be the change in energy through interaction with X_2^- [equation (5.144)]. Likewise the stronger the $C-X_2^-$ interaction, the greater is $\beta^{\ddagger}$ and so the smaller is β_{eff} [equation (5.148)]. The variation in $\Delta E^{\ddagger}$ with changes in a_{0r}, i.e., with changes in the aryl group, is then correspondingly less [equation (5.147)]. In the case of typical $S_N 2$ reactions, the $C-X_2^-$ interaction is likely to be strong. Indeed, we have seen that such species occur as stable entities in the gas phase (p. 234). The value of β_{eff} is then likely to be much less than $\alpha\beta$ in equation (5.141), accounting for the fact that the slope of the line in Fig. 5.21 is less than that of the line in Fig. 5.19. Furthermore, if the interaction between Ar and $CH_2 X_2^-$ is small, little will be lost by rotating the aryl group so as to avoid the unpleasant *peri* interaction in "α-naphthyl-type" transition states. This would account for the lack of distinction between "α-naphthyl" and "β-naphthyl" types in Fig. 5.21.

These results also throw further light on the relation between the $S_N 1$ and $S_N 2$ reactions. We have seen that positive ions tend to undergo covalent solvation by solvent molecules that have pairs of unshared electrons available for dative bonding. One would certainly expect this to be the case for a carbonium ion. One would therefore expect carbonium ions in polar solvents to be covalently bound in this fashion to pairs of solvent molecules, the resulting structure (25) being precisely analogous to that in an $S_N 2$ transition state. The available evidence shows that this is in fact the case, the only exceptions being carbonium ions with α heteroatoms (p. 254), such as (21), which are stabilized to such an extreme extent that they exist as stable ionic species in solution. These ions can hardly be classed as carbonium ions, since they are better represented by classical structures [cf. (21)] in which the positive charge is on a heteroatom, not carbon.

$$\underset{(25)}{\overset{a\ \ b}{\underset{c}{S\cdots C^+ \cdots S}}} \qquad \underset{(26)}{S:\overset{\frown}{R-}X} \longrightarrow \underset{(27)}{\overset{\delta^+}{S}\cdots R\cdots \overset{\delta^-}{X}}$$

Our analysis of the $S_N 2$ reaction assumed that the nucleophile and leaving group are the same. It is easily shown that similar conclusions hold if they are different, $\beta^{\ddagger}$ in equation (5.148) now being a measure of the net binding to both in the transition state. The analysis in the previous paragraph then suggests that the $S_N 1$ reaction is no different in kind from an $S_N 2$ reaction, involving a bimolecular attack by the solvent acting as a nucleophile, (26) $\longrightarrow$ (27).

Suppose now that we carry out the solvolysis of our arylmethyl chlorides in solvents of increasing nucleophilicity, i.e., tendency to combine covalently with positively charged carbon.* The transition states should then show a

---

* Nucleophilicity is discussed in detail in Section 5.13.

Chemical Reactivity

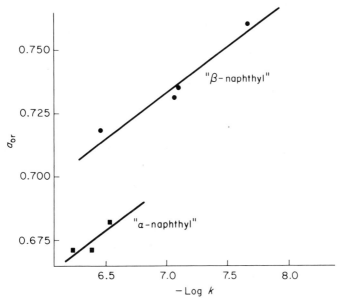

FIGURE 5.22. Plot of $\log_{10} k$ against a_{0r} for solvolysis of arylmethyl chlorides in formic acid/water/dioxane. Rate constants are from Refs. 1 and 2.

steady increase in the role of covalent bonding to the solvent. Therefore while plots of $\log k$ vs. a_{0r} should still be linear and show a distinction between "α-naphthyl" and "β-naphthyl" types (cf. Fig. 5.19), the slopes of the lines should decrease, and the α/β separation should also decrease, as the solvent becomes more nucleophilic. It should indeed be possible to find a complete spectrum, from plots corresponding to extreme $S_N 1$-type reactions (e.g., Fig. 5.19) to extreme $S_N 2$ type reactions (e.g., Fig. 5.21). This is the case and indeed it provides the best available evidence for the continuity of mechanism in S_N reactions. A typical example is shown in Fig. 5.22. Water and dioxane are both more nucleophilic than formic acid. The slopes of the plots for solvolysis in a formic acid/water/dioxane mixture are, as expected, less than in Fig. 5.21 and the spacing of the lines is also smaller.

As we stabilize the carbonium ion CH_3^+ by introducing substituents, we make its formation from $CH_3 X$ less endothermic. Since the $S_N 1$ reaction is a BEP reaction, the transition state must then become increasingly like the reactants in structure and the interaction between the nucleophile and substrate must correspondingly decrease. Changes in the nucleophile will then have correspondingly smaller effects on the rate of reaction. If the differences in reactivity between different nucleophiles become sufficiently small, reaction with the solvent may predominate since it is present in much higher concentration than any added nucleophile. The reaction of an alkyl derivative RX with a nucleophile Y^- will then be of zeroth order in Y^-, since the rate-determining step is a nucleophilic attack on RY by the solvent.

The product of the reaction may still be RY if the solvated carbonium ion (28) survives long enough for Y^- to replace one of the solvent molecules,

$$(S \cdots R \cdots S)^+ + Y^- \longrightarrow \overset{\delta^+}{S} \cdots R \cdots \overset{\delta^-}{Y} \longrightarrow S + RY$$
$$(28) \qquad\qquad\qquad (28a) \qquad\qquad (28b)$$

There should then be no hard and fast distinction between S_N1 and S_N2 reactions and none indeed is observed. At one extreme, we have reactions where participation by the nucleophile in the transition state is so small that no nucleophile can compete with the solvent; such reactions have been described as being of "limiting" S_N1 type. At the other extreme, we have reactions where the interactions with the nucleophile are so strong that the solvent can be outbid by nucleophiles only slightly stronger than itself. These are the purest S_N2 reactions. In between, we can find cases where the competition becomes progressively more even. These are the reactions of "mixed" type. A good example is the reaction of benzyl chloride in aqueous ethanol containing hydroxide ion. At low concentrations of hydroxide, the solvent wins, so the reactions are of zeroth order with respect to HO^-. As the concentration of hydroxide increases, an increasing amount of reaction takes place through bimolecular attack by HO^-. The reaction does not therefore show the characteristics either of a pure S_N1 process or a pure S_N2 process.

Since the dominant factor in S_N1 reactions is the stabilization of the intermediate carbonium ion by substituents, the fact that the ion is not really free does not alter the conclusions reached in Section 5.12A. Reactions of this type are facilitated by $-I$, $\pm E$, and $-E$ substituents and retarded by $+I$ or small $+E$ substituents.

In the case of S_N2 reactions, the effect of substituents depends on their ability to stabilize transition states in which the carbonium ion is strongly bonded to the nucleophile and leaving groups. We have seen that $\pm E$ groups can do this. It is easily seen that $+E$ groups should be even more effective, a conclusion drawn earlier (p. 259) on intuitive grounds. Just as a $\pm E$ substituent can be turned into a $+E$ substituent by introduction of heteroatoms that are more electronegative than carbon (p. 172), so likewise should it be turned into a $-E$ substituent by replacement of carbon with less electronegative atoms. As we have seen, the group $C-X_2^-$ in the transition state (22) is isoconjugate with vinyl [see (24)]. However, X_2 must be less electronegative than carbon, in view of the negative charge. Thus $C-X_2^-$ should therefore behave as a $-E$ group and the interaction between it and a $+E$ group should be stronger than between it and an isoconjugate $\pm E$ group, for the interaction between a $+E$ and a $-E$ group leads to mutual conjugation (p. 180). Thus $+E$ substituents should facilitate S_N2 reactions. We have already noted that compounds such as α-chloroketones (e.g., $PhCOCH_2Cl$) undergo S_N2 reactions very rapidly. Another example is the acceleration of the S_N2 reactions of benzyl chloride ($PhCH_2Cl$) by $+E$ substituents in the *para* position, *p*-nitrobenzyl chloride reacting much more readily than benzyl chloride itself.

This argument implies that $-E$ substituents should not accelerate S_N2

reactions. In fact, simple $-E$ substituents such as F or OCH_3 retard S_N2 reactions strongly; CH_3OCH_2Cl, for example, undergoes S_N2 reactions more sluggishly than methyl chloride itself. The reason for this is that in the S_N2 transition state (29) for such reactions, the AO of the nucleophile and leaving group cannot help interacting with the filled $2p$ AO of the substituent.

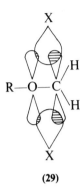

(29)

The transition state is consequently isoconjugate with cyclopropenium. Since the π-type system contains four electrons, two from the C—X_2 bond and two from the AO of the substituent, it is isoconjugate with the cyclopropenium anion and so is antiaromatic.

The balance between the S_N1 and S_N2 reactions can be altered by changes in the solvent or by suitable substituents. Thus we have already seen (Table 5.2) that polar solvents accelerate S_N1 reactions of neutral alkyl derivatives RX but retard their S_N2 reactions with anionic nucleophiles. In such cases, an increase in solvent polarity favors the S_N1 reaction, and a decrease favors the S_N2. This is seen nicely in the case of borderline reactions such as that of benzyl chloride. In aqueous alcohol, hydrolysis is of an intermediate character, hydroxide ion accelerating the reaction, which, however, is not of first order in HO^-. In the more ionizing solvent, formic acid, formolyis is of zeroth order in formate ion, the reaction being S_N1. In acetone, on the other hand, a poor ionizing solvent, benzyl chloride readily undergoes normal S_N2 reactions. Thus:

$PhCH_2Cl + HO^- \xrightarrow[\text{aq. ethanol}]{90\%} PhCH_2OH + Cl^-$; "mixed"

$PhCH_2Cl + HCOO^- \xrightarrow[\text{formic acid}]{\text{moist}} PhCH_2OOCH$; S_N1, zeroth order with respect to HCO_2^-

$PhCH_2Cl + I^- \xrightarrow{\text{dry acetone}} PhCH_2I + Cl^-$; S_N2 \hfill (5.149)

The balance can also be disturbed by substituent effects. Thus since $+E$ substituents accelerate S_N2 reactions but retard S_N1 reactions, p-nitrobenzyl chloride reacts with hydroxide ion in aqueous ethanol by an S_N2 reaction,

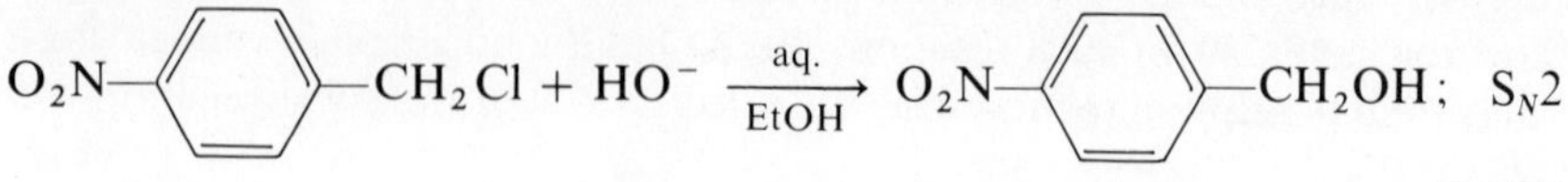

$$O_2N-C_6H_4-CH_2Cl + HO^- \xrightarrow[\text{EtOH}]{\text{aq.}} O_2N-C_6H_4-CH_2OH; \quad S_N2 \qquad (5.150)$$

On the other hand, p-methoxybenzyl chloride reacts by the S_N1 path, since ionization of Cl is accelerated by the $-E$ methoxy group,

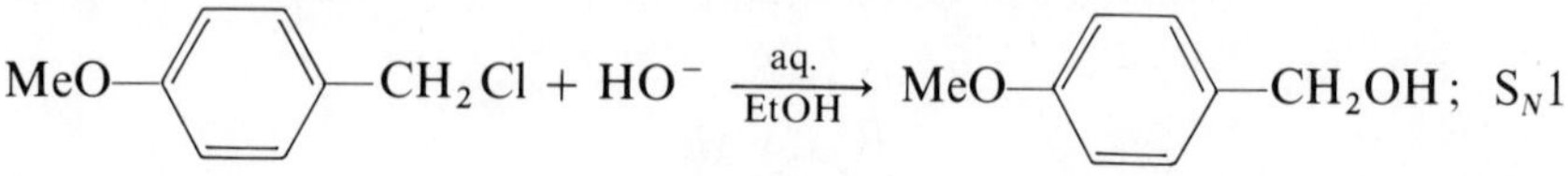

$$MeO-C_6H_4-CH_2Cl + HO^- \xrightarrow[\text{EtOH}]{\text{aq.}} MeO-C_6H_4-CH_2OH; \quad S_N1 \qquad (5.151)$$

A very nice example is provided by the arylmethyl chlorides $ArCH_2Cl$, where ArH is an even AH. As we have seen, both the S_N1 and S_N2 reactions are accelerated by an increase in the $\pm E$ activity of Ar (Figs. 5.19 and 5.21), but the acceleration is greater in the case of the S_N1 reaction. Therefore while benzyl chloride hydrolyzes by a "mixed" mechanism in aqueous ethanol, the corresponding hydrolyses of more reactive arylmethyl chlorides are S_N1.

The rates of S_N1 and S_N2 reactions can also be changed by altering the leaving group. The energy of ionization E_i of RX to $R^+ + X^-$ can be found in the same way as that of HX to $H^+ + X^-$ (p. 242),

$$E_i = E_{RX} + I_R - A_X \qquad (5.152)$$

This differs from the expression for HX only in the replacement of the HX bond energy E_{HX} by the corresponding CX bond energy E_{CX} and the ionization potential of H (I_n) by that of the radical R (I_R). If R is kept constant, the energy of ionization, and so the rate of the corresponding S_N1 reaction, will be determined by the difference $E_{RX} - A_X$.

The bond energies of bonds formed by X to carbon and to hydrogen show very similar variations with X (Table 5.4). The quantity $E_{RX} - A_X$ in equation (5.152) therefore varies with X in very much the same way as $E_{HX} - A_X$ in

TABLE 5.4. Comparison of HX and CX Bond Energies (kcal/mole)

X	E_{HX}	E_{CX}	$E_{HX} - E_{CX}$
C	99	83	16
N	93	73	20
O	111	86	25
S	83	65	18
F	135	116	19
Cl	103	81	22
Br	88	68	20
I	71	51	20

equation (5.117). The energies of ionization of HX and RX should therefore parallel one another, so RX should ionize more readily, and the corresponding S_N1 reaction should take place faster, the stronger an acid is HX.

The available data fit this prediction. Thus the electron affinity of the hydroxyl group HO· is increased if hydrogen is replaced by a $+I$ or $+E$ substituent Z since these stabilize the ion ZO^- formed by addition of an electron. In the series of acids

$$H_2O < HOOCCH_3 < HOOCCF_3 < HO_3SPh \qquad (5.153)$$

acid strength increases in the order indicated because the bond broken is the same in all cases (i.e., H—O), while the electron affinity of the radical ZO· increases along the series. For the same reason, the rates of S_N1 reactions of a given compound RX increase in the order

$$ROH < ROOCCH_3 < ROOCCF_3 < RO_3SPh \qquad (5.154)$$

Alcohols indeed undergo S_N1 reactions only in extreme cases where the ion R^+ is exceptionally stable. They can, however, be induced to do so by acid catalysts which convert them to oxonium ions, ROH_2^+. The electron affinity of the ion radical H_2O^+ (290.3 kcal/mole) is of course very much greater than that of neutral HO· (42.4 kcal/mole), so H_3O^+ is a very strong acid and H_2O^+ a correspondingly good leaving group.

The effect of changes in RX bond energy is seen in the alkyl halides. While the halogens have comparable electronegativities, the CX bond energy falls dramatically (Table 5.4) in the series

$$CF < CCl < CBr < CI \qquad (5.155)$$

The rates of S_N1 reactions therefore increase in the order

$$RF < RCl < RBr < RI \qquad (5.156)$$

just as acidity increases in the series $HF < HCl < HBr < HI$ (p. 242).

Comparison of the equations for S_N2 and S_N1 behavior (5.141), (5.148), and (5.152) shows that the rates of S_N2 reactions of RX vary in the same way with the CX bond energy and the electron affinity of X as do the rates of the corresponding S_N1 reactions. The rates of S_N1 and S_N2 reactions of derivatives RX of a given group R should therefore vary in the same way with the leaving group X. This again is the case, the S_N2 reactions of such compounds following, for example, equations (5.154) and (5.156).

5.13 Nucleophilicity and Basicity

The last factor influencing the course of S_N2 reactions is the nature of the nucleophile. Discussion of this has been left to a separate section because other reactions also involve nucleophilic attack and their rates likewise depend on the nucleophilicity of the nucleophile.

Since the combination of a base Y^- with H^+ to form HY can be regarded as a nucleophilic attack by Y^- on the proton, one might expect a general parallel between the nucleophilicity of nucleophiles and their basicity. Indeed, since protonation of a base Y^- is the reverse of the acid dissociation of the conjugate acid HY, one can see from equation (5.152) that the energy of protonation of Y^-, i.e., its *proton affinity* PA_{Y^-}, is given by

$$PA_{Y^-} = -E_{HY} + A_Y - I_H \qquad (5.157)$$

Therefore the stronger the HY bond or the smaller the electron affinity of Y, the stronger will be Y^- as a base.

We have already derived an expression for the activation energy of the simple S_N2 reaction of Y^- with RX [equation (5.97)]. From this, one can see that the rate of reaction should be greater, the greater is the CY bond energy and the lower is the electron affinity of Y. On this basis, one would expect nucleophilicity to run parallel to basicity, just as the efficacy of a leaving group X in RX runs parallel to the acid strength of HX.

This relation does indeed hold for a series of nucleophiles Y^- in which the R—Y bond in the product is the same. Thus since acid strength rises in the series of equation (5.153), the basicity of the conjugate bases falls in the corresponding order,

$$HO^- > CH_3COO^- > CF_3COO^- > PhSO_3^- \qquad (5.158)$$

This is also the order of decreasing nucleophilicity in S_N2 reactions. When, however, we compare nucleophiles in which a different bond is formed, the parallel fails. Thus the nucleophilicity of the halide ions increases in the series

$$F^- < Cl^- < Br^- < I^- \qquad (5.159)$$

This is exactly the opposite order to that of basicity. Likewise, the hydrosulfide ion HS^- is far more nucleophilic than HO^- even though it is a much weaker base. Indeed, hydroxide ion in protic solvents is a surprisingly poor nucleophile, lying between Br^- and I^-, although it is a strong base.

One very important factor is of course the need to desolvate the nucleophile so that it can react. When hydroxide ion reacts with an acid, it can do so via a hydrogen-bonded complex $HO^- \cdots HX$ whose heat of formation will be at least equal to that of a hydrogen bond between HO^- and a molecule HZ of a protic solvent, $HO^- \cdots HZ$. In an S_N2 reaction, however, at least one molecule of solvent must be removed from HO^- in order that it should be able to approach the substrate RX. Since approach is from the backside of the R—X bond, and since alkyl groups are weakly polar, the interaction between HO^- and RX will be small. This loss of solvation energy is indeed responsible for a major part of the activation energy of S_N2 reactions.

The importance of desolvation will depend on the strength of attachment of solvent groups to the nucleophile. In the case of anionic nucleophiles, solvation involves charge–dipole interactions (p. 224), which will be stronger, the smaller is the ion. Much of the difference between HO^- and HS^- is due to the smaller size and consequently stronger solvation of HO^-. Likewise, the abnormal order of nucleophilicity shown by the halides [equation (5.159)] is due to the increase in ionic radius along the series and the consequent decrease in the energy of desolvation.

Protic solvents are especially effective at solvating anions. If we use a less effective solvent such as dimethylsulfoxide (DMSO) or dimethylformamide (DMF), the contribution of desolvation should become less. Indeed, the nucleophilicity of F^- and HO^- is enormously increased in these solvents, the ratio of the rates of their S_N2 reactions to those of the large iodide ion being greater by several powers of ten than it is in a protic solvent such as water or methanol.

The relative importance of the bond energy and electronegativity terms on basicity and nucleophilicity is also likely to differ. In the symmetric S_N2 transition state $(X \cdots R \cdots X)^-$, the formal charge is shared equally between the two X groups. Consequently, half the charge on the entering nucleophile X^- is transferred to RX. However, the $R \cdots X$ bonds in the transition state will be weak both because two RX bonds are being formed by a single pair of electrons and because the central carbon atom is using a pure $2p$ AO for bonding instead of an sp^3 hybrid AO. The difference should be further enhanced in an asymmetric S_N2 transition state $(Y \cdots R \cdots X)^-$ if the ionization potential of the nucleophile Y^- (equal to the electron affinity A_Y of the radical $Y \cdot$) is less than the electron affinity of $X \cdot$. In this case, the transition state will be polarized so that the greater part of the formal negative charge is on X rather than Y. Since typical leaving groups are very electronegative, polarization of this kind should be very important in the case of nucleophiles of low electronegativity. The electronegativity of a nucleophile Y^- should therefore be more important in determining its nucleophilicity than the bond energy of the CY bond. This factor again would tend to make sulfur a better nucleophile than oxygen and would contribute to the opposite orders of basicity and nucleophilicity in the halide ions.

The S_N2 reaction involves a large desolvation energy, because there is no significant electrostatic interaction between the nucleophile and the substrate. Even if the latter contains a very polar $C^{\delta+}-X^{\delta-}$ bond, the charge–dipole interaction with the nucleophile will be weak because the backside of the CX bond is protected by hydrogen or substituent groups attached to carbon. In the case of other reactions, however, the nucleophile can closely approach the positive end of a dipole in the substrate. The resulting interaction with the substrate may then become comparable with that between the nucleophile and a solvent molecule, so that a nucleophile–substrate pair can be formed without much loss of energy. The role of desolvation then becomes correspondingly less important and the order of nucleophilicity and basicity more nearly equal. A good example is the addition of nucleophiles to carbonyl groups,

$$Y^- + R_2C{=}O \longrightarrow Y-CR_2-O^- \tag{5.160}$$

This reaction is really no different from an S_N2 reaction. In a normal S_N2 reaction, the leaving group X is singly bound to carbon and so departs as an anion, X^-. Here the "leaving group" O is doubly bound to carbon and so still remains attached to it by one bond. The similarity between the two reactions can be seen from their description in the curved arrow notation (p. 206),

$$Y^- \curvearrowright R \curvearrowright X \longrightarrow Y-R + X^- \quad (5.161)$$

$$Y^- \curvearrowright C \curvearrowright O \longrightarrow Y-C-O^- \quad (5.162)$$

There is, however, one important difference. Here the carbon at the reaction center is planar, so Y^- can approach the positive end of the $C^{\delta+}=O^{\delta-}$ dipole closely. The order of nucleophilicity (in a protic solvent) now corresponds to that of basicity, as the following comparisons show:

order of basicity $\qquad HO^- > CN^- > HS^- > F > Cl^- > Br^- > I^-$

order of nucleophilicity, $\quad HS^- > CN^- > I^- > HO^- > Br^- > Cl^- > F^-$
S_N2

order of nucleophilicity, $\quad HO^- > CN^- > HS^- > F^- > Cl^-, Br^- > I^-$
CO addition

Solvation also plays a major role in determining the products from reactions where the nucleophile is an ambident anion such as the ethyl acetoacetate anion (30). In the isoconjugate pentadienate anion (31), the formal charge is distributed evenly over the 1,3,5 positions, one-third on each. In (30), however, the π electrons will be polarized by the very electronegative oxygen atoms so that the formal charge will be almost entirely concentrated on them. Surrounding protic solvent molecules will therefore be very strongly attached to the oxygen atoms of (30), the central carbon being only weakly solvated.

$$\underset{(30)}{CH_3-\overset{\overset{O}{\|}}{C}\cdots \overset{\ominus}{CH}\cdots \overset{\overset{O}{\|}}{C}-OEt} \qquad \underset{(31)}{\overset{\frac{1}{3}-}{CH_2}\cdots CH\cdots \overset{\frac{1}{3}-}{CH_2}}$$

We have seen (p. 251) that protonation of (30) takes place much more rapidly on oxygen than on carbon. Attack on carbon would require specific activation because the carbon atom would have to change its hybridization to bind the proton. Since acid–base reactions in solution do not involve desolvation, protonation takes place more rapidly on oxygen.

The situation is very different in the case of an S_N2 reaction. Here desolvation *is* involved. The energy required to desolvate carbon in (30) is, moreover, very much less than that required to desolvate oxygen, since nearly all the formal charge is on oxygen. Consequently, (30) reacts with alkyl halides by C-alkylation rather that O-alkylation.

The situation is once more reversed on passing from an alkyl halide to an acyl halide. Nucleophilic displacements of halogen in the latter take place by preliminary addition of the nucleophile to the C=O bond (see Section 5.20). As we have seen, desolvation is relatively unimportant for this reaction. Acylation of (30) therefore takes place on oxygen, giving enol derivatives. Thus:

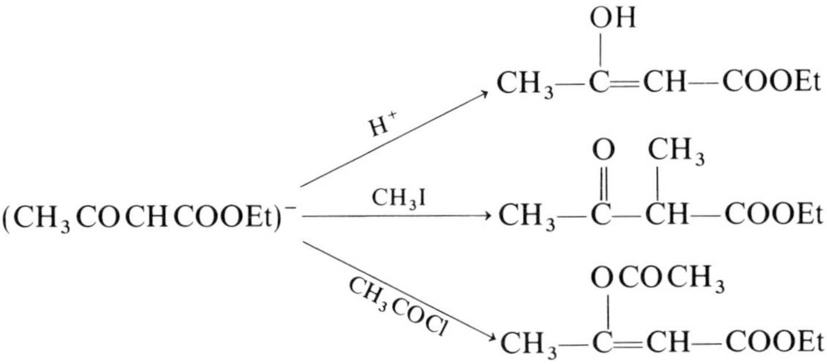

$$(CH_3COCHCOOEt)^- \begin{cases} \xrightarrow{H^+} CH_3-\underset{\underset{OH}{|}}{C}=CH-COOEt \\ \xrightarrow{CH_3I} CH_3-\underset{\underset{O}{\|}}{C}-\underset{\underset{CH_3}{|}}{CH}-COOEt \\ \xrightarrow{CH_3COCl} CH_3-\underset{\underset{OCOCH_3}{|}}{C}=CH-COOEt \end{cases} \quad (5.163)$$

Many examples of this kind are known, largely through the work of Kornblum and his collaborators. A very nice one is provided by the reactions of sodium β-naphthoxide (32) with benzyl chloride (33). In DMSO, the product consisted of 95% of the benzyl ether (34) and 5% of 1-benzyl-2-keto-1,2-dihydronaphthalene (35), while in trifluoroethanol (CF_3CH_2OH), the proportions were reversed. DMSO is a poor solvating agent for anions, as we have already remarked. Here oxygen can exert its inherent greater reactivity. Trifluoroethanol, on the other hand, is a particularly good solvent for anions since the $+I$ CF_3 group polarizes the OH bond and so makes it especially effective at hydrogen bonding. The formation of (35) as the major product is noteworthy since the equilibrium between (34) and (35) strongly favors (34), in which both aromatic rings are still intact.

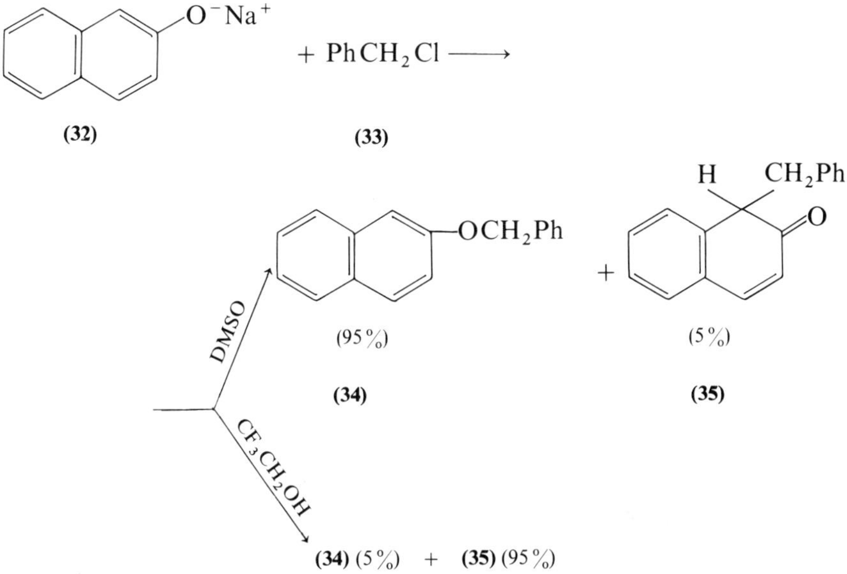

5.14 Electrophilic Aliphatic Substitution

In nucleophilic aliphatic substitution, an alkyl derivative RX is attacked by a nucleophile Y^-, i.e., a species with a filled orbital that can be used to bond Y to R through a vacant AO of R. The vacant AO is provided by heterolytic cleavage of the RX bond to $R^+ + X^-$. Electrophilic aliphatic substitution is an analogous process in which the attacking *electrophile* Y^+ has an empty orbital that can be used for bonding to R. Here R must have a filled AO to form the YR bond in the product; this filled AO is again derived by heterolytic cleavage of RX but in the opposite sense to that in S_N reactions.

These electrophilic processes can, like their S_N counterparts, be divided into one-step (S_E2) and two-step (S_E1) types; i.e.,

$$Y^+ \quad R\!-\!X \xrightarrow[\text{one-step}]{\text{concerted}} Y\!-\!R \quad X^+ \qquad (S_E2) \qquad (5.164)$$

$$R\!-\!X \longrightarrow R^- + X^+, \quad Y^+ \quad R^- \longrightarrow Y\!-\!R \qquad (S_E1) \qquad (5.165)$$

Some authors have regarded prototropic reactions as special kinds of S_E1 process, X being a hydrogen atom. However, hydrogen is not lost as the free ion H^+ and indeed such reactions occur only in the presence of a base that can combine with the proton. The course of the reaction depends critically on the nature of the base. It is therefore better to reserve the designation S_E1 for reactions where the group X is expelled as a neutral molecule [from an initial anion $(RX)^-$] or as a stable cation X^+.

The S_E1 and S_E2 reactions differ from their S_N1 and S_N2 counterparts in three important respects.

First, as we have already seen (p. 209), the most favorable geometry for the S_E2 reaction is one involving frontside attack by the reagent Y^+ on the RX bond instead of backside attack as in the S_N2 reaction. In the case of asymmetric alkyl derivatives, the S_E2 reaction should then lead to *retention* of configuration instead of inversion.

Second, in the S_E2 transition state (Fig. 5.3), the carbon atom at the reaction center retains its tetrahedral geometry and sp^3 hybridization. One sp^3 hybrid AO is used to bond the entering and leaving groups through a three-center, two-electron bond. Consequently E-type substituents should exert none of the specific stabilizing effects on the transition state that they do in an S_N2 reaction, where the carbon atom changes its hybridization to sp^2. The S_E2 reaction is nevertheless still an anti-BEP reaction, i.e., $EO_A^{\ddagger}$, because the effects of I-type substituents on the transition state will be different from those on the reactants or products. In the transition state, two electrons are shared between the carbon atom at the reaction center and the entering and leaving groups. The carbon atom is therefore likely to be more positive in the transition state than it is in the reactant or product, particularly if the reaction involves a cationic reagent Y^+, as is often the case. Therefore $-I$ substituents would be expected to accelerate S_E2 reactions and $+I$ substituents to retard them, although such substituents should have little effect on the heat of reaction.

Third, there is a possible fundamental difference in stereochemistry between the S_E1 and S_N1 reactions due to the fact that the S_E1 reaction involves an intermediate carbanion instead of a carbonium ion. Since the methyl cation (CH_3^+) is planar, the same should be true of its derivatives. If, then, an S_N1 reaction takes place at an asymmetric center in RX, the asymmetry will be lost in the intermediate cation R^+, so the reaction should lead to loss of configuration. Reactions of this type invariably lead to partial or complete racemization. The methyl anion (CH_3^-) is, however, pyramidal, being isoelectronic with ammonia, and the same should be true of its derivatives unless these are stabilized by $\pm E$ or $+E$ groups adjacent to the anionic carbon. When such a pyramidal anion R^- is formed by heterolysis of RX, the stereochemistry at the reaction center is retained. If the subsequent reaction of R^- with the electrophile X^+ is much faster than inversion of the pyramidal anion, the product will then be formed with complete retention of configuration.

While relatively little work has been done on the stereochemistry of S_E2 reactions, the evidence suggests that they do in fact lead to retention of configuration. A good example is provided by the reactions of cis- and trans-methylcyclohexylmercuric bromides with bromine and pyridine to form 1-bromo-4-methylcyclohexane, the *cis* isomer giving almost exclusively the *cis* product while the *trans* isomer gives 91% of the *trans* product. The reactions probably involve an electrophilic (S_E2) attack by bromine on the pyridine complexes, as indicated below.

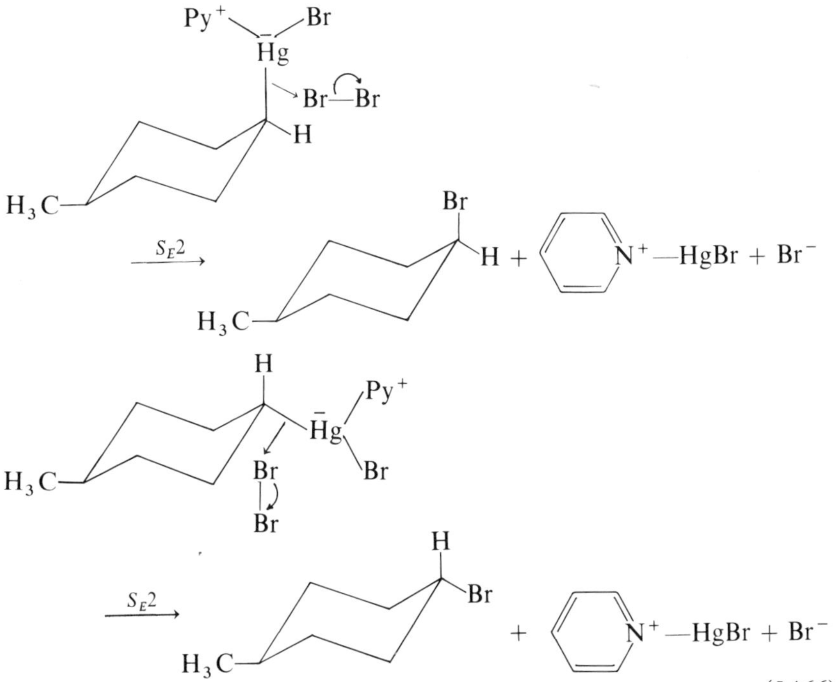

(5.166)

The effects of substituents on the rates of S_E2 reactions has been barely studied. We would expect them to depend only on the inductive and field effects of substituents, E-type substituents being unable to exert mesomeric effects. The S_E1 reaction should, like its S_N1 counterpart, be of BEP type, i.e., $EO_B^\ddagger$. It should be accelerated by substituents that can stabilize carbanions, i.e., those of $+I$, $\pm E$, or $+E$ type, the last named being particularly effective. Now $\pm E$ and $+E$ substituents also exert $+I$ effects.* Such substituents should therefore accelerate S_E1 reactions but retard S_E2 reactions.

The majority of S_E1 reactions involve the fragmentation of negative ions $(RX)^-$, X carrying the negative charge and being expelled as a neutral molecule. The most important reactions of this kind are of retro-aldol type, involving cleavage of a carbon–carbon bond to a C—O$^-$ group. Some examples follow.

Retro-aldol reaction:

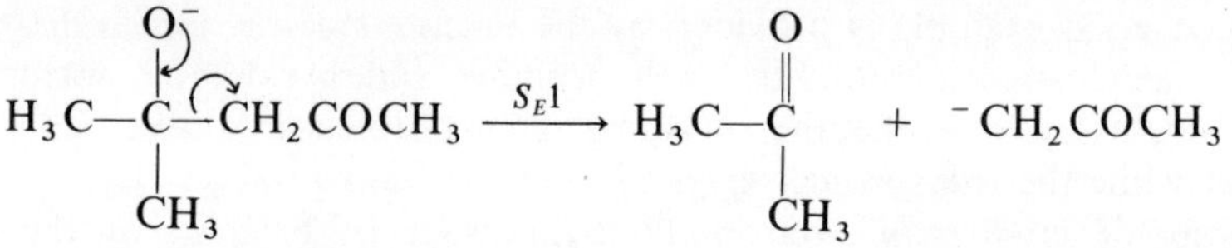

$$H_3C-\underset{CH_3}{\underset{|}{C}}(O^-)-CH_2COCH_3 \xrightarrow{S_E1} H_3C-\underset{CH_3}{\underset{|}{C}}=O + {}^-CH_2COCH_3 \quad (5.167)$$

Iodoform reaction:

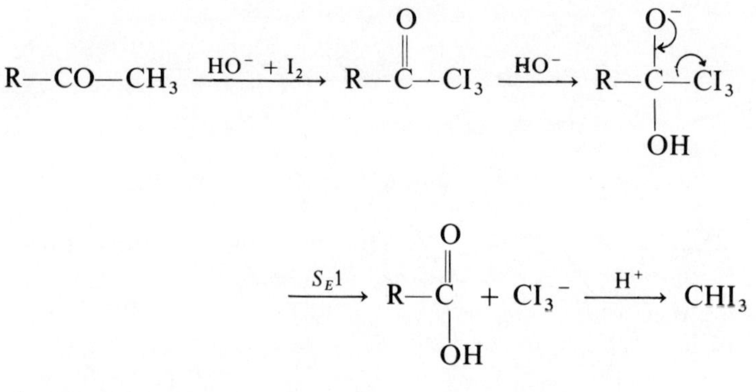

$$R-CO-CH_3 \xrightarrow{HO^- + I_2} R-\underset{}{\overset{O}{\overset{\|}{C}}}-CI_3 \xrightarrow{HO^-} R-\underset{OH}{\underset{|}{C}}(O^-)-CI_3$$

$$\xrightarrow{S_E1} R-\underset{OH}{\underset{|}{\overset{O}{\overset{\|}{C}}}} + CI_3^- \xrightarrow{H^+} CHI_3 \quad (5.168)$$

Decarboxylation:

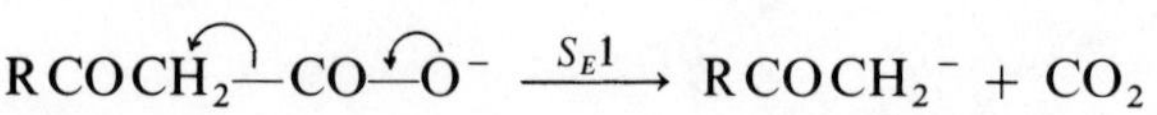

$$RCOCH_2-CO-O^- \xrightarrow{S_E1} RCOCH_2^- + CO_2 \quad (5.169)$$

---

* The $+I$ activity of $\pm E$ substituents such as phenyl is due to the fact that an sp^2 hybrid AO has a lower energy than an sp^3 AO so that a bond between sp^3 carbon and sp^2 carbon has a dipole moment in the sense $C(sp^3)^{\delta+}-C(sp^2)^{\delta-}$. In other words, a carbon atom using an sp^2 AO for bonding is effectively more electronegative than one using an sp^3 AO.

Haller–Bauer reaction (cleavage of ketones by sodamide in toluene):

$$\text{(cyclopropyl)}(CH_3)C(=O)Ph + NH_2^- \longrightarrow \text{(cyclopropyl)}(CH_3)\underset{O^-}{\overset{Ph}{C}}-NH_2$$

$$\xrightarrow{S_E 1} \text{(cyclopropyl)}(CH_3)C:^{-} + Ph-\underset{\|}{C}-NH_2 \xrightarrow{H^+} \text{(cyclopropyl)}(CH_3)H \qquad (5.170)$$
$$\qquad\qquad\qquad\qquad\qquad\qquad\qquad O$$

Reactions of this kind usually occur only if the resulting carbanion is stabilized, just as $S_N 1$ reactions usually require formation of a stabilized carbonium ion. The best groups for stabilizing carbanions are of course of $+E$ type; this is the case for the carbanions formed in equations (5.167) and (5.169). Equation (5.168) shows the formation of a carbanion stabilized by three $+I$ substituents. A third possible mode of stabilization is to have the lone pair of carbanionic electrons in an AO of unusually high s character, the binding energy of a hybrid AO being greater, the greater is its s character (p. 243). This is true of the exocyclic orbitals in cyclopropane (see p. 243 and Table 5.3; cyclopropane is for this reason a stronger acid than propane). Equation (5.170) shows the formation of such an anion by an $S_E 1$ process.

Stabilization of carbanions by $\pm E$ or $+E$ substituents tends to lead to planarity at the anionic center just as in the case of trivalent nitrogen (see p. 119). The effect of such substituents on carbanions is indeed greater than in the case of amines since the mesomeric interaction with C^- leads only to a dispersal of negative charge while interaction with N leads to a separation of positive and negative charges. Consider, for example, formamide (NH_2CHO) and the isoelectronic carbanion CH_2^-CHO derived from acetaldehyde, both of which are isoconjugate with the allyl anion $(CH_2\cdots CH\cdots CH_2)^-$. While the allylic resonance in $(CH_2CHO)^-$ leads only to a dispersal of negative charge, $^{\delta-}CH_2\cdots CH\cdots O^{\delta-}$, as in the allyl anion itself, in the case of formamide it leads to a separation of positive and negative charges, $^{\delta+}NH_2\cdots CH\cdots O^{\delta-}$. Since unlike charges can be separated only at the expense of their Coulomb energy of attraction, the resonance energy of formamide is much less than that of CH_2^-CHO. While therefore the nitrogen atom in formamide is pyramidal, the carbanionic carbon in anions stabilized by acyl groups is planar.

No such problems arise in the case of carbanions stabilized by $+I$ groups or by changes in hybridization. One would therefore expect such carbanions to be pyramidal. If an $S_E 1$ reaction involves such an intermediate carbanion, it may then, as we have already seen, lead to products formed with retention of configuration at the reaction center. This is the case for the reaction of equation (5.170), optically active ketone leading to optically active methyldiphenylcyclopropane. On the other hand, S_E1 reactions, involving resonance-stabilized planar carbanions as intermediates, lead to products in which any asymmetry at the reaction center is more or less lost.

5.15 Radical Substitution Reactions ($EO_B^{\ddagger}$)

Radical substitutions in paraffin derivatives do not take place by attack on carbon as in the S_N and S_E reactions. Instead, they involve abstraction of an atom, usually hydrogen, by a radical reagent to give an alkyl radical which then reacts in some way to form the product. This secondary reaction also usually involves transfer of an atom to the alkyl radical from some normal molecule which is thus in turn converted to a radical. The result is a radical chain reaction in which radicals effectively catalyze substitution. A good example is the photochemical bromination of paraffins, which takes place in the following way:

$$Br_2 + h\nu \longrightarrow 2Br\cdot \qquad \text{initiation} \qquad (5.171)$$

$$Br\cdot + RH \longrightarrow R\cdot + HBr \qquad (5.172)$$
$$R\cdot + Br_2 \longrightarrow RBr + Br\cdot \qquad \text{propagation} \qquad (5.173)$$

$$2Br\cdot \longrightarrow Br_2 \qquad \text{termination} \qquad (5.174)$$

The two propagation reactions have the overall effect of converting $RH + Br_2$ into $RBr + HBr$. Bromine atoms catalyze this reaction since the bromine atom used up in the first propagation step [equation (5.172)] is regenerated in the second [equation (5.173)]. Bromine atoms are removed by combination [equation (5.174)] but are generated by photochemical dissociation of Br_2 [equation (5.171)].

The propagation steps in reactions of this kind are usually very fast, so that one atom or radical brings about substitution in a large number of molecules before it is removed by termination. The number of molecules reacting per catalyzing atom or radical is called the *chain length*. Chain reactions are common in radical chemistry because since radicals contain unpaired electrons, the reaction of a radical with a normal molecule must give rise to another radical. Radicals can be destroyed only by reacting with other radicals, a pair of radicals containing between them an even number of electrons so that the product can be a normal molecule or pair of normal molecules.

In most of these reactions, the slowest step in the propagation cycle is the abstraction of an atom from the substrate, e.g., equation (5.172). If the rate of initiation is held constant, the overall rate is then determined by this step. Moreover the resulting radicals $R\cdot$ are removed very rapidly [e.g., equation (5.173)] so their concentration is very small. The termination reaction is therefore almost confined to radicals derived from the substituting agent [e.g., Br atoms, equation (5.174)]. The relative reactivities of different

Chemical Reactivity

alkyl derivatives and the relative reactivities of given atoms (e.g., H) in two or more different positions in the same alkyl derivative are therefore determined by the rates of such atom abstraction reactions.

Reactions of this kind are clearly analogous to S_E2 and S_N2 reactions except that the reaction center involves a single atom (e.g., H) instead of an alkyl group; thus,

$$Y{:} \quad R\text{—}X \longrightarrow (Y \cdots R \cdots X)^- \longrightarrow Y\text{—}R + {:}X^- \quad (5.175)$$

$$Y\cdot \quad H\text{—}R \longrightarrow (Y \cdots H \cdots R) \longrightarrow Y\text{—}H + \cdot R \quad (5.176)$$

In equation (5.176), the fishhook arrows denote formal displacements of single electrons. Thus the unpaired electron of $Y\cdot$ and one of the H—R bond electrons combine to form the Y—H bond in the product, while the other H—R bond electron remains unpaired in the radical $R\cdot$.

Such radical reactions are therefore termed S_R2. In the transition state, the central atom [e.g., H in equation (5.176)] uses a single AO to bind both the attacking radical $Y\cdot$ and the leaving group $R\cdot$ and three electrons are involved [see equation (5.176)]. The transition state is therefore isoconjugate with an allyl or cyclopropenyl radical (cf. Figs. 5.2 and 5.3), the distinction between the two being meaningless since both lead to similar products and since the cyclopropyl radical is nonaromatic (see Sec. 3.14). Since the reaction center in this case involves a single atom [e.g., H in equation (5.176)], the transition state contains no bonds that are not present either in the reactants or in the products. The reactions are therefore of BEP type and so are classed as $EO_B^{\ddagger}$.

For a given substitution reaction, Y and H in equation (5.176) are constant so the rate depends on the nature of the alkyl group R. Let us first consider the case where RH is an AH, e.g., an arylmethane $ArCH_3$, Ar being a $\pm E$ group. Equation (5.176) then becomes

$$Y\cdot + H\text{—}CH_2Ar \longrightarrow Y \cdots H \cdots CH_2Ar \longrightarrow YH + \cdot CH_2Ar \quad (5.177)$$

The overall energy of reaction ΔE is given by

$$\Delta E = E_{CH} - E_{HY} + \delta E_\pi \quad (5.178)$$

where E_{CH} and E_{HY} are the CH and HY bond energies and where δE_π is the difference in π energy between $ArCH_3$ and $ArCH_2\cdot$. For a given reagent $Y\cdot$, ΔE is therefore as usual of the form

$$\Delta E = \text{constant} + \delta E_\pi \quad (5.179)$$

From equation (3.44),

$$\delta E_\pi = 2\beta(1 - a_{0r}) \quad (5.180)$$

where a_{0r} is the NBMO coefficient of CH_2 in $ArCH_2\cdot$. Hence,

$$\Delta E = \text{constant} - 2a_{0r}\beta \quad (5.181)$$

The activation energy $\Delta E^{\ddagger}$, i.e., the difference in energy between $ArCH_3$ and the transition state of equation (5.177), can be found by a procedure exactly analogous to that used for the S_N2 reaction (p. 248). Both transition states are isoconjugate with 1-arylallyl. Assuming that the transition states have similar structures so that the resonance integrals of HY and HR bonds in them remain constant, we find that

$$\Delta E^{\ddagger} = \text{constant} - 2a_{0r}\beta' \qquad (5.182)$$

where β' is numerically less than β in equation (5.181), the ratio being smaller, the stronger is the CH bond in the transition state, i.e., the nearer the transition state is to the reactants in structure. This equation could have been derived directly from equation (5.181) by using the BEP principle, since the reaction is of BEP type.

A plot of the logarithms of the rate constants (or relative rate constants) for the reactions of a given radical $Y\cdot$ with a series of arylmethanes $ArCH_3$ against the NBMO coefficients a_{0r} should then be linear. Gleicher has shown this to be the case.

The energy required to break a given bond in a molecule is not equal to the bond energy, because the resulting radical, or pair of radicals, may be stabilized by mesomerism. Thus in the case of an arylmethane $ArCH_3$, where Ar is a $\pm E$ group, the energy required to break one of the methyl CH bonds is given by

$$BDE_{CH} = E_{CH} + \delta E_\pi \qquad (5.183)$$

where BDE_{CH} is the CH *bond dissociation energy*. The energy of reaction of $ArCH_3$ with a given radical $Y\cdot$ to form $HY + R\cdot$ is then [see equation (5.178)]

$$\Delta E = \text{constant} + BDE_{CH} \qquad (5.184)$$

It is easily seen that this relation must also hold for CH bonds in alkyl derivatives of all kinds. Since the reactions are of BEP type, the rate of abstraction of hydrogen should be greater, the smaller is the corresponding bond dissociation energy and so the more stable is the resulting radical. Abstraction of hydrogen is therefore facilitated by E-type substituents of all kinds at the reaction center (see Table 5.5), including alkyl groups, which act by hyperconjugation (being isoconjugate with vinyl and so of $\pm E$ type).

Table 5.5 shows the correspondence between rates of hydrogen abstraction by chlorine or bromine atoms and the corresponding CH bond dissociation energies. The parallel is good but not exact, alkyl groups accelerating hydrogen abstraction more than would be expected relative to phenyl.

The anomalous efficiency of alkyl groups can be attributed to mutual conjugation. Since the delocalized system in a transition state is topologically equivalent to a corresponding π system, the effects of substituents should be qualitatively similar in both cases. We have already seen that mutual conjugation should occur between substituents attached to positions of opposite parity in an even AH if one is of $-E$ type and the other is of $+I$, $\pm E$, or $+E$ type (p. 180). The same should be true for even alternant transition states. Here, however, we have a transition state that is isoconjugate with an odd

TABLE 5.5. Relative Reactivities in Radical Substitution Reactions[a]

Hydrocarbon substrate	Relative reaction rates		CH bond strength (kcal/mole)
	Br	Cl	
CH_3—H	7×10^{-4}	4×10^{-3}	104
XH_3—CH_2—H	1	1	98
$(CH_3)_2$—CH—H	220	4.3	95
$(CH_3)_3$—C—H	1.9×10^4	6.0	92
Ph_2—CH_2—H	6.4×10^4	1.3	85
Ph_2—CH—H	6.2×10^5	2.6	—
Ph_3C—H	1.14×10^6	9.5	—

[a] Data from Ref. 6.

AH radical, i.e., the allyl radical. Mutual conjugation also occurs in such odd systems but the rules are now different from those in even systems. It can be shown* that mutual conjugation occurs in an odd AH radical between two substituents provided that they are both attached to active (starred) positions and provided that one is of $-I$ or $-E$ type and the other is of $+I$ or $+E$ type. Here the halogen Y is much more electronegative than carbon, thus being equivalent to a carbon atom carrying a strong $+I$ substituent. If the substituents (a, b, c) in (36) are of $-I$ type (e.g., alkyl), mutual conjugation between them and Y will then selectively stabilize the transition state. The effect should be greater, the more electronegative is the halogen atom. This indeed is the case. Thus propane and isobutane are attacked less

$$Y \cdots H \cdots C \begin{array}{c} a \\ b \\ c \end{array}$$

(36)

rapidly than toluene by bromine atoms but more rapidly than toluene by chlorine atoms. Chlorine is more electronegative than bromine, so mutual conjugation is more effective in the case of attack by chlorine atoms.

Similar anomalies occur in other abstractions of hydrogen by electronegative radicals such as halogen atoms or alkoxyl; alkyl substituents again greatly accelerate the rate. This is the origin of the widespread idea that radical reactions have a certain polar character, "electrophilic" radicals such as alkoxyl tending to attack "nucleophilic" CH. The basis of this phenomenon is mutual conjugation in the transition state.

A very striking example is provided by *autooxidation*. Autooxidation is a radical chain oxidation of alkyl derivatives RH by oxygen to hydroperoxides ROOH, involving the following propagation steps:

$$RO_2\cdot + HR \longrightarrow RO_2H + \cdot R \qquad (5.185)$$

$$R\cdot + O_2 \longrightarrow RO_2\cdot \qquad (5.186)$$

* The proof of this is somewhat tortuous so it will not be given here; see Ref. 5, Theorem 65.

Except at very low oxygen pressures, the first step [equation (5.185)] is, as usual, rate determining. The reaction is accelerated by $\pm E$ and $-I$ substituents (e.g., aryl, methyl) but especially strongly by $-E$ substituents, in particular alkoxyl. Ethers therefore autooxidize very easily indeed, giving hydroperoxides, which are often dangerously explosive. Numerous explosions have occurred through the distillation of ethers from which the peroxides had not first been removed. Another example is chloroform, which autooxidizes rather easily to phosgene,

$$Cl_3CH + O_2 \longrightarrow Cl_3COOH \longrightarrow COCl_2 + HOCl \qquad (5.187)$$

Since phosgene is very toxic, a small amount of alcohol is usually added to chloroform to act as an inhibitor of the autooxidation. The easy oxidation of pure chloroform shows, incidentally, that the guiding principle in these cases is not one of simple nucleophilicity. Indeed, the hydrogen atom in chloroform is especially *positive*, due to the inductive effect of the three $+I$ chlorine atoms. Chloroform can consequently form hydrogen bonds with donors such as ethers and it also reacts easily with bases to form the CCl_3^- ion.

Complications due to mutual conjugation in the transition state are quite common in radical chemistry. They do not, however, lead to gross deviations from the BEP principle, mutual conjugation being a second-order effect. It is therefore convenient to regard such reactions as being of BEP type, mutual conjugation being regarded as a disturbing factor that must be taken explicitly into account, like steric hindrance in the transition state.

Since the reactions are of BEP type in an overall sense, the transition state should be nearer to the reactants in structure, and the proportionality factor α in the BEP relation [equation (5.69)] smaller, the more exothermic the reaction. Since the heat of formation of HCl is more negative than that of HBr, hydrogen abstraction by Cl· is more exothermic than abstraction by Br·. The spread of rates for bromination should therefore be greater than for chlorination. This effect is seen very clearly in Table 5.5, the ratio of the fastest to the slowest reactions being 2000 for chlorination, but 10^9 for bromination. Since the transition state for chlorination is nearer to the reactants than that for bromination, the central carbon atom should, moreover, be more nearly tetrahedral in the former and so less able to interact with $\pm E$ substituents. This may be an additional factor in explaining the relatively low efficacy of phenyl.

One final peculiarity of radical reactions may be noted. Since the reactants and products are neutral, the transition state must also be more or less neutral. The solvation energies of all three species must then be small so changes in them with changes in the solvent must likewise be small. Radical reactions are therefore relatively insensitive to solvent and can indeed take place easily in the gas phase.

5.16 Elimination Reactions

There are a number of bimolecular nucleophilic elimination reactions like that of equation (5.188) that take place in a single, concerted step. Since the transition state for such an E_N2 reaction is odd, being isoconjugate with the pentadienate anion $(CH_2 \cdots CH \cdots CH \cdots CH \cdots CH_2)^-$, while the reactants and products are even, E-type groups at the central carbon atoms can exert mesomeric effects in the transition state that are different from those in the reactants or products. Such eliminations are therefore anti-BEP reactions and are classed as $EO_A{}^\ddagger$.

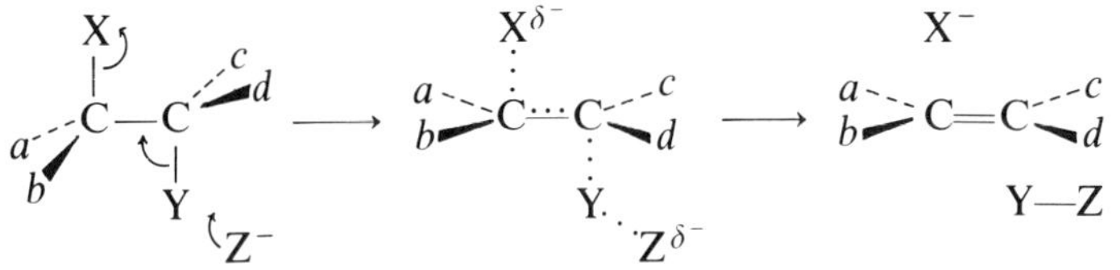

$$\begin{array}{c} X \\ | \\ a \diagdown \!\! \diagup C \!\! - \!\! C \diagdown \!\! \diagup d \\ b \diagup \quad | \quad \\ Y \\ \quad \quad Z^- \end{array} \longrightarrow \begin{array}{c} X^{\delta-} \\ \vdots \\ a \diagdown \!\! \cdot \!\! C \cdots C \diagdown \!\! \diagup c \\ b \diagup \quad \quad \diagdown d \\ \vdots Y \\ \quad \quad Z^{\delta-} \end{array} \longrightarrow \begin{array}{c} X^- \\ \\ a \diagdown \!\! C \!\! = \!\! C \diagdown \!\! \diagup c \\ b \diagup \quad \quad \diagdown d \\ Y \!\! - \!\! Z \end{array} \quad (5.188)$$

During such a reaction, the two central carbon atoms remain bonded throughout to two neighbors. Each therefore has a single AO to contribute to the delocalized MOs in the transition state. Since these AOs are initially sp^3 hybrids and end by being pure p AOs (in the C=C bond in the product), they overlap with one another in π fashion. For overlap to be a maximum, the AOs must be parallel and the X—C—C—Y system in the transition state consequently coplanar. The favored geometries of the transition states are consequently those indicated in Newman projections in Fig. 5.23.

In the *trans* transition state [Fig. 5.23(a)], the carbon atoms have a staggered conformation, while in the *cis* transition state [Fig. 5.23(b)] they are eclipsed. One would therefore expect reactions of this kind to take place by *trans* elimination and the available evidence shows that this is the case. Thus elimination of hydrogen bromide from *meso*-1,2-dibromo-1,2-diphenyl-ethane (37) with base gives exclusively α-bromo-*cis*-stilbene (38), while the corresponding reaction of the *dl*-dibromide (39) gives exclusively the *trans* isomer (40). Another classic example is provided by the β isomer (41) of 1,2,3,4,5,6-hexachlorocyclohexane, which undergoes E_N2 elimination 10,000 times more slowly than any of its isomers. There is no conformation of (41) in which hydrogen and chloride can have a dihedral angle of 180°; a 0° angle is possible, but only in strained conformations like (42).

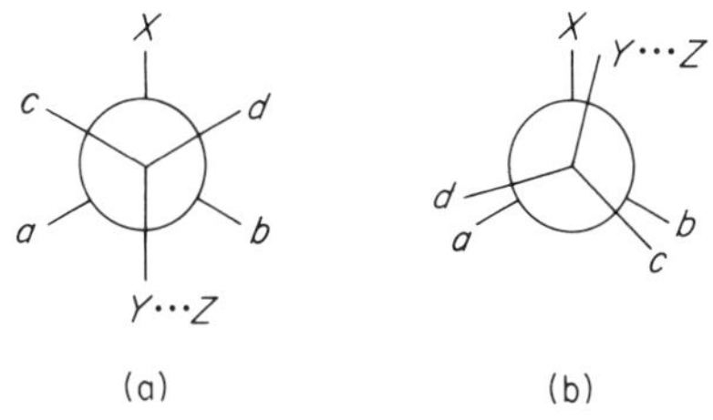

FIGURE 5.23. Newman projections for E_N2 eliminations; (a) *trans*; (b) *cis*.

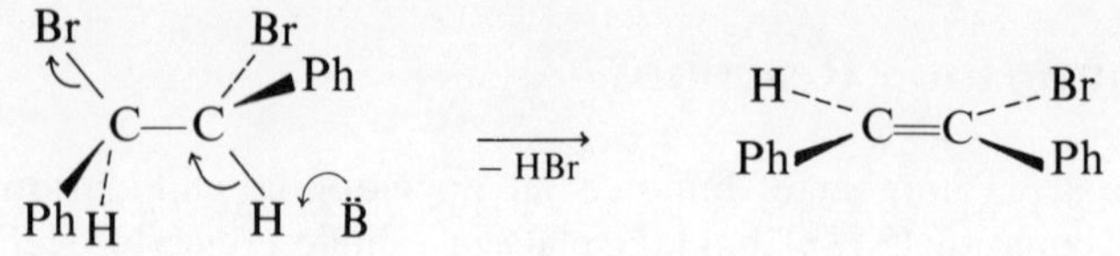

(37) →−HBr (38)

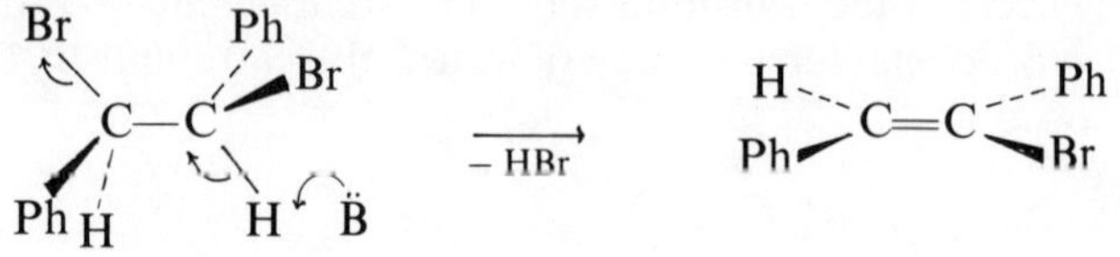

(39) →−HBr (40)

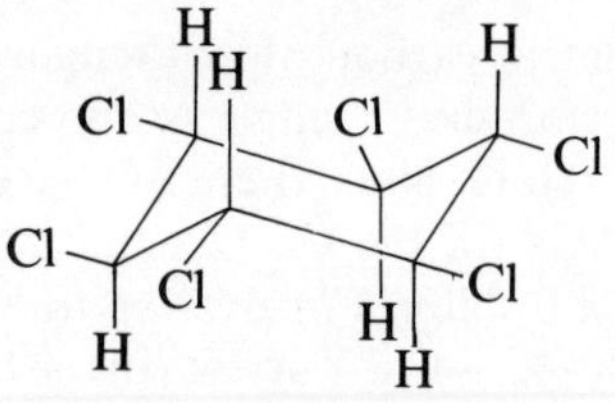

(41) 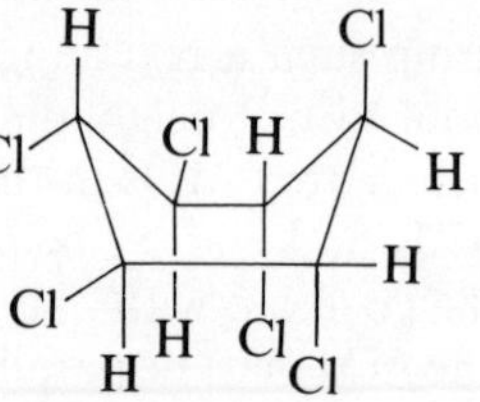 (42)

The E_N2 eliminations with leaving groups other than a proton are as stereoselective as the examples cited above. Thus elimination of bromine from 2,3-dibromobutane by iodide ion is stereospecifically *trans*, as is shown by the exclusive formation of *trans*-2-butene (44) from *meso*-dibromide (43) and of *cis*-2-butene (46) from *dl*-dibromide (45).

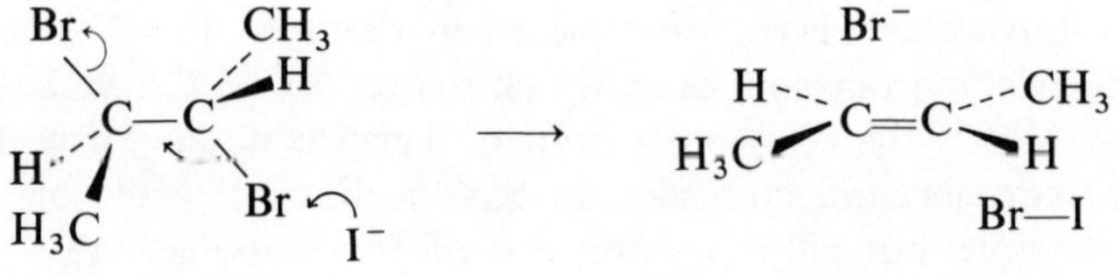

(43) → (44)

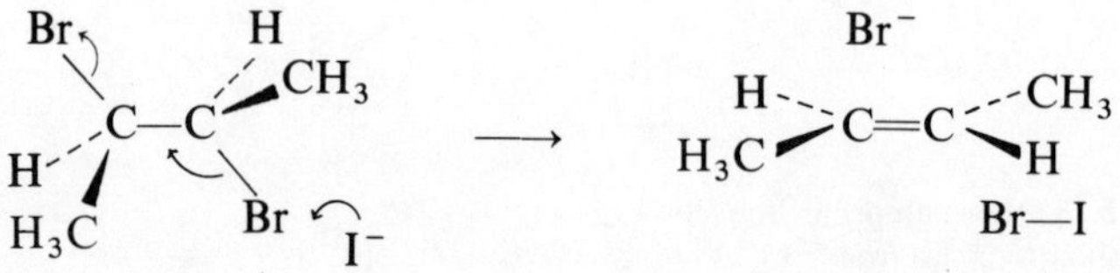

(45) → (46)

Chemical Reactivity

Cis E_N2 eliminations have been observed in rigid cyclic systems; e.g.,

$$\text{[norbornyl-Br,H,D,H]} \xrightarrow{\text{base}} \text{[alkene with H,H]} \ (94\%) + \text{[alkene with H,D]} \ (6\%) \quad (5.189)$$

Here the choice is between elimination of deuterium via a strictly planar *cis* transition state or elimination of hydrogen, which can only achieve a *trans*-planar configuration at the expense of distortion of the rigid ring system.

Elimination reactions can also take place by two two-step reactions in which the two groups are eliminated in turn rather than together. In the E_N1 process, X in equation (5.188) departs first as an anion, leaving a carbonium ion which is attacked by Z^-.

$$\underset{b}{\overset{a}{\diagdown}}\underset{Y}{\overset{X}{C}}-\underset{}{\overset{c}{C}}\diagdown d \longrightarrow \underset{b}{\overset{a}{\diagdown}}\overset{+}{C}-\underset{Y\ Z^-}{\overset{c}{C}}\diagdown d \longrightarrow \underset{b}{\overset{a}{\diagdown}}C=\underset{Y-Z}{\overset{c}{C}}\diagdown d \quad (5.190)$$

In the second ($E1cB$), Z^- first removes Y to form an intermediate anion which then expels X^-,

$$\underset{b}{\overset{a}{\diagdown}}\underset{Y\ Z^-}{\overset{X}{C}}-\overset{c}{C}\diagdown d \longrightarrow \underset{b}{\overset{a}{\diagdown}}\underset{Y-Z}{\overset{X}{C}}-\overset{c}{C}\diagdown d \longrightarrow \underset{b}{\overset{a}{\diagdown}}C=\underset{Y-Z}{\overset{X^-\ c}{C}}\diagdown d \quad (5.191)$$

The E_N1 reaction involves the same intermediate carbonium ion as an S_N1 replacement of Y and its rate is governed by similar considerations. Likewise, the $E1cB$ process involves an initial step analogous to a protoropic reaction (Y replacing hydrogen). Both reactions are of $EO_B^{\ddagger}$ type and the effect of structure on their rates can be predicted in the same way as the rates of S_N1 reactions (p. 237) or of deprotonations by base (p. 243). Thus the E_N1 reaction will be favored by $-I$, $\pm E$, and $-E$ substituents α to X and the $E1cB$ reaction by $+I$, $\pm E$, or $+E$ substituents α to Y. Indeed, elimination reactions involving a proton α to a powerful $+E$ group such as acyl always take place by the $E1cB$ mechanism, as in the conversion of β-chloroethyl ketones to vinyl ketones,

$$\text{RCO}-\text{CH}_2-\text{CH}_2\text{Cl} + \text{B} \longrightarrow \text{RCO}-\overset{\frown}{\text{CH}}-\text{CH}_2\overset{\frown}{\text{Cl}}$$
$$\longrightarrow \text{RCOCH}=\text{CH}_2 + \text{Cl}^- \quad (5.192)$$

Although E_N2 reactions must proceed through transition states where the bonds to both leaving groups are simultaneously weakened, there is no reason why the weakening should be symmetric. At one extreme, where the CY bond is barely weakened, the transition state (47) will resemble that for an E_N1 elimination. At the other extreme, where the CX bond is nearly intact, the transition state (48) will resemble that for an $E1cB$ process. One can therefore envisage a complete spectrum of mechanisms from one extreme to the other via E_N2 processes of varying symmetry.

$$
\begin{array}{cc}
X^{\delta-} & X \\
^{\delta+} & ^{\delta-} \\
a{-}\underset{b}{C}{-}{-}C{-\!\!\!-}{-}c_d & a{-}\underset{b}{C}{-}{-}C{-\!\!\!-}{-}c_d \\
Y\cdots Z^- & Y{-}{-}{-}Z^{\delta-} \\
(47) & (48)
\end{array}
$$

Such mechanistic details could be probed by using $\pm E$ substituents on the two carbon atoms. Introduction of such substituents should accelerate the reaction through favoring the progression toward a planar geometry, the substituent–carbon bonds changing from sp^2–sp^3 to sp^2–sp^2 during the the reaction and so becoming stronger. This effect should, however, be the same for all $\pm E$ substituents. On the other hand, if the transition state veers toward (47), a $\pm E$ group in place of a or b should also exert an electromeric accelerating effect that will depend on its $\pm E$ activity. The rate should increase greatly as we introduce more and more active $\pm E$ groups. A substituent on the right-hand carbon should, however, exert little conjugative effect. Here changing the substituent should have little effect on the rate. A converse effect of substituents should be seen if the transition state veers to the other extreme (48). Here the rate should be sensitive to the nature of the substituent on the right-hand carbon atom, not to that on the left. Since, moreover, the effects of such substituents on the S_N1 reaction and on deprotonation are known, it should be possible to get a semiquantitative indication of the relative extents to which the CX and CY bonds are weakened in the transition state.

While no measurements of this kind have as yet been reported, this argument shows the potential of this kind of technique for the study of reaction mechanisms. The effects of $\pm E$ substituents on the rate of a reaction will depend on the structure of the transition state and these effects can be predicted for various possible structures by the PMO method. Most work in this area has used substituted phenyl groups as probes, the results being analyzed in terms of the Hammett relation. This is a much less satisfactory procedure, for three reasons. First, the introduction of polar substituents introduces possibly large and certainly unpredictable changes in the energies of solvation. Second, the effects of substituents of other types are much harder to treat theoretically than are those of neutral, nonpolar $\pm E$ substituents. Third, the introduction of such substituents is much more likely to perturb the balance between alternative transition states such as (47) and (48). It is therefore difficult to disentangle the true course of the unperturbed reaction from such measurements.

Many elimination reactions can lead to more than one possible product, the double bond produced by elimination appearing in different positions. Thus dehydrobromination of a secondary alkyl bromide R_1R_2CHBr may involve loss of hydrogen from either R_1 or R_2, while Hofmann elimination from a quaternary ammonium hydroxide $R_1R_2R_3R_4N^+OH^-$ may lead to loss of any one of the four alkyl groups. Early studies of these reactions led to the formulation of two conflicting empirical rules:

Hofmann's Rule. When the leaving group is positively charged, elimination leads mainly to the least alkylated olefin.

Saytzeff's Rule. When the leaving group is neutral, elimination leads mainly to the most alkylated olefin.

Typical examples illustrating these rules are

$$CH_3CH_2\underset{Br}{\overset{|}{C}}HCH(CH_3)_2 \xrightarrow{EtO^- \text{ in EtOH}} \underset{85\%}{CH_3CH_2CH=C(CH_3)_2} +$$

$$\underset{18\%}{CH_3CH=CHCH(CH_3)_2} \qquad (5.193)$$

$$CH_3CH_2\underset{\overset{|}{SMe_2}}{\overset{+}{C}}HCH_3 \xrightarrow{EtO^- \text{ in EtOH}} \underset{26\%}{CH_3CH=CH-CH_3} +$$

$$\underset{74\%}{CH_3CH_2CH=CH_2} \qquad (5.194)$$

Since sp^3–sp^2 bonds are stronger than sp^3–sp^3 bonds, reactions of this kind are generally more exothermic, the more alkyl groups there are attached to the double bond in the resulting olefin. If the overall elimination followed the BEP principle, Saytzeff's rule would then hold.

In an E_N1 elimination of HX from RX, the first (and rate determining) step is loss of X^- to form an intermediate carbonium ion R^+. The product is formed from this by removal of a proton by base. If two or more different protons can be lost, the product will be determined by the relative rates of the corresponding proton transfer reactions. Since proton transfer reactions are of BEP type (p. 240), the most stable olefin should then be favored. Indeed E_N1 reactions always follow Saytzeff's rule except in cases where steric hindrance favors an alternative route.

Conversely, in an $E1cB$ reaction, both the rate of reaction and the position of the double bond in the product are determined by the initial attack by base. The product formed will then depend on the ease of the initial deprotonation and hence on the stability of the resulting carbanion. Since alkyl groups are of $-I$ type and so destabilize anions, such reactions should tend to lead to olefins with as few alkyl groups as possible on the double

TABLE 5.6. Leaving Group Effects on Orientation in Elimination Reactions[a,b]

Alkyl halide	% 1-olefin (Hofmann)
$CH_3-CH_2-CH_2-CH(CH_3)-F$	82
$CH_3-CH_2-CH_2-CH(CH_3)-Cl$	35
$CH_3-CH_2-CH_2-CH(CH_3)-Br$	25
$CH_3-CH_2-CH_2-CH(CH_3)-I$	20
$CH_3-CH_2-C(CH_3)_2-F$	71
$CH_3-CH_2-C(CH_3)_2-Cl$	45, 32[c]
$CH_3-CH_2-C(CH_3)_2-Br$	38, 26[c]

[a] Date from Ref. 7.
[b] EtO$^-$ in EtOH at 78°.
[c] Ph—S$^-$ in EtOH

bond, particularly on the end of the double bond to which the proton was originally attached. Such a reaction would follow Hofmann's rule.

In the case of E_N2 reactions, the transition state can have structures ranging all the way from those very similar to that for an E_N1 reaction at one extreme to structures very similar to $E1cB$ at the other. The course of the reaction should depend on the position of the transition state in this range, reactions closer to the $E1cB$ end following Hofmann's rule and reactions closer to the E_N1 end following Saytzeff's rule.

Chemical Reactivity

The more easily the leaving group X in RX can separate as a negative ion, the closer should a corresponding E_N2 transition state (for elimination of HX from RX) approximate to the E_N1 extreme and the more it should tend to follow Saytzeff's rule rather than Hofmann's rule. Since quarternary ammonium ions such as $RNMe_3^+$ undergo S_N1 reactions much less easily than do analogous halides, it is not surprising to find that the Hofmann elimination follows Hofmann's rule. Likewise, since SR_2^+ is a much better leaving group in this connection than NR_3^+, it is also not surprising that elimination from sulfonium ions gives more Saytzeff-type products than do the corresponding reactions of ammonium ions.

Since, moreover, the ease of S_N1 reactions increases in the order $RF < RCl < RBr < RI$, one might equally expect E_N2 reactions to show an increasing tendency toward Saytzeff-type products in the same order. This is the case (Table 5.6), alkyl fluorides often differing from other halides in giving, as here, mainly Hofmann-type products.

The effects of solvents on the course of these reactions follow a pattern similar to that of S_N1/S_N2 reactions. Thus E_N2 reactions are favored by solvents of low ionizing power, E_N1 reactions by ionizing solvents. The choice between substitution and elimination can also be altered in a predictable way. Thus the E_N1 and S_N1 reactions involve a common intermediate carbonium ion. Substitution is favored if good nucleophiles are present, elimination is favored by bases. Hydroxide ions in water or aqueous ethanol are therefore good reagents for bringing about E_N1 reactions. Likewise, E_N2 reactions can be favored over S_N2 by using reagents which are better bases than nucleophiles. Alkoxide ion in a dry alcohol is a good choice. Dry alcohols are not very good ionizing solvents and so do not promote S_N1 and E_N1 reactions, while RO^- in ROH is a strong base but a poor nucleophile. Finally, since the transition state for an E_N2 reaction is more spread out than that for an analogous S_N2 reaction, the dispersal of charge is greater in the former. Therefore E_N2 reactions are favored over S_N2 reactions by a reduction in the solvating power of the medium.

5.17 π-Complex Reactions ($E\pi_B^+$)

A large number of molecular rearrangements are now known in which aliphatic substitution or elimination is accompanied by a 1,2 migration. The classic example is the Wagner–Meerwein rearrangement in which 1,2 migration of an alkyl group accompanies an S_N1 or E_N1 reaction; e.g.,*

---

* Primary alkyl chlorides do not easily undergo S_N1 reactions because primary alkyl cations are unstable and because primary alkyl derivatives easily undergo S_N2 reactions. Neopentyl chloride is exceptional in that the S_N2 reaction is sterically hindered. Moreover, Ag^+ promotes S_N1 reactions of chlorides by effectively "solvating" the resulting chloride ion, AgCl having a very negative heat of formation.

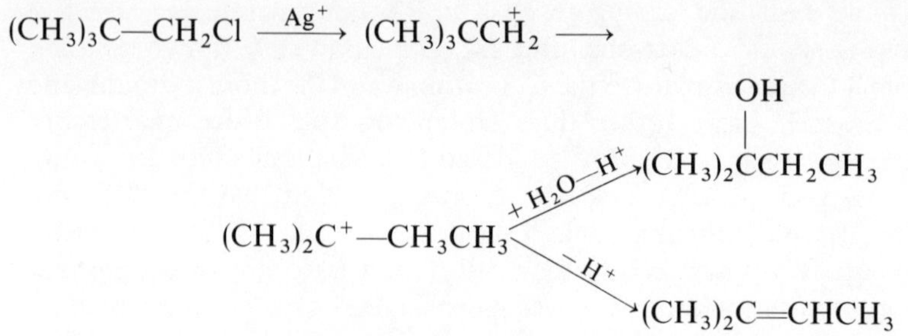

$$(CH_3)_3C\text{---}CH_2Cl \xrightarrow{Ag^+} (CH_3)_3CC\overset{+}{H}_2 \longrightarrow$$

$$(CH_3)_2\overset{+}{C}\text{---}CH_3CH_3 \begin{array}{c} \xrightarrow{+H_2O, -H^+} (CH_3)_2CCH_2CH_3 \\ | \\ OH \\ \xrightarrow{-H^+} (CH_3)_2C\text{=}CHCH_3 \end{array} \qquad (5.195)$$

Reactions of this kind formally involve rearrangements of intermediate carbonium ions by a 1,2 migration of alkyl. In this case the carbonium ion initially formed is an unstable primary ion. The rearrangement leads to its conversion to a tertiary carbonium ion in which three $-I$ alkyl groups are attached to the cationic center. The products are then derived from the resulting *tert*-amyl cation rather than neopentyl.

During such a reaction, the rearranging cation must pass through an intermediate where the migrating alkyl group R is half-way between the carbon atom it is leaving and the carbon atom to which it is going. It must then be attached to both carbon atoms by a three-center bond, formed from one AO of each carbon atom and one of R. Thus,

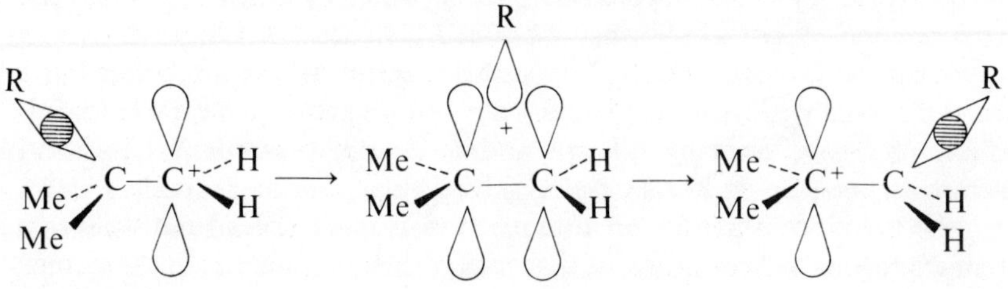

$$(5.196)$$

Let us then consider the structure of this intermediate.

For maximum overlap of the AO of R with the other two AOs, it will clearly be best for the latter to be of p type. Any hybridization will tend to tip the orbital away from R. The three-center bond, moreover, contains just two electrons, i.e., those in the original CR bond; for the $2p$ AO of the second carbon atom is initially empty. If then R were removed as R^+, we would be left with an olefin (here $Me_2C\text{=}CH_2$), the two $2p$ AOs combining to form two π MOs, one bonding and the other antibonding, and the two electrons then occupying the bonding π MO.

Now let us bring up R^+. The antibonding π MO will have a node on which R will more or less lie in the resulting adduct (Fig. 5.24a). The empty AO of R lies more or less on this node. Thus while the antibonding MO is antisymmetric with respect to reflection in the nodal plane, the AO of R is symmetric. The two orbitals do not therefore interact with one another, so the antibonding MO survives unchanged in the adduct.

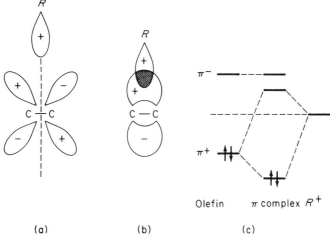

FIGURE 5.24. Formation of a π complex; (a) the antibonding π MO (π⁻) of the olefin does not interact with the empty AO of the acceptor R⁺; (b) interaction between the bonding π MO (π⁺) and the AO of R⁺; (c) changes in orbital energies on forming the π complex.

The bonding π MO, on the other hand, is symmetric and so can interact with the AO of R^+ (Fig. 5.24b). The situation is exactly the same as in the formation of a normal dative bond by interaction of two AOs, one full and the other empty (p. 17). Here the donor component is a filled MO but this makes no difference. An orbital represents a possible mode of motion of an electron in a given field of force. There is therefore no essential distinction between an AO and a MO and either can be used equally well to form further bonds. Indeed, we have already used this concept in our treatment of union. The union of two odd AH radicals is entirely analogous to the combination of a pair of hydrogen atoms. In the present case, the interaction depresses the energy of the bonding π MO, making it more strongly bonding, while raising the energy of the AO of R^+, making it antibonding (Fig. 5.24c).

Compounds of this type, containing dative bonds in which the donor is a filled π MO, are called π *complexes*. The dative bond is represented by an arrow just as in the case of analogous atom–atom bonds. Thus,

$$\underset{C=C}{\overset{R^+}{\uparrow}}$$

Compare

$$Me_3N \longrightarrow O \quad \text{or} \quad Me_3N \longrightarrow CH_3^+ \; (\equiv Me_3N^+ - CH_3) \quad (5.197)$$

As the last comparison emphasizes, the bond linking the olefin to R^+ should be a normal covalent bond. Although represented for convenience as a dative bond, it involves no separation of charge. It should be entirely analogous to the covalent NC bonds in NMe_4^+ which, as (5.197) indicates, can be represented in a π-complex-like way as $Me_3N \longrightarrow CH_3^+$.

$$\begin{array}{c} CH_3^+ \\ \uparrow \\ CH_2 = CH_2 \end{array} \qquad\qquad CH_3-CH_2-CH_2^+$$

(49)   (50)

Let us consider the two isomeric ions (49) and (50) that can be formed by addition of CH_3^+ to ethylene.

Formation of the π complex (49) is exothermic by E_π, the bond energy of the olefin–CH_3 covalent bond.

The heat of reaction in forming $n\text{-Pr}^+$ (50) can be estimated from the following thermocycle:

$$CH_3\cdot \quad CH_2=CH_2 \xrightarrow{-E_{C-C}-E'_{C-C}+E_{C=C}} CH_3-CH_2-CH_2$$

$$-e \Big| +I_{Me} \qquad\qquad\qquad\qquad\qquad\qquad -e \Big| +I_{Pr} \qquad (5.198)$$

$$CH_3^+ \quad CH_2=CH_2 \xrightarrow{\Delta H} CH_3-CH_2-CH_2^+$$

In the addition of a methyl radical to ethylene, a $C=C$ bond (bond energy $E_{C=C}$) is replaced by one sp^3–sp^3 C—C bond (bond energy E_{C-C}) and one sp^3–sp^2 C—C bond (bond energy E'_{C-C}). Substituting values for the three bond energies ($E_{C=C}$ 146, E_{C-C} 83, E'_{C-C} 86 kcal/mole) and the ionization potentials of the methyl and n-propyl radicals (I_{Me} 227, I_{Pr} 187 kcal/mole), we find

$$\begin{aligned} \Delta H &= -E_{C-C} - E'_{C-C} + E_{C=C} + I_{Pr} - I_{Me} \\ &= -63 \quad \text{kcal/mole} \end{aligned} \qquad (5.199)$$

This is no more than the bond energy of a rather weak covalent bond. The heat of formation of the π complex (49) from $CH_3^+ + CH_2=CH_2$ should therefore be at least comparable with that of the $n\text{-Pr}^+$ cation (50). Indeed, recent detailed MO calculations have indicated that n-alkyl cations should be less stable in the gas phase than π complexes analogous to (49). In solution, the situation will be different because the solvation energy of cations such as (50) should be greater than that of an isomeric π complex [e.g., (49)]. The cation has an empty AO that can interact strongly with molecules of solvent while empty orbitals in the π complex are antibonding.

Nevertheless, it is not surprising to find that rearrangements of this kind proceed very easily in solution and that in cases where the balance between the classical ion and the π complex is disturbed, the latter may exist as a stable entity even in solution. The classic example of this is the norbornyl cation (51), which rearranges on formation to the π complex (52). The con-

Chemical Reactivity

version of (51) to (52) involves a decrease in strain, both because the geometry of the dative bond in (52) is more favorable* than that of the C—C bond in (51) and because it can be deformed at less cost in energy, being weaker.

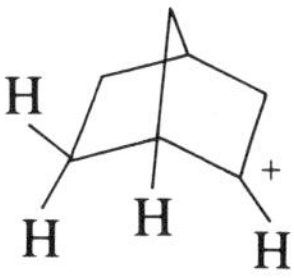

(51) 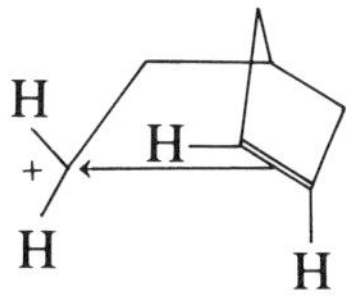   (52)   i.e.,   (53)

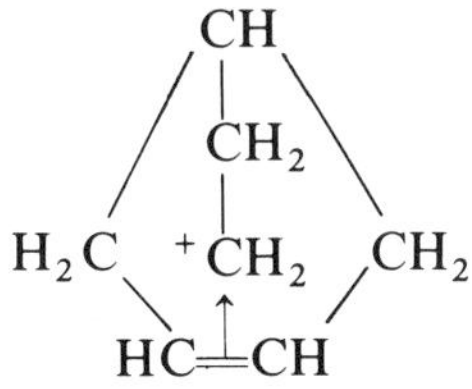

Since all the bonding MOs in a π-complex cation are full, the corresponding radical would have one electron in an antibonding MO while the corresponding anion would have two antibonding electrons. Wagner–Meerwein-type rearrangements involving migration from one carbon atom to another are therefore rarely observed in the case of radicals and never with carbanions because the intermediate π complexes would have too high energies.

The effect of substituents on the equilibrium between a carbonium ion and an isomeric π complex can easily be predicted. In the carbonium ion, one atom carries a full unit of formal charge while the others are neutral. In the π complex, the charge is shared between the *apical* carbon atom in the acceptor and the two *basal* carbon atoms in the olefin moiety. The apical carbon atom and one basal carbon atom are therefore more positive in the π complex than in the carbonium ion, while the other basal atom is much more positive in the carbonium ion. If then we replace one of both hydrogen atoms at a basal position in (49) by $-I$ groups, e.g., alkyl, we will stabilize one of the classical ions very strongly relative to the π complex. There is little doubt that in the rearrangement of neopentyl cation [equation (5.195)], the intermediate π complex immediately rearranges to the more stable *t*-amyl cation,

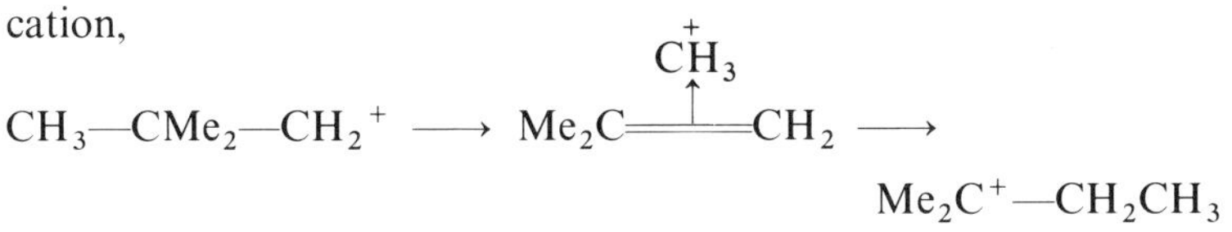

$$CH_3-CMe_2-CH_2^+ \longrightarrow Me_2C\overset{\overset{+}{C}H_3}{=\!=\!=}CH_2 \longrightarrow \\ Me_2C^+-CH_2CH_3 \quad (5.200)$$

Likewise in the case of the norbornyl cation (52), a methyl substituent is sufficient to displace the equilibrium in favor of the "classical" ion [see (54)].

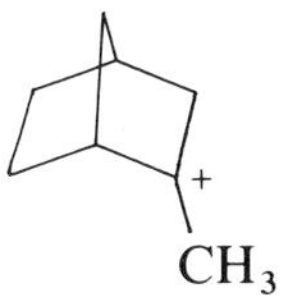

(54) 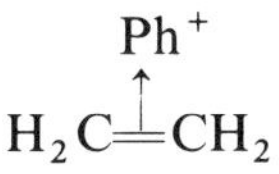   (55)   (56) $PhCH_2CH_2^+$

---

* This can be seen very easily from molecular models.

Apical − *I* substituents should have the converse effect, the apical carbon atom becoming more positive on forming the π complex. The tendency of alkyl groups to migrate ("migratory aptitude") therefore increases in the order CH_3 < primary alkyl < secondary alkyl < tertiary alkyl.

An entirely different mode of stabilization of π complexes is seen in cases where the apical group is unsaturated. Consider, for example, the π complex (55) formed from (phenyl)$^+$ + ethylene. If the benzene ring is oriented perpendicular to the CC bond of the ethylene, the nodal plane of the phenyl π MOs coincides with the nodal plane of the antibonding ethylenic π MO (Fig. 5.25a). Since the filled π MOs of phenyl are antisymmetric with respect to reflection in that plane, they can interact with with the antibonding ethylenic π MO. Since the latter is empty, the result is a second dative bond, this time of π type, opposite in direction to the normal σ-type dative bond linking ethylene to Ph$^+$. The effect of course is to greatly stabilize the π complex relative to the corresponding "classical" carbonium ion (56), phenyl in the latter being attached to a saturated carbon atom. The double dative bonding in the π complex is conveniently indicated by the symbol in Fig. 5.25(b).

Furthermore, as a comparison of Figs. 5.25(a) and 5.25(c) shows very clearly, the π complex (55) is isoconjugate with the benzyl cation $PhCH_2{}^+$. In each case, the π MOs of the benzene ring interact with a single empty adjacent orbital. The formation of the π complex from an isomeric carbonium ion is therefore a typical $EO_B{}^\ddagger$ process, entirely analogous to the S_N1 ionization of a benzyl ester. The stabilities of ions with apical groups R should therefore run parallel to those of the corresponding $RCH_2{}^+$ ions (p. 255). In cases where R is a $\pm E$ group, the stability of the π complex should parallel the $\pm E$ activity of the group. Likewise, $-I$ or $-E$ substituents at positions of like parity to CH_2 in RCH_2 should stabilize the π complex, while $+I$ or $+E$ substituents should destabilize it. Thus *p*-nitrophenyl is less effective, and *p*-anisyl, more effective, than phenyl.

A huge literature now exists concerning reactions of this kind; reviews are listed at the end of this chapter. In many cases, the π complex is not only more stable than the classical carbonium ion, but can be formed from it without activation. If the carbonium ion is produced by an S_N1 process, it starts to rearrange to the π complex while it is being formed. The rearrangement consequently acts as a driving force in the ionization, so S_N1 reactions of this kind take place faster than analogous ones where no concerted rearrangement of the cation to a π complex occurs. The subsequent reaction of the π complex at one of its basal positions with a nucleophile involves a concerted reversion of the apical group to the other basal position; thus,

$$\begin{array}{c} R \\ \diagdown \\ CH_2\text{---}CH_2 \\ \diagdown_X \end{array} \longrightarrow \begin{array}{c} R^+ \\ \downarrow \\ CH_2\text{=}CH_2 \\ \diagdown_{Y^-} \end{array} \longrightarrow \begin{array}{c} R \\ \diagdown \\ CH_2\text{---}CH_2 \\ \diagdown \\ Y \end{array} \qquad (5.201)$$

Chemical Reactivity　　　　　　　　　　　　　　　　　　　　　　　　　　　291

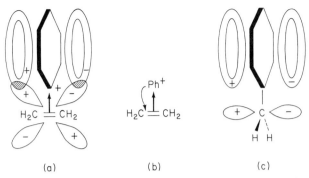

FIGURE 5.25. (a, b) π complex formed from ethylene + Ph^+; (c) the benzyl cation.

The second step is clearly analogous to an S_N2 reaction, the $2p$ AO of the central carbon atom being replaced by the bonding π MO of the olefin. The reaction should therefore involve attack from the backside of the R–olefin bond, as indicated in equation (5.201). In cases where X is replaced by Y without rearrangement but via such a concerted π-complex mechanism, the replacement is then stereospecific, leading to complete retention of configuration. Reactions of this type can therefore be distinguished by the fact that they take place faster than would be expected for the simple S_N1 mechanism and lead to complete retention of configuration. They are classed as $E\pi_B^{\ddagger}$, the rate-determining formation of the π complex following the BEP principle since it is analogous to the S_N1 reaction.

A classic example of an $E\pi_B^{\ddagger}$ reaction is the solvolysis of *exo*norbornyl esters [(57) → (58)]. These take place faster than the solvolyses of analogous cyclopentyl esters, lead exclusively to *exo* derivatives (i.e., configuration is retained), and in the case of optically active norbornyl derivatives, lead to completely racemic products. The intermediate here is the π complex (52), which has a plane of symmetry, as is seen if the ion is written in the form (53). Recently Olah *et al.* have been able to obtain stable solutions of (52) in fluorosulfonic acid mixtures and to establish its structure.

Numerous other examples will be found in the reviews listed at the end of this chapter.

An interesting confirmation of these ideas is provided by the rates of solvolysis of β-arylethyl esters, $ArCH_2CH_2X$, with ArH being an alternant aromatic hydrocarbon. Primary chlorides normally react by the S_N2 mechanism. The rates of such reactions should not vary with the aryl group since this is neutral and cannot conjugate with the transition state. On the other hand, an $E\pi_B^\ddagger$ reaction will lead to an intermediate π complex isoconjugate with $ArCH_2^+$. The rates of such reactions should therefore vary with Ar in the same way as the S_N1 reaction of $ArCH_2X$, a plot of $\log k$ against a_{0r} (the NBMO coefficient of CH_2 in $ArCH_2$) being linear (see Figure 5.19). Consequently, a plot of $\log k$ against a_{0r} for the solvolysis of $ArCH_2CH_2X$ might be expected to consist of two parts. For the less reactive compounds (i.e., a_{0r} large) the S_N2 reaction will predominate, so the plot should be horizontal, the rate being independent of a_{0r}. At some point, the $E\pi_B^\ddagger$ mechanism should, however, take over, so the plot should then change to a line of finite slope. Moreover, the formation of the ionic π complex from neutral ester should be strongly accelerated by ionizing solvents. The crossover from S_N2 to $E\pi_B^\ddagger$ should therefore occur at larger values of a_{0r} in good ionizing solvents than in poor ones.

Furthermore, the $S_N1/E\pi_B^\ddagger$ mechanism should be favored relative to S_N2 by a decrease in the ability of the solvent to solvate carbonium ions and an increase in its tendency to solvate anions. Such a change in the solvent should therefore move the crossover point from S_N2 to $E\pi_B^\ddagger$ to larger values of a_{0r}.

Figure 5.26 shows plots of $\log k$ vs. a_{0r} for the solvolyses of 2-arylethyl tosylates in acetic acid, formic acid, and trifluoracetic acid. Along this series, acidity increases, so nucleophilicity decreases, while the solvating power for anions by hydrogen bonding increases. In acetic acid, the plot consists, as predicted, of two lines, the crossover point being near α-naphthyl. In trifluoracetic acid, all the points lie on a single line, even phenylethyl tosylate reacting by the $E\pi_B^\ddagger$ mechanism. In formic acid, phenylethyl tosylate is just beginning to deviate.

In the reactions so far considered, the intermediate π complex was formed by rearrangement of a carbonium ion. In certain cases, however, a π complex can be formed by direct combination of an olefin with a carbonium ion. The classic example is 7-*anti*-norbornenyl chloride (59), which solvolyses one hundred billion times faster than the saturated analog (61). The acceleration is due to internal attack by the double bond, leading to the π complex (60). The reaction is indeed a typical S_N2 process in which the double bond serves as a nucleophile, emphasizing once again the complete equivalence of a filled π MO and a filled AO as donors for forming dative bonds.

(59)   (60)   (61)   (62)

This example raises a further point. The bond in a π complex linking the apical group to the olefin moiety is a normal covalent bond so far as the

Chemical Reactivity

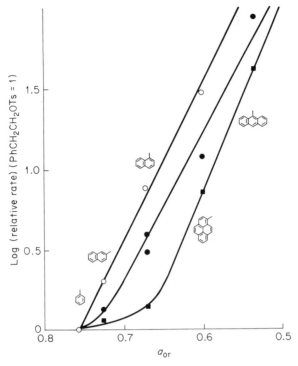

FIGURE 5.26. Plots of log k vs. a_{or} for solvolyses of β-arylethyl tosylates (ArCH$_2$CH$_2$OTs) in acetic acid (■), formic acid (●), and trifluoroacetic acid (○). Data from Ref. 8.

apical group itself is concerned. Thus if the apical group is alkyl, the carbon atom adjacent to the olefin will be tetrahedral, forming three normal σ bonds and one π bond, as one may term a dative bond in which the donor is an MO instead of an AO. If the alkyl group was initially asymmetric, it will retain its configuration in the π complex. Consequently, optically active groups migrate with complete retention of configuration in Wagner–Meerwein rearrangements. Moreover, the reaction of (60) with a nucleophile will itself be an S_N2 process, the double bond now acting as leaving group,

Comparison with (59) and (60) shows that the overall effect is to replace the chlorine in (59) by Y with complete retention of configuration at the 7-position. These reactions do indeed take place in this way, providing further evidence that the π complex (60) is an intermediate.

A further prediction of the π-complex theory is that the π complex should have the geometry indicated in (60), the hydrogen at the 7-position being tilted out of the CCC plane because the adjacent carbon atom is quadricovalent and should therefore have a tetrahedral geometry. This carbon should also be tilted over toward the double bond to which it is covalently bound. Both these predictions have been confirmed in the case of the analogous 7-norbornadienyl cation (62), which can be obtained as a stable species in super acids.

If olefins can act as donors, one might expect them to act as ligands to metal ions. Complexes of this kind have indeed been known for many years, the classic examples being the complexes formed from olefins and silver cations (63) and platinum derivatives such as Zeise's salt (64).

$$\begin{array}{cccc} \text{Ag}^+ & \left[\begin{array}{c}\text{H}_2\text{C}\\ \text{H}_2\text{C}\end{array}\Vdash \text{PtCl}_3\right]^- \text{K}^+ & & \\ (63) & (64) & (65) & (66) \end{array}$$

These complexes are stable only when the acceptor is a transition metal with filled d AOs that can be used for back-coordination. As indicated in (65), a filled d AO of the metal has the correct symmetry to interact with the empty antibonding MO of the olefin, just like the π MOs of phenyl in the π complex of Fig. 5.25 (a). In this case, the double dative bonding [see (66)] is particularly effective because it enables bonding of the ligand to the metal to occur without producing a large positive charge on the olefin. The structures of a number of these π complexes have now been determined and conform nicely to the predictions of the π-complex theory. Thus the CC bond in the olefin is still partly double, being shorter than a C—C single bond, but at the same time, it is longer and weaker than a normal C=C double bond. This is because the bonding π electrons are now diluted in the CC region since they are shared with a third atom and electrons are also fed back from the metal into the antibonding π MO. The carbon atoms still have nearly planar geometries, though some distortion from planarity is observed due to repulsions between the adjacent groups and the rest of the complex.

So far, π complexes have been regarded as dative complexes in which an olefin acts as the donor. One can, however, also regard a π complex as a "nonclassical" structure in which the two basal atoms and the apical atom are linked by a three-center bond. Bonds of this type are well known in the electron-deficient compounds of group III elements, e.g., the boron hydrides and trimethylaluminum (see p. 48). These bonds are usually formed by the mutual overlap of three orbitals on the atoms, the resulting structure being consequently isoconjugate with cyclopropenium; e.g., compare

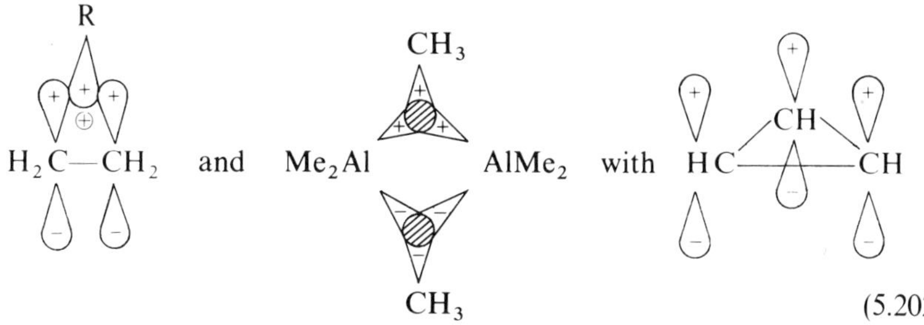

Chemical Reactivity

$$\text{(5.202)}$$

The special stability of compounds containing such bonds can be attributed to their being isoconjugate with the cyclopropenium cation and consequently aromatic. Likewise, the failure of radicals and carbanions to undergo Wagner–Meerwein rearrangements could be attributed to the fact that the intermediate π complexes would be isoconjugate with the nonaromatic cyclopropenium radical or the antiaromatic cyclopropenium anion.

These arguments suggest that π complexes should be merely a special case of a general class of nonclassical carbonium ions containing three-center bonds. In practice, however, very few reactions have been found that involve nonclassical ions other than π complexes. Two exceptions are the trishomocyclopropenium ion (68), which seems to be an intermediate in $S_N 1$ reactions of bicyclohexyl derivatives (67), and edge-protonated cyclopropane (69), which seems to be an intermediate in reactions of cyclopropane with acid*

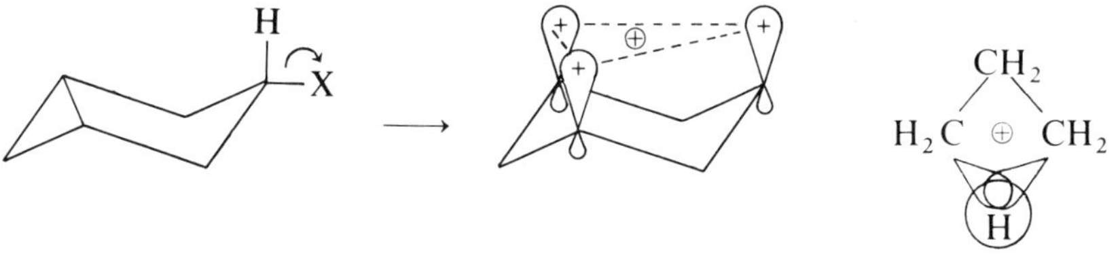

(67) (68) (69)

This representation does, however, lead to an important conclusion concerning the effects of E-type substituents on the stability of π complexes. A $\pm E$ substituent stabilizes an odd AH very strongly because the latter has an NBMO (see p. 167). The interactions between the NBMO and the filled, bonding MOs of the $\pm E$ substituent are much greater than the bonding–antibonding interactions that arise when such a substituent is attached to an even system. Now while cyclopropenium is odd, it is *not* alternant. The MOs in it are consequently all bonding or antibonding. The interaction between a $\pm E$ substituent such as phenyl and cyclopropenium is therefore

---

* Recent calculations indicate that (69) should be a stable species, intermediate in stability between the isomeric $CH_3CH_2CH_2^+$ and $CH_3\overset{+}{C}HCH_3$. Franklin *et al.* have shown that the ion formed by protonating cyclopropane in the gas phase has these properties.

no greater than that between a $\pm E$ substituent and an even AH. Since this second-order interaction is taken into account in the sp^2–sp^2 bond energy, phenyl has *no* mesomeric stabilizing effect on cyclopropenium. Since phenyl is also a much weaker $-I$ substituent than alkyl, due to the greater electronegativity of sp^2 carbon, phenyl groups are actually *less* effective in stabilizing cyclopropenium than are alkyl groups. This observation, which surprised Breslow, the discoverer of cyclopropenium, is seen to follow as an immediate consequence of PMO theory.

The analogy between cyclopropenium and π complexes suggests that the same situation should also hold in the latter. Phenyl groups at basal positions in π complexes should exert no mesomeric stabilizing effect. Solvolysis of benzyl derivatives should therefore take place by normal S_N1 mechanisms rather than by the $E\pi_B^{\ddagger}$ process because the phenyl group can stabilize the carbonium ion in the former but not the π complex in the latter. This indeed is the case. If ions of this type undergo Wagner–Meerwein rearrangement, they do so only subsequently in a distinct step involving activation. There is never any question of a concerted rearrangement to a π complex during the initial ionization. This is true even when a very good group (e.g., phenyl) is available for the apical position in the π complex. The stable form is still the classical carbonium ion. A classic example is the scrambling of phenyl groups in the solvolysis of 1,2,2-triphenylethyl derivatives,

$$Ph_2CH\!-\!CHPh\!-\!X \xrightarrow{\text{slow}} Ph_2CH\!-\!\overset{+}{C}HPh \xrightarrow{\text{slow}}$$

$$\underset{PhCH=\!=\!=CHPh}{\overset{Ph^+ \uparrow}{}} \longrightarrow Ph\overset{+}{C}H\!-\!CHPh_2 \quad \text{etc.} \quad (5.203)$$

Scrambling does not take place by concerted phenyl migration during the initial ionization.

While the π-complex theory was proposed before any definite examples were known and while the term π complex and the associated arrow symbolism have been generally adopted in inorganic chemistry, organic chemists have mostly preferred to use the term "nonclassical carbonium ion" and to represent the corresponding three-center bonds by dotted lines; e.g.,

$$\underset{CH_2\!-\!-\!-\!-\!-\!-CH_2}{\overset{CH_3^+}{\triangle}} \quad \equiv \quad \underset{CH_2=\!=\!CH_2}{\overset{\overset{+}{CH_3} \uparrow}{}} \quad (5.204)$$

The trouble with this "neutral" notation is that it conveys none of the definite orbital and stereochemical implications that the π-complex notation does and as a result it can prove, and indeed has proved, very misleading, as numerous examples in the recent literature show.

5.18 Electrophilic Addition ($E\pi_B^{\ddagger}$ and $EO_B^{\ddagger}$)

Addition reactions are processes in which a reagent XY adds across a double bond,

$$C=C + X-Y \longrightarrow \underset{\underset{C-C}{|\ \ |}}{\overset{X\ \ Y}{|\ \ |}} \qquad (5.205)$$

For reasons that will become apparent later in this chapter (Section 5.26), reactions of this type never take place by a single concerted process. The groups X and Y add in two separate steps, either as X^+ followed by Y^- (*electrophilic addition*) or as X^- followed by Y^+ (*nucleophilic addition*) or as $X\cdot$ followed by $Y\cdot$ (*radical addition*). Here we will consider electrophilic additions in which the rate-determining step is a bimolecular reaction between the olefin and XY to form an intermediate cation and Y^-,

$$(R_2C=CR_2) \overset{\frown}{\ } X\overset{\frown}{-}Y \longrightarrow C_2R_4X^+ + Y^- \qquad (5.206)$$

The reaction is completed by a rapid reaction between the cation and Y^- to form the product. As one might expect, the intermediate cation is free to react with other anions or nucleophiles that may be present. Thus ethylene reacts with chlorine in water to form a chlorohydrin,

$$C_2H_4 \overset{\frown}{\ } Cl\overset{\frown}{-}Cl \longrightarrow C_2H_4Cl^+ + Cl^- \qquad (5.207)$$

$$C_2H_4Cl^+ \overset{\frown}{\ } OH_2 \longrightarrow CH_2ClCH_2OH + H^+ \qquad (5.208)$$

Clearly, X^+ is an electrophile, the AO used by X to form the X—Y bond being empty. Additions of X^+ to an olefin can therefore give rise either to a carbonium ion (71) or to a π complex (72):

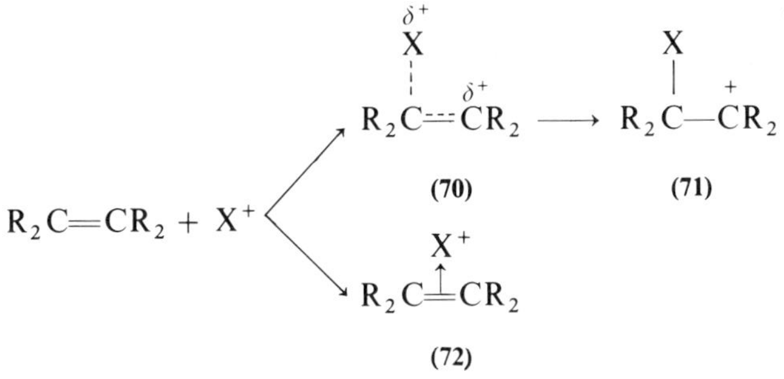

$$R_2C=CR_2 + X^+ \begin{cases} \overset{\overset{\delta^+}{X}}{\underset{R_2C\overset{\delta^+}{=\!=\!=}CR_2}{|}} \longrightarrow \underset{R_2C-CR_2}{\overset{X}{\underset{|}{|}}}\overset{+}{\ } \\ (70) \qquad\qquad (71) \\ \\ \underset{R_2C\overset{X^+}{\overset{\uparrow}{=\!=\!=}}CR_2}{\ } \\ (72) \end{cases} \qquad (5.209)$$

The latter reaction is clearly of $E\pi_B^{\ddagger}$ type. In the former, the transition state equally clearly contains no bonds that are not present either in the reactants or in the products. The reaction is therefore of BEP type and is classed $EO_B^{\ddagger}$.

Formation of the π complex (72) takes place without any possibility of rotation about the double bond. Likewise, the reaction of (72) with Y^- will involve attack *trans* to the apical group X (see p. 290). Addition by the π-complex mechanism should therefore be stereospecifically *trans*. Thus in cyclopentene,

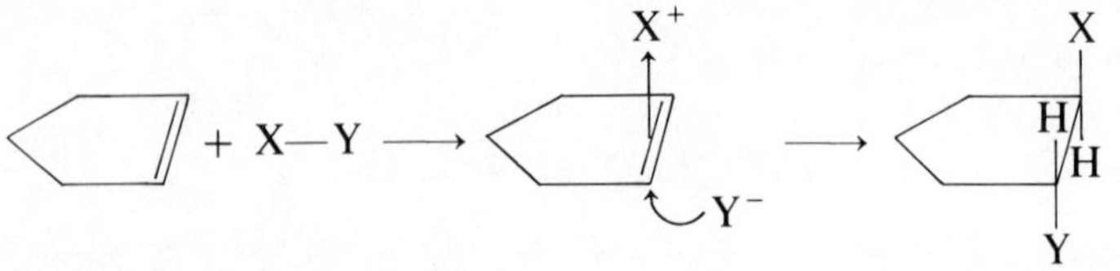

$$\begin{array}{c}\text{(diagram)}\end{array} \quad (5.210)$$

Addition via a carbonium ion can, on the other hand, lead equally well to *cis* or *trans* adducts.

Next let us consider how the relative stabilities of the π complex and the carbonium ion will vary with changes in X.

In forming the carbonium ion, there is a change in net bond energy equal [see equation (5.198)] to:

$$E_{CX} + E_{C-C} - E_{C=C} \qquad (5.211)$$

In forming the π complex, the change in bond energy is just E_π. The difference in total bond energy between the classical ion and the π complex is then

$$(E_{CX} - E_\pi) + (E_{C-C} - E_{C=C}) \qquad (5.212)$$

The strengths of the CX and π bonds should run parallel to one another since both are bonds from X to carbon. Since, however, the CX bond is stronger, variations in its bond energy with X should be greater than those in the π-bond energy. Therefore the stronger the CX bond, the more the carbonium ion should be favored.

A second difference between the carbonium ion and the π complex is that the former is derived from a nonbonding carbon AO and the latter from a bonding π MO. Since the π MO has a lower energy than the AO, the olefin in the π complex behaves like an atom more electronegative than carbon. Consequently, X in the π complex is more positive than in the carbonium ion. Thus the conversion of the carbonium ion to the π complex involves a transfer of δq electrons from X to carbon,

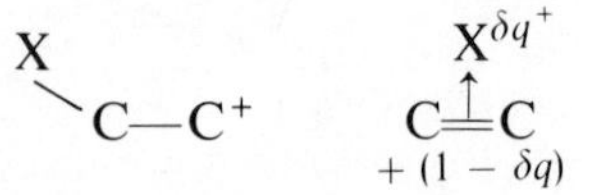

$$\begin{array}{cc} X & X^{\delta q^+} \\ \diagdown C-C^+ & C\!\!=\!\!C \\ & +(1-\delta q) \end{array} \qquad (5.213)$$

The corresponding change in energy due to this electron transfer is then

$$(\alpha_C - \alpha_X)\,\delta q \qquad (5.214)$$

where α_C and α_X are the Coulomb integrals of carbon and X. Thus the more electronegative X and so the more negative α_X, the more the classical ion is favored.

A final difference between the carbonium ion and the π complex arises when X has unshared p electrons. In this case, back-coordination can occur in the π complex as in the π complex of Fig. 5.25(a); i.e.,

$$\text{(5.215)}$$

The interaction between the AO of X and the antibonding π MO will be smaller, the further apart they are in energy. Since the AO of X is lower in energy than the antibonding π MO, the interaction will be smaller, the lower is the energy of the AO, i.e., the more electronegative is X.

While these three factors cannot be assessed quantitatively, we can account qualitatively for the available evidence in terms of them. Consider, for example, electrophilic addition of an acid to an olefin,

$$R_2C=CR_2 + HY \longrightarrow C_2HR_4{}^+X^- \longrightarrow R_2CHCR_2Y \quad (5.216)$$

The proton has no p electrons, so back-coordination cannot occur. Hydrogen is also approximately equal in electronegativity to carbon. However, the CH bond is much stronger than C—C. Since simple carbonium ions are more stable in solution than are isomeric π complexes, the same should be even more true for π complexes with an apical proton. Addition of acids to olefins should therefore take place by the carbonium ion mechanism—as indeed it does.

Next let us consider addition of ICl to an olefin. Since iodine is less electronegative than chlorine, the iodine will add first as I^+. Iodine has p electrons and is not very strongly electronegative. Also, the CI bond is very weak. These reactions therefore invariably take place by the π-complex mechanism giving *trans* adducts; e.g.,

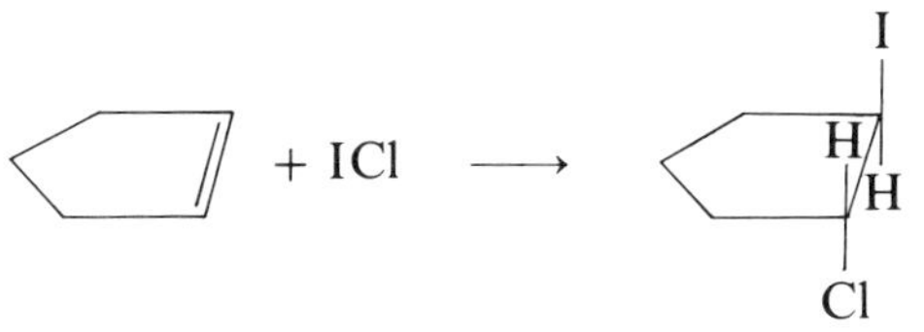

$$\text{(5.217)}$$

Bromine is more electronegative than iodine and the CBr bond is stronger than CI. Nevertheless, electrophilic addition of bromine usually takes place by the $E\pi_B{}^{\ddagger}$ mechanism,* giving only *trans* adducts. One can however, bring about addition of Br^+ via the carbonium ion route $(EO_B{}^{\ddagger})$ by introducing into the olefin $-E$ substituents which can stabilize the intermediate carbonium ion. An extreme example is (73), where the intermediate ion (74) can be isolated as a salt.

---

* Provided that the addition is electrophilic.

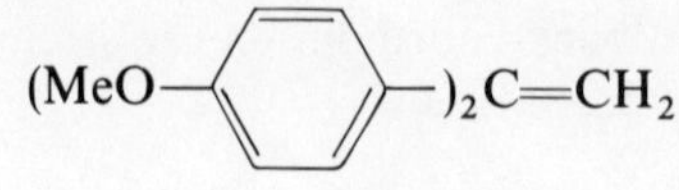

(MeO—⟨◯⟩—)$_2$C=CH$_2$ (MeO—⟨◯⟩—)$_2$C$^+$CH$_2$Br

(73) 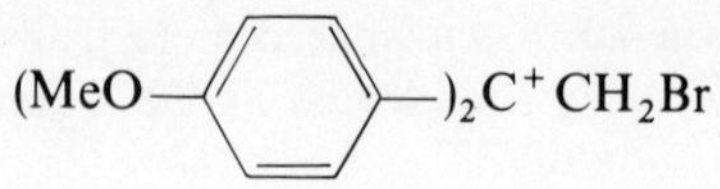   (74)

Chlorine is even more electronegative than bromine and the CCl bond is still stronger. Chlorine still adds *trans* to simple olefins by the $E\pi_B^{\ddagger}$ mechanism, but even a $\pm E$ group such as phenyl is sufficient to divert the reaction to $EO_B^{\ddagger}$. Thus *trans*-stilbene (75) reacts with chlorine to give both *meso* (76) and *dl* (77) dichlorides in similar amounts:

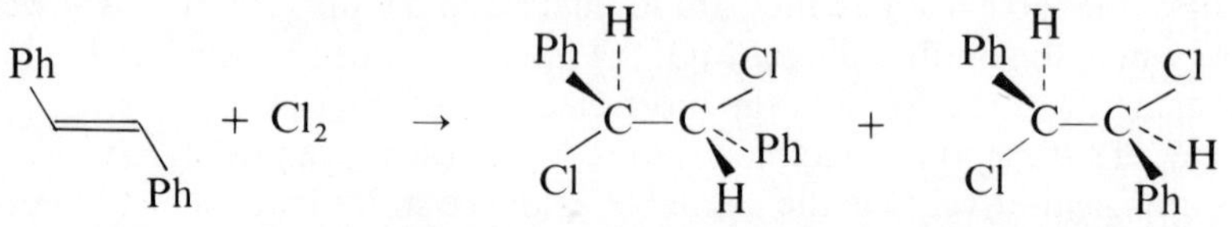

(75) Ph–CH=CH–Ph + Cl$_2$ → (76) + (77)

Fluorine would undoubtedly add by the $EO_B^{\ddagger}$ mechanism rather than $E\pi_B^{\ddagger}$. However the F—F bond is so very weak that fluorine reacts by radical mechanisms, leading to both addition and substitution. Simple adducts therefore cannot be obtained.

5.19 π Complexes vs. Three-Membered Rings

An oxygen atom in the ^1D state has the configuration $(1s)^2(2s)^2(2p_x)^2(2p_y)^2$. It therefore has an empty $2p$ AO and two filled $2p$ AOs. Such an atom can therefore combine with ethylene to form a π complex, using one of its filled p AOs for back-coordination [cf. equation (5.215)],

(78)   (79)

The product is clearly ethylene oxide, for which one can thus write a π-complex structure. On the other hand, one can also write a "normal" structure for ethylene oxide containing bent ("banana") bonds formed by the overlapping of sp^3 hybrid AOs,

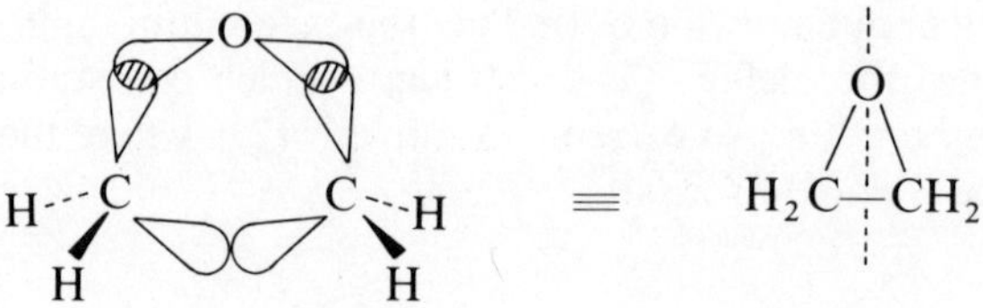

(80)   (81)

Which of these two representations is correct?

Since ethylene oxide has a plane of symmetry bisecting the CC bond [marked by a dotted line in (81)], its MOs must be either symmetric or antisymmetric with respect to reflection in this plane. The MOs in the π-complex representation (78) satisfy this criterion but the bent-bond orbitals in (80) do not. We can, however, construct from the two bent-bond orbitals ϕ_1 and ϕ_2 two linear combinations $\phi_1 + \phi_2$ and $\phi_1 - \phi_2$, one of which is symmetric and the other antisymmetric. Since the orbitals ϕ_1 and ϕ_2 are filled with electrons, an equivalent representation is obtained by replacing the orbitals ϕ_1 and ϕ_2 by the combinations $\phi_1 \pm \phi_2$. It can be shown that the latter representation is in fact identical with the MO or π-complex representation. The relationship between the two representations is indeed just just like that between the MO and localized bond descriptions of methane (p. 28). The answer to our question is therefore "either or both, because they are equivalent."

There is, however, one important factor to be considered, namely the hybridization of the carbon atoms. In a "pure π complex," the carbon atoms should have sp^2 hybridization, while in a "pure classical structure," their hybridization is sp^3.

Hybrid AOs form stronger σ bonds than do plane p AOs because the greater part of the hybrid is concentrated in a single lobe. This of course is no advantage in the case of a π bond, so π bonds formed by hybrid AOs are, if anything, weaker than ordinary $p\pi:p\pi$ bonds. Hybridization will similarly strengthen the CX bonding in (78) by making the lobes of the π MOs of the olefin unequal in size, the larger lobes pointing toward X. At the same time it will tend to weaken the other σ bonds formed by the carbon atoms by diverting some of the s character of the relevant carbon AOs. The extent to which hybridization will be favorable in (78) therefore depends on the strength of the CX interactions. If these are large in the case where the carbon atoms use $2p$ AOs for bonding, the increase in their strength when the carbon atoms use hybrid AOs will be proportionately large. Hybridization will then be extensive. If, on the other hand, the CX interactions are small, the gain in CX bonding through the use of hybrid AOs will not be sufficient to offset the loss in the bond energies of the other bonds.

In the case of a simple π complex where the apical atom X has no filled AOs available for back-coordination, the olefin $\rightarrow$ X bond will be relatively weak because it is of dative type. Donation of electrons from the olefin to X makes the olefin more electronegative and X less electronegative. These changes limit the extent to which electrons can be donated to X. If, however, X has an AO available for back-coordination, the reverse dative bonding feeds electrons back from X to the olefin. This reduces the electronegativity of the olefin and raises that of X, thus enabling an additional transfer of charge in the σ-type dative bond olefin $\rightarrow$ X. The result is a kind of positive feedback in which the charge transfer produced by each dative bond reinforces the other. Since the olefin $\rightarrow$ X bond reduces the electron density in the bonding π MO of the olefin and since the X $\rightarrow$ olefin dative bond introduces electrons into the antibonding π MO, the result is a great increase in the strength of the CX bonding combined with a decrease in the strength of of the CC bond.

Because of the positive feedback effect, the resulting structure will tend to one or the other of two extremes. If back-coordination is not strong enough to lead to significant charge transfer, the CX bonding will remain weak and the CC bonding strong. In this case the carbon atoms should use AOs near to $2p$ in type. Such a structure will behave like a π complex and is best represented as such. If, however, back-coordination is strong, forward and backward dative bonding will reinforce one another, so that the CX bonds become strong and the CC bonds weak. In this case, the carbon atoms should use hybrid AOs probably not far from sp^3 in type, so the resulting structure is best represented by a three-membered ring.

There should also be a corresponding change in the length of the CC bond. In the π complex, where the population of the bonding π MO is not much less than in the corresponding olefin and the antibonding π MO is barely populated, the CC bond should be near to double, in agreement with the π complex representation. If, however, forward and backward coordination are both strong, the CC bonding π MO is depopulated while electrons are fed into the antibonding π MO. The net effect will be virtually to eliminate the CC π bond, the bond between the carbon atoms being best represented as single, i.e., as in the classical ring structure.

The crucial factor is then the strength of the interaction between the filled AO of X and the antibonding π MO of the olefin. This is a second-order perturbation and will therefore depend on how far the orbitals are apart in energy. If X is comparable with carbon in electronegativity, the interaction will be strong. If, however, X is very much more electronegative than carbon, the filled AO of X will lie so far below the antibonding π MO in energy that the interaction will be very weak.

Figure 5.27 sums up these conclusions. In Fig. 5.27(a), the AOs of X are comparable with those of carbon in energy, so the interactions are strong. The structure should then correspond to a classical ring. In Fig. 5.27(b), on the the other hand, X is extremely electronegative, so the separation of its AO from the antibonding π MO is very large and the interaction between them correspondingly weak. The structure will now correspond to that of a π complex.

While it should in principle be possible to have a whole range of intermediate structures, changing from one extreme to the other over a range of electronegativity of X, the transition is in fact likely to occur rather suddenly because of the positive feedback. The moment back-coordination exceeds a critical value, the structure will tend to flip over from one best represented by a π complex to one approximating to a classical three-membered ring.

The difference in geometry has a very important effect on the reaction of (78) or (80) with anions. Reactions of this type occur in the second phase of electrophilic addition [equation (5.210)] and are analogous to S_N2 reac-

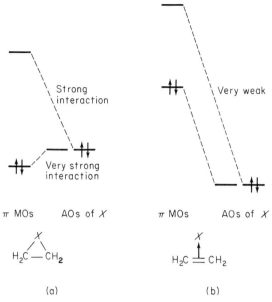

FIGURE 5.27. (a) When X differs little from carbon in electronegativity, the compound C_2H_2X is best represented by a classical structure; (b) when X is far more electronegative than carbon, C_2H_4X is best represented by a π-complex structure.

tions. Similar processes also lead to ring opening in epoxides, azirines, and thioepoxides; e.g.,

$$CH_2\overset{O}{\underset{\underset{\displaystyle -OMe}{\diagup}}{\diagdown}}CH_2 \longrightarrow \underset{\displaystyle CH_2-CH_2OMe}{\overset{\displaystyle O^-}{|}} \xrightarrow{H^+} HOCH_2CH_2OMe \quad (5.218)$$

There is, however, an important difference in cases where the ethylene moiety is asymmetrically substituted. This can be seen from a comparison of the products formed by addition of iodine chloride to propene and by reaction of propylene oxide with methoxide ion.

In the first case, an intermediate π complex is formed which reacts with Cl^- as follows:

$$CH_3-CH\overset{I^+}{=\!=\!=}CH_2 \qquad CH_3-\underset{\underset{\displaystyle Cl}{|}}{CH}-CH_2I \quad (5.219)$$
$$Cl^-$$

The nucleophile attacks the *more* alkylated carbon atom. The product is therefore formed by addition of the electrophile to the *less* substituted end of the double bond and by subsequent addition of the nucleophile to all electrophilic addition reactions including those of $EO_B{}^{\ddagger}$ type,* a statement known as *Markownikov's rule*.

On the other hand, nucleophilic attack on propylene oxide takes place at the *less* substituted carbon atom, e.g.,

$$\underset{\underset{\displaystyle {}^-OCH_3}{\big\uparrow}}{CH_3\!-\!\overset{\displaystyle O}{\overset{\displaystyle /\,\backslash}{CH\!-\!CH_2}}} \longrightarrow CH_3\!-\!\underset{\displaystyle |}{\overset{\displaystyle O^-}{CH}}\!-\!CH_2OCH_3$$

$$\longrightarrow CH_3\!-\!\underset{\displaystyle |}{\overset{\displaystyle OH}{CH}}\!-\!CH_2OCH_3 \quad (5.220)$$

This in fact is the orientation that would be expected if the reaction were a simple S_N2 process, as is implied in equation (5.219). Primary alkyl derivatives react faster than secondary ones and secondary faster than tertiary. It is therefore the addition reactions that are apparently anomalous.

The reason for this difference is probably mainly stereochemical. The decrease in reactivity in the S_N2 reaction in the series Me > Et > iPr > tBu is due to repulsion between the alkyl group and the solvation sphere of the nucleophile. If the basal atoms in (78) or (80) have approximately sp^3 hybridization, their geometry will be similar to that in normal alkyl derivatives, so similar steric effects should be observed (Fig. 5.28a). If, however, the carbon atoms use $2p$ AOs to bond X, the situation will be as represented in Fig. 5.28(b). Not only is the alkyl group no longer in the line of fire of the approaching nucleophile, but the carbon atom also has a nice large backside lobe to interact with the nucleophile. Both basal positions are therefore accessible. If this is the case, one would then expect attack to occur preferentially at the more alkylated carbon atom for the following reasons. Since alkyl groups have a $-I$ effect, they will alter the relative electronegativities of the carbon atoms so that they are no longer equal. The π MO will then be polarized toward the less alkylated carbon as indicated in Fig. 5.28(c). The apical group X should therefore be attached off center, being nearer to the less alkylated carbon atom. Attack by a nucleophile at the more alkylated carbon should then be facilitated both by the lower electron density there and by the fact that X is initially nearer the other carbon atom.

The electronegativities of neutral oxygen, neutral nitrogen, and neutral sulfur are not very much greater than that of carbon. Nucleophilic ring

---

* Since the $EO_B{}^{\ddagger}$ reactions are of BEP type, they must proceed via the more stable possible intermediate classical carbonium ion. This will clearly be the one with most alkyl ($-I$) groups attached to the cationic center. The electrophile must therefore add to the less alkylated carbon atom, thus following Markownikov's rule.

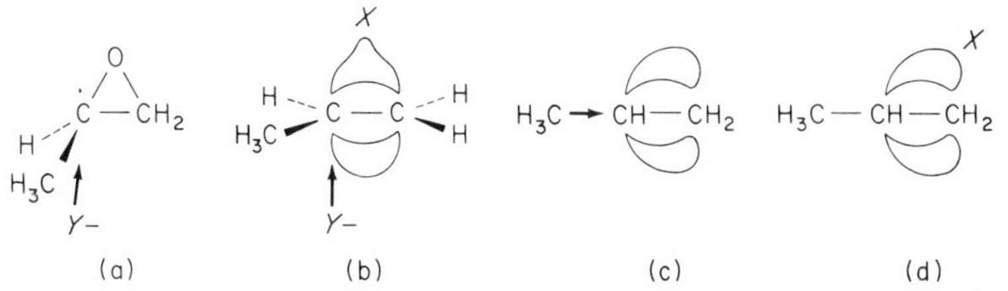

Chemical Reactivity

FIGURE 5.28. (a) Steric hindrance to approach of a nucleophile in propene oxide; (b) lack of steric hindrance in analogous approach to a propene π complex; (c) polarization of the π MO in propene by the inductive effect of methyl; (d) a symmetric structure of a propene π complex.

opening of epoxides, aziridines, and thioepoxides consequently follows the S_N2 pattern. Even positively charged nitrogen and sulfur are not excessively electronegative; "onium" salts derived from aziridines and thioepoxides therefore also follow the S_N2 pattern. All these compounds are best represented by classical structures with three-membered rings. For example,

$$CH_3\text{—}\overset{\overset{S}{\diagup\diagdown}}{CH\text{—}CH_2} + HO^- \longrightarrow CH_3\overset{\overset{S^-}{|}}{CH}\text{—}CH_2OH \qquad (5.221)$$

$$CH_3\text{—}\overset{\overset{\overset{+}{N}Me_2}{\diagup\diagdown}}{CH\text{—}CH_3} + HO^- \longrightarrow CH_3\text{—}\overset{\overset{NMe_2}{|}}{CH}\text{—}CH_2OH \qquad (5.222)$$

On the other hand, Cl^+, Br^+, and I^+ are all extremely electronegative. The corresponding cyclic compounds consequently react with nucleophiles according to Markownikov's rule, i.e., in an anti-S_N2 manner. These compounds are best represented as π complexes [see equation (5.219)]. A very nice final touch is provided by oxygen. As we have seen, propylene oxide reacts with nucleophiles in S_N2 manner [e.g., equation (5.220)]. However, acid-catalyzed ring opening leads to Markownikov-type products formed by nucleophilic attack by solvent on protonated epoxide,

$$\underset{CH_3OH}{CH_3CH\overset{\overset{\overset{+}{OH}}{\diagup\diagdown}}{\text{——}}CH_2} \longrightarrow CH_3\overset{|}{CH}\text{—}CH_2OH \qquad (5.223)$$
$$\phantom{CH_3CH\overset{\overset{+}{OH}}{\text{——}}CH_2 \longrightarrow CH_3CH} OCH_3$$

It was at one time thought that the difference might be due to a change of mechanism, the protonated epoxide undergoing S_N1 ring opening to a carbonium ion,

$$CH_3CH\overset{\overset{+}{O}H}{\underset{}{\diagup\!\!\!\!\diagdown}}CH_2 \longrightarrow CH_3\overset{+}{C}H\text{—}CH_2OH \qquad (5.224)$$

This, however, is not the case since the ring opening is still completely stereospecific. It is due simply to a change in geometry with a change in electronegativity of the apical group. Thus while epoxides are best represented by classical structures, their salts should be written as π complexes, e.g.,

$$CH_3\text{—}CH\overset{\overset{+}{O}H}{\underset{\uparrow}{=\!=}}CH_2$$

(82)

5.20 Nucleophilic Addition and Related Reactions ($EO_B^\ddagger$)

In the addition of a nucleophile to a C=C bond, four electrons are involved, i.e., the pair of electrons in the relevant orbital of the nucleophile and the π electrons of the double bond. Since a π complex has only one bonding orbital, nucleophilic addition cannot take place by a π-complex mechanism. The addition therefore follows a path analogous to $EO_B^\ddagger$ electrophilic addition, i.e.,

$$Y:\!\!\diagdown\overset{}{\underset{}{>}}\!C\!\!=\!\!C\!\!<\quad \longrightarrow \quad \overset{Y^{\delta-}}{\underset{}{>}}\!C\!\cdots\!C\!\!<^{\delta-}\quad \longrightarrow \quad \overset{Y}{\underset{}{>}}\!C\!-\!\bar{C}\!\!< \qquad (5.225)$$

Comparison of equations (5.45) and (5.225) shows that this process is analogous to a normal S_N2 reaction. The only difference is that the leaving group (here the right-hand carbon atom) is initially doubly bound to the central carbon so that instead of separating as a free, negative ion, it still remains bound to the central carbon by a single bond. There is, however, one important difference from normal S_N2 reactions. Here the central atom is initially unsaturated, with sp^2 hybridization, and its hybridization in the transition state is intermediate between that in the reactant (sp^2) and that in the product (sp^3). The effect of E-type substituents there on the transition state is therefore intermediate between their effects on the reactants and on the products. The reactions are consequently of BEP type, being classified as $EO_B^\ddagger$.

Since carbon is only weakly electronegative, carbanions are extremely poor leaving groups, so S_N2 reactions do not lead to the cleavage of C—C σ bonds. The situation in equation (5.225) is more favorable for an S_N2-type reaction since the C—C π bond is much weaker than a C—C σ bond. Never-

Chemical Reactivity

theless, simple olefins still fail to react in this way. Addition occurs only if the resulting carbanion is strongly stabilized by positive substituents, $+E$ being naturally the best. Nucleophiles readily add to compounds containing such groups adjacent to a double bond, the so-called Michael reaction, e.g.,

$$NC^- \; CH_2{=}CH{-}\underset{\underset{R}{|}}{C}{=}O \longrightarrow N{\equiv}C{-}CH_2{-}CH{=}\underset{\underset{R}{|}}{C}{-}O^-$$

$$\xrightarrow{H^+} N{\equiv}C{-}CH_2{-}CH_2{-}CO{-}R \quad (5.226)$$

$$MeS^- \; CH_2{=}CH{-}C{\equiv}N \longrightarrow MeSCH_2{-}CH{=}C{=}N^-$$

$$\xrightarrow{H^+} MeSCH_2CH_2CN \quad (5.227)$$

$$(EtOOC)_2CH^- \; CH_2{=}CH{-}\underset{\underset{OEt}{|}}{C}{=}O$$

$$\longrightarrow (EtOOC)_2CH{-}CH_2{-}CH{=}\underset{\underset{OEt}{|}}{C}{-}O^-$$

$$\xrightarrow{H^+} (EtOOC)_2CH{-}CH_2{-}CH_2COOEt \quad (5.228)$$

Analogy with the S_N2 reaction suggests that the rates of nucleophilic additions should be greater, the more nucleophilic is the attacking nucleophile and the less basic is the resulting carbanion. This is the case. Thus compounds of the type $H_2C{=}CXY$, where X and Y are both $+E$ substituents, are extremely reactive. Likewise, $-I$ and $-E$ substituents at the reaction center again retard the reaction, though this time for a different reason.

The S_N2 reaction is retarded by $-I$ substituents because of steric effects in the transition state and by $-E$ substituents because of unfavorable electronic interactions in the transition state. Neither of these factors appears in the reactants or products, so the decrease in rate is not accompanied by a corresponding change in the energy of reaction. However in the case of an olefin $RCH{=}CHX$, where R is a $-E$ or $-I$ group and X a $+E$ group, mutual conjugation will occur between R and Y. The resulting stabilization disappears in the product, where R and X are now separated by a C—C single bond, and it will be correspondingly attenuated in the transition state. The retarding effect of $-I$ and $-E$ substituents is due to this decrease in mesomeric stabilization on passing from the reactants to the transition state, a decrease which is reflected by a corresponding change in the energy of reaction. Nucleophilic additions are therefore always of BEP type.

In the Michael reaction, addition to an otherwise inert C=C bond is facilitated by $+E$ substituents. An alternative course is to replace one of the carbon atoms by a more electronegative atom. Thus the carbonyl group readily adds nucleophiles; e.g.,

$$N\equiv C^- \curvearrowright CR_2 = O \curvearrowright \longrightarrow N\equiv C-CR_2O^- \xrightarrow{H^+} N\equiv C-CR_2OH$$
(cyanhydrin reaction)
(5.229)

$$BrMg-CH_3 \curvearrowright CR_2 = O \curvearrowright \longrightarrow CH_3CR_2O^-MgBr^+ \quad (5.230)$$
(Grignard reaction)

$$H_2N \curvearrowright CR_2 = O \curvearrowright \longrightarrow \underset{NHPh}{H_2\overset{+}{N}-CR_2-O^-}$$

$$\longrightarrow \underset{NHPh}{HN-CR_2OH} \longrightarrow \underset{NHPh}{N=CR_2}$$
(phenylhydrazone formation)
(5.231)

The last step in equation (5.231) is an elimination of water, brought about here either by an $E_N 1$ or by an $E1cB$ mechanism; i.e.,

$$PhNHNHCR_2OH \xrightarrow{H^+} PhNH-\overset{H}{\underset{\curvearrowleft}{N}}-CR_2\overset{+}{\underset{\curvearrowright}{-}OH_2} \quad (5.232)$$

$$\searrow PhNHN=CR_2$$

$$PhNHNHCR_2OH \xrightarrow{B} PhNH\overset{\curvearrowleft}{\underset{}{N}}-CR_2\overset{\curvearrowright}{\underset{}{-}}OH \nearrow \quad (5.233)$$

A similar series of reactions is involved in the *aldol condensation*, where a stabilized carbanion adds to a carbonyl group; e.g.,

$$CH_3CO\bar{C}H_2 \curvearrowright \underset{Ph}{CH=O} \curvearrowright \longrightarrow CH_3COCH_2\underset{Ph}{CHO^-}$$

$$\xrightarrow[-H_2O]{+H^+} CH_3COCH=CHPh \quad (5.234)$$

Since nitrogen is less electronegative than oxygen, Schiff bases react less readily with nucleophiles than the analogous carbonyl compounds. However they do still add strong nucleophiles such as Grignard reagents. Nucleophilic addition can also be catalyzed by acids, the Schiff base being

Chemical Reactivity

converted to a salt in which neutral nitrogen is replaced by the more electronegative N^+. Since nitrogen is quite basic, such acid-catalyzed reactions can occur in solutions which are neutral or even weakly alkaline. The Strecker amino acid synthesis follows this pattern, involving two nucleophilic additions with an intervening elimination,

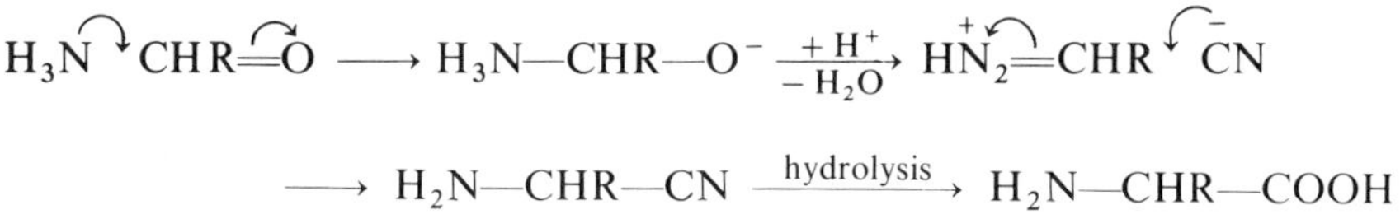

$$H_3N{:}\,CHR{=}O \longrightarrow H_3N-CHR-O^- \xrightarrow[-H_2O]{+H^+} \overset{+}{H}N_2{=}CHR\,{:}CN$$

$$\longrightarrow H_2N-CHR-CN \xrightarrow{\text{hydrolysis}} H_2N-CHR-COOH$$

(5.235)

Since nucleophilic addition is a two-step reaction, the intermediate anion may be diverted in directions other than that leading to simple addition. The most important of these involve ejection of an anion from a position adjacent to the anionic center, or addition to a second molecule of olefin. The first of these leads to an overall substitution, the latter to polymerization.

Substitution occurs in compounds of the type $YCH{=}CHX$, where X is a $+E$ substituent and Y is a suitable leaving group. Thus Michael reactions with β-chlorovinyl ketones lead to replacement of chlorine by the nucleophile, e.g.,

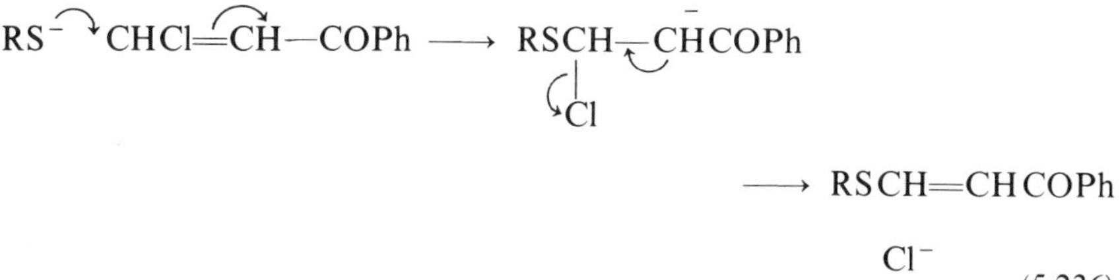

$$RS^-{:}\,CHCl{=}CH-COPh \longrightarrow RSCH-CHCOPh$$
$$\qquad\qquad\qquad\qquad\qquad\qquad\quad |$$
$$\qquad\qquad\qquad\qquad\qquad\qquad\;\,Cl$$

$$\longrightarrow RSCH{=}CHCOPh$$
$$\qquad\qquad\qquad Cl^-$$

(5.236)

The most important reactions of this type occur with derivatives of carboxylic acids, e.g.,

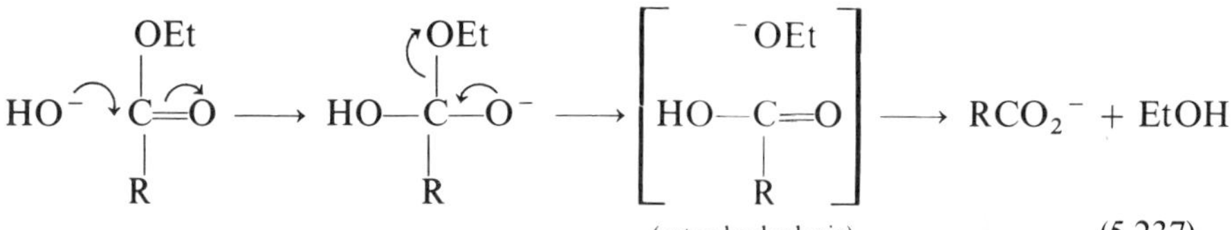

$$HO^-{:}\,\overset{\overset{\displaystyle OEt}{|}}{\underset{\underset{\displaystyle R}{|}}{C}}{=}O \longrightarrow HO-\overset{\overset{\displaystyle OEt}{|}}{\underset{\underset{\displaystyle R}{|}}{C}}-O^- \longrightarrow \left[HO-\overset{\overset{\displaystyle ^-OEt}{|}}{\underset{\underset{\displaystyle R}{|}}{C}}{=}O\right] \longrightarrow RCO_2^- + EtOH$$

(ester hydrolysis) (5.237)

$$EtOOC\bar{C}H_2{:}\,\overset{\overset{\displaystyle OEt}{|}}{\underset{\underset{\displaystyle CH_3}{|}}{C}}{=}O \longrightarrow EtOOCCH_2-\overset{\overset{\displaystyle OEt}{|}}{\underset{\underset{\displaystyle CH_3}{|}}{C}}-O^- \longrightarrow EtOOCHCH_2\overset{\overset{\displaystyle ^-OEt}{|}}{\underset{\underset{\displaystyle CH_3}{|}}{C}}{=}O$$

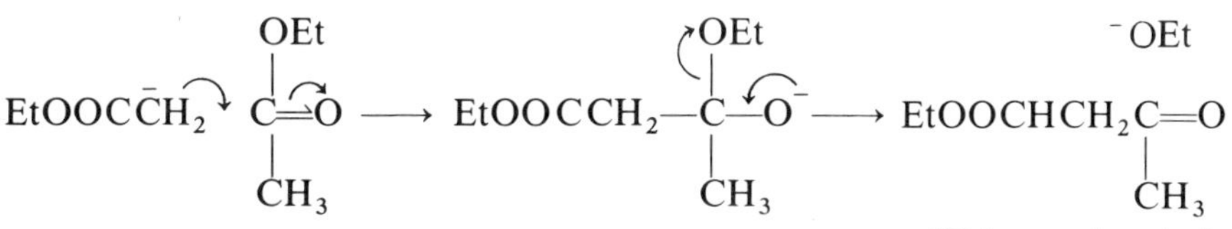

(Claisen condensation)

(5.238)

Nucleophilic polymerization takes place when vinyl ketones or acrylic esters are treated with base, e.g.,

$$\text{RO}^- \quad \text{CH}_2{=}\text{CHCOR} \longrightarrow \text{ROCH}_2\overset{-}{\text{CH}}\text{COR}$$

$$\text{ROCH}_2\overset{-}{\text{CH}} \quad \text{CH}_2{=}\text{CHCOR} \longrightarrow \text{ROCH}_2\text{CHCH}_2\overset{-}{\text{CH}}\text{COR}$$
$$\quad\;\;|\qquad\qquad\qquad\qquad\qquad\qquad\quad\;|$$
$$\text{COR}\qquad\qquad\qquad\qquad\qquad\;\;\text{COR}$$

$$\text{ROCH}_2\text{CHCH}_2\overset{-}{\text{CH}} \quad \text{CH}_2{=}\text{CHCOR} \qquad\qquad (5.239)$$
$$\;\;\;|\qquad\;\;\;|$$
$$\text{COR}\;\;\text{COR}$$

$$\longrightarrow \text{ROCH}_2\text{CHCH}_2\text{CHCH}_2\overset{-}{\text{CH}}\text{COR} \quad \text{etc.}$$
$$\qquad\qquad\;\;\;|\qquad\;\;\;|$$
$$\qquad\qquad\text{COR}\;\;\text{COR}$$

5.21 Radical Addition and Polymerization ($EO_B^{\ddagger}$)

Radical addition, like nucleophilic addition, is barred from a π-complex mechanism by the presence of an extra electron which would have to go into an antibonding MO in the π complex. Reactions of this kind therefore take place by $EO_B^{\ddagger}$ mechanisms exactly analogous to those in nucleophilic and $EO_B^{\ddagger}$ electrophilic addition, except that the intermediates are radicals instead of negative or positive ions.

Reactions of this kind differ from ionic additions in four respects.

First, since radicals are neutral, their solvation energies are small. No expensive desolvation of the reagents is necessary for them to come together. The activation energies of radical reactions in general, and of radical addition reactions in particular, therefore tend to be much less than those of analogous ionic reactions.

Second, since radical reactions lead to no displacements of charge, the electronegativities of the atoms involved play no role. The ease of addition to a double bond depends solely on the strength of its π component. Since CC π bonds are weaker than those involving heteroatoms, radical addition takes place most readily to C=C bonds.

Third, since radicals usually react very exothermically with one another in pairs to form "normal" closed shell molecules, it is usually impossible to obtain good yields of products by simple addition of radicals in pairs.

Fourth, on the other hand, radicals are durable in the sense that the reaction of a radical with a normal molecule necessarily gives rise to another radical, since the total number of electrons in the product must still be odd. Reactive ions, on the other hand, can be rapidly inactivated by combining

Chemical Reactivity

with, e.g., protons or anions derived from a protic solvent. Radical reactions can therefore take place by chain mechanisms in which a radical, once formed, can lead to the reaction of thousands or even millions of normal molecules. Reactions of this kind can take place rapidly, even though the concentrations of radicals is extremely small, because radical reactions tend to be very fast for the reasons indicated above.

A radical chain reaction consists of one or more repeating reactions constituting the propagation cycle (see p. 274). In the case of radical addition of X—Y to an olefin, the propagation steps are

$$\underset{}{\overset{X}{\underset{|}{>}C}}-\dot{C}< \;+\; X-Y \;\longrightarrow\; \underset{}{\overset{X}{\underset{|}{>}C}}-\underset{}{\overset{Y}{\underset{|}{C}}}< \;+\; X\cdot \qquad (5.240)$$

$$>C=C< \;+\; X\cdot \;\longrightarrow\; \underset{}{\overset{X}{\underset{|}{>}C}}-\dot{C}< \qquad (5.241)$$

Since a chain reaction of this kind can occur only if both steps are fast, and since even thermoneutral reactions involve activation, it is essential that both steps be exothermic, or at least not endothermic. In the first step [equation (5.240)], a CY bond is formed at the expense of the XY bond and any resonance stabilization of the initial radical. This reaction will be exothermic only if the CY bond energy is greater than the XY bond strength *plus* the resonance stabilization. Likewise, the second step will be exothermic if the CX bond energy is greater than the strength of the CC π bond, i.e., the CC π-bond energy *less* the resonance energy of the resulting radical.

Since the CCl and CBr bonds are strong, both chlorine and bromine add readily to olefins by radical chain mechanisms. The reactions are usually initiated photochemically, blue light dissociating Cl_2 or Br_2 into atoms. Iodine adds much less readily because the CI bond is rather weak.

Addition of Br_2 or Cl_2 can of course also take place by the electrophilic mechanism. Since this involves ionic intermediates, it occurs much faster in polar solvents. Ionic addition therefore prevails in solvents such as water or acetic acid. Radical reactions take place almost equally well in solvents of all kinds. Radical addition is therefore favored by nonpolar solvents such as carbon tetrachloride and by the presence of radicals or atoms produced, e.g., by light.

Compounds of the type HZ can add only if H plays the role of Y in equation (5.240), because bonds to hydrogen (HZ) are invariably stronger than corresponding bonds to carbon (CZ). Hydrogen chloride and hydroxyl compounds (ROH) are also unable to add because the HO and HCl bonds are stronger than CH. Compounds of the type HCR_3 can add if one of the groups R can stabilize the radical R_3C resulting from hydrogen abstraction [equation (5.240), Y = H]. This is true (see Table 4.3) if R is an *E*-type substituent. The first step [equation (5.240)] is then exothermic. Reactions

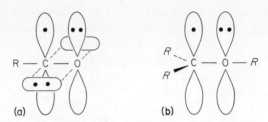

FIGURE 5.29. Comparison of the three-electron π bonds in (a) an acyl radical; (b) an alkoxyl radical. The dots represent π electrons.

of this kind do not, however, take place very easily, because abstraction of hydrogen from HCR_3 involves a change in hybridization of carbon and so requires extra activation (see p. 245). An interesting example is the addition of aldehydes to double bonds,

$$\underset{\underset{\text{COR}}{|}}{>}\!\!C\!\!-\!\!\dot{C}\!\!<\ +\ RCHO\ \longrightarrow\ \underset{\underset{\text{COR}}{|}}{>}\!\!C\!\!-\!\!CH\!\!<\ +\ R\dot{C}O \qquad (5.242)$$

$$>\!\!C\!\!=\!\!C\!\!<\ +\ R\dot{C}O\ \longrightarrow\ \underset{\underset{\text{COR}}{|}}{>}\!\!C\!\!-\!\!\dot{C}\!\!< \qquad (5.243)$$

The radical RĊO is stabilized by an interaction of a lone-pair AO of oxygen with the singly occupied $2p$ AO of carbon, forming a three-center π bond, while the original CO π bond remains untouched. Oxygen thus behaves as a $-E$ group. See Fig. 5.29.

Hydrogen iodide also fails to add by a radical chain mechanism, partly because the CI bond is weak and partly because it can add extremely easily by the electrophilic path. Hydrogen bromide can, however, add by either mechanism, the choice depending on the conditions. Electrophilic addition is favored by polar solvents (e.g., alcohol) and the presence of *radical inhibitors*.* Radical addition is favored by nonpolar solvents such as hexane or carbon tetrachloride and the presence of radical initiators such as peroxides.† Hydrogen sulfide and thiols (RSK) also undergo radical addition to olefins very easily. In this case, electrophilic addition cannot occur because thiols are very weak acids.

---

* An inhibitor is a compound that can combine with radicals to form adducts which, while still radicals, are too highly stabilized by mesomerism to react with normal molecules. An inhibitor thus effectively removes radicals and so suppresses radical reactions.
† Since the OO bond is very weak, peroxides easily decompose to form radicals; ROOR → 2RȮ. A radical initiator is a compound which easily gives rise to radicals, usually by rupture of a weak bond.

Chemical Reactivity

Alkyl radicals react very rapidly with Cl_2, Br_2, HBr, or HSR, according to equation (5.240). The rate-determining step is consequently attack by the resulting atom or radical on the olefin [equation (5.241)]. Since the reaction is of BEP type, it will be facilitated by any substituent that can stabilize the resulting radical, i.e., $\pm E$, $+ E$, or $- E$ substituents (Table 4.3). Alkyl groups act as weak $\pm E$ groups in this connection, being isoconjugate with vinyl. Thus styrene ($PhCH\!=\!CH_2$), methyl acrylate ($CH_2\!=\!CHOOCH_3$), vinyl acetate ($CH_2\!=\!CHOAc$), and propene ($CH_2\!=\!CHCH_3$) all react more easily with radicals than does ethylene. Addition to the double bond will moreover take place at the unsubstituted carbon in order that the resulting radical be stabilized by the substituent. Thus in the addition of hydrogen bromide to propene, the propagation steps are

$$BrCH_2\dot{C}HCH_3 + HBr \longrightarrow BrCH_2CH_2CH_3 + Br\cdot \quad (5.244)$$

$$Br\cdot + CH_2\!=\!CHCH_3 \longrightarrow BrCH_2\dot{C}HCH_3 \quad (5.245)$$

Note that addition of HX by this mechanism gives products violating Markownikov's rule. Thus electrophilic addition of HBr to propene gives isopropyl bromide, $CH_3CHBrCH_3$, not n-propyl bromide. The reversal of addition of HBr under conditions where the radical chain mechanism operates is synthetically useful since it allows one to convert terminal olefins into primary alkyl derivatives.

As in the case of nucleophilic addition, the intermediate radical in a radical addition can add to another molecule of olefin and repetition of this step can give rise to a polymer. Such radical chain polymerization of vinyl derivatives, commonly termed vinyl polymerization, is of very great technical importance, being the procedure used to make the majority of industrial polymers. The most important *monomers* used in such polymerizations are ethylene, styrene ($PhCH\!=\!CH_2$), methyl methacrylate ($CH_2\!=\!CMeCOOMe$), vinyl acetate ($CH_2\!=\!CHOAc$), acrylonitrile ($CH_2\!=\!CHCN$), vinyl chloride ($CH_2\!=\!CHCl$), butadiene ($CH_2\!=\!CH\!-\!CH\!=\!CH_2$), and isoprene ($CH_2\!=\!CH\!-\!CMe\!=\!CH_2$).

The radicals formed at each step in such a polymerization clearly have identical resonance energies. Thus in the polymerization of styrene, each radical in the growing chain is an α-phenylalkyl radical, namely

$$\sim\!\!\sim\!\!CH_2\underset{Ph}{\dot{C}H}\ \curvearrowright\ CH_2\!=\!\dot{C}HPh \longrightarrow\ \sim\!\!\sim CH_2\underset{Ph}{CH}\!-\!CH_2\underset{Ph}{\dot{C}H}$$

$$\xrightarrow{+\ CH_2=CHPh}\ \sim\!\!\sim CH_2\underset{Ph}{CH}CH_2\underset{Ph}{CH}CH_2\underset{Ph}{\dot{C}H}\ \text{etc.}$$

$$(5.246)$$

The exothermicity of the polymerization is therefore much the same in all cases,* being equal to the difference in bond energy between two C—C single bonds and a C=C double bond, i.e., ~ 20 kcal/mole of monomer. Substituents do not therefore greatly alter the rate of polymerization, the extra reactivity of the olefin being offset by a corresponding stabilization of the attacking radical. A very large variety of polymers can therefore be made by radical chain polymerization of the corresponding ethylene derivatives.

There are two special features of these reactions which can be explained rather simply in terms of the PMO method.

First, the rate of the propagation step, e.g., equation (5.246), is less, the more stable is the intermediate radicals. This of course represents a deviation from the BEP principle since the heats of polymerization of different vinyl compounds are similar. However, the deviations are relatively small, so the reaction can still be regarded as of $EO_B^\ddagger$ type.

We will consider the polymerization of CH_2=CHR where R is a $\pm E$ group. The argument can easily be extended to groups of other kinds. The radical is effectively an odd AH, $R\dot{C}H_2$. Denote the NBMO coefficient at the methylene group in this by a_{0r}. The transition state (83) for addition is is clearly isoconjugate with the 1,3-disubstituted allyl radical (84). If we neglect variations in β, the NBMO coefficients of the atoms adjacent to R will be equal and opposite ($\pm a'_{0r}$). The same must be true for all pairs of corresponding atoms in the two R groups. Each coefficient occurs twice; hence in order that the NBMO of (84) be normalized, we must have

$$a'_{0r} = (1/\sqrt{2})a_{0i} \qquad (5.247)$$

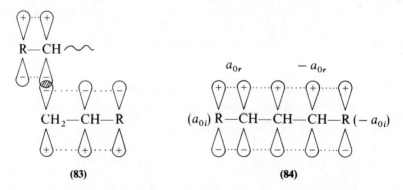

(83)   (84)

We can now construct the reactants and transition state by union of RH,

---

* This of course is true only if no steric effects are present. Polymerization of a compound of the type CH_2=CR_1R_2 gives a polymer in which alternate carbon atoms are doubly substituted. As one can see very easily from a model, there are severe 1,3 repulsions between the R groups in such a polymer, so the heats of polymerization are correspondingly reduced. That of methyl methacrylate, for example, is only 12 kcal/mole.

Chemical Reactivity

CH_3, and $RCH=CH_2$, as follows:

$$RH \xleftarrow{u} \dot{C}H_3 \quad CH_2=CHR \longrightarrow R\dot{C}H_2 \quad CH_2=CHR,$$

$$\delta E_{\text{deloc}} = 2\beta(1 - a_{0r}) \quad (5.248)$$

$$RH \xleftarrow{u} \dot{C}H_3 \xleftarrow{u} CH_2=CHR \longrightarrow R\dot{C}HCH=CHR$$

$$\delta E_{\text{deloc}} = 2\beta(1 - a'_{0r}) \quad (5.249)$$

Thus the difference in delocalization energy between the reactants and the transition state, i.e., the delocalization energy of activation $\Delta E^{\ddagger}_{\text{deloc}}$, is given by

$$\Delta E^{\ddagger}_{\text{deloc}} = 2\beta(1 - a'_{0r}) - 2\beta(1 - a_{0r})$$
$$= 2\beta(a_{0r} - a'_{0r}) = 2a_{0r}\beta[1 - (1/\sqrt{2})] \simeq 0.6 a_{0r}\beta \quad (5.250)$$

If the transition states of such reactions have similar structures, any variations in activation energy are as usual due to changes in $\Delta E^{\ddagger}_{\text{deloc}}$. Equation (5.250) shows that this is less negative, the smaller is a_{0r}. But the smaller a_{0r}, the more stable is the radical $R\dot{C}H_2$ (see p. 104). Thus the more stable the radical RCH_2, the greater is the activation energy for its addition to $RCH=CH_2$.

If therefore we want polymers of high molecular weight, we must be careful not to overstabilize the intermediate radicals.

The second problem is concerned with *copolymerization*, i.e., the polymerization of mixtures of two or more monomers. The problem here is to predict the extent to which mixed polymers will result.

Consider a two-component system of two monomers A and B and denote by $\sim\sim\dot{A}$ and $\sim\sim\dot{B}$ polymer radicals ending in units of A and B, respectively. There are then four possible propagation reactions:

$\sim\sim\dot{A} + A \longrightarrow \sim\sim\dot{A}$	rate constant	k_{AA}	(5.251)
$\sim\sim\dot{A} + B \longrightarrow \sim\sim\dot{B}$	rate constant	k_{AB}	(5.252)
$\sim\sim\dot{B} + A \longrightarrow \sim\sim\dot{A}$	rate constant	k_{BA}	(5.253)
$\sim\sim\dot{B} + B \longrightarrow \sim\sim\dot{B}$	rate constant	k_{BB}	(5.254)

Suppose now that $\sim\sim\dot{A}$ is much more stable than $\sim\sim\dot{B}$. The cross-propagation step (5.252) will then convert a more stable radical into a less stable one; it will therefore be much less exothermic than either of the homo-propagation steps (5.251) or (5.254). Conversely, the other cross-propagation step (5.253) converts a less stable radical into a more stable one; it should therefore be much more exothermic than the homopropagation steps. Since the reactions are of BEP type,

$$k_{BA} \gg k_{AA}, k_{BB} \gg k_{AB} \quad (5.255)$$

Thus any radicals $\sim\sim\dot{B}$ will be rapidly converted to $\sim\sim\dot{A}$ and these will react with A rather than B. Polymerization will therefore give an almost pure

A-polymer until almost all the A is used up; after this, almost pure B-polymer will be formed from the residue of B. In order to get genuine copolymers, we must therefore choose pairs of monomers that lead to radicals of comparable stability.

There is, however, a further point to be considered. In the transition states (85a) and (85b) for cross-propagation between two different monomers $R_1CH=CH_2$ and $R_2CH=CH_2$, R_1 and R_2 are attached to active positions in a delocalized system isoconjugate with allyl. If one is of $-I$ or $-E$ type and the other of $+I$ or $+E$ type, mutual conjugation will occur and will stabilize the transition states for cross-propagation relative to those for homopropagation. This effect will favor the formaton of genuine copolymers between monomers that might otherwise have been expected not to copolymerize. The resulting copolymer tends to have an alternating structure, —A—B—A—B—A—B— etc. Indeed, in some cases, pairs of monomers that will not polymerize alone, because the propagation step is too slow, will give alternating copolymers of this type. This effect is of practical importance in the production of copolymers.

$$\begin{array}{cc} \sim\!\!\sim\!\!\sim\text{CH}\!=\!\!=\!\text{R}_1 & \sim\!\!\sim\!\!\sim\text{CH}\!=\!\!=\!\text{R}_2 \\ | & | \\ \text{CH}_2\!=\!\!=\!\text{CH}\!=\!\!=\!\text{R}_2 & \text{CH}_2\!=\!\!=\!\text{CH}\!=\!\!=\!\text{R}_1 \\ (85a) & (85b) \end{array}$$

5.22 Aromatic Substitution in Even Systems $(EO_B^\ddagger)$

$EO_B^\ddagger$ addition is a two-step process in which part of the reagent adds to a double bond to form an intermediate ion or radical, this then combining with the rest of the reagent to form the final product. There is, however, the possibility that the intermediate may be diverted in some way, either by combining with an extraneous reagent [cf. equation (5.207)] or by eliminating some group to regenerate a double bond. The latter possibility occurs in nucleophilic substitution by addition/elimination as in equation (5.236). Similar substitutions can take place in the case of radical and electrophilic substitution in olefins, a classic example being the reactions of chlorine and bromine with cinnamate ion,

$$\text{PhCH}=\text{CH}-\text{CO}_2^- + \text{Br}_2 \longrightarrow \text{PhCH}\overset{+}{-}\text{CH}-\text{CO}_2^- \overset{\text{Br}^-}{\underset{\text{Br}}{|}}$$

$$\longrightarrow \text{PhCH}=\text{CHBr} + \text{CO}_2 + \text{Br}^- \quad (5.256)$$

Aromatic substitution is an analogous addition/elimination process in which an electrophilic, nucleophilic, or radical reagent attaches itself to one position (t) in an aromatic ring. If the aromatic compound is even, the result is an odd ion or radical. This then eliminates the atom or group originally attached to atom t, to regenerate the original aromatic system. In the case of olefins, such eliminations occur only in cases where exceptionally good leaving groups are present. In the case of aromatic compounds, the driving force is the tendency to regenerate the original aromatic structure, since addition would destroy the cyclic conjugation with consequent loss of the corresponding aromatic energy.

The most important substitution reactions are electrophilic reactions in which hydrogen is replaced. Thus nitration of benzene in sulfuric acid involves attack by the electrophilic cation NO_2^+ as follows:

$$C_6H_6 + NO_2^+ \longrightarrow [C_6H_6\cdot NO_2]^+ \longrightarrow C_6H_6\cdot NO_2^+$$
$$\qquad\qquad\qquad\qquad (86) \qquad\qquad (87)$$

$$\longrightarrow [C_6H_6\cdot NO_2]^+ \longrightarrow C_6H_5NO_2 + H^+ \qquad (5.257)$$
$$\qquad\qquad (88)$$

The intermediate ion in reactions of this kind, e.g., (87), is clearly isoconjugate with the arenonium ion formed by protonation of the hydrocarbon. Indeed, they differ only by the replacement of one localized σ bond (CH) by another. Thus (87) differs from the benzenonium ion only by replacement of C—H by C—NO_2.

As in the case of electrophilic addition, the transition state for formation of the intermediate [e.g., (86)] contains no bonds that are not present either on it or the reactants. The same is true for the transition state [e.g., (88)] for the elimination reaction, which indeed is the converse of a reverse substitution in which H^+ displaces, e.g., NO_2 from the ring. Both reactions should therefore be of BEP type. Moreover, since electrophilic substitution must be exothermic overall in order to occur, the conversion of the intermediate to the product must be more exothermic than its reversion to the reactants. The reaction path for the system should then have the form indicated in Fig. 5.30, the first maximum being higher than the second. Consequently the first step in such reactions, i.e., addition of the electrophile, should normally be rate determining. This indeed is usually the case, replacement of deuterium taking place as easily as that of hydrogen. The absence of a deuterium isotope effect shows that the CH or CD bond must be intact in the transition state.

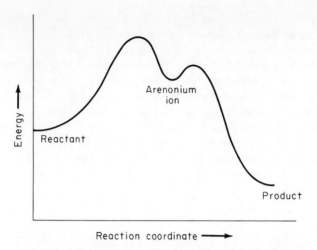

FIGURE 5.30. Reaction path for a typical electrophilic substitution reaction.

First let us consider the case where the substrate is an even AH, i.e., a benzenoid hydrocarbon such as benzene, naphthalene, phenanthrene, etc., substitution being by an electrophile Y^+ at position t,

$$\text{(structures)} \tag{5.258}$$

The π energy of reaction is then given [see equation (3.24)] by

$$\Delta E_\pi = 2\beta(a_{0r} + a_{0s}) = \beta N_t \tag{5.259}$$

where a_{0r} and a_{0s} are the NBMO coefficients at the positions adjacent to atom t in the arenonium* intermediate and N_t is the corresponding reactivity number.

As usual, variations in the energy of reaction ΔE arise from variations in ΔE_π. Indeed, the formation of an arenonium ion has been used as an example to demonstrate this (p. 140). Hence

$$\Delta E = \text{constant} + \Delta E_\pi = \text{constant} + \beta N_t \tag{5.260}$$

Since the reaction is of BEP type, the activation energy $\Delta E^\ddagger$ is given by

$$\Delta E^\ddagger = \text{constant} + \alpha\,\Delta E = \text{constant} + \alpha\beta N_t \tag{5.261}$$

---

* These intermediates have also been termed *Wheland intermediates* or σ *complexes*.

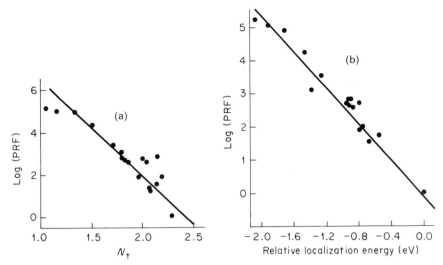

FIGURE 5.31. Plot of logarithms of partial rate factors for nitration of even AHs in acetic anhydride (a) against reactivity numbers N_t; (a) against energies of reaction calculated by an SCF MO method [data from (a) Ref. 10; (b) Ref. 3].

Hence a plot of the logarithms of the rate constants for substitution of even AHs by Y^+ against N_t should be linear.

The absolute rates of electrophilic substitution reactions have been measured only in a few cases. The relative rates of reaction at different positions in a given compound can, however, be found from the proportions of the various isomers formed by substitution in a given compound, while the reactivities of positions in two different compounds Ar_1 and Ar_2 can be estimated from the amounts of the products formed when a mixture of Ar_1 and Ar_2 undergoes substitution.* The rate of reaction at a given position in an aromatic compound relative to the rate of substitution of a single position† in benzene under the same conditions is called the *partial rate factor* (PRF). From equation (5.261), we have

$$\text{PRF}_i = \alpha\beta(N_i - N_B) \tag{5.262}$$

where PRF_i is the partial rate factor for substitution at position i in some even AH, N_i is the corresponding reactivity number, and N_B is the reactivity number of an atom in benzene. A plot of $\log(\text{PRF}_t)$ against N_t should therefore be linear.

Figure 5.31(a) shows this plot for the nitration of a number of benzenoid hydrocarbons of widely varying reactivity in acetic anhydride. The points lie quite close to a straight line. This indeed was the first case where the PMO

---

* This is the competitive method for finding relative rate constants. For details of its application to aromatic substitution, see Ref. 9.
† The rate of substitution in benzene is six times the rate of substitution at a single position because all six carbon atoms are equivalent.

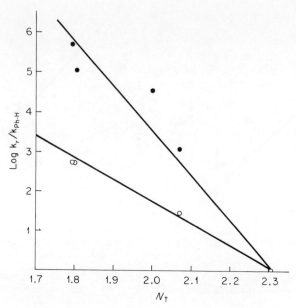

FIGURE 5.32. Plots of log k_r/k_{Ph-H} vs. reactivity number N_t for nitration of AHs in acetic anhydride (○) and chlorination in acetic acid (●). Data from Ref. 11.

method was used in a quantitative way. In view of the relationship between rates of substitution and the stability of the intermediate arenonium ions, one might expect an even better linear relation if the π-energy differences were calculated by the more elaborate MO procedures referred to earlier (see p. 10 and Fig. 4.3). This is the case (Fig. 5.31b), but the gain is again relatively small considering the simplicity of the PMO procedure and the fact that it invariably predicts correctly the most reactive position in any given AH.

Let us now consider the value of α in equation (5.262). The nearer the transition state is to the arenonium ion, the larger will be α. Now the transition state will be nearer to the product (i.e., the arenonium ion), the more endothermic is the formation of the arenonium ion and consequently the slower is the substitution reaction. The slope of the plot in Fig. 5.31 should therefore be greater, the slower is the substitution reaction, i.e., the less active is the attacking electrophile.

Nitration in acetic anhydride is a much more facile reaction than chlorination by molecular chlorine. A plot analogous to Fig. 5.31 for chlorination should therefore be linear but with a steeper slope than that for nitration. Figure 5.32 shows that this is indeed the case.

While Fig. 5.32 refers to one specific reaction, the arguments leading to it apply equally well to other BEP processes. In the general case, we have a series of compounds $X_1, X_2, \ldots, X_m$ that undergo analogous reactions with one or another of a set of reagents $Y_1, Y_2, \ldots, Y_n$. The reactions of each Y_i with the X_j will follow the BEP relation [equation (5.69)] with a corresponding proportionality factor α_i. From the BEP principle, the activation energies

Chemical Reactivity

will be greater, the less exothermic (or more endothermic) are the reactions. Furthermore, as the reactions become less exothermic, the transition states become closer to the products in structure. The proportionality factor α_i therefore becomes greater. This means that the difference in activation energy between two compounds of the series, X_r and X_s, becomes correspondingly greater and so does the ratio of their rates. Thus the spread of rates of reaction over the X_j for reaction with some reagent Y_i will be greater, the greater are the absolute values of the activation energies, i.e., the more slowly the reactions take place. If, then, Y_i has a choice between reacting with two or more different X_j, the selectivity will be greater the less reactive is Y_i. This aspect of the BEP principle was well recognized forty years ago but has been overlooked and consequently rediscovered from time to time. A good example is Brown's "selectivity rule" for aromatic substitution which was based on studies of electrophilic substitution in toluene. As we shall see presently, toluene undergoes such substitution *para* to the methyl group faster than *meta*. Brown found that the *para/meta* ratio varied with the electrophilic reagent, being greater, the slower the reaction, i.e., the less reactive the reagent. Some examples are shown in Table 5.7.

The rates of electrophilic substitution on other even aromatic systems, and the effects of substituents, can be deduced in the usual way. Thus since $+I$ substituents destabilize odd AH cations while $-I$ substituents stabilize them, substitution should tend to take place in such a way as to ensure that $+I$ substituents are at inactive positions in the arenonium intermediate and $-I$ substituents at active positions. For example, in benzene, where the positions *ortho* and *para* to the point of substitution are active in the intermediate benzonium ion (89), $+I$ substituents (e.g., CF_3) are best placed in the *meta* position, while $-I$ substituents (e.g., CH_3) should be *ortho* or *para*. Benzotrifluoride ($PhCF_3$) consequently substitutes mainly *meta*, nitration, for example, giving (90), while toluene ($PhCH_3$) substitutes mainly *ortho* and *para*, giving, e.g., (91) and (92) on nitration.

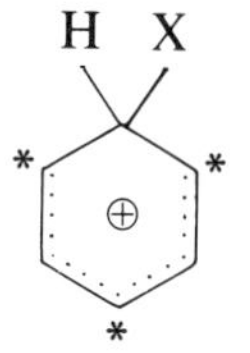

(89) 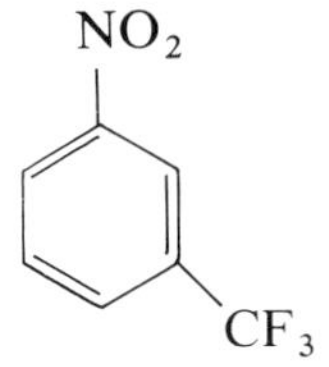 (90) 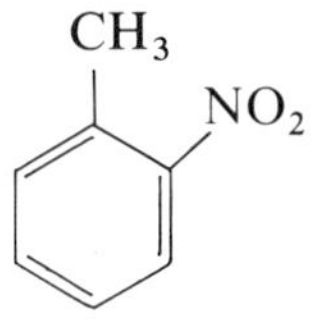 (91)

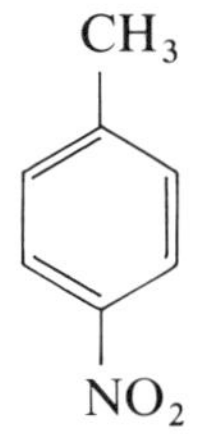

(92) 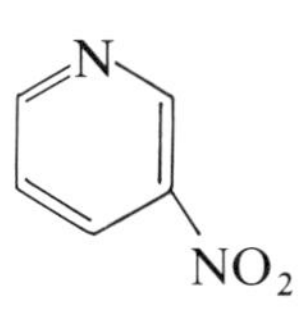 (93) 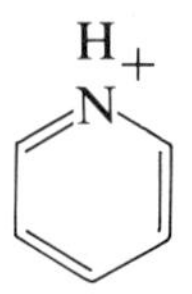 (94)

TABLE 5.7. Selectivity Effects in Electrophilic Substitution in Toluene[a]

Reaction	Relative rate $k_{\text{toluene}}/k_{\text{benzene}}$	Toluene product distribution	
		% meta	% para
Bromination	605	0.3	66.8
Chlorination	350	0.5	39.7
Benzoylation	110	1.5	89.3
Nitration	23	2.8	33.9
Mercuration	7.9	9.5	69.5
Isopropylation	1.8	25.9	46.2

[a] Data from Ref. 14.

Since nitrogen is more electronegative than carbon, replacement of carbon in an AH by nitrogen should have the same effect as introduction of a $+I$ substituent (Table 4.3). Pyridine therefore substitutes *meta*, nitration, for example, giving (93). However, although nitrogen occupies an inactive position in the transition state leading to (93), the reaction nevertheless takes place only with difficulty. Indeed, whereas benzene is nitrated rapidly and quantitatively by nitric acid in sulfuric acid at 0°C, (93) is obtained only in low yield after heating pyridine with concentrated sulfuric acid and potassium nitrate to 300°C for a week.

There are two reasons for this lack of reactivity. The first, and more important, is the fact that pyridine is a base, so nitration involves attack not on pyridine itself but on the conjugate acid (94). There is a very large repulsion (field effect) in the transition state between the positive charges on the intermediate arenonium ion and the positive charge on nitrogen. Second, even if the nitrogen atom is neutral, and even if it occupies an inactive position in the transition state, it will still deactivate it through the π-inductive effect (see p. 164). The CN bonds in pyridine are polar in the sense $C^{\delta+}$—$N^{\delta-}$. The carbon atoms adjacent to nitrogen consequently have positive charges and so are more electronegative than carbon atoms in a neutral AH. If the nitrogen atom is at an inactive position in the transition state, the two adjacent carbon atoms will be at active positions and will exert a deactivating effect.

The same situation arises in the case of $+I$ and $-I$ substituents. The $+I$ substituents have an overall deactivating effect on electrophilic substitution, partly through the field effect due to the polarity ($C^{\delta+}$—$X^{\delta-}$) of the bond linking such a substituent to the ring and partly through the π-inductive effect, the carbon atom adjacent to X being more electronegative than usual (because of the positive charge) and so in turn attracting electrons from the neighboring carbon atoms ($C^{\delta+}$—$C^{\delta+}$—$X^{\delta-}$). Conversely, $-I$ substituents have an overall activating effect. Thus the partial rate factor for *meta* substitution in benzotrifluoride is less than unity while that for *meta* substitution in toluene is greater than unity.

The application of the PMO method to electrophilic substitution in

Chemical Reactivity

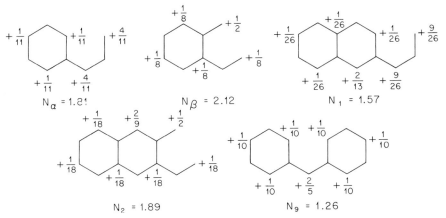

FIGURE 5.33. Distributions of formal charge in transition states for electrophilic substitution in naphthalene and anthracene and corresponding reactivity numbers.

heteroaromatic compounds may be illustrated by its application to quinoline (95a), isoquinoline (95b), and acridine (95c):

First we examine the isoconjugate AHs naphthalene and anthracene. The distributions of formal charge in the various transition states and the corresponding reactivity numbers are shown in Fig. 5.33. The formal charge is largely concentrated in the ring where substitution is taking place. The nitrogen atom will therefore have a much larger deactivating effect if it is in that ring. Substitution should therefore take place in a benzenoid ring of (95a), (95b), or (95c), not in a ring containing nitrogen.

Consider isoquinoline. The reactivity numbers of the 5 and 8 positions [see 95b] are lower than at 6 and 7. Substitution at the 5 position also leads to an intermediate where nitrogen occupies an inactive position. This then should be the preferred point of attack. Nitration of isoquinoline does indeed give mainly 5-nitroisoquinoline (96). The next most reactive position might be either 7 (higher reactivity number but with nitrogen at an inactive position) or 8 (lower reactivity number but nitrogen at an active position). The difference in reactivity numbers wins here since the positive charge on nitrogen is only $+ 1/11$. The main byproduct in the nitration is therefore 8-nitroisoquinoline (97).

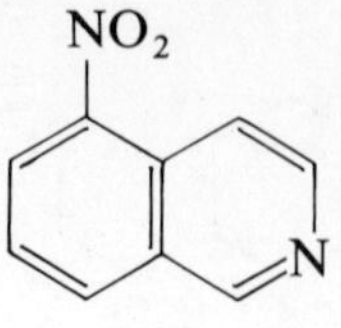

(96) 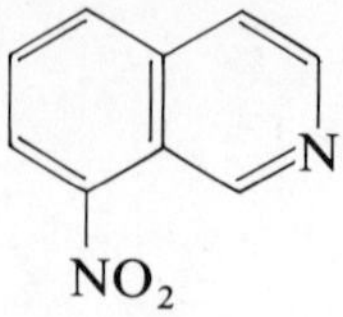 (97) 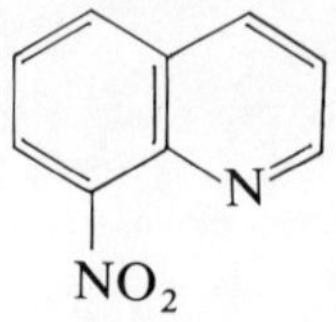 (98) 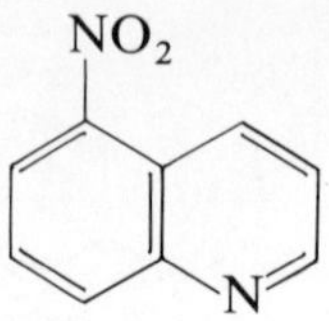 (99)

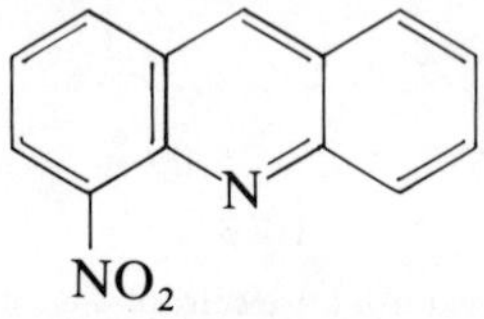

(100) 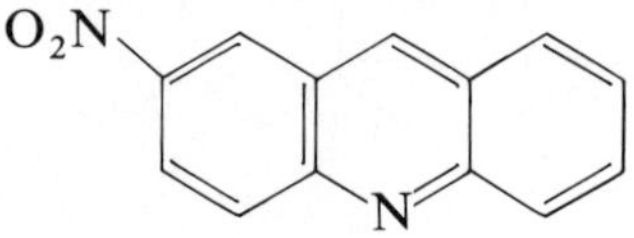 (101)

Since the reactivity numbers and charges are the same, we would similarly expect quinoline to give mainly 8-nitroquinoline (98) with 5-nitroquinoline (99) as the main byproduct. Nitration does indeed give (98) and (99), but in approximately equal quantities. The reason for this is that the reactions are carried out in sulfuric acid, where the nitrogen atom is protonated. In the transition state for substitution, the CN bond between the entering electrophile NO_2^+ and the ring is not completely formed, so a considerable positive charge remains on the NO_2 group. The electrostatic repulsion between this and the NH^+ group in the ring has a selective deactivating effect on the 8 position since the NO_2 and NH groups are then very close together.

In acridine one would expect the 4 position to be the most reactive, having a low reactivity number with nitrogen in an inactive position. The main product from nitration of acridine is indeed (100). The next choice is again between a position of lower reactivity number but deactivated by nitrogen (i.e., 1), and a position of higher reactivity number but not deactivated (i.e., 2). The difference in reactivity number between the two positions (0.32) is almost identical with that in quinoline or isoquinoline (0.31), but the formal charge on nitrogen ($+2/13$) is almost double. The resulting increase in deactivation of the 1 position is so great that the main byproduct is (101), not the 1 isomer. Since this is so much less reactive than the 4 position, (100) is the major product in spite of deactivation by the field effect of NH^+ as in the formation of (98).

The π energy of reaction ΔE_π to form the arenonium intermediate is given by

$$\Delta E_\pi = \beta N_t - Q_N \delta\alpha_N - \sum Q_r \delta\alpha_N' \qquad (5.263)$$

where Q_i is the formal charge at position i in the odd AH intermediate ($i = N$ is the position occupied by the nitrogen atom), $\delta\alpha_N$ is the Coulomb integral (relative to carbon) of nitrogen and $\delta\alpha_N'$ is that of a carbon atom adjacent to nitrogen, and the sum is over positions r adjacent to nitrogen. The first

Chemical Reactivity

term indicates the reactivity of position t in the isoconjugate AH, the second gives the direct deactivating effect of nitrogen, and the final sum represents the π-inductive effect.

While it is possible to use equation (5.263) in a quantitative way, and while one can in fact account surprisingly well in this way for the available data on substitution in nitrogen heterocycles, rather a lot of parameters are involved, so the main interest lies in more qualitative applications. An interesting example is provided by the nitration of 10-methyl-10,9-borazarophenanthrene (102). This compound is isoconjugate with phenanthrene (103), the B^- and N^+ being isoelectronic with carbon. In deriving (102) from (103), one carbon atom is replaced by a more electronegative atom (N^+) and another by a less electronegative atom (B^-). One can see from equation (5.263) that nitrogen should deactivate all positions of opposite parity to itself; conversely, boron should activate all positions of opposite parity. The positions starred in (102) are consequently activated by boron and not deactivated by nitrogen, while the converse is true for the unstarred positions. The reactivity numbers are shown in (103). It can be seen that the 8 position is predicted to be the most reactive, having the lowest reactivity number of any starred position. Nitration of (102) indeed gives mainly (104). The next most reactive position might be expected to be 4; however substitution here is subject to severe steric hindrance and indeed the 4 position in (103) is consequently less reactive than 3. The main byproduct in the nitration of (102) is consequently the 6 isomer (105).

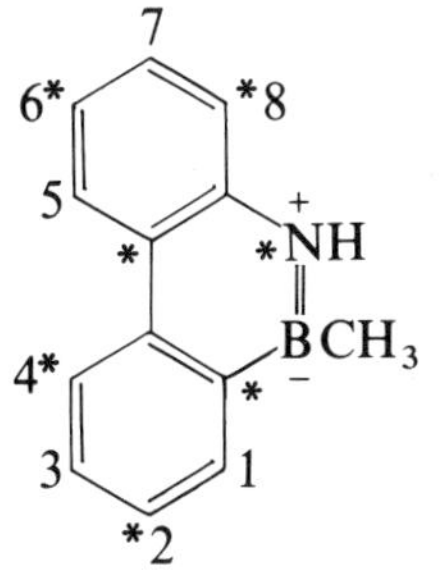

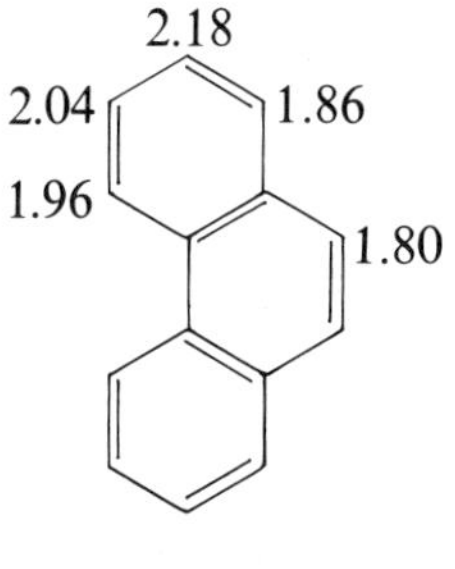

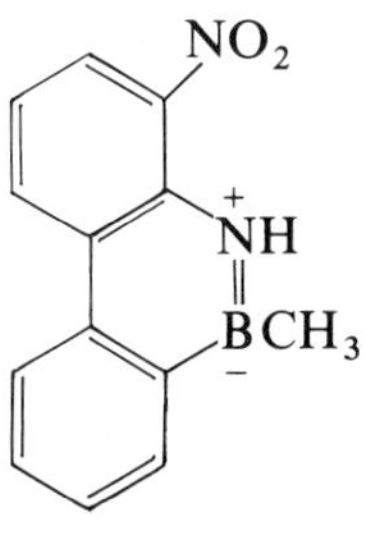

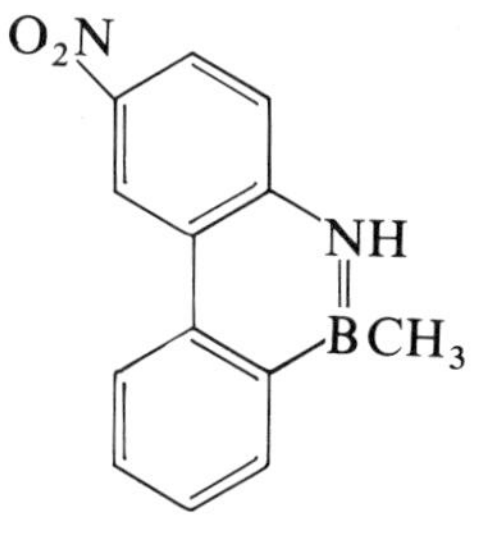

The effect of electromeric substituents on electrophilic substitution can be deduced immediately from the results in Chapter 4 (cf. Table 4.3). A $\pm E$ or $-E$ substituent at an active position stabilizes an odd AH cation, the effect being proportional to the square of the NBMO coefficient, i.e. the positive formal charge. Thus substituents of this type in benzene promote *ortho* and *para* substitution. In polycyclic systems, they will promote substitution most strongly in the ring carrying the substituent since the charges in the intermediate arenonium ion are largest in the ring where substitution is taking place. Thus 1-methoxynaphthalene substitutes 2 and 4 (see Fig. 5.33), while 2-methoxynaphthalene substitutes 1; e.g.,

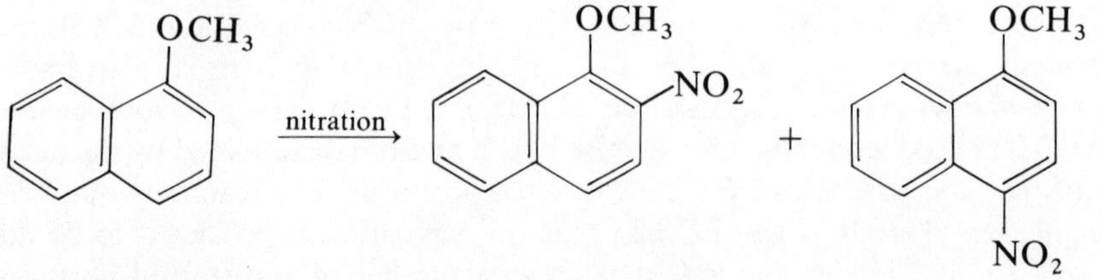

(5.264)

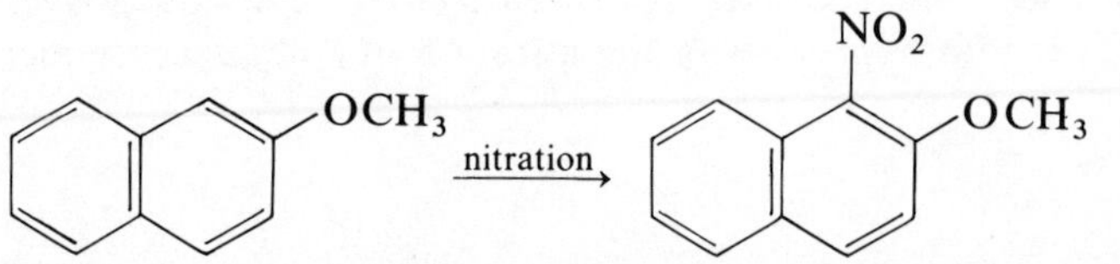

(5.265)

Note that in the latter, only one of the positions *ortho* to methoxy is reactive. The reason for this is evident from Fig. 5.33.

The effect of $+E$ substituents is ambiguous (see Table 4.3). Small $+E$ substituents have a deactivating effect at active positions in the intermediate arenonium ion, combined with a general overall deactivation due to their $+I$ effect (cf. CF_3, p. 321). Thus nitrobenzene ($PhNO_2$) and acetophenone ($PhCOCH_3$) nitrate *meta*, while further nitration of 1-nitronaphthalene (106) gives a mixture of (107) and (108), and that of 2-nitronaphthalene (109) gives mainly 2,5-dinitronaphthalene (110) with some 2,8 isomer (111), nitration in the ring containing the nitro group being inhibited, as in the case of (95a) or (95b), by the field and π-inductive effects of the substituent.

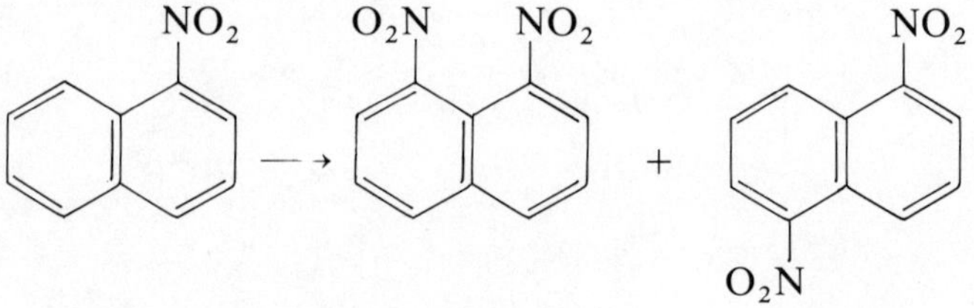

(106)   (107)   (108)

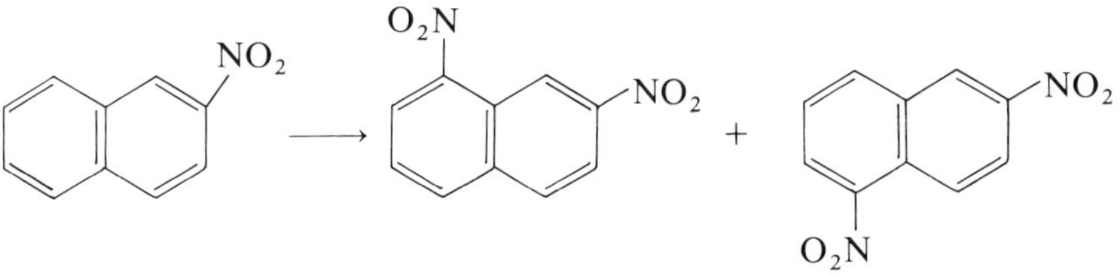

Chemical Reactivity

(109) → **(110)** + **(111)**

[structures: 2-nitronaphthalene → 1,6-dinitro and 5,6-dinitro naphthalenes]

Large $+E$ substituents are, however, *ortho and para directing*, since they exert an overall activating effect on odd AH cations (see p. 175). Thus nitration of cinnamic acid (112) gives a mixture of (113) and (114), and that of 4-phenyl-pyridine (115) gives a mixture of (116) and (117). The charges on the heteroatoms in these substituents (isoconjugate with CH_2=CH—CH=CH— and Ph—) are much smaller than in the case of substituents derived from vinyl.

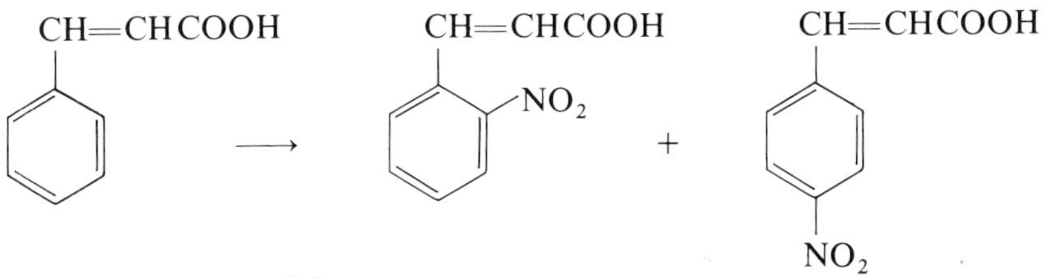

(112) → **(113)** + **(114)**

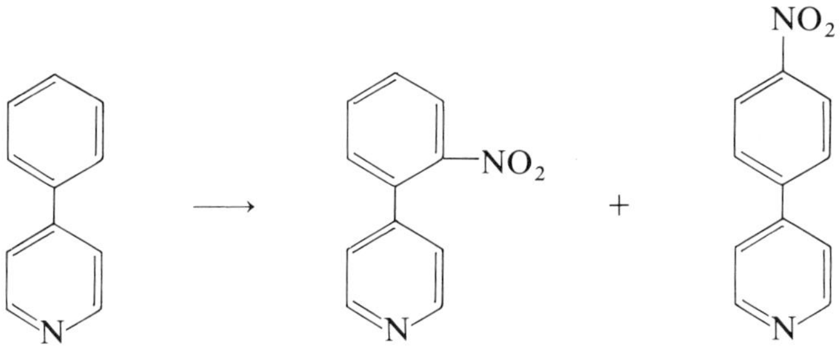

[cinnamic acid → ortho- and para-nitro cinnamic acids]

(115) → **(116)** + **(117)**

[4-phenylpyridine → ortho- and para-nitro derivatives]

One final structural change in an even AH may be considered, i.e., intramolecular union to a nonalternant system. Thus azulene (119) can be derived (see p. 94) by intramolecular union from cyclodecapentaene (118). Since union is between atoms of like parity, it has no first-order effect on the π energy. This, however, is no longer true in the arenonium intermediate for electrophilic substitution, the bond order between two starred positions r and s being $-a_{0r}a_{0s}$ (p. 75). Union will then stabilize the transition state if atoms r and s are active and if a_{0r} and a_{0s} have opposite signs. The NBMO coefficients for substitution in (118) are shown in (120). It will be seen that the conditions are met only if substitution takes place in the five-membered ring at positions adjacent to the bridge, to give, e.g., (121).

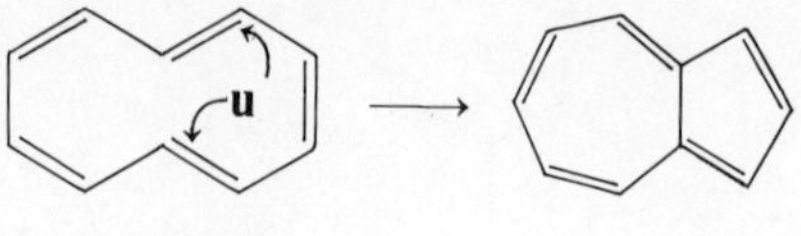

(118)   (119)

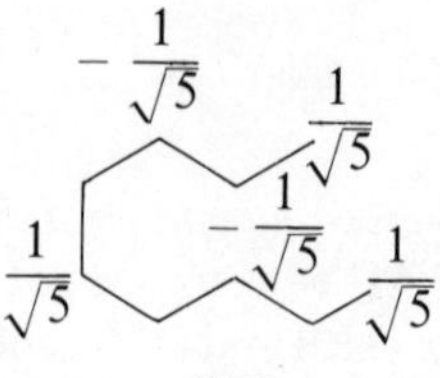

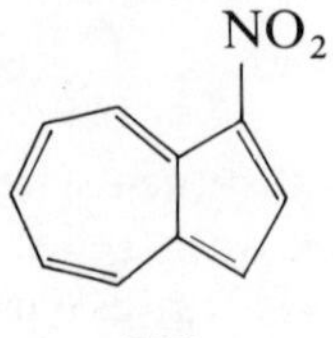

(120)   (121)

The π energy of reaction with an electrophile ΔE_π is given by

$$\Delta E_\pi = -\beta N_t - 2a_{0r}a_{0s}\beta \qquad (5.266)$$

where N_t is the reactivity number of any position in (118) ($= 4/\sqrt{5} = 1.79$) and a_{0r} and a_{0s} are the NBMO coefficients at the point of union ($a_{0r} = -a_{0s} = 1/\sqrt{5}$). Substituting in equation (5.266), we find that $\Delta E_\pi = 0$! Azulene is in fact extraordinarily reactive to electrophiles, giving remarkably stable arenonium ions. It is indeed quite basic, being readily protonated to azulenium ion.

Radical and nucleophilic substitution reactions are entirely analogous to electrophilic reactions, so only a brief discussion will be given. The main difference is that hydrogen cannot easily be expelled as a free atom (because the CH bond is so strong) or as an anion (H^- being an extremely powerful nucleophile). Substitution of hydrogen by radicals can take place only through attack by a second radical on the arenonium intermediate or by disproportionation; e.g.,

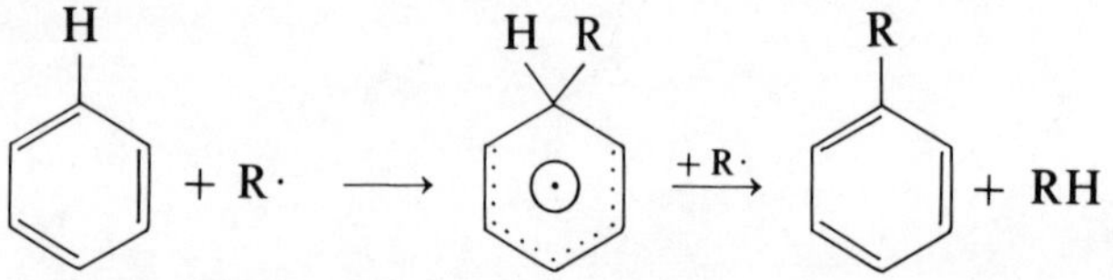

$$\text{PhH} + R\cdot \longrightarrow [\text{arenonium}] \xrightarrow{+R\cdot} \text{PhR} + RH \qquad (5.267)$$

Substitution of hydrogen by nucleophiles is possible only if the nucleophile is very powerful or if an H^- acceptor is present (e.g., oxygen, which can react with H^- to give HO_2^-). A classic example is the Chichibabin amination reaction; e.g.,

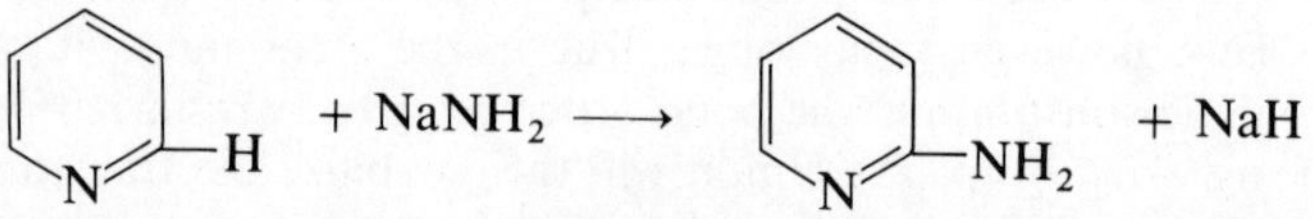

$$\text{pyridine-H} + NaNH_2 \longrightarrow \text{pyridine-}NH_2 + NaH \qquad (5.268)$$

In the case of nucleophilic substitution, this difficulty can be avoided by using a better leaving group. The majority of nucleophilic substitution reactions consequently involve replacement of halogen or alkoxyl rather than hydrogen.

Chemical Reactivity

In radical substitution, an even AH is converted to an intermediate odd AH radical and in nucleophilic substitution to the corresponding anion. Since these differ from the corresponding electrophilic intermediate only in the number of nonbonding electrons, the energies of reaction should be the same in all three cases. The relative reactivities of different even AHs and the orientation of substitution in a given AH, should therefore be the same for all reagents. This is the case. Thus naphthalene substitutes mainly in the 1 position both with electrophiles (e.g., nitration) and with radicals (e.g., phenylation by phenyl radicals, as in the Gomberg reaction) and with nucleophiles (e.g., amination by sodamide, the Chichibabin reaction); i.e.,

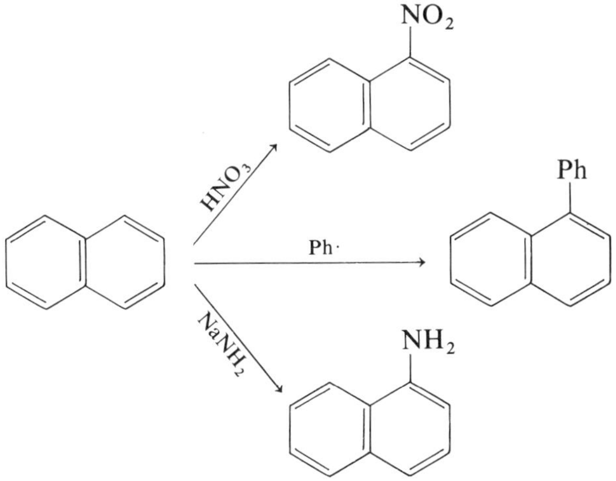

(5.269)

All these substitution reactions should consequently follow equation (5.262), so plots of PRF vs. N_t should be linear. This is true for methylation by $CH_3\cdot$ radicals and trifluoromethylation by $CF_3\cdot$ radicals, as Fig. 5.34 shows. Note that the slope of the line for trifluoromethylation is slightly less than that for methylation. The $CF_3\cdot$ radicals are somewhat more reactive than $CH_3\cdot$ radicals.

The effects of substituents on the rates of these reactions can also be deduced in the same way as their effect on electrophilic substitution, using the results obtained in Chapter 4 (see Table 4.3). Thus heteroatoms or $+I$ or $-I$ substituents have little effect on the rates of radical reactions since the π-electron density at each position in the intermediate AH radical is unity. Heteroatoms do, however, exert small second-order effects. It can be shown by second-order perturbation theory that the self-polarizability of a position in an odd AH radical is greater, the greater the NBMO coefficient. If a heteroatom such as nitrogen replaces a carbon atom in a neutral AH, it will acquire a negative charge at the expense of the rest of the system. The resulting charge transfer from carbon atoms to the more electronegative heteroatom leads of course to stabilization. Since the charge transfer will be greater, the greater the self-polarizability of the position occupied by nitrogen, the

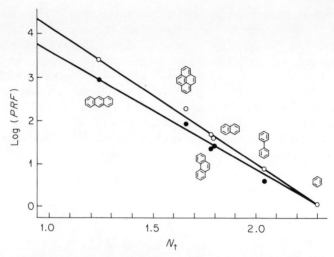

FIGURE 5.34. Plot of log (PRF) for methyl (○) and trifluoromethyl (●) substitution reactions of aromatic hydrocarbon. Data from Ref. 12.

stabilization will be greater, the greater the NBMO coefficient at that point. Radical substitution consequently takes place more easily in the α and γ positions in pyridine than in the β position, while quinoline reacts most easily in the 2,4 positions and isoquinoline in the 1 position. Similar directing effects are shown by $+I$ and $-I$ substituents for the same reason.

Electromeric substituents of all kinds stabilize radicals if they are attached to active positions. All such substituents are consequently *ortho* and *para* directing, regardless of whether they are $+E$, $-E$, or $\pm E$. The rules for orientation, etc., are identical with those for electrophilic substitution and therefore need not be elaborated.

Nucleophilic substitution is accelerated by substituents that stabilize carbanions. Thus electronegative heteroatoms promote nucleophilic substitution very strongly, the effect being greatest when the heteroatom is in a position with a high NBMO coefficient (and hence a large negative charge) in the intermediate anion. Thus, whereas chlorobenzene fails to undergo normal nucleophilic substitution of chlorine and the chloronaphthalenes react only with great difficulty if at all, 2- and 4-chloropyridine or chloroquinoline hydrolyze easily in water, as does 1-chloroisoquinoline;

(5.270)

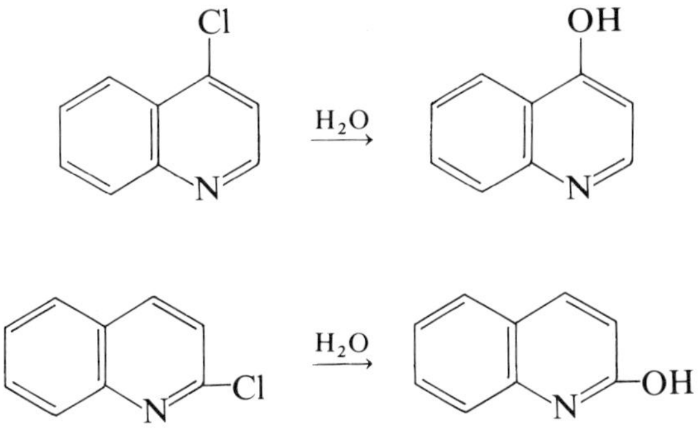

Chemical Reactivity

$$\text{4-chloroquinoline} \xrightarrow{H_2O} \text{4-hydroxyquinoline}$$

$$\text{2-chloroquinoline} \xrightarrow{H_2O} \text{2-hydroxyquinoline} \quad (5.271)$$

$$\text{1-chloroisoquinoline} \xrightarrow{H_2O} \text{1-hydroxyisoquinoline} \quad (5.272)$$

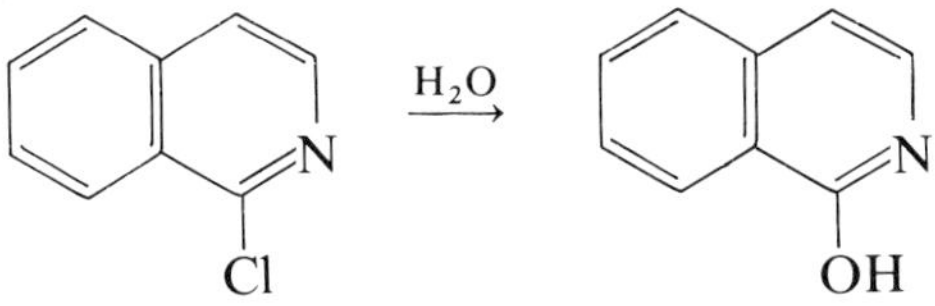

The other isomers of chloropyridine, chloroquinoline, and chloroisoquinoline are much less reactive, as would be expected.

Nucleophilic substitution is also assisted by $\pm E$ and $+E$ substituents, the latter being especially effective. Indeed, in many cases the intermediate anions can actually be isolated as stable salts (Meisenheimer complexes), e.g.,

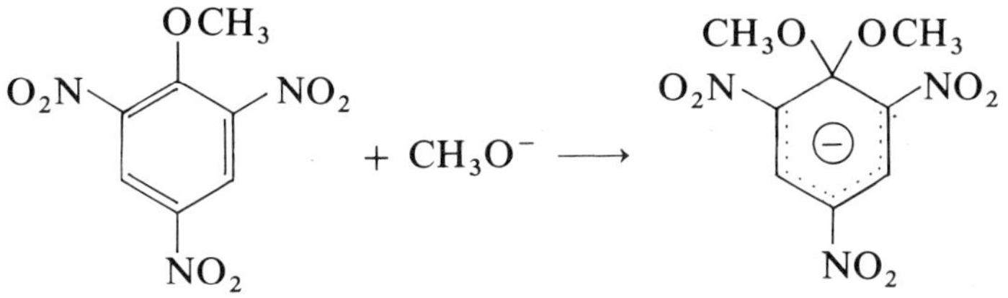

$$\text{2,4,6-trinitroanisole} + CH_3O^- \longrightarrow \text{Meisenheimer complex} \quad (5.273)$$

Since the rules for orientation, etc., are identical with those derived for electrophilic substitution, no detailed discussion is necessary.

5.23 Substitution vs. Addition

The classic criterion for an aromatic compound was its ability to undergo substitution rather than addition. Confusion was therefore caused when ostensibly aromatic compounds underwent addition. Phenanthrene (122), for example, easily adds chlorine to form (123) and even benzene will add chlorine under radical conditions to form hexachlorocyclohexane* (124).

---

* This reaction is of technical importance since one of the isomers of (124) is a useful insecticide (gammexane).

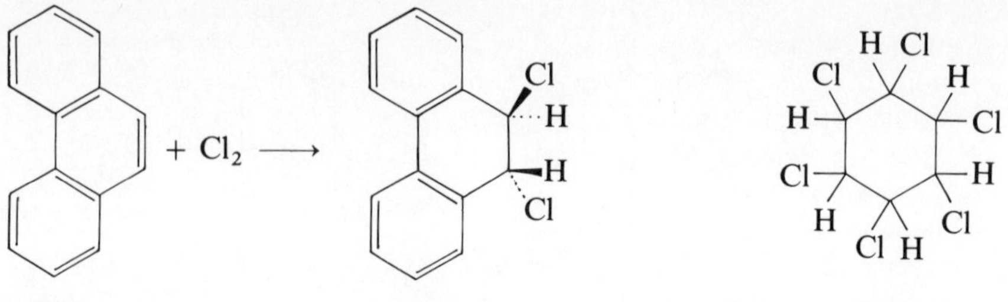

(122) + Cl₂ ⟶ (123)   (124)

As we have seen, the first step in aromatic substitution is the same as in $EO_B^{\ddagger}$ addition, converting an even conjugated system into an odd ion or radical. The distinction comes at the second step, when the intermediate can either eliminate some atom or group to regenerate the parent even conjugated system or undergo addition to form a new even conjugated system with two atoms less.

The majority of substitution and addition reactions are of electrophilic type. Here aromatic compounds tend to undergo substitution for three reasons.

First, the loss of resonance energy on addition is sufficient to make many such reactions (e.g., addition of Br_2) endothermic.

Second, aromatic compounds do not usually form good π complexes. It can be shown that the heat of formation of a π complex by addition of an acceptor to a CC π bond in a conjugated molecule is proportional to the π bond order. In an aromatic compound, this is usually a good deal less than unity. The $E\pi_B^{\ddagger}$ mechanism naturally favors addition, since direct elimination from a π complex involves a rather tortuous reorganization of bonds.

Third, aromatic compounds tend to react less readily than olefins and so more drastic conditions have to be used. This usually means that there are fewer nucleophiles around to add to the intermediate odd AH cation. This intermediate, however, behaves as a very strong acid, so even very weakly basic solvents can remove a proton from it.

(125)   (126)

There are three situations where aromatic compounds undergo addition rather than substitution.

First, when addition takes place in a ring whose aromatic energy (p. 89) is small. Thus the π energy of anthracene (125) is very little greater than that of two molecules of benzene—or equivalently 9,10-dihydroanthracene (126). Anthracene therefore undergoes addition reactions quite easily.

Chemical Reactivity

Second, if a bond in an aromatic compound is of sufficiently high order, addition may take place by the $E\pi_B^{\ddagger}$ mechanism. This is probably true for addition of chlorine to phenanthrene, the product (123) being the *trans* isomer.

Third, if no suitable leaving group is available, addition may take place. This situation is most common in the case of radical reactions. As we have seen, radical substitution can occur only by a radical attack on the intermediate arenonium radical or by its disproportionation. It is impossible for a hydrogen atom to be ejected as such because the CH bond is much too strong. If some compound is present that can react easily with radicals, addition may then occur. This is what happens when benzene is chlorinated under nonpolar conditions in presence of light to produce chlorine atoms. The first step is an addition of a chlorine atom to benzene to form the radical (127). The concentration of atoms and radicals is necessarily low under such conditions because the rate at which radicals combine with one another to form ordinary molecules is very high. Therefore the rate at which (127) can lose its extra hydrogen atom is quite small. On the other hand, chlorine reacts very easily with alkyl radicals to give alkyl chlorides. Since chlorine is present in relatively high concentration, (127) reacts with it to form the dichloride (128). This is no longer aromatic, so it can undergo further addition of two more molecules of chlorine very rapidly, just like an olefin.

5.24 Neighboring Group Participation

In a bimolecular reaction, each of the two reactant molecules can move about freely, having three degrees of translational freedom. When they unite to form the transition state, three degrees of freedom are lost since the molecules are now locked together. As a result there is a large decrease in the entropy, reflected by a negative contribution to $\Delta S^{\ddagger}$, the entropy of activation, and a corresponding decrease in rate.

If now we had a similar reaction in which the reactant molecules were joined together, the reaction taking place between different parts of the same molecule rather than two different molecules, no translational entropy would be lost on forming the transition state. The entropy of activation should then be much more positive, and the rate consequently much greater, than for the corresponding bimolecular reaction.

This effect is seen very clearly in the intramolecular reactions of ω-chloroalkylamines to form rings; e.g.,

$$\text{CH}_2\begin{array}{c}\diagup\text{CH}_2\text{—NH}_2\\ \diagdown\text{CH}_2\text{—CH}_2\diagup\end{array}\text{CH}_2\text{—Cl} \longrightarrow \text{CH}_2\begin{array}{c}\diagup\text{CH}_2\text{—}\overset{+}{\text{NH}}_2\diagdown\\ \diagdown\text{CH}_2\text{—CH}_2\diagup\end{array}\text{CH}_2 + \text{Cl}^-$$

(5.274)

In the cases where five- or six-membered rings are formed, the rates are far greater than those of corresponding bimolecular reactions of primary amines (RNH_2) with primary alkyl chlorides (RCH_2Cl). The rate is less for reactions of γ-chloropropylamines, leading to trimethyleneimines, because of the strain in the resulting four-membered rings. The rate also decreases with increasing size of the ring being formed because the longer the chain separating the CH_2NH_2 and CH_2Cl groups, the more entropy is lost when they are tied together in the transition state for cyclization.

The same effect is seen in chemical equilibria. Thus the reaction of two molecules of a carboxylic acid to form the anhydride,

$$2\text{RCOOH} \longrightarrow (\text{RCO})_2\text{O} + \text{H}_2\text{O} \qquad (5.275)$$

leads to two molecules of product. There is therefore no change in translational entropy. On the other hand, in cyclization of a dicarboxylic acid to form a cyclic anhydride, one molecule of reactant leads to *two* molecules of product; e.g.,

$$\begin{array}{c}\text{CH}_2\text{—COOH}\\ |\\ \text{CH}_2\text{—COOH}\end{array} \longrightarrow \begin{array}{c}\text{CH}_2\text{—CO}\diagdown\\ |\qquad\qquad\text{O} + \text{H}_2\text{O}\\ \text{CH}_2\text{—CO}\diagup\end{array} \qquad (5.276)$$

Cyclic anhydrides are therefore formed much more easily than acyclic anhydrides and are less easily hydrolyzed, particularly if the resulting anhydride has a five- or six-membered ring. Anhydrides of this kind are formed very easily just by heating the dicarboxylic acid, whereas acyclic anhydrides must usually be obtained by other procedures.

The facility of ring formation allows reactions to proceed that do not occur at all in the acyclic case. Thus while carboxylic acid chlorides do not react with ethers, δ-methoxyvaleryl chloride (129) rearranges on heating to methyl δ-chlorovalerate (130) through nucleophilic attack on the —COCl group by the ether moiety; i.e.,

$$\begin{array}{ccc}
\begin{array}{c}\quad\text{CH}_3\\ \diagup\\ \text{CH}_2\text{—O}\\ \text{CH}_2\quad\quad\quad\text{CO—Cl}\\ \diagdown\quad\diagup\\ \text{CH}_2\text{—CH}_2\\ (129)\end{array} &
\begin{array}{c}\text{Cl}^-\\ \quad\text{CH}_3\\ \diagup\\ \text{CH}_2\text{—}\overset{+}{\text{O}}\\ \text{CH}_2\quad\quad\quad\text{CO}\\ \diagdown\quad\diagup\\ \text{CH}_2\text{—CH}_2\end{array} \longrightarrow &
\begin{array}{c}\text{Cl}\quad\text{CH}_3\\ |\quad\quad|\\ \text{CH}_2\quad\text{O}\\ \diagup\quad\diagdown\\ \text{CH}_2\quad\quad\quad\text{CO}\\ \diagdown\quad\diagup\\ \text{CH}_2\text{—CH}_2\\ (130)\end{array}
\end{array}$$

(5.277)

Chemical Reactivity

While such *neighboring group participation* occurs readily when the intermediate rings have five or six members and much less readily when they have four, analogous S_N2-type reactions involving intermediate three-membered rings occur very easily indeed. The reason for this is that the intermediates are in fact analogous to the structures considered in the previous section. Thus β-bromoalkyl esters can undergo solvolysis very readily by mechanisms following S_N1-type kinetics because the intermediate carbonium ion rearranges during its formation to a more stable π complex; e.g.,

$$\underset{X}{\overset{Br}{\underset{|}{-C-C-}}} \longrightarrow \overset{Br^+}{\underset{}{>C\doteq C<}} + X^- \qquad (5.278)$$

$$\overset{Br^+}{\underset{Y^-}{>C\doteq C<}} \longrightarrow \underset{Y}{\overset{Br}{-C-C-}} \qquad (5.279)$$

The rearrangement accompanies separation of X as X^- and so acts as a driving force for the ionization, the reactions being faster than analogous unassisted S_N1 reactions. Since, moreover, the migration of Br accompanies ionization of X, the reaction is analogous to an S_N2 substitution, so X must approach from the backside of the CX bond. The intermediate π complex is therefore formed stereospecifically. Since the subsequent reaction with some nucleophile (Y^- in equation (5.279)] is also stereospecific, taking place *trans* to Br, the overall reaction takes place with complete retention of configuration, as indicated in equations (5.278) and (5.279).

Obviously the ability of a given group Z to act in this way when β to a suitable leaving group X, i.e.,

$$\underset{X}{\overset{Z}{-C-C-}} \longrightarrow \overset{Z^+}{>C\doteq C<} + X^- \qquad (5.280)$$

$$(131)$$

will depend on the stability of the resulting π complex (131) relative to that of the corresponding classical carbonium ion $>CZ-\overset{+}{C}<$ (see Section 5.21). The Z group will be more effective, the lower its electronegativity and the weaker the CZ σ bond. Thus the acceleration due to neighboring group participation in compounds of this type rises in the series $Z = F < Cl < Br < I$. No participation is seen in β-fluoroethyl derivatives and only weak participation in β-chloroethyl derivatives, while strong participation is seen in the solvolysis of β-bromoethyl esters and very strong participation in β-iodoethyl esters. Similarly, alkoxy is a better neighboring group than chloro, oxygen being less electronegative than chlorine, while amino and thioether groups are better still. Also while neutral oxygen (as in $ROCH_2CH_2Cl$) is a poor neighboring group, O^- is excellent. One of the standard procedures for making epoxides is to treat β-hydroxylethyl halides or esters with base, e.g.,

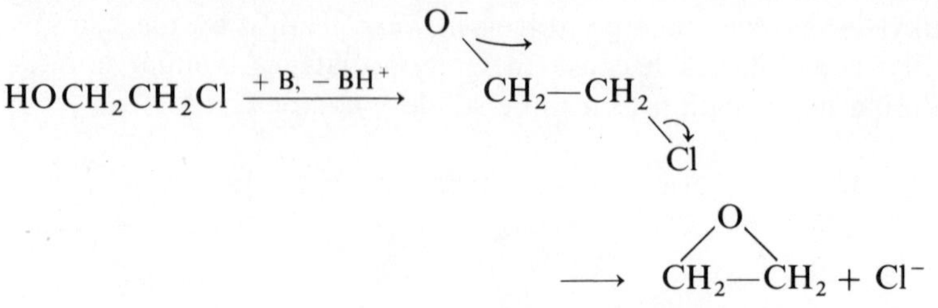

$$HOCH_2CH_2Cl \xrightarrow{+B, -BH^+} \begin{array}{c} O^- \\ \diagdown \\ CH_2\text{---}CH_2 \\ \diagdown \\ Cl \end{array} \longrightarrow \begin{array}{c} O \\ \diagup \diagdown \\ CH_2\text{---}CH_2 \end{array} + Cl^- \quad (5.281)$$

One special case of neighboring group participation has already been noted, i.e., that occurring in the Wagner–Meerwein rearrangement (p. 289). Compounds of the type

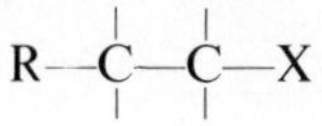

$$R\text{---}\underset{|}{\overset{|}{C}}\text{---}\underset{|}{\overset{|}{C}}\text{---}X$$

can undergo ionization of X with concerted formation of a π complex, i.e.,

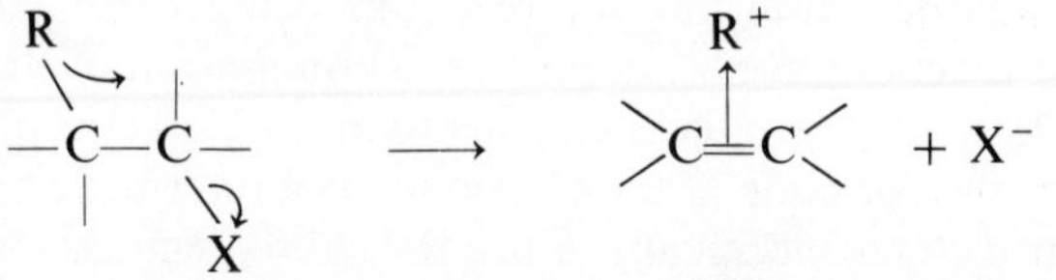

$$\begin{array}{c} R \\ \diagdown \\ \text{---}C\text{---}C\text{---} \\ | \quad \diagdown \\ \quad X \end{array} \longrightarrow \begin{array}{c} R^+ \\ \uparrow \\ \diagup C\text{=}C \diagdown \end{array} + X^- \quad (5.282)$$

(132)

only if the classical carbonium ion is destabilized by steric strain (as in the norbornyl cation; see p. 289) or if R is a $\pm E$ group (e.g., phenyl, naphthyl, etc.) or a $-E$ group (e.g., p-anisyl). In cases of this kind, R occupies an apical or bridging position between the two basal carbon atoms. Since a nucleophile can attack either to form the final product, the overall reaction can take place with rearrangement (i.e., a Wagner–Meerwein-type rearrangement) or by an aliphatic substitution following S_N1 kinetics but leading to a product where X is replaced with complete retention of configuration.

5.25 Some $OE^{\ddagger}$ Reactions

The reader may have noted that the reactions so far considered have all been of $EO^{\ddagger}$ type and may consequently have wondered if the EO/OE classification has any point to it. The reason for this bias is that the majority of odd neutral conjugated molecules are conventionally regarded as derivatives of even ones containing a $-E$ substituent. Thus since aniline (133) is isoconjugate with an odd AH anion (134), electrophilic substitution of (133) could be classed as $OE_B^{\ddagger}$, the rate-determining step being conversion of (133) to an even intermediate [e.g., (135)]. However the conventions of organic chemistry lead us to regard this rather as an $EO_B^{\ddagger}$ reaction of an even AH (i.e., benzene) carrying a $-E$ substituent (i.e., NH_2).

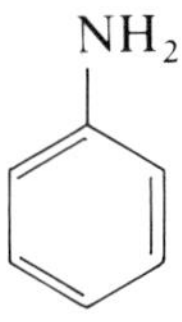

(133) 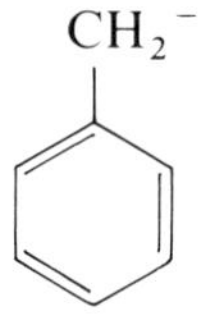   (134) 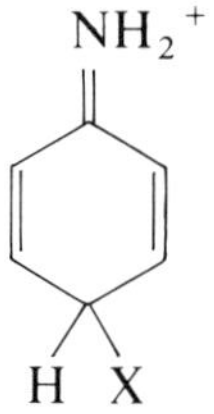   (135)

There are, however, a number of reactions that are properly classed as $OE^{\ddagger}$, for example, the reactions of ambident anions with electrophiles. The examples we have already cited (p. 245) show very clearly that this is of anti-BEP type, i.e., $OE_A^{\ddagger}$, the protonation of enolate ions giving enols rather than ketones even when the latter are more stable. Furthermore, the course of such reactions can often be changed by altering the solvent (see, e.g., p. 269) in a manner unrelated to the resulting changes in the heats of reaction.

The reactions of enamines with electrophiles follow a similar course and are also of $OE_A^{\ddagger}$ type. Enamines are isoconjugate with enolate ions and can likewise react with electrophiles in two different ways, either on nitrogen to form an enammonium ion or on carbon to form a methyleneammonium ion. As in the case of enolate ions, protonation takes place more rapidly on the heteroatom (here nitrogen) rather than carbon while allyl halides, e.g., methyl iodide, give C-alkylated products; e.g.,

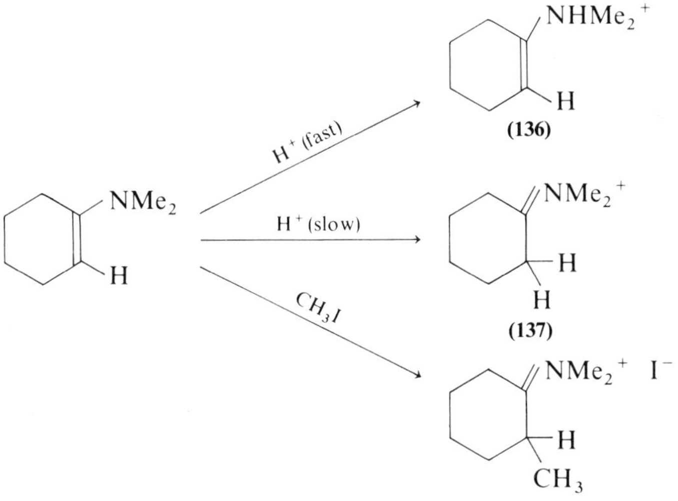

(5.283)

In the case of enolate ions, the choice between attack on oxygen or on carbon was attributed to solvent effects (p. 269). Here solvation is unlikely to be responsible since enamines are neutral. There is, however, another factor that could contribute to the distinction between O attack and C attack in enolate ions and which could also operate in the case of enamines.

Methyleneammonium ions [e.g. (137)] are more stable than the isomeric enammonium ions [e.g., (136)], just as simple ketones are more stable than the corresponding enols, because the sum of the C=N and C—C bond energies in, e.g., (137) is greater than that of the C—N and C=C bond energies in, e.g., (136). The heat of reaction for attack on carbon will be correspondingly more negative than that for attack on nitrogen by an amount δE which will be independent of the electrophile:

$$\delta E = (E_{\text{C-N}} + E_{\text{C=C}}) - (E_{\text{C=N}} + E_{\text{C-C}}) \qquad (5.284)$$

If the reaction obeyed the BEP principle, there would be a corresponding difference $\delta E^\ddagger$ between the activation energies for attack on carbon and on nitrogen given [see equation (5.69)] by

$$\delta E^\ddagger = \alpha \, \delta E \qquad (5.285)$$

However, whereas attack on nitrogen involves no reorganization of bonds, attack on carbon involves both a change in the lengths of the CC and CN bonds and a change in hybridization of carbon. This will lead to an extra contribution $\delta E_A^\ddagger$ to the activation energy. Which point of attack is preferred will then depend on whether or not $\delta E_A^\ddagger$ is greater or less than $\delta E^\ddagger$, i.e., than $\alpha \, \delta E$.

Now according to the BEP principle, the more exothermic a reaction, the smaller is the activation energy, the nearer is the transition state to the reactants in structure, and the smaller is the BEP factor α. The selectivity between C attack and N attack is therefore less for reactions with low activation energies than for those with high energies, i.e., is less, the faster is the reaction. For very fast reactions (e.g., protonation), α will then be small, so the term $\delta E_A^\ddagger$ dominates the situation and attack consequently takes place on nitrogen. For slow reactions, e.g., alkylation by alkyl halides, the reverse is true, so attack on carbon predominates.

5.26 Thermal Pericyclic Reactions ($EE_A^\ddagger$ and $OO_A^\ddagger$)

Early in the history of organic chemistry, benzene was regarded as undergoing a rapid oscillation between the two Kekulé structures, a process represented by Lapworth and Robinson as a corresponding concerted cyclic electromeric displacement of pairs of electrons, namely

$$(5.286)$$

This led to the idea that a similar cyclic interchange of bonds might occur in chemical reactions and a number of such mechanisms were suggested from time to time. One example was the rearrangement of iminoethers to amides,

$$R-C\underset{O}{\overset{NH}{\diagup}}R' \longrightarrow R-C\underset{O}{\overset{NHR'}{\diagup}} \quad (5.287a)$$

and another the Cope rearrangement,

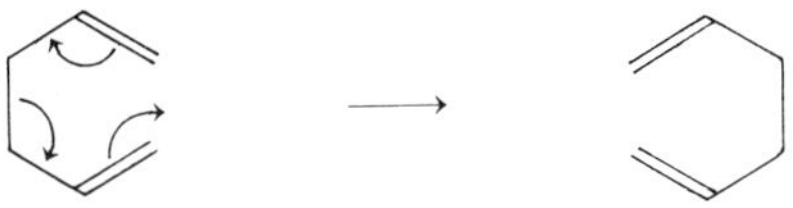

$$\text{(structure)} \longrightarrow \text{(structure)} \quad (5.287b)$$

However, it is only during the last decade that reactions of this kind, for which Woodward and Hoffman have coined the convenient term *pericyclic*, have become a matter of major concern to organic chemists. There are three reasons for this.

First, pericyclic reactions present awkward mechanistic problems in the sense that alternative two-step mechanisms are usually possible and that the distinction between these and true concerted pericyclic processes is extremely difficult. It is only in recent years that techniques have been developed for solving this problem. Thus of the two examples cited above, only the latter [equation (5.287b)] is a true pericyclic reaction.

Second, the majority of these reactions are thermal rearrangements in which mixtures of products are formed. It was impossible to analyze the products before the development of gas chromatography for their separation and IR and NMR spectroscopy for their identification.

Third, just when the development of the necessary techniques had initiated studies in this field, an enormous impetus was given to them by the enunciation of the Woodward–Hoffmann rules. These not only explained various unexpected results that had already been obtained, but also led to a number of rather startling and surprisingly successful predictions.

Woodward and Hoffmann based their original derivation of their rules on Fukui's frontier orbital theory and have subsequently relied on arguments based on the conservation of orbital symmetry. Both these approaches suffer from deficiencies (which will be discussed at the end of this chapter) so we will rely here on an earlier treatment by M. G. Evans, which is both simpler and better.

In the Diels–Alder reaction, an ethylene derivative (*dienophile*) reacts with a 1,3-diene to form a cyclohexene, the simplest case being the reaction of ethylene with 1,3-butadiene,

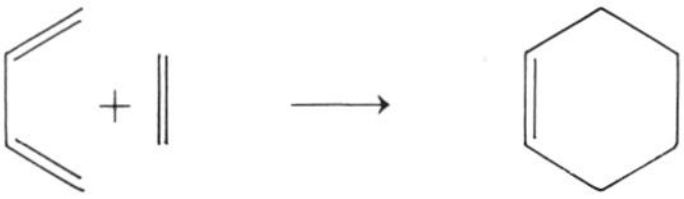

$$\text{(diene)} + \text{(ethylene)} \longrightarrow \text{(cyclohexene)} \quad (5.288)$$

In 1938, there was still much discussion concerning the mechanism. Two possibilities were recognized, the first a one-step pericyclic process involving a cyclic transition state (138) and the second a two-step process involving an intermediate biradical (140); i.e.,

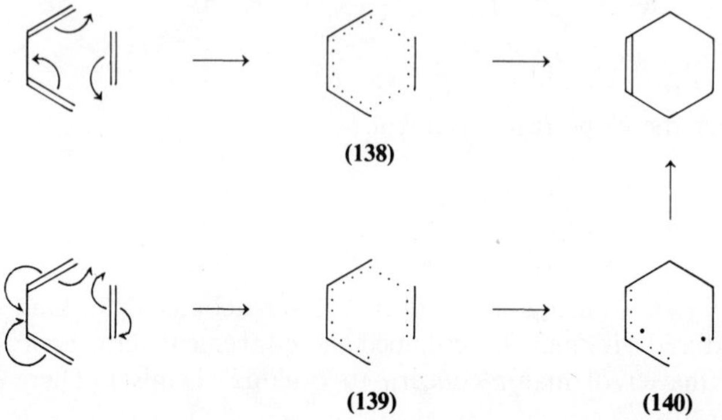

(138)

(5.289)

(139)   (140)

Let us consider the structure of the cyclic transition state (138). In it, each carbon atom remains linked by σ bonds to the same three atoms as in the reactants. Three of its AOs will be used in this way. Each carbon atom has just one AO left over; these will be used to form the delocalized MOs indicated by dotted lines in (138).

Figure 5.35(a) shows the corresponding structure. It can be seen that the six $2p$ AOs can overlap in a cyclic fashion, the AO of each carbon atom overlapping with the AOs of the atoms on each side of it in the six-membered ring. Moreover the phases of the AOs can be chosen so that all the overlaps are in phase. The situation is thus topologically equivalent to, or isoconjugate with, that in benzene (Fig. 5.35b), the only difference being in the geometry of overlap of the AOs (which is not important; see p. 50). Similarly, the transition state (139) leading to the biradical (140) is isoconjugate with hexatriene [cf. Figs. 5.35(c, d)]. Consequently, the energy of the delocalized electrons in the cyclic transition state (138) should be less than in the linear transition state (139) for the same reason that the π energy of benzene is less than that of hexatriene. In other words, the cyclic transition state should be specifically stabilized by its cyclic structure because it is *aromatic*.

Evans pointed out that one would therefore expect the Diels–Alder reaction to be a concerted one-step process because the activation energy for this should be less than that involving an intermediate biradical. This prediction has proved to be correct. He also pointed out that a one-step Diels–Alder-like dimerization of ethylene to cyclobutane,

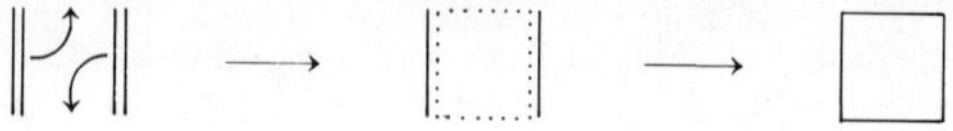

(5.290)

Chemical Reactivity

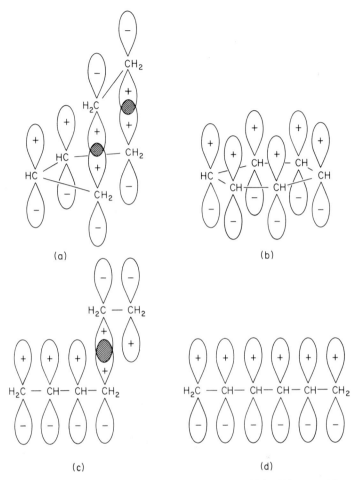

FIGURE 5.35. Overlap of $2p$ AOs in (a) the cyclic Diels–Alder transition state (138); (b) benzene; (c) the linear transition state (139); (d) 1,3,5-hexatriene.

would involve a transition state isoconjugate with cyclobutadiene and so not aromatic* (Fig. 5.36). Such a reaction, if it occurs at all, should therefore take place by a much less facile two-step biradical mechanism,

$$\text{[diagram]} \quad (5.291)$$

Reactions of this kind are now known to take place in this way and are observed only if the intermediate biradical is stabilized by suitable substituents (Section 5.25).

---

* In 1938, chemical theory had progressed to a point at which it could be predicted that cyclobutadiene would not be aromatic but not that it would be antiaromatic. The destabilization of cyclobutadiene was first predicted by the PMO method fourteen years later.[13]

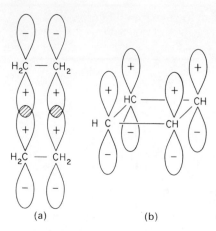

FIGURE 5.36. Overlap of $2p$ AOs in (a) the cyclic transition state for concerted pericyclic dimerization of ethylene; (b) cyclobutadiene.

In a pericyclic reaction, a cyclic permutation of bonds takes place around a ring of atoms, e.g.,

$$\text{(141)} \longrightarrow \text{(142)} \longrightarrow \text{(143)} \quad (5.292)$$

The remaining bonds in the system remain intact throughout. Thus if atoms a and b in (141) are initially linked by a double bond, the bond represented in (141) will be the π component of this, so atoms a and b remain singly linked in the product (143). Each bond in (141) or (143) is formed by the interaction of two AOs, one on each atom. In the reaction, the AO of a given atom (say a) switches from forming a bond to one partner in the ring (b) to another (j). In the transition state (142), the n AOs of the n atoms therefore overlap in cyclic fashion, like the overlap of $2p$ AOs in the π MOs of a cyclic polyene (e.g., benzene). The transition state is therefore isoconjugate with a cyclic polyene and is consequently either aromatic or antiaromatic. This result depends solely on the topology of overlap of the AOs involved, not on the shapes of the AOs or the way in which they overlap. Thus while the arguments embodied in Figs. 5.35 and 5.36 are based on the overlapping of $2p$ AOs, they are not in any way affected if the AOs in fact have some s character and are consequently of hybrid type.

There is, however, one very important difference between the delocalized MOs in a pericyclic transition state and the delocalized π MOs in a cyclic polyene. In the latter, the phases of the $2p$ AOs can be chosen so that they always overlap in phase; the resulting π systems are consequently of Hückel type (p. 106). While it is possible to envisage anti-Hückel π systems (p. 108), none has yet been prepared. Pericyclic transition states, on the other hand,

Chemical Reactivity

have three-dimensional structures in which the participating AOs can overlap in σ fashion as well as in π fashion. Here it is quite possible for anti-Hückel systems to occur.

This possibility is best illustrated by an example, the conversion of cyclobutene (144) to 1,3-butadiene (145) on heating*:

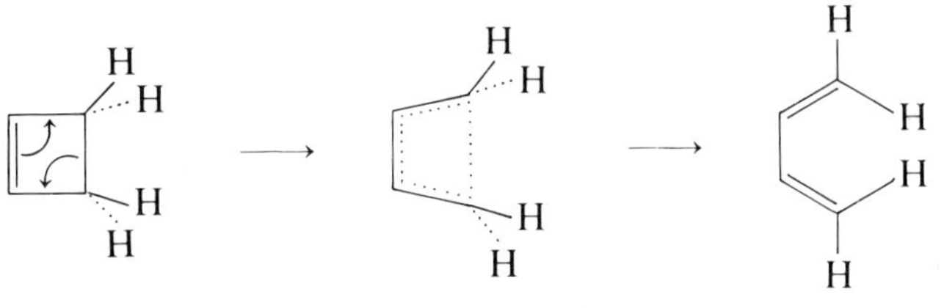

$$(144) \longrightarrow \longrightarrow (145) \tag{5.293}$$

In (144), the methylene groups lie in planes perpendicular to the ring, while in (145), all the atoms can lie in a plane. Each methylene group must therefore rotate through 90° during the reaction. There are then two distinguishable paths by which the reaction can occur, the methylene groups rotating in the same sense, both clockwise or both counterclockwise (146), or in opposite senses like a pair of meshed gearwheels (147). The former mode of opening is termed *conrotatory*, the latter *disrotatory*. Our problem is to determine which, if either, will be preferred.

conrotatory (146)   disrotatory (147)

In (144), the AOs forming the CH_2—CH_2 σ bond lie in the nodal plane of the C=C π MO. There is therefore no interaction between them. However, as soon as the methylene groups begin to rotate, each hybrid AO begins to overlap and interact with the $2p$ AO of the adjacent carbon atom in the C=C bond. We now have a typical pericyclic situation, each carbon atom having an AO which interacts with AOs of the two carbon atoms adjacent to it in the ring.

Figure 5.37 shows the resulting situation in the conrotatory and disrotatory transition states. It is immediately obvious that in the latter (Fig. 5.37b), the phases of the orbitals can be chosen so that all the overlaps are

---

* The thermal conversion of cyclobutenes to butadienes is of especial historical importance since it was a reaction of this kind, encountered during work directed to the synthesis of vitamin B12, that first interested Woodward in what are now termed pericyclic processes.

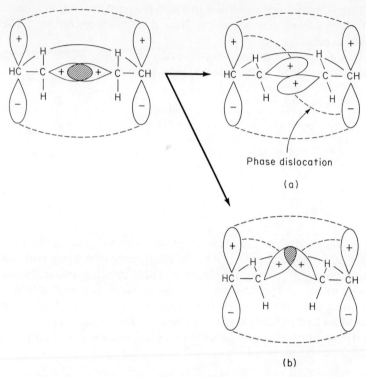

FIGURE 5.37. Conversion of cyclobutene to (a) the conrotatory and (b) the disrotatory transition states for electrocyclic opening to butadiene. The C=C bond has been sketched in the diagram to show the orbital overlap more clearly. Dashed lines indicate the overlap of AOs.

in phase. The disrotatory transition state is therefore a Hückel-type conjugated system. In the conrotatory transition state, on the other hand, this is not possible. However the phases are chosen, there will be at least one *phase discontinuity* where two AOs overlap out of phase with one another. The conrotatory transition state is therefore of anti-Hückel type.

We are dealing here with four-atom conjugated systems containing four electrons, i.e., the two pairs of electrons that form the C=C π bond and the CH_2—CH_2 σ bond in (144). The disrotatory transition state (145), being of Hückel type, will then be isoconjugate with normal cyclobutadiene and so will be antiaromatic, whereas the conrotatory transition state will be isoconjugate with an anti-Hückel analog of cyclobutadiene and so will be aromatic (see Table 4.2).

The conrotatory transition state should therefore be the more stable, being stabilized by the cyclic conjugation instead of destabilized. The reaction is therefore predicted to take place by conrotatory opening rather than disrotatory, as indeed it does. For example, the 3,4-dimethylcyclobutenes open

Chemical Reactivity

stereospecifically in the following way:

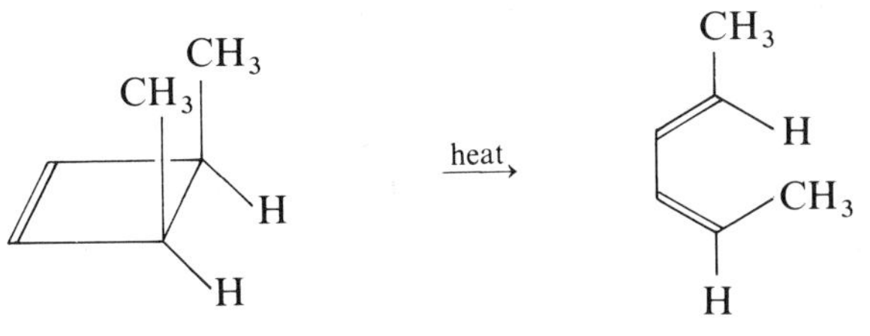

(5.294)

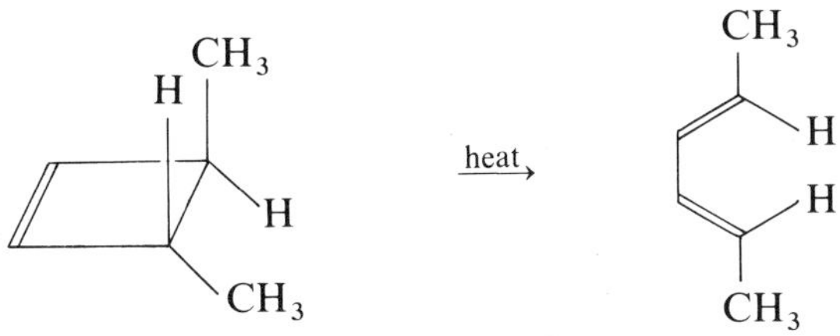

(5.295)

This argument can obviously be extended to concerted pericyclic reactions of all kinds. The transition state for any such reaction will be isoconjugate with a normal Hückel-type cyclic polyene or an anti-Hückel analog of one. If the transition state is aromatic, the resulting stabilization will lower its energy and so accelerate the reaction. If it is antiaromatic, the converse will be true. Since, moreover, the rules for aromaticity in Hückel-type and anti-Hückel-type systems are diametrically opposite, in each case one will be aromatic and the other antiaromatic. If, then, a reaction can follow one of two alternative pericyclic paths, one involving a Hückel-type transition state and the other an anti-Hückel-type transition, the reaction will prefer to follow the path in which the transition state is aromatic. If, on the other hand, only one of the two alternatives is sterically possible, the reaction will take place relatively easily if the corresponding transition state is aromatic and with relative difficulty if it is antiaromatic. In the latter case, the antiaromatic transition state will, if possible, be bypassed by a two-step mechanism in which the transition state is linear instead of cyclic [e.g., equation (5.291)].

This obvious extension of Evans' ideas is sufficient to explain all the known experimental results for pericyclic reactions. It can be summed up in a single simple rule which can therefore conveniently take the place of the Woodward–Hoffmann rules:

Evan's Principle. Thermal pericyclic reactions take place preferentially via aromatic transition states.

Whether or not a given transition state is aromatic can be found in the following way.

1. Construct a three-dimensional model or drawing of the transition state with the relevant AOs (e.g., Figs. 5.35–5.37).

2. Assign phases to the AOs, choosing them so as to avoid phase dislocations as far as possible (see again Figs. 5.35–5.37).
3. Is there an unavoidable phase dislocation? (There cannot be more than one, see p. 106). If not, the transition state is of Hückel type. If there is, it is of anti-Hückel type.
4. Use Section 3.14 to deduce whether or not the transition state is aromatic. If it is, the reaction is "allowed" in the Woodward–Hoffmann sense. If not, it is "forbidden."

In order to use Evans' principle, one must of course have some way of telling whether or not a given cyclic conjugated system is aromatic. The PMO method provides a simple procedure for this purpose. Since the concept of aromaticity is an essential one in organic chemistry, Evans' principle reduces the problem of pericyclic reactions to a special case of this more general theme. Indeed, it greatly enriches the range of aromatic structures by introducing anti-Hückel types that are unknown in normal π systems.

If the pericyclic ring has an even number of atoms, both the reactants and the transition state will have even conjugated systems. Such reactions are of $EE^{\ddagger}$ type. Furthermore, the course of the reaction depends on the topology of the overlap of AOs in the transition state and this is unrelated to the structures of the reactants or products. The reactions are of anti-BEP type and are therefore classed as $EE_A^{\ddagger}$. This is evident from the examples given in equation (5.294) or (5.295). The stereospecificity of the reactions is clearly unrelated to their heats of reaction since both cis- and trans-3,4-dimethylcyclobutene can rearrange without steric restraints to any of the isomeric 2,4-hexadienes.

So far we have considered only reactions in which the pericyclic ring contains an even number of atoms. Reactions of this kind are, however, known in which an odd-numbered ring is involved. A simple example is the Diels–Alder-like addition of 2-methylallyl cation (148) to cyclopentadiene (149) to form the methylbicyclooctyl cation (150). The transition state for this reaction is easily seen to be of Hückel type (151) and so isoconjugate with tropylium. Since the allyl cation contains only two π electrons, we are dealing here with a six-electron system isoconjugate with the tropylium cation (147) and hence aromatic. In reactions of this kind, both the reactants and the transition state are odd. The reactions are therefore of $OO^{\ddagger}$ type. Since, moreover, the aromaticity or antiaromaticity of the transition state is again unrelated to the structures of the reactants or products, the reactions are of anti-BEP type and are consequently classed as $OO_A^{\ddagger}$.

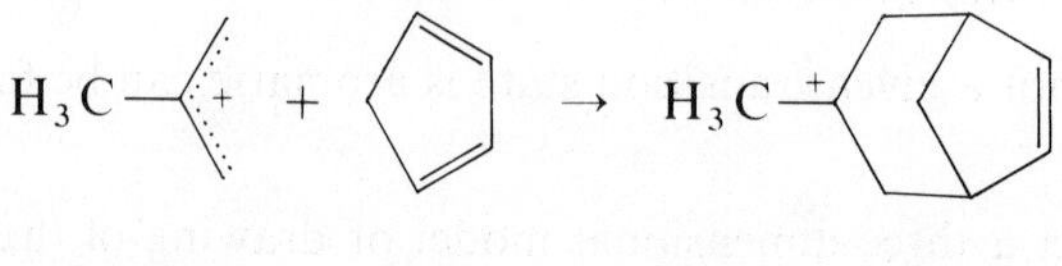

(148)   (149)   (150)

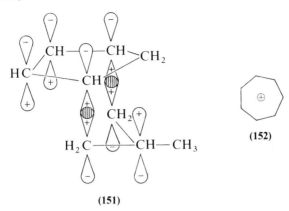

5.27 Examples of Pericyclic Reactions

In this section, we will briefly review the various kinds of pericyclic reactions so far discovered in order to show how Evans' principle can be used to interpret them. Many additional examples will be found in the reviews listed at the end of this chapter.

A. Cycloaddition Reactions ($EE_A{}^{\ddagger}$ and $OO_A{}^{\ddagger}$)

The majority of cycloaddition reactions involve interactions between two (occasionally three) π systems as in the Diels–Alder reaction [equation (5.284)] or the analogous additions of allyl cations to dienes (148)–(150). Such π cycloadditions usually take place by *cis* addition to each conjugated system because the corresponding transition states are less strained. The reactions are then of Hückel type (see, e.g., Figs. 5.35a and 5.36a) and follow the same rules for aromaticity as do ordinary conjugated hydrocarbons. Some examples follow:

(5.296)

This is one of the rare cases of a triple addition. The interactions between the 2p AOs in the transition state are indicated by dotted lines. It is easily seen that addition to each double bond is *cis*, so the six-membered ring in the transition state is isoconjugate with benzene.

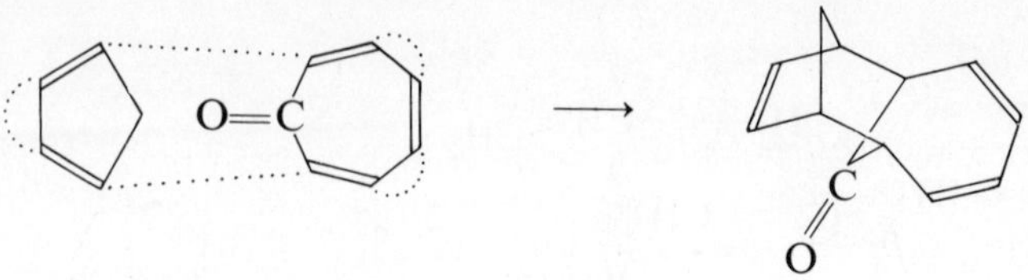

(5.297)

In cases such as this, where more than one conjugated ring seems to be involved, it is usually best to examine the conjugated hydrocarbon analog of the transition state. Here the analog is (153). The bonds ringed by a dotted line in (153) are essential single or double bonds. Deleting them, we have (152), which is aromatic according to the rules in Sections 3.14 and 3.17 (pp. 96 and 106) (Hückel-type system with $4 \times 2 + 2$ atoms).

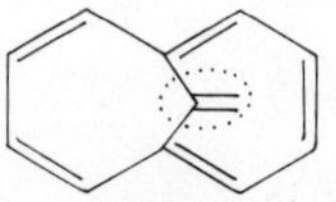

(153) 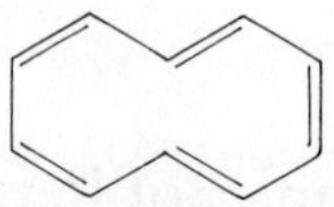   (154)

This example illustrates another point. Cyclodecapentaene (154) is in fact rather unstable because it cannot exist in a planar form, for steric reasons. These steric problems arise, however, entirely from the framework of σ bonds. It has nothing to do with aromaticity, this being a function of the π electrons. In applying the rules for aromaticity to diagrams such as (153) or (154), we should therefore ignore all steric effects since these are clearly irrelevant to a consideration of the delocalized MOs in the corresponding transition state.

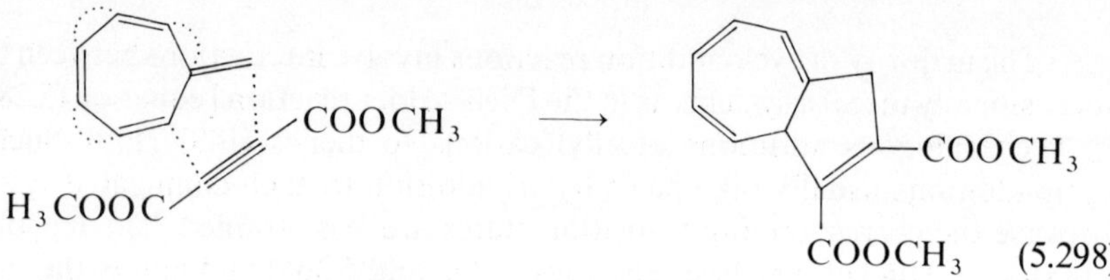

(5.298)

The transition state is isoconjugate with azulene.

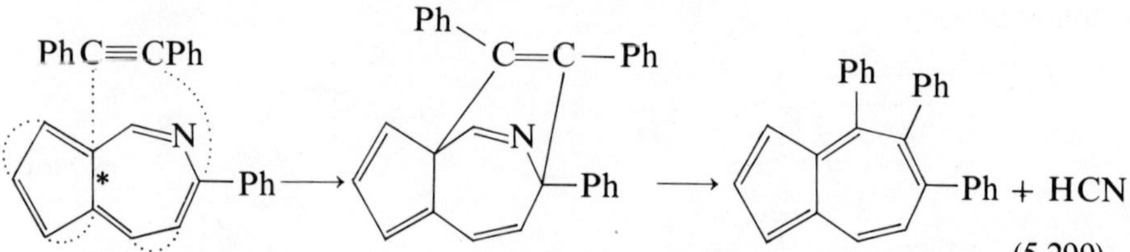

(5.299)

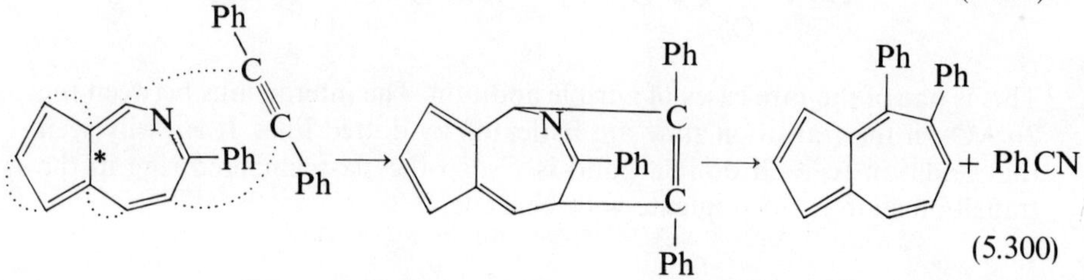

(5.300)

Chemical Reactivity

The bonds marked with asterisks are easily seen to be essential single bonds in the two transition states (cf. azulene; p. 100). Deleting them, the transition states are seen to be isoconjugate with (155), which contains six- and ten-membered rings, both aromatic. The second steps are the reverse of π cycloadditions, taking place via transition states analogous to those in the initial additions. This can be seen very easily if the intermediate adducts are written as (156) or (157).

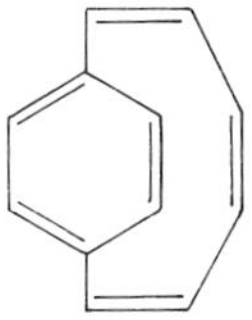

(155) 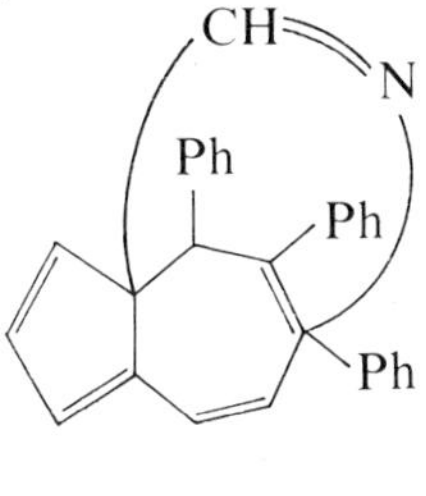   (156) 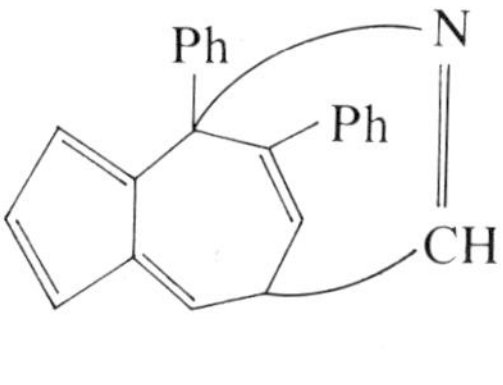   (157)

If a π cycloaddition takes place by *cis* addition to the π systems, the transition state is of Hückel type. If, however, one of them adds *trans*, the transition state has a phase dislocation and so is anti-Hückel. This is shown in (158) and (159). The ribbons indicate the π MOs of the two components, the signs along them indicating the phase of the component 2p AOs. In *trans–trans* addition (160), the transition state is once more of Hückel type.

(158)   (159)   (160)

The classic example of a *cis–trans* π cycloaddition is that of tetracyanoethylene (161) to heptafulvalene (162):

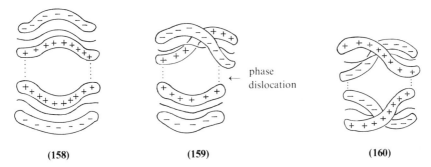

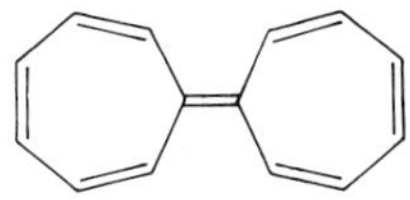

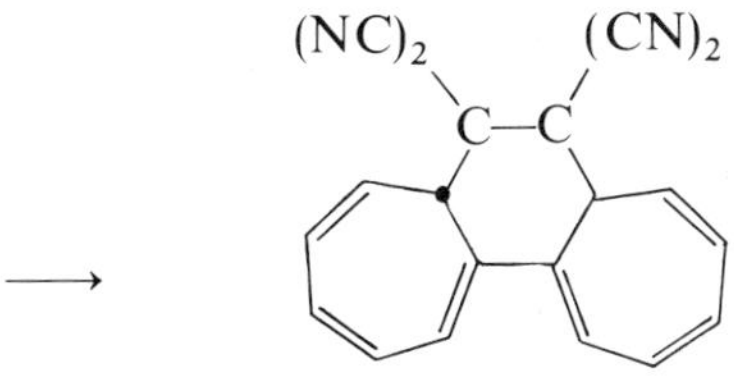

(5.301)

As indicated, addition takes place *trans* to the heptafulvalene. The transition state is isoconjugate with the tricyclic hydrocarbon (163). The bonds marked with asterisks in this are essential single bonds. Deleting them, we have (164), a cyclic polyene with 4 × 4 = 16 atoms. The corresponding Hückel system is consequently antiaromatic, so the reaction should prefer to take place via an anti-Hückel transition state. Of the two possibilities [*trans* addition to (161) and *cis* to (162), or *trans* to (162) and *cis* to (161)], only the former is sterically feasible.

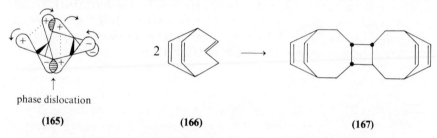

(163)   (164)

Even the concerted pericyclic dimerization of an olefin to a cyclobutane [equation (5.290)] can occur if the two double bonds are twisted sufficiently. It may then be possible for the top lobes of the $2p$ AOs in one olefin moiety to overlap with a top and a bottom lobe of the $2p$ AOs of the other (165). The transition state is then of anti-Hückel type and so is aromatic, being an anti-Hückel analog of cyclobutadiene. The arrows in (165) indicate the relative directions in which the terminal atoms of the C=C bonds are twisted.

↑
phase dislocation

(165)   (166)   (167)

The spontaneous dimerization of (166) to (167) probably takes place in this way. The *trans* double bond in (166) is severely twisted, as one can see from models. Addition via the transition state (165) leads to *cis* addition to one double bond and *trans* addition to the other. As indicated in (167) the stereochemistry of the product conforms to this.

B. Some Special Features of the Diels–Alder Reaction

The Diels–Alder reaction is the oldest and best studied pericyclic reaction. Here we will consider some special features in the light of PMO theory.

(a) In the case of dienophiles carrying unsaturated substituents, the reaction leads almost exclusively to *endo–cis* adducts, although the *exo–cis* isomers are the more stable and can be formed eventually by isomerization of the *endo–cis* products initially formed.* Thus cyclopentadiene (168) reacts with maleic anhydride (169) to form the *endo* adduct (170), although the *exo* isomer (171) is more stable.

* This represents a further deviation from the BEP principle.

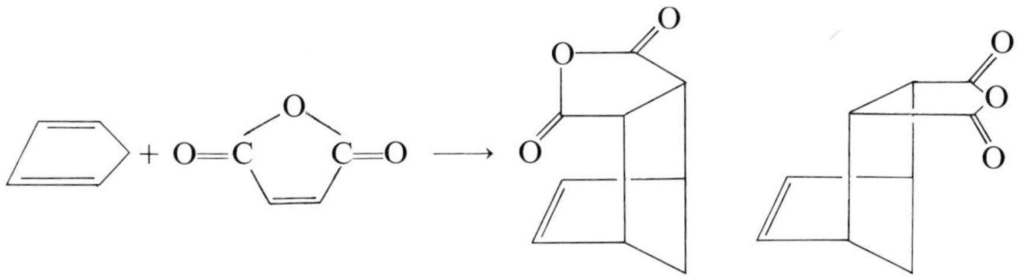

Chemical Reactivity   351

(168)   (169)   (170)   (171)

The reason for this can be seen from an examination of the transition states for *endo* (172) and *exo* (173) dimerization of 1,3-butadiene,

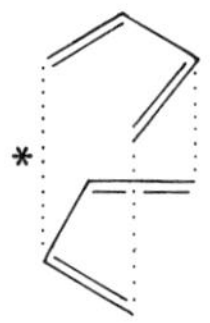

(172) 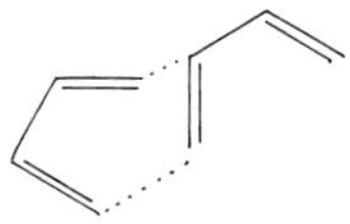   (173) 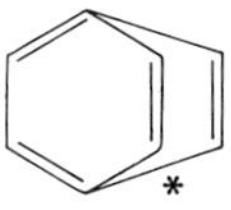   (174a) 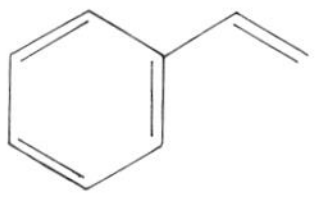   (174b)

All the interactions between the two molecules of butadiene [indicated by dotted lines in (172) and (173)] are in phase, so both transition states are of Hückel type. They differ only by the possibility of an extra interaction in (172) (indicated by an asterisk). The AHs isoconjugate with (172) and (173) are (174a) and (174b), respectively, the extra interaction in (172) corresponding to the asterisked bond in (174). Since both rings in (174a) are six-membered, both are aromatic. The extra interaction is therefore energetically favorable and is sufficient to tip the balance in favor of (172). The predominance of *endo–cis* addition in other cases [e.g., (168) + (169) ⟶ (170)] can be likewise attributed to analogous "secondary interactions" in the *endo–cis* transition states.

(b) In cases where the diene and dienophile are asymmetrically substituted, one of the two possible adducts usually predominates. The effect is particularly marked in cases where one of the reactants carries a $+E$ substituent (CHO, CN, etc.) and the other a $-I$ or $-E$ substituent (e.g., CH_3 or CH_3O). The preferred product is then the one in which the substituents are 1,2 or 1,4 to each other. Thus,

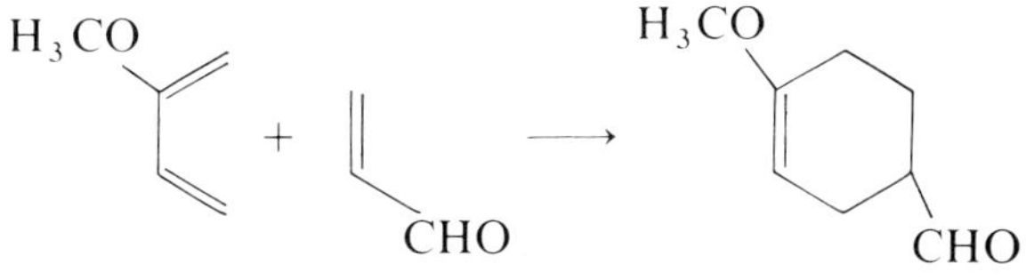

rather than

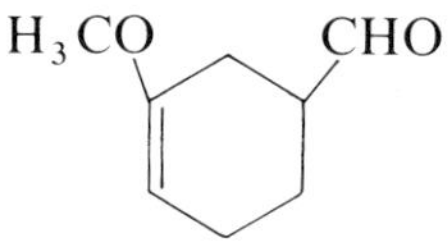

(5.302)

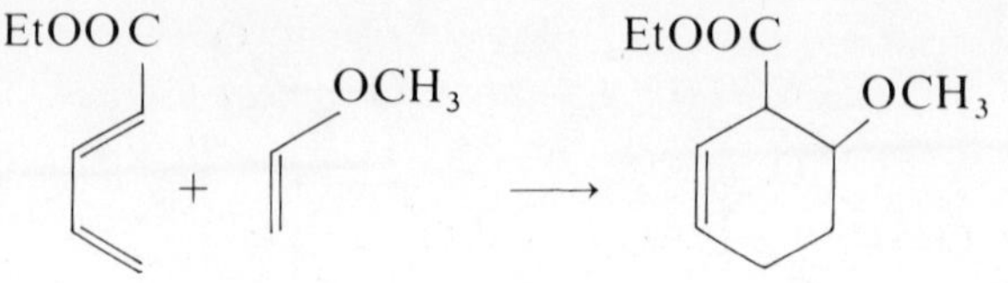

$$\text{EtOOC-butadiene} + \text{methoxyethylene} \longrightarrow \text{4-EtOOC-5-OCH}_3\text{-cyclohexene} \quad$$

rather than

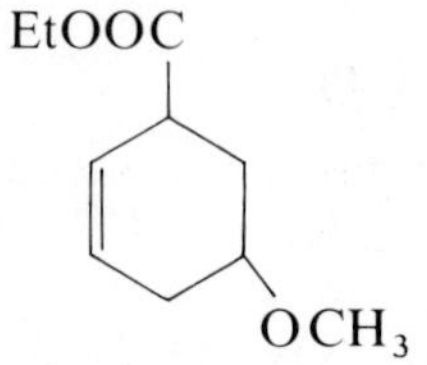

$$\text{4-EtOOC-3-OCH}_3\text{-cyclohexene} \quad (5.303)$$

Since the transition state of the Diels–Alder reaction is isoconjugate with benzene, E-type substituents should be able to conjugate with it. If then we have a $+E$ substituent, and a $-I$ or $-E$ substituent, attached to atoms taking part in the transition state, mutual conjugation can occur between them if they are *ortho* or *para* to one another. The resulting stabilization leads to the preferred formation of the product formed via a mutually conjugated transition state.

C. Sigmatropic Reactions ($EE_A^{\ddagger}$ and $OO_A^{\ddagger}$)

The term *sigmatropic rearrangement* was introduced by Woodward and Hoffmann to describe reactions in which σ bond flanked by one or two π systems migrates to another position along the π system(s). A classic example would be a 1,3-allylic migration,

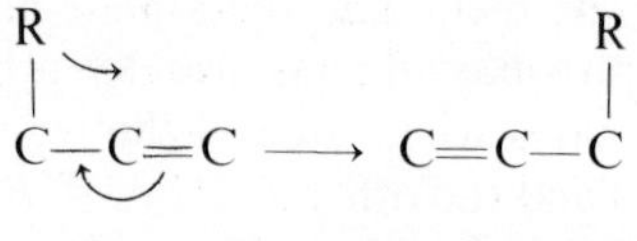

$$\underset{|}{\overset{R}{C}}-C=C \longrightarrow C=C-\underset{|}{\overset{R}{C}} \quad (5.304)$$

Another classic example, this time involving two π systems, is the Cope rearrangement,

$$\text{1,5-hexadiene} \longrightarrow \text{1,5-hexadiene} \quad (5.305)$$

The reactant and product in a reaction of this type are isomers formed by alternative possible modes of combination of two conjugated systems. Thus the reactant and product in equation (5.304) can both be formed by combination of $(R\cdot + \text{allyl}\cdot)$ and those in equation (5.305) by alternative modes of dimerization of allyl. We will first consider the case where both the systems in question are odd, so that the reactant and transition state are even. Pericyclic reactions of this kind will be of $EE_A^{\ddagger}$ type.

Let us consider a reaction of this type involving two odd AHs R and S linked initially through atom r in R to atom s in S and finally through atom t in R to atom u in S; i.e.,

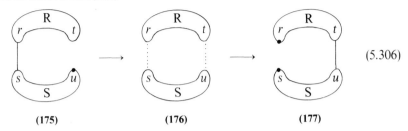

(5.306)

In the transition state (176), the two conjugated systems interact simultaneously through rs and tu, leading to a single cyclic conjugated system. We have to determine whether or not this is aromatic.

The transition state is precisely analogous to that in a π cycloaddition, the only difference being that the reactants in the π cycloaddition are both even, while here R and S are both odd. We can therefore see at once that if the π systems in (176) both interact in *cis* fashion or both in *trans* fashion, the transition state will be of Hückel type, while a *cis*–*trans* interaction will lead to an anti-Hückel transition state.

The transition state for a *cis-cis* allylic migration [equation (5.304)] will therefore be isoconjugate with cyclobutadiene and consequently antiaromatic. No such rearrangement has as yet been observed. For a concerted rearrangement of this kind to be at all facile, it is necessary that one or other component should interact in *trans* fashion. Migration from one side to the other of the allylic system (*antarafacial* migration) is virtually impossible on steric grounds, so the allyl moiety must interact *cis* (*suprafacial* migration). Therefore the only way in which a rearrangement of this kind can take place by an aromatic pericyclic process is by the mechanism indicated in the following equation, the migrating group R using different lobes of a p AO for bonding to the end of the allyl; effectively, the reaction involves migration of allyl from one side to the other of the p AO in R:

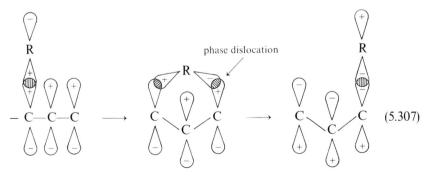

(5.307)

If the migrating group R is an asymmetric alkyl group, the reaction will lead to inversion of its configuration since the reaction, so far as R is concerned, is in effect an S_N2 substitution, allyl acting simultaneously as a nucleophile and a leaving group. Several rearrangements of this kind have been reported, the classic example, due to Berson, being

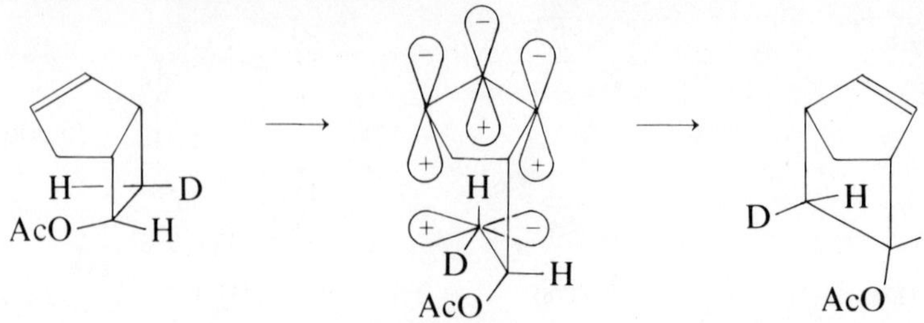

$$ \qquad (5.308) $$

Note that the CHD group undergoes inversion during the reaction, D being *trans* to acetoxy in the reactant but *cis* in the product.

While 1,3 migrations of this kind can occur only if the migrating group undergoes inversion, 1,5 migrations can take place without inversion. Thus hydrogen, which has only a single *s*-type AO, cannot undergo allylic migration but can migrate 1,5 in dienes; e.g.,

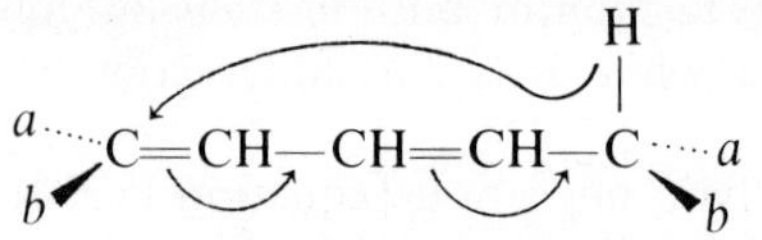

(178)

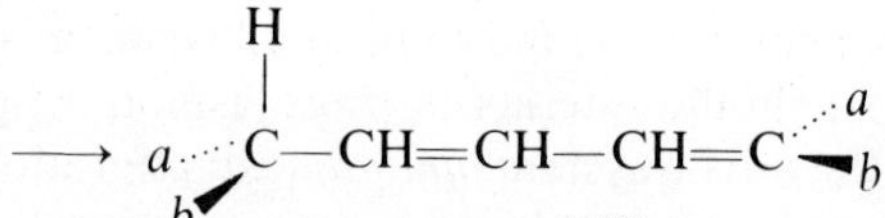

$$ \longrightarrow \qquad (5.309) $$

(179)

If one examines the orbitals in the transition state (180), one can see that it is isoconjugate with benzene* and so is aromatic:

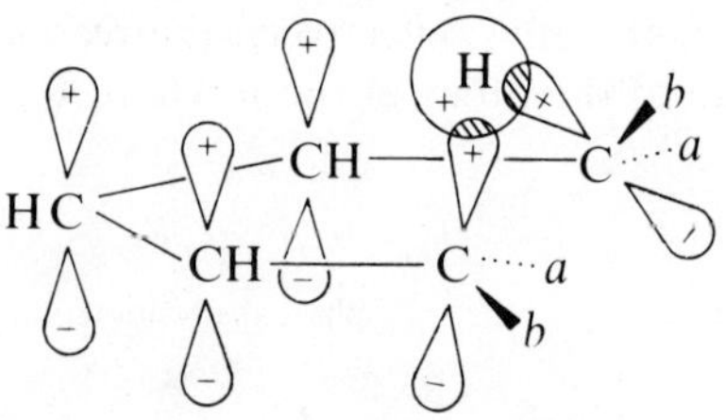

(180)

---

* Since the 1*s* AO of hydrogen is so small, the AOs of the two terminal carbon atoms cannot overlap with it without also overlapping with one another. The transition state is therefore strictly isoconjugate not with benzene but with bicyclohexatriene

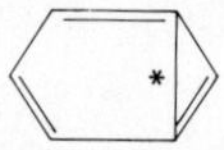

Since, however, the transannular bond (asterisked) is an essential single bond, it can be ignored. Thus the stability of the transition state is unaffected by mutual overlap of the carbon AOs.

Chemical Reactivity

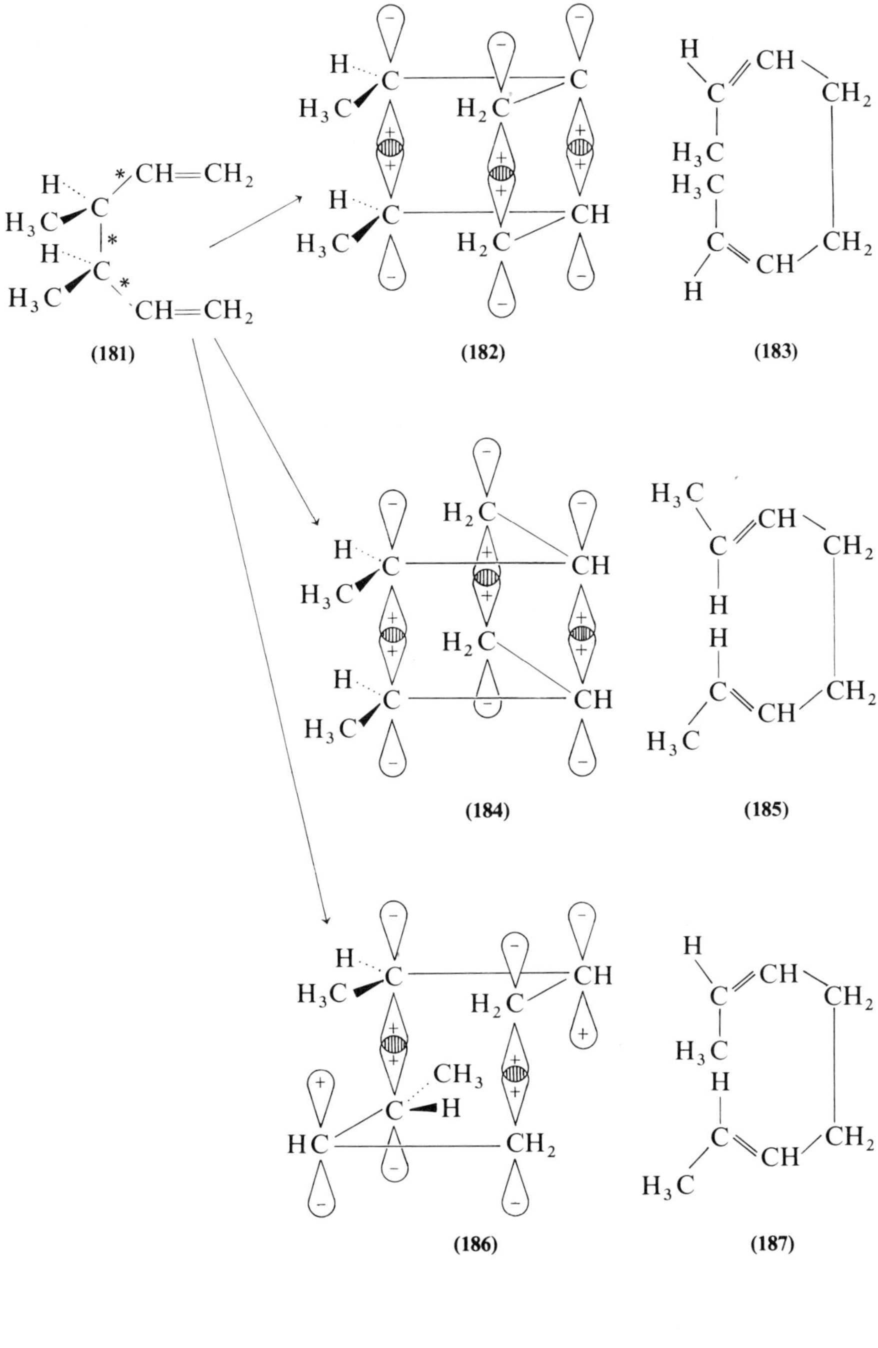

If the terminal carbon atoms are substituted, the stereochemistry of the reaction can be deduced from that of the products. Thus if the terminal atoms are substituted in the manner indicated in equation (5.309), rearrangement converts the reactant (178) into its mirror image (179). If the starting material was optically active, the product will then be racemized. By experiments of this kind, Roth *et al.* have confirmed that the reaction does take place by suprafacial migration of hydrogen.

Turning now to the Cope rearrangement [equation (5.305)], one can see at once that the *cis–cis* transition state will be aromatic, being of Hückel type and isoconjugate with benzene. However, there is still a further interesting stereochemical point which can be seen from a simple example, the rearrangement of *meso*-3,4-dimethyl-1,5-hexadiene (181) into 2,7-octadiene. By rotating the vinyl groups in (181) about the asterisked bonds, we can arrange for the transition state to have any of three different geometries (182), (184), (186), leading to different geometric isomers (183), (185), (187) of the octadiene. Which if any of these will be favored?

The transition states (182) and (184) are similar in shape to the boat isomer (188) of cyclohexane, while (186) is similar to the chair isomer (189). One cannot, however, draw any conclusions from the fact that (186) is less stable than (189), because this difference is due to eclipsing of saturated carbon atoms in the boat. One can also see that the transition state (186) is isoconjugate with benzene, having six AOs in a ring, each AO overlapping in phase with the two AOs on either side of it. The boat transition states (182) and (184) differ from (186) in having an additional overlap between the AOs of the central atoms in each allyl residue. They are consequently isoconjugate not with benzene but with [2,2,0]bicyclohexatriene (190). By forming this through union of methyl and pentadienyl (191), one can see at once that the extra interaction is antibonding, the NBMO coefficient at the central atom of pentadienyl differing in sign from the two terminal ones. Therefore (190) has a higher π energy than benzene and so (186) should be more stable than (182) or (184). Doering and Roth have shown that rearrangement of (181) does indeed give (185). This result again violates the BEP principle because the most stable of the three octadienes is (187), in which both double bonds are substituted *trans*.

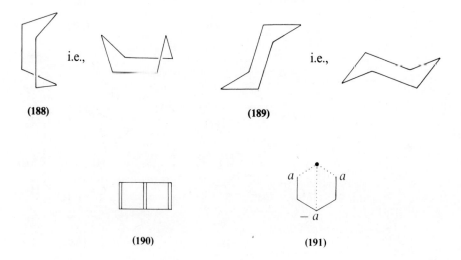

(188) (189)

(190) (191)

Chemical Reactivity

The energy of union of methyl with pentadienyl (191) to form (190) is equal to

$$2\beta(a + a - a) = 2a\beta \qquad (5.310)$$

This is the same as that for union to hexatriene. Thus (190), and so likewise the boat transition state, is neither aromatic or antiaromatic but nonaromatic. The difference in energy between it and the aromatic chair (186) should therefore be about one-half that between an aromatic transition state and an analogous antiaromatic one. Recently Goldstein has measured the difference in activation energy between the rearrangements of an analog of (181) with deuterium in place of methyl (to avoid complications from steric effects). They found it to be 10 kcal/mole, rather more than half the corresponding difference (14 kcal/mole; p. 345) between the aromatic conrotatory and antiaromatic disrotatory transition states for conversion of cyclobutene to butadiene (Fig. 5.37).

These results are interesting because neither the frontier orbital method nor arguments based on the conservation of orbital symmetry during reactions can account for the difference between the two modes of the Cope rearrangement. According to these theories, rearrangement via boat and chair transition states should be equally allowed.

The Claisen rearrangement of phenyl allyl ethers [e.g., (192)] is analogous to the Cope rearrangement and could likewise take place via chair (193) or boat (194) transition states, namely

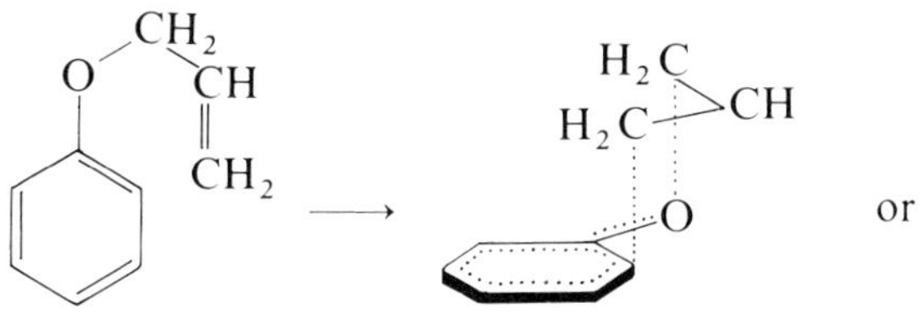

(192)   (193)   or

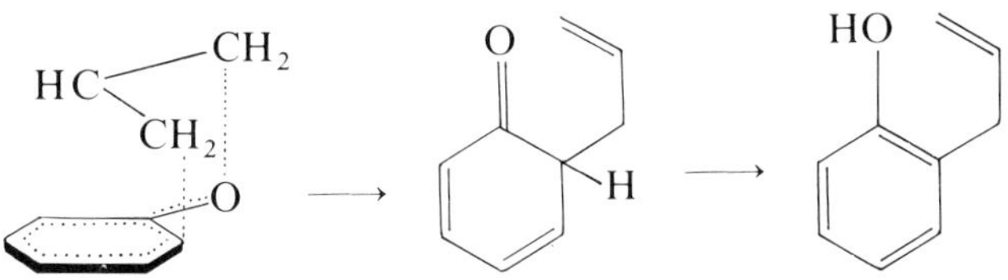

(194)   (195)   (196)   (5.311)

As in the Cope rearrangement, the extra interaction in (194) should be antibonding, so rearrangement via the chair transition state (193) should be favored. Schmidt has shown this to be so.

Our last example, also due to Schmidt, is of an unusual concerted *para* Claisen rearrangement. Direct *para* rearrangement does not occur with phenyl allyl ethers [e.g., (192)] because the transition state (197) would be isoconjugate with (198), which contains an eight- (= 2 × 4) membered ring and would consequently be antiaromatic. However, the phenyl butadienyl ether (199) does rearrange directly to (200) via a transition state (201) which is aromatic, being isoconjugate with the bicyclic hydrocarbon (155) considered earlier in connection with π cycloaddition to an azaazulene. Rearrangement of allyl phenyl ethers can give *para* products, but only by secondary Cope rearrangements of ketonic intermediates [e.g., (195)] formed first by normal *ortho* rearrangement. This results in a product where allyl is linked to carbon by the atom through which it was linked to oxygen in the original ether; cf. (202) ⟶ (203) ⟶ (204).

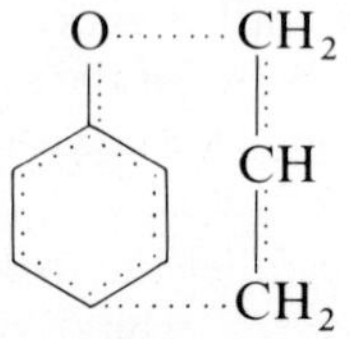

(197) 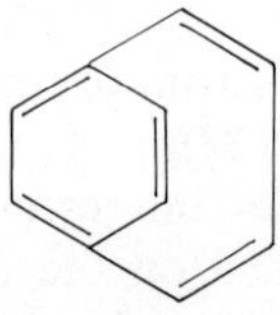   (198)

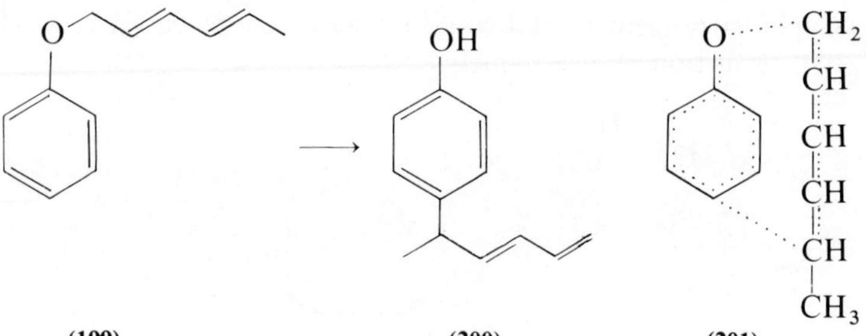

(199)   (200)   (201)

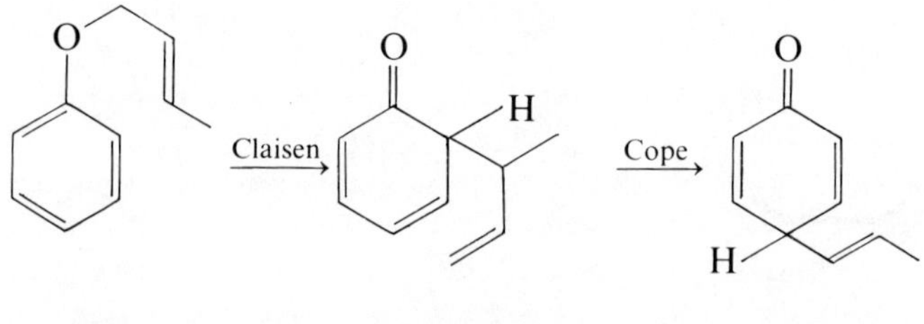

(202)   (203)   (204)

Sigmatropic rearrangements can also occur in odd systems, being then of $OO_A^{\ddagger}$ type. A good example is that shown in equation (5.312). The *cis–cis* transition state (205) is of Hückel type, isoconjugate with the $C_5H_5^-$ anion (206) and so aromatic.

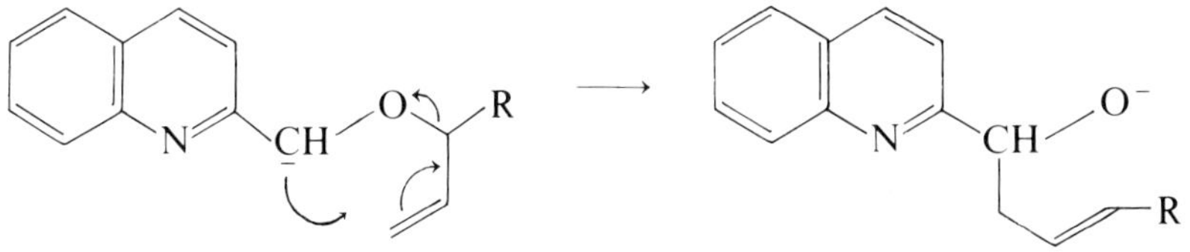

Chemical Reactivity

(diagram of quinoline-CH-O-R with allyl rearrangement)

(5.312)

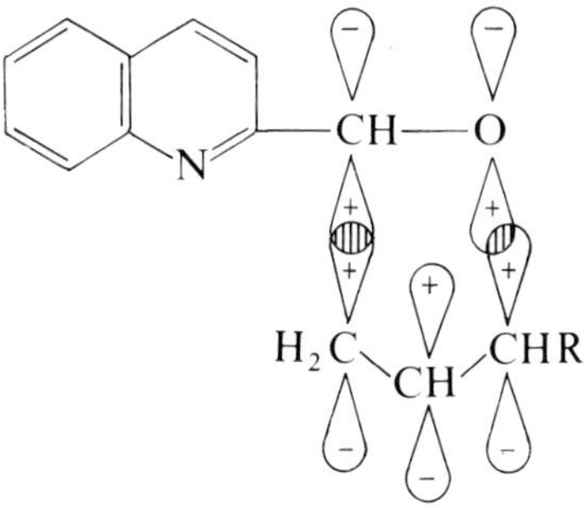

(205) 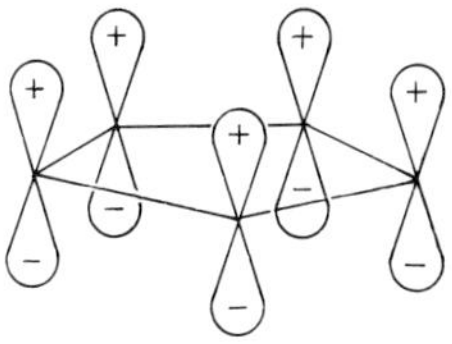   (206)

Another example, this time involving rearrangement of a cation, is the following:

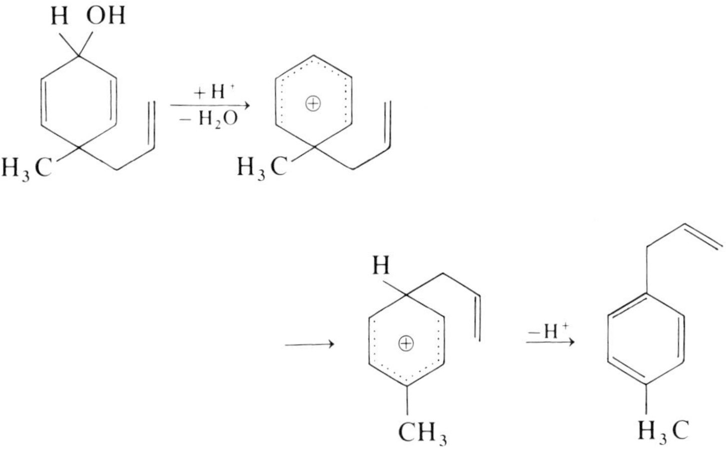

(5.313)

Here the transition state (207) is isoconjugate with (208), containing two fused aromatic rings (i.e., benzene + tropylium):

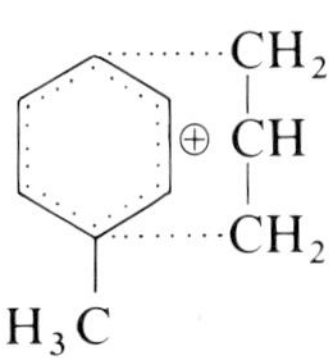

(207) 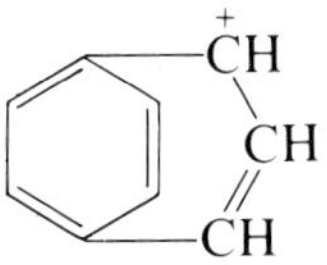   (208)

D. Electrocyclic Reactions

The term *electrocyclic reaction* was introduced by Woodward and Hoffmann to describe reactions in which a ring, which is unsaturated except for two atoms, opens by breaking at the single bond linking those atoms to form an open-chain conjugated system with two additional atoms. For example,

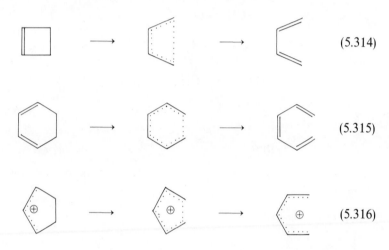

(5.314)

(5.315)

(5.316)

Reactions of this type are of course reversible, an open-chain conjugated system then undergoing cyclization to a ring in which two of the atoms in the original conjugated system are now saturated.

We have already considered in detail (p. 343) one reaction of this kind. i.e., the opening of cyclobutene to butadiene [equation (5.294) or (5.295)]. The general case can be visualized from Fig. 5.37. In the case of cyclobutene, the two $2p$ AOs in Fig. 5.37 are those forming the CC π bond in cyclobutene. In the general case, they are the $2p$ AOs of the terminal atoms in a conjugated system of arbitrary length. Evidently in all cases the electrocyclic ring opening can follow either of two stereochemically distinct paths, one disrotatory and the other conrotatory. Equally clearly, the disrotatory opening leads to a cyclic conjugated system in which there is no phase dislocation, whereas conrotatory opening leads to a phase dislocation. The transition states for electrocyclic reactions in which a single bond is broken are therefore of Hückel type if the opening is disrotatory and of anti-Hückel type if it is conrotatory.

The course of such reactions can therefore be predicted immediately from the rules for aromaticity in Section 3.14. All we have to do is to count the number of atoms in the ring and the total number of delocalized electrons. We can then tell at once if the transition state is isoconjugate with an even cyclic polyene, with an odd polymethine cation, or an odd polymethine anion. If the resulting structure is aromatic in the Hückel series, ring opening (or

Chemical Reactivity

closure) is disrotatory. If it is antiaromatic, ring opening (or closure) is conrotatory. Thus in the reactions of equations (5.314)–(5.316):

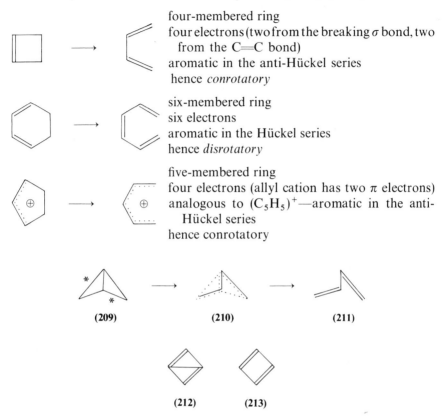

An interesting further example is the conversion of bicyclobutane (209) to butadiene (211). This involves a double electrocyclic opening by simultaneous breaking of the bonds marked with asterisks in (209). The corresponding transition state (210) is isoconjugate with bicyclobutadiene (212). The transannular bond in this is an essential single bond, just like that in azulene (p. 100). Omitting this, we find that the transition state is an analog of cyclobutadiene (213). The transition state (214) should therefore be of anti-Hückel type. Now the two bond-breaking processes could take place in three different ways.

(a) *Disrotatory–disrotatory.* Neither opening introduces a phase dislocation, so the corresponding transition state will be of Hückel type.
(b) *Conrotatory–conrotatory.* Each opening introduces a phase dislocation, making two in all. However, phase dislocations can be reduced or increased in number two at a time by suitable changes in the phases of the contributing AOs. The number of phase dislocations can thus be reduced to zero by suitable choice of the

phases of the AOs. This transition state is therefore also of Hückel type.

(c) *Disrotatory–conrotatory.* This combination at last gives a single-phase dislocation and so leads to an anti-Hückel, and hence aromatic, transition state.

The rearrangement should therefore involve disrotatory opening of one ring but conrotatory opening of the other, a curious conclusion which has been shown to be correct by Closs. This reaction is again of interest in that it cannot be discussed either in terms of the frontier orbital method (since it is unimolecular; see p. 367) or in terms of the conservation of orbital symmetry (since all symmetry is lost in forming the transition state).

E. Chelotropic Reactions

The term chelotropic reaction refers to reactions in which an odd-numbered ring is formed by addition of a monocentric reagent across the end of an even, conjugated system. The classic example is the formation of the sulfone (214) by addition of sulfur dioxide to butadiene,

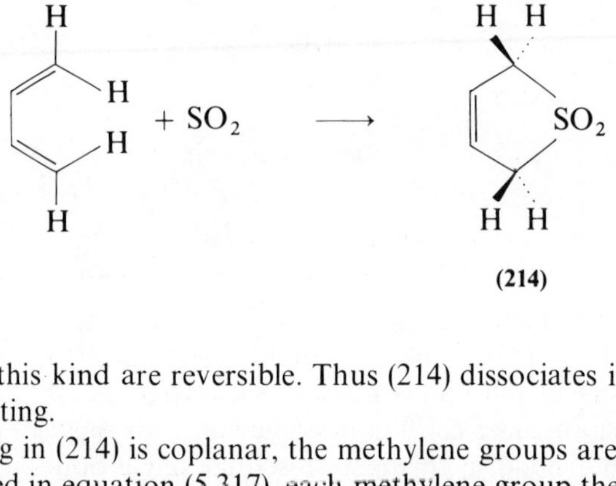

$$\text{butadiene} + SO_2 \longrightarrow \text{(214)} \tag{5.317}$$

(214)

Reactions of this kind are reversible. Thus (214) dissociates into its components on heating.

If the ring in (214) is coplanar, the methylene groups are orthogonal to it. As indicated in equation (5.317), each methylene group then has to rotate through 90 during the reaction. Consequently, chelotropic reactions, like electrocyclic reactions, fall into two stereochemically distinct categories. In one, the rotations of the methylene groups are conrotatory, in the other, they are disrotatory.

The following treatment of reactions of this kind is clearer and simpler than those given in the reviews listed at the end of this chapter.

In a chelotropic reaction, the reagent has to have two AOs available to form the new bonds. These AOs must also contain just two electrons between them in order that the new bonds be covalent rather than dative. Thus we can write SO_2 in the form (215) with a filled hybrid AO (ϕ_1) and an empty $3p$ AO (ϕ_2).

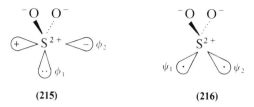

(215)   (216)

We can therefore replace these two AOs by two equivalent hybrid AOs ψ_1 and ψ_2 (216), given by

$$\psi_1 = (1/\sqrt{2})(\phi_1 + \phi_2); \quad \psi_2 = (1/\sqrt{2})(\phi_1 - \phi_2) \quad (5.318)$$

Indeed, in the usual localized bond picture, these are the AOs used to form the CS bonds in the product (214).

The AOs ψ_1 and ψ_2 will be orthogonal, the overlap integral between them vanishing. However, this does *not* mean that the resonance integral between them will also vanish. The resonance integral between two "correct" AOs of an atom (s, p, d, etc.) vanishes as a result of symmetry. The symmetry is lost when "correct" AOs are replaced by hybrids. Thus the resonance integral between two sp^3 AOs of a carbon atom is equal to $\sim 2\,\text{eV}$, not so very much less than that between the $2p$ AOs forming a C=C π bond.

In the atom, ψ_1 and ψ_2 will consequently interact with one another, the interaction leading to a splitting into a "bonding" orbital $\psi_1 + \psi_2$ and an "antibonding" orbital $\psi_1 - \psi_2$. From equation (5.318), these are seen to be the "correct" AOs ϕ_1 and ϕ_2, respectively. These relations are shown in Fig. 5.38(a), while Fig. 5.38(b) shows the corresponding interactions of the $2p$ AOs in ethylene to form bonding and antibonding π MOs. The two descriptions are exactly analogous. Thus we can treat our atomic reagent as an analog of ethylene, the two hybrid AOs in (216) playing exactly the same role as the $2p$ AOs in ethylene. The reaction of equation (5.317) is therefore topologically equivalent to, or isoconjugate with, the Diels–Alder reaction of butadiene with ethylene, as is indicated in Fig. 5.38(c, d).

Chelotropic reactions are therefore analogs of cycloaddition reactions in which ethylene adds to the ends of an even conjugated system. Since addition to the ethylene analog has to be *cis* (since the hybrid AOs are asymmetric and since the reaction must clearly involve their larger lobes), the transition state will be of Hückel type if addition to the conjugated system is *cis* (as in Fig. 5.38d) and of anti-Hückel type if it is *trans*. Thus addition will be *cis* if the conjugated system contains $4n$ atoms, so that the transition state contains effectively $4n + 2$ atoms, the reagent behaving as a pseudopair of atoms, and *trans* if the conjugated system contains $4n - 2$ atoms, so that the transition state contains $4n$ atoms.

It is easily seen (Fig. 5.39) that *cis* addition, involving approach of the the reagent in a direction more or less perpendicular to the conjugated system (Figs. 5.38d and 5.39), leads to disrotation of the terminal methylene groups. On the other hand, *trans* addition involves approach of the reagent in the plane of the conjugated system and takes place conrotatorily (Fig. 5.40).

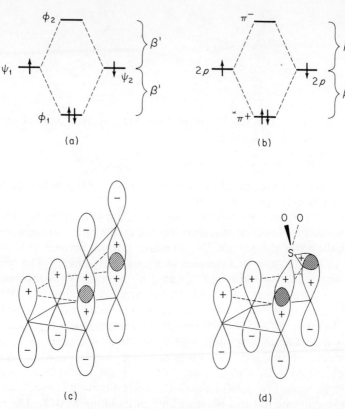

FIGURE 5.38. (a) Interaction of hybrid AOs of (216) to form the "correct" AOs of (215), β' being the resonance integral between the hybrid AOs; (b) analogous interaction between $2p$ AOs of the two carbon atoms in ethylene to form bonding (π^+) and antibonding (π^-) MOs; (c) transition state for the Diels–Alder reaction between butadiene and ethylene (cf. Fig. 5.35); (d) transition state for the chelotropic reaction of equation (5.317).

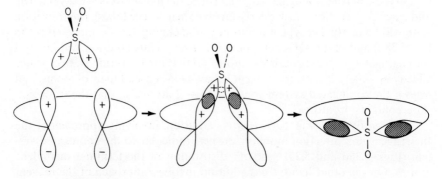

FIGURE 5.39. Disrotatory chelotropic reaction of SO_2 with a $4n$-polyene.

Chemical Reactivity

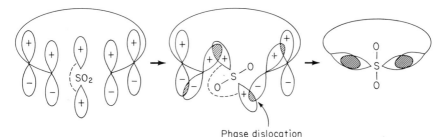

FIGURE 5.40. Conrotatory chelotropic reaction of SO_2 with a $(4n + 2)$-polyene. The orbitals of SO_2 are the two hybrid AOs of (216), the dotted line indicating the interaction between them.

These predictions are again supported by the available evidence. Some examples follow. It should be noted that the geometry deduced here for the disrotatory reactions (e.g., Fig. 5.39) differs from that proposed by Woodward and Hoffmann, who assume that the reagent approaches in the plane of the conjugated system as in Fig. 5.40. They did this in order to simplify their analysis of the reaction in terms of the conservation of orbital symmetry.

Example 1.

$$\text{(structure with } SO_2\text{)} \rightleftharpoons \text{(diene)} + SO_2 \qquad (5.319)$$

This is disrotatory since the polyene has $4 \times 1 = 4$ atoms.

Example 2.

$$\text{(structure with CO)} \longrightarrow \text{(benzene derivative)} + CO \qquad (5.320)$$

This reaction has to be disrotatory for steric reasons. It occurs easily because the transition state is isoconjugate with (217), which has two six-membered rings and is consequently aromatic.

(217)   (218)   (219)

Example 3.

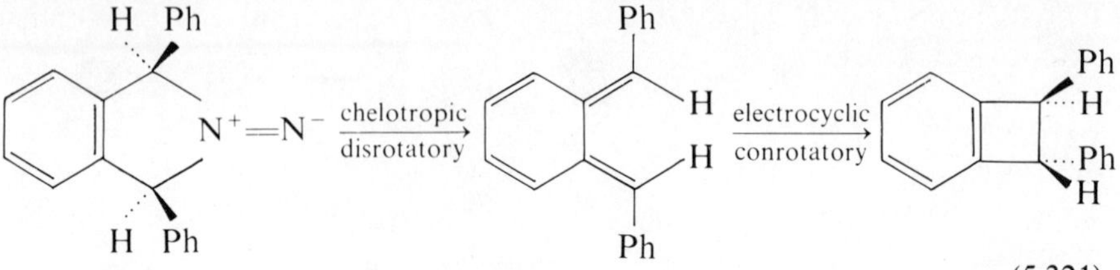

$$\text{(structure)} \xrightarrow[\text{disrotatory}]{\text{chelotropic}} \text{(structure)} \xrightarrow[\text{conrotatory}]{\text{electrocyclic}} \text{(structure)} \quad (5.321)$$

The first step is a disrotatory chelotropic reaction, the transition state being isoconjugate with naphthalene (218). The second step is a conrotatory electrocyclization, the transition state being isoconjugate with an anti-Hückel analog of (219).

The addition of carbene to an olefin could be regarded as the simplest possible chelotropic reaction and indeed Woodward and Hoffmann class it as such; e.g.,

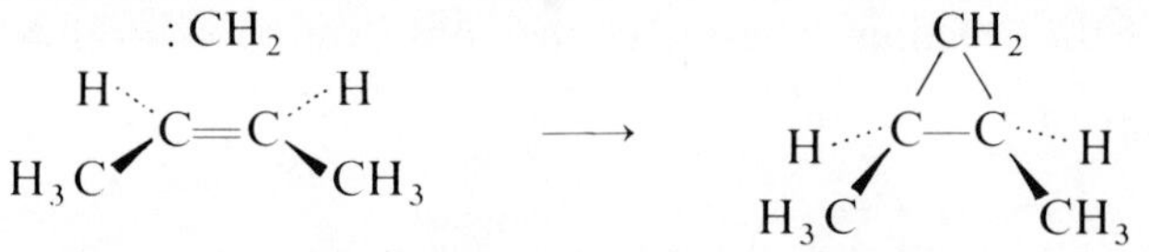

$$\text{(structure)} \longrightarrow \text{(structure)} \quad (5.322)$$

Carbene certainly fulfills the requirements for a chelotropic reagent, having two unused AOs containing two electrons. On the other hand, ethylene differs from all other even systems in that it cannot for steric reasons undergo conrotatory chelotropic additions. Since disrotatory *cis* addition by the path of Fig. 5.39 would lead to an antiaromatic transition state isoconjugate with cyclobutadiene, the reaction has to follow a different course. If therefore reactions of this kind are classed as chelotropic, they have to be regarded as unique exceptions.

As Skell pointed out some years ago, carbene is the conjugate base of the methyl cation CH_3^+ ($CH_2: + H^+ \to CH_3^+$). One might therefore expect it to add to an olefin in an analogous manner to form a π complex; namely

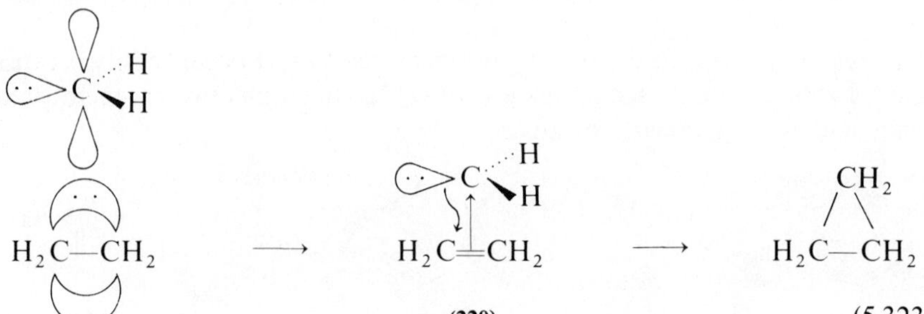

$$\text{(structure)} \longrightarrow \text{(220)} \longrightarrow \text{(structure)} \quad (5.323)$$

Addition is assisted by back-coordination of the pair of electrons in the hybrid AO of the carbene into the empty, antibonding π MO of the ethylene. Once the π complex (220) has been formed, it can rearrange into cyclopropane without any difficulty.

Skell's geometry of approach is the same as that deduced later by Woodward and Hoffmann, i.e., sideways approach of the carbene in a "non-least-motion" fashion. Since this would represent a unique exception if the reaction were classed as pericyclic, it seems better to treat it as a typical π-complex process, $E\pi_A{}^{\ddagger}$.

5.28 Alternative Derivations of the Woodward–Hoffmann Rules. "Allowed" and "Forbidden" Pericyclic Reactions

In their original investigation, Woodward and Hoffmann based their analysis of pericyclic reactions on Fukui's frontier orbital (FO) method.

The FO method treats a bimolecular reaction in terms of perturbation theory, the MOs of the transition state being regarded as perturbed forms of those in the reactants. As we have seen, the only interactions that alter the total energy of the system will be those between filled bonding MOs of one reactant and empty antibonding MOs of the other. Since these are second-order perturbations, they will in general be the greater, the closer together are the two MOs. The largest interactions will therefore tend to be those between the highest occupied MO (HOMO) or *frontier orbital* of one component and the lowest unoccupied MO (LUMO) of the other. The FO method assumes that the remaining interactions can be neglected in comparison with these, the course of a reaction being determined by the HOMOs and LUMOs of the reactants.

The FO method works well for AHs, this being another consequence of the pseudosymmetry of their wave functions, embodied in the pairing theorem. It is much less satisfactory for compounds of other kinds and indeed sometimes fails drastically. A nice example is the nitration of 10-methyl-10,9-borazarophenanthrene [(102); p. 325] mentioned earlier. According to the FO method, reaction should occur most easily at the point of maximum HOMO density. In fact, (102) reacts most easily at the 8 position, where the HOMO density is least.

The FO method is also unable to treat unimolecular reactions except in a purely intuitive way. During a unimolecular reaction, the perturbations involve changes in Coulomb and resonance integrals of atoms and bonds in the reactant molecule. These lead to *first*-order perturbations which affect *all* the MOs in the system. The FO method can be applied only by the artificial and unsatisfactory expedient of treating such processes as bimolecular reactions between different parts of a single molecule.

An alternative treatment of pericyclic reactions was introduced by

Longuet-Higgins and Abrahamson and has been rather generally adopted. This is based on the idea that if a molecule has symmetry and if the symmetry is conserved during a reaction, the wave function will tend to retain similar symmetry. Thus if the molecule has a plane of symmetry, and if the wave function is initially symmetric with respect to reflection in that plane, and if the plane of symmetry is retained throughout a reaction, then the wave function must remain symmetric. In the MO approximation, it can be shown that similar restrictions apply to MOs. This has led to the interpretation of pericyclic reactions in terms of the conservation of orbital symmetry during them.

The trouble with this approach is that very few organic molecules have any symmetry and even those that do need not retain it during a reaction. Consider, for example, a classic example of this kind of argument, the cis-π-cyclodimerization of ethylene,

$$\cdots\|\cdots\|\cdots x \longrightarrow \cdots[\cdots]\cdots x \longrightarrow \cdots[\ \ ]\cdots x \quad (5.324)$$

If the reacting system retains two planes of symmetry (x, y) throughout [marked with dotted lines in equation (5.324)], it can be shown that the reaction must lead to a doubly excited form of the product if the symmetry of the wave function remains unchanged. The reaction would then lead to a highly excited form of cyclobutane and so be very endothermic. However, there is no need for the reaction to follow this course. There is no reason at all why the transition state should have the symmetric structure indicated in equation (5.324), in which the two new bonds have been formed to equal extents. It could very well have an asymmetric trapezoidal structure

in which one new bond has been formed to a greater extent than the other and in which one of the planes of symmetry is lost. Arguments based on conservation of symmetry could not then be applied. There are also cases where this kind of argument is inherently inapplicable because the reaction is by its nature asymmetric (cf. the electrocyclic opening of bicyclobutane, p. 361), and others where it fails to account for observed differences between apparently equally "allowed" reactions (cf. the Cope rearrangement; p. 355).

A further criticism of this approach is its purely qualitative nature. There is no way of telling how large or how small will be the factors preventing "violations" of the rules derived from it. This has led to an exaggerated estimate of the potency of the factors distinguishing "allowed" pericyclic reactions from "forbidden" ones and the consequent belief that "forbidden" reactions cannot occur as concerted pericyclic processes.

This is quite incorrect. The factor involved is simply the stabilization or destabilization of cyclic conjugated systems relative to analogous open-chain analogs, i.e., the factor that distinguishes aromatic systems from antiaromatic ones. Antiaromatic systems can be, and have been, prepared. There is no reason at all why pericyclic reactions should not equally take place via antiaromatic transition states. The intramolecular Stevens rearrangement is a reaction which proceeds through an antiaromatic transition state analogous to the cyclopropenate anion.

The aromatic and antiaromatic energies of pericyclic transition states are indeed likely to be smaller than those of analogous π systems. In the latter, all the overlapping pairs of $2p$ AOs are close together because the atoms in question are linked by σ bonds. This is not the case in a typical pericyclic transition state. Thus recent detailed calculations indicate that the methylene groups in the transition states of Fig. 5.37 are over 2 Å apart. The orbital overlap is correspondingly reduced and so is the effect of the cyclic conjugation. It has indeed recently been found possible to isolate the minute amounts of by-products formed in the opening of cyclobutenes via the "forbidden" antiaromatic disrotatory route. The corresponding difference in activation energy between the conrotatory and disrotatory paths is only 14 kcal/mole, corresponding to aromatic and antiaromatic energies of ± 7 kcal/mole. This is less than one-third the aromatic energy of benzene or the antiaromatic energy of square cyclobutadiene. Indeed, it is little more than the difference (10 kcal/mole) between the activation energies for rearrangement of 2,5-hexadiene via boat and chair transition states (p. 355), two reactions that are both "allowed" in terms of conservation of orbital symmetry. As we have seen, Evans' principle and the PMO approach provide an entirely satisfactory explanation of these results.

It is therefore inaccurate and misleading to talk about "allowed" and "forbidden" pericyclic reactions. The terms "aromatic" and "antiaromatic" pericyclic reaction are much more appropriate. It is also clear that the distinction between them has nothing to do with symmetry. It depends on the topology of overlap of the AOs in pericyclic transition states, not on the symmetries of MOs. If symmetry were involved, the distinction between "allowed" and "forbidden" reactions would be attenuated as symmetry was lost. This is not the case. The Woodward–Hoffmann rules, or the equivalent statement embodied in Evans' principle, hold just as strongly in systems lacking symmetry as in symmetric systems. Indeed, if this were not the case, they would be far less useful and important.

5.29 Catalysis of Pericyclic Reactions by Transition Metals

The ring opening of the codimer (221) of cyclobutadiene and benzocyclobutadiene rearranges on heating to benzocyclooctatetraene (224). This reaction inevitably involves, as indicated, the successive disrotatory opening of two cyclobutene rings via antiaromatic transition states. The reaction is is therefore quite recalcitrant, rearrangement being complete only after 3 hr at 150°C.

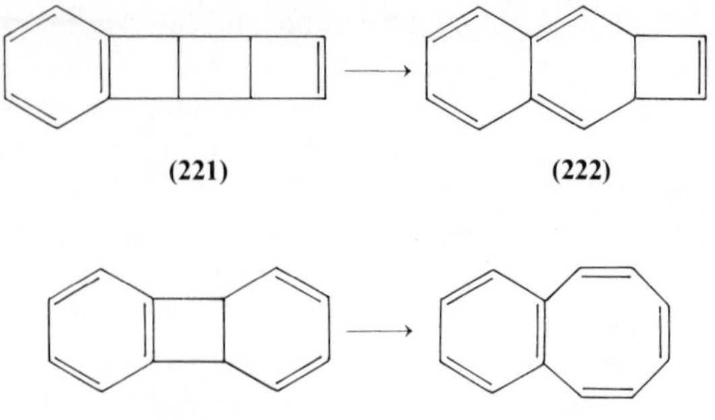

(221) → (222) or

(223) → (224) (5.325)

Pettit and his collaborators found that the rearrangement is very strongly catalyzed by salts of silver or platinum. Thus addition of a solution of silver nitrate to a solution of (222) in alcohol at room temperature leads to complete conversion to (224) in less than 10 sec. This startling observation led to extensive studies of the effects of transition metals on other forbidden pericyclic reactions and many of them have indeed been found to undergo similar accelerations.

The mechanisms of these reactions are still a matter of controversy and some at least proceed via organometallic intermediates formed by replacement of hydrogen atoms in the reactant by metals. However it seems very likely that the original examples [equation (5.325)] represent a true catalysis by a transition metal of a pericyclic reaction involving an antiaromatic transition state.

Transition metals, particularly those at the end of each transition series, form π complexes to olefins in which the olefin acts as a donor and in which d electrons of the metal are simultaneously used for back-coordination (p. 294). Similar complexes are also formed by conjugated hydrocarbons and by aromatic hydrocarbons such as benzene.

The formation of such a π complex is indicated in Fig. 5.41. The dative bond formed by the π electrons of the hydrocarbon arises mainly from an interaction between the HOMO of the hydrocarbon and an empty valence shell AO of the metal, while the reverse dative bond involves an analogous interaction between a filled d AO of the metal and the LUMO of the hydrocarbon. The nearer together these pairs of orbitals are, the stronger is the interaction and the more stable is the π complex.

Consider now the formation of an even AH by union of two odd AH radicals. The π energy of union arises mainly from a first-order interaction between the two NBMOs. This interaction is stronger for formation of an aromatic AH and weaker for an antiaromatic AH than for a nonaromatic AH. Indeed, our theory of aromaticity and antiaromaticity was based on this.

Now the two MOs arising by interaction of the NBMOs will correspond

Chemical Reactivity

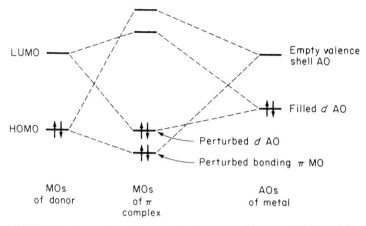

FIGURE 5.41. Formation of a π complex from a transition metal with an olefin or conjugated hydrocarbon.

to the HOMO and LUMO of the resulting even AH. It follows (Fig. 5.42) that their separation will be greater in an aromatic AH and less in an antiaromatic AH than in an analogous nonaromatic AH. Thus the HOMO of an antiaromatic AH is higher in energy than that of an analogous nonaromatic polyene, while that of an analogous aromatic AH is lower. Likewise, the LUMO of an antiaromatic AH is lower and that of an aromatic AH is higher than that of an analogous nonaromatic AH.

It follows that antiaromatic AHs should form exceptionally strong π complexes with transition metals because they have HOMOs of unusually high energy and also LUMOs of unusually low energy. Both the interactions in Fig. 5.41 should therefore be unusually strong. This prediction was made by Longuet-Higgins and Orgel before any such complex was known. It has been

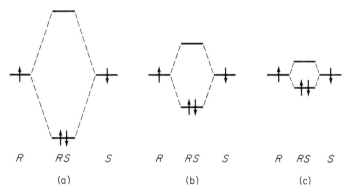

FIGURE 5.42. Interactions between NBMOs of two odd AHs R and S on union (a) to an aromatic AH', (b) to a nonaromatic AH' (c) to an antiaromatic AH.

dramatically confirmed by recent work, notably that by Pettit *et al.* on cyclobutadiene complexes.

Now since antiaromatic transition states are isoconjugate with antiaromatic hydrocarbons, similar arguments should apply to them. If, then, we form a metal π complex from some compound that can undergo an antiaromatic pericyclic reaction, the binding of the metal should increase on passing from the reactant to the transition state. The result should be a lowering of the activation energy, which may be sufficient to enable the reaction to proceed under mild conditions. This of course accounts well for the catalysis of cyclobutene ring openings observed by Pettit *et al.*

The hydrocarbon (221) [equation (5.325)] has two unsaturated groups that could bind metals, i.e., the benzene ring and the terminal double bond. Silver ions form better π complexes with benzene than with ethylene; silver should therefore attach itself to the benzene ring in (221) and so catalyze the opening of the adjacent ring via a transition state isoconjugate with benzocyclobutadiene to form (222). Conversely, platinum has a greater affinity for olefins than for benzene, so it should attach itself to (221) at the isolated double bond. Platinum should therefore catalyze the opening of the terminal four-membered ring in (221) to give (223). Thus

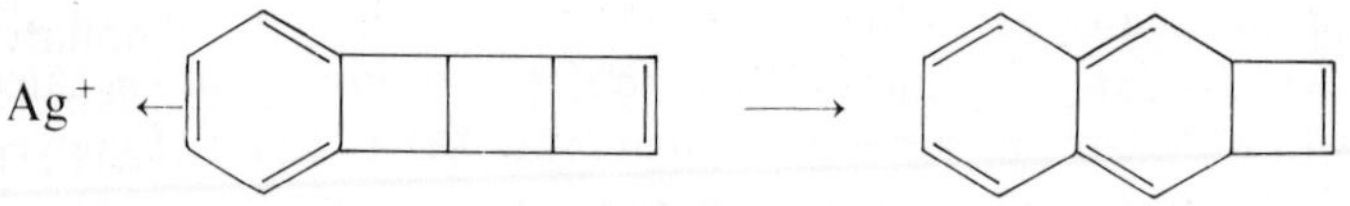

$$Ag^+ \quad \text{(structure)} \longrightarrow \text{(structure)} \quad (5.326a)$$

$$\text{(structure)} \quad Pt^{II} \longrightarrow \text{(structure)} \quad (5.326b)$$

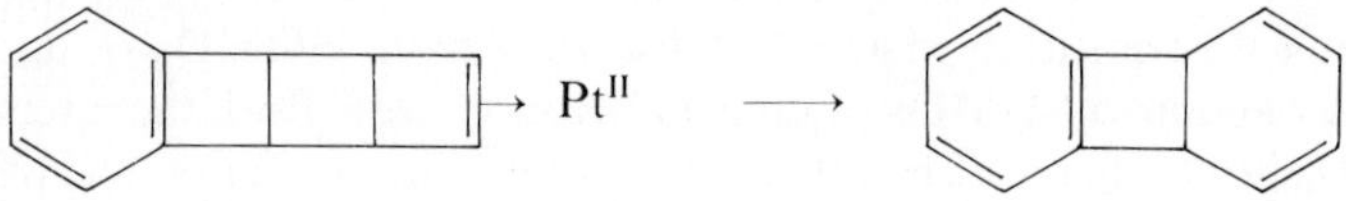

Now the subsequent rearrangement of (222) or (223) to (224) involves transition states isoconjugate with naphthalocyclobutadiene (225) or biphenylene (226). Each of these has two aromatic benzene rings and one antiaromatic cyclobutadiene ring. The latter will cancel the effect of one of the benzenes, so (225) and (226) should have resonance energies similar to that of benzene. The same should be true of the corresponding transition states for (222) and (223).

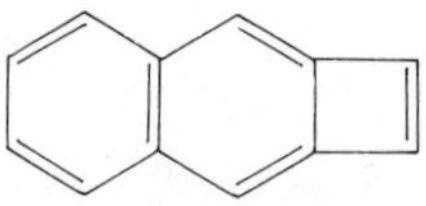

(225) 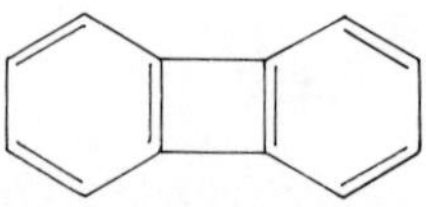   (226)

Since (222) is nonaromatic, all the bonds in it being essential single or double bonds, rearrangement to (223) will be in effect an aromatic pericyclic reaction since it leads from nonaromatic starting material to a transition state that is, on balance, aromatic. The reaction therefore takes place readily, so no intermediate can be isolated in the silver-catalyzed reaction [equation (5.326a)]. On the other hand, (223) is itself aromatic, having an intact benzene ring. The rearrangement of (223) to (224) is therefore a *nonaromatic* pericyclic reaction and so is less facile than that of (222). Indeed, (223) could be isolated as an intermediate in the platinum-catalyzed rearrangement [equation (5.326b)].

5.30 Reactions Involving Biradical Intermediates ($ER_A{}^\ddagger$)

To break the C—C bond in ethane requires a great deal of energy (88 kcal/mole), so the reaction

$$H_3C\text{—}CH_3 \longrightarrow H_3C\cdot + \cdot CH_3 \qquad (5.327)$$

takes place only at very high temperatures. If, however, we introduce substituents that can stabilize the resulting radicals (i.e., $\pm E, +E, -E$), the reaction can be greatly facilitated and may take place at an appreciable rate under quite mild conditions. Indeed, some substituted methyl radicals are so stable that they show no tendency to dimerize, (227) being a good example.

(227)   (228)   (229)

If now we introduce similar substituents at the ends of a C—C bond forming part of a ring, we will equally facilitate its cleavage. In this case, however, the product will be a *biradical*,* since the two fragments still remain bound together by the other half of the ring. A good example is (228), which could undergo cleavage of the PhC—CPh bond under moderate conditions to give the biradical (229). Biradicals, like ordinary radicals, are very reactive. However, they differ from ordinary radicals in that they contain two radical centers in the same molecule. These can interact very easily in various different ways to regenerate a normal molecule, either the starting material or an isomer of it, and they can also undergo addition reactions, again giving "normal" products. A number of reactions are known which involve processes of this kind, the reactants and products being "normal" molecules, while the rate-determining step is the formation of an intermediate biradical. Since the

---

* A biradical is a species in whose ground state there are two unpaired electrons. This situation can occur either because the radical centers are separated by one or more saturated carbon atoms and so cannot interact with one another, or in special cases, such as the triplet ground state of O_2 (p. 36), where the unpaired electrons occupy orbitals of different symmetry and so again are unable to interact. The term should not be used for excited states, in particular triplet excited states, since their behavior is different from that of true biradicals (see Chapter 6, p. 397).

stability of the latter depends on resonance interactions that are present neither in the reactants nor in the products, the rate of the overall reaction is unrelated to the heat of reaction. Processes of this kind are therefore of anti-BEP type and are termed $ER_A^\ddagger$.

The biradicals in $ER_A^\ddagger$ reactions are usually formed either by bond cleavage or by the combination of two unsaturated molecules. Thus two molecules of ethylene can combine to form a butadiyl biradical,

$$H_2C=CH_2 + H_2C=CH_2 \longrightarrow H_2\dot{C}CH_2CH_2\dot{C}H_2 \qquad (5.328)$$

This process is less endothermic than simple C—C bond fission, the calculated heat of reaction being +62 kcal/mole.

The intermediate biradicals can undergo further reaction in one of four ways:

(a) Recombination, e.g., (229) → (228).
(b) Fission, e.g., the reverse of equation (5.328).
(c) Disproportionation; thus (229) might react as follows:

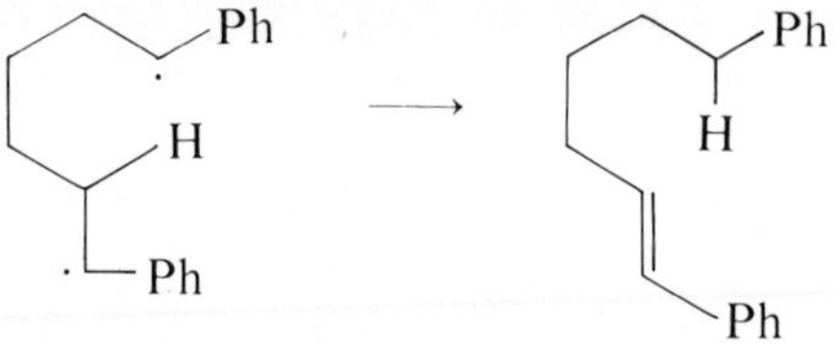

$$(5.329)$$

(d) Addition to double bonds, a 1,4 biradical being equivalent to a diene with zero 1,4 localization energy;

$$(5.330)$$

One of the first $ER_A^\ddagger$ reactions to be discovered was that involved in the initiation step in the uncatalyzed thermal polymerization of styrene,

$$PhCH=CH_2 + CH_2=CHPh \longrightarrow Ph\dot{C}HCH_2CH_2\dot{C}HPh \qquad (5.331)$$

The polymerization is inhibited by quinone through a Diels–Alder-type addition to the biradical to give adducts which can be isolated.

The calculated heat of reaction for formation of the biradical from styrene [equation (5.331)] is 24 kcal/mole. The observed activation energy for the initiation step in the thermal polymerization of styrene is considerably greater (37 kcal/mole). The dissociation of the biradical into styrene therefore requires activation (~ 13 kcal/mole). This is because it involves a change in the hybridization of two carbon atoms (see p. 245). Since the same situation should hold quite generally, biradicals of this type should have appreciable stability and so survive long enough to undergo other reactions. In particular they may undergo recombination to form cyclobutanes.

Chemical Reactivity

A number of cases are known where ethylene derivatives undergo π cycloaddition to form cyclobutanes. Tetracycanoethylene [TCNE; $(NC)_2C{=}C(CN)_2$] and fluorinated ethylenes are particularly prone to behave in this way. These reactions were discovered at the laboratories of duPont de Nemours & Co. and have been studied in great detail by Bartlett and his collaborators. As we have already seen, concerted *cis* dimerizations of this kind should be less favorable than an $ER_A^{\ddagger}$ path via an intermediate 1,4-butadiyl biradical; all these reactions in fact take place in this way.* An interesting example is the reaction of TCNE with bismethylenecyclobutene (230). The normal Diels–Alder reaction to form (231) is inhibited because this would be an antiaromatic cyclobutadiene derivative and because the transition state leading to it would be isoconjugate with benzocyclobutadiene. The product is therefore the spiran (232), formed as indicated by an $ER_A^{\ddagger}$ process.

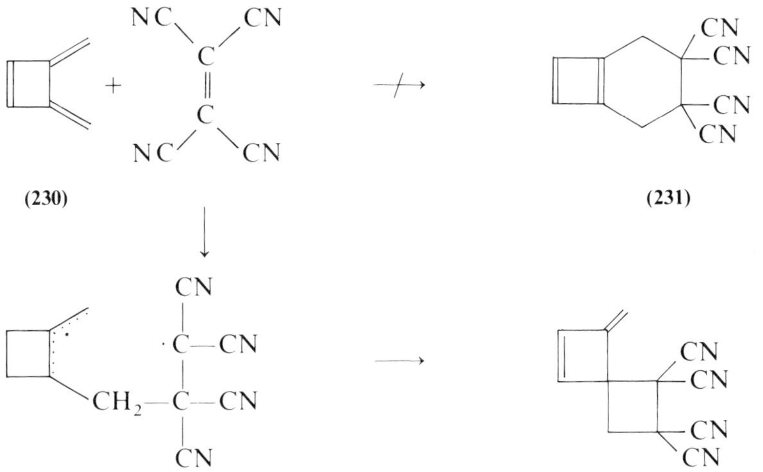

(230) + TCNE $\not\to$ (231)

↓

(232)

The use of the terms "allowed" and "forbidden" has had an unfortunate psychological effect, giving rise to the impression that the allowed reactions are necessarily favorable and that the "forbidden" ones cannot take place at all.

Thus it is generally believed that if a reaction leads to products that could have been formed by an antiaromatic pericyclic process, it must in fact have taken place by a nonconcerted $ER_A^{\ddagger}$ mechanism. As we have already seen, this is not the case. Thus the electrocyclic opening of bicyclopentane (233) to cyclopentadiene (234) must be a pericyclic process because there is no way

---

* The reactions could take place by an aromatic pericyclic process via twisted transition states of the type indicated in (165). The energy required to twist the two double bonds is, however, so high that no reactions of this kind have been reported except in the case of molecules where the double bonds are twisted initially [e.g., (166)]. Indeed, recent calculations suggest that this path to cyclobutanes should be even less favorable, in the case of unstrained olefins, than the antiaromatic pericyclic route involving *cis–cis* addition.

in which the AOs of the breaking σ bond can avoid overlapping with the $2p$ AOs of the double bond. It is impossible for the biradical to exist in a configuration (235), where the $2p$ AOs lie in the plane of the ring.

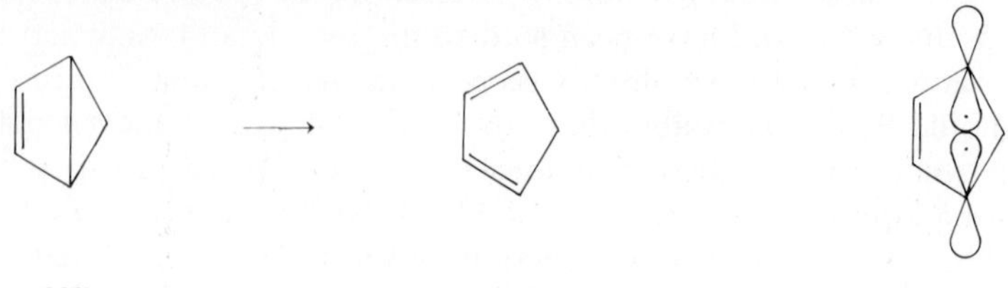

(233)   (234)   (235)

Conversely, cases are known where π cycloadditions take place by the $ER_A^{\ddagger}$ mechanism in preference to an alternative "allowed" Diels–Alder process. A good example is the reaction of butadiene with trifluoroethylene to give the adducts (236)–(239). The fact that (236) and (239) are formed in equal amounts shows that the reaction is not a pericyclic cycloaddition but must take place via the biradical (240). In this the $F_2C\text{—}CFH$ bond is single, so rotation is possible, the configuration of the original reactants being lost. Even the 13% of "normal" Diels–Alder product may be, and probably is, formed by cyclization of the same intermediate biradical, either (240) or the precursor (241) of (238).

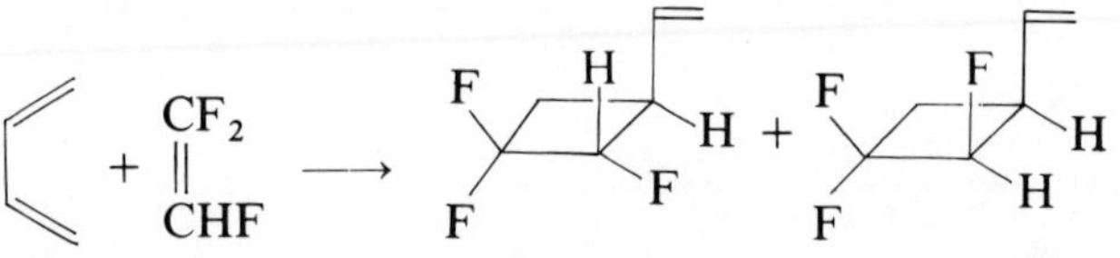

$$\text{butadiene} + \begin{array}{c} CF_2 \\ \| \\ CHF \end{array} \longrightarrow$$

(236) (29%)   (237) (30%)

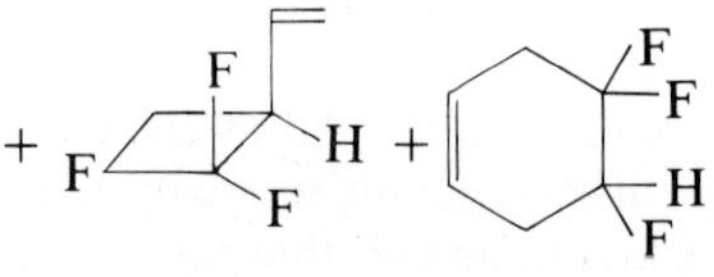

+ (238) (28%) + (239) (13%)   (5.332)

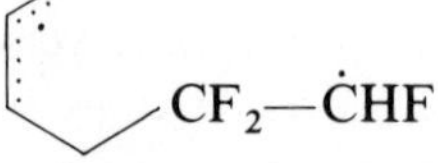

(240) $CF_2\text{—}\dot{C}HF$ (241) $CHF\text{—}\dot{C}F_2$

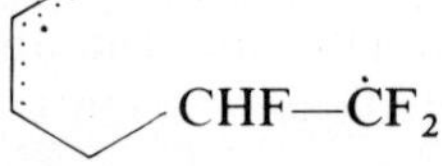

The intermediate biradical (240) or (241) could cyclize equally well to cyclobutane (236)–(239) or cyclohexene (239) derivatives. Indeed one might have expected the latter to be favored because it is much more stable, the strain energy in a cyclobutane ring being ~25 kcal/mole. The formation of cyclobutane derivatives is therefore a violation of the BEP principle. The reason for this violation is that the combination of radicals requires no energy of activation. The relative rates of formation of the various possible products by internal combination of a biradical therefore depends solely on the corresponding entropies of activation. The transition state leading to (230) has a rather rigid six-membered ring in which no internal rotation is possible. On the other hand, the transition states leading to (237)–(239) have a vinyl group that is free to rotate about the bond linking it to the four-membered ring. The entropies of the latter are therefore more positive than that of the transition state leading to (239). This is quite a general phenomenon. When a biradical can cyclize to form rings of different sizes, the smallest is always preferred (unless of course, special steric effects are present to prevent it).

An $ER_A^{\ddagger}$ π cycloaddition can lead to nonstereospecific *cis–trans* addition. If a π cycloaddition is not stereospecific, one can be sure that it has taken place by an $ER_A^{\ddagger}$ mechanism. The converse is not, however, true because the intermediate biradical in an $ER_A^{\ddagger}$ cycloaddition may collapse before it has had time to undergo internal rotation. Thus the additions of benzyne to *cis*- and *trans*-1,2-dichloroethylene are stereoselective because the intermediate biradicals collapse before they have had time to equilibrate,

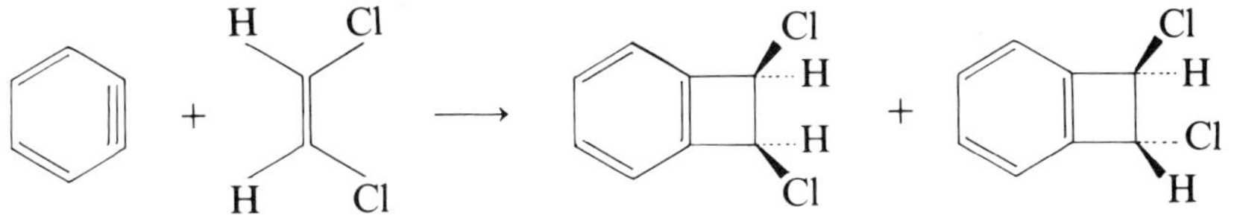

$$\text{(242)}(68\%) \qquad \text{(243)}(32\%) \qquad (5.333)$$

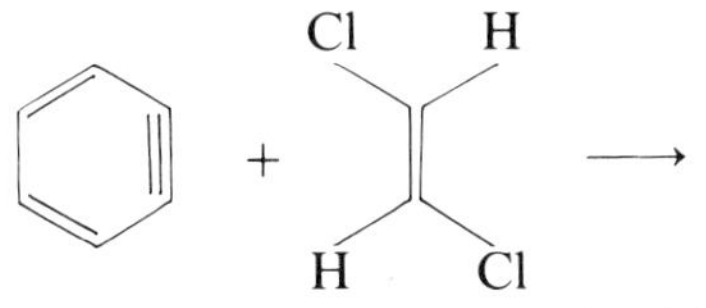

$$\text{(242)}(20\%) + \text{(243)}(80\%) \qquad (5.334)$$

Even stronger stereoselectivity is observed in the addition of dichlorodifluoroethylene to *trans*-cyclooctene,

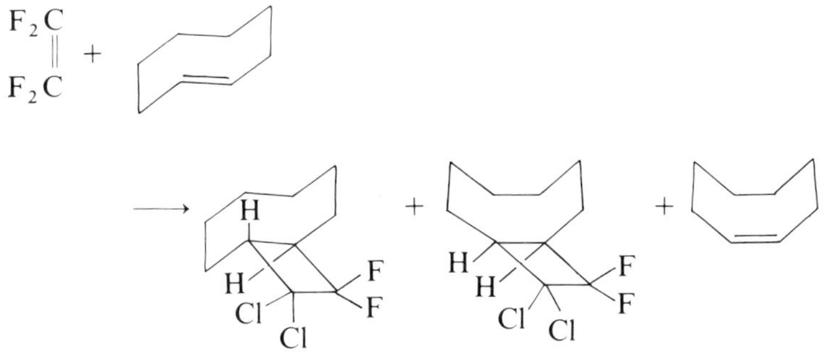

$$(5.335)$$

(244) (91%)   (245) (1%)   (246) (8%)

Under the conditions of the reaction, *cis*-cyclooctene (246) does not react. Evidently, therefore, the biradical that would be formed from (246) prefers to cleave into its components rather than cyclize to (245). The selectivity due to the tendency of the initially formed biradical to collapse immediately to form (244), rather than isomerize to the precursor of (245), is enhanced in the product by extensive (88%) dissociation of the latter into (246) + $F_2C=CCl_2$.

Another example of a reaction which follows an $ER_A^{\ddagger}$ path in preference to a normal aromatic pericyclic path is the Cope rearrangement of 2,5-diphenyl-2,5-hexadiene (247) to (249). It has recently been shown that this is not a normal, concerted, pericyclic, sigmatropic reaction but an $ER_A^{\ddagger}$ reaction involving the intermediate biradical (248). Indeed, Doering has recently shown that the homologous diene (250), which cannot undergo a Cope rearrangement, cyclizes to (252) by an analogous $ER_A^{\ddagger}$ mechanism via the biradical (251).

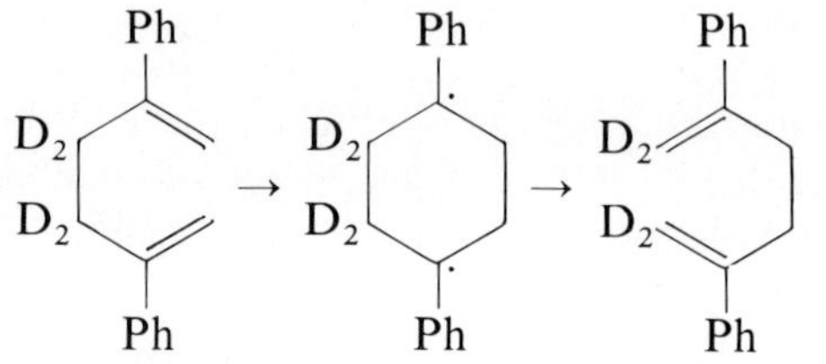

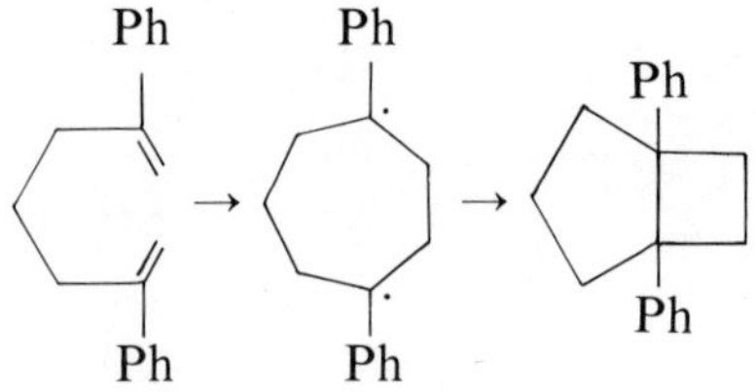

5.31 The $\pm E$ Substituent Technique

As we saw in the previous section, reactions that are formally of pericyclic type may in fact take place by an alternative two-step $ER_A^{\ddagger}$ mechanism. One can, moreover, envisage a continuous transition between these two extremes. Thus the Diels–Alder reaction between butadiene and ethylene might take place in one step via a symmetric pericyclic transition state (253) or it might take place in two steps via a biradical (254). In the latter case, formation of the biradical should involve passage over an intermediate transition state (255) since the formation of (254) involves a change in hybridization of two carbon atoms [see the discussion following equation (5.331)]. A third alternative would be a pericyclic reaction via an asymmetric transition state (256) in which one of the new bonds is stronger than the other. By progressive

Chemical Reactivity

weakening of one of the new bonds in (253), we can clearly pass continuously to the extreme represented by (255). Consequently, it is not sufficient merely to distinguish between the pericyclic and $ER_A^{\ddagger}$ mechanisms. In the case of a pericyclic process, one also wants to know the timing of the various bond-forming and bond-breaking processes.

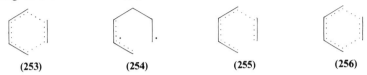

(253)　　　　(254)　　　　(255)　　　　(256)

Problems of this kind are often exceedingly difficult to solve and one promising technique for the purpose is that indicated earlier (p. 282) in our discussion of elimination reactions. If we have a reaction in which the reactants and possible transition states are AHs and if we introduce $\pm E$ substituents into the reactants, they and the corresponding transition states will still be AHs. Since the properties of AHs can be calculated simply and reliably by the PMO method, we can calculate the way the rate should vary for various possible models of the transition state and compare the results with experiment. In this way, we can deduce the structure of the actual transition state.

This procedure, which was first used to study aromatic substitution in 1956, may be termed the $\pm E$ *substituent technique.*

The objective in the original application of this technique was to see if the arenonium intermediates in aromatic substitution are stable intermediates or, as Wheland had suggested, transition states. A study (see p. 316) of the rates of substitution in a number of aromatic AHs, combined with a PMO analysis, showed very clearly that arenonium ions or radicals are stable intermediates and also led to an explanation of the relation between the reactivity of reagents and their selectivity in substitution reactions.

Another application, reported in 1958, was to another current problem, that of the relationship between the S_N1 and S_N2 reactions. As we have seen (p. 256), a study of the rates of reaction of arylmethyl derivatives $ArCH_2X$, Ar being a $\pm E$ group, showed that the S_N1 and S_N2 reactions are not qualitatively distinct but rather represent the extremes in a continuous range of possible mechanisms.

The $\pm E$ substituent technique has recently been applied to the Diels–Alder reaction.

Consider the reaction of a dienophile with a series of polycyclic hydrocarbons such as anthracene (257). These can be regarded as derivatives of 1,3-butadiene carrying $\pm E$ substituents, as indicated in the following equation:

(257)　　　　　　　　　　　　　　　　　　　　　(5.336)

Compare this with

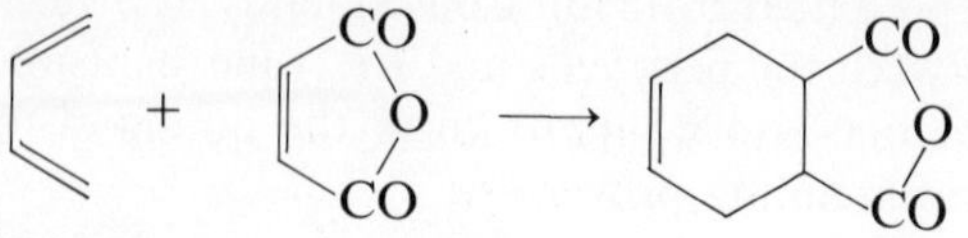

$$\text{[diene]} + \text{[maleic anhydride]} \longrightarrow \text{[adduct]} \quad (5.337)$$

If the dienophile remains unchanged, the part of the transition state containing it should be similar in all cases. Differences in rate between different dienophiles will then be due to changes in the π energy of activation associated with the diene.

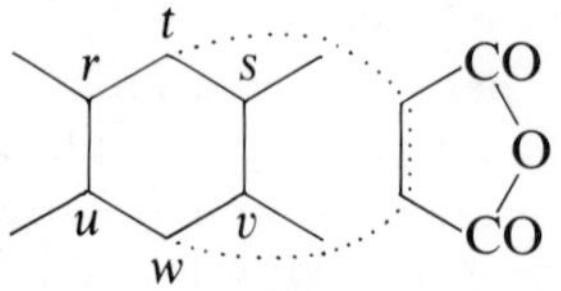

(258)

Consider the ring containing the diene moiety (258). Association of the dienophile with atom t in the transition state will reduce the rt and st π interactions just as they would in the case of substitution at atom t. If the corresponding changes in the π resonance integrals are $\delta\beta_t$, the corresponding change in π energy $\delta E_{\pi t}$ will be

$$\delta E_{\pi t} = 2(p_{rt} + p_{st})\delta\beta_t \quad (5.338)$$

where p_{rt} and p_{st} are the corresponding π bond orders in the original hydrocarbon. Now we have already shown (p. 145) that p_{rt} and p_{st} are approximately equal to a_{0r} and a_{0s}, the NBMO coefficients in the odd AH obtained by removing atom t. Thus,

$$\delta E_{\pi t} = 2(p_{rt} + p_{st})\delta\beta_t = 2(a_{0r} + a_{0s})\delta\beta_t = N_t\delta\beta_t \quad (5.339)$$

where N_t is the reactivity number of atom t. Likewise, if association of the dienophile with atom w in the transition state lowers the uw and vw resonance integrals by $\delta\beta_w$, the corresponding charge $\delta E_{\pi w}$ in the π energy will be given by

$$\delta E_{\pi w} = N_w\delta\beta_w \quad (5.340)$$

The values of $\delta\beta_t$ and $\delta\beta_w$ will be greater, the greater is the binding of the dienophile to the corresponding atom in the transition state. If the transition state is symmetric [cf. (253)], $\delta\beta_t = \delta\beta_w$; otherwise, let us assume that $\delta\beta_t > \delta\beta_w$. The extreme case of an $ER_A^\ddagger$ reaction will then correspond to $\delta\beta_w = 0$.
The π energy of activation ΔE_π is then given by

$$\Delta E^\ddagger = \delta E_{\pi t} + \delta E_{\pi w} = N_t\delta\beta_t + N_w\delta\beta_w$$

$$= (N_t + N_w)\delta\beta_w + N_t(\delta\beta_t - \delta\beta_w) \quad (5.341)$$

Now the 1,4-bislocalization energy (paralocalization energy; PLE_{tw}) for double addition to atoms t and w is given (see p. 145) by

$$\text{PLE}_{tw} = (N_t + N_w)\beta \quad (5.342)$$

TABLE 5.8. Rates of Diels–Alder Reactions[a] and Reactivity Numbers

Compound	$10^4 k$ (liters/mole sec)[b]	N_t	N_w	$N_t + N_w$
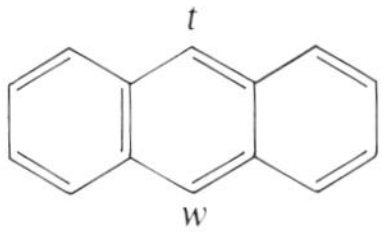 (structure 1)	79.9	1.265	1.265	2.530
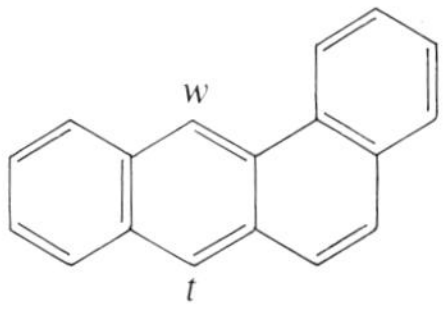 (structure 2)	6.92	1.353	1.436	2.789
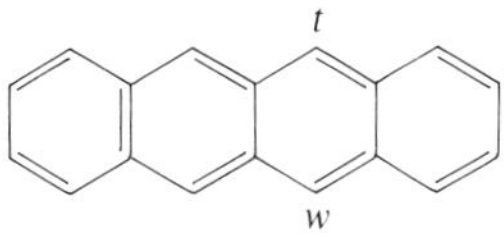 (structure 3)	1380	1.026	1.026	2.052
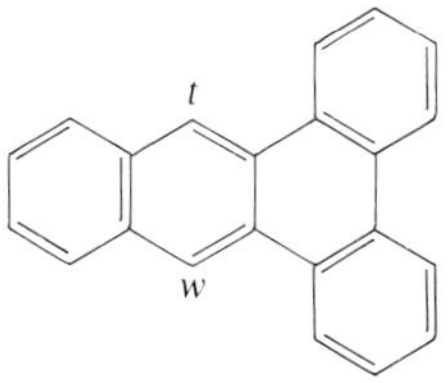 (structure 4)	3.58	1.504	1.504	3.008
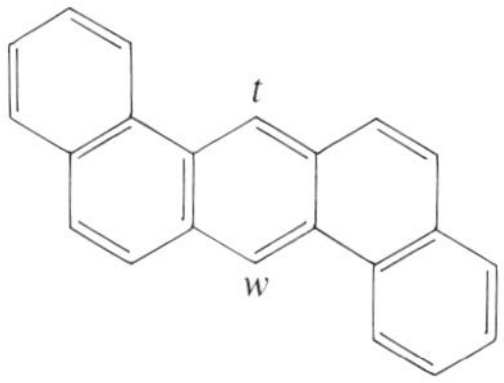 (structure 5)	0.747	1.512	1.512	3.024
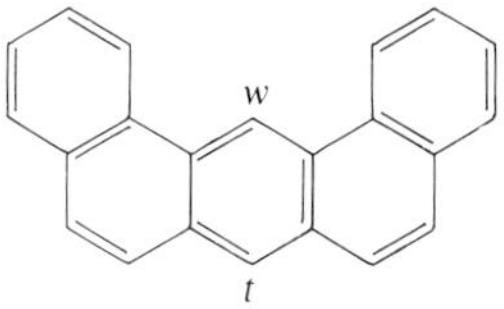 (structure 6)	0.675	1.437	1.600	3.037

[a] Data from Ref. 15.
[b] In diethyl succinate at 129.7°C.

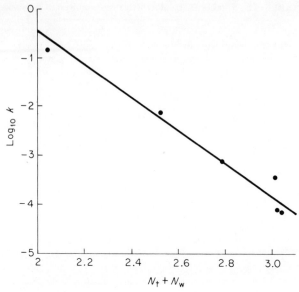

FIGURE 5.43. Plot of $\log_{10} k$ against $N_t + N_w$ for the Diels–Alder reactions of aromatic hydrocarbons; data from Table 5.8.

Likewise, the localization energy LE_t for attack of atom t is given by

$$LE_t = N_t \beta \tag{5.343}$$

Hence,

$$\Delta E_\pi^\ddagger = A(PLE_{wt}) + B(LE_t) \tag{5.344}$$

where $A\, (\equiv \delta\beta_w/\beta)$ is a measure of the extent to which the weaker bond linking the diene and dienophile has been formed in the transition state, while $B\, [\equiv (\delta\beta_t - \delta\beta_w)/\beta]$ is a measure of the asymmetry of the transition state. Thus if $B = 0$, the reaction is a symmetric pericyclic one, while if $A = 0$, it involves two-step $ER_A^\ddagger$ processes.

Table 5.8 shows the measured rates of reaction of maleic anhydride with several hydrocarbons of this type, together with the values of N_t and N_w. The points fit best to equation (5.344) with $B = 0$ (Fig. 5.43), implying that the reactions are of pericyclic type with symmetric transition states.*

---

* Dewar and Pyron[15] analyzed the reaction in a similar way but with localization energies and 1,4-bislocalization energies given by the more accurate MO treatment mentioned earlier. Their plot is somewhat better than Fig. 5.43.

Problems

5.1. Explain why the order of stability of carbanions in the gas phase reverses itself in solution, while the order of stability of the corresponding cations is the same in the gas phase as in solution.

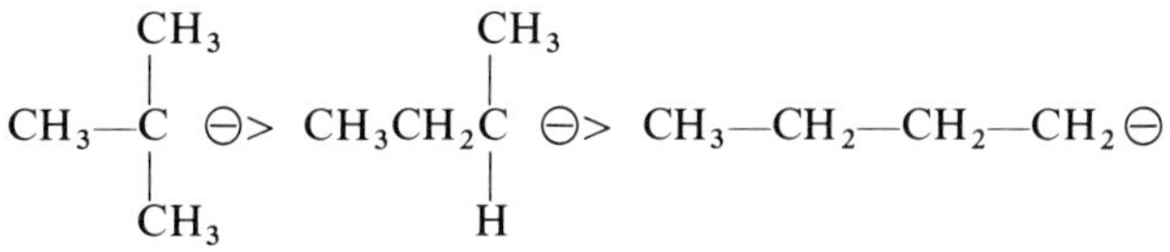

$$CH_3-\underset{\underset{CH_3}{|}}{\overset{\overset{CH_3}{|}}{C}}{}^{\ominus} > CH_3CH_2\underset{\underset{H}{|}}{\overset{\overset{CH_3}{|}}{C}}{}^{\ominus} > CH_3-CH_2-CH_2-CH_2{}^{\ominus}$$

5.2. In the Stevens rearrangement below, would you expect product (A) or product (B) to be favored by increasing the electron affinity (solvating power) of the solvent? Explain your answer.

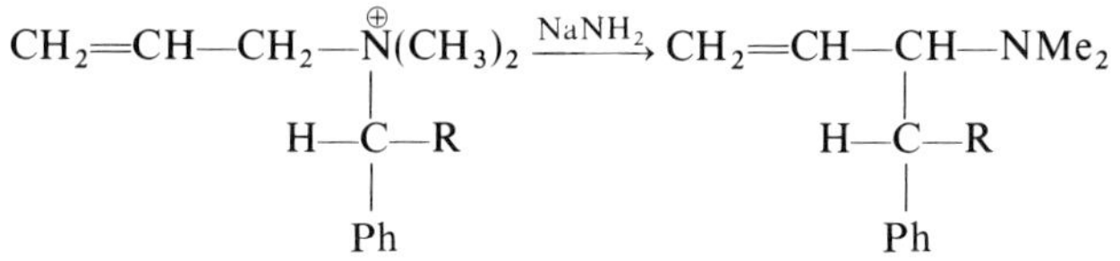

$$CH_2=CH-CH_2-\overset{\oplus}{N}(CH_3)_2 \xrightarrow{NaNH_2} CH_2=CH-CH-NMe_2$$
$$\underset{\underset{Ph}{|}}{\overset{\overset{|}{H-C-R}}{}} \qquad\qquad \underset{\underset{Ph}{|}}{\overset{\overset{|}{H-C-R}}{}} \quad (A)$$

or

$$\underset{\underset{Ph}{|}}{\overset{\overset{CH_2-CH=CH-NMe_2}{|}}{H-C-R}} \quad (B)$$

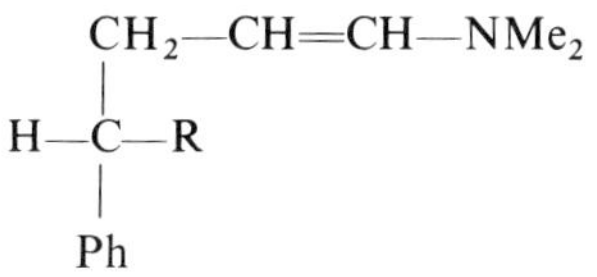

5.3. In competition between $E2$ and $E1cB$ elimination reactions, which would be favored by increasing the solvents' ability to solvate anions? Explain.

5.4. Arrange the following in order of increasing rate for deprotonation reactions:

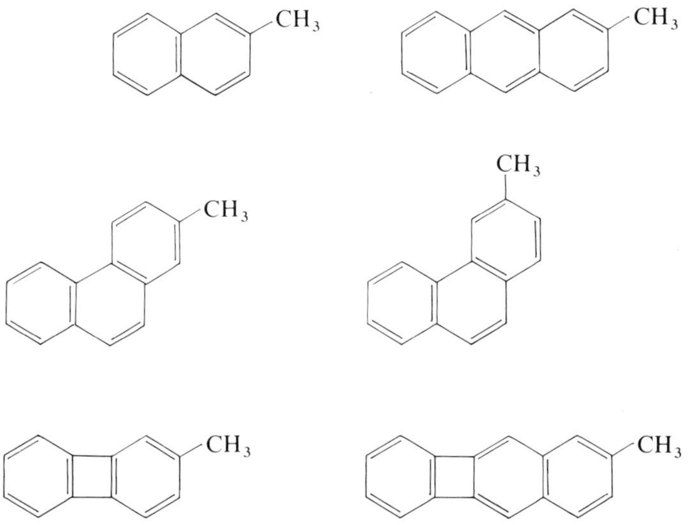

5.5. Predict the most reactive sites in the following hydrocarbons for electrophilic nitration:

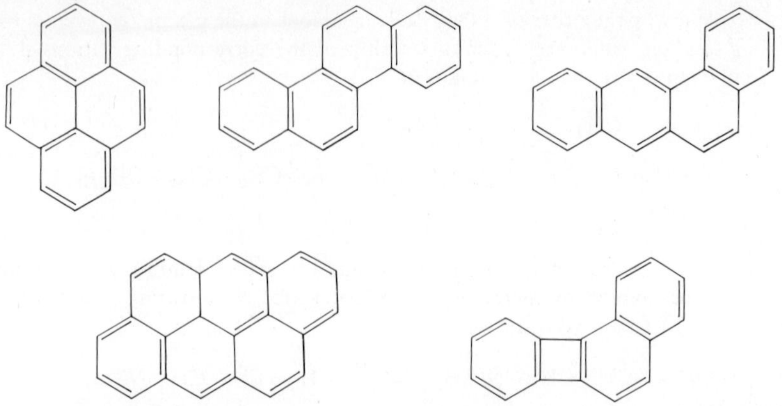

5.6. a. Arrange the following in order of increasing rates for S_N1 solvolysis reactions.

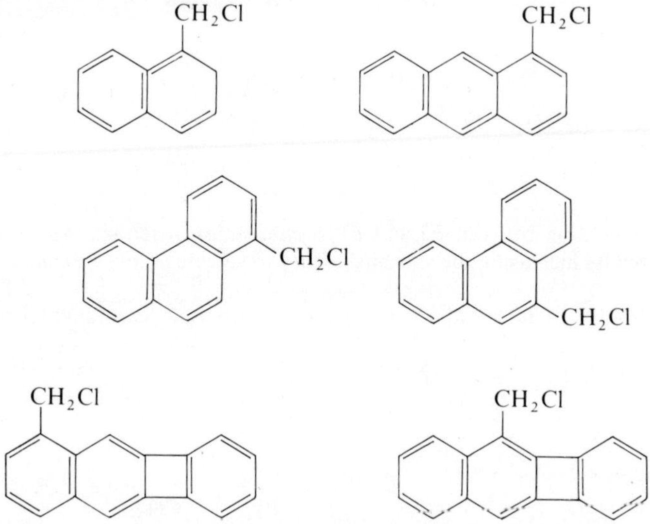

  b. Arrange the same systems in order of increasing rate difference between a given S_N1 and S_N2 reaction.

5.7. Write reasonable mechanisms for the following multistep pericyclic reactions based on Evans' principle:

  a.

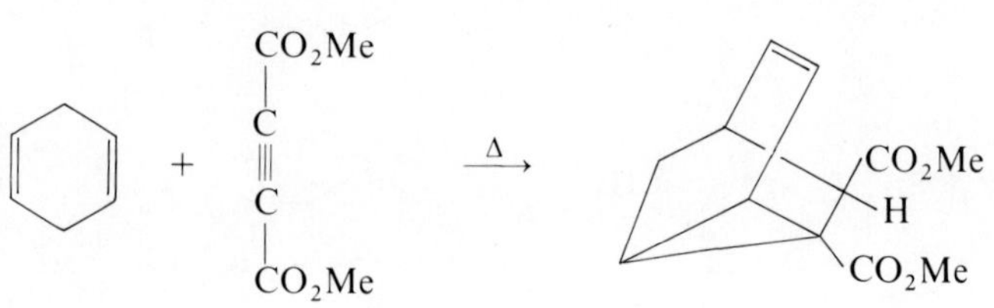

b. 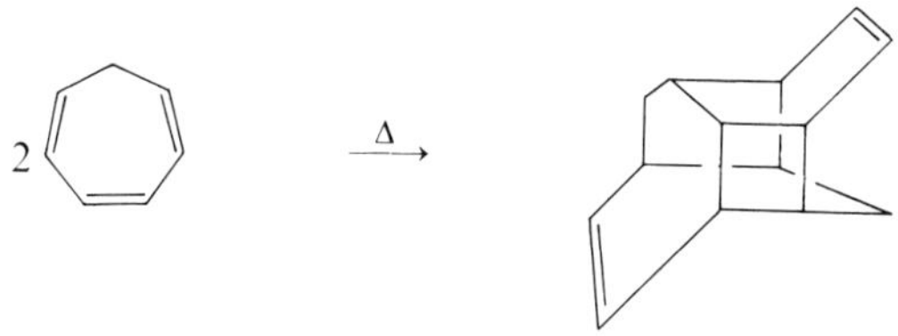
2
Δ

c. 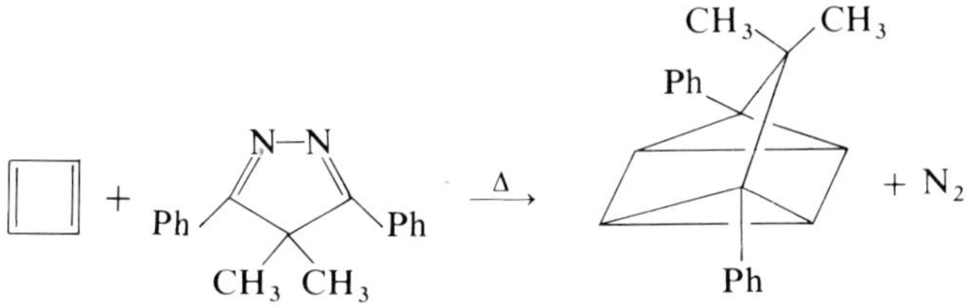
N—N
Ph
Ph
CH3 CH3
Δ
CH3
CH3
Ph
Ph
+ N2

d. 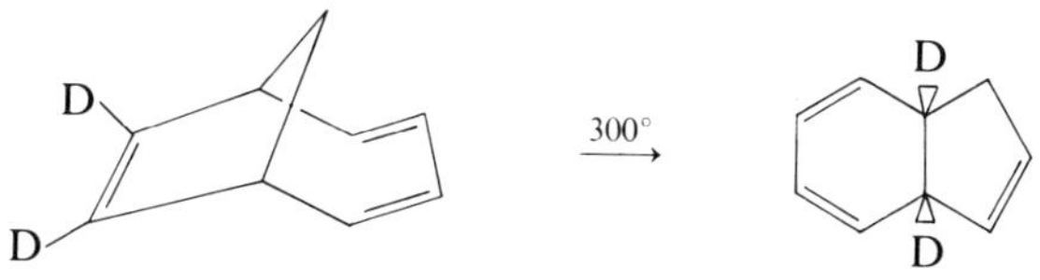
D
D
300°
D
D

e. 
D
D
500°
D
D

f. 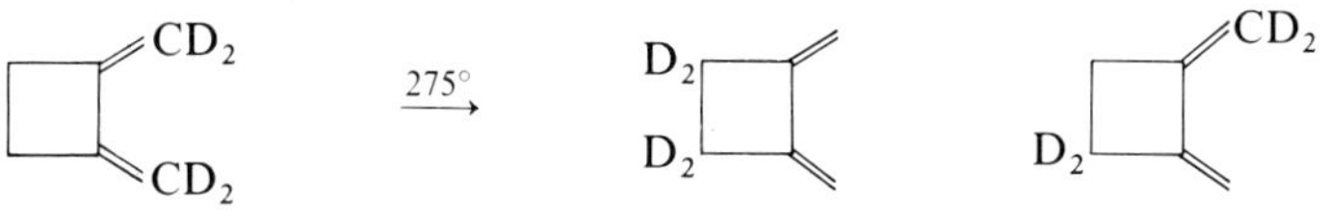
CD2
CD2
275°
D2
D2
D2
CD2

g. 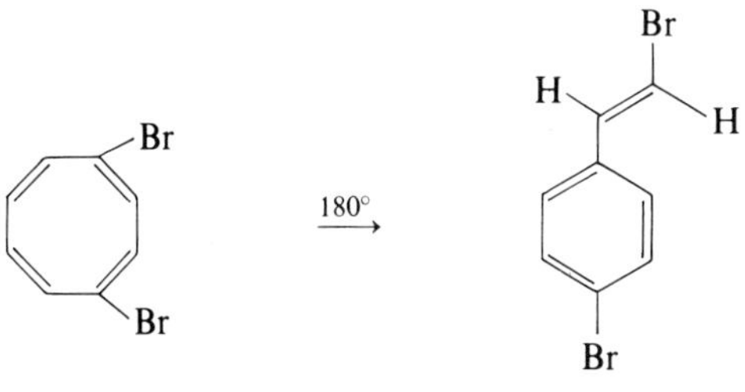
Br
Br
180°
Br
H
H
Br

(reaction rate increases dramatically on changing to a polar solvent).

h. 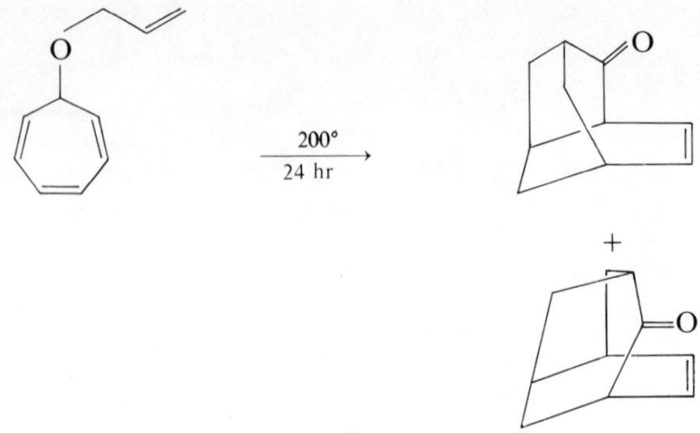 [allyl phenyl ether of cycloheptatriene] $\xrightarrow{200°\ 24\ hr}$ [tricyclic ketone] + [tricyclic ketone isomer]

i. 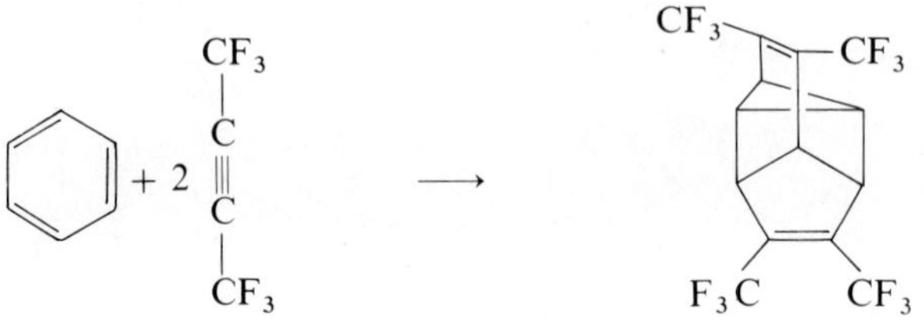 [benzene] + 2 [hexafluorobutyne: CF$_3$–C≡C–CF$_3$] → [hexakis(trifluoromethyl) adduct]

j. 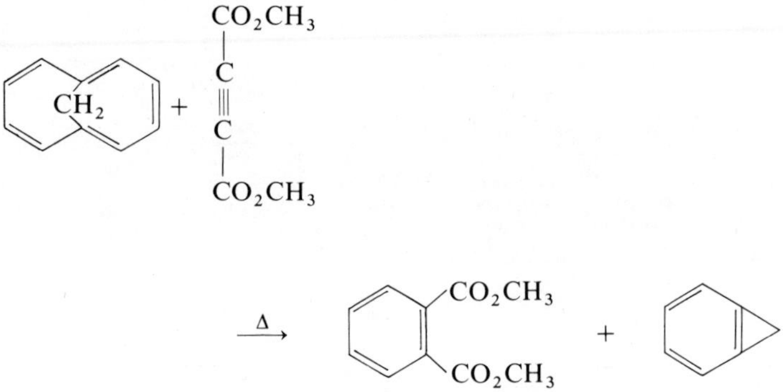 [2-methylene-naphthalene derivative] + [dimethyl acetylenedicarboxylate: CH$_3$O$_2$C–C≡C–CO$_2$CH$_3$] $\xrightarrow{\Delta}$ [dimethyl phthalate derivative with CO$_2$CH$_3$ groups] + [bicyclic alkene]

5.8. a. Given the fact that the chloride (C) shows no tendency to revert to (B), indicate the stereochemistry of (A) and (B).

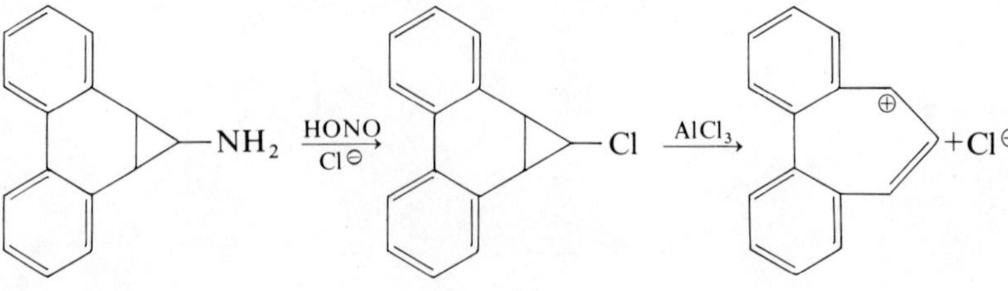

(A) [dibenzo cyclopropane-NH$_2$] $\xrightarrow[\text{Cl}^\ominus]{\text{HONO}}$ (B) [dibenzo cyclopropane-Cl] $\xrightarrow{\text{AlCl}_3}$ (C) [dibenzotropylium cation] + Cl$^\ominus$

b. Solvolysis of the bromides (E) and (F) results in the formation of two distinct 3-hydroxycyclooctenes, one *cis* and one *trans*. Indicate the stereochemistry of the product in each case.

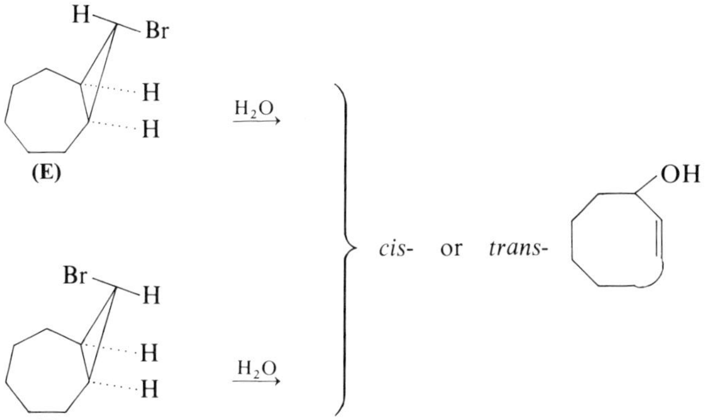

(E)

(F)

$\xrightarrow{H_2O}$

$\xrightarrow{H_2O}$

} *cis-* or *trans-*

5.9. Nucleophilic aromatic substitution reactions can proceed by the following mechanism in potassium amide–potassium–liquid ammonia mixtures:

initiation $(K \cdot) + ArI \longrightarrow (K^+) + ArI^{-\cdot}$

$ArI^{-\cdot} \longrightarrow Ar\cdot + I^-$

$Ar\cdot + NH_2^- \longrightarrow ArNH_2^{-\cdot}$

$ArNH_2^{-\cdot} + ArI \longrightarrow ArNH_2 + ArI^{-\cdot}$

termination steps

Comment on the requirements that would have to be met to obtain electrophilic substitution by a similar radical cation mechanism.

5.10. B_2Cl_4 adds stereospecifically to olefins,

$$\begin{matrix} R_2C & BCl_2 \\ \| & + & | \\ R_2C & BCl_2 \end{matrix} \longrightarrow \begin{matrix} R_2C-BCl_2 \\ | \\ R_2C-BCl_2 \end{matrix}$$

Discuss the probable mechanism for this reaction.

5.11. Dicobaltoctacarbonyl catalyzes the condensation of acetylenes to give benzenes. Comment on the mechanisms of this reaction. Would you expect that cyclooctatetraenes would be side products of this reaction?

5.12. On treatment of (A) with alcoholic KOH, which of the two chlorides should be replaced first?

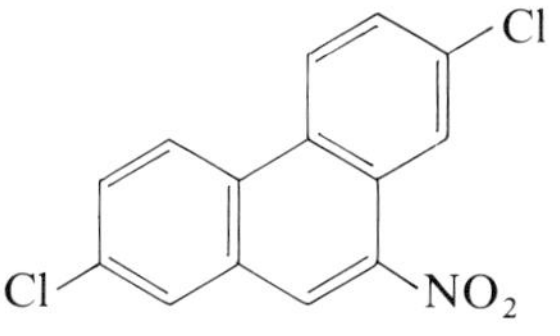

(A)

5.13. Show that the stability of a π complex formed by adding CH_3^+ to one of the π bonds in a conjugated hydrocarbon is approximately proportional to the bond order of the π bond.

5.14. When one mole of maleic anhydride reacts with one mole of pentaphene (A) only the bis adduct (B) is formed. Explain this.

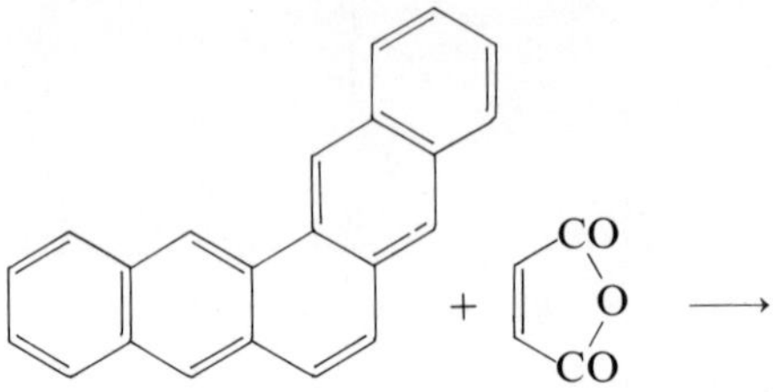

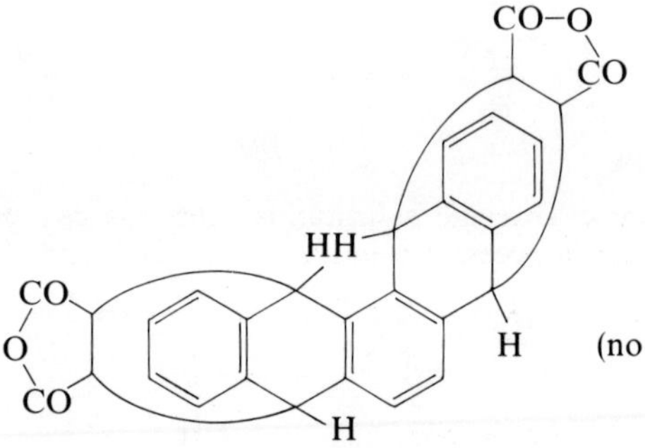

Selected Reading

Reaction rates, solvation effects:

L. P. HAMMETT, *Physical Organic Chemistry,* 2nd ed., McGraw-Hill, New York, 1970.

E. M. KOSOWER, *An Introduction to Physical Organic Chemistry,* John Wiley, New York, 1968.

PMO treatment of reactivity:

M. J. S. DEWAR, *The Molecular Orbital Theory of Organic Chemistry,* McGraw-Hill, New York, 1969.

Organic reaction mechanisms:

J. MARCH, *Advanced Organic Chemistry; Reactions, Mechanisms, Structure,* McGraw-Hill, New York, 1968.

Pericyclic reactions:

R. B. WOODWARD and R. HOFFMANN, *The Conservation of Orbital Symmetry,* Academic Press, New York, 1970.

R. E. LEHR and A. P. MARCHAND, *Orbital Symmetry—A Problem Solving Approach,* Academic Press, New York, 1972.

References

1. A. Streitweiser, Jr., and W. C. Longworthy, *J. Am. Chem. Soc.* **85,** 1757 (1963).
2. M. J. S. Dewar and R. J. Sampson, *J. Chem. Soc.* **1956,** 2789; **1957,** 2946.
3. M. J. S. Dewar and C. C. Thompson, Jr., *J. Am. Chem. Soc.* **87,** 4414 (1956).
4. P. J. C. Fierens, H. Hannaert, J. V. Rysselberge, and R. H. Martin, *Helv. Chim. Acta* **38,** 2009 (1955).
5. M. J. S. Dewar, *J. Am. Chem. Soc.* **74,** 3354 (1952).
6. G. A. Russell and C. DeBoer, *J. Am. Chem. Soc.* **85,** 3136 (1963).
7. W. H. Saunders, Jr., S. R. Farenholtz, E. A. Caress, J. P. Lowe, and M. Schreiber, *J. Am. Chem. Soc.* **87,** 3401 (1965).
8. M. D. Bentley and M. J. S. Dewar, *J. Am. Chem. Soc.* **92,** 3996 (1970).
9. M. J. S. Dewar, T. Mole, and E. W. T. Warford, *J. Chem. Soc.* **1956,** 3576.
10. M. J. S. Dewar, T. Mole, and E. W. T. Warford, *J. Chem. Soc.* **1956,** 3581.
11. M. J. S. Dewar and T. Mole, *J. Chem. Soc.* **1957,** 342.
12. A. P. Stefani and M. Szwarc, *J. Am. Chem. Soc.* **84,** 3661 (1962).
13. M. J. S. Dewar, *J. Am. Chem. Soc.* **74,** 3347 (1952).
14. L. M. Stock and H. C. Brown, *Adv. Phys. Org. Chem.* **1,** 34 (1963).
15. M. J. S. Dewar and R. S. Pyron, *J. Am. Chem. Soc.* **92,** 3098 (1970).

CHAPTER 6

Light Absorption and Photochemistry

6.1 Introduction

When a molecule absorbs light, it is converted to one of its *excited states*. In the orbital approximation such a process is represented by a jump or *transition* of an electron from one of the orbitals that was originally filled or partly filled to one that was empty or partly empty (Fig. 6.1). The energy required to bring this about is equal, in this approximation, to the difference $\mathscr{E}_1 - \mathscr{E}_2$ in energy between the two orbitals. The transition involves absorption of a photon of light of frequency v, given by

$$hv = \mathscr{E}_1 - \mathscr{E}_2 \qquad (6.1)$$

where h is Planck's constant.

Since transitions can take place between any pair of filled and empty orbitals, a molecule can exist in any of a number of excited states. The transi-

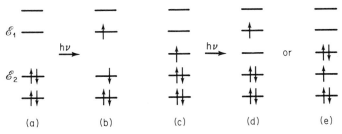

FIGURE 6.1. (a, b) Lowest transition for a closed shell molecule where all the electrons are paired; (c–e) two lowest transitions for a radical.

tion from the ground state to each excited state will involve absorption of light of a different frequency [Equation (6.1)]. These absorptions are studied by passing light of known and variable frequency and known intensity through the substance in question, usually in solution, and measuring the intensity of the light emerging. If we vary the frequency of the light and plot the fraction absorbed against frequency (or equivalently wavelength), we will get a typical absorption spectrum (Fig. 6.2). The process is in fact carried out automatically in a *spectrometer*.

A molecule contains vibrational and rotational (i.e., *vibronic*) energy as well as electronic energy and these may also change during an electronic transition. Vibronic energy is quantized so that a molecule exists in definite states, each with a fixed amount of vibronic energy. Each electronic transition therefore gives rise to a *band* in the spectrum composed of innumerable lines, each line corresponding to the same electronic transition but a different vibronic one. The rotational lines are not resolved in solution, so each vibrational transition appears as a broad band. Frequently these are so broad and numerous that they merge together in the absorption spectrum. In Fig. 6.2, one band has vibrational structure, while the other does not.

The intensity of a given absorption should properly be measured by the total area under a band in a plot such as that of Fig. 6.2. However, these areas are usually fairly closely proportional to the maximum extinction of the band. The amount of absorption depends on the number of molecules through which the light photons pass and on the probability that a photon will be absorbed when it goes through, or near, a molecule. The former will be proportional to a product of the concentration c (moles/liter) of the material and the thickness d (cm) of the layer of solution through which the light passes. The latter is of course characteristic of the material and is measured by the *extinction coefficient* ε. Since the probability of absorption is a molecular characteristic and does not depend on the concentration,

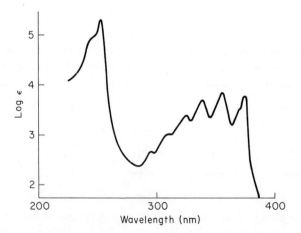

FIGURE 6.2. Absorption spectrum of anthracene.

the fraction of light absorbed in a given distance will be independent of the intensity. One could give a quantitative estimate of ε in terms of the fraction of light absorbed by a given thickness of, say, a 1 M solution ($c = 1$), or by the distance that light travels before its intensity is reduced by some fixed fraction. The latter procedure is the one commonly used, the extinction coefficient being defined as the reciprocal of the thickness (in cm) of a 1 M solution required to reduce the intensity by a factor of ten. It follows that if light of intensity I_0 passes through d cm of a solution of concentration c, the emerging intensity I will be given by

$$I = I_0 \times 10^{-\varepsilon dc} \quad \text{or} \quad \log(I/I_0) = -\varepsilon dc \tag{6.2}$$

We will be mainly concerned in this chapter with the special chemical reactions that excited states of molecules undergo. The study of such reactions is called *photochemistry*. Photochemical reactions differ from ordinary thermal or ground-state reactions because excited states have different electronic structures from, and much more energy than, the ground state. The absorption of light in the middle of the visible region, at 500 nm, is equivalent to the absorption of about 60 kcal/mole and of course absorption of UV light is even more energetic. This energy may become available to bring about chemical reactions that do not take place thermally because they would need too much activation.

In order to deal with reactions of this kind, we need to develop a theory of excited states and we also need to know the conditions under which the excitation energy can be made available for chemical purposes.

6.2 The Nature of Electronically Excited States

For reasons which will appear presently, only the lower excited states of molecules are of importance in photochemistry. These involve transitions from the highest occupied orbitals in molecules into the lowest unoccupied orbitals. Transitions involving inner shell electrons, outer shell AOs, or σ MOs are not important because their energies are as a rule all either too low or too high. Molecules containing only σ bonds usually have extremely high excitation energies and so absorb only in the *vacuum* ultraviolet, so termed because air absorbs strongly below 200 nm and experiments in that region consequently have to be carried out in the absence of air. Such work is difficult and so little has been done in the field of vacuum ultraviolet photochemistry. Besides, the amount of energy becomes so extreme that almost anything can happen to the resulting photoexcited molecules. It is as though in studying an unknown machine, instead of taking it apart piece by piece and studying each piece separately, we were to blow it to pieces with a charge of dynamite. The results would be dramatic but difficult to interpret.

The filled orbitals of importance in organic photochemistry are therefore of two kinds; bonding π MOs (π), and nonbonding AOs occupied by lone-pair electrons (n). The empty MOs into which the electrons move are nearly always antibonding π MOs (π^*). Rare cases are known, however, in which transitions involving filled, bonding (σ) or empty, antibonding (σ^*) MOs are involved. If we ignore these, the types of transition of importance then fall into three categories:

1. **$\pi \rightarrow \pi^*$ Transitions.** These involve a simple transfer of an electron from a filled π MO into an empty π MO (or a partly empty π MO in the case of a radical; see Fig. 6.1c–e). Note the self-explanatory symbol used to describe the transition.

2. **$n \rightarrow \pi^*$ Transitions.** These likewise involve a transition between an electron occupying a lone-pair AO n and an empty π MO. As we shall see presently, transitions occur with reasonable ease only between orbitals that overlap in space. An $n \rightarrow \pi^*$ transition can therefore occur only if the atom with the AO n forms part of the conjugated system that contributes the π^* MO.

3. **Charge Transfer (CT) Transitions.** If two molecules A and B are close together so that their MOs overlap, an electron in a filled orbital of A may undergo a transition into an empty orbital of B (Fig. 6.3). The result is an ion pair A^+B^-. Transitions of this type are called *charge transfer transitions*, abbreviated to CT transitions.

According to a naive orbital picture, one might expect the lowest CT transition between two identical molecules A, giving an ion pair A^+A^-, to have the same transition energy as the corresponding lowest "locally excited" transition in one of the molecules, giving AA* (Fig. 6.3b). This, however, is not the case, as the following argument shows. We can generate AA* or A^+A^- by adding an electron to a LUMO in the pair A^+A (see Fig. 6.3b), namely

$$A^+ + e^- \longrightarrow A^* \xrightarrow{+A} AA^* \qquad (6.3)$$

$$A + e^- \longrightarrow A^- \xrightarrow{+A^+} A^+A^- \qquad (6.4)$$

In both cases the electron enters the LUMO, but in the case of A, the HOMO has two electrons in it, whereas in A^+, it has one only. In forming A^- from

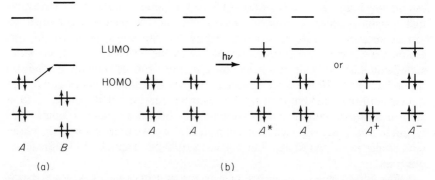

FIGURE 6.3. Charge transfer transitions (a) between dissimilar molecules A and B; (b) between two similar molecules A.

A, one has to overcome the extra repulsion between the additional HOMO electron and the electron we are trying to put into the LUMO. Consequently, the CT transition energy is greater than the corresponding locally excited energy by this repulsion, which typically amounts to 60–100 kcal/mole.

The CT transitions generally have low extinction coefficients since the mutual overlap of MOs of two different molecules A and B is relatively poor even when the molecules are in contact, nonbonded atoms being unable to approach one another closely. Since the locally excited transitions of A and B are still available, the CT transition can usually be seen only if it occurs at longer wavelengths than any of the locally excited transitions. Since the CT transition is further handicapped by the transfer of charge involved in it, such transitions are usually seen only if they involve a "donor" A with a HOMO of very high energy and an "acceptor" B with a very low-energy LUMO.

6.3 The Franck–Condon Principle

Light absorption is an extremely rapid process. Therefore when a molecule absorbs light, it does so without any change in its instantaneous geometry. This is known as the *Franck–Condon principle*. Molecules do not have fixed geometries, since they can undergo a variety of vibrations and internal rotations. The geometry of a molecule consequently oscillates around that corresponding to a true minimum in its potential energy. This would be true even at the absolute zero because vibration is never completely stilled. A molecule always has *residual energy* (*nullpunktsenergie, zero-point energy*) in each of its possible modes of vibration. The geometry in which a given excited state is formed depends on the instantaneous geometry of the ground state at the instant when it absorbs a photon.

The consequences of this can be seen most simply from the case of a diatomic molecule. If we plot its energy E as a function of bond length r, we will get a curve like that in Fig. 6.4. The energy rises very rapidly if we try to compress the bond, while as we stretch it, the energy will approach asymptotically the sum of the energies of the two atoms. This will lie above the minimum by an amount equal to the bond energy of the bond. The molecule can contain only fixed amounts of vibrational energy. It will therefore exist in a number of states represented by the horizontal lines in Fig. 6.4. The length of the line represents the amplitude of the corresponding vibration. The amplitude of course increases as the amount of vibrational energy increases. Note that even in its ground state, the molecule has residual energy $\mathscr{E}_0$.

There is no reason why the equilibrium bond length in an excited state should be the same as that in the ground state. Indeed, in the case of a diatomic molecule, it should be greater because one of the bonding electrons has moved into an antibonding MO. Excitation will then involve a transition between states with different geometries (Fig. 6.5). Now molecules at

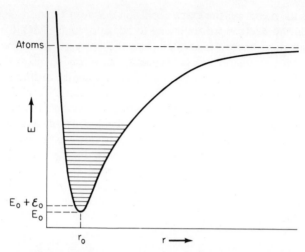

FIGURE 6.4. Plot of energy vs. bond length for a diatomic molecule XY, $\mathscr{E}_0$ being its residual energy of vibration and r_0 the equilibrium bond length.

room temperature are nearly all in their vibrational ground states. Excitation from the vibrational/electronic ground state should lead to the excited state in the same geometry. One can see from Fig. 6.5 that this will be one of the vibrationally excited forms of the electronically excited state. The argument can obviously be extended to more complex molecules. Unless the excited state and ground state have the same geometry, excitation will usually lead to a vibrationally excited form of the excited state. This is why electronic absorptions appear as wide bands, representing transitions to various vibrationally excited excited states.

While vibrational excitation is important in spectroscopy, it is usually unimportant in photochemistry because the vibrationally excited excited state is really just a "hot" excited state containing a lot of excess thermal energy. This is lost extremely rapidly by collisions with surrounding molecules, so that chemical reactions in solution involve only the vibrational ground state.

The process of excitation is reversible. An excited molecule can revert to the ground state by emission of light, the frequency of which is of course given by equation (6.1). The lifetime of the excited state is always long enough for its excess thermal energy to be dissipated. The emission of light therefore takes place from the vibrational ground state. The argument given above then shows (see Fig. 6.5) that the deexcitation will probably take place not to the vibrational/electronic ground state, but to a vibrationally excited ground state. The emitted light will therefore also cover a band of frequencies.

One can also see from Fig. 6.5 that the absorption band will lie to the high-frequency or short-wavelength side of the *zero–zero* transition, i.e., that between the two vibrational ground states, whereas the corresponding emission will lie to the low-frequency or long-wavelength side. The two

Light Absorption and Photochemistry

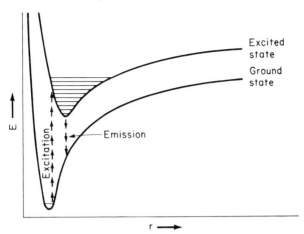

Figure 6.5. Excitation (upward arrows) from the vibrational/electronic ground state tends to lead to a vibrationally excited state. Downward arrows correspond to emission.

bands in the spectrum will therefore appear approximately as mirror images of one another (Fig. 6.6).

6.4 Singlet and Triplet States

In a normal molecule, where all the MOs are either doubly occupied or empty, the electrons all have to appear in pairs with opposite spin because of the Pauli principle. This is not the case in an excited state (see Fig. 6.1b), because there are now two singly occupied MOs. The electrons in these may or may not retain their opposite spins (Fig. 6.7). States in which they do are called *singlet* states and those in which they do not are called *triplet*

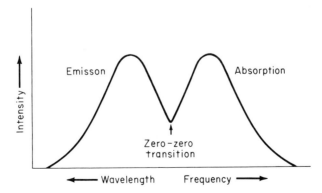

FIGURE 6.6. Relation between absorption and emission bands corresponding to the same electronic transition.

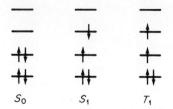

FIGURE 6.7. Ground state S_0 and the singlet S_1 and triplet T_1 states arising from the first electronic transition.

states.* The singlet states are denoted in order of energy by S_0, S_1, S_2, etc., S_0 being the ground state. Since all the triplet states are excited, the lowest is denoted by T_1, followed by T_2, T_3, etc.

Since a given pair of singlet and triplet states differ only in the spins of the electrons in the two single occupied MOs, it follows from Hund's rule that the triplet must be lower in energy. The order of the states in Fig. 6.7 is therefore $S_0 < T_1 < S_1$.

6.5 Extinction Coefficients and Transition Moments

During a transition, an electron initially in one orbital suddenly finds itself in another. Intuitively, one can see that such a process should be unfavorable unless the electron is initially in a reasonable part of the first orbital ϕ_1, i.e., one where the probability $\phi_1{}^2$ of finding it is significant, and ends in correspondingly reasonable part of the second orbital ϕ_2. For the transition to occur easily, the two orbitals must therefore overlap in space, i.e., there must be regions where their product $\phi_1\phi_2$ does not vanish.

Since absorption involves an interaction of the charged electron with the fluctuating electric field associated with electromagnetic radiation (i.e., light), one might also expect the absorption to depend on the relation of ϕ_1 and ϕ_2 to the direction x along which the electric vector is oscillating. It is therefore not surprising to find that the ease of a given transition, as measured by the extinction coefficient ε, is proportional to the value of the triple product $\phi_1\phi_2 x$ summed over the whole region occupied by the two orbitals. The value of this sum is given by the so-called *transition moment*

---

* The terms are derived from the number of different ways the molecule can orient itself in a magnetic field. Each electron has an angular momentum of $\pm h/4\pi$, due to its spin, and as a result it behaves like a magnet. In a singlet, all the spins are paired, so there is no net spin or magnetic moment. The molecule therefore does not orient itself in any particular way in a magnetic field and its energy does not depend on its orientation. In a triplet, however, there will be a net spin of $h/2\pi$, due to the two unpaired electrons. Quantum theory tells us that the electrons in such a molecule must be oriented so that the component of spin momentum is either $+h/2\pi$ or $-h/2\pi$ (corresponding to the axis of spin lying along the field) or zero (with the axis perpendicular to the field). Since the spin leads to a magnetic moment, the three states correspond to three different orientations of a magnet in a magnetic field and so have different energies. Hence the term *triplet state*. Likewise, the odd electron in a radical must orient itself with its spin parallel to ($+h/4\pi$) or antiparallel to ($-h/4\pi$) the field. States with single unpaired electrons are therefore called *doublet states*.

integral $\int \phi_1 x \phi_2 \, d\tau$. This can be illustrated with a pictorial representation which will do quite well for our purpose.

The total overlap between two AOs of an atom, say the 1s and $2p_x$ AOs of the hydrogen atom, vanishes because although the two AOs do overlap in space, the regions where the product $(1s)(2p_x)$ is positive are balanced by those where it is negative (Fig. 6.8a). This is because the yz plane is a node for the $2p_x$ AO, which is positive to the right of it and negative to the left of it, whereas the 1s AO has the same sign all over. Now the variable x has the same nodal properties as the $2p_x$ AO (Fig. 6.8a). It follows that the product $(2p_x)x$ is positive everywhere, as is the 1s AO. The triple product $(1s)(2p_x)x$ is therefore also positive everywhere. There is therefore a finite transition moment for the transition $1s \to 2p_x$, which should be "allowed."

The reason the transition moment does not vanish is that one of the AOs has an additional angular node, perpendicular to the direction along which the electronic vector is oscillating, and that both AOs have the same number of nodes (here zero) in other directions. One can see that this result must be quite general. Since the number of angular nodes is equal to the azimuthal quantum number l, a transition between two AOs is allowed, i.e., has a nonvanishing transition moment and so a measurable extinction coefficient, only if l changes by one unit. In particular, the transition moment integral for two different $2p$ AOs vanishes.

Note that molecules in solution have random orientations, so that the plane of polarization of incident light is effectively random. Thus while the transition $1s \to 2p_x$ could not be brought about by light polarized along the y axis (Fig. 6.8c), nor the transition $1s \to 2p_y$ by light polarized along the x axis, there are always photons around with all possible polarizations so it is only necessary that *one* of the transition moments differ from zero.

Similar arguments apply to transitions in molecules. For example, a $\pi \to \pi^*$ transition will be possible only if the incident light is polarized parallel to the π MOs and if one of the MOs has an additional node perpendicular to the plane of polarization. Thus a transition between the bonding and antibonding MOs of ethylene is possible if, and only if, the light is polarized along the C—C bond (Fig. 6.9a). In the case of such a $\pi \to \pi^*$ transition, the product $\psi_1 \psi_2 x$ will usually be large because the π MOs will

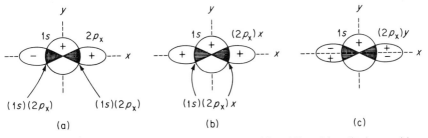

FIGURE 6.8. (a) The total overlap between the AOs 1s and $2p_x$ of H vanishes; (b) the transition moment along the x axis does not, but (c) that along the y axis does.

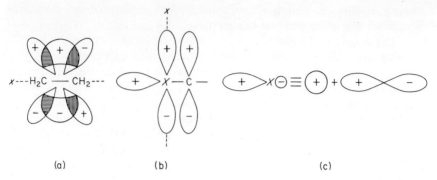

FIGURE 6.9. (a) Bonding and antibonding MOs of ethylene; the $\pi \to \pi^*$ transition is allowed for light polarized along the C—C bond (cf. Figure 6.8a, b); (b) n and p AOs of atom X; the $n \to \pi^*$ transition is allowed only from light polarized along the p AO; (c) dissection of the hybrid n AO into s and p components.

overlap efficiently with one another. The situation is different in the case of an $n \to \pi^*$ transition (Fig. 6.9b). Here the π MO has a node passing through the nuclei, while the AO does not. This time the overall value of the product $n\pi^*x$ will vanish unless the direction of polarization is perpendicular to the π nodal plane, i.e., parallel to the $2p$ AO of the n atom. Since the AO (n) overlaps only with that one $2p$ AO, only a small part of the π^* is involved. Furthermore, only the s component of the hybrid n AO (Fig. 6.9c) can take part since the overall value of the product $\psi_1\psi_2 x$ vanishes if ψ_1 and ψ_2 are two different p AOs of a given atom. Even the sp contribution is poor because here the s AO is a $2s$ AO and this has an additional spherical node. The overall value of the product $(2s)(2p_x)x$ indeed vanishes for AOs of hydrogen and it remains small for AOs of other atoms.

The extinction coefficients for $n \to \pi^*$ transitions are therefore small, ranging from less than ten to a few hundred. The extinction coefficients of $\pi \to \pi^*$ transitions, on the other hand, are usually large, in the range 10^4–10^5. It is of course possible that a $\pi \to \pi^*$ transition may be weak, due to poor overlap of the MOs. This happens quite often as a result of symmetry. Thus the extinction coefficient for the first $\pi \to \pi^*$ transition of benzene (at 254 nm) is only 204.

These considerations apply to all transitions, but in the case of a singlet–triplet transition there is a further factor that reduces the extinction coefficient. In a singlet–singlet transition, an electron simply shifts from one MO to another, but in a singlet–triplet transition it also has to reverse its spin (see Fig. 6.7). Now a spinning electron behaves like a tiny gyroscope and so resists having its spin inverted. Such transitions are therefore normally forbidden. They occur only in the presence of strong electromagnetic fields which can, as it were, get a grip on the electron and help to upend it. Such fields exist near the nuclei of atoms, particularly those of high atomic number and consequently large nuclear charge. The presence of such an atom facilitates singlet–triplet transitions. The heavy atom need not even be in the molecule itself. Collisions with solvent molecules containing heavy atoms can suffice. Indeed, collisions generally help things along because of the disturbances caused by the resulting interpenetration of electronic atmospheres.

The usual "organic" elements, C, H, O, N, are too light to be of much help, so the extinction coefficients for singlet–triplet transitions in organic molecules are often less than unity. The "heavy atom" effect is seen very clearly in the halonaphthalenes. Naphthalene has a singlet–triplet transition on the violet–ultraviolet border but it is too weak to give rise to color. In chloronaphthalene, the transition is stronger but still not enough for chloronaphthalene to be visibly colored. In bromonaphthalene, however, the transition becomes quite strong and in iodonaphthalene very strong indeed. Bromonaphthalene is pale yellow and iodonaphthalene is yellow. The solvent effect is seen very nicely when colorless chloronaphthalene is dissolved in colorless methyl iodide. The resulting solution is deep yellow, not because of any chemical reaction but solely as a result of a huge increase in the extinction coefficient of the singlet–triplet band.

A transition between a singlet state and a triplet state is called an *intersystem crossing*. The most important process of this kind is the conversion of a singlet excited state into the corresponding triplet (see Fig. 6.7). As we have seen, such a process is always energetically possible since the singlet must lie above the triplet. Whether it in fact occurs will depend on whether the rate of this formally forbidden process can compete with other modes of destruction of the singlet.

6.6 Excitation and Deexcitation; Lifetimes of States, Fluorescence, and Phosphorescence

We have so far only considered excitation produced by the absorption of light. Excitation can, however, also occur if a molecule can be given enough energy in some other way. The most important process of this kind is *sensitization*, in which the energy is transferred to a molecule A from some excited atom or molecule B*,

$$A + B^* \longrightarrow A^* + B \tag{6.5}$$

For this to occur, the excitation energy of B must be greater than that of A.

Since B* is normally produced by photoexcitation of B, it is not immediately obvious that anything is gained by using it as an intermediate instead of exciting A directly. This is certainly true for singlet–singlet transitions, but not for singlet–triplet transitions. The $S_0 \to T_1$ extinction coefficients are extremely small, so that direct photoexcitation of triplets is difficult. Sensitization by a triplet excited state B* (↑↑) of B is, however, easy because the *total* spin angular momentum of A and B remains unchanged if one electron in each flips; i.e.,

$$A(\uparrow\downarrow) + B^*(\downarrow\downarrow) \longrightarrow A^*(\uparrow\uparrow) + B(\uparrow\downarrow) \tag{6.6}$$

We are, however, still left with the problem of getting triplet B. This can be done if B undergoes easy intersystem crossing from the S_1 state to T_1. We can easily excite B to S_1, this being a normal singlet–singlet transition, and intersystem crossing then leads to the T_1 state. Photochemical reactions involving triplet states are commonly studied in this way. For example, biacetyl ($CH_3COCOCH_3$) undergoes intersystem crossing very efficiently and its lowest triplet T_1 has a higher energy than the lowest triplet state of butadiene (BD). Photodimerization of butadiene via its triplet can therefore be sensitized by biacetyl, which incidentally absorbs at wavelengths (400 nm) where butadiene is completely transparent. In presence of small amounts of biacetyl, butadiene dimerizes rapidly in visible light.

Whether or not intersystem crossing occurs will of course depend on the rate of intersystem crossing and the lifetime of the excited singlet. The lifetime in turn depends on the rates of the processes that remove it. There are five of these possible processes.

1. Emission of Light. The simplest way in which an excited state can disappear is by emitting light and reverting to the ground state, a process considered in Section 6.3. Such emission of light by a singlet–singlet transition is called *fluorescence*. From the principle of microscopic reversibility, the ease of such a process should parallel the ease of the corresponding photoexcitation. The lifetime for fluorescence of an excited singlet state should therefore vary inversely as the extinction coefficient* of the corresponding transition from the ground state. Fluorescence lifetimes usually lie in the range 10^{-7}–10^{-10} sec.

It is of course equally possible for a triplet state to revert to the ground state by emission of light, a process known as *phosphorescence*.† Since, however, extinction coefficients for $S_0 \rightarrow T_1$ transitions are so very small, the lifetimes for phosphorescence are very much longer than those for fluorescence, usually lying in the range 10–10^{-3} sec. Indeed, it is difficult to measure such lifetimes in solution or at room temperature because the triplets are removed too rapidly by the other processes considered below. It is usually necessary to generate the triplet in a solid glass at low temperatures when it has little alternative to phosphorescence.

2. Sensitization. This process, the transfer of excitation energy from one molecule to another, has already been considered [equations (6.5) and (6.6)].

3. Internal Conversion. This is a process in which the electronically excited molecule passes over into a vibrationally excited ground state, a single large quantum of electronic excitation being replaced by a very large number of small vibrational quanta. It can be shown that such processes are rather improbable, the difficulty increasing with the amount of electronic energy that has to be dissipated (see Section 6.13). Now the gap between the ground state S_0 and the first excited state S_1 is always much greater than

---

* Strictly, the lifetime varies inversely as the total area under the absorption band, but this is usually proportional to the extinction coefficient; see Section 6.1.
† It is easy to confuse the terms "fluorescence" and "phosphorescence." A useful mnemonic is that fluorescence is the faster process.

that between any two successive excited states. Internal conversion between the latter is therefore very much faster than that between S_1 and S_0. If, therefore, we excite a molecule to one of its higher singlet states S_n, internal conversion via S_{n-1}, S_{n-2}, etc., will take place very rapidly until the first excited state is reached. The lifetimes of the higher excited states are consequently so short that they rarely play a role in photochemistry. Exceptions occur only in cases where a given step $S_m \rightarrow S_{m-1}$ involves a transition with a very low transition moment so that there is an additional factor hindering it.

Internal conversion from T_1 to the ground state can also occur, but its rate will be less than that for S_1 because the $S_0 \rightarrow T_1$ transition is spin forbidden. On the other hand, phosphorescence is a correspondingly slower process than fluorescence, so the competition between light emission and internal conversion can be much the same for both S_1 and T_1.

4. Intersystem Crossing. This is a special type of internal conversion in which S_1 is converted to T_1 instead of to the ground state. The difference in energy between S_1 and T_1 is much less than that between T_1 and S_0, so intersystem crossing is a much easier process than internal conversion of the triplet T_1 to the ground state S_0. Nevertheless, it is by no means easy, so special conditions must hold if it is to compete effectively with fluorescence. First, the fluorescence lifetime of the singlet must be long. This means that the extinction coefficient for the $S_0 \rightarrow S_1$ process must be small. Second, the rate of intersystem crossing must be as high as possible; this means that the S_1-T_1 separation should be small. Both these conditions are met particularly well by $n \rightarrow \pi^*$ excited states. As we have seen, the extinction coefficients for $n \rightarrow \pi^*$ transitions are small because of poor overlap between the corresponding orbitals. Furthermore, the singlet–triplet separation depends on an interaction between the pair of electrons in question and this is also small if they occupy orbitals that overlap poorly with one another. The singlet–triplet separations for $n \rightarrow \pi^*$ states are therefore unusually small. Intersystem crossing is consequently especially effective in the case of $n \rightarrow \pi^*$ excited states, in particular those for carbonyl compounds. Ketones and α-diketones are therefore commonly used as sensitizers in the study of triplet reactions.

5. Chemical Reactions. The removal of excited states by chemical reactions is another special case of internal conversion. Such processes can be of two kinds, corresponding to internal conversion of an excited state to the ground state, and to internal conversion between two different excited states. In other words, the chemical reaction may lead to an excited state of some other species or to that species directly.

6.7 Excitation Energies of Even AHs

Having established the ground rules, our next task is to apply PMO theory to the calculation of excitation energies. We will start with the simplest case, $\pi \rightarrow \pi^*$ transitions of even AHs. We are interested here only in the first $\pi \to \pi^*$ states because the higher states rarely play any role in photochemistry.

The frequency ν_0 of the first absorption band (FFAB) is given by equation (6.1), i.e.,

$$h\nu_0 = \mathscr{E}_{\pi^*} - \mathscr{E}_\pi \tag{6.7}$$

where $\mathscr{E}_{\pi^*}$ is the energy of the lowest unoccupied MO (LUMO) and $\mathscr{E}_\pi$ is that of the highest occupied MO (HOMO). The corresponding wavelength λ_0 is given by

$$\lambda_0 = \frac{c}{\nu_0} = \frac{hc}{\mathscr{E}_{\pi^*} - \mathscr{E}_\pi} \tag{6.8}$$

where c is the velocity of light.

In our PMO treatment of even AHs by union of pairs of odd AHs (Section 3.8), it is clear that the perturbed NBMOs represent approximations to the HOMO and LUMO of the resulting even AH (see Fig. 3.9). The separation between them is thus given [see equation (3.23)] by

$$\mathscr{E}_{\pi^*} - \mathscr{E}_\pi = 2\beta \left| \sum_r a_{0r} b_{0s} \right| \tag{6.9}$$

There is, however, an ambiguity in using this expression because an even AH can be divided into two odd AHs in more than one way and the resulting values for $\mathscr{E}_{\pi^*} - \mathscr{E}_\pi$ are not the same. This is because second-order perturbations, particularly those between the filled MOs, and between the empty MOs, significantly affect the energies of individual MOs even though they do not seriously affect the total energy (see the discussion of collective and one-electron properties in Section 1.12). Since the second-order perturbations between pairs of bonding MOs are generally greater than those between bonding and antibonding MOs, and since the HOMO is by definition the highest occupied MO, the second-order perturbations will on balance push it up in energy. A similar argument shows that second-order interactions between the empty MOs should push the LUMO down. The excitation energy given by first-order theory alone [equation (6.9)] is therefore too large, so the best first-order estimate will come from that division of the even AH into a pair of odd AHs that minimizes the NBMO interaction. This point is illustrated in Fig. 6.10 by two different divisions of benzene.

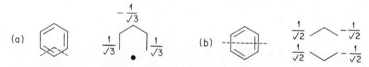

FIGURE 6.10. Two different dissections of benzene into a pair of odd AHs, showing the corresponding NBMO coefficients. The resulting estimates of the excitation energy are: (a) $\mathscr{E}_{\pi^*} - \mathscr{E}_\pi = 2\beta[(1/\sqrt{3}) + (1/\sqrt{3})] = 2.31\beta$. (b) $\mathscr{E}_{\pi^*} - \mathscr{E}_\pi = 2\beta(\frac{1}{2} + \frac{1}{2}) = 2\beta$.

TABLE 6.1. Light Absorption of Aromatic Hydrocarbons

Compound (and method of division)	λ_0, nm		
	PMO	HMO	obs
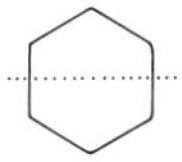	(210)	(210)	210
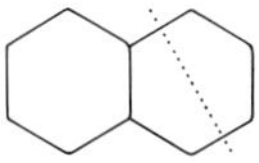	262	340	275
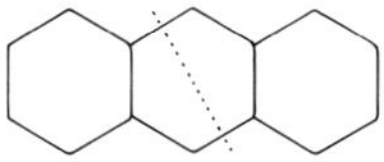	368	507	375
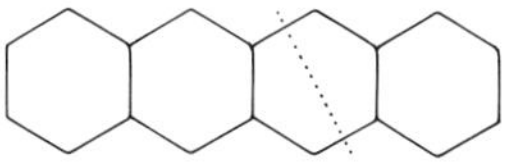	458	713	474
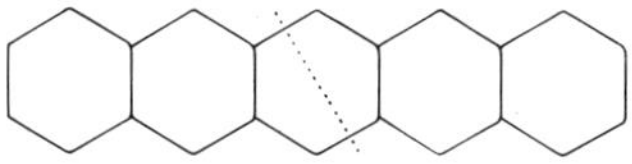	595	955	581
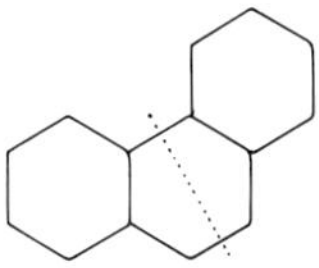	331	347	293
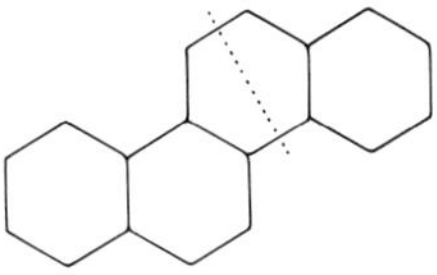	311	404	319
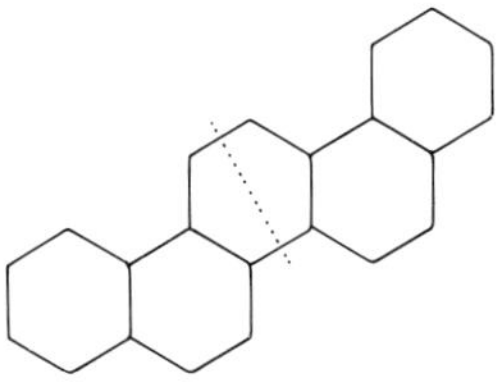	323	419	329
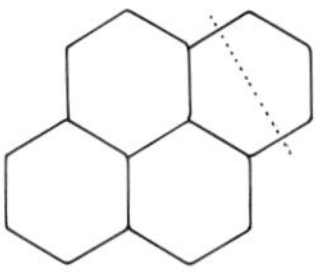	364	472	364

TABLE 6.1 (cont.)

Compound (and method of division)	λ_0, nm		
	PMO	HMO	obs
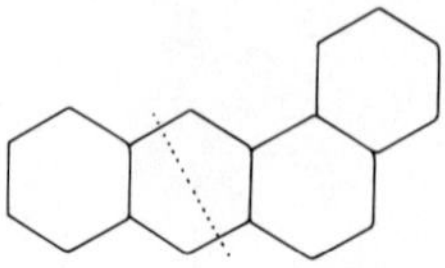	351	465	359
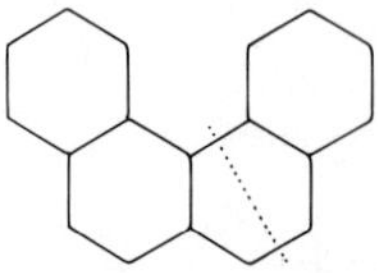	284	370	317
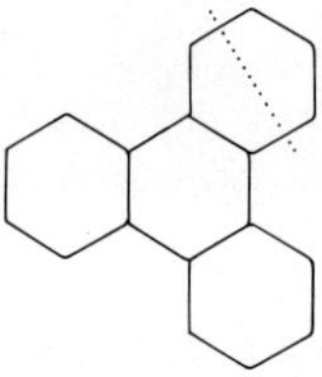	245	307	257
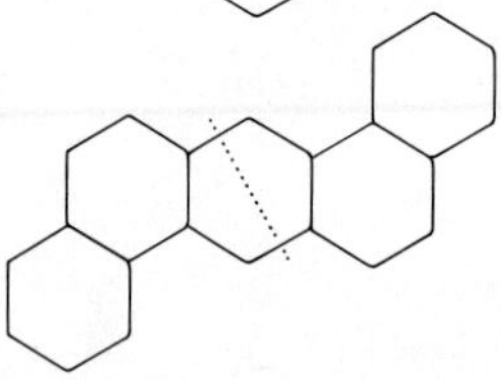	346	444	351
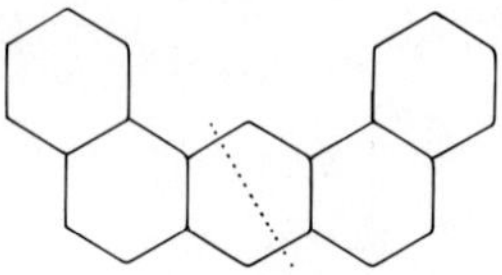	319	428	338
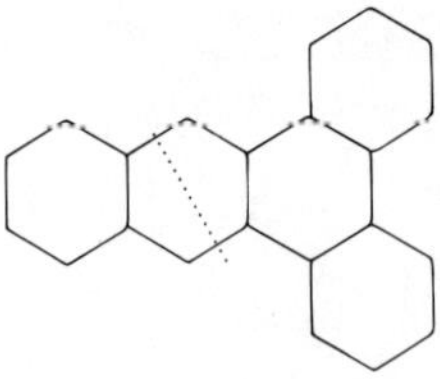	317	420	336
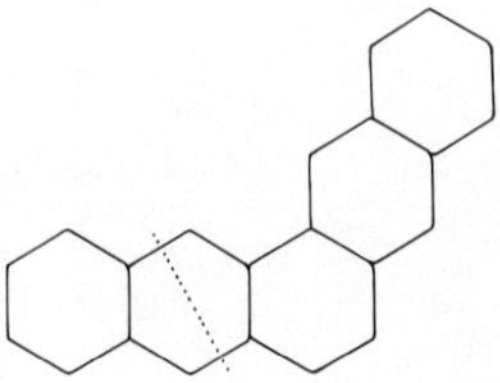	357	480	348

Light Absorption and Photochemistry

TABLE 6.1 (cont.)

Compound (and method of division)	λ_0, nm		
	PMO	HMO	obs
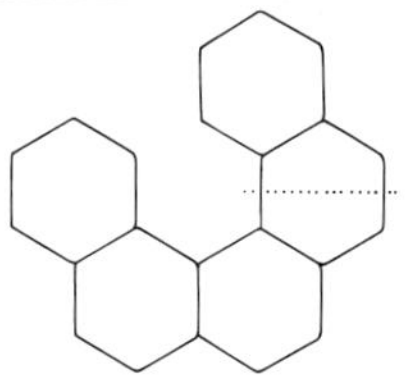	296	392	330
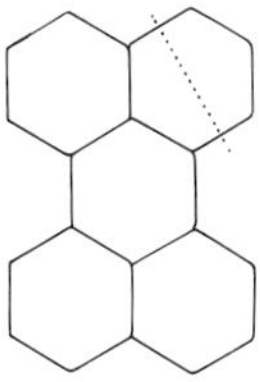	395	605	434
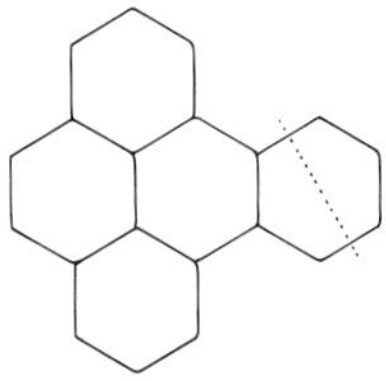	323	423	331
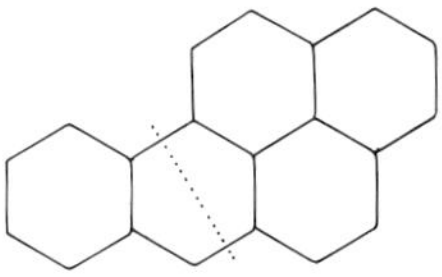	449	565	385
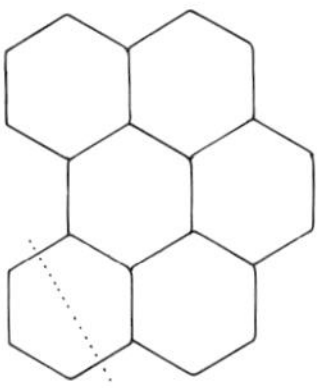	342	478	388
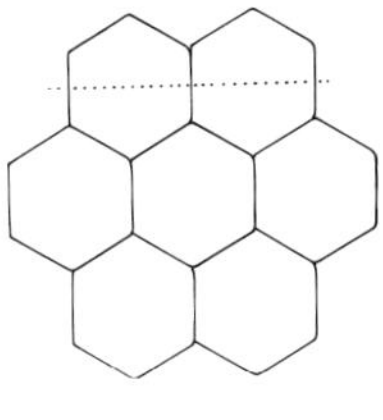	293	389	341

Of these, the second (Fig. 6.10b) is the better, leading to a lower value for $\mathscr{E}_{\pi^*} - \mathscr{E}_{\pi}$.

When this convention is followed, remarkably good predictions of spectra are obtained for even AHs. Indeed, the results are usually much better than those given by the standard Hückel method. Table 6.1 lists structures and calculated and observed wavelengths for the first $\pi \to \pi^*$ transitions for a wide range of benzenoid hydrocarbons. Values calculated by the Hückel method are included for comparison.

6.8 Excitation Energies of Odd AHs

Odd AHs differ from even AHs in that they can exist in three distinct states, cation, radical, and anion, depending on the number of electrons in the NBMO. Thus in the cation, the NBMO is the LUMO, while in the anion, it is the HOMO (Fig. 6.11).

The lowest $\pi \to \pi^*$ transition corresponds to one between the HOMO and LUMO. From the pairing theorem, the corresponding differences $(\delta E^+, \delta E^-)$ are equal. In the radical, however, transitions are possible of an electron both from the highest bonding MO into the half-empty NBMO ($\delta E_1^{\cdot}$ in Fig. 6.11b) and also from the NBMO into the lowest antibonding MO ($\delta E_2^{\cdot}$ in Fig. 6.11b). Both these transitions should, moreover, have similar energies, both being equal to the lowest transition energies for the cation and anion.

The first absorption band of an odd AH should therefore be the same for the cation, radical, and anion. A good example is the triphenylmethyl system, Ph_3C. The first absorption bands of the cation and anion are at almost identical wavelengths (~ 450 nm). It is difficult to compare the

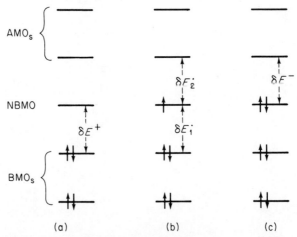

FIGURE 6.11. MOs and lowest transitions in an odd AH
(a) cation; (b) radical; (c) anion.

absorptions of the ions with that of the radical since the former are greatly affected by solvation and by the nature of the gegenion; however, the first absorption band of the radical (519 nm) is reasonably close to those of the ions, and solutions of all three species have similar red colors.

6.9 $\pi \to \pi^*$ and $n \to \pi^*$ in Even, Heteroconjugated Systems

When a carbon atom r in a conjugated hydrocarbon is replaced by a heteroatom, the main effect, so far as the π electrons is concerned, arises from the corresponding change $\delta\alpha_\mu$ in the Coulomb integral. This leads to a change δE_μ in the energy of the MO given by

$$\delta E_\mu = a_{\mu r}^2 \, \delta\alpha_i \qquad (6.10)$$

The usual heteroatoms (N, O, F, etc.) are more electronegative than carbon, so $\delta\alpha$ for them is negative; introducing such an atom in place of carbon therefore lowers the energy of each MO by an amount proportional to the density of the MO at the position in question. The effect on the energies of the HOMO (Φ_μ) and LUMO (Φ_ν), and so on the lowest $\pi \to \pi^*$ transition energy δE, is indicated in Fig. 6.12.

In the case where the hydrocarbon is an even AH, the coefficients $a_{\mu r}$ and $a_{\nu r}$ will differ at most in sign (pairing theorem). Consequently, $a_{\mu r}^2 = a_{\nu r}^2$, so both MOs are depressed equally in energy by the heteroatom. The lowest $\pi \to \pi^*$ transition should therefore remain unchanged. The first $\pi \to \pi^*$ absorption bands of even, alternant, heteroconjugated molecules should therefore appear in the same place as those of the isoconjugate AHs—and they do. The UV spectra of benzene and pyridine (Fig. 6.13) illustrate this well, the first absorption maxima being at 254 and 258 nm, respectively.

Heteroatoms, however, also possess lone pairs of electrons which, in the case of a heteroconjugated molecule, can undergo transitions ($n \to \pi^*$) into the lowest empty π MOs. The AO occupied by such a lone pair is usually higher in energy than even the highest occupied π MO, so the first $n \to \pi^*$ transition usually appears at lower energies (longer wavelengths) than the

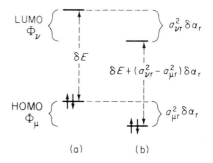

FIGURE 6.12. Effect of replacing atom r in a conjugated hydrocarbon by a heteroatom; (a) HOMO and LUMO in the hydrocarbon; (b) HOMO and LUMO in the heteroconjugated system.

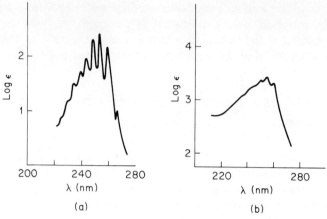

FIGURE 6.13. Ultraviolet spectra of (a) benzene and (b) pyridine.

first $\pi \to \pi^*$ transition (Figure 6.14). In such cases, the photochemistry of the molecule is usually dominated by the $n \to \pi^*$ transition. Pyridine is one of the exceptions, the $n \to \pi^*$ transition occurring at almost the same wavelength as the first $\pi \to \pi^*$ transition and being hidden by it. Benzaldehyde (PhCHO) is more typical. The first (and very intense) $\pi \to \pi^*$ transition occurs at the same wavelength (244 nm) as that of the isoconjugate AH (styrene, PhCH=CH$_2$) but there is also a weaker band at 328 nm due to an $n \to \pi^*$ transition involving lone-pair electrons of the oxygen atom.

6.10 $\pi \to \pi^*$ Transitions in Odd, Heteroconjugated Systems

The majority of odd, heteroconjugated systems are isoconjugate with carbanions. Here introduction of heteroatoms has very drastic effects on the spectrum, effects which have been used extensively in the dye industry in general, and the photographic dye industry in particular, to modify the

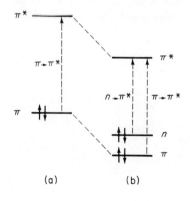

FIGURE 6.14. Orbital energies and transitions in (a) an even AH, and (b) an isoconjugated system containing a heteroatom.

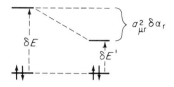
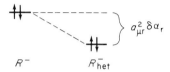

FIGURE 6.15. MO energies in an odd AH anion R^- and an isoconjugate system R^-_{het} containing a heteroatom at an unstarred position.

light absorption of dyes. As we shall now see, these effects can be accounted for in a very satisfactory manner in terms of two simple rules that follow at once from the PMO method.

Rule 1. Replacement of an unstarred atom in an odd AH anion by a heteroatom lowers the energy of the first absorption band.

The first absorption band of an odd AH anion is due (Fig. 6.11c) to a transition between the NBMO and the lowest unoccupied MO (LUMO). The NBMO vanishes at unstarred atoms, so the introduction of a heteroatom at those locations will have no effect on the NBMO but it will lower the energy of the LUMO (Φ_μ) by $-a^2_{\mu r}\,\delta\alpha_r$. The result (Fig. 6.15) is a decrease in the energy of the first absorption band. In the case of dyes, there are a number of comparisons which illustrate this point; e.g.,

Michler's Hydrol Blue
$\lambda_0 = 610$ nm

Bindschedler's Green
$\lambda_0 = 740$ nm

Acridine Orange, $\lambda_0 = 491$ nm

$\lambda_0 = 546$ nm

[Structure: quinolinium-CH=CH-C6H4-NMe2 with starred atoms]

$\lambda_0 = 530$ nm

[Structure: quinolinium-CH=N-C6H4-NMe2]

$\lambda_0 = 580$ nm

In each case, replacement of an unstarred atom by a heteroatom increases the wavelength of the first absorption band. In each of these comparisons, the compound on the left is isoconjugate with an odd AH anion and the compound to its right is derived from it by replacement of an unstarred carbon atom by a heteroatom. In each case, the latter absorbs at longer wavelengths. Introduction of the heteroatom thus has a *bathochromic* effect, the term bathochromic implying displacement of the first absorption band to longer wavelengths.

Rule 2. Replacement of a starred atom in an odd AH anion by a heteroatom generally increases the energy of the first absorption band.

Here again, the heteroatom will alter the energy of the individual orbitals by $-a_{\mu r}^2 \delta\alpha_r$. This effect will be largest for the NBMO because the NBMO is confined to only half the atoms and thus a_{0r}^2 should tend to be larger than any other $a_{\mu r}^2$, the orbital Φ_μ extending over all the atoms. The net result will be an increase in the energy gap between the NBMO and the LUMO; see Fig. 6.16.

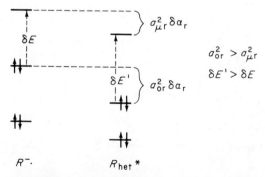

FIGURE 6.16. MO energies of an odd AH anion R^- and of an isoconjugate system R_{het}, containing a heteroatom of a starred position.

Analogs of carbanions containing heteroatoms at starred positions are usually neutral; cf. aniline, $PhNH_2$, with the isoconjugate benzyl anion, $PhCH_2^-$. Comparisons of the spectra of such pairs of compounds would be of little significance because those of the ions must be affected greatly by solvation and the nature of the gegenion.* We can, however, study the effects of heteroatoms in isoconjugate neutral systems, or cationic dyes, derived from the carbanions by the introduction of heteroatoms. Thus (the numbers are values of λ_0 in nm)

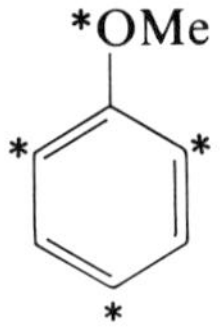

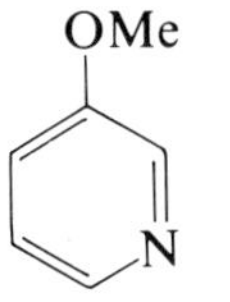

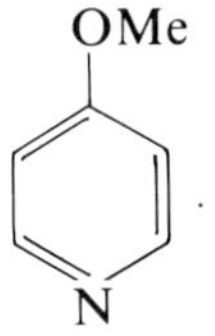

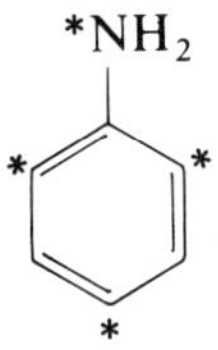

6.11 Effect of Substituents on Light Absorption

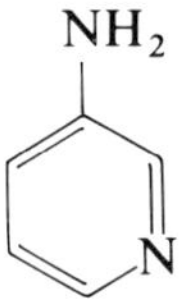

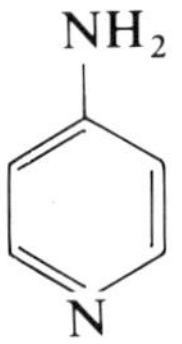

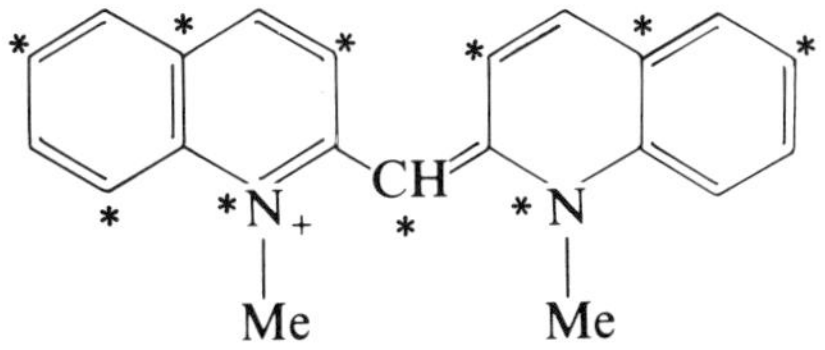

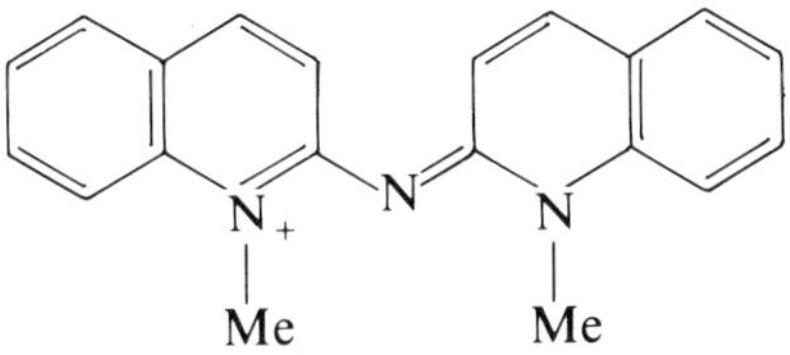

As we have seen, substituents can be divided into two types, inductive and electromeric. Substituents of the former type do not alter the size of a conjugated system but act only by changing the electronegativity of the atom to which they are attached. The corresponding changes in α will bring about changes in light absorption qualitatively similar to those produced by the replacement of carbon atoms by heteroatoms. In the case of a $+I$ substituent (e.g., CF_3), the changes in light absorption will be exactly parallel

---

* Our calculations refer to isolated molecules or ions in the gas phase. In solution, the excitation energy will be altered by the difference in solvation energy between the ground state and the excited state. In the case of an ion, this may well be large, the solvation energies of ions being of the order of 50 kcal/mole.

to those produced by a heteroatom such as N or O. In the case of a $-I$ substituent (e.g., CH_3), the changes will be in the opposite direction.

These effects can be seen very clearly in the cyanine dyes, where subtle changes in light absorption can be brought about by inductive substituents at the terminal positions. Since these are starred positions, Rule 2 applies. A $+I$ substituent there should make the adjacent starred atom more electronegative and so have a *hypsochromic* effect (displacement to shorter wavelengths, higher energies), while a $-I$ substituent should have a bathochromic effect. A typical example is the following (numbers are values of λ_0 in nm):

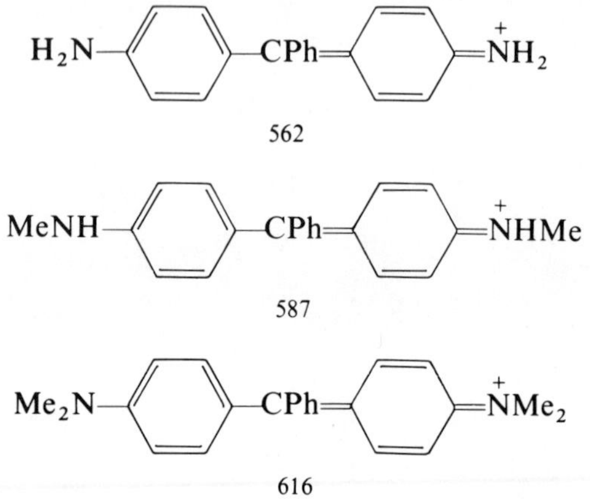

H_2N—⟨⟩—CPh=⟨⟩=$\overset{+}{N}H_2$

562

MeNH—⟨⟩—CPh=⟨⟩=$\overset{+}{N}HMe$

587

Me_2N—⟨⟩—CPh=⟨⟩=$\overset{+}{N}Me_2$

616

Introduction of methyl substituents into the starred NH_3 groups produces a bathochromic effect. By using groups with varying inductive activity, the light absorption of such a dye can be controlled very closely. This is important in color photography, where the accuracy of color reproduction depends on the use of dyes with exactly the right shades.

These conclusions can be summarized in our third rule:

Rule 3. Introduction of a $+I$ substituent has qualitatively the same effect on light absorption as replacing the corresponding carbon atom by a heteroatom, while a $-I$ substituent has the opposite effect.

For the purpose of light absorption, electromeric substituents can be divided into two types. Those ($\pm E$, $+E$) with both filled and empty π MOs that can interact with π MOs of the substrate, and simple $-E$ substituents (NH_3, OCH_3, etc.) that have a pair of p electrons.

Since the interactions between MOs are greater, the closer together the MOs are in energy, it is clear that the interactions between a pair of bonding MOs, or a pair of antibonding MOs, will be greater than that between a bonding MO and an antibonding MO. If then the HOMO and LUMO of the substituent lie outside the HOMO and LUMO of the substrate (Fig. 6.17a), the interactions will squeeze the latter MOs together and so lower the transition energy. If, on the other hand, the HOMO or LUMO of the substituent lies between the HOMO and LUMO of the substrate

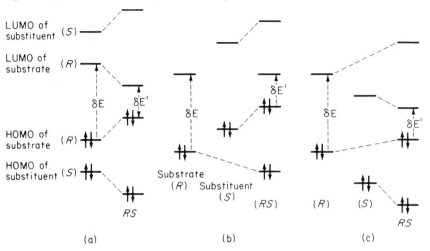

FIGURE 6.17. Interaction between the HOMO and LUMO of a substituent S and the HOMO and LUMO of a substrate R; (a) LUMO of substituent above LUMO of substrate and HOMO of substituent below HOMO of substrate; (b) HOMO of substituent between HOMO and LUMO of substrate; (c) LUMO of substituent between HOMO and LUMO of substrate. In all cases the transition energy $\delta E'$ for the substituted system is less than that δE for the unsubstituted system.

(Fig. 6.17b,c), then there will be a low-energy transition between the perturbed HOMO of the substituent and the perturbed LUMO of the substrate (Fig. 6.17b), or between the perturbed HOMO of the substrate and the perturbed LUMO of the substituent (Fig. 6.17c). In either case, the transition energy will be lowered by the substituent. Hence we have the following rule:

Rule 4. A $\pm E$ or $+E$ substituent lowers the transition energy. Some examples follow (numbers are values of λ_0 in nm):

311   358   254   269

608

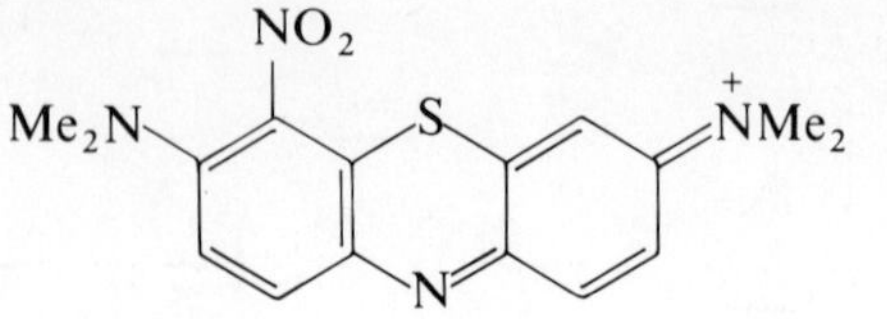

658 (structure)

The only place where this rule fails is in the case of simple derivatives of benzene such as aniline (PhNH$_2$) or phenol (PhOH) that are isoconjugate with the benzyl anion (PhCH$_2^-$). Thus while the nitrophenols do follow the rule, absorbing at longer wavelengths than phenol, the cyanoanilines do not; namely (λ_0 in nm):

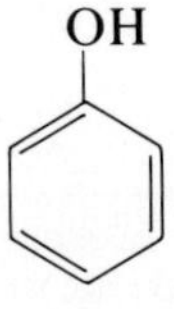

PhOH: 270
o-nitrophenol: 345
m-nitrophenol: 330
p-nitrophenol: 311

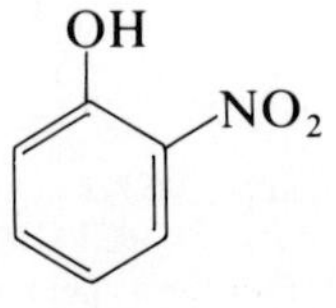

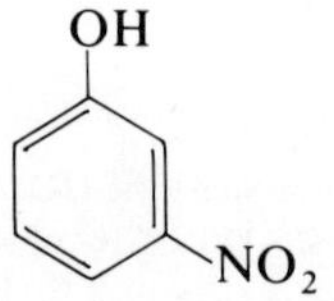

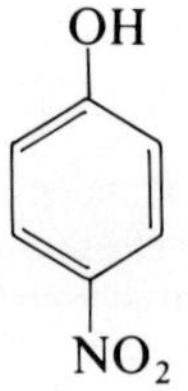

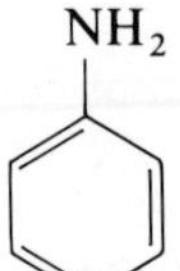

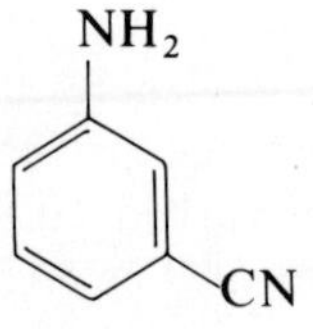

PhNH$_2$: 280
m-cyanoaniline: 236
p-cyanoaniline: 270

This is because replacement of a starred atom in a carbanion greatly perturbs the π system, the resulting structure having a pair of electrons more or less localized on the heteroatom (cf. p. 119). Usually this perturbation does not qualitatively affect our conclusions concerning the effect of an added $\pm E$ or $+E$ substituent; however, benzene is exceptional in having a degenerate LUMO and this gives rise, in disubstituted benzenes, to unusually large higher-order perturbations which are neglected in the present treatment.

We are left with simple $-E$ substituents where the substituent has a filled p AO that interacts with an adjacent π system but where there are no empty orbitals on the substituent.

It is immediately obvious that such a substituent will normally have a bathochromic effect on an adjacent π system since either the AO of the substituent lies between the HOMO and LUMO of the substrate (Fig. 6.18a; cf. Figure 6.17b), so that there will be a low-energy transition between it and the LUMO, or the AO lies below the HOMO of the substrate, in which case the interaction between it and the HOMO will be greater than between

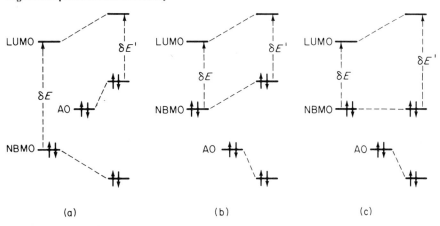

FIGURE 6.18. Effect of a simple $-E$ substituent on the energy δE of the first transition of an odd AH anion; (a) when the AO of the substituent lies between the LUMO and NBMO (HOMO); (b) when the AO lies below the NBMO and the substituent is attached to a starred atom; (c) when the AO lies below the NBMO and the substituent is attached to an unstarred atom. In (a, b) the transition energy is lowered ($\delta E > \delta E'$), while in (c) it is raised ($\delta E < \delta E'$).

it and the LUMO (Fig. 6.18b; cf. Fig. 6.17a). The only exception occurs when the substituent is attached to an inactive position in an odd AH anion. In this case, it will have no effect on the energy of the NBMO but will still raise that of the LUMO (Fig. 6.18c). Hence we have the following rules.

Rule 5. A $-E$ substituent in an even AH or at a starred position in an odd AH anion has a bathochromic effect.

Rule 6. A $-E$ substituent at an unstarred position in an odd AH anion has a hypsochromic effect.

Some examples follow (values of λ_0 in nm):

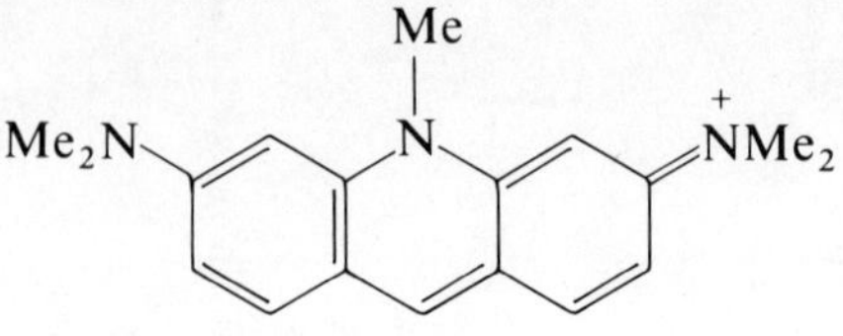

491 — structure with N-Me bridge, Me₂N and =NMe₂⁺ groups

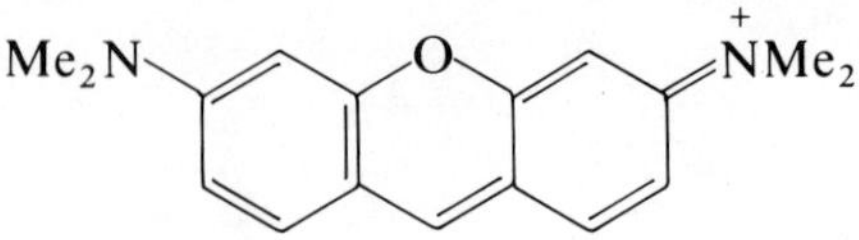

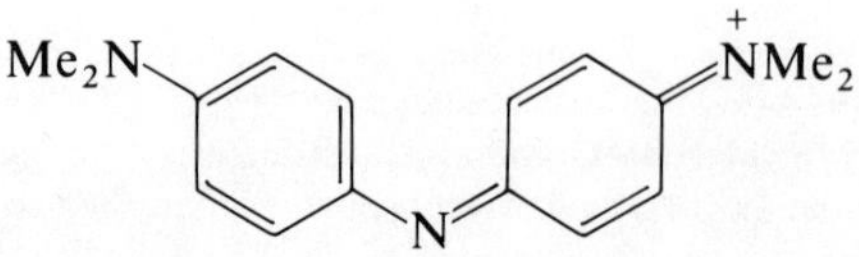

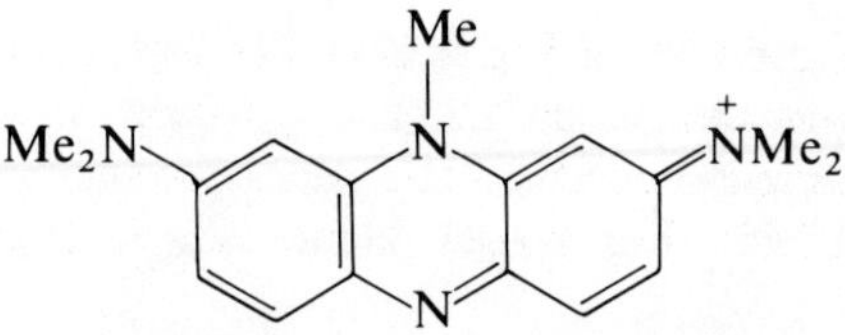

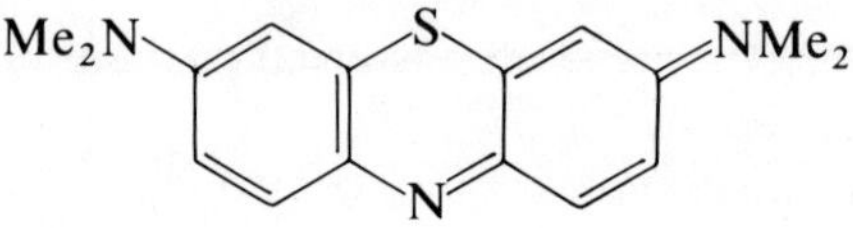

548 — structure with O bridge

700 — structure with N bridge (open)

564 — structure with N-Me and N bridges

608 — structure with S and N bridges

Here again the rules may fail in disubstituted benzenes; e.g.,

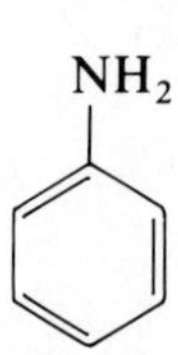

- 280: PhNH₂
- 293: 1,3-diaminobenzene
- 286: 3-aminophenol
- 270: phenol

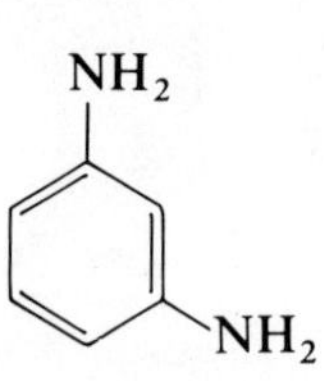

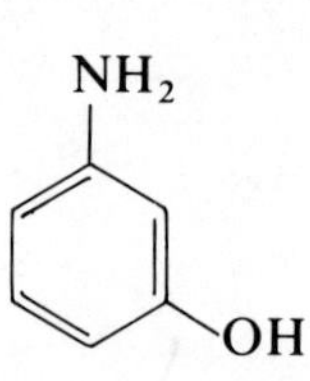

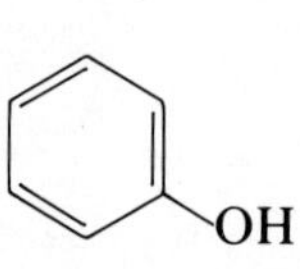

6.12 Basic Principles of Photochemistry; Types of Photochemical Processes

Photochemical reactions represent processes in which light energy is converted to chemical energy, either leading to reactions that could not occur thermally because too much activation would be required, or bringing about thermal-type reactions in an alternative and more expeditious way. There is of course a basic problem, i.e., the short lifetimes of excited states. Unless an excited state reacts very rapidly indeed, it will simply lose its excess energy by fluorescence or internal conversion and so revert to the original reactant. This restriction is particularly severe in the case of singlet excited states since their lifetimes are usually so very short. Photochemical reactions involving them occur only if they involve little or no activation. The lifetimes of triplet states, though longer, are still short by chemical standards. Here again photochemical reactions will be observed only if their activation energies are low.

Reactions of excited states, i.e., photochemical reactions, can be divided into three main classes.

1. Photosensitized Electron Transfer and Excitation Transfer. For an electron transfer to occur between two species A and B,

$$A + B \longrightarrow A^+ + B^- \tag{6.11}$$

it is necessary that the first ionization potential of A should be less than the first electron affinity of B. In orbital terms, these can be linked to orbital energies (Section 4.4) since the first ionization potential is approximately equal to *minus* the energy of the HOMO and the first electron affinity to *minus* the energy of the LUMO. The feasibility of the reaction of equation (6.11) there depends on the relative energies of the HOMO of A and the LUMO of B.

The situation is, however, drastically changed if we photoexcite A or B (Fig. 6.19). In the lowest singlet (S_1) and triplet (T_1) states, A now has an electron in what originally was its LUMO; this electron can be removed very much more easily than any of the electrons in the ground state (S_0).

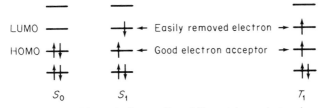

FIGURE 6.19. The excited states S_1 and T_1 contain an electron in a high-energy MO and so more easily removed than one from the HOMO of S_0; they also contain a low-energy hole that acts as a better acceptor than the LUMO of S_0.

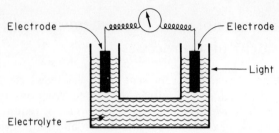

FIGURE 6.20. The photovoltaic effect. Irradiation of one electrode by light absorbed by the dissolved, colored redox agent leads to a potential difference between the electrodes, due to the change in redox potential on photoexcitation.

Equally, the original HOMO of B will be only singly occupied in the corresponding states S_1 or T_1; the electron affinity corresponding to filling of this hole will clearly be much greater than that for the LUMO in the ground state. Thus even if the reaction of equation (6.11) fails to proceed thermally, it may very well do so if either A or B is photoexcited. Electron transfers of this kind are of course very fast, being in effect CT transitions of a special kind.

Photoexcitation can therefore alter the effective redox potential of an oxidizing or reducing agent. Effects of this kind are now well established. The simplest example is provided by the so-called *photovoltaic effect* (Fig. 6.20). If we place two platinum electrodes in a solution of a colored compound that can undergo redox reactions and shine light on one of the electrodes, a potential is set up due to the change in redox potential of the solute near the irradiated electrode.

Another example is provided by the reversible redox reaction between Methylene Blue (A) and ferrous ion, forming the colorless leuco base (B) and ferric ion,

$$\text{(A)} + 2\text{Fe}^{2+} + \text{H}^+ \rightleftharpoons \text{(B)} + 2\text{Fe}^{3+} \quad (6.12)$$

The reaction does not go to completion, so adding a ferrous salt to a solution of Methylene Blue weakens the blue color of the dye but does not destroy it. If, however, we shine light on the solution, the dye becomes a stronger oxidizing agent and the equilibrium is pushed to the right. Sufficiently strong illumination bleaches the solution completely. When the light is turned off, the equilibrium reasserts itself and the blue color returns.

A third example of such a reaction, and one of considerable practical importance, is the so-called tendering of fabrics by light. Fabrics dyed with certain dyes disintegrate on exposure to light, a phenomenon which caused some concern to manufacturers of curtains when it was first observed. The effect seems to be due to a photosensitized oxidation of water or hydroxide ion by electron transfer, leading to hydroxyl radicals and hydrogen peroxide which attack the fabric.

It is evident from Fig. 6.19 that the change in effective redox potential will be greater, the greater is the separation of the HOMO and LUMO and so the greater is the energy of the first $\pi \rightarrow \pi^*$ transition. The dyes that show tendering action all absorb at the short-wavelength of the visible spectrum, the worst being those which absorb in the extreme violet and are consequently lemon yellow in color.

2. Reactions Taking Place in the Excited State. Excited states differ in their electronic structure from ground states because one of the electrons is now in a different MO. Their chemical behavior should be correspondingly different. When a molecule is photoexcited, it may undergo reactions which would not occur in the ground state and at the same time it may retain an electron in a high-energy MO. The product of such a reaction will be the excited form of some molecule or molecules that differ from the reactants. Deexcitation can then occur in the usual way by fluorescence, phosphorescence, or internal conversion. This type of reaction does not make effective use of light energy as such; it depends rather on the change in electronic structure brought about by photoexcitation. Most of the excitation energy is thrown away at the end in the form of light or heat. Excitation, reaction, and deexcitation are distinct and take place independently in successive steps. A good example of such a process is the photochemical ketonization of methyl salicylate that occurs on irradiation with UV light,

$$\text{(6.13)}$$

The reaction is endothermic in the ground state but exothermic in the excited state. The product is certainly first formed in its excited state because fluorescence is observed, the fluorescence being different from that for the ground state (which can be deduced from the absorption spectrum; see Fig. 6.6).

3. Reactions Leading Directly to the Product in Its Ground State. The third type of reaction is one in which deexcitation occurs during the reaction and forms an essential part of it. A simple example is provided by the photochemical dimerization of an olefin

$$\| + \|^* \longrightarrow \square \qquad (6.14)$$

It is very easily shown that the reaction cannot possibly lead to photoexcited cyclobutane because such a process would be strongly endothermic. Deactivation must occur during the reaction. One can also rule out an internal conversion of the photoexcited olefin to a "hot," vibrationally excited olefin followed by a normal thermal dimerization, because such dimerizations do not occur thermally. The reaction must therefore involve attack by an electronically excited molecule of olefin on another molecule of olefin, deexcitation occurring during the reorganization of bonds to form the cyclobutane.

4. **Photothermal Reactions.** The argument in the previous paragraph indicates another possible way in which light energy could be used, i.e., as a source of thermal energy to overcome activation barriers. A good example of this is provided by the reactions of carbene, formed by photolysis of diazomethane or ketene in the gas phase,

$$CH_2N_2 + h\nu \longrightarrow CH_2 + N_2 \qquad (6.15)$$

$$CH_2CO + h\nu \longrightarrow CH_2 + CO \qquad (6.16)$$

The initial reaction forms a nonclassical molecule (carbene) and it can be carried out over a range of wavelengths. The reactions are exothermic and the excess energy ends up in the form of kinetic energy of the products. The reactions of the resulting carbene consequently depend on the frequency of the light used in its generation, since this determines how much excess energy it possesses.

While the secondary reactions are of interest, they do not concern us here because they are essentially thermal ground-state reactions. Moreover, the phenomenon is of importance only in the gas phase because in solution, the "hot" molecules lose their excess energy extremely rapidly.

6.13 The Role of the Born–Oppenheimer (BO) Approximation

The foregoing discussion has described reactions that occur in real systems. It has assumed that the Born–Oppenheimer (BO) approximation (p. 16) holds throughout, i.e., that one can treat electronic energy and vibrational energy separately. If this were rigorously true, no interconversion of electronic and vibronic energy would be possible and the various kinds of conversion of electronic to chemical potential listed above would not occur. The fact that they occur is evidence that the BO approximation does not hold rigorously; however, it does hold to a very good approximation because excited states are generally distinct states. This is seen from the fact that so many molecules exhibit fluorescence. An excited molecule can fluoresce only if it survives long enough without deexcitation. Now the lifetimes of excited states, while apparently short by conventional standards ($10^{-7} - 10^{-10}$ sec), are very long compared with the time involved in a molecular vibration ($\sim 10^{-13}$ sec). An enormous number of molecular vibrations must therefore take place before the electronic energy is finally converted into vibronic quanta. The mechanism of the conversion is indeed the subject of photochemistry and it cannot be dismissed with wavy lines under the title of "internal conversion."

The distinctness of excited electronic states is due to the enormous disparity between the energy usually involved in an electronic transition and a single vibrational quantum. Internal conversion from the first excited state (S_1 or T_1) to the ground state involves replacement of one large (electronic quantum) by many smaller (vibronic) ones and it must involve a cascade of specific electronic states to be efficient. Internal conversion becomes progressively easier the smaller the amount of electronic energy that has to be converted. We have already used this argument to explain why higher excited states usually slide down to S_1 to T_1 too rapidly for them to play a role in photochemistry.

A thermal reaction is controlled by the shape of the potential surface representing the energy of the ground state as a function of its geometry. A photochemical reaction is controlled by the corresponding excited-state potential surface, representing the energy of the excited state as a function of its geometry, as well as the ground-state surface to which the system must eventually return. If the two surfaces remain far apart, return to the ground-state surface by internal conversion will be relatively slow at every point. The photochemical reaction will then be analogous to a ground-state reaction, the initially excited molecule passing over to a lower minimum on the excited-state surface, perhaps by surmounting an intervening potential barrier. Eventually, fluorescence or internal conversion will occur. The excited state then reverts to the ground state without change in geometry (Franck–Condon principle) so the product may be formed in a geometry which, for the ground state, is very peculiar. The product may therefore be entirely different from that formed in a thermal reaction because it corresponds to a minimum in the excited state, not in the ground state. The situation is illustrated pictorially in Fig. 6.21. Here the reactant A would normally be converted to B, via the low-lying transition state X. If, however, we photoexcite A, the corresponding point A' is on the side of an incline leading down to C'. The excited state therefore runs downhill to C', where it is trapped long enough for internal conversion or fluorescence to occur. It then falls vertically to C. This happens to be a minimum on the ground-state surface but one of high energy and with a low barrier Y separating it from a second deep minimum D. The thermal conversion of C to D will therefore be rapid. Thus the photochemical reaction converts A to D instead of B, even though D is less stable than B and is separated from A by a much higher energy barrier.

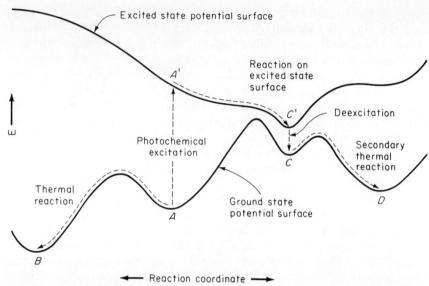

FIGURE 6.21. A photochemical reaction taking place on the excited-state potential surface can lead to a product different from that of a corresponding thermal reaction (recall that this is a two-dimensional representation of a multidimensional surface).

The point of this argument is that the potential surfaces for the ground state and for the excited state may be different in shape because one of the electrons in the molecule occupies very different MOs. In certain cases, the difference in shape may become so extreme that the surfaces touch at some point. This will happen if the HOMO and LUMO become identical in energy so that the difference in energy between them i.e., the excitation energy, vanishes (Fig. 6.22). When the excitation energy becomes very small, the probability of internal conversion becomes very large because we are no longer dealing with quanta of different sizes. Interchange of vibronic and electronic energy can then become very rapid so that we can no longer distinguish between them, and the BO approximation consequently fails. The rate of interchange will become comparable with the rate of interchange of energy between different vibrational modes. This is extremely rapid; indeed, the rapidity of such processes is responsible for the fact that vibrationally excited excited states usually cascade down to their vibronic ground states before any other process, e.g., emission, can occur.

This argument shows that an entirely new situation can occur when a photochemical reaction takes place in a system where the excited-state surface and the ground-state surface come very close together at some point, the separation being comparable with the energy of a single vibrational quantum. At this point, the reaction can pass over from one surface to the other in a time of the order of a single vibration, i.e., in a time of the order of that involved in a typical chemical reaction. Such a point, where the surfaces come close together and the BO approximation consequently breaks

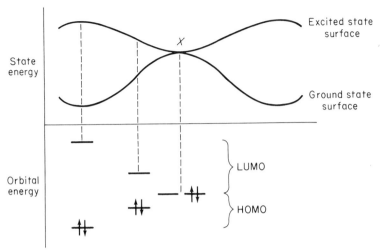

FIGURE 6.22. As the HOMO and LUMO move together in energy in one dimension, so do the potential surfaces for the ground state and excited state. At X, the MOs become degenerate, so the surfaces touch.

down, will therefore act as a flaw or hole in the upper excited-state surface through which the reaction can leak through to the ground-state surface below. Such a reaction can lead directly from the excited state of the reactant to the ground state of the product in a single step without any need for separate deexcitation. Indeed, deexcitation forms an essential part of the reaction and the reaction coordinate must be explicitly involved with the deexcitation. This is illustrated in Fig. 6.23. The *BO hole* (X) acts as a trap for the unwary reactant, through which it falls to the product B. If the hole were

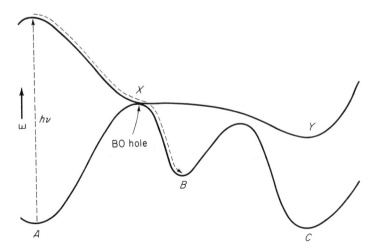

FIGURE 6.23. Trapping of the excited state in a reaction by a "BO hole" where the excited-state and ground-state potential surfaces touch or nearly touch.

not there, the reactant would have gone on to the lower minimum Y in the potential surface and so given a different product C.

The condition for the existence of such a BO hole is that the HOMO and LUMO should become degenerate, or very nearly degenerate. There is therefore an important qualitative distinction between reactions where the surfaces remain well apart, so that the BO approximation holds throughout, and those where they touch or nearly touch at some point, so that the BO approximation may fail there. Note the phrase "may fail." It is of course entirely possible that the transition from the excited state to the ground state at a point such as X in Fig. 6.23 may be forbidden for other reasons. In this case, the reaction will behave "normally," leading to a minimum in the excited-state potential surface from which the product is eventually formed by deexcitation.

We can sum up these conclusions in two simple rules:

Rule 1. If a photochemical reaction takes place in a system where the potential surfaces for the excited state and ground state remain well separated throughout, or where transitions between them are forbidden, the reaction will take place entirely on the excited state surface and will lead to excited products.

Rule 2. If a photochemical reaction takes place in a system where the potential surfaces for the excited state and ground state come very close together at a point which is accessible and corresponds to an allowed transition, the reaction will pass from the excited-state surface to the ground-state surface at that point and so will lead directly to ground-state products.

Note a further qualification in Rule 2, the word "accessible." A BO hole will obviously play no part in a reaction if the entry to it lies too far uphill from the point on the excited-state surface corresponding to the original excitation. In practice, however, BO holes, when they occur, are usually accessible.

6.14 The Role of Antibonding Electrons

In our simple MO treatment, the bond in a simple diatomic molecule X—X arises from an interaction between a pair of AOs when the atoms are close together. This leads to replacement of the two AOs by two MOs, one bonding and one antibonding, separated in energy by $\pm \beta$ (the resonance integral) from the AOs (Fig. 1.10). According to this picture, the bond energy of the bond in X_2, and the transition energy for excitation of one of the bonding electrons into the antibonding MO, should both be 2β. This is not in fact the case.

Consider, for example, the π bond in ethylene, which is formed by just such a two-center interaction of AOs (Section 1.13). The bond energy of the π bond in ethylene is about 2 eV, but the transition energy for the $\pi \to \pi^*$ transition is 7.6 eV, nearly four times as great.

The reason for this can be seen if we consider the distribution of the electrons in the molecule. Each electron occupies a MO ψ given by

$$\psi = (1/\sqrt{2})(\phi_1 + \phi_2) \tag{6.17}$$

where ϕ_1 and ϕ_2 are the two $2p$ AOs. Since the MO is symmetric, the electron spends half its time near nucleus 1 and half near nucleus 2. If we were to pull the atoms apart without altering the electron distribution, β would decrease and eventually become very small because ϕ_1 and ϕ_2 would no longer overlap. We would then in effect have two separated carbon atoms, so the change in energy would be 2β. However, the atoms would not be normal carbon atoms, each with a whole electron in its $2p$ AO, but queer pseudoatoms each with two "half-electrons" in its $2p$ AO, because the wave function of equation (6.17) represents a state in which each electron spends half its time in each, so in each case each electron is half in ϕ_1 and half in ϕ_2. The repulsion between the two "half-electrons" in each AO will be $\frac{1}{4}\gamma$, where γ represents the total repulsion between two whole electrons in a carbon $2p$ AO. The energy of each pseudoatom is therefore $\frac{1}{4}\gamma$ greater than that of an ordinary carbon atom (Fig. 6.24). As a result, the effective bond energy of the π bond in ethylene is not 2β, but $2\beta - \frac{1}{2}\gamma$. The value of γ (11 eV) can be estimated from spectroscopic data for carbon; in this way one can construct the energy level diagram of Fig. 6.24. It will be seen that the bonding MO, while lying β (i.e., 3.8 eV) below the pseudoatom level, lies only about 1 eV below the level corresponding to normal carbon atoms. The bond energy found in this way (2 eV) thus corresponds to experiment.

It will be seen from this argument that the antibonding MO is far more antibonding than the bonding MO is bonding. A corollary of this is that the

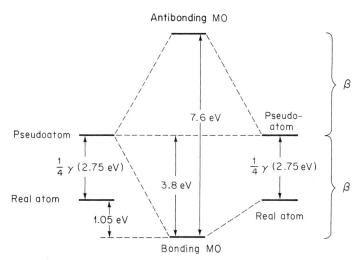

FIGURE 6.24. Energy level diagram for π MOs of ethylene.

interaction between two filled AOs on adjacent atoms is not merely nonbonding, as the argument of Section 1.7 implied, but antibonding. Two helium atoms do not merely fail to attract one another; on the contrary, they resist strongly attempts to bring them together. Likewise the three-electron bond in HeH is extremely weak, the third electron, in an antibonding MO, nearly neutralizing the bonding effect of the other two. The repulsion between lone-pair electrons is also seen in the bond energies (kcal/mole) of homopolar single bonds between second row elements,

$$C-C, 85; \quad N-N, 60; \quad O-O, 50; \quad F-F, 30$$

Other things being equal, one would expect the bond energy of a bond to increase with increasing electronegativity of the bonded atoms because the attraction between the overlap cloud (see Section 1.5) and the nuclei, and so β, should be greater, the greater is the affinity of the two nuclei for electrons. Indeed, bond energy *increases* in the series of bonds to hydrogen,

$$C-H, 100; \quad N-H, 103; \quad O-H, 119; \quad F-H, 153$$

The decrease in bond energy for homopolar bonds is due to an increase in the number of repulsive lone-pair interactions, carbon having no lone pairs, nitrogen one, oxygen two, and fluorine three.

A molecule with an electron in an antibonding MO is therefore in a very unhappy situation and will be anxious to undergo any reaction that will allow the electron to revert into an orbital of lower energy. This tendency of electrons to try to escape from antibonding MOs plays an important role in determining the course of photochemical reactions.

6.15 Classification of Photochemical Reactions

The discussion in Section 6.14 showed that photochemical reactions can be divided into three main categories. First, there are reactions involving electron transfer. Second, there are reactions that take place on the excited-state surface and lead initially to excited products. Third, there are reactions that lead directly from excited reactants to products in their ground states, deexcitation occurring during the reaction and forming an integral part of it. In this section, we will consider this classification in detail and the factors that control the course of reactions of each type.

E-Type reactions involve electron transfer. Electron transfer reactions generally produce ground-state radical cations and radical anions which subsequently react on the ground-state surface. Thermal reactions of radical ions are discussed in detail in Chapter 7. The requirement for an electron transfer reaction is the existence of a vacant acceptor MO at a lower energy than the HOMO of the donor.

X-Type reactions take place entirely on the excited-state surface. The PMO description of such a process is analogous to that of ground-state reaction (Chapter 5), with one very important exception. During an X-type reaction, one electron remains throughout in an antibonding MO of high energy. Now, as we have already seen (Section 6.14), antibonding MOs are more antibonding than corresponding bonding MOs. Furthermore, this difference is due to the contributions of the electron repulsion integrals γ (Fig. 6.24), which are neglected in the PMO approach. In discussing X-type reactions, one must allow for these neglected effects. In particular, processes that lead to a decrease in the antibondingness of the antibonding electron will be very strongly favored in comparison with analogous ground-state reactions.

Consider, for example, the way the energies of the ground state and of the first ($\pi \to \pi^*$) excited state of ethylene change as we twist the molecule about the CC bond. According to our simple MO theory, the π bond in ethylene arises from an interaction between the $2p$ AOs of the two carbon atoms. The AOs split into a bonding MO and an antibonding MO, spaced in energy from the $2p$ AOs by β (Fig. 6.25a), β_0 being the π resonance integral.

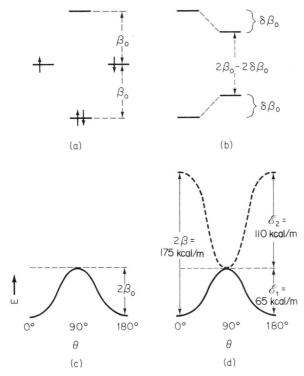

FIGURE 6.25. (a) Formation of π MOs by interactions of $2p$ AOs in ethylene; (b) effect of a change $\delta\beta_0$ in β_0 on the energies of the π MOs; (c, d) plots of energies of ground (———) and $\pi \to \pi^*$ excited (– – –) states of ethylene as a function of the angle of twist θ about the CC bond (c) with energies calculated by simple MO theory and (d) with experimental values for the energies.

One would therefore expect both the bond energy and the $\pi \to \pi^*$ excitation energy to be $2\beta_0$. If the decrease in β_0 is $\delta\beta_0$, the change in the energy of the ground state will then be $2\delta\beta_0$. However, the excitation energy will also change by $2\delta\beta_0$ (Fig. 6.25b). Thus the energy of the $\pi \to \pi^*$ state should remain unchanged. The variation of the energies of the ground state and excited ($\pi \to \pi^*$) state with the angle of twist θ should be as indicated in Fig. 6.25(c). This is not the case. The excitation energy is very much greater than the bond energy and decreases rapidly with θ. The energy of the excited state therefore also decreases with θ, the ground state and excited state surfaces touching at $\theta = 90°$ (Fig. 6.25d).

We have seen (Section 6.14) that the PMO resonance integral (β_0 in Fig. 6.25) is not the true carbon–carbon resonance integral (β in Figure 6.24), being equal to $\beta + \frac{1}{4}\gamma$. The true resonance integral β is much larger than β_0, being equal in this case (see Fig. 6.24) to 3.8 eV or 87.5 kcal/mole. This error does not, however, explain the difference between the potential curves of Fig. 6.23c,d, because if the electronic repulsions remained the same during the rotation, the excitation energy would still change in such a way as to compensate the changes in ground-state energy. The only effect would be to replace β_0 in Fig. 6.25(c) by β. The fact that the energy of the excited state decreases very rapidly with θ (Fig. 6.25d) must then imply that twisting reduces the electronic repulsions. In the 90° configuration, the two electrons that originally took part in the π bond are more or less confined to separate carbon $2p$ AOs. The repulsion between them is then much less than in ethylene itself. In the terminology of Section 6.14, breaking the π bond effectively replaces the two carbon "pseudoatoms" by more or less normal atoms of much lower energy. One can see that the effect is very large, from estimates of the energy differences in Fig. 6.25(d). The π-bond energy in ethylene is about 35 kcal/mole ($= 2\beta_0$). The difference in energy between normal ethylene in its ground state and ethylene twisted through 90° ($\mathscr{E}_1$ in Fig. 6.25d) is equal to the barrier to rotation about the C=C bond, i.e., to the activation energy for *cis–trans* isomerization. One cannot of course observe *cis–trans* isomerization in ethylene itself, but one can do so in the isotopically labeled 1,2-dideuterioethylene, CHD=CHD. The observed activation energy is 65 kcal/mole. Since $2\beta = 175$ kcal/mole (Fig. 6.24), the decrease ($\mathscr{E}_2$ in Fig. 6.25d) in the electronic repulsions on twisting the excited state through 90° is $(2\beta - \mathscr{E}_1)$, or 110 kcal/mole. This is a huge amount of energy, chemically speaking. As a result, photoexcitation of ethylene leads immediately to rotation through 90° about the double bond.

It is unfortunately impossible to estimate these changes in the electronic repulsions in any simple way. However, there seems to be a general tendency for them to follow bond energies in the same way that they do in ethylene. In other words, as the bonds in the ground state become weaker and its energy consequently increases, a corresponding decrease in the electronic repulsions tends to lower the energy of the excited state. *The excited-state and ground-state surfaces therefore tend to resemble mirror images of one another,* a maximum in one corresponding to a minimum in the other.

This of course means that reactions in the excited state will be very different from those in the ground state. Many processes which are unknown or occur with extreme difficulty as normal thermal reactions on the ground-state surface take place easily on the excited-state surface as photochemical reactions.

G-Type reactions are those that lead directly to products in their ground state. This can happen only if the ground-state and excited-state surfaces touch, or very nearly touch, at some point, thus providing a "BO hole" for direct passage from the excited state to the ground state (Section 6.13). At the BO hole, the excitation energy vanishes and this can happen only in one of five very specific situations. It is therefore convenient to subdivide G-type reactions into five classes, depending on the type of structure responsible for the BO hole.

G_R *Reactions.* In a G_R reaction, the BO hole corresponds to a pair of radicals or a biradical. The separation of the bonding and antibonding MOs in a diatomic molecule is proportional to the resonance integral β between their two AOs (Section 1.5). This in turn is proportional to the extent to which they overlap in space (Section 1.5). As the atoms move apart, β will therefore decrease, becoming zero when the molecule has dissociated completely. In the limit of complete dissociation, the ground-state and excited-state surfaces for the molecule consequently coincide (Fig. 6.26).

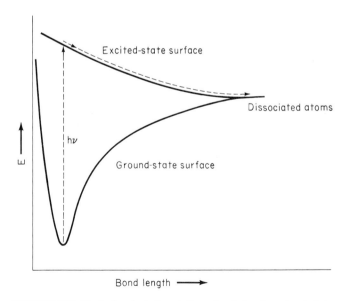

FIGURE 6.26. Photochemical dissociation of a diatomic molecule into atoms.

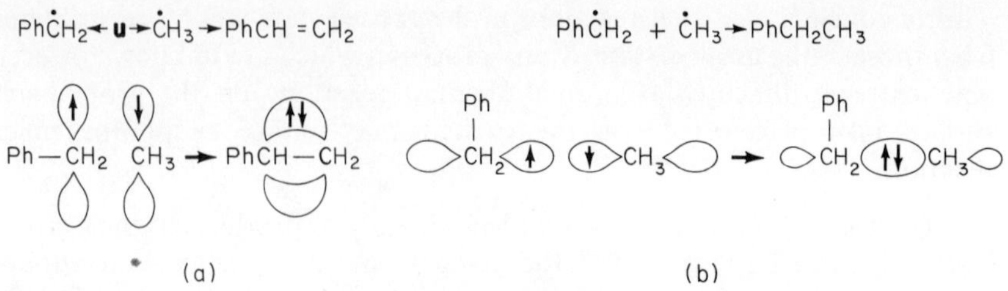

PhĊH₂ ← **u** → ĊH₃ → PhCH=CH₂   PhĊH₂ + ĊH₃ → PhCH₂CH₃

Ph—CH₂ CH₃ → PhCH—CH₂   >CH₂(↓) (↓)CH₃< → >CH₂(↑↓)CH₃<
                                  |Ph |Ph

(a)   (b)

FIGURE 6.27. The union (a) of benzyl and methyl is analogous to their combination (b).

If, therefore, we excite the molecule photochemically, it can dissociate into ground-state atoms whose subsequent reactions will take place on the ground-state surface. A very simple example of this is provided by the halogens, which dissociate into atoms on exposure to light. If the reaction takes place in the presence of other materials, one can observe ground-state reactions between them and the photochemically produced halogen atoms.

The same argument will apply to a "localized" σ bond in a saturated molecule. In this case dissociation gives rise to a localized radical or a pair of such radicals. The photochemical dissociation of an alkyl iodide RI into an alkyl radical R· and an iodine atom is a simple example. A further extension is seen if one or both of the new radical centers is adjacent to an even AH. Such a radical will be an odd AH in which the odd electron occupies a NBMO, i.e., an orbital of the same energy as an AO. In this case, dissociation is the reverse of an analog of union (Fig. 6.27). The situation differs from normal union only in the mode of overlap of the two interacting orbitals. Union of an odd AH with methyl gives rise to a product $>$C$=$CH₂ with a localized CC π bond; combination of the same radical with methyl gives rise to a product $>$CH—CH₃ with a localized CC σ bond.

The combination of radicals can therefore be treated, like union, in terms of an interaction between a pair of orbitals, either NBMOs or AOs. In the case of hydrocarbons, the orbitals will be degenerate; in the case of compounds containing heteroatoms, they will not be. However, the same situation will still hold, the relationship being similar to that between the homopolar molecule H_2 and the heteropolar ion HHe^+ (Sections 1.5 and 1.7).

The bond that is being broken may form part of a ring. If the rest of the ring is composed of saturated atoms, these will "insulate" the resulting radical centers from each other so that the product will behave just like a pair of separate radicals. Such a *biradical* can be produced photochemically, so this is another process in which excitation leads to products in their ground state. Indeed, since a pair of radicals, or a biradical, represents a point at which the ground-state and excited-state surfaces touch, any process by which an excited state leads to such a pair of radicals, or biradical, will provide a direct path from the excited state to the ground state. Two examples will illustrate this.

If we rotate stilbene through 90° about the central bond, the net overlap between the $2p$ AOs of the central atom vanishes because the positive and

Light Absorption and Photochemistry 433

negative regions of overlap cancel [as indicated in equation (6.18), the product contains two uncoupled benzyl systems]*

$$\begin{array}{c}\text{Ph}\diagdown\diagup\text{H}\\\text{C}=\text{C}\\\text{H}\diagup\diagdown\text{Ph}\end{array} \longrightarrow \begin{array}{c}\text{Ph}\diagdown\diagup\text{H}\\\text{C}-\text{C}\\\text{H}\diagup\diagdown\text{Ph}\end{array} \longleftrightarrow \left[\text{H}-\underset{\text{H}}{\overset{\text{Ph}}{\bigoplus}}-\text{Ph}\right] \quad (6.18)$$

Photoexcitation of acetophenone by an $n \longrightarrow \pi^*$ transition gives an excited state in which one of the oxygen AOs is singly occupied

$$\begin{array}{c}\text{Ph}\diagdown\\\text{C}=\text{O}\\\text{H}_3\text{C}\diagup\end{array} \xrightarrow{h\nu} \begin{array}{c}\text{Ph}\diagdown\\\text{C}-\text{O}\\\text{H}_3\text{C}\diagup\end{array} \xrightarrow[G_R]{HR} \begin{array}{c}\text{Ph}\diagdown\diagup\text{H}\quad\dot{\text{R}}\\\text{C}-\text{O}\\\text{H}_3\text{C}\diagup\end{array}$$

This radical center can abstract a hydrogen atom from some adjacent molecule RH to give a pair of radicals, both in their ground states.

The basic theme underlying all these reactions is that the excited state and ground state of a pair of radicals, or of a biradical, are similar in energy. Such a radical pair or biradical can therefore form a direct link between the excited-state surface and the ground-state surface.

G_I *Reactions.* A G_I reaction is an analog of a G_R reaction in which the product is a pair of ions (M^+N^-) instead of a pair of radicals ($M\cdot\cdot N$). A G_I reaction can therefore be regarded as a superposition of a G_R reaction (leading to radicals $M\cdot + \cdot N$) and an E reaction (leading to transfer of an electron from $M\cdot$ to $N\cdot$). The principles guiding such reactions are therefore exactly the same as those for G_R reactions. The distinction lies in the possibility of the electron transfer. Since the conversion of a pair of radicals to a pair of ions is usually endothermic, G_I reactions are rather rare.

G_N *Reactions.* A G_N reaction is one in which the BO hole corresponds to a nonclassical conjugated molecule, i.e., one for which no classical structure can be written. This is the case for even AHs where there are more starred than unstarred atoms (Section 3.14); e.g.,

$$\begin{array}{c}\overset{*}{\text{CH}}_2\\|\\\overset{*}{\text{H}_2\text{C}}\diagdown\overset{*}{\text{C}}\diagup\overset{*}{\text{CH}}_2\end{array} \qquad \begin{array}{c}\overset{*}{\text{CH}}_2\\\diagup\diagdown\\\text{(ring)}\\\diagdown\\\text{CH}_2\end{array}$$

---

* In the rest of this chapter, photochemical processes will be indicated by an arrow with $h\nu$ and the classification (X, G_R, etc) of the type of photochemical process involved.

As we have seen, such an AH possesses two NBMOs. Its HOMO and LUMO therefore have identical energies.

If heteroatoms are present, the degeneracy of the HOMO and LUMO is removed. However, the relative energies of the two resulting MOs will still be similar and since, moreover, they can be altered by geometric distortions, there will still be an accessible point on the potential surface where their energies coincide and where the ground-state and excited-state surfaces consequently touch.

As yet, few processes of this kind have been discovered. The simplest is the production of trimethylenemethane by photolysis of its cyclic azo derivative,

$$\text{[cyclic azo compound]} \xrightarrow[G_N]{hv} \text{[trimethylenemethane diradical]} + N_2 \qquad (6.19)$$

G_A *Reactions.* A G_A reaction is the photochemical analog of an antiaromatic pericyclic reaction (Section 5.28), the BO hole corresponding to the antiaromatic transition state. Since it is not immediately obvious that such a structure will be a BO hole, i.e., a point where the ground-state and excited-state surfaces touch, we must first establish that this is in fact the case.

The delocalized MOs of pericyclic transition states are analogous to the π MOs of cyclic, conjugated systems. We will therefore start by looking at the simplest antiaromatic even AH, i.e., cyclobutadiene. This can be constructed by union of allyl with methyl,

$$\text{allyl} \cdot + \text{methyl} \longrightarrow \text{cyclobutadiene} \qquad (6.20)$$

There is no change in first-order energy during union, because the net interaction between the NBMO of allyl and the AO of methyl vanishes. Since this interaction vanishes, the allyl MO must remain unchanged, so cyclobutadiene must have a NBMO. But cyclobutadiene is an even AH in which all the MOs are paired. Each NBMO must be paired with an MO of equal and opposite energy, i.e., another NBMO. Cyclobutadiene must therefore contain a pair of NBMOs. It is easily seen that the same argument will apply to any antiaromatic even AH with a single, symmetric $4n$-membered ring. The HOMO and LUMO of such a system are degenerate.

As we have seen (Section 3.11), antiaromatic rings show bond alternation because this reduces the antiaromatic energy. During union of methyl with an alternating odd AH radical to form a cyclic polyene with alternating bonds, the positive and negative interactions no longer cancel. The NBMOs now interact to form a bonding–antibonding pair of MOs. The distortion thus removes the degeneracy present in the symmetric ring.

We can represent this in another way. The reason there is no net interaction during the union of methyl and allyl in equation (6.20) is that one of

FIGURE 6.28. NBMOs of cyclobutadiene.

the new bonds involves a bonding interaction between the NBMO and AO, the other an antibonding interaction. Since the same argument applies to each atom in the ring, the NBMO must be alternately bonding and antibonding for successive bonds around the ring. The mirror image NBMO will then be one in which the bonding–antibonding roles are reversed (Fig. 6.28).

When we introduce bond alternation, we strengthen alternate π interactions and weaken the intervening ones. This will make the net effect bonding for one of the MOs Φ_1 and Φ_2 (Fig. 6.28) and antibonding for the other. Since there are only two electrons to be accommodated, the distortion is favorable energetically because both electrons can be housed in the bonding MO.

Now there are two alternate modes of distortion, i.e.,

$$\underset{A}{\square} \rightleftarrows \underset{B}{\square} \rightleftarrows \underset{C}{\square} \qquad (6.21)$$

In one of these (A), Φ_1 will be bonding and Φ_2 antibonding; in the other, Φ_2 will be bonding and Φ_1 antibonding. If, then, we convert A into C by a continuous distortion via the symmetric intermediate B, the MOs will cross (Fig. 6.29).

It is very easily seen that the same argument must apply in all cases

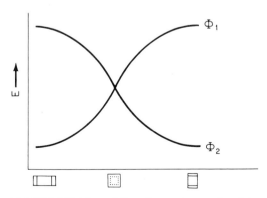

FIGURE 6.29. Effect of distortion on the energies of the two central π MOs in cyclobutadiene.

where there is an antiaromatic ring. There are always two classical structures for such a ring that are both more stable than the degenerate symmetric intermediate, and in passing from one to the other, the HOMO and LUMO must change places. There must therefore be some intermediate structure in which the two MOs are degenerate. At this point, the HOMO and LUMO are identical in energy, so the excitation energy vanishes. This structure therefore represents a point at which the ground-state surface and the excited-state surface touch. The argument does not depend on symmetry and it will hold equally well if heteroatoms are present. It is a basic consequence of antiaromaticity.

Now, as we have seen (Section 5.26), a pericyclic reaction involves a cyclic interchange of bonds which is precisely analogous to the interconversion of two classical structures for a cyclic polyene. In the case of an antiaromatic ("forbidden") pericyclic reaction, the transition state is an antiaromatic delocalized system. The argument given above shows that at some point during such a reaction, the HOMO and LUMO must cross, so that the potential surfaces for the ground state and for the excited state touch. One can go further. Suppose we make a BEP plot for the process indicated in equation (6.21). One line will represent the energy for a situation corresponding to A, i.e., where Φ_1 is filled and Φ_2 empty, and the other to B, where Φ_1 is empty and Φ_2 filled (Fig. 6.30). It is obvious that the crossing point, corresponding to the transition state of the thermal reaction, will be the point at which Φ_1 and Φ_2 cross. This will also be the point at which the excitation energy vanishes (Fig. 6.30) and where the reaction can con-

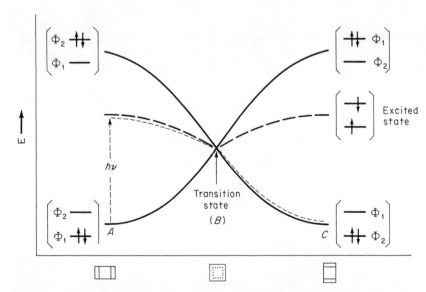

FIGURE 6.30. The transition state for an antiaromatic pericyclic reaction is a point where the excited-state and ground-state potential surfaces touch. The dotted line indicates a photochemical route from A to C.

Light Absorption and Photochemistry 437

sequently pass from the excited-state directly to the ground-state surface. The antiaromatic transition state therefore serves as a BO hole.

It should therefore be possible to bring about such processes photochemically by direct conversion of the excited form of the reactant to the ground state of the product via the BO hole corresponding to the antiaromatic transition state. A classic example is the photochemical formation of cyclobutenes from 1,3-butadienes by disrotatory ring closure; e.g.,

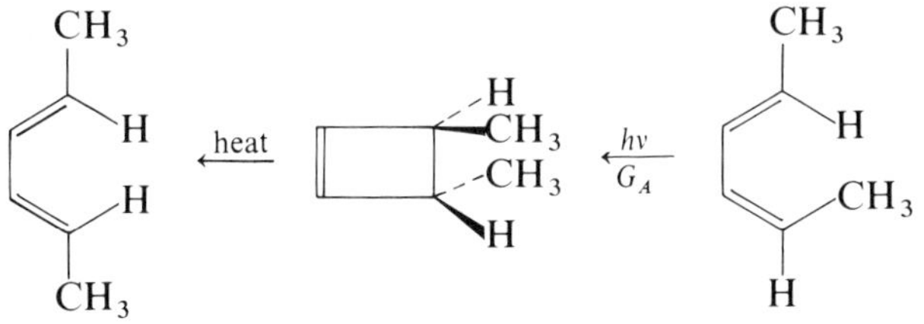

[structure: cis-1,3-pentadiene with CH₃ groups] ⇌(heat) [cyclobutene intermediate with H, CH₃, CH₃, H] ⇌($\frac{hv}{G_A}$) [diene product with CH₃, H, CH₃, H]

The (reverse) thermal reaction is of course conrotatory (Section 5.26).

G_S *Reactions.* A G_S reaction is one in which the degeneracy of the ground-state and excited-state surfaces arises through solvation. If the excited state is more polar than the ground state, its energy will be lowered correspondingly more by a polar solvent. A polar solvent will therefore lead to a decrease in the separation between the two surfaces and if the initial separation was sufficiently small, the result may be a BO hole.

Although there is a strong physical basis for expecting solvation-generated BO holes, the number of examples of reactions of this type that have been clearly identified as such is extremely limited.[1a-c] Reactions (6.22a)[1a] and (6.22b)[1b] are both bimolecular processes in which the excited forms of the transitions states should be more polar than the corresponding transitions state for reaction on the ground-state surface:

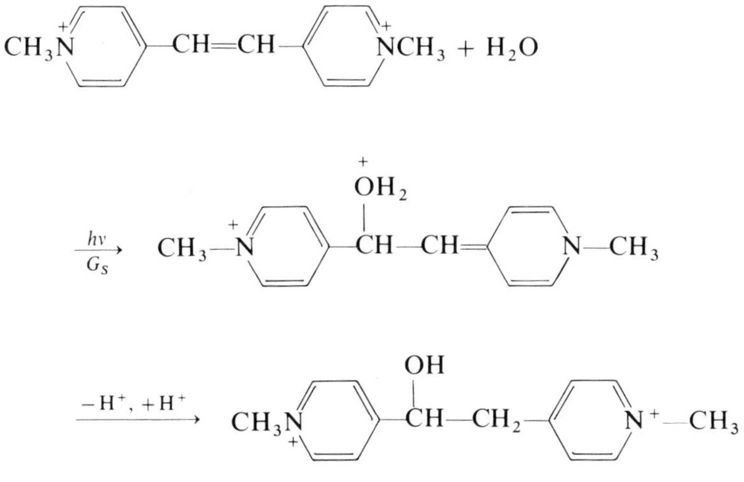

$CH_3\overset{+}{N}$⟨⟩$-CH=CH-$⟨⟩$\overset{+}{N}CH_3 + H_2O$

$\xrightarrow[G_S]{hv}$ $CH_3-\overset{+}{N}$⟨⟩$-\overset{\overset{+}{OH_2}}{CH}-CH=$⟨⟩$N-CH_3$

$\xrightarrow{-H^+, +H^+}$ $CH_3\underset{+}{N}$⟨⟩$-\overset{\overset{OH}{|}}{CH}-CH_2-$⟨⟩$\overset{+}{N}-CH_3$ (6.22a)

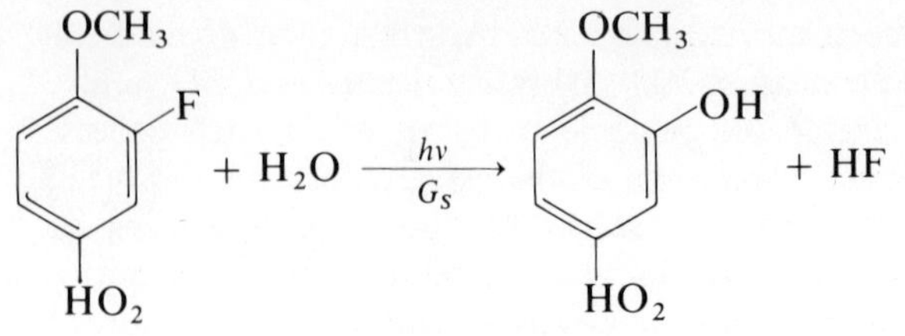

$$\text{2-methoxy-4-peroxy-fluorobenzene} + H_2O \xrightarrow[G_S]{h\nu} \text{2-methoxy-4-peroxy-phenol} + HF \qquad (6.22b)$$

$$\text{norbornyl-I} \xrightarrow[G_S]{h\nu/Et_2O} [\text{norbornyl cation} \cdot I^-] \xrightarrow{-H^+} \begin{array}{c} 81\% \text{ norbornene} \\ \\ 19\% \text{ nortricyclene} \end{array} \qquad (6.22c)$$

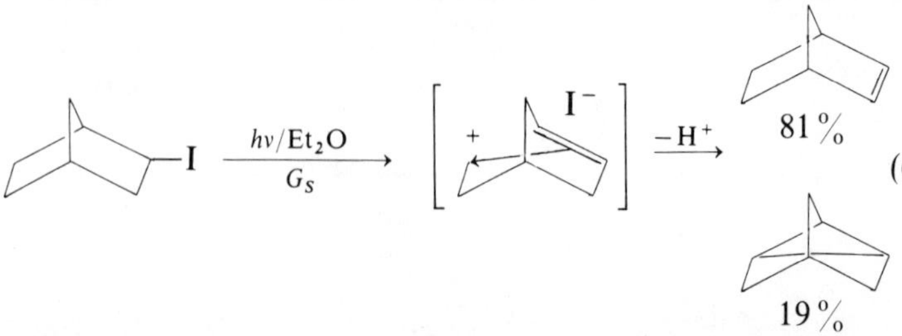

The evidence certainly suggests that these are both G_S processes; however, further work is needed to establish their detailed mechanism. Reaction (6.22c)[1c] has been clearly shown to be a G_S process. In nonpolar solvents or with the corresponding bromides, photolysis results in the formation of radicals. Because of the limited number of examples of this kind of reactivity and the continuing uncertainty about the mechanisms of many of the processes shown, further discussion of G_S photochemical reactions is beyond the scope of this book.

6.16 Examples of Photochemical Reactions

A. X-Type Reactions

X-Type reactions take place entirely on the excited-state surface, leading to products in excited states. The most conclusive evidence that a given reaction is of X type is the observation of fluorescence or phosphorescence corresponding to the product. This of course involves spectroscopic examination of the light emitted during the reaction, because the excited reactants may themselves fluoresce or phosphoresce. Unfortunately, few observations of this kind have been made. There are at least three reasons for this lack of data: (1) special equipment is needed; (2) most organic photochemists have been primarily concerned with the products of reactions; and (3) the distinction between X-type reactions and G-type reactions has not been generally appreciated.

Photochemical reactions can occur only if they are exceedingly fast, so that they can compete with deexcitation of the reactants. In the case of X-type reactions, there are three general situations where this condition is met.

1. Association reactions of excited molecules, made possible by the fact that such a system contains two unpaired electrons. The best example

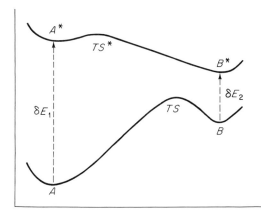

FIGURE 6.31. If the excitation energies δE_1 and δE_2 of reactant A and product B in a reversible reaction differ, the equilibrium between them will be different in the ground state and excited state and the transition state will move accordingly to a different value of the reaction coordinate.

of this (considered in detail below) is the formation of excited dimers or *excimers* by combination of two identical molecules, one of which is excited. The reactions are fast because the association needs no activation, being analogous to the combination of atoms or radicals.

2. The position of equilibrium in a reversible reaction will be changed on passing from the ground state to the excited state if the excitation energies of the reactant and product are different (see Fig. 6.31). If the activation energy in the excited state is low enough, the excited molecules may exist long enough for a new equilibrium to be set up. Equally, in the case of a reversible reaction $A \rightleftharpoons B$, where A is normally the more stable species, one may observe the formation of excited B from excited A due to a reversal of their relative stabilities. In such cases, the structure of the transition state will be different in the ground and excited states, lying nearer the less stable species in each case (BEP principle; Section 5.5). This again is illustrated in Fig. 6.31.

3. In an $n \rightarrow \pi^*$ transition, the resulting excited molecule has an electron in a high-energy antibonding MO. A molecular rearrangement which enables this unpaired electron to move to a lower-energy orbital, in particular an AO or an NBMO, will therefore be strongly exothermic. This release of energy may be a sufficient driving force to bring about some reaction that would not otherwise occur.

The following paragraphs give examples of these three X-type processes.

1. Excimer Formation.[†] Dilute solutions of pyrene (1) show a normal fluorescence, corresponding (see Fig. 6.6) to the absorption spectrum of (1). In concentrated solutions, however, the normal fluorescence is *quenched*,

---

[†] See Refs. 2a and 2b for recent reviews of excimer fluorescence.

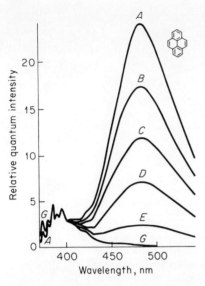

FIGURE 6.32. Fluorescence spectra of solutions of pyrene in cyclohexane; A, 10^{-2} M; B, 7.75×10^{-3} M; C, 5.5×10^{-3} M; D, 3.25×10^{-3} M; E, 10^{-3} M; G, 10^{-4} M (after Birks and Christophoron[2c]).

i.e., disappears, and is replaced by a new emission at longer wavelengths (Fig. 6.32). The new fluorescence appears at the expense of the monomer emission, though this is not apparent in Fig. 6.32 because the concentration of monomer, and so the total light absorbed, is increasing from (G) to (A). The absorption spectrum does not change, so the light is still being absorbed by isolated pyrene molecules and no chemical reaction is observed.

Light Absorption and Photochemistry 441

It is now established that the phenomenon is due to the formation of an excited dimer or *excimer* (3) from combination of pyrene (1) with excited pyrene (2). The new fluorescence comes from the excimer, which in this way is converted to two molecules of (1). Since pyrene has no tendency to dimerize, (4) is unstable and breaks apart into two molecules of (1).

Reactions of this type can be studied by measuring the quantum yield of monomer fluorescence as a function of concentration. A simple kinetic analysis shows that the reaction scheme indicated above leads to the so-called *Stern–Volmer relationship*

$$\frac{\Phi}{\Phi_0} = \frac{1}{1 + [M]/[M]_{1/2}} \qquad (6.23)$$

Here Φ is the quantum yield for a solution of concentration $[M]$, Φ_0 is the quantum extrapolated to infinite dilution, and $[M]_{\frac{1}{2}}$ is the concentration at which the quantum yield has fallen to half its limiting value Φ_0. (Note that the quantum yield refers to the *monomer* fluorescence, not to that of the excimer.) The Stern–Volmer relation holds only when the fluorescence of an excited molecule M* is quenched by reaction of M* with another molecule. If in this case the dimer were formed in the ground state and were undergoing fluorescence of the light it had absorbed as a dimer, the quantum yield of the monomer fluorescence would not follow equation (6.23).

One can see at once from PMO theory why excimers form. Figure 6.33 shows the first-order interactions between the HOMOs and LUMOs of two molecules M when they come close together so that the MOs overlap and interact. If both molecules are in their ground states, there is no first-order gain in energy because the interactions are between a pair of filled MOs and a pair of empty MOs (cf. Section 2.5). If, however, one of the molecules is excited, *three* electrons are reduced in energy on forming the excited dimer (i.e., excimer) and only *one* is increased. There is therefore a net stabilization.

As a result of this stabilization, the molecules in the dimer are pulled close together, to a separation which would be uncomfortably intimate for the corresponding ground state. Figure 6.34 indicates this situation

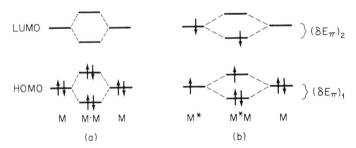

FIGURE 6.33. (a) No first-order stabilization of the dimer M·M formed by two molecules in their ground state; (b) first-order stabilization of the corresponding excimer M*M.

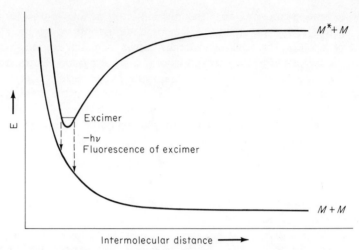

FIGURE 6.34. Formation of and fluorescence from an excimer.

diagrammatically. Deexcitation of the excimer produces the ground-state system M + M in a compressed high-energy state. The fluorescence covers a range of wavelengths because the range of geometry covered in the lowest state of the excimer (indicated by the horizontal line) corresponds to a range of different energies of the ground state. There is no vibrational structure because the ground state is formed in an unstable form MM which breaks up into two molecules M.

Nearly all polycyclic aromatic hycrocarbons form excimers and show typical excimer fluorescence. Exceptions occur only in the case of certain extremely reactive hydrocarbons such as pentacene, where chemical dimers are formed on exposure to light, e.g.,

(6.24)

The corresponding thermal reaction would be an antiaromatic ("forbidden") pericyclic reaction (Section 5.26) since it involves a *cis–cis* π cycloaddition involving a 4*n*-membered ring. The transition state for such a reaction is a point at which the BO approximation breaks down (see Fig. 6.29), so the dimerization is not an *X*-type process but is of G_A type (see Section 6.17).

Polynuclear aromatic hydrocarbons also form analogous *exiplexes* with other molecules. For example, anthracene will form exiplexes with tertiary

Light Absorption and Photochemistry

amines which exhibit exiplex emission in nonpolar solvents. With secondary amines, exiplex formation still occurs, but the exiplexes evidently decompose to radicals with relaxation of the excitation[2b]—a G_R reaction (see Section 6.16). The photochemical reaction of anthracene with diethylamine is illustrated as follows[2b]:

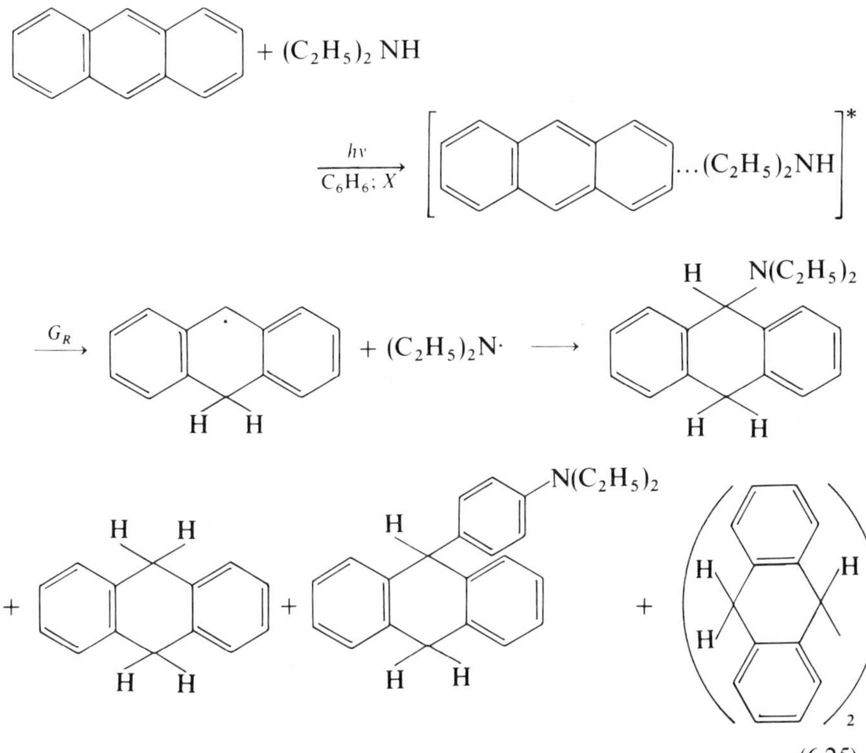

$$\text{anthracene} + (C_2H_5)_2NH \xrightarrow[C_6H_6; X]{hv} \left[\text{anthracene}\cdots(C_2H_5)_2NH\right]^*$$

$$\xrightarrow{G_R} \text{(9-H anthracenyl radical)} + (C_2H_5)_2N\cdot \longrightarrow \text{9-N(C}_2\text{H}_5)_2\text{-9,10-dihydroanthracene}$$

$$+ \text{9,10-dihydroanthracene} + \text{9-(4-N(C}_2\text{H}_5)_2\text{-phenyl)-9,10-dihydroanthracene} + \text{(9,10-dihydroanthracene dimer)}_2 \qquad (6.25)$$

In this reaction, both the reaction products and the quantum yields were strongly solvent dependent, which suggested that at least some of the decomposition was occurring through an ionic, G_S, pathway.

2. Proton Transfer Reactions in the Excited State.[3] Proton transfer reactions involving aromatic bases and acids can take place in the excited state as well as the ground state because proton transfer in such systems is an extremely fast process. Since there is no reason why the excited-state and ground-state surfaces should touch, these are X-type reactions. In other words, we are dealing with an equilibrium between an excited acid HX* and a base B to form an excited salt BH$^+$(X*)$^-$. The equilibrium constant for this process, and so the pK_A of the excited acid (pK_A*), will normally differ from that of the acid in its ground state.

A simple example is provided by the proton transfer reactions of aromatic amines,

$$\text{ArNH}_2 + \text{H}^+ \rightleftharpoons \text{ArNH}_3^+ \qquad (6.26)$$

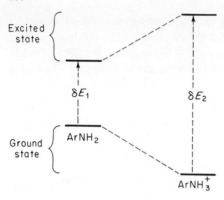

FIGURE 6.35. Since the excitation energy δE_1 of an amine $ArNH_2$ is less than that δE_2 of the corresponding ion $ArNH_3^+$, the latter is destabilized relative to the base in the excited state.

The relative equilibrium constants for different aromatic amines are determined by the relative changes in π energy during protonation (see Section 4.6). Now the excitation energy of the amine will be less than that of the ion $ArNH_3^+$ because in the latter, the nitrogen atom is saturated and so does not conjugate with the aryl group. The group NH_3^+ is thus a $+I$ substituent and will have no effect on the first absorption band of an adjacent, even AH ARH, whereas the $-I$ group NH_2 should lower the transition energy (Section 6.11, Rule 5). The situation is indicated in Fig. 6.35 (cf. Figure 6.30). Evidently, the excited base should be stronger, and the excited acid weaker, than their ground-state counterparts.

The same situation should apply in the equilibrium between a phenol and phenate ion,

$$ArOH \rightleftharpoons ArO^- + H^+ \qquad (6.26)$$

Here again the acid should be stronger in the excited state than in the ground state. The same conclusion holds, though for a different reason, in the case of heterocyclic bases such as quinoline (5).

(5)　　　(6)

Here the base and conjugate acid [e.g., (6)] are isoconjugate, the change in π energy and in $\pi \to \pi^*$ transition energy being due solely to a change in electronegativity of the nitrogen atom on conversion to $\geqslant NH^+$. This change should not affect the energy of the first $\pi \to \pi^*$ transition (Section 6.11). However, the bases can also undergo $n \to \pi^*$ transitions, which are usually lower than the first $\pi \to \pi^*$ transition, while this possibility is excluded in the salt $\geqslant N^+H$, where the nitrogen no longer has unshared electrons. Therefore the excitation energy of the base is again less than that of the salt, though for a different reason. Table 6.2 shows some examples of the

TABLE 6.2. Acid–Base Properties of Ground States and Excited States[4a]

Acid	$pK_A(S_0)$	$pK_A(S_1)$	$pK_A(T_1)$
2-Naphthol	9.5	3.0	8.1
(2-Naphthylamine) H^+	4.1	-2	3.3
$PhNHMe_2^+$	4.9	—	2.7
Quinoline · H^+	4.9	—	6.0
Acridine · H^+	5.5	10.6	5.6

change in pK_A on passing from the ground state S_0 to the first excited singlet S_1. The measurements can be made quite easily by flash photolysis, virtually all the acid or base being converted in this way to the excited state.†

An interesting difference is observed between the pK_A for the lowest excited singlet S_1 and triplet T_1 states. In the case of quinoline and acridine, the lowest singlet–singlet transitions are of $n \to \pi^*$ type. Since, however, the singlet–triplet separation is much less for such a transition than the lowest $\pi \to \pi^*$ transition (see Section 6.6), the lowest triplet state T_1 is of $\pi \to \pi^*$ type and may lie below S_1. In this case, our arguments are reversed because the excitation energy of the acid is now less than that of the base. Consequently, the pK_A of the excited acid may be greater than that of the unexcited acid; see the case of acridine in Table 6.2.

A rather similar situation occurs in the other bases, e.g., $ArNH_2$. Here the HOMO is a perturbed NBMO which in fact is largely concentrated on the nitrogen atom, while the LUMO has a low density there. The singlet–triplet separation for the base is therefore unusually small. That of the salt $ArNH_3^+$, however, is quite normal, corresponding to the (almost unperturbed) absorption of an even AH (ArH). The change in pK_A with excitation is therefore much less for the triplet state T_1 than for the singlet S_1 (see Table 6.2).

It should be added that the X-type nature of these reactions has been confirmed by observation of fluorescence from the excited states.

The nature of the solvent can have important effects on acid–base reactions in the excited state. In the photoinduced hydrogen–deuterium exchange in naphthalene, the quantum yield for β-hydrogen exchange is larger than that for α-hydrogen exchange by roughly 15% in the range of 54–64% sulfuric acid.[4a,b] With lower acid concentrations (30–50%), the relative α- and β-exchange quantum yields are reversed, with that for α being almost two times the β value at 30% D_2SO_4.[4a,b] The reversal may be partly due to competing X- and G_S-type reactions in this case.

The inversion of the relative energies of tautomers on excitation is quite a general phenomenon. A rather nice example is provided by the

---

† Flash photolysis is a technique in which an exceedingly intense light flash of very short duration is produced by discharging a bank of capacitors through a tube containing a suitable gas. The intensity is so high that the molecules in a solution exposed to the flash can all be converted to their excited states.

azaindole (7).[5] This is more stable than the isomer (8). In the excited state, however, the stabilities are reversed. In fact, (7) exists as the hydrogen-bonded dimer (9), which is less stable than the isomeric dimer (10), of (8) as indicated by the relative lengths of the arrows in equation (6.27). In the excited state, however, the relative stabilities of excited (9) [i.e., (9*)] and (10) [i.e., (10*)] are reversed [equation (6.27)].[5]

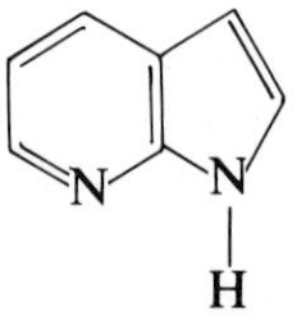

$$(9) \rightleftarrows (10)$$
$$\downarrow h\nu \quad \uparrow -h\nu$$
$$(9^*) \rightleftarrows (10^*)$$

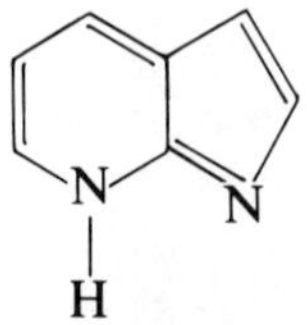

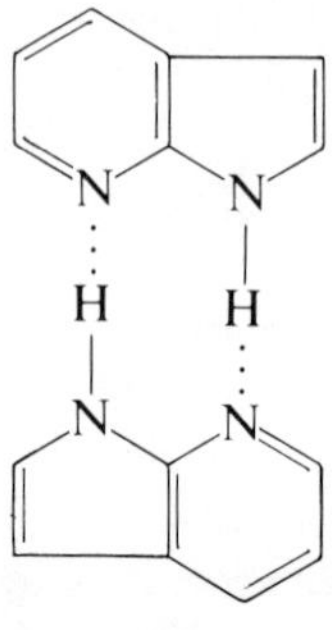

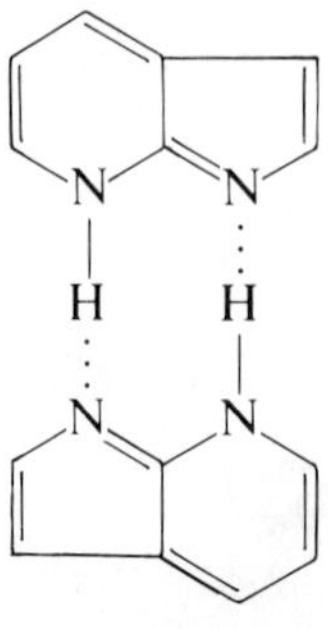

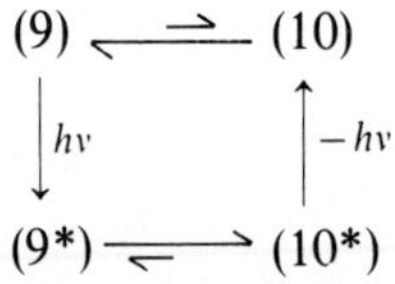

(6.27)

3. X-Type Reactions Involving $n \to \pi^*$ Transitions. A number of very interesting photochemical rearrangements of ketones[6a–d] involve $n \to \pi^*$ excited states. Such an excited state contains two unpaired electrons, one in a high-energy antibonding π MO, the other in a nonbonding oxygen AO (n). The driving force in most of these reactions is the tendency of the antibonding electron to get, if possible, into an orbital of lower energy. Any rearrangement which allows the odd π electron to pass into an orbital of lower energy (i.e., an AO or NBMO) will be greatly accelerated by the consequent gain in energy. Such a rearrangement gives a product which is in effect the $n \to \pi^*$ excited form of a molecule isomeric with, and usually much less stable than, the original reactant. The excited form of this unstable product is, however, more stable than the original molecule because it does not have an antibonding π electron. Therefore, rearrangement on the excited-state surface followed by $n \leftarrow \pi^*$ deactivation converts the original molecule into an unstable and highly reactive isomer. The final products are usually formed by further thermal rearrangement of this primary product (cf. Figure 6.21).

A very good example of such a process is the rearrangement[6a–d] of 4,4-disubstituted cyclohexadienones [e.g., (11)] to bicyclohexenones [e.g., (15)]. It is fairly certain that the reaction follows the general path indicated below. Excitation ($n \to \pi^*$) of (11) gives (12), which then undergoes disrotatory electrocyclization to (13), which is formally an $n \to \pi^*$ excited form

Light Absorption and Photochemistry

of (14). Deexcitation of (13) to (14) followed by a ground-state rearrangement gives the product (15). This mechanism is supported by the fact that "nonclassical" compounds such as (14) do rearrange as indicated (as in the Favorskii rearrangement).

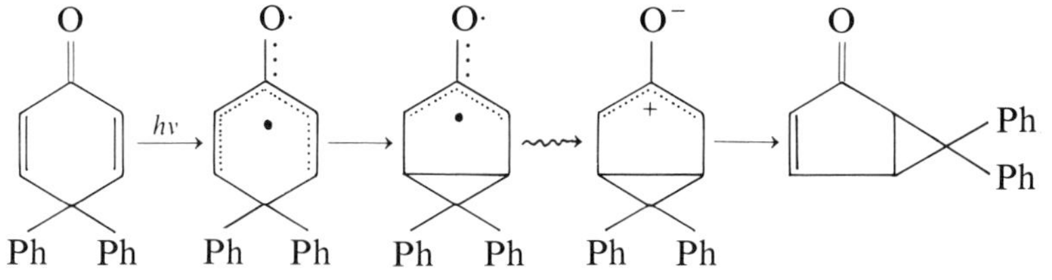

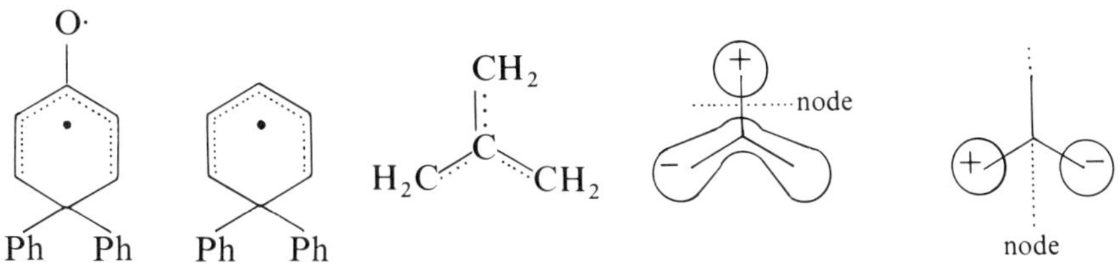

This mechanism was suggested by Zimmerman, who, however, formulated the initial excited state (12) as a simple biradical (16). At this point it should be emphasized that the use of conventional valence structures for excited states is incorrect in principle and can prove very misleading in practice. Indeed, much of the current confusion in organic photochemistry can be attributed to this error. The arguments on which the localized bond model is based (see Section 1.12) rest on the assumption that we are dealing with a system in which all the orbitals are either doubly occupied or empty. Since this condition is not met in the case of excited states, they cannot be represented by classical valence structures.

In the present case, the next step in the reaction is a disrotatory electrocyclization of (12) to the intermediate (13). Now such an electrocyclization of a radical [e.g., (17)] would not be expected to occur easily, because it would involve a nonaromatic transition state (see Section 5.26). Indeed, no such reactions have as yet been observed because radicals, being very reactive, have short lifetimes. Since the lifetimes of excited states are shorter still, it is very unlikely that such a cyclization could occur as part of a photochemical process. However, (12) is *not* in fact analogous to (17), because the unpaired π electron is *not* in an NBMO but in an antibonding MO of consequently high energy. The conjugated system in the product, on the other hand, is isoconjugate with the nonclassical hydrocarbon trimethylenemethane (18) in which there are two NBMOs. Introduction of oxygen will lower the energy of one of these (19) and make it bonding relative to a carbon 2p AO, but the other (20) will remain unchanged since it has a node passing through the carbonyl group. Of the five π electrons in (13), two can then occupy the strongly bonding MO of the trimethylenemethane system, two the bonding MO analogous to (19), and one the NBMO analogous to (20). Thus in the conversion of (12) to (13), the unpaired electron passes from a high-energy antibonding MO into an NBMO. The cyclization therefore occurs much more easily than it would in the case of a simple radical such as (17) (see Section 6.14).

The next step is a deexcitation of (14) by a reverse $n \to \pi^*$ transition. This cannot easily take place directly because the AO of oxygen and the singly occupied π MO [see (20)] do not overlap well. However, the deexcitation can take place in steps, by an $n \to \pi^*$ process removing an electron from the π MO analogous to (19) and then a $\pi \to \pi^*$ transfer of an electron between the two singly occupied π MOs. Whether these deexcitations occur entirely by internal conversion or are accompanied by fluorescence or phosphorescence is not known. Knowledge that the product is formed in its excited state is essential to the X-type classification for this reaction. The unpaired electron in the n AO plays no part in the X-type reaction, so this should occur just as easily in the triplet excited state (12) as in the singlet. Indeed, sensitized excitation of (11) by a triplet sensitizer led to rearrangement.

The presumed deexcitation leads to the nonclassical zwitterion (14) which then undergoes a thermal, ground-state rearrangement to the product (15). This may involve fission of a bond in the three-membered ring to form the zwitterion (21) as an intermediate, but more probably it involves a concerted sigmatropic rearrangement via the transition state (22). This transition state is aromatic, being of anti-Hückel type with four delocalized electrons. Note that the overall effect of the X-type rearrangement is to convert (11) into the much less stable and highly reactive isomeric species (14). On the excited-state surface, (13) is more stable than (12) because it does not contain an antibonding electron.

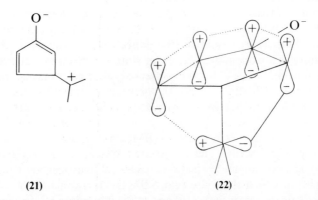

(21) (22)

A number of apparently related rearrangements of cyclohexenones are known, i.e., compounds analogous to (11) but with only one double bond; e.g.,

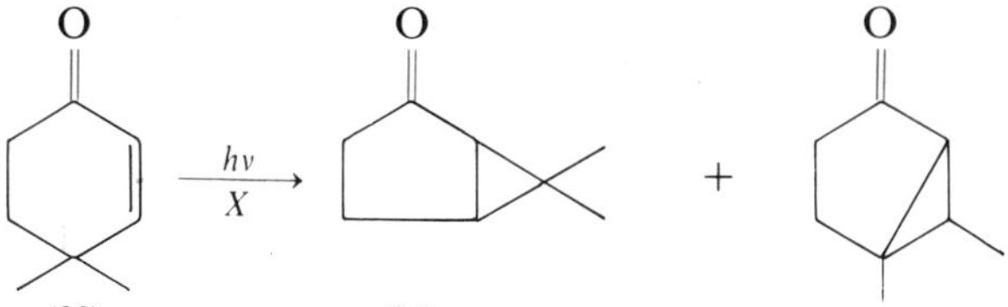

Light Absorption and Photochemistry

$$\text{(23)} \xrightarrow{hv} \text{(24)} + \text{(25)} \quad (6.28)$$

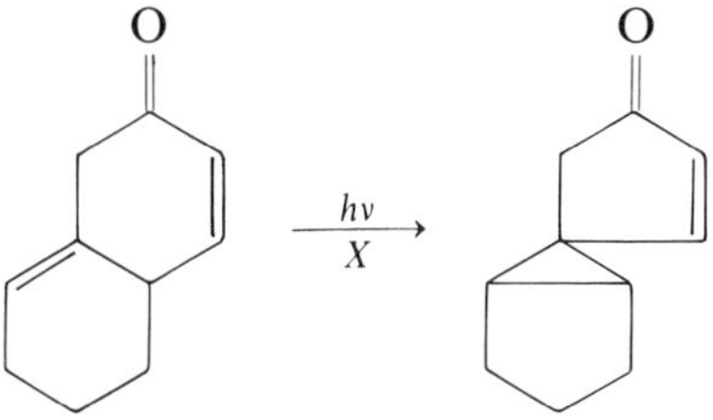

$$\text{(26)} \xrightarrow{hv} \text{(27)} \quad (6.29)$$

The mechanisms of these reactions have presented problems mainly because of the confusion caused by writing $n \to \pi^*$ transition states as biradicals. It seems very likely that they all follow a common pattern, i.e., the facilitation of otherwise rather unfavorable processes by the gain in energy when the unpaired antibonding electron in the π^* MO of the reactant passes into an AO or NBMO in the product (see Section 6.14).

In the case of (11), the reaction in question was an analog of the electrocyclization of a radical. In the reactions of equations (6.28), it is probably a similar analog of the Wagner–Meerwein rearrangement of a radical. Such processes are known in the ground state, e.g.,

$$\underset{\underset{CH_3}{|}}{\overset{\overset{CH_3}{|}}{Ph-C-CH_2\cdot}} \longrightarrow \underset{\underset{CH_3}{|}}{\overset{\overset{CH_3}{|}}{\cdot C-CH_2-Ph}}$$

However, they are much less facile than the corresponding rearrangement of carbonium ions, because in the intervening π complex one electron has to occupy an antibonding MO (Section 5.17). Here the reactant already has an antibonding electron, so the rearrangement can take place with the same ease as the ground-state rearrangement of a carbonium ion, i.e., very easily indeed. The product in each case is an excited radical in which one unpaired electron occupies a nonbonding carbon $2p$ AO, the other the oxygen n orbital. Thus,

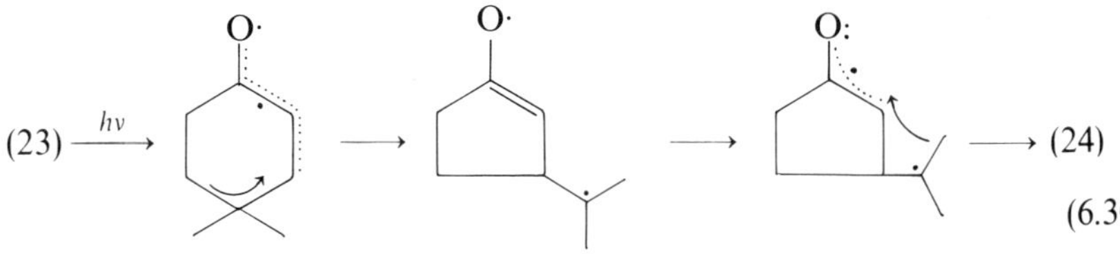

$$(23) \xrightarrow{hv} (23^*) \longrightarrow (28) \longrightarrow (29) \longrightarrow (24) \quad (6.30)$$

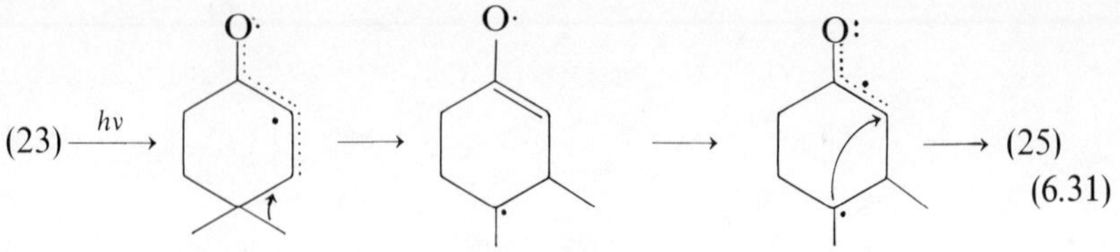

(23) $\xrightarrow{h\nu}$ [structure (23*)] → [structure (30)] → [structure (31)] → (25)

(6.31)

A reverse $n \to \pi^*$ transition now converts the product into a ground-state biradical (29) or (31) in which one of the unpaired electrons occupies an orbital analogous to an allyl NBMO, the other a carbon $2p$ AO. Ground-state combination of the radicals gives the final product. The net effect of the X-type reaction is therefore again to convert the reactant to the excited form of a species [(29) or (31)] which is much less stable in the ground state, but more stable in the excited state, than the original molecule.

The reaction in equation (6.29) probably involves a radical-type addition to the isolated double bond; i.e.,

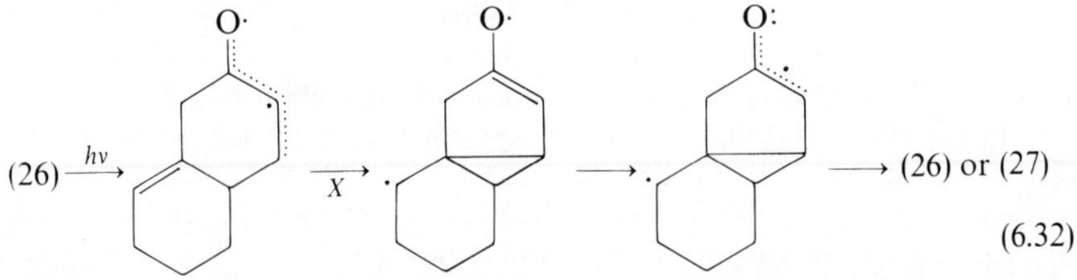

(26) $\xrightarrow{h\nu}$ [structure (26*)] $\xrightarrow{X}$ [structure (32*)] → [structure (32)] → (26) or (27)

(6.32)

In this case, analogous intramolecular addition reactions of radicals are known. Here addition will be facilitated by the transfer of an electron from an antibonding π MO into a nonbonding carbon $2p$ AO. The product (32*) then undergoes $n \to \pi^*$ deexcitation to (32), which now is a simple biradical, one of the unpaired electrons occupying an MO analogous to the NBMO of allyl. The biradical undergoes a concerted bond-forming/bond-breaking process either to regenerate (26) or to form the product (27).

One final reaction in this series is interesting in that the zwiterionic intermediate (33) can be generated either photochemically or by a ground-state path[6d] (6.33b). In both cases the ratios of products show the same solvent dependence:

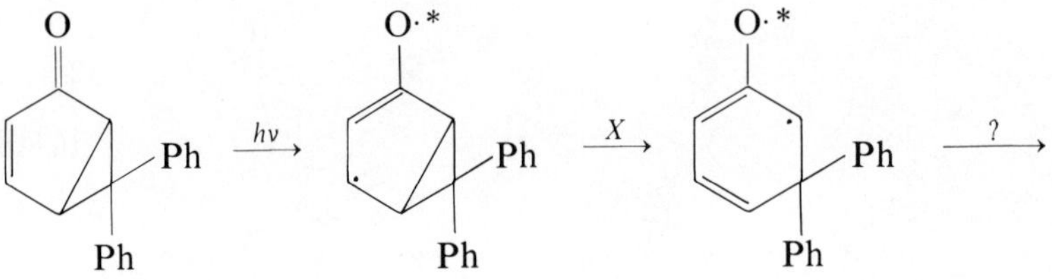

[structure with O, Ph, Ph] $\xrightarrow{h\nu}$ [structure with O·*, Ph, Ph] $\xrightarrow{X}$ [structure with O·*, Ph, Ph] $\xrightarrow{?}$

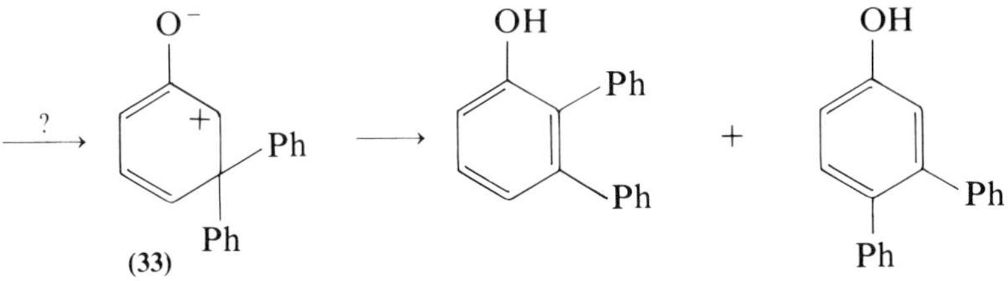

Light Absorption and Photochemistry

$$\text{(33)} \xrightarrow{?} \cdots \longrightarrow \cdots + \cdots \quad (6.33a)$$

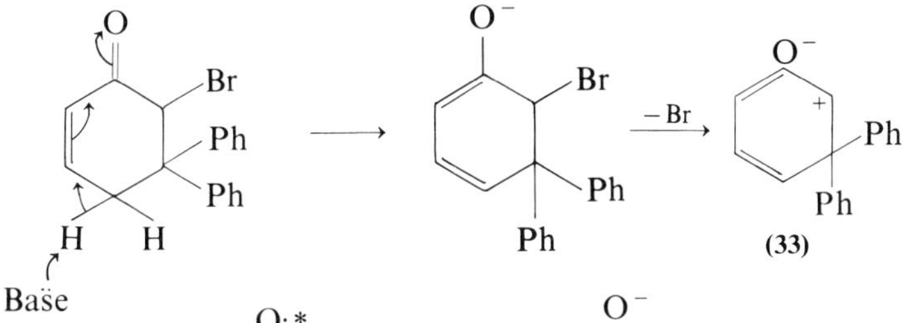

$$\cdots \longrightarrow \cdots \xrightarrow{-\text{Br}} \text{(33)} \quad (6.33b)$$

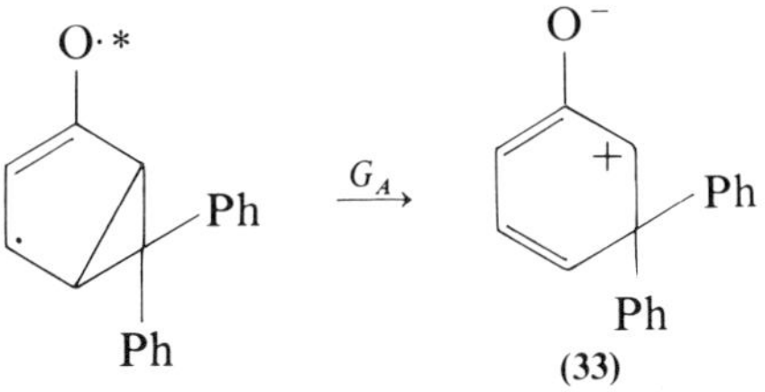

$$\cdots \xrightarrow{G_A} \text{(33)} \quad (6.33c)$$

The evidence leaves little doubt that the product-forming steps occur from (33) in its ground state. The nature of the photochemical rearrangement, X type as shown in (6.33a) or G_A type as suggested in (6.33c), is not yet settled; however, the available evidence seems to favor the X-type path.

The reactions so far considered have depended on primary reactions involving the antibonding electron resulting from an $n \to \pi^*$ transition. The other unpaired electron, occupying an oxygen AO (n), is analogous to that in an alkoxy radical $R\dot{O}$. Since such radicals are extremely reactive, it is possible for the primary reaction of the $n \to \pi^*$ excited state to involve this electron rather than the one in the antibonding π MO. The result of such a process is usually a biradical rather than an excited state and reactions involving such biradicals are treated separately as G_R reactions (see below). However, one or two exceptions are known, an interesting example being the photochemical conversion[6a-d] of methyl salicylate (34) to the enol (37). The reaction involves the $n \to \pi^*$ excited state (35*) of (34), which undergoes an internal hydrogen transfer to (36*). This is the $n \to \pi^*$ excited state of the product (37), into which it is converted by deexcitation. The reaction is certainly of X type, since the final step is accompanied by observable fluorescence.

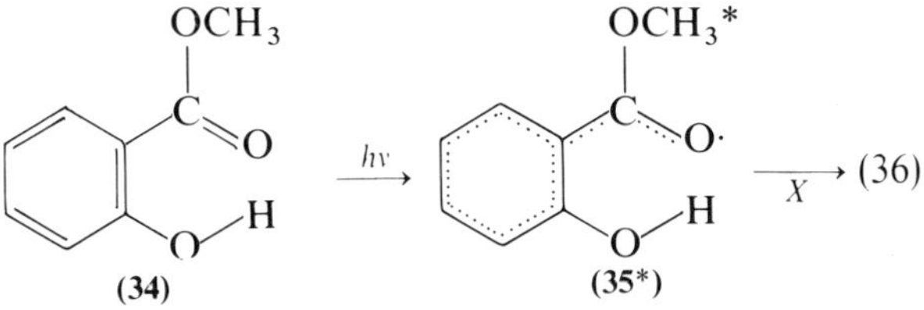

$$(34) \xrightarrow{h\nu} (35^*) \xrightarrow{X} (36) \quad (6.34a)$$

[see overleaf for (36)]

$$\text{(36*)} \xrightarrow{-h\nu} \text{(37)} \quad (6.34b)$$

B. G_R Reactions

The majority of G reactions involve radical intermediates, i.e., those classified here as G_R. A G_R-type reaction involves the direct conversion of an excited reactant molecule into ground-state products via the BO hole corresponding to a pair of radicals, or to a biradical in which the radical centers are isolated from one another by an intervening chain of saturated atoms. Here again three different processes of this kind can be distinguished:

1. Reactions leading to cleavage of π bonds by rotation [see the discussion of the ring opening of cyclobutene, p. 343].
2. Cleavage of σ bonds to give pairs of atoms or radicals.
3. Reactions of radical type involving $n \to \pi^*$ excited states, the singly occupied n AO acting as a radical center.

We now illustrate these possibilities.

1. Photochemical Cleavage of π Bonds; Cis–Trans Isomerization. As we have seen (Section 6.14), interaction of two AOs on adjacent atoms gives rise to a bonding MO which is relatively weakly bonding and an antibonding MO which is very strongly antibonding. Excitation of an electron from the filled, bonding MO in a diatomic molecule X_2 into the antibonding MO should then lead to a net antibonding interaction and the excited molecule should consequently dissociate into atoms. The halogens do in fact dissociate in this way on irradiation, e.g.,

$$Cl_2 \xrightarrow[G_R]{h\nu} 2Cl \cdot \quad (6.35)$$

These were indeed almost the first photochemical reactions to be studied in detail.

The same argument applies to the π bond in ethylene (see Figs. 6.24 and 6.25). In this case, the carbon atoms still remain attached by a σ bond, so dissociation does not occur, but the same effect can be achieved, so far as the π bond is concerned, by a rotation through 90° [equation (6.17)]. This rotation destroys the net overlap between the carbon $2p$ AOs and so makes the corresponding β equal to zero. The bonding and antibonding MOs thus become nonbonding. One can see from this argument that the ground-state and $\pi \to \pi^*$ excited-state potential surfaces for ethylene must have the form indicated in Figs. 6.25 and 6.36, touching at the point where the CH_2 groups are perpendicular to one another.

The point where the surfaces touch is thus a BO hole. It also corresponds to the transition state for thermal *cis–trans* isomerization of ethylene. Photo-

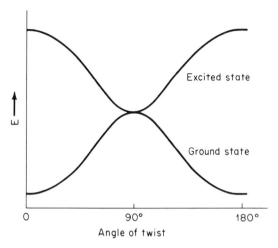

FIGURE 6.36. Energy of the ground state and $\pi \to \pi^*$ excited state of ethylene as a function of the angle of twist about the CC bond.

chemical excitation should—and does—bring about a G_R-type photochemical *cis–trans* isomerization. One can see, moreover, that the same thing should happen in ethylene derivatives carrying conjugated substituents. Consider, for example, stilbene (38). Stilbene can be formed as indicated by union of two benzyl radicals. The strength of the central π bond in (38) arises from consequent interaction between the two benzyl NBMOs (Section 3.9). Now the MOs arising from this interaction are the HOMO and LUMO of (38). Consequently, excitation of (38) transfers an electron which is strongly bonding between the two central carbon atoms into an orbital which is strongly antibonding between them. Excitation should therefore lead to rotation about the central bond, the potential energy diagram being analogous to Fig. 6.36.

There is, however, an interesting additional factor to be considered. *Cis*-stilbene (39) is less stable than *trans*-stilbene (40) because in the *cis* isomer, two of the *ortho* hydrogen atoms get in each other's way. Consequently, (39) is forced out the planar configuration that is most favorable for a conjugated system.† This distortion both alters the wavelength of the first $\pi \to \pi^*$ absorption and reduces its extinction coefficient. If we irradiate a mixture of *cis*-stilbene and *trans*-stilbene, the latter absorbs more strongly and so is converted to *cis* faster than *cis* absorbs and is converted to *trans*. Photochemical isomerizations of this type provide a convenient practical route to less stable geometric isomers.

2. Photochemical Cleavage of σ Bonds. As remarked above, $\sigma \to \sigma^*$ photochemical excitation of the halogens leads to cleavage into atoms. Similar cleavages can be observed in polyatomic molecules, the reactions falling into three groups.

(*a*) *Cleavage of Weak σ Bonds.* The S—S bond is weak, so its β is small. If we treat a disulfide in the way we treated paraffins in Section 1.12, i.e., starting with a set of localized two-center bonds and then letting them interact, the S—S bonding MO, with its low β, will be at the top of the filled MOs in energy and the interactions between the filled MOs (cf. Fig. 1.27) will leave it there. The HOMO of a disulfide is therefore a perturbed S—S σ MO, concentrated mainly between the sulfur atoms, and the same argument shows that the LUMO is also primarily a S—S antibonding MO. Excitation ($\sigma \to \sigma^*$) of an S—S bonding electron, or $n \to \sigma^*$ excitation of a sulfur lone-pair electron, will therefore transfer an electron which is bonding or nonbonding between the sulfur atoms into an MO which is strongly antibonding there. Disulfides on irradiation consequently dissociate into radicals,

$$\text{RS—SR} \xrightarrow[G_R]{hv} \text{RS}\cdot + \text{SR} \qquad (6.36)$$

Similar reactions are observed in other molecules containing weak σ bonds, e.g., alkyl iodides which dissociate on irradiation with UV light into alkyl radicals and iodine atoms,

$$\text{RI} \xrightarrow[G_R]{hv} \text{R}\cdot + \cdot\text{I} \qquad (6.37)$$

(*b*) *Cleavage of Bonds $\beta\gamma$ to a Conjugated System.* The discussion of *cis*–*trans* isomerization above (p. 452) shows that $\pi \to \pi^*$ excitation in a conjugated molecule containing an essential double bond adjacent to an even, conjugated system [e.g., stilbene, (38)] leads to a π antibonding interaction in that bond. Now the argument indicated in Fig. 6.2 shows that a σ bond $\beta\gamma$ to a conjugated molecule is analogous to an isolated π bond in that position. Consequently, $\pi \to \pi^*$ excitation of a conjugated system should tend to weaken σ bonds $\beta\gamma$ to it.

This effect is seen very clearly in a number of benzyl derivatives PhCH$_2$X which dissociate on irradiation with light absorbed by the π electrons of the phenyl group,

$$\text{PhCH}_2\text{X} \xrightarrow[G_R]{hv} \text{PhCH}_2\cdot + \cdot\text{X} \qquad (6.38)$$

† Note that while the single bonds in conjugated polyenes are localized, they do nevertheless have π components; see Section 3.6. The corresponding π-bond energy is hidden because it is lumped in with the σ-bond energy in the overall "polyene single-bond energy." Rotation out of planarity consequently reduces the π component and so weakens the bond.

An entertaining example of such $\beta\gamma$ cleavage seems to be provided by a photo-Cope rearrangement studied by Houk and Northington,[7]

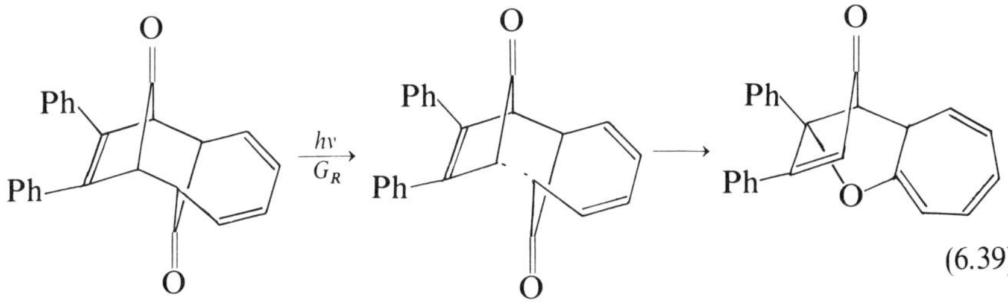

$$(6.39)$$

The transition state for the concerted reaction would be a six-electron Hückel-type transition state. Such a process should not take place photochemically since only antiaromatic pericyclic reactions occur in this way. The reaction moreover seems to involve a triplet excited state since it occurs only in the presence of a ketone as sensitizer. Therefore there seems little doubt that it is, as indicated in equation (6.39), a G_R-type process in which the first step is cleavage of a bond $\beta\gamma$ to the excited diene system to give a pair of mesomeric biradicals. These combine on the ground-state surface to form the product.

3. Production of Radicals from $n \to \pi^*$ Excited States. [8a-c] An $n \to \pi^*$ transition leaves the AO n of a heteroatom singly occupied. The situation will be similar to that in a corresponding radical, so the excited state may undergo typical intermolecular or intramolecular radical reactions. Since the product is a biradical or pair of radicals, these are typical G_R processes.

The best investigated reactions of this kind are those involving $n \to \pi^*$ excited states of ketones and other carbonyl compounds. It is well known that alkoxy radicals can very easily undergo cleavage and can also very readily abstract hydrogen from CH bonds; e.g.,

$$CH_3-\underset{\underset{CH_3}{|}}{\overset{\overset{CH_3}{|}}{C}}-O\cdot \longrightarrow CH_3-\underset{\cdot CH_3}{\overset{\overset{CH_3}{|}}{C}}=O \quad (6.40)$$

$$RO\cdot + H-\underset{|}{\overset{|}{C}}- \longrightarrow ROH + \cdot \underset{|}{\overset{|}{C}}- \quad (6.41)$$

Both these reactions are observed with $n \to \pi^*$ excited states. The cleavage and intramolecular hydrogen abstraction were first studied in detail by Norrish and are accordingly called *Norrish type I* and *Norrish type II* processes. In the type I process, the excited ketone (41*) undergoes cleavage into acyl and alkyl radicals (42). In the type II process, the n radical center is used to abstract a γ hydrogen atom to form a biradical (45), which may either cleave into the enol (46) of a lower ketone (47) or collapse by radical combination into a cyclobutanol (48).

Norrish type I:

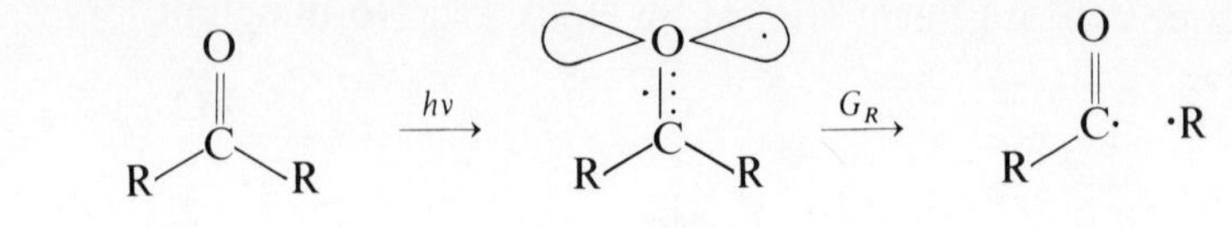

$$R-\overset{O}{\underset{\|}{C}}-R \xrightarrow{h\nu} (41^*) \xrightarrow{G_R} R-\overset{O}{\underset{\|}{C}}\cdot \;\; \cdot R \;\; (42)$$

Norrish type II:

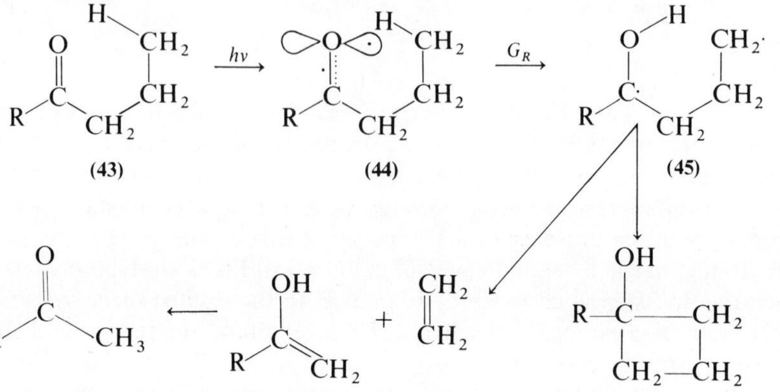

(43) → (44) → (45) → (46) + (48)

(46) → (47)

(6.43)

The selective abstraction of γ hydrogen is a purely geometric phenomenon, due to the fact that in normal conformations of the alkyl chain it is the γ hydrogen that is nearest to the oxygen atom. The overriding importance of geometry in these reactions is exemplified by the fact that variations in the photochemical reactivity in a series of straight-chain and polycyclic alkyl ketones could be attributed almost entirely to variations in $\Delta S^{\dagger}$ for the reactions.[8b]

The balance between the two types of processes is quite subtle, depending on the nature of the alkyl group expelled in a type I process, the stability of the radical formed in a type II process, and on the geometry of the alkyl groups. Similar reactions are also observed in aryl alkyl ketones, though here the type II process usually leads to cyclobutanol rather than cleavage to an olefin. Ring substitution in aryl ketones can completely suppress the type II process.[8c] If the lowest triplet state of the ketone is a $\pi \to \pi^*$ state rather than an $n \to \pi^*$ state and if the energy gap between the two states is greater than ~ 2 kcal/mole, the type II process will be effectively suppressed.

In addition to these intramolecular processes, $n \to \pi^*$ excited ketones can undergo typical intermolecular radical reactions. Thus irradiation of benzophenone (49) in isopropanol (50) gives the pinacol (53) and acetone as follows:

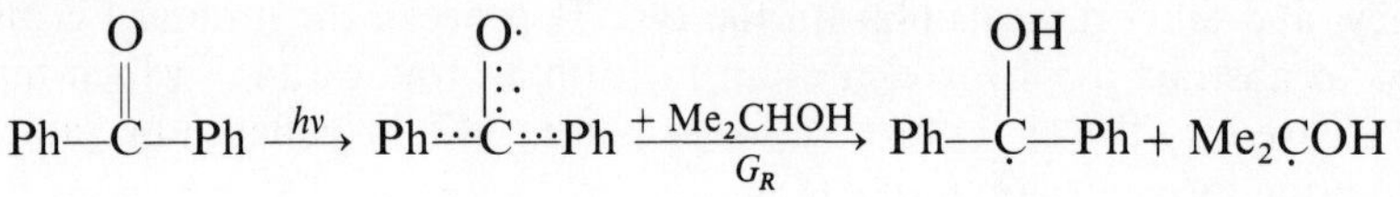

$$\underset{(49)}{Ph-\overset{O}{\underset{\|}{C}}-Ph} \xrightarrow{h\nu} \underset{(49^*)}{Ph\cdots\overset{O\cdot}{C}\cdots Ph} \xrightarrow[G_R]{+Me_2CHOH} \underset{(51)}{Ph-\overset{OH}{\underset{|}{C}}-Ph} + \underset{(52)}{Me_2\dot{C}OH} \quad (6.44)$$

Light Absorption and Photochemistry

$$Me_2\dot{C}OH + Ph_2CO \longrightarrow Me_2CO + Ph-\underset{|}{\overset{OH}{C}}-Ph \qquad (6.45)$$
$$(52) \qquad (49) \qquad\qquad\qquad (51)$$

$$2\ (Ph-\underset{|}{\overset{OH}{\dot{C}}}-Ph) \longrightarrow HO-\underset{\underset{Ph}{|}}{\overset{\overset{Ph}{|}}{C}}-\underset{\underset{Ph}{|}}{\overset{\overset{Ph}{|}}{C}}-OH \qquad (6.46)$$
$$\qquad\qquad\qquad\qquad (53)$$

The transfer of hydrogen from the radical (52) to (49) to form (51) [equation (6.45)] occurs readily because phenyl ($\pm E$) groups stabilize radicals much better than do alkyl groups (Section 4.13). There is therefore a gain in delocalization energy during the reaction.

Another intermolecular reaction of $n \to \pi^*$ excited ketones is their addition to olefins to form oxetanes, a very nice example being the photochemical addition of acetone to *trans*- (54) and *cis*- (61) 1-methoxy-1-butene.[9a]

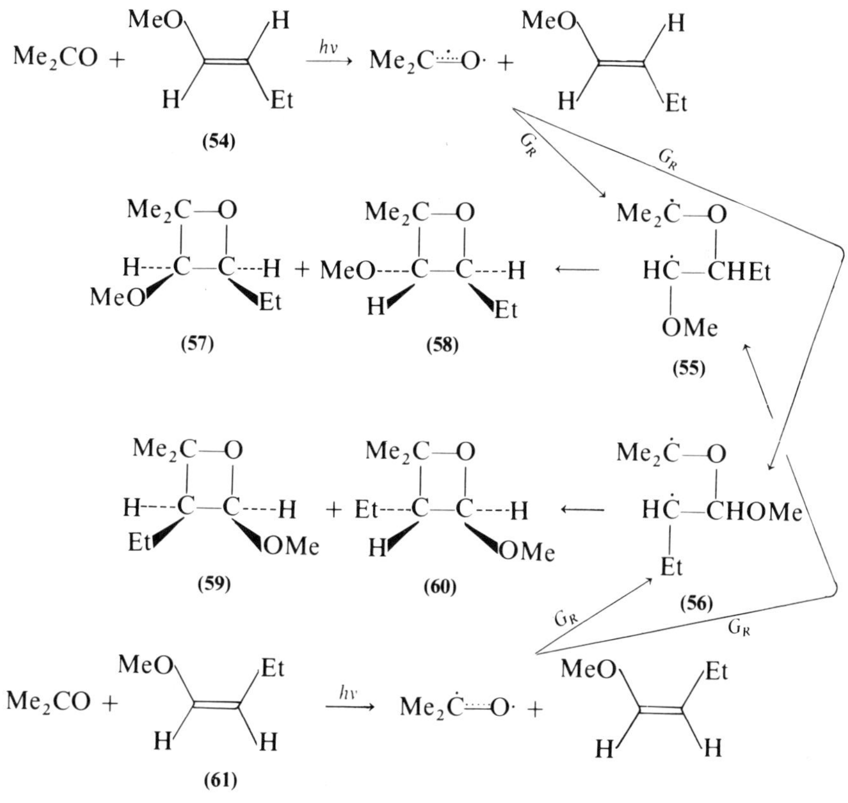

As indicated, the reaction is completely nonstereospecific, both (54) and (61) giving mixtures of all four possible products (57)–(60). There is therefore no question of concerted cycloaddition, the reaction taking place via intermediate adducts (55) and (56) where the double bond of (54) or (61) has become single so that rotation about it can occur freely.

Another interesting example is provided by the photochemical conversion of *o*-methylbenzophenone (62) to the enol (63). The evidence suggests very strongly that (63) is formed directly and not by a secondary photochemical *cis–trans* isomerization of the enol (64) that would be formed by a simple Norrish type II reaction, by collapse of the intermediate (65*). The reason is that the reaction involves the triplet $n \to \pi^*$ excited state of (62), not the singlet. The intermediate (65*) is therefore in a triplet state, the two unpaired electrons having parallel spins. Now the arguments concerning *cis–trans* isomerization (see Fig. 6.36) apply equally well to triplet states. Here, too, the stable configuration of a substituted ethylene will be one in which the double bond has been twisted 90°. The initial product of the G_R reaction is therefore the triplet twisted form (65*) of (64*), with two benzyl radicals in which the unpaired electrons have parallel spins. This eventually undergoes intersystem crossing and can then revert to (64) or (63). Since (64) will undergo a rapid conversion back to (62) by an aromatic pericyclic reaction, the product observed is (63).

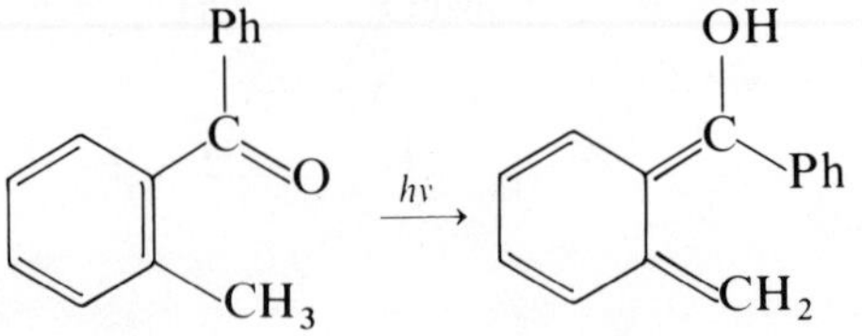

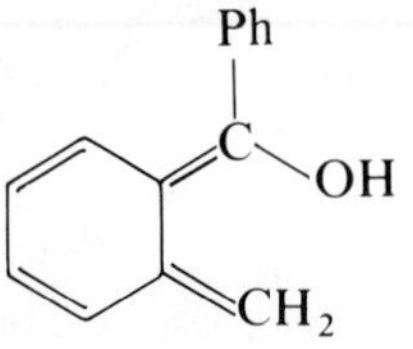

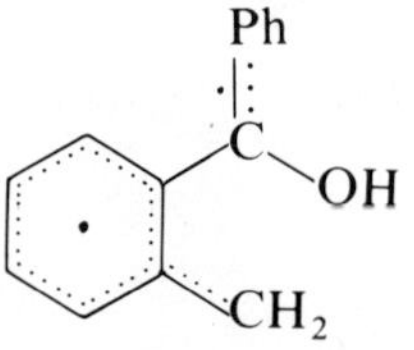

Another fairly general class of G_R photochemical isomerizations are the di-π-methane photoisomerizations. These reactions involve the photochemical conversion of a divinyl methane unit to a corresponding vinyl cyclopropane through intermediate biradical states. The example shown below[9b] is interesting in that the naphthalene chromophore in (66) absorbs the light, undergoes intersystem crossing, and sensitizes the benzo-bicyclo-[2.22]-octatriene for reaction.

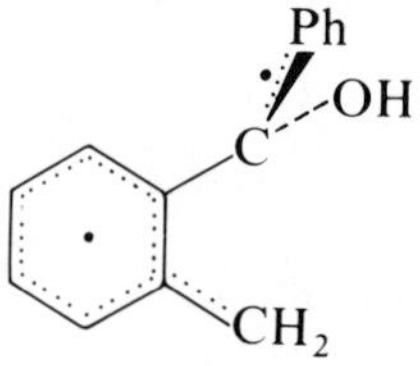

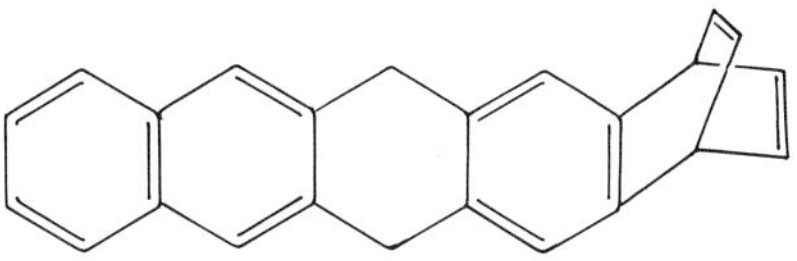

(66)

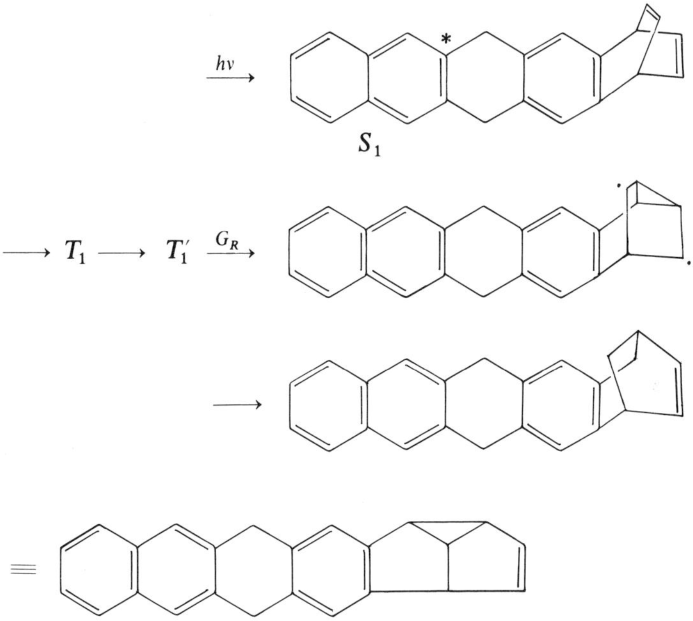

$$\xrightarrow{h\nu} S_1 \longrightarrow T_1 \longrightarrow T_1' \xrightarrow{G_R}$$

As one would expect, the di-π-methane rearrangement is regiospecific in asymmetric cases, e.g.,

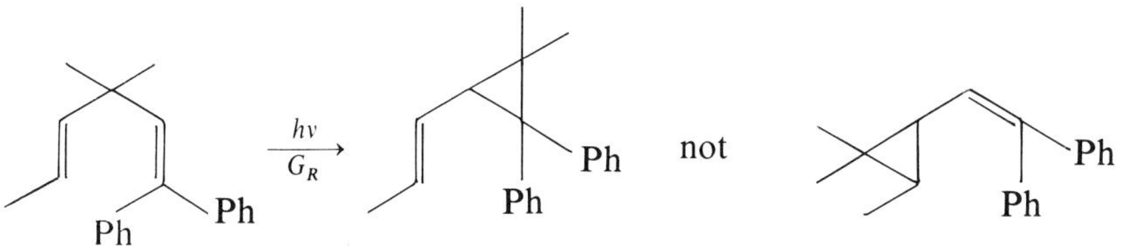

The reaction is also regiospecific when the directing groups are not bonded to any of the atoms directly involved in the rearrangement[9c]; this is presumably due to interactions through space with the unpaired spins.

C. G_I Reactions

G_I reactions are in effect G_R processes in which the two radical sites have very different electronegativities. A typical example is the photochemical conversion of colorless leuco Malachite Green cyanide to the

cationic dye Malachite Green,

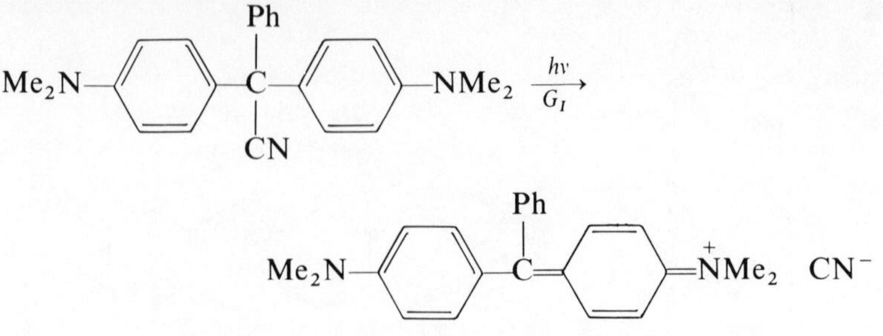

$$Me_2N-C_6H_4-C(Ph)(CN)-C_6H_4-NMe_2 \xrightarrow{h\nu}_{G_I}$$

$$Me_2N-C_6H_4-C(Ph)=C_6H_4=\overset{+}{N}Me_2 \quad CN^- \quad (6.47)$$

In this case, solvation of the ions makes them more stable than the corresponding radicals, so dissociation of the C—CN bond takes place heterolytically.

The photochemical reactions of pyrole with benzene and naphthalene involve both X-type (6.48) and bimolecular G_I-type reactions (6.49)[10]:

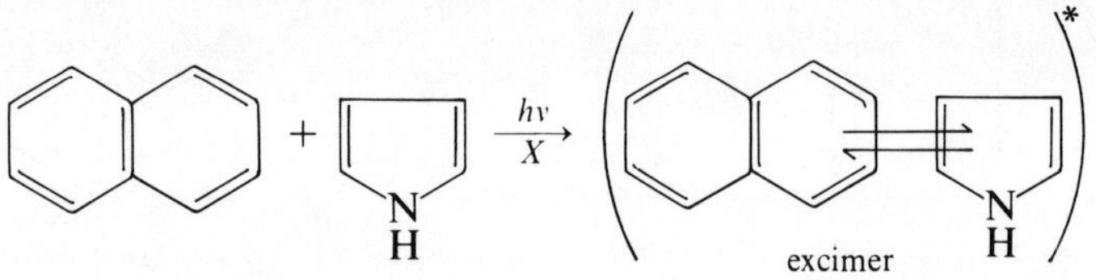

$$\text{naphthalene} + \text{pyrrole} \xrightarrow{h\nu}_{X} (\text{naphthalene}\cdots\text{pyrrole})^* \quad \text{excimer} \quad (6.48)$$

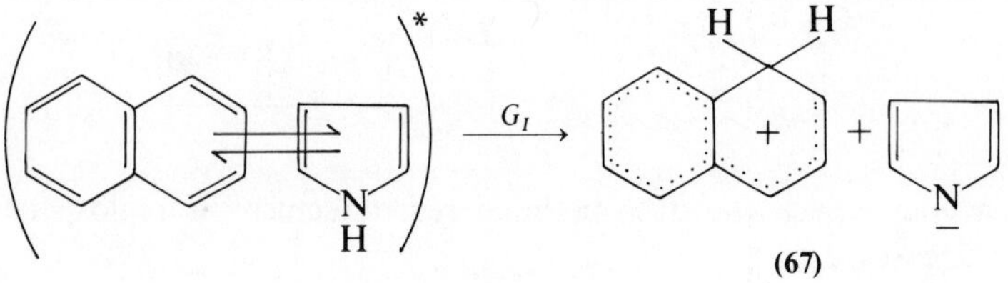

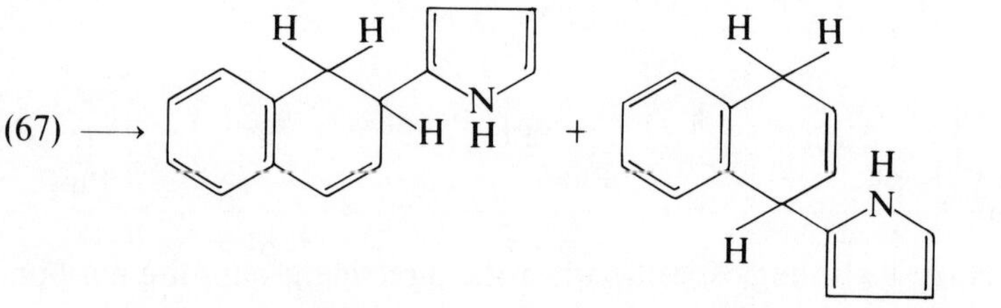

$$(\text{naphthalene}\cdots\text{pyrrole})^* \xrightarrow{G_I} (67) + \text{pyrrole}^- \quad (6.49)$$

$$(67) \longrightarrow (68) + (69) \quad (6.50)$$

The quenching of the fluorescence of naphthalene by pyrroles depends strongly on the polarity of the solvent, being much more efficient in solvents of high polarity. This is consistent with the mechanism of equations (6.48)–(6.50), in which the interaction of excited naphthalene with pyrrole leads to the formation of an ion pair. The ratio of 1,2 (68) and 1,4 (69) adducts also corresponds to that expected for combination of the ions (67). Finally,

Light Absorption and Photochemistry 461

the transfer of a proton [equation (6.49)] was indicated by deuterium labeling and by the fact that N-methylpyrrole quenches the fluorescence of naphthalene but does not react.

Virtually all of these reactions depend on solvation of the ion pair to make the reactions exothermic. As such, G_I reactions are properly a subclass of G_S processes (see Section 6.15 and Ref. 1c). We have maintained the distinction here to emphasize the relationship between radical and ion pair processes.

D. G_N Reactions

G_N reactions lead either to a nonclassical, even AH, i.e., one for which no classical structure can be written and which therefore has two NBMOs (Section 3.14), or to an analogous heteroconjugated system in which there is again degeneracy between the HOMO and LUMO. Few reactions of this type have been reported, partly because structures of this kind are uncommon and partly because introduction of heteroatoms into a nonclassical, even AH usually destroys the necessary orbital degeneracy [cf. Section 6.16A(3); the conjugated system (14) is isoconjugate with (18), but the orbitals in it all have different energies].

Haselbach and Eberbach have reported an intriguing example of a G_N reaction of the tricyclic system (70).[11a,b] Photoelectron spectra as well as theoretical studies supported the intermediacy of the nonclassical molecule (71) in this reaction.

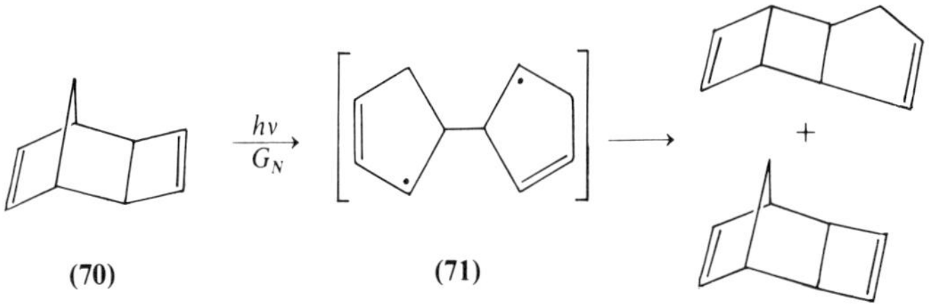

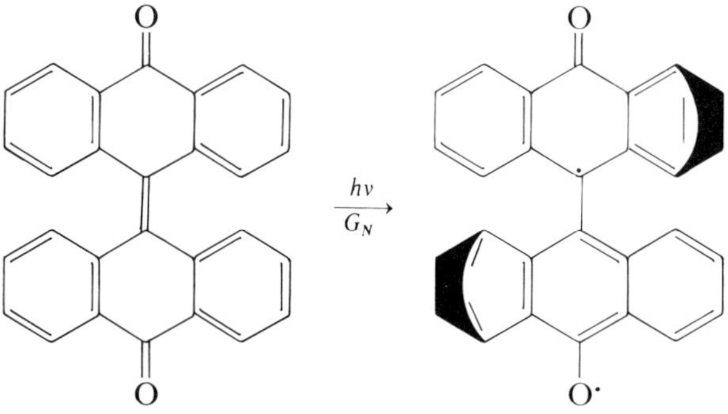

The photochromism of bianthrone (72) is another example of a formal G_N reaction. The structure of the colored form (73) has long been a subject of interest. The bond-twisted structure[11c] (73) was suggested after extensive NMR, UV–visible, and theoretical studies by Muszkat and his co-workers. Although the photochromism of bianthrone is formally similar to G_R photoreactions like the isomerization of *trans*-stilbene, the bond-twisted form in this case is a relatively stable nonclassical molecule.

E. G_A Reactions

An aromatic ("allowed") ground-state pericyclic process involves a transition state in which the conjugated system is aromatic (Section 5.26). The ground-state potential surface for such a system will remain separated throughout from the excited-state surface, so the corresponding photochemical reaction would be X type. Now pericyclic processes usually require considerable activation because they involve an extensive rearrangement of bonds. A process requiring such activation could not occur in the excited state. Photochemical reactions of this kind can therefore be expected to occur only when the excitation energy of the product is much less than that of the reactant so that the reaction is very exothermic in the excited state and so has a low activation energy. One might, for example, see a photochemical reversal of the Diels–Alder reaction (Section 5.27A) due to the fact that the excitation energy of cyclohexene is much greater than that of butadiene (cf. Figure 6.29). Reactions of this type would, however, involve no new principles and there are, moreover, relatively few cases where they have been observed. Even these may be of G_R type.

A much more interesting case is provided by antiaromatic pericyclic reactions where the transition state is a BO hole at which the excited-state surface and the ground-state surface touch (Section 6.13). Here photoexcitation can lead directly to the ground-state product by passage through the BO hole. This, as we have seen, is the basic requirement for a G_A photochemical reaction. Such a process will lead to products which are qualitatively different from those given by corresponding thermal processes and the formation of these products is often the only evidence that a G_A reaction is involved.

Sometimes, however, one can be sure that the reaction has involved a BO violation and is not of X type since an X-type reaction would be endothermic; it has already been pointed out that the photochemical dimerization of olefins is such a case (p. 422).

Therefore G_A photochemical reactions follow the same general course as ground-state pericyclic reactions except that the rules are diametrically opposite. The reactions that occur easily thermally, having aromatic transition states, do not occur as G-type photochemical processes and so are not usually facilitated by light. On the other hand, the antiaromatic reactions, which do not normally occur thermally, commonly take place very easily as G_A photochemical processes. We will consider a number of examples, grouping them by the number of delocalized electrons involved in the transition state.

1. **Four-Electron Hückel Systems.** The literature abounds with examples of photochemical olefin addition reactions which proceed through transition states that are analogous to Hückel cyclobutadiene; e.g.,

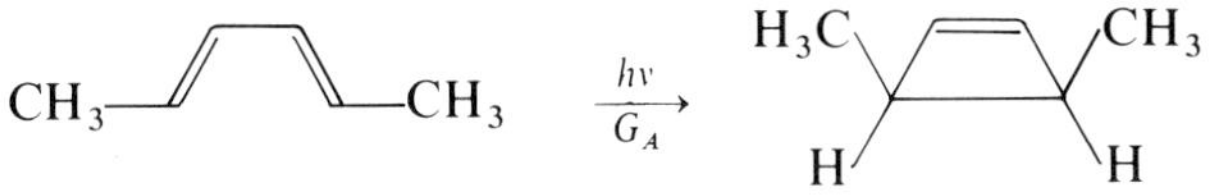

$$CH_3-\text{\_/\_}-CH_3 \xrightarrow[G_A]{h\nu} \begin{array}{c} H_3C \quad CH_3 \\ \text{\_/\_} \\ H \quad H \end{array} \quad \text{(Ref. 12)} \quad (6.51)$$

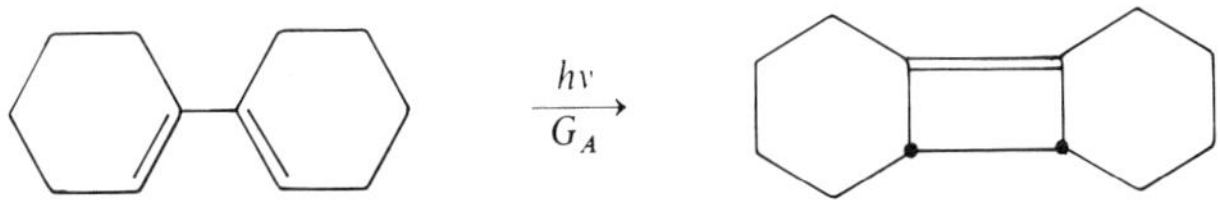

$$\xrightarrow[G_A]{h\nu} \quad \text{(Ref. 13)} \quad (6.52)$$

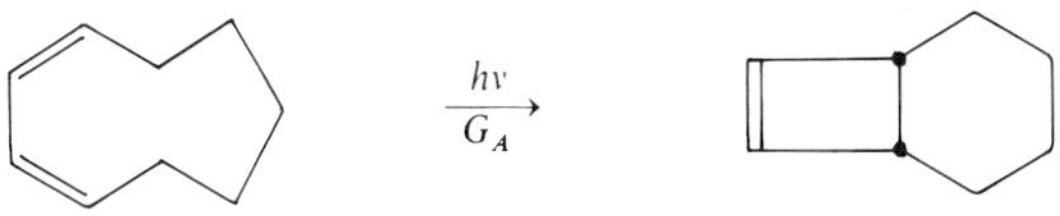

$$\xrightarrow[G_A]{h\nu} \quad \text{(Refs. 14a, b)} \quad (6.53)$$

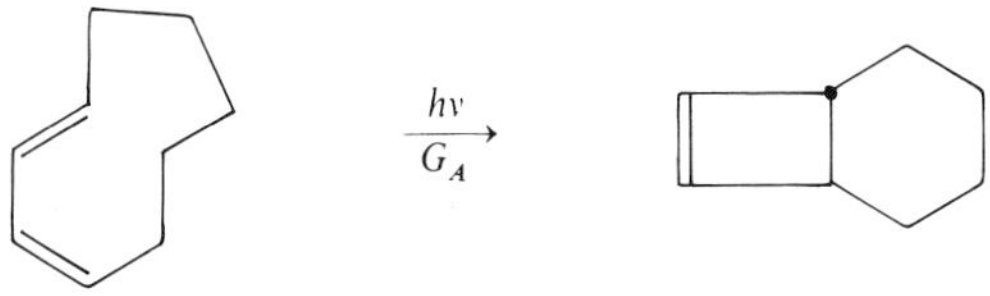

$$\xrightarrow[G_A]{h\nu} \quad \text{(Refs. 14a, b)} \quad (6.54)$$

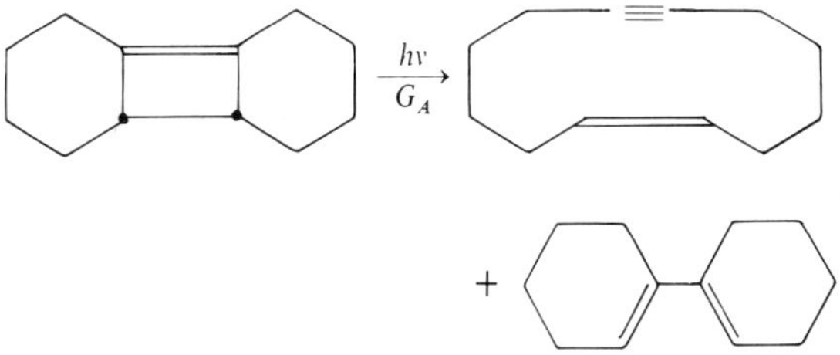

$$\xrightarrow[G_A]{h\nu} \quad + \quad \text{(Ref. 15)} \quad (6.55)$$

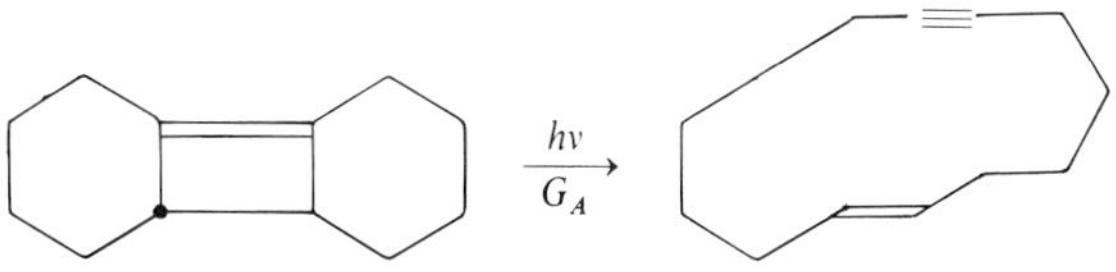

$$\xrightarrow[G_A]{h\nu} \quad \text{(Ref. 15)} \quad (6.56)$$

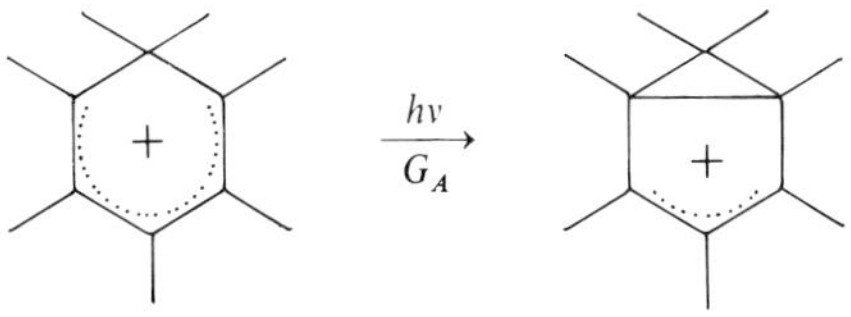

$$\xrightarrow[G_A]{h\nu} \quad \text{(Ref. 16)} \quad (6.57)$$

All these reactions involve Hückel-type, antiaromatic transition states, the cyclizations in equations (6.51)–(6.54) and (6.57) being disrotatory and the ring fissions in equations (6.55) and (6.56) being the reverse of *cis–cis* $2 + 2\pi$ cycloadditions. In the case of equation (6.56), one can also be certain that the reaction must have involved formation of the product in its ground state because otherwise *cis–trans* isomerization would have taken place to give the more stable isomer [cf. equation (6.55)].

2. Six-Electron Anti-Hückel Systems. G_A-type photochemical reactions in six-electron systems involve anti-Hückel transition states. Equations (6.58)[17] and (6.59)[18] show two typical examples:

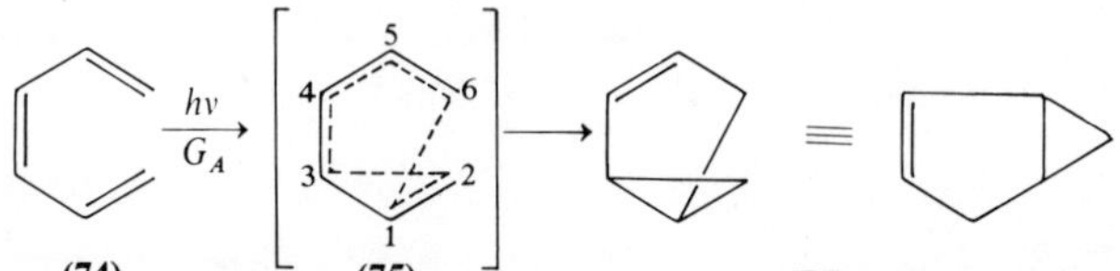

$$\text{(74)} \xrightarrow{h\nu}_{G_A} \text{[(75)]} \longrightarrow \text{(76)} \equiv \quad (6.58)$$

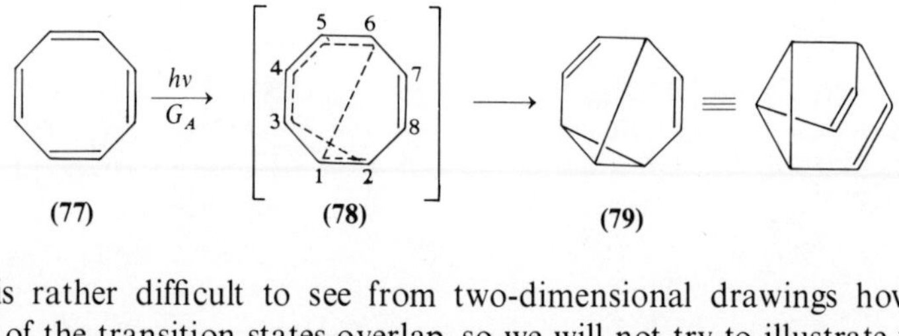

$$\text{(77)} \xrightarrow{h\nu}_{G_A} \text{[(78)]} \longrightarrow \text{(79)} \equiv \quad (6.59)$$

It is rather difficult to see from two-dimensional drawings how the orbitals of the transition states overlap, so we will not try to illustrate them. Examination of models shows that the intermediates (75) and (78) are in fact of anti-Hückel type. If the orbital phases are chosen so that the overlaps in (74) and (77) are all positive, the 1,6 overlaps in (75) and (78) are both negative, the rest positive. The transition states are isoconjugate with (80) and (81) but with one β inverted as indicated.

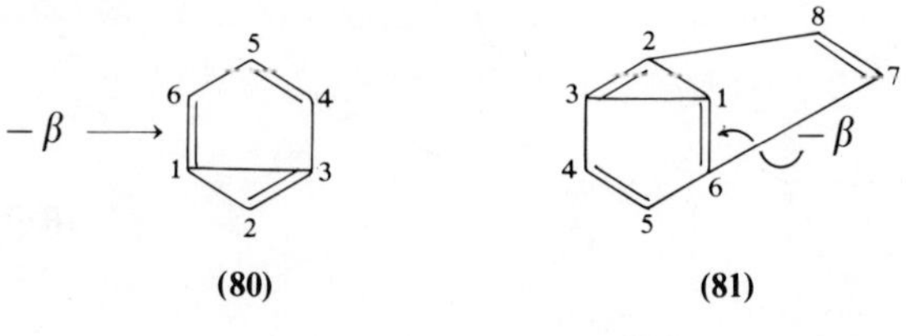

$$-\beta \longrightarrow \text{(80)} \qquad \text{(81)} \; -\beta$$

Note that in (81), the 2,8,7,6 segment contains only essential single and essential double bonds and so does not take part in the delocalized system.

These reactions must unquestionably lead directly to ground-state products and so be of G type since the corresponding reactions on the excited-state surfaces would be highly endothermic.

Hexatrienes can also cyclize photochemically through anti-Hückel transition states which do not involve an extra *meta* bond[19] [equations (6.60) and (6.61)]:

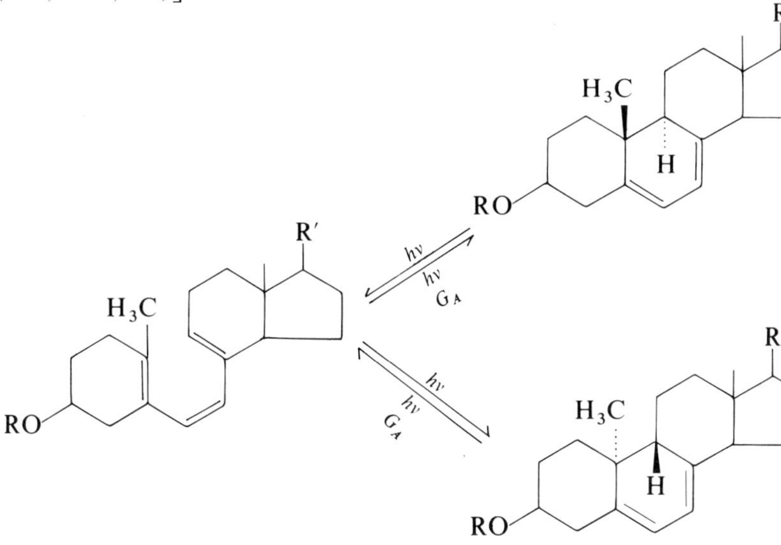

$$(6.60)$$

$$(6.61)$$

These reactions are highly stereospecific and reversible photochemically, indicating that the ground states of both the dienes and the triene are accessible from the antiaromatic intermediate structure.

A third type of photocyclization of hexatrienes occurs through a transition state which is analogous to bicyclo-[2.2.0]-hexatriene, the cyclobutadiene analog of naphthalene; e.g.,

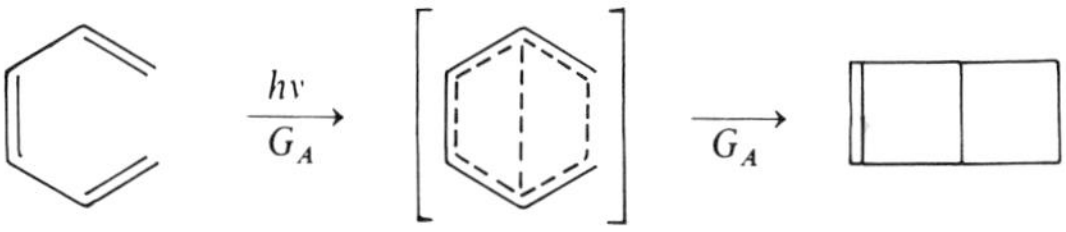

$$(6.62)$$

Benzene and its derivatives undergo photoisomerizations which may be of G type. Thus the formation of benzvalene (82)[20] may be a G_A photoreaction involving a transition state isoconjugate with (83). It is, however, possible that the reaction may be of G_R type, involving the intermediate biradical (84):

(82)   (83)   (84)

Dewar benzene[21] (85) is very probably formed by a G_A reaction via a transition state isoconjugate with (86) [cf. equation (6.62)]. The formation of Dewar pyridine (87)[22] isolated as its hydrate (88) probably involves a similar G_A process.

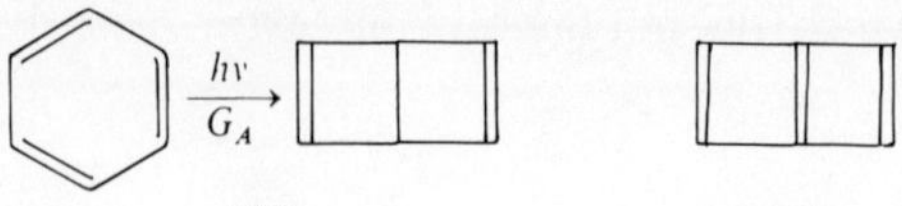

(schemes showing reactions 85→86, 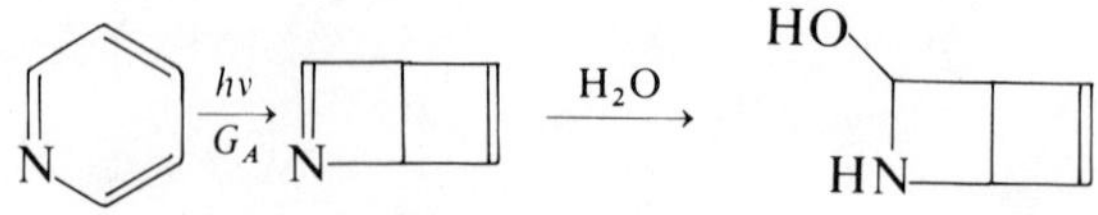 87→88, and structures)

Wilzbach and Kaplan[23] have recently reported the results of a study of the stereospecific addition of olefins to photoexcited benzene. The stereospecificity of these reactions, and the relative energies of the first excited states of the reactants and products, are clear indications that these reactions must be concerted processes. The kinetics and quantum yields of the reactions suggest that all three of the addition processes involve a common exciplex intermediate (89). The formation of the exciplex is an X-type process.

The conversion of the exciplex to (90) or (91) involves transition states that are isoconjugate with the hydrocarbons (93) and (94). The former is antiaromatic, differing only by two essential single bonds [marked a in (93)] from cyclooctatetraene, while (94) contains an antiaromatic cyclobutadiene ring. These are G_A photoreactions, the exciplex reverting to the ground states of (90) and (91) via the corresponding BO holes.

The conversion of (89) to (92), however, involves an aromatic transition state isoconjugate with (95). It seems very likely that either this is a G_R reaction or that it is not a true photochemical reaction but rather a thermal Diels–Alder reaction. The yield of (92) was much less than that of (90) or (91).

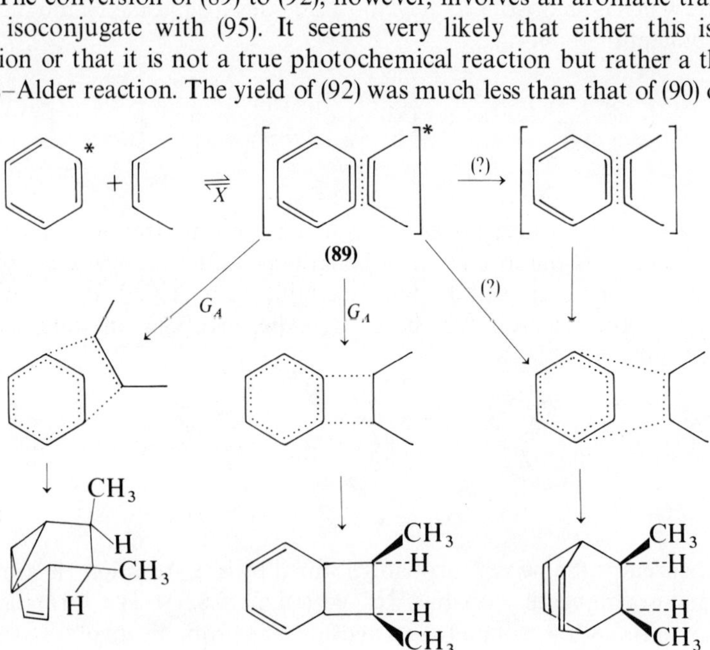

(6.63)

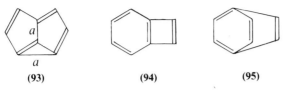

(93)   (94)   (95)

3. Systems with Eight or More Delocalized Electrons. The photochemical additions of acetylenes to 1,4-cyclohexadienes provide interesting examples of reactions involving Hückel-type antiaromatic transition states isoconjugate with cyclooctatetraene; e.g.,[24]

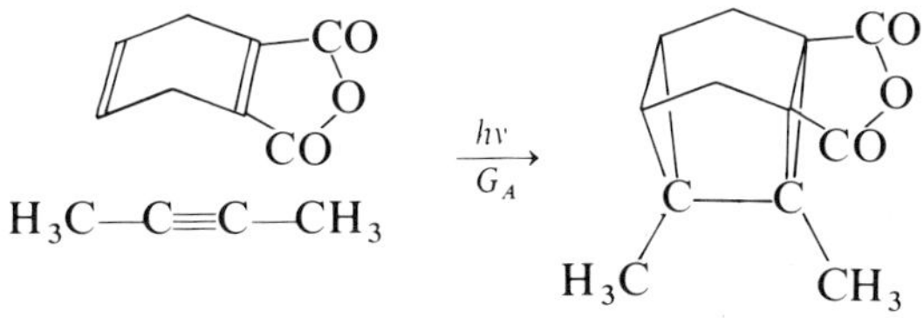

(6.64)

and[25]

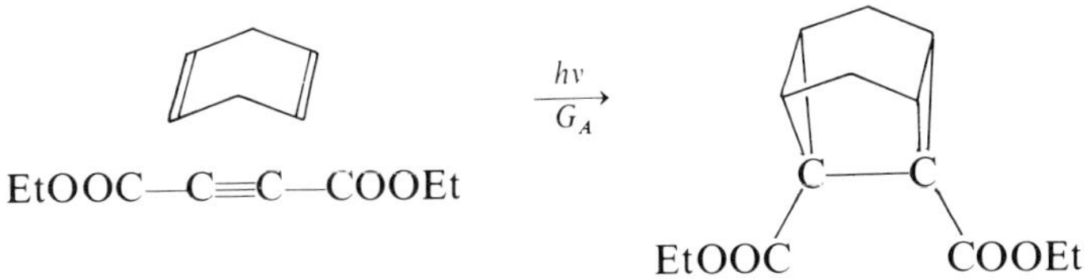

(6.65)

Here the two independent π MOs of the acetylene both take part in the eight-membered ring. Since they are orthogonal, there is no net overlap between them, so the effect is the same as if they constituted two independent π bonds (Fig. 6.37; the AOs overlap in the cyclic order 1,2,3,4,5,6,7,8 and back to 1).

A more interesting reaction involving a transition state isoconjugate with cyclooctatetraene is the photochemical racemization[26] of (96),

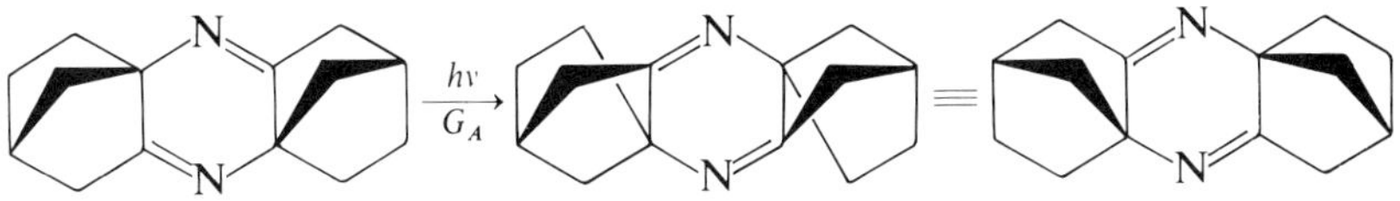

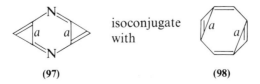

(96)

The transition state for this process is isoconjugate with (97). This in turn is equivalent to cyclooctatetraene (98) since the transannular bonds (a) are essential single bonds.

(97)   isoconjugate with   (98)

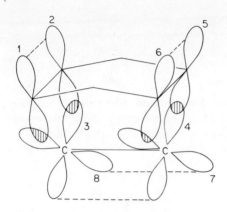

FIGURE 6.37. Orbital overlap in the reactions of equations (6.64) and (6.65).

cis-9,10-Dihydronaphthalene (99) photocyclizes[27a] through an eight-electron transition state to give a tetracyclic intermediate (100) which undergoes a retro-Diels–Alder reaction,

$$\text{(99)} \xrightarrow[G_A]{h\nu} \text{(100)} \longrightarrow \qquad (6.66)$$

Once again the transition state resembles planar cyclooctatetraene.

The photochemical interconversion of *cis*- (99) and *trans*- (103) 9,10-dihydronaphthalene involves both G_A and G_R reactions.[27b]

$$(6.67)$$

(99) (101) (102) (103) (104)

The conversion of (99) to (101) involves a conrotatory ring opening via a transition state isoconjugate with anti-Hückel naphthalene (104). The conversion of (101) to (102) is a G_N *trans–cis* photoisomerization. The final conrotatory ring closure (102)–(103) again occurs through a transition state isoconjugate with anti-Hückel naphthalene. The stereochemical selectivity of the first and third processes shows them to be G_A reactions rather than G_R reactions involving intermediate radicals.

Light Absorption and Photochemistry

There are a large number of photocyclizations reactions of polycyclic aromatic systems that all proceed through antiaromatic transitions states. The cyclization[28a] of *cis*-stilbene (105) and the related diarylethylene (106)[28b] are classical examples of this reaction,

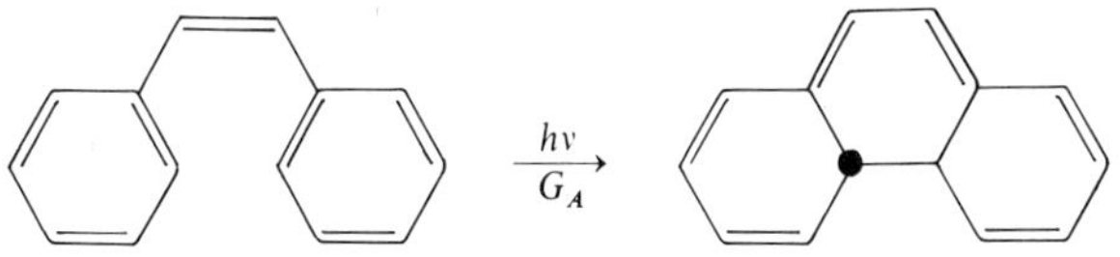

$$\text{(105)} \xrightarrow[G_A]{hv} \text{product} \qquad (6.68a)$$

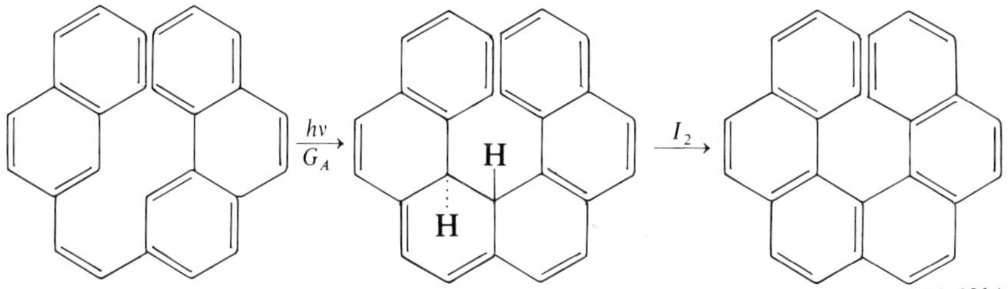

$$\text{(106)} \xrightarrow[G_A]{hv} \text{dihydro intermediate} \xrightarrow{I_2} \text{hexahelicene} \qquad (6.68b)$$

When reaction (6.68b) is conducted with circularly polarized light, the product hexahelicene is found to be optically active,[28b] a result of the circular dichroism of the enantiomorphic conformations of the starting material.

The cyclization of stilbene occurs through an anti-Hückel transition state analogous to anti-Hückel phenanthrene. Photodimerizations of polycyclic aromatic systems are also very common. The dimerization of anthracene derivatives is a classical example:

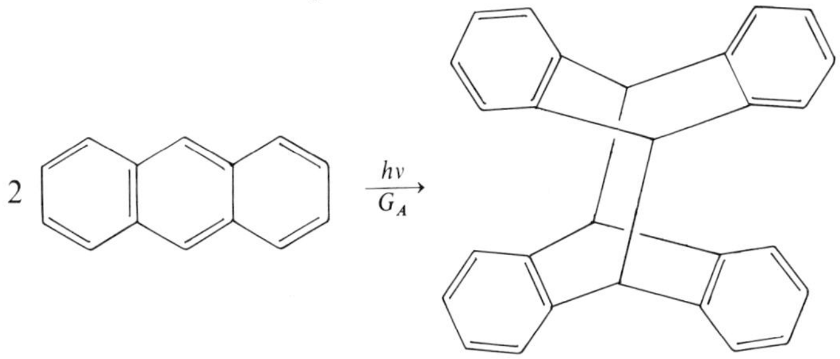

$$2 \, \text{anthracene} \xrightarrow[G_A]{hv} \text{dimer} \qquad (6.69)$$

The core of the transition state for the anthracene dimerization is the tricyclic relative (107) of cyclooctatetraene, which can easily be shown to be antiaromatic.

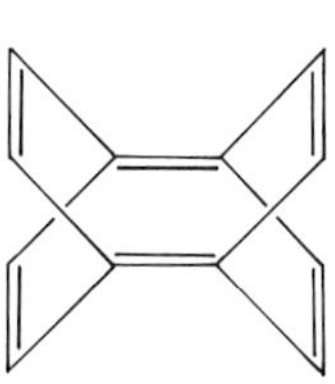

(107)

The photocyclization[29] of (108) to (110) is another ten-electron process, this time involving a system isoconjugate with an odd hydrocarbon anion. The first step is probably a G_A photochemical conrotatory closure to (109). The transition state, being of anti-Hückel type in a fused 6–5 system with ten delocalized electrons, should be antiaromatic. The intermediate (109), isoconjugate with a linear C_7 carbanion, was not isolated but underwent rearrangement, probably by a 1,5 sigmatropic shift via a transition state isoconjugate with the indenyl anion (111).

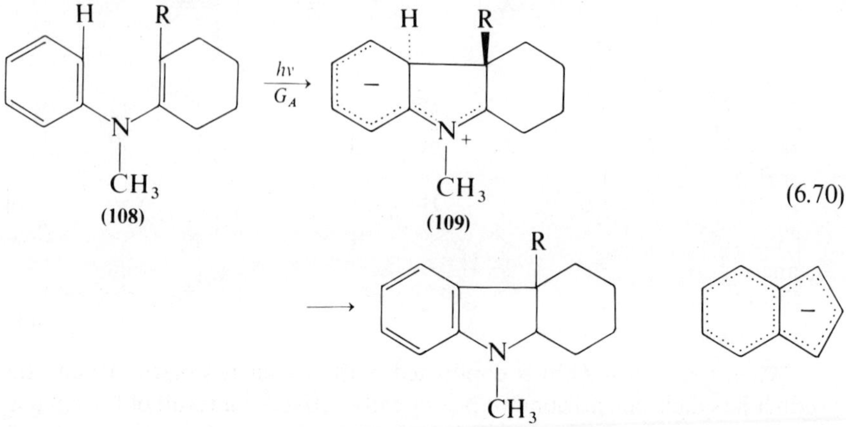

$$(6.70)$$

Another curious reaction is the photochemical conversion[30] of (112) to (113) which occurs only from higher excited states of (112), excitation to the lowest excited state being ineffective:

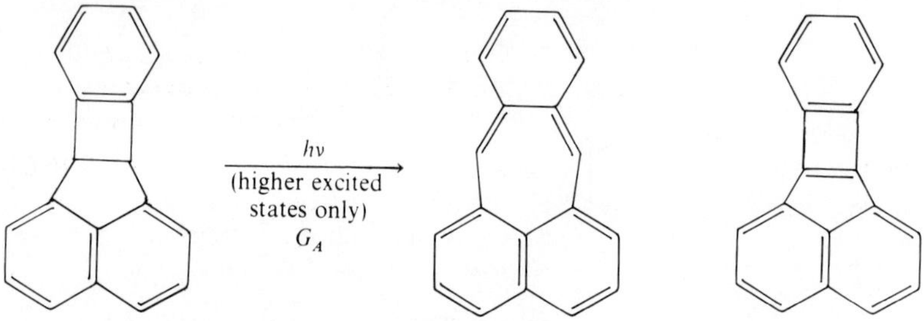

$$(6.71)$$

The reason for this is that the benzene and naphthalene moieties in (112) are "insulated" from one another by the two central saturated atoms. The light absorption of (112) therefore corresponds to transitions inside the naphthalene moiety, or the benzene moiety. Since naphthalene absorbs at longer wavelengths than benzene, the first absorption band of (112) corresponds to a transition localized in the naphthalene system.

The conversion of (112) to (113) involves a transition state which is isoconjugate with (114). In this molecule, the bonds linking the naphthalene

Light Absorption and Photochemistry

and benzocyclobutadiene units are essential single bonds, so the units are essentially uncoupled. Excitation of the naphthalene moiety in (112) therefore fails to lead to reaction because the excitation energy is in the wrong place.

It is of course impossible for the excitation energy in (112) to be transferred from the naphthalene unit to the benzene unit, because this would be uphill. However, during the antiaromatic rearrangement of (112) to (113), the transition energy at some point must fall to zero (at the BO hole). Much earlier in the reaction, the excitation energy of the nascent benzocyclobutadiene unit must fall below that of naphthalene; could not, then, transfer of excitation energy from naphthalene to this nascent benzocyclobutadiene unit lead to reaction?

The answer is that the corresponding crossing point would occur only after the reaction had proceeded to a considerable extent. Up to this point, the reaction is in effect a thermal antiaromatic pericyclic rearrangement and would be proceeding uphill in energy. Thus the transfer of excitation occurs too late for the reaction to occur without appreciable activation and this means that deexcitation of the reactants will take place before the reaction has a chance to occur. Excitation to the T_2 state of the naphthalene, which can directly sensitize the benzene moiety, solves this problem.

A 16-electron photo-double-cyclization[31] is our last example of photochemical G_A pericyclic reaction,

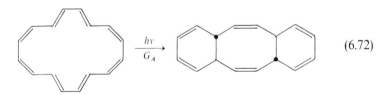

(6.72)

The transition state for this reaction is a superantiaromatic (anti-Hückel-dibenzo)–normal-Hückel-cyclooctatetraene. If the phase relationships in these reactions are examined, it is clear that there will be a phase inversion at *both* ring junctions. By proper choice of the basis set of AOs, these phase inversions can be eliminated, so the transition state is of Hückel type.

6.17 Chemiluminescent Reactions

From the principle of microscopic reversibility, all chemical processes must be reversible and this principle must apply to photochemical processes as well as others. Since a photochemical reaction is one brought about by absorption of light, its reverse will be a reaction in which light is emitted. Such a process is called *chemiluminescence*. It is of course an essential feature of these reactions that chemical energy is converted into light energy; the reverse of an X-type process in which absorbed light was merely reemitted in the form of light of higher energy would not be very significant. The reactions that concern us are therefore reversals of *G*-type photochemical reactions in which a reaction between molecules in their ground states leads to emission of light.

In order that light be emitted in the ultraviolet or visible region, it is obviously essential that the reaction in question should be extremely exothermic. As we have seen, emission of a quantum in the middle of the visible region (500 nm) corresponds to the emission of energy equivalent to 50 kcal/mole; a chemiluminescent reaction can therefore lead to emission of such light only if it is exothermic by this amount.

The simplest way to get chemical energy on this scale is by the combination of radicals, by the recombination of ion pairs, or by extremely exothermic electron transfer processes. All these three types of reaction can give rise to chemiluminescence under suitable conditions. A simple example is the oxidation of extremely energetic anion radicals formed by reduction of neutral molecules. As Fig. 6.38 indicates, an electron transfer from such an ion to a strong electron acceptor, i.e., one with an AO of low energy, can leave the resulting molecule in an excited state. A good example[32] is the reaction of sodium 9,10-diphenylanthracene with radicals. This is a − *E* (reverse *E*-type) reaction.

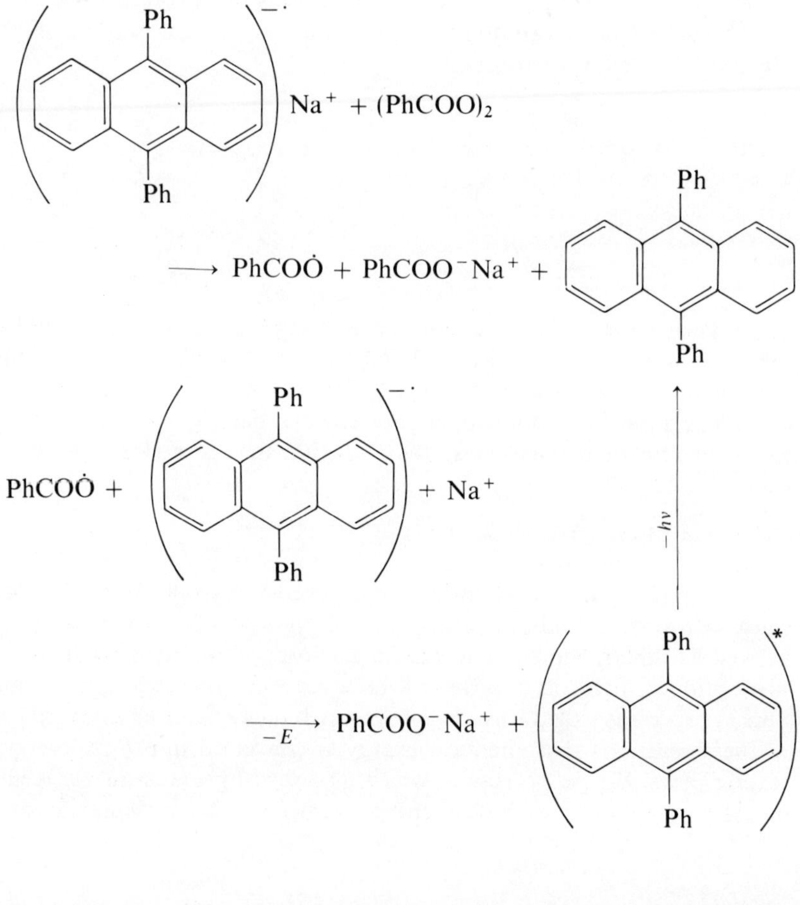

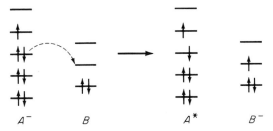

FIGURE 6.38. Electron transfer between high-energy anion radical A^- with an oxidizing agent B, leading to an excited state A^* of A.

It is evident that reaction of a molecule in its ground state to give a product in its excited state can only occur if the process involves a breakdown in the BO approximation. Just as the excited state can leak through a BO hole to the ground state, so can the ground state in a reaction, if it has sufficient energy, burrow upward through the BO hole into the excited state and so give rise to chemiluminescence. Otherwise the conversion of chemical (i.e., thermal) energy into electronic energy would be far too inefficient to compete with dissipation of the thermal energy of collisions with surrounding solvent molecules.

While $-E$-type chemiluminescent processes are of some interest, the most important reactions of this kind are of $-G$-type. A $-G$-type chemiluminescent reaction proceeds through a transition state or intermediate where the ground-state and excited-state surfaces touch or almost touch. This point serves as a BO hole, permitting the system to pass directly from the ground state of the reactants to electronically excited product. It goes without saying that the overall reaction must be very exothermic for visible

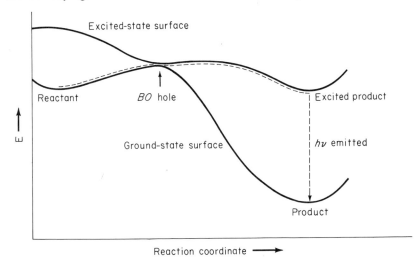

FIGURE 6.39. The course of a $-G_A$-type chemiluminescent reaction; the reaction path is indicated by the broken line.

light to be emitted because light in the visible region corresponds to large amounts of energy, chemically speaking.

One possibility would be a very exothermic $-G_A$-type reaction, i.e., a pericyclic reaction involving an antiaromatic transition state which, as we have seen, can act as a BO hole. The essential features of such a process are indicated in Fig. 6.39. Because the ground-state reaction is very exothermic, the activation energy is relatively low in spite of the antiaromatic nature of the transition state. At this point there is a BO hole, so the reaction can pass on to the excited state surface and so end up in excited products; emission of light completes the reaction.

Another possibility is a $-G_R$-type reaction in which fission of one bond in a ring gives rise to a biradical. If the biradical can rearrange to give a "normal" molecule or pair of molecules in which all the electrons are paired, the latter can be formed in an electronically excited state; for a biradical can again act as a BO hole, its excitation energy being zero.

A number of chemiluminescent reactions are known which seem to follow one or the other of these two possible paths. The simplest example is the thermal decomposition of dioxetanes into pairs of carbonyl compounds, one of which may be excited,

$$\begin{array}{c} R_2C-O \\ | \quad | \\ R_2C-O \end{array} \xrightarrow{G_A} R_2C{=}O + (R_2C{=}O)^* \tag{6.74}$$

This could take place by a concerted $-G_A$-type process which would be the reverse of a *cis–cis* $(2 + 2)\pi$ cycloaddition and so involve an antiaromatic transition state isoconjugate with cyclobutadiene. Alternatively, it might be of $-G_R$ type, taking place in two steps via the rate-determining formation of an intermediate biradical,

$$\begin{array}{c} R_2C-O \\ | \quad | \\ R_2C-O \end{array} \xrightarrow{\text{slow}} \begin{array}{c} R_2C-O\cdot \\ | \\ R_2C-O\cdot \end{array} \xrightarrow[-G_R]{\text{fast}} \begin{array}{c} R_2C{=}O \\ \\ (R_2C{=}O)^* \end{array} \tag{6.75}$$

The choice between these two possibilities ($-G_A$ or $-G_R$) is still uncertain. Note that if the reaction is concerted, then it must take place by the reverse of a face-to-face π cycloaddition. If the reaction took place by the reverse of an "allowed" *cis–trans* π cycloaddition (see p. 350), it could not be chemiluminescent because in an aromatic ("allowed") pericyclic reaction, the ground-state and excited-state surfaces remain far apart throughout. Similar chemiluminescent reactions are also observed in the case of bicyclic oxetanes, a good example being the chemiluminescent oxidation[33a] of lophine (115) by alkaline hydrogen peroxide. Here again it is uncertain whether the final step is a concerted $-G_A$-type process or a $-G_R$-type process involving an intermediate biradical.

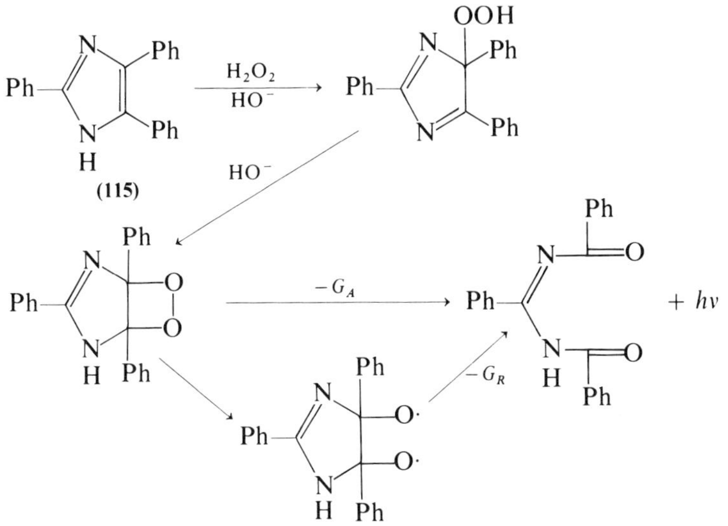

[Scheme showing reactions of compound (115): Ph-substituted imidazoline reacting with H_2O_2/HO^- to form OOH-substituted intermediate, then via HO^- to a dioxetane, which loses G_A to give the diimine product Ph-C(=N-)-C(=N-H)-Ph + $h\nu$, with an alternative pathway through a biradical intermediate losing G_R.]

In the case of tetra-methyl-1,2-dioxetane [(6.75)], the luminescence originates from triplet acetone which is formed directly from the starting material. In the presence of oxygen, the decomposition reaction is first order in dioxetane; in the absence of oxygen in hydrocarbon solvents, the destruction of dioxetane is second order in dioxetane and appears to be autocatalytic in triplet acetone.[33b] Impurities in hydroxylic solvents (presumably transition metal ions) can catalytically quench the luminescence; however, in the presence of lanthanide ions, the chemiluminescence is quite efficient, with emission occurring from lanthanide excited states.[33f]

Potential surface calculations[33g] for the ground state, first excited singlet, and first excited triplet indicate that the triplet excited state surface intersects the ground-state surface *before* the antiaromatic transition state is reached, while the singlet surface lies higher in energy. The easiest path for the reaction thus involves intersystem crossing to the triplet surface, leading to one molecule of unexcited ketone and one of triplet excited ketone. In the autocatalytic reaction, energy and spin transfer from excited ketone to the dioxetane excites the latter to its triplet state, which at once dissociates. The frequency factors for these reactions are large and typical of spin-allowed processes and not singlet–triplet conversions. A likely explanation involves the relative allowedness of spin inversions in orbitally degenerate systems.

Our final example is one of the classic cases of chemiluminescence, the oxidation of luminol [aminophthalhydrazide (116)] by alkaline hydrogen peroxide in the presence of a ferrous salt (i.e., alkaline Fenton's reagent).[33a,d] This is of particular interest because the autooxidation of lumiferin by air, the process which fireflies use to generate light, probably follows a similar course. The first step in the oxidation of (116) probably leads to the cyclic axo derivative (117). Base-catalyzed addition of hydrogen peroxide to (117) then gives (118), which decomposes very exothermically to biphthalate ion (120) and nitrogen.

(116) (117) (118) (120) (119)

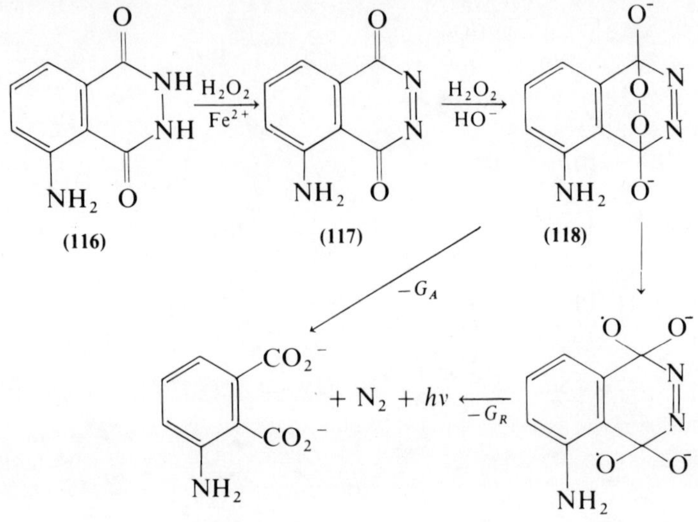

In this case, the final reaction can be most simply interpreted as a $-G_R$-type process involving the intermediate biradical (119). If it were of $-G_A$-type, the transition state (121) would be isoconjugate with the *p*-xylylene dianion (122), which is not antiaromatic. It is admittedly impossible to predict the MO energies in (121) on the basis of PMO theory because there are so many heteroatoms in it and because (122) is not in any case a "normal" molecule, being derived from an even AH [i.e., *p*-quinodimethane (123)] by the introduction of two electrons into the lowest antibonding MO (LUMO). It is quite possible that the HOMO and LUMO of (121) are degenerate and that the reaction is an unusual *G*-type process in which the BO hole arises from this "accidental" degeneracy.

(118)

(121)

(122)

(123)

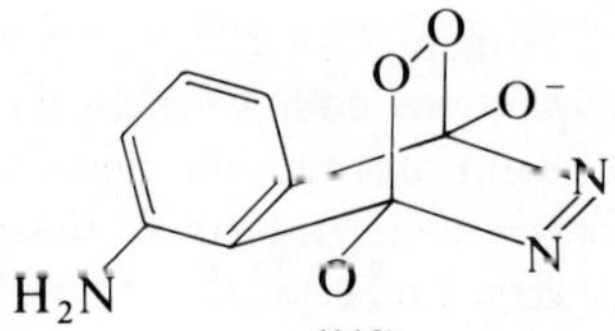

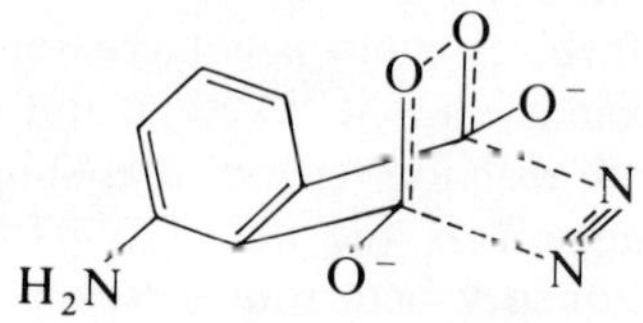

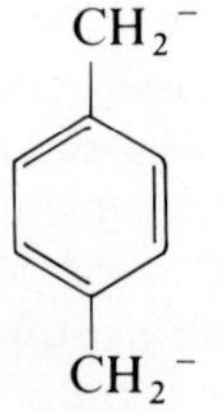

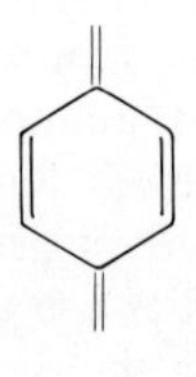

One interesting feature of the luminol chemiluminescence is the recently discovered fact that the chemiluminescence of aminophthalate ions can be significantly different from luminescence generated by light absorption.[33d,e] The basis of this difference appears to be greater ion pairing, particularly with sodium ions, by the excited aminophthalate ions produced by light absorption. These results suggest that the transition state for the chemiluminescent decomposition is rotationally restricted as compared to the ground state of the aminophthalate ion. Unfortunately, this added information does not seem to probe the distinction between the $-G_R$ and $-G_A$ paths for this reaction.

6.18 Summary

While this discussion has shown how the PMO method can be extended to cover the main phenomena of photochemistry, it is by no means exhaustive. We have deliberately omitted reference to several types of photochemical processes because although it is not difficult to extend the PMO treatment to them, such an extension would take an undue amount of space. One example which we previously mentioned is the photochemical conversion of ketene or diazomethane to carbene,

$$CH_2{=}N{=}N \xrightarrow{h\nu} CH_2 + N_2$$
$$CH_2{=}C{=}O \xrightarrow{h\nu} CH_2 + CO$$

(6.76)

The driving force in these reactions is of course the extremely high heat of formation of N_2 or CO, the bonds in these molecules being the strongest known. The details of the G-type processes involved, while not difficult, would take rather a lot of space to explain.

The large majority of photochemical processes can, however, be easily seen to fall into the categories enumerated here and can consequently be explained in a very simple manner in terms of the PMO approach. This is an important achievement because the development of organic photochemistry has been bedeviled in recent years by an enormous outpouring of unexplained observations, most of which can be interpreted very simply in terms of the ideas presented here, including the following:

1. The distinction between X-type reactions where the BO approximation holds and those where it does not; PMO theory can be used to follow the course of X-type reactions and to enumerate the situations where the BO approximation fails.
2. Reactions (G type) where the reaction path is determined by the presence of a "BO hole"; PMO theory can be used to follow the course of such reactions.

Finally, the use of "internal conversion" as a cosmetic term for vague mechanisms should be avoided. In the cases where radiationless internal conversion is definitely known to occur, the cascade of states can often be discovered,[34] or the processes may be pericyclic pseudo reactions.

Problems

6.1. Prove that:
Antiaromatic ground states of nonalternant monocyclic hydrocarbons will be aromatic with reference to the excited singlet states of their open-chain analogs; nonaromatic (radical), nonalternant systems will be nonaromatic with reference to the excited singlet states of their open-chain analogs.
Complete the proof that the rules for aromaticity in anti-Hückel systems hold for the excited states of monocyclic Hückel systems.

6.2. Prove that

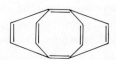

would be antiaromatic on the ground-state surface.

6.3. Deduce a reasonable G-type mechanism for the following:

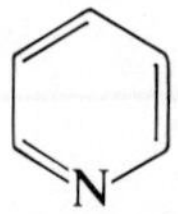

$\xrightarrow[H_2O]{h\nu,\ 2537\text{Å}}$ $H_2N—CH{=}CH—CH{=}CH—CHO$

6.4. 1-Iodonaphthalene is yellow and its triplet T_1 state lifetime τ is 3×10^{-3} sec, although the frequencies of its main absorptions are virtually identical with those of naphthalene. Naphthalene is colorless and its T_1 lifetime is 20 sec. What does this imply about the effect of heavy atoms on singlet–triplet and singlet–singlet processes?

6.5. Pyridine shows no luminescence. Kasha and co-workers have explained this by observing that the $n \to \pi^*$ state of pyridine is lower in energy than the $\pi \to \pi^*$ state, although the extinction coefficient for absorption is extremely small. Pyridine radiationlessly crosses to the $n \to \pi^*$ state rapidly and this state does not emit. Propose a mechanism for the rapid, radiationless relaxation of an $n \to \pi^*$ excitation in pyridine.

6.6. Select the best compound from the following pairs for the stated purpose.

a.

$E_t = 60$ kcal/mole

$E_t = 63$ kcal/mole

to quench phosphorescence of

$E_t = 69$ kcal/mole

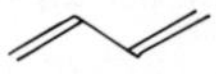

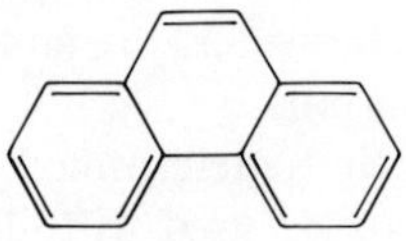

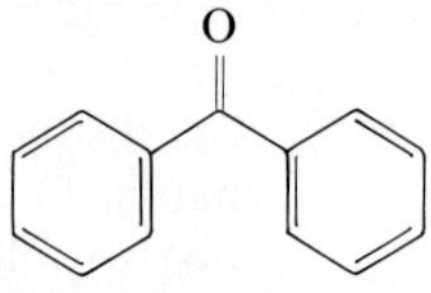

b.

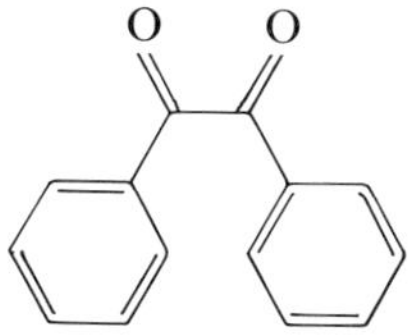

benzil
$E_{t, n \to \pi^*} = 60$ kcal/mole

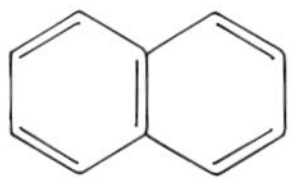

naphthalene
$E_{t, \pi \to \pi^*} = 60$ kcal/mole

to sensitize phosphorescence of

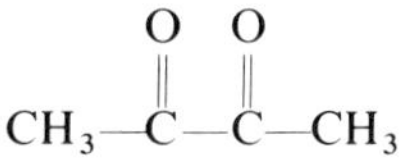

$CH_3-\overset{O}{\underset{\|}{C}}-\overset{O}{\underset{\|}{C}}-CH_3$
$E_{t, n \to \pi^*} = 54$ kcal/mole

6.7. Arrange the following sets of compounds in order of increasing wavelength of their first absorption:

a.

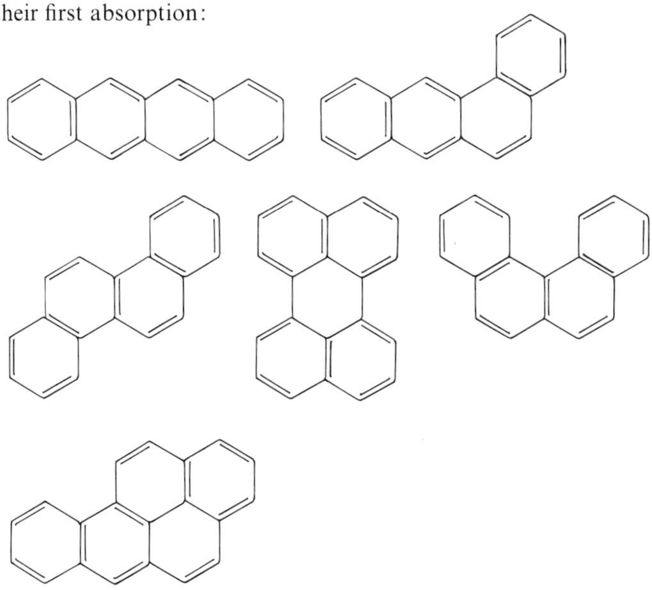

b.

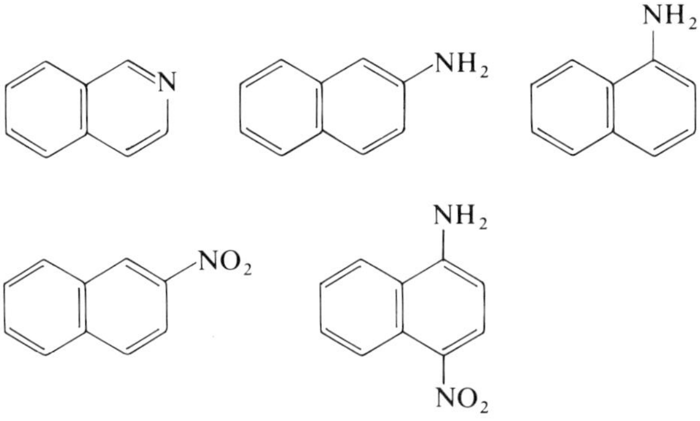

6.8. Suggest mechanisms for the following reactions. Indicate whether the reactions are *X*- or *G*-type processes by explicitly indicating the transition state for return to the ground-state surface.

a. $CH_3C(O)-C(O)-OCH_2CH_3 \xrightarrow[C_6H_6]{h\nu} 2CH_3CHO + CO$

b.

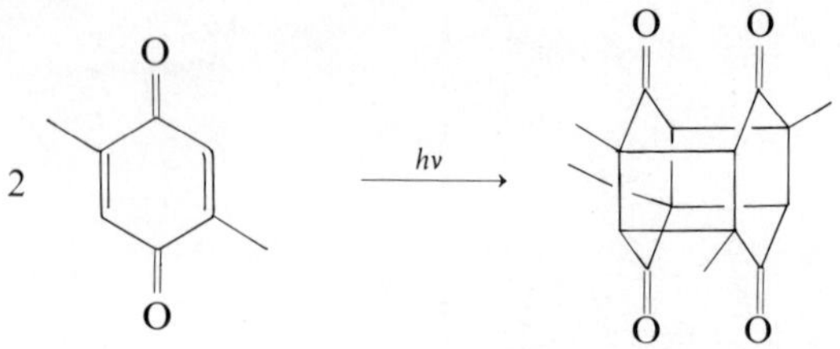

c.

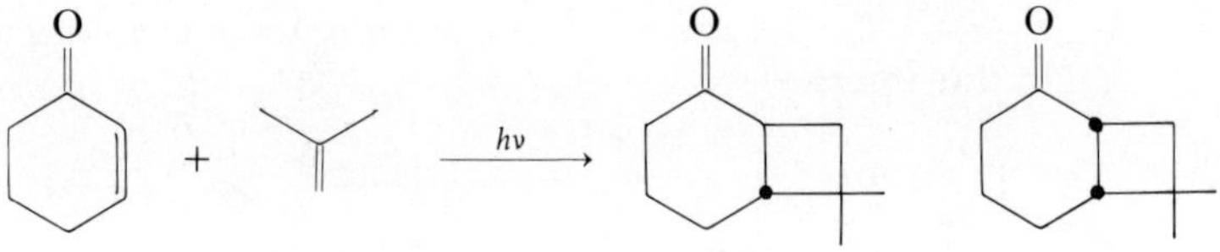

d.

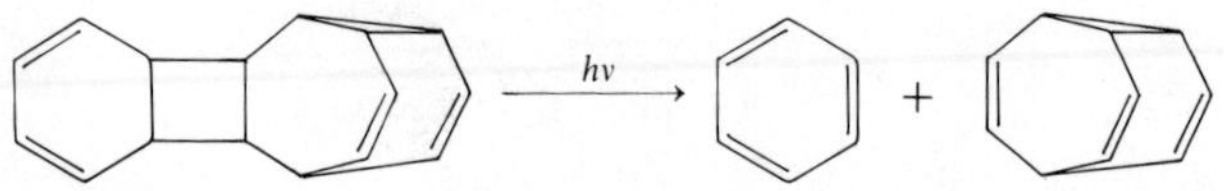

e.

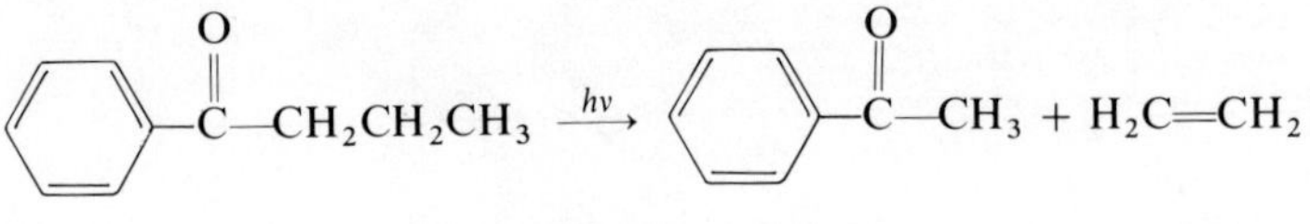

6.9. Discuss the electronic mechanism of the following reaction sequence. Is the last thermal step potentially chemiluminescent?

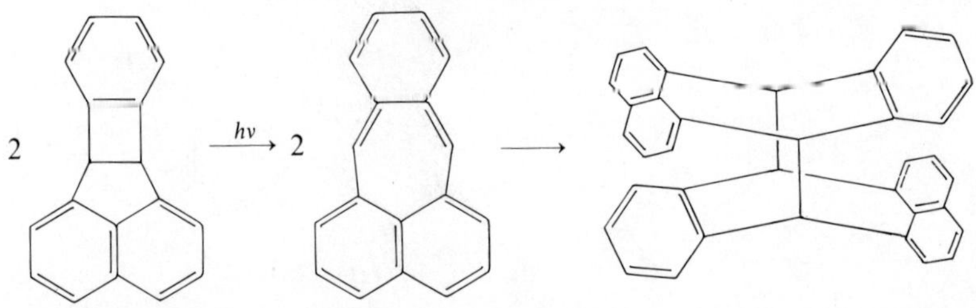

6.10. Explain the following observations.
   a. The quantum yield for photochemical solvolysis of 3-methoxy-benzyl acetate in aqueous dioxane is an order of magnitude greater than that for 4-methoxy-benzyl acetate.

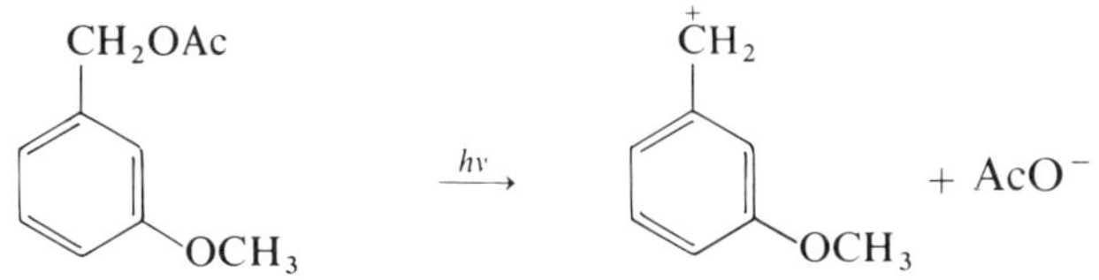

b. *m*-Nitrophenyl trityl ether undergoes a substantially enhanced photochemical solvolysis to *m*-nitrophenolate in aqueous dioxane, whereas photolysis does not significantly speed the rate of solvolysis of the *p*-nitro isomer.

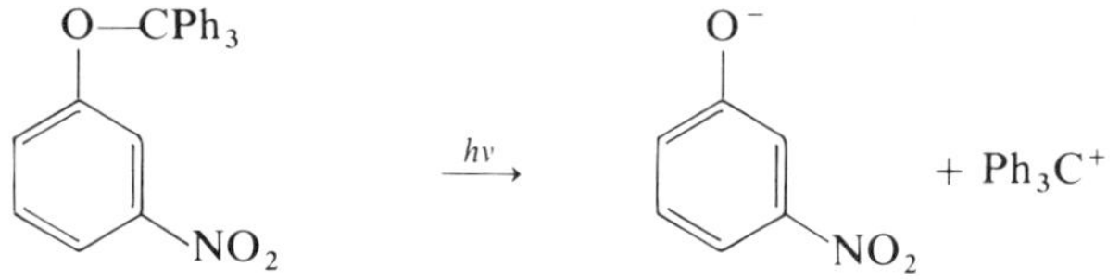

6.11. By heating the diol (A) with aluminum chloride, Clar obtained a hydrocarbon which could have been (B) or (C). The first absorption band was at 643 nm. Which was it?

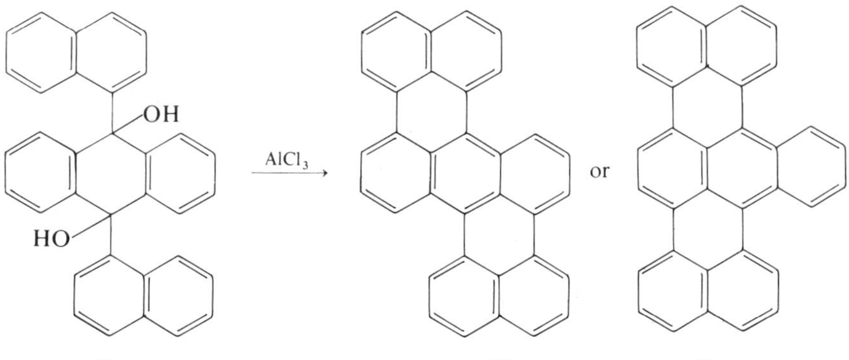

Selected Reading

The following books deal with organic spectroscopy and photochemistry and are recommended to the reader who wishes to delve further into the subject:

- H. H. JAFFE AND M. ORCHIN, *Theory and Applications of Ultraviolet Spectroscopy*, John Wiley and Sons, New York, 1962.
- J. G. CLAVERT AND J. N. PITTS, Jr., *Photochemistry*, John Wiley and Sons, New York, 1966.
- N. S. TURRO, *Molecular Photochemistry*, W. A. Benjamin, New York, 1965.
- R. O. KAY, *Organic Photochemistry*, McGraw-Hill Book Co., New York, 1966.
- M. J. CARMIER, D. M. HERCULES, AND J. LEE, eds., *Chemiluminescence and Bioluminescence*, Plenum Press, New York, 1973.

Numerous reviews will be found in:

Advances in Photochemistry, 1963–present.

A detailed discussion of the treatment outlined here will be found in the following articles:

M. J. S. Dewar, *J. Chem. Soc.* **1950**, 2329; **1952**, 3532, 3544.
R. C. Dougherty, *J. Am. Chem. Soc.* **93**, 7187 (1971).

General discussions of photochemical reactivity can be found in the following articles:

N. D. Epiotis, *J. Am. Chem. Soc.* **94**, 1941, 1946 (1972).
J. Michl, *Mol. Photochem.* **4**, 243, 256, 287 (1972).

Both the treatments above depend strongly on symmetry arguments. The series of papers by Michl explicitly discusses singlet and triplet energy and reactivity differences. However, neither of the above treatments emphasizes the phenomenological and theoretical distinctions among X-, G_R-, G_N-, and G_A-type photochemical reactions.

References

1. (a) J. W. Happ, M. T. McCall, and D. G. Whitten, *J. Am. Chem. Soc.* **93**, 5496 (1971).
   (b) P. Brasem, J. G. Zammers, J. Cornelisse, J. Lagtenburg, and E. Havinga, *Tetrahedron Letters* 685 (1972); see also R. R. Hawtala and R. L. Letsinger, *J. Org. Chem.* **36**, 3762 (1971).
   (c) P. J. Kropp, T. H. Jones, and G. S. Poindexter, *J. Am. Chem. Soc.* **95**, 5420 (1973).
2. (a) J. B. Birks, *Progress in Reaction Kinetics* **8**, 181 (1969).
   (b) N. C. Yang and J. Libman, *J. Am. Chem. Soc.* **95**, 5783 (1973).
3. E. Vander Dorekt, *Progress in Reaction Kinetics* **8**, 274 (1969).
4. (a) G. Jackson and G. Porter, *Proc. Roy. Soc.* **A2000**, 13 (1961).
   (b) C. G. Stevens and S. J. Strickler, *J. Am. Chem. Soc.* **95**, 3922 (1973).
5. C. A. Taylor, M. A. El-Bayoomi, and M. Kasha, *Proc. Nat. Acad. Sci. U.S.* **63**, 253 (1969).
6. (a) K. Schaffner, *Adv. Photochem.* **4**, 81 (1966).
   (b) P. J. Knapp, *Org. Photochem.* **1**, 1 (1967).
   (c) H. E. Zimmerman, *Adv. Photochem.* **1**, 183 (1963).
   (d) H. E. Zimmerman and G. A. Epling, *J. Am. Chem. Soc.* **94**, 3245 (1972).
7. K. N. Houk and D. J. Northington, *J. Am. Chem. Soc.* **93**, 6693 (1971).
8. (a) J. C. Dalton and M. J. Turro, *Ann. Rev. Phys. Chem.* **21**, 499 (1970).
   (b) F. D. Lewis, R. W. Johnson, and D. R. Kory, *J. Am. Chem. Soc.* **95**, 6470 (1973).
   (c) P. J. Wagner, A. E. Kemppainan, and H. N. Schott, *J. Am. Chem. Soc.* **95**, 5604 (1973).
9. (a) N. J. Turro and P. A. Wriede, *J. Am. Chem. Soc.* **92**, 320 (1970).
   (b) H. E. Zimmerman, D. R. Amil, and H. Hemetsberger, *J. Am. Chem. Soc.* **95**, 4606 (1973).
   (c) H. Hart and G. M. Love, *J. Am. Chem. Soc.* **95**, 4592 (1973).
10. J. J. McCullough and W. S. Wu, *Tetrahedron Letters* 3951 (1971).
11. (a) E. Haselbach and W. Ebarbach, *Helv. Chim. Acta* **56**, 1944 (1973);
    (b) W. Eberbach, *Chimia* **25**, 248 (1971).
    (c) R. Korenstein, K. A. Muszkat, and S. Sharafy-Ozeri, *J. Am. Chem. Soc.* **95**, 6177 (1973).
12. R. Srinivasan, *J. Am. Chem. Soc.* **90**, 4498 (1968).

13. W. G. Dauben, R. G. Cargill, R. M. Coates, and J. Saltiel, *J. Am. Chem. Soc.* **88**, 2742 (1966).
14. (a) K. M. Schumate, R. N. Neuman, and G. J. Fonken, *J. Am. Chem. Soc.* **87**, (1965).
    (b) W. G. Dauben and R. L. Cargill, *J. Org. Chem.* **27**, 1910 (1962).
15. J. Saltiel and L. N. Lim, *J. Am. Chem. Soc.* **91**, 5404 (1969).
16. R. F. Childs and S. Winstein, *J. Am. Chem. Soc.* **90**, 7146 (1968).
17. J. Meinwald and P. H. Mazzocchi, *J. Am. Chem. Soc.* **88**, 2850 (1966).
18. H. E. Zimmerman and H. Iwamura, *J. Am. Chem. Soc.* **92**, 2015 (1970).
19. G. M. Sanders and E. Havinga, *Rec. Trav. Chim.* **83**, 665 (1964) and papers cited therein.
20. K. E. Wilzbach, J. S. Ritscher, and L. Kaplan, *J. Am. Chem. Soc.* **89**, 1031 (1967).
21. E. E. van Tamelen and S. P. Pappas, *J. Am. Chem. Soc.* **84**, 3789 (1962).
22. K. E. Wilzbach and D. J. Rausch, *J. Am. Chem. Soc.* **92**, 2178 (1970).
23. K. E. Wilzbach and L. Kaplan, *J. Am. Chem. Soc.* **93**, 2073 (1971).
24. R. Askani, *Chem. Ber.* **98**, 3618 (1965).
25. M. Takahashi, Y. Kitahova, I. Murata, T. Nitta, and M. C. Woods, *Tetrahedron Letters* 3387 (1968).
26. D. G. Farnom and G. R. Carlson, Abstracts of papers presented at the 160th ACS meeting.
27. (a) S. Masamuni, R. T. Seidnur, H. Zenda, M. Weisel, N. Nakatsuka, and G. Bigam, *J. Am. Chem. Soc.* **90**, 5286 (1968).
    (b) E. E. van Tamelen and J. H. Bukath, *J. Am. Chem. Soc.* **89**, 151 (1967).
28. (a) K. A. Muszkat and E. Fisher, *J. Chem. Soc.* B **1967**, 662.
    (b) A. Moradpour, J. F. Nicoud, G. Balavoine, H. Kagan, and G. Tsoucaris, *J. Am. Chem. Soc.* **93**, 2353 (1971).
29. O. L. Chapman and G. L. Eian, *J. Am. Chem. Soc.* **90**, 5329 (1968).
30. J. Michl, *J. Am. Chem. Soc.* **93**, 523 (1971).
31. G. S. Schroder, W. Martin, and J. R. M. Oth, *Angew. Chem. Int. Ed.* **6**, 879 (1967).
32. (a) E. Chandross and F. I. Sonntag, *J. Am. Chem. Soc.* **88**, 1089 (1966).
    (b) R. Bezman and L. R. Faulkner, *J. Am. Chem. Soc.*, **94**, 6317, 6324, 6331 (1972).
33. (a) J. McCaptra, *Quart. Rev.* **20**, 485 (1966).
    (b) P. Lechten, A. Yekta, and N. J. Turro, *J. Am. Chem. Soc.* **95**, 3027 (1973).
    (c) T. Wilson, M. E. Landis, A. L. Baumstark, and P. D. Bartlett, *J. Am. Chem. Soc.* **95**, 4765 (1973).
    (d) P. D. Wildes and E. H. White, *J. Am. Chem. Soc.* **95**, 2610 (1973).
    (e) E. H. White, D. F. Roseell, C. C. Wei, and P. Wildes, *J. Am. Chem. Soc.* **95**, 6223 (1973).
    (f) P. D. Wildes and E. H. White, *J. Am. Chem. Soc.* **93**, 6286 (1971).
    (g) M. J. S. Dewar and S. Kirschner, *J. Am. Chem. Soc.*, in press.
34. (a) R. J. Hoover and M. Kasha, *J. Am. Chem. Soc.* **91**, 6508 (1969).
    (b) S. Japar and D. A. Ramsay, *J. Chem. Phys.* **58**, 5832 (1973).

CHAPTER 7

Reactions of Transient Ions

7.1 Ions in the Gas Phase; the Mass Spectrometer and Ion Cyclotron Spectroscopy

In the three previous chapters, we have considered a variety of reactions involving ions in solution. Under these conditions, the ions are solvated and the corresponding energies of solvation can be very large. Solvation consequently plays a major role in determining the rates of such reactions and it is indeed a matter of some difficulty to dissect the overall effects into contributions by solvation and by the inherent reactivities of the species involved.

Recently, however, it has become possible to study ionic reactions in the gas phase, where of course there can be no interference by solvation. This work has already led to the development of a whole new area of chemistry and one in which a variety of novel phenomena occur. Before discussing these in terms of the PMO approach, we may first consider briefly the techniques used in their study.

Most of the work in this area has been carried out with various types of *mass spectrometer*. In a mass spectrometer, a beam of gaseous ions is sorted out according to their masses and the result displayed as a plot of number of ions vs. mass. Such a *mass spectrum* is shown in Fig. 7.1.

Since the masses of isotopes are almost whole numbers on the atomic weight scale, the masses of individual molecules are also close to whole numbers. The *mass number* of a molecule is its molecular weight to the nearest integer. Early mass spectrometers could not determine the masses of ions to much better than 1 amu and so could not distinguish between ions of

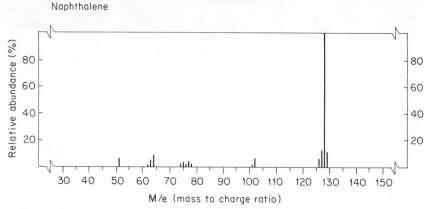

FIGURE 7.1. The mass spectrum of naphthalene (M/e is the mass to charge ratio).

different chemical composition but similar mass number. Thus the species $^{12}CH_4$, $^{13}CH_3$, $^{14}NH_2$, ^{15}NH, and ^{16}O all have mass number 16 but their exact molecular weights differ significantly (16.0309, 16.0273, 16.0190, 16.0141, and 15.9949, respectively). Recently, high-resolution mass spectrometers have been developed that enable ionic masses to be determined to one or two units in the fourth decimal place. Such an instrument allows one to deduce the exact chemical composition of each ion.

The ions are produced by bombarding molecules of a gas at very low pressure with a beam of electrons. If a sufficiently energetic electron hits a molecule, it can knock an electron out of it,

$$M + e \longrightarrow M^{+\cdot} + 2e \tag{7.1}$$

The positive ions so produced are accelerated by an electric field. If the potential through which they are accelerated is V, each ion with unit charge $+e$ will acquire kinetic energy eV. Since the kinetic energy of a particle of mass M and velocity u is $\frac{1}{2}Mu^2$, the velocity of an ion depends on its mass, being given by $(2eV/M)^{1/2}$.

The beam of ions is analyzed by passing it through an electric or magnetic field. A charged particle moving through such a field is deflected. Since the force acting depends on the magnitudes of the charge and of the field and since these are the same for all the ions, the deflection depends on the mass of the ion and its velocity. In this way, the beam of ions is spread out into a number of different beams, each composed of ions of a single mass. In the earlier mass spectrometers, only a single analyzer was used. In modern high-resolution instruments, the ions are passed through two analyzing fields in succession, one electrostatic and one magnetic. Figure 7.2 shows such a spectrometer. The ions are produced by bombardment by a stream of electrons in the source (a) and are accelerated by an electric field, usually of about 8 kV. A narrow beam of the ions, selected by a slit, then passes first through an electrostatic analyzer (b) and then through a magnetic analyzer (c). The geometry is chosen so that all ions of a certain mass are focused on a second slit through which they pass to the detector (d). By varying the field, we can select the mass of the ions that are detected, so a plot of field vs. detector response will give us a typical mass spectrum.

Such a mass spectrometer has to operate at very low pressures to avoid collisions between the ions and gas molecules (which would of course defocus the ion beam). In order to get a sufficient number of ions, it is necessary to use fairly energetic electrons to bombard the molecules of gas in the source chamber. In this case, the ions need not be produced in their ground states (cf. the discussion of photoelectron spectroscopy in Section 1.9). The ions generated in this way can indeed contain a great deal of excess energy, sufficient to shatter them into fragments. The mass spectrum of an organic molecule M therefore contains not only a *parent peak* corresponding to the ion $M^{+\cdot}$, but also a series of additional peaks corresponding to the fragments produced by its fragmentation. If the electrons in the source have energies greater than ~ 50 eV, the probability of a given excited state of the ion being formed becomes independent of the electron energy and so consequently does the resulting mass spectrum. Under these conditions, the mass spectrum becomes an easily reproducible characteristic of the compound, a circumstance which makes the mass spectrometer a useful analytical tool.

Apart from fragmentation, the ions can also undergo internal rearrangements. These can be deduced from the fragmentation patterns from the resulting rearranged ions. The interpretation of mass spectra therefore involves an understanding of the chemistry of ions in the gas phase.

Reactions of this kind can normally be observed only if they occur in the source because if an ion disintegrates in the middle of the analyzer, the

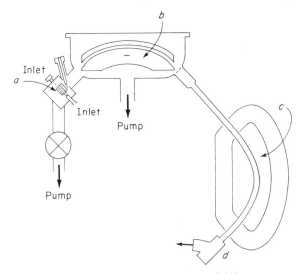

FIGURE 7.2. Block drawing of an AEI MS-903 mass spectrometer; (a) source; (b) electrostatic analyzer; (c) magnetic analyzer; (d) detector (an electron multiplier).

fragments no longer have the velocities appropriate to their mass [equation (7.1)] and so are not focused properly. The reactions must therefore take place before the ion is swept out of the source by the accelerating field, typically in 1–5 μsec. Consequently, only very fast reactions can be observed. The only exception to this occurs if an ion disintegrates after passing through the electrostatic analyzer and before passing through the magnetic one. In this case, a rather diffuse peak occurs at an apparent mass M^* given by

$$M^* = M_2{}^2/M_1 \qquad (7.2)$$

where M_1 is the mass of the ion leaving the source and M_2 is that of the fragment formed by the fission reaction. Such peaks are described as being due to *metastable* ions and may be observed if the half-life of the original ion $M_1{}^+$ lies in the range 1–100 μsec.

Mass spectra usually provide information only concerning unimolecular reactions of ions since the pressure in the source is normally too low for bimolecular processes to occur in the very short time that the ions are in the source. The only exception concerns uptake of hydrogen atoms and/or ions, since the concentration of hydrogen is usually relatively high.

Recently, two different mass spectrometric procedures have been developed for studying bimolecular reactions of ions in the gas phase. In the first, an ordinary mass spectrometer is modified by running the source chamber at high pressure (up to several mm of mercury). The pressure in the spectrometer is kept sufficiently low by making the first slit extremely narrow and by very rapid pumping. The ions produced in the source chamber can undergo bimolecular reactions and the resulting ions can at the same time escape through the slit into the mass spectrometer. Such an instrument is called a *chemical ionization* (CI) mass spectrometer.

The second approach uses a mass spectrometer of entirely different type. When an ion moves in a magnetic field, it is deflected by the field so that its path is a circle. For an ion with unit charge $+e$ and of given mass, the radius of the circle is proportional to the velocity of the ion. If, therefore, we double the velocity of the ion, the circumference of the circle will double, so the time it takes to make one revolution stays the same. In a magnetic field of given strength, each such ion will then be characterized by the time it takes to make one revolution, this time depending only on its mass. Suppose now we apply an alternating electric field to the ion, in the plane in which it is moving. It is easily seen (Fig. 7.3) that if the period of oscillation of the field is the same as the period of revolution of the ion, the ion will be continually accelerated because the period of revolution depends only on the mass of the ion and not on its velocity. This principle has been known for a long time in the case of particle accelerators, being the basis of the cyclotron.

The extra energy of the ion has to come from somewhere. It comes in fact from the accelerating field. If, therefore, we vary the frequency of the applied field, we will observe absorption of energy when the frequency coincides with the *cyclotron resonance* frequency of the ion. If we plot the absorption of energy as a function of frequency, we will then get a typical

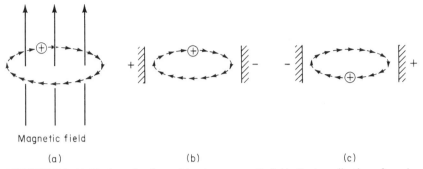

Magnetic field

(a)   (b)   (c)

FIGURE 7.3. (a) Motion of a charged ion in a magnetic field; (b, c) application of an alternating electric field will continuously accelerate the ion if the field switches in step with the revolution of the ion.

mass spectrum, energy being absorbed whenever the frequency corresponds to the cyclotron resonance frequency of one of the ions present. Such a device (originally called the *omegatron*) can therefore serve as a very simple mass spectrometer. Ion cyclotron resonance (ICR) can be used to study bimolecular reactions of ions and neutrals because an ICR spectrometer does not require the pressure to be low. Since the cyclotron resonance frequency of an ion does not depend on its velocity, the frequency is not affected by occasional collisions with gas molecules. In an ICR experiment, a given ion can be selectively accelerated by applying the appropriate accelerating frequency. This in effect means that we can selectively "cook" a given ion A^+ and so make it react faster by raising its effective temperature. If we then observe an increase in the cyclotron resonance signal corresponding to some ion B^+, we know that B^+ must have been produced by some reaction involving A^+. Arguments of this kind provide a powerful tool for studying reactions of ions.

An ion cyclotron resonance spectrometer is illustrated in Fig. 7.4. The ions are produced by a beam of electrons from a filament and drift through the cell under the influence of an applied field. In the central section of the cell, the ions are exposed to alternating electric fields produced by applying

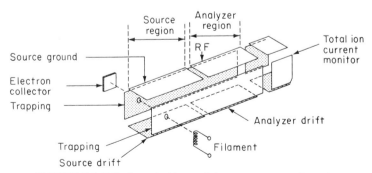

FIGURE 7.4. Cell of a typical ion cyclotron resonance spectrometer.

appropriate ac potentials to the top and bottom plates. The whole cell is filled with the reactant gas at low pressure and placed in the uniform magnetic field of a large magnet.

In the discussion so far, we have assumed that the ions with which we are dealing are positively charged. This is the case for the ions formed by bombarding neutral molecules with energetic (70 eV) electrons and also of course for any secondary ions formed by their dissociation or by their reaction with neutral molecules. It is, however, also possible to form negative ions in the gas phase by allowing slow electrons to combine with neutral molecules,

$$M + e^- \longrightarrow M^{-\cdot *} \xrightarrow{N} M^{-\cdot} \tag{7.3}$$

The energy released in such a process is called the *electron affinity* of the molecule. The negative ions can be studied by any of the usual mass spectrometric procedures. Thus in a standard mass spectrometer, negative ions can be analyzed if the electric and magnetic fields in the analyzer are reversed. In an ICR spectrometer, only the trapping potential must be changed since negative and positive ions of a given mass resonate at the same frequency, the direction of revolution being simply reversed.

7.2 The Structure of Radical Ions

The species initially produced in the mass spectrometer are ions derived from normal molecules by loss of an electron. Such an ion will normally contain an odd number of electrons and is consequently described as a *radical ion*. The same is true of the negative radical ion formed from a neutral molecule by capture of an electron [equation (7.3)]. The relationships between a neutral molecule M, the derived radical ions $M^{+\cdot}$ and $M^{-\cdot}$, and the lowest singlet excited state M^* are shown in Fig. 7.5.

Excitation of a normal molecule involves transfer of an electron from a bonding MO to an antibonding MO. The minimum energy required for such a process is then equal to the difference in energy between the HOMO and LUMO, generally of the order of 4 eV. In the case of a radical ion, however, excitation can take place between two bonding MOs or two antibonding MOs (see Fig. 7.6). Now the separation of the bonding MOs can be estimated from photoelectron spectroscopy (see Section 1.9). It is typically of the order of 1 eV. Since the arguments of Section 6.14 show that the spectrum

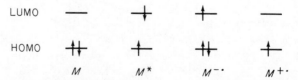

FIGURE 7.5. Relations between a neutral molecule M and the corresponding lowest excited singlet state M^* and radical ions M^+ and M^-.

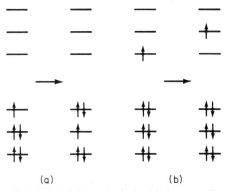

FIGURE 7.6. Hypothetical orbital energy diagram for the lowest singlet transition in (a) a radical cation; (b) a radical anion.

of antibonding MOs should parallel that of the bonding MOs, the separation of antibonding MOs should also be of the order of 1 eV. The excitation energy of a radical ion should therefore be small, several times less than that of the neutral molecule from which it is derived.

These arguments suggest that the chemical behavior of radical ions should resemble both that of the ground state and that of the excited state of the corresponding neutral molecule and also that the difference in reactivity between the ground state and excited state should be less for a radical ion than for a neutral molecule. Furthermore, since the excitation energy of the radical ion is less than that of the neutral molecule, the probability of internal conversion to the ground state should be correspondingly greater (see Section (6.13). In the case of an ion in the gas phase at low pressure, the vibrational energy resulting from such internal conversion cannot be dissipated rapidly by collisions. Internal conversion of the excited ions produced by electron impact in the mass spectrometer should therefore give "hot," vibrationally excited species which may very well undergo unusual chemical reactions as a result of their excess energy.

In the majority of cases that concern us, radical cations are formed by loss of a π electron or a lone-pair electron, and radical anions by capture of an electron into an antibonding π^* MO. The relative stabilities of the π radical ions can be estimated by considering their formation through union of an AH radical with an AH cation or anion (Fig. 7.7).

The first-order stabilization in each case is just half what it would be for union of the two corresponding neutral AH radicals (R· and S·) to form the neutral molecule RS. The radical cation differs from the corresponding neutral molecule in having one bonding electron less, while the radical anion has its extra electron in an antibonding MO. In the PMO approximation, both changes have an equal net antibonding effect. However, as we saw in the previous chapter (Section 6.14), antibonding electrons are in general more antibonding than analogous bonding electrons are bonding. On this

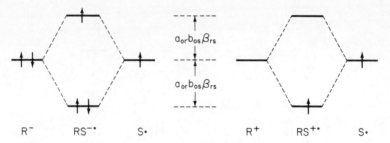

FIGURE 7.7. First-order interactions between an AH radical S· and an AH anion R^- or cation R^+.

basis, one might expect radical cations to be more stable than the corresponding radical anions. One might also expect the reactions of a radical cation to be similar to those of the molecule in its ground state rather than in its first excited state, while the reactions of radical anions should tend to show an opposite trend.

7.3 Reactions of Cation Radicals

A. Cleavage Reactions[1]

The simplest type of process undergone by radical cations in the mass spectrometer is cleavage into a radical and a cation. Thus the mass spectrum of an n-paraffin, $C_nH_{2n+2}^{+\cdot}$, normally shows no peak at the mass number $M = (14n + 2)$ corresponding to the initially formed cation radical $C_nH_{2n+2}^{+\cdot}$. The radical ion is unstable to dissociation into alkyl cation and a hydrogen atom or alkyl radical, namely

$$C_nH_{2n+2}^{+\cdot} \longrightarrow C_mH_{2m+1}^{+} + C_{n-m}H_{2(n-m)+1}^{\cdot} \qquad (7.4)$$

Here m can have any value from unity to n. Thus the mass spectrum shows a series of peaks at mass number $M - 1$, $M - 15$, $M - 29$, etc., corresponding to the various alkyl cations derived from the parent radical cation by breaking a CH or CC bond.

In the case of the paraffins, the initially formed radical cation is unstable with respect to dissociation. One can, however, also observe similar cleavage of bonds in cases where the cleavage is endothermic. This is because the radical cation is generated by bombardment with very energetic electrons and so may be formed initially in an electronically excited state. Internal conversion will then lead to a vibrationally excited form of the radical cation in which the excess vibrational energy may be enough to counter the endothermicity of bond rupture.

The amount of energy made available in this way can indeed be very large, more than enough to break any bond. One might therefore expect a chaotic mixture of products to be formed by random bond cleavage. If this were so, the results would be both difficult to interpret and of little interest—but fortunately it is not. Strong selectivity is observed between the various possible cleavage reactions, the easiest modes of cleavage being favored.

This is due partly to the fact that the ion radicals are formed with a whole spectrum of energies. However, an additional factor favors the easier modes of cleavage even when excess energy is available. When electronic excitation energy is converted into vibrational energy, the latter is normally distributed over a number of bonds. As the molecule vibrates, the vibrational energy is continually redistributed over its various modes of vibration and so the partitioning between the various bonds changes as well. For a given bond to break, enough vibrational energy must be concentrated at some given instant in that particular bond to break it. Clearly this will occur more readily as the fraction of the whole vibrational energy that has to be concentrated in this way becomes smaller. Other things being equal, it will then be the weakest bond in the molecule that ruptures most readily even if ample energy is available to break other, stronger bonds. The latter will of course break too, but less readily, so the peaks in the mass spectrum corresponding to the products formed by their rupture will be correspondingly less intense. The same arguments apply to other possible reactions. Even though very large amounts of excess energy may be available, the reactions taking place can be interpreted as normal ground-state reactions of the species in question. The mass spectrometer therefore enables us to deduce the ground-state reactions of ions in the gas phase in spite of the rather drastic procedure used to produce them. One simple example is provided by the mass spectra of n-alkanes [equation (7.4)]. Each rupture of a bond gives rise to a cation A^+ and a radical $B\cdot$. It could, of course, equally well give rise to the pair $A\cdot + B^+$ formed by alternate fission of the A—B bond. The heat of the reaction

$$A^+ + B\cdot \longrightarrow A\cdot + B^+ \tag{7.5}$$

is simply equal to $I_B - I_A$, where I_B and I_A are the ionization potentials of the two radicals. One would therefore expect to see in the mass spectrum a much larger peak for the species with the lower ionization potential. This is the case. Thus cleavage of a CH bond gives mainly an alkyl cation and a hydrogen atom, rather than an alkyl radical and a proton, because the ionization potentials of alkyl radicals (8–10 eV) are much less than that (13.6 eV) of the hydrogen atom. The peak for CH_3^+ is also weak because the ionization potential of methyl (9.84 eV) is greater than the values (8.0–8.5 eV) for other primary alkyl radicals. Consequently, cleavage of $RCH_3^{+\cdot}$ gives $R^+ + CH_3\cdot$ rather than $R\cdot + CH_3^+$.

Similarly, the effects of substituents on such cleavage can be well interpreted in terms of the principles deduced in Chapter 5. Thus compounds of type $ArCH_2R$, where ArH is an even AH, readily undergo the following cleavage, assisted by stabilization of the resulting carbonium ion by the $\pm E$ aryl group:

$$ArCH_2R \longrightarrow ArCH_2R^{+\cdot} \longrightarrow ArCH_2^+ + R\cdot \tag{7.6}$$

Cleavage of this kind is assisted even more strongly by $-E$ groups such as alkoxyl. Similar considerations apply to the converse cleavage of ArXR into ArX· + R⁺. Thus aryl alkyl ethers ArOR undergo cleavage in this way, the radical ArO· being stabilized,

$$\text{ArOR} \longrightarrow \text{ArOR}^{+\cdot} \longrightarrow \text{ArO}\cdot + \text{R}^+ \qquad (7.7)$$

This type of argument can be put on a more quantitative basis by using the PMO approach of Chapters 3–5. One simple example will suffice. Compounds of the type $ArCH_3$, where ArH is an aromatic compound, commonly show peaks at mass number $M - 1$, M being the mass number of $ArCH_3$, due to loss of hydrogen from the methyl group of the parent radical cation,

$$\text{ArCH}_3{}^{+\cdot} \longrightarrow \text{ArCH}_2{}^+ + \text{H}\cdot \qquad (7.8)$$

Let us examine the probability of this process for various methyl derivatives $ArCH_3$ of a given alternant aromatic compound ArH.

First consider the corresponding isoconjugate hydrocarbon $Ar'CH_3$. The energy ΔE_1 for fission of this to the radical $Ar'CH_2\cdot$,

$$\text{Ar'CH}_3 \longrightarrow \text{Ar'CH}_2\cdot + \text{H}\cdot \qquad (7.9)$$

is given (see p. 104) by

$$\Delta E_1 = E_{\text{CH}} + 2\beta(1 - a_{0r}) \qquad (7.10)$$

where E_{CH} is the CH bond energy and a_{0r} is the NBMO coefficient at the methylene carbon in $ArCH_2$. Since $ArCH_3$ and $ArCH_2\cdot$ are both neutral AHs, the energy change in this process will be unaffected if carbon atoms in Ar' are replaced by heteroatoms (see p. 109). The corresponding heat of reaction ΔE_2 is thus equal to ΔE_1; i.e.,

$$\text{ArCH}_3 \longrightarrow \text{ArCH}_2\cdot + \text{H}\cdot; \qquad \Delta E_2 = E_{\text{CH}} + 2\beta(1 - a_{0r}) \qquad (7.11)$$

The change in energy ΔE_3 for equation (7.6) can now be found from the following thermocycle:

$$\begin{array}{ccc} \text{ArCH}_3 & \xrightarrow{E_{\text{CH}} + 2\beta(1 - a_{0r})} & \text{ArCH}_3\cdot + \text{H}\cdot \\ \downarrow I_1 & & \downarrow I_2 \\ \text{ArCH}_3{}^{+\cdot} & \xrightarrow{\Delta E_3} & \text{ArCH}_2{}^+ + \text{H}\cdot \end{array} \qquad (7.12)$$

The ionization potentials I_1 of different isomers $ArCH_3$ will be almost the same since the $-I$ effect of methyl is relatively small. As a first approximation, we can set them equal ($= I_0$). The ionization potential of the odd AH radical $Ar'CH_2$, isoconjugate with $ArCH_3$, is of course just $-\alpha_c$, since the electron lost is in an NBMO. The ionization potential I_2 of $ArCH_2\cdot$ is then

$$I_2 = -\alpha_c - \sum_i a_{0i}^2 \delta\alpha_i \qquad (7.13)$$

where a_{0i} is the NBMO coefficient of atom i in Ar'CH$_2$ and $\delta\alpha_i$ is the Coulomb integral (relative to carbon) of the corresponding atom in ArCH$_2$. Combining equations (7.12) and (7.13), we have

$$E_3 = E_{CH} + 2\beta(1 - a_{0r}) + I_2 - I_1$$
$$= C + 2\beta(1 - a_{0r}) - \sum_i a_{0i}^2 \, \delta\alpha_i \qquad (7.14)$$

where C is the same for different isomers ArCH$_3$, being given by

$$C = E_{CH} - I_0 - \alpha_c \qquad (7.15)$$

Since β is quite small (~ 1 eV), the effects of changes in the corresponding term in equation (7.14) are negligible compared with those of changes in the last term. The ease of the reaction of equation (7.6), and so the intensity of the $M - 1$ line in the mass spectrum, should therefore run parallel to the final sum in equation (7.14). This is the case, as the examples in Table 7.1 show.

TABLE 7.1. Ease of Fission in Isomeric Ions[1b] ArCH$_3^{+\cdot}$

Parent compound	$\sum_i a_{0i}^2 \, \delta\alpha_i$	Intensity at $M - 1$ peak / Intensity at M peak
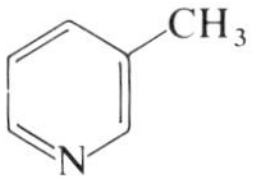 3-methylpyridine	0	30
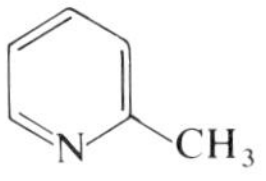 2-methylpyridine	$0.15\delta\alpha_N$	20
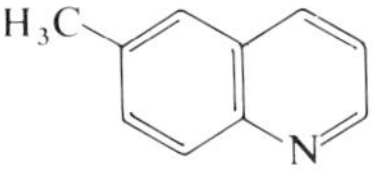 8-methylquinoline	0	45
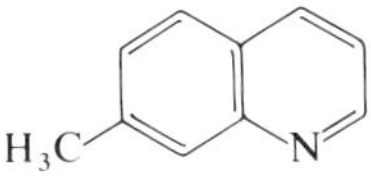 6-methylquinoline	$0.06\delta\alpha_N$	43
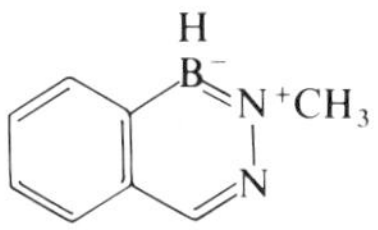 cinnoline N-methyl (N+CH$_3$, B−)	$0.24\delta\alpha_B + 0.06\delta\alpha_N$	60
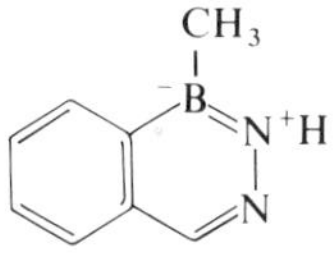 cinnoline CH$_3$-B, N+H	$0.20\delta\alpha_N$	0

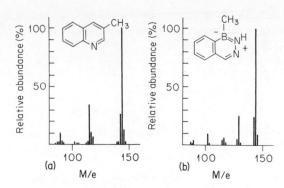

FIGURE 7.8. Low-resolution mass spectra of (a) 3-methylquinoline and (b) 4-methyl-4,3-borazaroisoquinoline.

Figure 7.8 shows two typical mass spectra. In that of 3-methylquinoline, there is a strong $M - 1$ peak. Here the final sum in equation (7.14) is zero. At first sight, one might suppose that the spectrum of 4-methyl-4,3-borazaroisoquinoline (Fig. 7.8b) shows an equally strong $M - 1$ peak, although here the final sum in equation (7.14) is large (Table 7.1). This peak, however, is in fact due to the parent molecular ion containing the isotope ^{10}B instead of ^{11}B. That the $M - 1$ peak is in fact absent is shown by the absence of the corresponding peak for the ^{10}B isomer (which would occur at $M - 2$).

B. Internal Displacement Reactions

The second type of reaction to be considered involves processes where the radical ion center in a radical ion attacks some other part of the molecule, thus leading to an internal reorganization, i.e., a molecular rearrangement. Numerous reactions of this kind have been discovered from studies of fragmentation patterns in the mass spectrometer.

It might be pointed out in this connection that mass spectrometry provides an extraordinarily quick and easy way to study reactions of ions. With a modern mass spectrometer system, a mass spectrum can be recorded and the intensities and elemental compositions of all the mass peaks assigned, all within 10 min. The detailed analysis of the products of a single thermal or photochemical reaction can easily take as many days. In view of the correlation between reactions in the mass spectrometer and the thermal ground-state reactions of ions, experiments of this kind can clearly act as a major source of chemical information.

Returning to internal displacement reactions, these normally involve ions where the radical ion center is localized and where a reaction takes place between it and some other part of the parent system. The classic example of this is the so-called McLafferty rearrangement, which is precisely analogous to the Norrish type II photochemical reactions of ketones; e.g.,

$$\underset{\underset{CH_2}{|}}{\overset{H}{\underset{|}{CH_2}}}\overset{O}{\underset{||}{C}}-R \xrightarrow[G_R]{h\nu} \underset{\underset{CH_2}{|}}{\overset{\overset{H}{\cdot}}{\underset{|}{CH_2}}}\overset{\overset{\cdot}{O}}{\underset{|}{C}}-R \longrightarrow \underset{CH_2}{\overset{CH_2}{||}} + \underset{CH_2}{\overset{OH}{\underset{|}{C}-R}} \quad (7.16)$$

$$\Big\downarrow -e$$

$$\underset{\underset{CH_2}{|}}{\overset{H}{\underset{|}{CH_2}}}\overset{O^{+\cdot}}{\underset{||}{C}}-R \longrightarrow \underset{\underset{CH_2}{|}}{\overset{\overset{H}{\cdot}}{\underset{|}{CH_2}}}\overset{\overset{\cdot}{O}}{\underset{|}{{}^+C}}-R \longrightarrow \underset{CH_2}{\overset{CH_2}{||}} + \left[\underset{CH_2}{\overset{OH}{\underset{|}{C}-R}}\right]^{+\cdot} \quad (7.17)$$

A large number of analogous reactions are now known which involve similar internal abstraction reactions, ususally involving six-membered transition states.

Discovery of rearrangements in mass spectrometry depends on the fact that reactive ions contain excess vibrational energy and this energy will not be entirely dissipated by rearrangement reactions. Such a reaction will take place as soon as enough energy has accumulated in the appropriate vibrational mode. Therefore even if activation is required, only an amount of energy equal to the activation energy is lost during the reaction. Any excess vibrational energy remains in the reaction products and can lead to secondary reactions, in particular cleavage of bonds. The mass spectrum will therefore contain peaks corresponding to the fragmentation products of the ions produced by the reaction in question. The structure of these ions can in turn be deduced from the fragmentation products to which they give rise. Further information can, if necessary, be obtained by isotopic labeling. Thus in the reaction of equation (7.17), labeling the terminal methyl of the *n*-propyl group with deuterium would give rise to a peak corresponding to $(CH_2\!=\!CR\!-\!OD)^+$.

C. Pericyclic Reactions

In Chapters 5 and 6 we have discussed thermal and photochemical pericyclic reactions. We will now consider the possibility of analogous reactions of radical ions.

The discussion of radical cations (see Fig. 7.7) indicates that the energy of union of an odd AH radical and an odd AH cation to form a radical cation is just half the energy of union of the two corresponding odd AH radicals to form a neutral AH. As long as the resulting AH is alternant, the rules for aromaticity for it and for the radical cation derived from it will be the same and it is easily shown that this is also true for isoconjugate molecules containing heteroatoms. The course of pericyclic reactions where the transition states are alternant should therefore be the same for radical cations as for the parent neutral molecules.

In the case of nonalternant systems, however, the parallel between neutral molecules and the corresponding radical cations breaks down. Consider, for example, heptafulvene (Fig. 7.9b). Heptafulvene can be derived by union of tropylium with CH_3^- (Fig. 7.9b). Likewise, the open-chain analog of heptafulvene, octatetraene (Fig. 7.9d) can be derived by union of the odd AH cation heptatrienylium with CH_3^- (Fig. 7.9c). Since both heptafulvene and octatetraene are classical conjugated polyenes, their total π energies should be the same.

In the union of heptatrienate cation with CH_3^- (Fig. 7.9e), the NBMO of the cation and the $2p$ AO of CH_3^- are degenerate. The interaction between the orbitals is consequently large, so the HOMO of the product, octatetraene, lies well below the nonbonding level. In tropylium, on the other hand, the LUMO is antibonding (see Section 3.13). The interaction of this antibonding MO with the $2p$ AO of CH_3^- on union of the two (Fig. 7.9f) is consequently much less. The HOMO of heptafulvene therefore lies much closer to the nonbonding level than does the HOMO of octatetraene. The ionization potential of heptafulvene must therefore be much less than that of octatetraene.

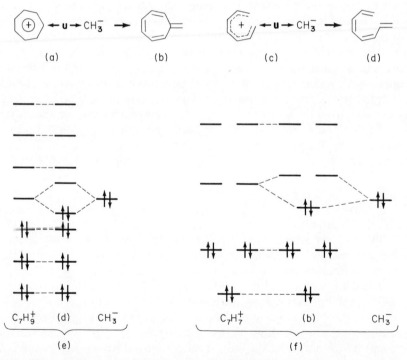

FIGURE 7.9. (a) Union of tropylium with CH_3^- to form heptafulvene (b); (c) union of heptatrienate cation with CH_3^- to form octatetraene (d); (e) orbital interactions in the union (c); (f) orbital interactions in the union (a).

Since both hydrocarbons are nonaromatic, the radical cation from heptafulvene must therefore be more stable than that from octatetraene, i.e., heptafulvene radical cation is aromatic, as can be emphasized by the representation in (1). The same argument can be applied to other even, nonalternant hydrocarbons containing $(4n + 3)$-membered rings, e.g., triafulvene (2), which should give an aromatic radical cation (3). An analogous argument, left to the reader, shows that fulvene (4) and analogous compounds with $(4n + 1)$-membered rings should give aromatic radical anions [e.g., (5)]. It is also easily shown that the compounds with $(4n + 3)$-membered rings should give antiaromatic radical anions and the compounds with $(4n + 1)$-membered rings should give antiaromatic radical cations. The aromaticity or antiaromaticity of such structures must be kept in mind in applying Evans' rule to pericyclic reactions involving radical ions isoconjugate with them.

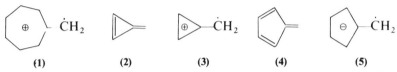

(1)   (2)   (3)   (4)   (5)

A major problem in interpreting pericyclic reactions of radical ions is the difficulty of distinguishing between concerted and nonconcerted paths. The same difficulty occurs in the case of thermal reactions, but the problem here is more acute since there is less scope for the application of subtle experimental techniques. Thus cyclohexene (6) undergoes a retro-Diels–Alder reaction to (7) in the mass spectrometer but the products formed do not enable us to distinguish between a two-step process via the intermediate radical ion (8) and a pericyclic process involving the cyclic transition state (9).

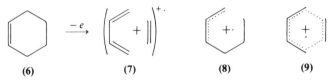

(6)   (7)   (8)   (9)

The difficulty is illustrated rather clearly by a recent study of the retro-Diels–Alder reaction of vinylcyclohexene (10) by Smith and Thornton.[2a] Here two molecules of butadiene are formed, one of which is ionized (11). Smith and Thornton reasoned that if the reaction took place via the intermediate open-chain ion radical (12), the distinction between the two butadiene moieties would be lost since the two allyl systems would have identical orbital energies. A one-step reaction, on the other hand, could involve either of the cyclic transition states (13) or (14). While (13) is again symmetric, fragmenting to two molecules of cis-butadiene, (14) should retain the distinction between them. Smith and Thornton therefore studied the mass spectrum of the isotopically labeled analog (15) of (10). They found that the positive charge was predominantly (1.85:1) retained by the butadiene moiety comprising the vinyl group in (15), i.e., lacking in deuterium. Theoretical calculations and photoelectron spectroscopy[26] suggest that the ionization potential of *trans*-butadiene is in fact lower than that of the *cis* isomer. Thornton and Smith's result is therefore consistent with a one-step mechanism via a mixture of (13) and (14). However, during the conversion of (10) to (12), the *cis* configuration of the butadiene moiety in the ring would be retained because rotation about the CC bonds in allyl is hindered by the loss in π energy that such rotation would entail. The ion radical (12) would therefore exist as a mixture of *cis–cis* (12) and *cis–trans* (16) isomers. The former would fragment to two molecules of *cis*-butadiene, one ionized, but the latter would fragment to one of *cis* and one of *trans*. The retention of charge in the *trans* isomer in the latter could equally well account for Thornton and Smith's results.

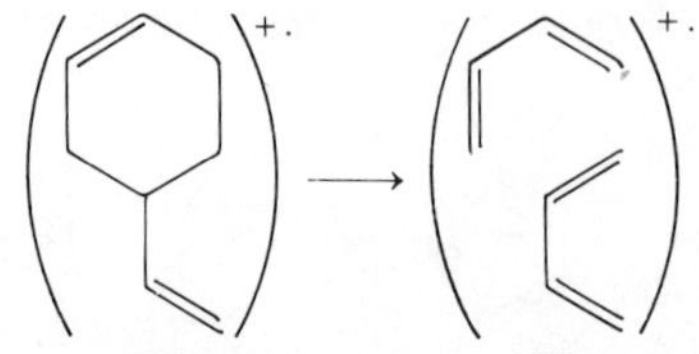

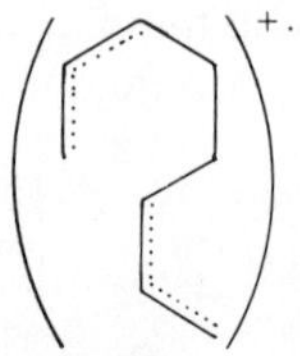

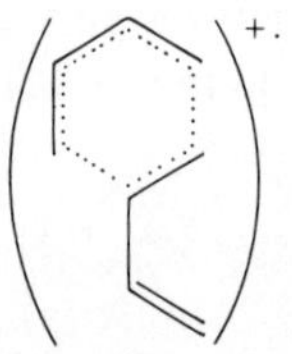

Mandelbaum and his co-workers have presented convincing evidence that electron-impact RDA reactions are highly stereospecific in three systems of multicyclic diketones.[3a] These results must be contrasted with those obtained by Hammerum and Djerassi, who found little stereochemical dependence in the RDA reactions of the molecule ions of seven hexahydronaphthalenes.[3b] The fact that metastable ions were not observable in the latter studies is significant inasmuch as the frequency factors for the RDA reactions are low enough to virtually guarantee the existence of metastable ions if the reactions had occurred from the electronic ground states of the ions.

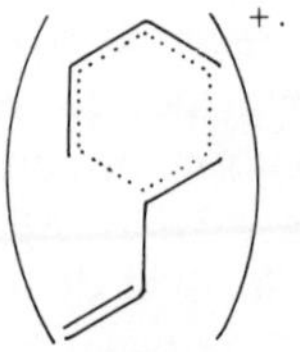

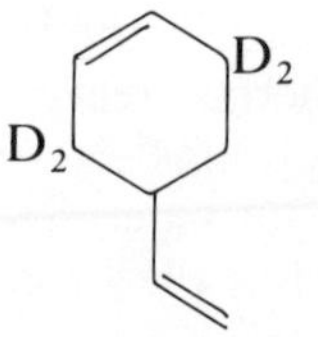

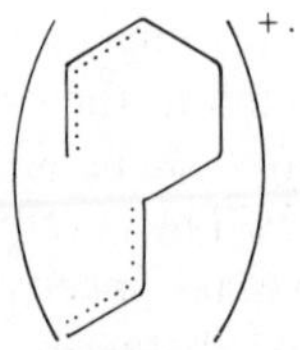

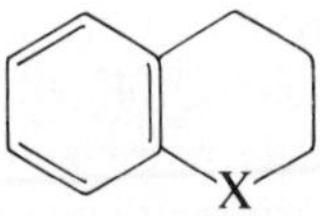

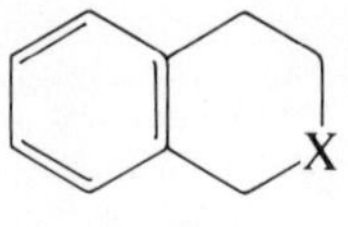

The parallel between retro-Diels–Alder reactions in the mass spectrometer and the corresponding thermal reactions of neutral cyclohexenes does, however, suggest that both proceed by a common mechanism, i.e., a one-step pericyclic process. This conclusion has been strongly supported by Loudon *et al.*[4] in a very detailed comparison of the thermal and mass spectrometric reactions in the case of several analogs (17) and (18) of tetrahydronaphthalene (X = O, S, or NH). They deduced the course of the reactions of the radical ions from a careful study of the mass spectra and then showed that the high-temperature thermolysis of the parent compounds led to similar products. In several cases, they were able to trap the o-quinonoid product of the initial retro-Diels–Alder reaction with a suitable dienophile.

The hexahelicene rearrangement is another pericyclic reaction that was anticipated on the basis of mass spectrometric studies.[1a,b] Analysis of the metastable ions from hexahelicene (19) suggested that C_2H_4 is eliminated in one step, presumably as ethylene, leaving the coronene cation (20). If the reaction is concerted, the transition state (21) is isoconjugate with the internal benzocoronene (22). This can be shown to be aromatic by the procedures of Chapters 3 and 4 (see pp. 99 and 179). The transition state for a one-step conversion of hexahelicene radical cation to ethylene and (20) should therefore be "allowed" and the same should be true for the analogous reaction of hexahelicene itself. Pyrolysis of (19) at high temperatures did in fact give small quantities of coronene and ethylene, while no detectable trace of acetylene was formed. The formation of ethylene must therefore have involved a concerted cycloelimination via the neutral equivalent of the transition state (21). The dominant product from the pyrolysis was, however, a dimer of hexahelicene, formed by an alternative type of pericyclic reaction (π cycloaddition). This of course cannot take place in the mass spectrometer since bimolecular processes are too unfavorable at the very low pressures used. It is, however, interesting that the most intense ion in the mass spectrum of the dimer corresponded to hexahelicene itself. If the dimer is formed by an "allowed" pericyclic process, the dissociation of the dimer should of course also be an "allowed" process.

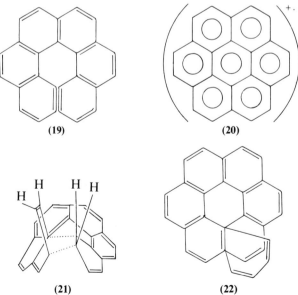

While the reactions of ions in the mass spectrometer tend to follow the reactions of corresponding neutral molecules, it must be remembered that the ions can contain enough excess energy to undergo reactions with photochemical analogs. A typical process of this kind is fragmentation via a four-membered ring,

$$\begin{pmatrix} A-B \\ | \; | \\ D-C \end{pmatrix}^{+\cdot} \longrightarrow \begin{pmatrix} A & B \\ | & || \\ D & C \end{pmatrix}^{+\cdot}$$

Processes of this kind do not usually occur as thermal reactions since the transition states would be antiaromatic, but they can take place in the mass spectrometer. A good example is the elimination of hydrogen to form aromatic rings, a reaction which takes place if the hydrogen atoms are *cis*. A typical example is represented by (23)–(25) (see Fig. 7.10). That H_2 is eliminated in one step is indicated by the presence of a single metastable peak corresponding to decomposition of the parent ion (M, 167) to the product with two hydrogen atoms less. While the loss of hydrogen is slow, as indicated by the metastable peak in Fig. 7.10, it takes place in a single step.

[Structures (23), (24), (25) showing conversion of 2-biphenylmethyl cation to fluorenyl cation + H_2]

(23)   (24)   (25)

In each of the two pericyclic processes so far considered, the course of the reaction was unambiguous. An interesting distinction arises in cases where there is a choice between two stereochemistries for a given pericyclic reaction, one corresponding to a thermally "allowed" process with a transition state which is aromatic in the ground state, the other to a photochemically "allowed" process with a transition state which is antiaromatic in the ground

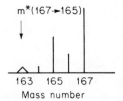

FIGURE 7.10. The mass spectrum for conversion of 2-biphenylmethyl cation to the fluorenyl cation [(23)–(25)] contains a metastable peak m^* for loss of H_2 but not for successive loss of 2H·.[4b]

state but aromatic in the excited state. Ions in the mass spectrometer are usually produced in excited states. If such an excited-state ion reacts as such, before internal conversion to a vibrationally excited ground state, its reactions should parallel photochemical reactions of the parent molecule rather than thermal reactions. A very nice example of such a process has in fact been reported by Johnstone and Ward.[4b]

This involves the cyclization of *cis*-stilbene (26) to phenanthrene (29) (Fig. 7.11), a process reminiscent of (23)–(25). In this case, the mass spectrum contains *two* metastable peaks, one corresponding to conversion of the original ion ($M^{+\cdot}$, 180) to an ion with one mass unit less (m/e 179), the other to conversion of this secondary ion to the radical ion of phenanthrene (m/e 178). Here the hydrogen atoms are lost successively, not simultaneously. The reaction must presumably involve first an electrocyclic cyclization of (26) to the dihydrophenanthrene radical ion (27), which is then converted to (29) via the ion (28). Since *cis* elimination of H_2 in such cases is a one-step process (cf. Fig. 7.9), the bridgehead hydrogen atoms in (27) must be *trans* to one another. Now the conversion of (26) to (27) is a typical electrocyclic process. In order to lead to a *trans* geometry, the cyclization must be conrotatory and so involve an anti-Hückel transition state. Since the ring being formed is six-membered, this is therefore an antiaromatic process in the ground state. Since (27) could equally well have been formed by disrotatory closure leading to *cis* hydrogens, it is clear that the reaction of the radical ion deliberately chooses a path which is unfavorable in the ground state. The reaction must therefore involve an electronically excited form of (26). It is interesting that the same ring closure can be brought about photochemically

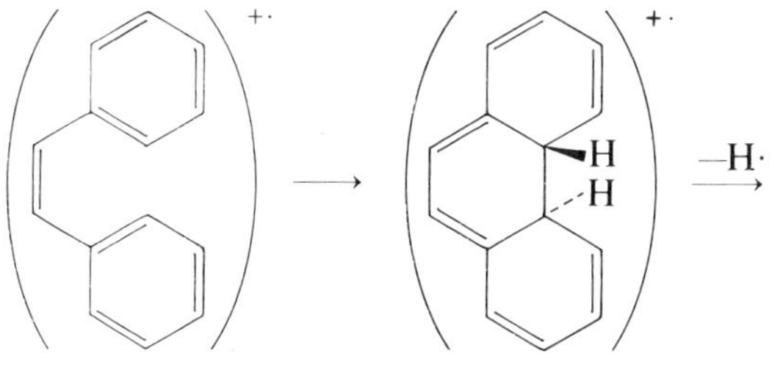

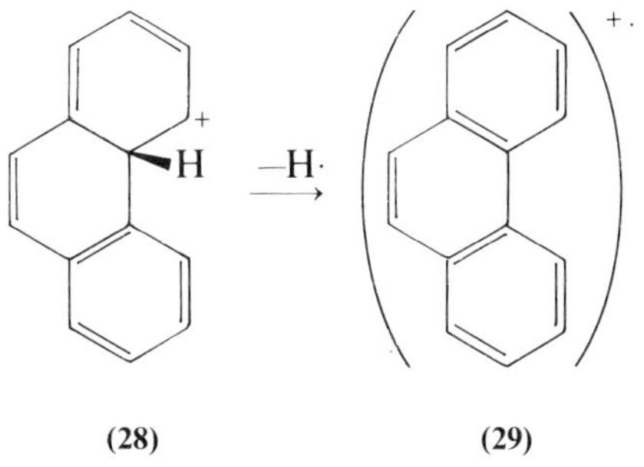

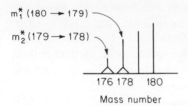

FIGURE 7.11. In the conversion of *cis*-stilbene to phenanthrene [(26)–(29)] in the mass spectrometer; the presence of two metastable peaks (m_1^*, m_2^*) shows that the two hydrogen atoms are lost successively, not simultaneously.[4b]

in the case of *cis*-stilbene itself, phenanthrene being formed via the *trans*-dihydro derivative corresponding to (27).

The general problem of the relationship between fragmentation reactions in a mass spectrometer and photochemical or thermochemical reactivity has fascinated mass spectroscopists for many years. The correlation in both cases is quite general and scores of examples are known. The scope and limits of these correlations have recently been reviewed[5] from the point of view of PMO theory, and we will not discuss them further here.

7.4 Radical Anions in the Gas Phase

Positive radical ions can be formed rather easily by bombarding neutral molecules with energetic electrons. The formation of negative radical ions, i.e., radical anions, is more difficult since radical anions have very low ionization potentials and easily lose their extra electrons to regenerate the parent molecules. Such a species can therefore be formed by electron capture, i.e., by the process

$$M + e \longrightarrow M^{-\cdot *} \xrightarrow{N} M^{-\cdot} \qquad (7.18)$$

only if very low-energy electrons are used. This can be achieved[6] most easily be adding an inert gas (xenon or nitrogen) to slow down the electrons so that they can be captured. This of course means that we must operate at reasonably high pressures so that the incident electrons *are* slowed down. We are therefore concerned here with processes which must be studied by chemical ionization (CI) mass spectrometry or by ion cyclotron resonance (ICR) spectroscopy (see Section 7.1). It also follows that the radical anions produced in this way must be formed in their ground states and with negligible thermal energies. Thus, whereas almost *any* positive ion undergoes fragmentation in the conventional mass spectrometer because the ions are inherently endowed with large amounts of vibrational energy and so effectively are very hot, the corresponding reactions of radical anions normally take place only if the ion is unstable with respect to dissociation, e.g.,

$$AB + e^- \longrightarrow AB^{-\cdot} \longrightarrow A^- + B\cdot \qquad (7.19)$$

The cardinal feature of a molecule, insofar as reactions of this kind are concerned, is its electron affinity, i.e., the energy liberated when it combines with an electron [equation (7.18)]. The electron affinity of a neutral molecule M is of course equal to the ionization potential of the corresponding radical anion M⁻·. Because of the negative charge, the ionization potentials of such radical anions are much less (0–1 eV) than those of normal molecules (~10 eV).

The electron affinities of conjugated molecules can be found by a PMO procedure very similar to that used to find excitation energies (see Section 6.7). It is evident from Fig. 3.9 that the energy $\mathscr{E}^*$ of the LUMO of an even AH, R is given by

$$\mathscr{E}^* = \alpha - a_{0r}b_{0s}\beta \tag{7.20}$$

where the dissection of R into two odd AHs S and T is such as to minimize the product of NBMO coefficients $a_{0r}b_{0s}$. From Koopmans' theorem (p. 26), the electron affinity of R should be equal to $-\mathscr{E}^*$. In order to estimate this, we must of course know the Coulomb integral α for carbon as well as the CC resonance integral β. This was not necessary in our discussion of light absorption because in estimating the differences in energy between the HOMO and LUMO, the term involving α cancelled.

We can use the measured ionization potentials I of any two even AHs to determine α and β in equation (7.20). Fig. 7.12 shows this calculation, taking naphthalene (I, 0.15 eV) and phenanthrene (I, 0.31 eV) as the AHs.

Using these values, one can calculate electron affinities of other even AHs. It is unfortunately rather difficult to compare the calculated values with experiment because the published values for electron affinities vary greatly. The values in Table 7.2 are from a recent survey. The agreement is

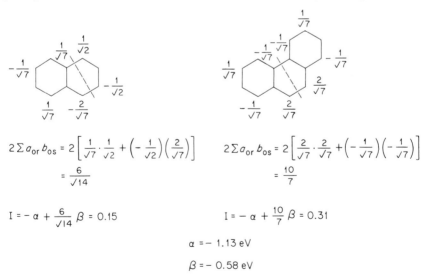

FIGURE 7.12. Calculations of α and β from the electron affinities of naphthalene and phenanthrene.

TABLE 7.2. Calculated and Observed Electron Affinities I of Even AHs

Compound	$2a_{0r}b_{0s}{}^a$	I(calc), eV	I(obs),[29] eV	
			Best value	Range of observed values
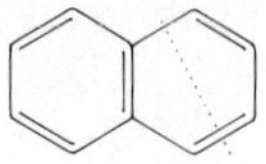 (naphthalene)	1.60	$(0.15)^b$	0.15	$-0.20, 0.15$
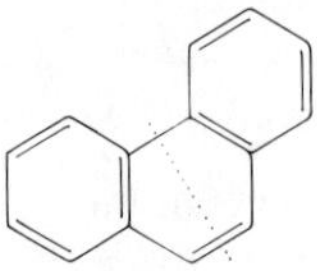 (phenanthrene)	1.43	$(0.31)^b$	0.31	$-0.20, 0.31$
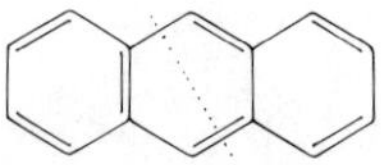 (anthracene)	0.92	0.57	0.82	0.82, 1.15
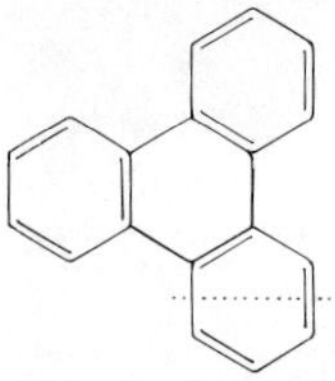 (triphenylene)	1.72	0.17	0.14	$-0.28, 0.14$
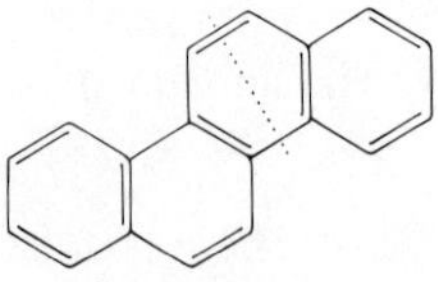 (chrysene)	1.35	0.37	0.33	0.33, 0.42
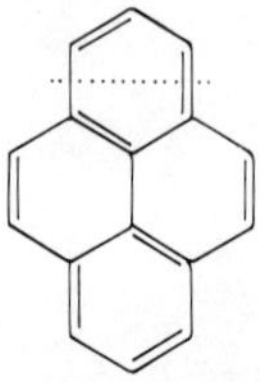 (pyrene)	1.15	0.48	0.50	0.42, 0.53

^{a} Product of NBMO coefficients for optimum dissection of the even AH, indicated in the first column.
^{b} Assumed (see Fig. 7.12).

reasonably satisfactory given the uncertainty in the experimental values.

Radical anions can be prepared in solution by reducing conjugated molecules with alkali metals or by cathodic reduction; namely

$$M + Na\cdot \longrightarrow M^-\cdot Na^+ \qquad (7.21)$$

$$M + e^-_{solv} \longrightarrow M^-_{solv} \qquad (7.22)$$

The stabilities and reactions of radical anions under such conditions are, however, often very different from what they are in the gas phase because the energies of solvation, even in such apparently innocuous solvents as ethers, are large. Thus, while benzene fails to combine with electrons in the gas phase, it does so readily in solution, giving a radical anion ($C_6H_6^{-\cdot}$) that can be detected by ESR spectroscopy. This in turn can react reversibly with other aromatic hydrocarbons,

$$C_6H_6^{-\cdot} + A \longrightarrow C_6H_6 + A^{-\cdot} \qquad (7.23)$$

Measurement of the equilibrium constants for such processes allows one to estimate the relative electron affinities (in solution) of different aromatic compounds. Measurements[7] of this kind for the equilibria between benzene and various monoalkyl and dialkyl benzenes, i.e.,

$$C_6H_6^{-\cdot} + C_6H_4R_1R_2 \longrightarrow C_6H_6 + C_6H_4R_1R_2^{-\cdot} \qquad (7.24)$$

are shown in Table 7.3. Note that the equilibrium constants are all less than unity and less for dialkylbenzenes than monoalkylbenzenes, implying that alkyl groups destabilize the negative ions. This of course might be expected from our theory of substituent effects (Chapter 4), but in fact it represents another difference between the behavior of ions in solution and in the gas phase. As we shall see presently, anions in the gas phase are *stabilized* by alkyl substituents. The last column of Table 7.3 shows the standard free energies of reaction for equation (7.24). These should be equal to the differences in electron affinity between benzene and the alkyl derivative.

In spite of the obvious importance of solvation, a number of correlations nevertheless exist between the reactivities of anions in the gas phase and in solution. Many more are likely to be discovered as the study of negative-ion mass spectra develops. Some examples follow.

TABLE 7.3. Relative Stabilities of Substituted Benzene Anion Radicals in Solution[a]

R_1	R_2	$10^3 \times k$	ΔG_{173}, eV
CH_3	H	230	0.02
C_2H_5	H	45	0.05
$CH(CH_3)_2$	H	25	0.06
$C(CH_2)_3$	H	9.1	0.07
CH_3	o-CH_3	1.1	0.10
CH_3	m-CH_3	20	0.06
CH_3	p-CH_3	110	0.03
C_2H_5	p-C_2H_5	8.3	0.07
$CH(CH_3)_2$	p-$CH(CH_3)_2$	1.6	0.06
$C(CH_3)_3$	p-$C(CH_3)_3$	< 0.2	0.12

[a] In dimethoxyethane, $CH_3OCH_2CH_2OCH_3$, at 173°K; Ref. 7. See equation (7.24). The equilibrium constant k is for the reaction of equation (7.24).

The reactions of radical anions in the gas phase often parallel those of analogous anions in solution. For example, *p*-nitrocumyl halides [(30); X = halogen] in solution react with anions A⁻ in the following way:

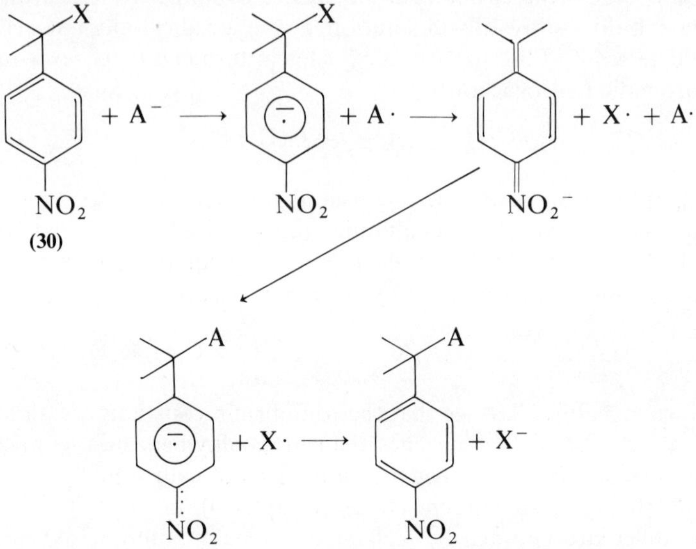

$$\text{(30)} + A^- \longrightarrow [\text{radical anion}] + A\cdot \longrightarrow [\text{quinoid}] + X\cdot + A\cdot$$

$$\longrightarrow [\text{A-substituted radical anion}] + X\cdot \longrightarrow [\text{A-substituted}] + X^- \tag{7.25}$$

In the mass spectrometer, one can observe the following reaction of the corresponding radical anion of (30) (this reaction could also occur by dissociative resonance capture, i.e., direct formation of the nitrobenzyl anion and a halide radical):

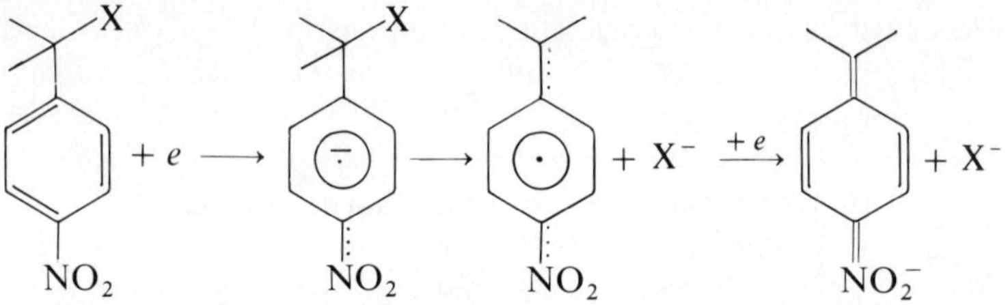

$$\text{ArX} + e \longrightarrow [\text{ArX}]^{-\cdot} \longrightarrow [\text{Ar}\cdot]^- + X^- \xrightarrow{+e} [\text{quinoid}] + X^- \tag{7.26}$$

In all the cases studied, the intensity of the benzyl anion in the mass spectrum parallels the rate of the substitution reaction in solution.

Another, similar parallel is shown by a comparison[8] of the electron affinities of polycylic aromatic hydrocarbons in the gas phase with their polarographic reduction potentials.

The reduction potential is a measure of the change in energy in the process

$$A + e \longrightarrow A^{-\cdot} \tag{7.27}$$

There is a very good linear relation between the reduction potentials and the electron affinities in the gas phase.

Another example is the reduction of alkyl halides by electron transfer.

The dominant negative ions in the mass spectra of alkyl halides are halide ions, produced by the process

$$RX + e \longrightarrow RX^{-\cdot} \longrightarrow R\cdot + X^- \qquad (7.28)$$

The same reaction is observed in the reduction of solutions of alkyl halides either by alkali metals or electrochemically.

7.5 Ion–Molecule Reactions in the Gas Phase

So far we have mostly considered first-order reactions of radical ions and ions in the gas phase. Here we will consider bimolecular reactions between ions and neutral molecules. Such reactions can be studied by CI mass spectrometry or ion cyclotron resonance (ICR) spectrometry (see Section 7.1).

A simple and intriguing reaction of this type is observed on ionizing methane. The initally formed radical ions $CH_4^{+\cdot}$ react as follows:

$$CH_4^{+\cdot} + CH_4 \longrightarrow CH_5^+ + CH_3\cdot \qquad (7.29)$$

Since CH_5^+ contains only four pairs of electrons, two of the hydrogen atoms must be linked to carbon by a three-center bond. The reaction is exothermic because the bond energy of such a bond is greater than the sum of a two-center, two-electron bond and a two-center, one-electron bond. Under conditions where the reaction of equation (7.29) can take place, an intense peak is observed corresponding to CH_5^+. It might be added that Olah et al.[9] have recently found evidence for the formation of CH_5^+ by protonation of CH_4 in very strong acids (e.g., $FSO_3H:SbF_5$, see p. 291).

In the case of other paraffins RH, where the corresponding carbonium ion R^+ is stabilized by $-I$ or $-E$ substituents, an alternative process is observed; e.g.,

$$\left(CH_3\!\!-\!\!\underset{\underset{CH_3}{|}}{\overset{\overset{CH_3}{|}}{C}}\!\!-\!\!H \right)^{+\cdot} + CH_3\!\!-\!\!\underset{\underset{CH_3}{|}}{\overset{\overset{CH_3}{|}}{C}}\!\!-\!\!H \longrightarrow CH_3\!\!-\!\!\underset{\underset{CH_3}{|}}{\overset{\overset{CH_3}{|}}{C^+}} + H\!\!-\!\!H + \cdot\underset{\underset{CH_3}{|}}{\overset{\overset{CH_3}{|}}{C}}\!\!-\!\!CH_3$$

$$(7.30)$$

Here the stabilization of the ion $(CH_3)_3C^+$ by the three $-I$ methyl groups is sufficient to make the decomposition of the $(CH_3)_3CH_2^+$ ion into $(CH_3)_3C^+ + H_2$ exothermic.

Both ICR spectroscopy[10] and CI mass spectrometry[11] can be used to measure the relative basicities of compounds by studying exchange reactions of the type

$$AH^+ + B \rightleftharpoons A + BH^+ \qquad (7.31)$$

The relative orders of basicity of various amines in the gas phase found in this way are

$$\text{C}_6\text{H}_{11}\text{NH} > (\text{CH}_3)_3\text{N} > \text{C}_5\text{H}_9\text{NH} > \text{C}_4\text{H}_7\text{NH} \qquad (7.32)$$

$$\text{Me}_3\text{CNH}_2 > \text{Me}_3\text{CCH}_2\text{NH}_2 > \text{Me}_2\text{CHNH}_2$$
$$> n\text{-BuNH}_2 > \text{EtNH}_2 > \text{MeNH}_2 > \text{NH}_3 \qquad (7.33)$$

$$\text{Et}_2\text{NH} > \text{Me}_2\text{NH} \qquad (7.34)$$

$$\text{Et}_3\text{N} > \text{Me}_3\text{N} \qquad (7.35)$$

$$\text{Me}_3\text{N} > \text{Me}_2\text{NH} > \text{MeNH}_2 \qquad (7.36)$$

$$\text{Et}_3\text{N} > \text{Et}_2\text{NH} > \text{EtNH}_2 \qquad (7.37)$$

$$\text{Me}_3\text{N} > \text{Me}_3\text{CNH}_2 \qquad (7.38)$$

$$\text{Me}_2\text{NH} \simeq \text{Me}_2\text{CHNH}_2 \qquad (7.39)$$

$$\text{Me}_3\text{N} \simeq \text{Et}_2\text{NH} \qquad (7.40)$$

These results are exactly what would be predicted on the basis of the theory presented in Chapter 4. One would expect the protonation of ammonia,

$$\text{NH}_3 + \text{H}^+ \longrightarrow \text{NH}_4^+ \qquad (7.41)$$

to be assisted by replacing the hydrogen atom by $-I$ alkyl groups and one would expect the $-I$ activity of methyl to be in turn increased by replacing hydrogen atoms in it by alkyl. The order of the effects of different alkyl groups in equations (7.32)–(7.34) and the effect of increasing alkyl substitution in equations (7.35) and (7.36) follow the predicted patterns.

The order of basicity in the gas phase differs significantly from that in solution. In water, the order of basicity of the methylamines is

$$\text{Me}_2\text{NH} > \text{Me}_3\text{N} > \text{MeNH}_2 > \text{NH}_3 \qquad (7.42)$$

The reason for this is that cations of the type R_3NH^+ are more strongly solvated in hydroxylic solvents than are amines of the type R_2NH. This is because the hydrogen atom in R_3NH^+ is more positive than that in R_2NH and so able to form stronger hydrogen bonds (see Section 5.7); the ΔE_{solv} values are in the order

$$\text{R}_3\text{N}^+\!\!-\!\!\text{H}\cdots\text{O}(\text{H})_2 > \text{R}_2\text{N}\!-\!\text{H}\cdots\text{O}(\text{H})_2 \qquad (7.43)$$

In the series of ions NH_4^+, MeNH_3^+, Me_2NH_2^+, Me_3NH^+, the number of positive hydrogen atoms decreases as the number of methyl groups increases. The stabilization of the resulting cation by solvation decreases correspondingly and eventually, at Me_3N, this decrease overtakes the increase in stability of the cation due to the increase in the number of $-I$ (i.e., methyl) substituents.

Reactions of Transient Ions

The relative stabilities of alkyl cations in the gas phase are the same as in solution, e. g.,

$$Me_3C^+ > Me_2CH^+ > MeCH_2^+ > CH_3^+ \qquad (7.44)$$

In this case, the solvation of the various ions in solution is similar, involving the empty $2p$ AO:

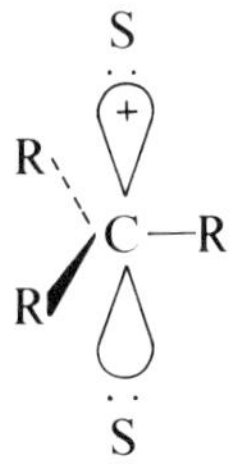

$$\begin{array}{c} \ddot{S} \\ R \diagdown \\ \diagdown C-R \\ R\diagup \\ \ddot{S} \end{array} \qquad (7.45)$$

Differences occur, however, in the case of carbonium ions that are stabilized by $\pm E$ or $-E$ substituents. Here the positive charge is delocalized over the adjacent conjugated system so the solvation energy is less. Thus Field[12a] has shown that t-amyl cations are formed only slightly more rapidly than benzyl in the gas phase by fission of the cations formed by protonation of the corresponding acetates,

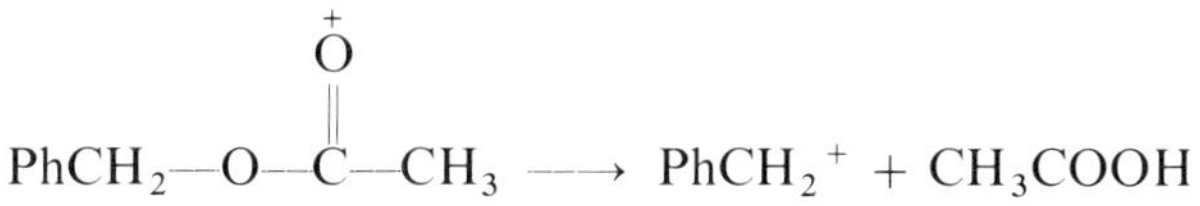

$$PhCH_2-O-\overset{+}{\underset{\parallel}{C}}-CH_3 \longrightarrow PhCH_2^+ + CH_3COOH \qquad (7.46)$$

$$\underset{\underset{Me}{|}}{\overset{\underset{Me}{|}}{Et-C}}-O-\overset{+}{\underset{\parallel}{C}}-CH_3 \longrightarrow \underset{\underset{Me}{|}}{\overset{\underset{Me}{|}}{Et-C^+}} + AcOH \qquad (7.47)$$

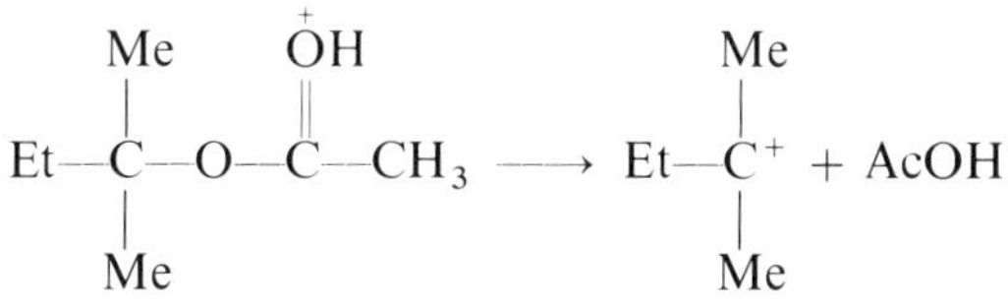

In solution, t-amyl cation is formed very much more readily than benzyl because its solvation energy is greater, the positive charge in benzyl being dispersed over the adjacent benzene ring.

Solvation-based discrepancies in reactivity are not limited to gas phase–solution comparisons. In alcohol solutions, a cyclopropyl group will generally increase the rate of an S_N1 solvolysis more than a phenyl. By use of ^{13}C NMR, Olah and co-workers have clearly shown that in superacid solutions, benzyl cations have higher electron densities on the α carbon than their cyclopropyl carbinyl analogs.[12b]

So far we have considered only cations. Here there is usually a close parallel between reactions in the gas phase and those in solution. The situation in the case of anions is, however, very different.

For example, data are available[13,14] for the relative gas-phase acidities of a number of protic acids, i.e., for the process

$$XH \longrightarrow X^- + H^+ \qquad (7.48)$$

Some of the results obtained in this way are as follows:

$$Me_3CCH_2OH > Me_3COH > Me_2CHOH > MeCH_2OH > CH_3OH > H_2O \quad (7.49)$$

$$CH_3CH_2OH > PhCH_3 > CH_3OH > H_3C-\!\!\!\!\bigcirc\!\!\!\!-CH_3 > H_2O \quad (7.50)$$

$$Me_3COH > n\text{-}C_5H_{11}OH > n\text{-}C_4H_9OH > n\text{-}C_3H_7OH > C_2H_5OH \quad (7.51)$$

$$HC\!\!\equiv\!\!CH > n\text{-}BuC\!\!\equiv\!\!CH > CH_3C\!\!\equiv\!\!CH > H_2O \quad (7.52)$$

$$Et_2NH > Me_3CCH_2NH_2 > Me_3CNH_2 > Me_2NH > Me_2CHNH_2 \quad (7.53)$$

$$n\text{-}BuNH_2 > n\text{-}PrNH_2 > EtNH_2 > MeNH_2 > NH_3 \quad (7.54)$$

The very first series [equation (7.49)] shows a shocking departure from the order that would be expected on the basis of the simple theory of substituents in Chapter 4. According to it, alkyl substituents, having $-I$ activity, should hinder the development of a negative charge at an adjacent atom. Yet here we find alkyl groups *stabilizing* such a negative charge! The same effect is seen in the amines [equation (7.54)]. Even the effects of different alkyl substituents are in the "wrong" order. The $-I$ activity of n-alkyl groups should increase in the series Me > Et > n-Pr > n-Bu. Yet equation (7.54) shows their acidifying power to increase in this order instead of decreasing. Again, toluene is an extremely weak acid in solution, far weaker than any alcohol, while in the gas phase, it is a stronger acid than methanol or water [equation (7.50)]. The same is true of acetylene [equation (7.52)]. In this case, alkyl substituents do decrease the acidity, though methyl is again anomalously more effective than n-butyl.

The discrepancy between the orders of basicity in solution and in the gas phase must be due to (1) solvation factors that act to produce the "expected order" of Chapter 4, and (2) factors that we did not consider in Chapter 4 that influence the relative stabilities of anions. The discrepancies here should be compared to the discrepancy in the case of amines which we considered earlier in this section. The anions, however, differ from the amines in two notable respects. First, the effects of solvation are far more extreme, completely inverting the order of basicity. Second, whereas the basicities of amines in the gas phase follow the pattern predicted by the theory of Chapter 4, as one would expect since solvation was neglected in developing it, in the case of the anions, it is the basicities *in solution* that follow the predicted pattern, the order in the gas phase being almost completely inverted.

The effects of solvation can be explained fairly reasonably in the following way. The values for basicities in solution usually refer to protic solvents (e.g., water, alcohols, formamide) in which anions are stabilized by hydrogen bonding. The strength of such hydrogen bonds must be especially large since they involve a charge–dipole interaction between a small anion

(RO^-) and a dipole (H^+X^-) which can approach the anion very closely. The total energy of solvation of a given anion, e.g., RO^-, will depend on the negative charge on the central atom and on the number of solvent molecules attached to it. Since the negative charge on the central atom is concentrated in its lone pairs of electrons, the number of hydrogen-bonded solvent molecules will depend on the number of such lone pairs. It may also be restricted by steric effects, due to the difficulty of packing solvent molecules around the anion. The basicity of the anion in the gas phase, on the other hand, will depend only on the availability of the most available lone pair of electrons, these being used to combine with a proton to form the conjugate acid. There is therefore no reason why the solvation energies of anions should run parallel to their intrinsic (i.e., gas-phase) basicities.

A good example is provided by toluene. In solution, toluene is a far weaker acid than any alcohol, having a pK_A of ~ 30, whereas the values for alcohols lie in the range 15–18. In the gas phase, however, toluene is a stronger acid than any normal alcohol and indeed stronger even than water (pK_A 14). The reason for this is that a carbanion contains only a single pair of unshared electrons and in the benzyl anion, the negative charge, instead of being concentrated at the carbanionoid center, is dispersed over a large conjugated system. In an alkoxide ion, the negative charge is concentrated on the oxygen atom, which, moreover, has *three* lone pairs of electrons. The solvation energy of an alkoxide ion is therefore enormously greater than that of the benzyl anion.

While this argument accounts for the inversion of relative basicities of oxygen anions and carbanions on passing from solution to the gas phase, it does not account for the equally dramatic inversion of the order of basicity of the alkoxide ions themselves. This inversion is due to changes in the strengths of individual hydrogen bonds rather than in their number.

Consider, for example, an equilibrium between an alcohol R_1OH and an alkoxide ion R_2O^-,

$$R_1OH + {}^-OR_2 \rightleftharpoons R_1O^- + HOR_2 \qquad (7.55)$$

If R_1 and R_2 are $-I$ groups with R_1 the more active, R_1O^- should be an intrinsically stronger base than R_2O^-, so the equilibrium should lie to the left in the gas phase. In solution in an alcohol ROH, each of the anions will be solvated by hydrogen bonding. The strength of each hydrogen bond should be greater, the greater is the negative charge on the oxygen of the alkoxide ion. Each hydrogen bond will therefore be stronger for R_1O^- than for R_2O^-. Furthermore, the oxygen atoms in the solvating molecules of ROH will become more negative as a result of the bonding to the anion. These oxygen atoms will in turn form secondary hydrogen bonds to other solvent molecules and the strengths of these bonds will parallel those of the primary hydrogen bonds and so the basicity of the central anion. Now the primary association involves hydrogen bonds to three molecules of solvent, and the secondary associations to six more, e.g.,

$$\begin{array}{c}
\text{ROH} \cdots \quad \cdots \text{HOR} \\
\text{O} \\
\text{H} \diagup \quad \diagdown \text{R}
\end{array}$$

$$\begin{array}{c}
\text{R}_1\text{—O}^- \cdots \text{H} \diagdown \quad \cdots \text{HOR} \\
\text{O} \\
\text{R} \diagup \quad \cdots \text{HOR}
\end{array} \qquad (7.56)$$

$$\begin{array}{c}
\text{R} \quad \text{H} \\
\diagdown \text{O} \diagup \\
\text{ROH} \cdots \quad \cdots \text{HOR}
\end{array}$$

An increase in basicity of the central alkoxide can therefore lead to a very large increase in the total energy of solvation. If this effect is great enough, it may outweigh the effect of primary basicity and so displace the equilibrium of equation (7.55) to the *right*.

An interesting confirmation of this argument has recently been found by McAdams and Bone,[13d] who studied the reaction of equation (7.55) with R_1 = methyl and R_2 = ethyl, using intermediate pressure (< 100 Torr) radiolysis. At low pressures, where the ions and molecules are separated, the equilibrium

$$CH_3O^- + C_2H_5OH \rightleftharpoons CH_3OH + C_2H_5O^- \qquad (7.57)$$

lies to the right [cf. equation (7.49)], the free energy of reaction being about -5 kcal/mole. At higher pressures, where each anion is solvated by about three molecules of alcohol, the free energy charge falls to -0.8 kcal/mole. In solution, where solvation is still more extensive [cf. (7.56)], the free energy of solution is *positive*, so the equilibrium now lies to the left.

These arguments explain the inversion in order of basicity of anions on passing from the gas phase to solution. We are still, however, left with the problem of explaining the anomalous effect of alkyl substituents in the gas phase. It has commonly been assumed that alkyl groups are $-I$ substituents, with a further potential of $-E$ activity by hyperconjugation in special circumstances (e.g., carbonium ions). The order of basicity of amines in the gas phase follows that predicted on this basis, as does the order of basicity of alkoxide ions in solution. However, we now see that the last result is spurious, being due to an inversion of the order of basicity by solvation. The gas-phase data show that alkyl groups must in fact *stabilize* adjacent, negatively charged centers, not destabilize them, so they must be acting by some process other than $-I$ or $-E$ effects.

A clue to the nature of this additional interaction seems to be provided by the data for acetylenes [equation (7.52)]. Here replacement of H by alkyl in H—C≡C— increases the basicity, as would be expected if the alkyl group exerts a $-I$ effect. However, *n*-butyl is less effective than methyl, contrary to expectations based on the $-I$ effect. This suggests that we are dealing with a charge–induced-dipole interaction. The electric field due to the negative charge polarizes the alkyl group and the resulting dipole and negative charge then attract one another, i.e.,

$$\langle alkyl \rangle - O^- \longrightarrow \langle -+ \rangle - O^- \qquad (7.58)$$

Forces of this kind are important in the solvation of anions in aprotic solvents and they vary as the fifth power of the distance between the charge and the polarizable group. Similar interactions may of course also occur in the case of positive ions, but there they could not be distinguished from the parallel consequences due to the $-I$ and $-E$ effects of alkyl. In the case of anions, the charge–induced-dipole interactions oppose the $-I$ and $-E$ effects of alkyl and apparently may outweigh them.

The competition between polarizability effects and inductive effects is probably clearest in the gas-phase acidities of carboxylic acids. A number of these acidities were quantitatively determined in an elegant set of high-pressure mass spectrometry experiments by Kebarle and co-workers.[14a,b] The gas-phase acidity order[14a,b]

$$HI > CF_3CO_2H > HBr > Cl_2CHCO_2H > F_2CHCO_2H >$$
$$BrCH_2CO_2H > ClCH_2CO_2H > HCl > FCH_2CO_2H >$$
$$HCO_2H > CH_3CH_2CH_2CO_2H > CH_3CH_2CO_2H >$$
$$CH_3CO_2H > HF \qquad (7.59)$$

corresponds to the order one would expect from solution, with three exceptions: (1) the placements of HCl, HF; (2) the order in the n-alkyl acid series; and (3) the order in the halo alkyl acid series. High solvation energies for the small Cl^- and F^- ions will easily account for the discrepancy there. The polarizability effects mentioned above would account for the fact that butyric acid is a stronger acid than acetic acid in the gas phase, while the reverse is true in solution. The operation of normal $-I$ inductive effects in the gas phase is obvious from the placement of formic acid in the alkyl acid series; however, the importance of polarizability effects on the stabilities of gas-phase ions should not be underestimated. Polarizability effects are almost certain to account for the order of acidity in the haloacetic acids. The observed acidity increase in the monohaloacetic acids F, Cl, Br is the reverse of the order in solution. In solution, the environment of an anion is exceptionally polarizable as compared to the gas phase, and this probably accounts for the general success in neglecting polarizability considerations in studies in solution.

Another striking—and indeed startling—confirmation of the stabilizing effect of alkyl groups on anions, and also of the dramatic effects of solvation, is provided by recent studies[15a-c] of the association of alkyl bromides with bromide ion in the gas phase,

$$RBr + Br^- \longrightarrow RBr_2^-$$

In solution, the product of this reaction is the transition state for S_N2 bromide exchange (see p. 208),

$$Br^- + R-Br \longrightarrow [Br \cdots R \cdots Br]^- \longrightarrow Br-R + Br^- \qquad (7.60)$$

In the gas phase, it is formed exothermically as a stable species, while in solution, it is only an unstable intermediate. This is because the energy of solvation of RBr_2^-, in which the negative charge is dispersed over two bromine atoms, is less than that of the monatomic anion Br^-. Table 7.4 shows the heats of reaction ΔH for alkylbromide–bromide association in the gas phase and the corresponding activation energies $\Delta E^\ddagger$ of bromide exchange in solution. Since the two are related by

$$\Delta E^\ddagger = \Delta H + \Delta E^\ddagger_{solv} \qquad (7.61)$$

where $\Delta E^\ddagger_{solv}$ is the solvation energy of activation (i.e., the difference in energy of solvation between the reactants and the transition state), one can immediately find the values of $\Delta E^\ddagger_{solv}$ (last column of Table 7.4).

The arguments on p. 235 indicate that the species $CH_3Br_2^-$ is effectively carbanionoid in nature. This is supported by the stabilization of this intermediate by $+E$ substituents, indicated by, e.g., the high S_N2 reactivity of α-haloketones. If so, $CH_3Br_2^-$ should be destabilized by $-I$ substituents such as alkyl. The results in the first column of Table 7.4 show that in the gas phase the reverse is true. Introduction of an alkyl group into CH_3Br makes ΔH much more negative. The effect, moreover, increases with the size of the alkyl group, being greatest with t-butyl (i.e., for neopentyl bromide). Introduction of further alkyl groups has little effect, presumably because the steric repulsions between them and the bromine atoms outweigh the electrostatic interactions (cf. p. 222). The order of reactivity of the bromides in solution is seen to arise entirely from the solvation terms. This is a particularly striking conclusion in the case of neopentyl, where the low S_N2 reactivity has always been attributed to steric hindrance to the primary association with the nucleophile. The heat of association of neopentyl bromide with bromide ion was in fact more negative than for any other bromide studied, so steric effects of this kind cannot be important if the symmetric dibromide is in fact the transition state. Either the transition state must occur early in the reaction, the symmetric dibromide being a stable intermediate, or the low S_N2 reactivity must be due to steric hindrance of solvation by the bulky t-butyl group in t-$BuCH_2Br_2^-$.

TABLE 7.4. Heat of Reaction ΔH for $RBr + Br^- \rightarrow RBr_2^-$ and Activation Energy $\Delta E^\ddagger$ and Solvation Energy of Activation $\Delta E^\ddagger_{solv}$ for Bromide Exchange ($Br^- + RBr \rightarrow BrR + Br^-$) in Acetone[15]

R	ΔH, kcal/mole	$\Delta E^\ddagger$,[30] kcal/mole	$\Delta E^\ddagger_{solv}$, kcal/mole
Me	−9.2	15.8	25.0
Et	−11.6	17.5	29.1
n-Pr	−11.6	17.5	29.1
i-Bu	−12.9	18.9	31.8
Me_3CCH_2	−14.4	22.1	36.3
i-Pr	−12.2	19.7	31.9
t-Bu	−12.4	21.8	34.2

The importance of solvation in determining the rates of S_N2 reactions has been well established in recent years. Thus the relatively low nucleophilicity of the fluoride and hydroxyl ions in hydroxylic solvents is due solely to their strong solvation by hydrogen bonding. In aprotic solvents such as dimethylsulfoxide (DMSO) or hexamethylphosphoramide (HMPA), both these ions are extremely active nucleophiles, their activity relative to, e.g., iodide ion increasing by four or five orders of magnitude in comparison with solutions in alcohols.

The gas-phase reactions of ions so far considered have all had counterparts in solution. It is, however, possible to prepare and study ions in the gas phase that could not easily be involved in reactions in solution. A good example is the oxygen radical anion $O^{-\cdot}$, which can be produced very easily in gaseous plasmas by combination of electrons with oxygen atoms. Since $O^{-\cdot}$ is isoelectronic with the fluorine atom, one might expect it to be extraordinarily reactive—and it certainly is. Thus, whereas "ordinary" radicals react with paraffins and alkyl halides by abstracting hydrogen or halogen atoms (see p. 274), $O^{-\cdot}$ can bring about radical substitution on carbon; e.g.,

$$RCH_2R + O^{-\cdot} \longrightarrow RCH_2O^- + R\cdot \qquad (7.62)$$

$$RCH_2Cl + O_2^{-\cdot} \longrightarrow RCH_2O^- + ClO\cdot \qquad (7.63)$$

The principles involved in these reactions are no different from those controlling the course of other radical processes (see Chapter 5), so we need not elaborate on them here. Thus the reaction of equation (7.62) tends to give the most stable possible alkyl radicals, while cleavage of CCl bonds in preference to CC [equation (7.63)] is due to the lower bond energy of the former.

Reactions of this type, involving highly reactive ionic species in gaseous plasmas, seem likely to prove of practical value in organic synthesis and possible commercial applications are being studied in a number of laboratories (though mostly under conditions of industrial secrecy).

7.6 Radical Cations in Solution

The majority of reactions in solution involve neutral molecules, radicals, and even-electron ions. Such processes were discussed in Chapters 4 and 5. There are, however, cases where radical ions play a role in solution chemistry. We have deferred discussing them until now because they can be understood most easily in terms of the gas-phase reactions examined in the earlier parts of this chapter. Here we will consider such reactions involving radical cations.

In order to prepare a radical cation in solution, an electron must in some way be abstracted from a "normal" neutral molecule. There are three main ways in which this can be achieved. In the first, the electron is transferred to a powerful electron acceptor such as a metal in a high valence state (e.g., Mn^{III}, Co^{III}) or a "high-energy" quinone (e.g., DDQ, dichlorodicyanoquinone). In the second, the electron is transferred to the positive electrode (anode) in an electrochemical cell (anodic oxidation). In the third, the electron is ejected photochemically, usually by X rays or γ-radiation (the basic step in radiation chemistry). We will consider these three processes in turn.

A. Oxidation by Electron Transfer

Reactions of this kind usually involve molecules with relatively low ionization potentials because the oxidizing power of convenient chemical oxidizing agents is limited. The majority of such reactions therefore take place by abstraction either of π electrons or of the lone-pair electrons of the nitrogen atoms. The ionization potentials of σ electrons, or of the lone-pair electrons of more electronegative atoms (O, F, Cl, etc.), are too high for them to be removed easily. The radical cation formed in the primary oxidation usually reacts by loss of a proton, gain of an anion, or fragmentation, to form a neutral radical and an even-electron species. Since the odd electrons in radicals normally occupy orbitals that are essentially nonbonding and have correspondingly low ionization potentials, the radicals formed in this way are in turn rapidly oxidized to cations by the electron acceptor. The final reaction product is derived from this cation by conventional processes. Oxidations proceeding by mechanisms of this general type have been termed ECE reactions (electron transfer/chemical reaction/electron transfer) by electrochemists.

A good example is the oxidation of aromatic compounds by manganic or cobaltic acetates in acetic acid, in particular the oxidation of arylmethanes to arylmethyl acetates, which takes place[16,17] as follows:

$$\begin{aligned} ArCH_3 + M^{III} &\longrightarrow (ArCH_3)^{+\cdot} + M^{II} \\ (ArCH_3)^{+\cdot} &\longrightarrow ArCH_2\cdot + H^+ \text{(solvated)} \\ ArCH_2\cdot + M^{III} &\longrightarrow ArCH_2^+ + M^{II} \\ ArCH_2^+ + AcOH &\longrightarrow ArCH_2OAc + H^+ \text{(solvated)} \end{aligned} \quad (7.64)$$

Manganic acetate is not a sufficiently powerful oxidizing agent to attack toluene in this way, but it will attack derivatives of toluene in which the ionization potential has been lowered by introduction of $-E$ substituents (e.g., p-methoxytoluene). Cobaltic acetate, being a more powerful oxidizing agent than manganic acetate, will also attack toluene.

An interesting example seems to be provided by the Shell process for

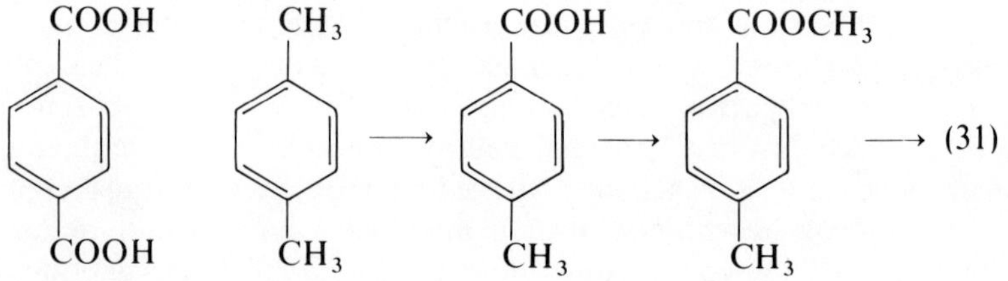

(31)   (32)   (33)   (34)

terephthalic acid (31), the key intermediate for the manufacture of dacron (terylene). Here p-xylene (32) is oxidized to p-toluic acid (33) by air in hot acetic acid containing cobalt acetate as catalyst. The acid is then converted to the methyl ester (34), which undergoes further oxidation under similar conditions to (31).

The final step is undoubtedly a typical radical chain autooxidation (p. 274), involving the propagation steps

$$\text{ArCH} + \text{ArC}-\text{O}_2\cdot \longrightarrow \text{ArC}\cdot + \text{ArC}-\text{O}_2\text{H} \qquad (7.65)$$

$$\text{ArC}\cdot + \text{O}_2 \longrightarrow \text{ArC}-\text{O}_2\cdot \qquad (7.66)$$

The fact that this takes place with (34) but not with (33) can be attributed to the fact the —COOCH$_3$ is a more powerful $+E$ group than is —COOH and so facilitates the hydrogen abstraction step in the autooxidation [equation (7.65)] by stabilizing the resulting benzyl radical. Since —COOH is a better radical stabilizer than CH$_3$, and since (33) apparently fails to undergo autooxidation under the conditions of the reaction, it is extremely unlikely that (32) could do so. The first step in the oxidation of p-xylene therefore very probably[16a] involves electron transfer oxidation of (32) to p-methylbenzyl acetate (35) by Co^{III}, following equation (7.64). The acetate can now autooxidize [equations (7.65) and (7.66)] since the intermediate radical (ArĊHOAc) is stabilized by the additional $-E$ acetoxy group. Autooxidation gives a hydroperoxide (36) which can dehydrate to the mixed anhydride (37) or react with Co^{II} to give Co^{III} and the aldehyde (38). Aldehydes autooxidize very easily under the conditions of the reaction to carboxylic acids.

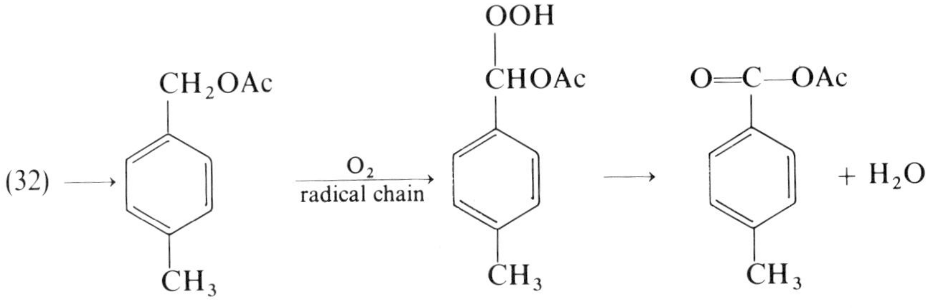

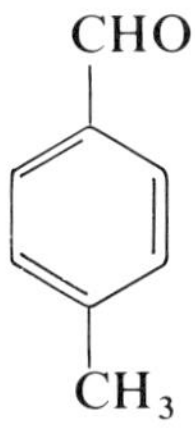

The oxidation of arylmethanes to carboxylic acids can also be brought about by alkaline ferricyanide, probably by a similar electron transfer mechanism; i.e.,

$$ArCH_3 + Fe(CN)_6^{3-} \longrightarrow (ArCH_3)^{+\cdot} + Fe(CN)_6^{4-}$$
$$(ArCH_3)^{+\cdot} + HO^- \longrightarrow ArCH_2\cdot + H_2O$$
$$ArCH_2\cdot + Fe(CN)_6^{3-} \longrightarrow ArCH_2^+ + Fe(CN)_6^{4-}$$
$$ArCH_2^+ + HO^- \longrightarrow ArCH_2OH$$

(7.67)

The alcohol is then oxidized in a similar way to the aldehyde and this in turn to the acid. In the case of α-methylnaphthalene, the alcohol and aldehyde can in fact be isolated as intermediates.

In these reactions, the key step is loss of a proton from the initially formed radical cation to form an arylmethyl radical. The driving force is the production of a highly stabilized radical. In other cases, the radical cation can combine with an anion to form a neutral radical which then undergoes further oxidation. Thus the first step in the oxidation of anthracene involves a process of this kind,

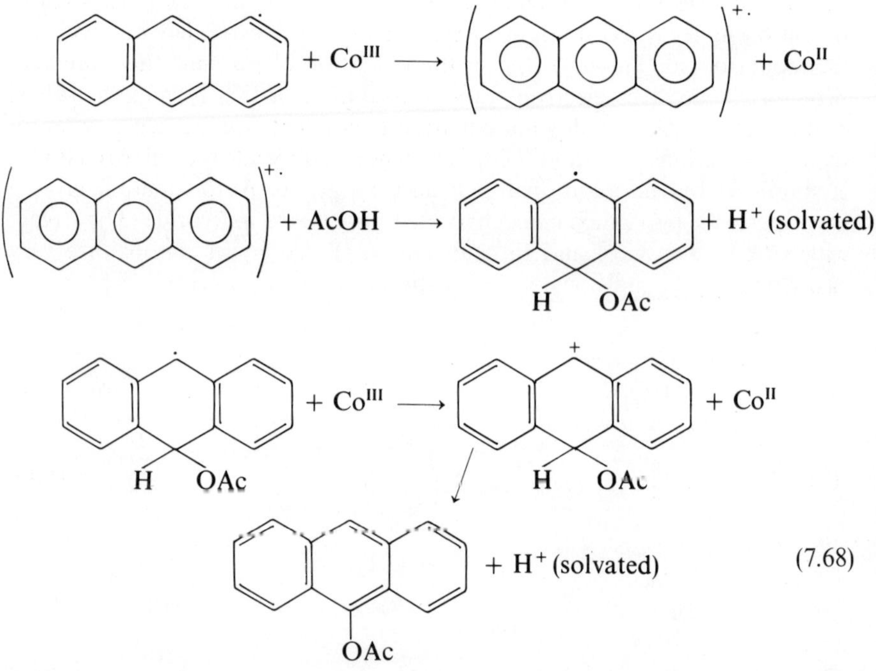

(7.68)

Here the product from the second step is the same as that from addition of an acetate radical to anthracene. The orientations of both reactions are therefore similar (see aromatic substitution by radical; p. 328).

The number of electrophilic aromatic substitution reactions that proceed with radical cation mechanisms analogous to (7.68) has increased dramatically in recent years. Even benzene may be converted to its trifluoroacetate derivative by treatment with 2 equiv of $(CF_3CO_2)_3Co^{III}$ in TFA.[17a] Thallation of aromatic compounds by thallium tris-(trifluoroacetate) has been generally thought to proceed by a two-electron transfer mechanism[17b]; however, direct observation of aromatic radical cations in solution under thallation conditions strongly suggests that one-electron processes may be important in thallation reactions.[17c] Considering the energetics of one-electron vs. two-electron transfer (second ionization potentials are usually two to three times those of first ionization potentials), this result is hardly surprising.

Oxidations involving loss of a lone-pair (nonbonding) electron rather than a π electron are observed in the reactions of amines with electron acceptors (potassium ferricyanide, chlorine dioxide). In the case of a tertiary aliphatic amine, the result is cleavage of an alkyl group; i.e.,

$$\begin{aligned} RCH_2NR_2 + ClO_2 &\longrightarrow RCH_2\overset{+}{\overset{..}{N}}R_2 + ClO_2^- \\ RCH_2\overset{+}{\overset{..}{N}}R_2 &\longrightarrow R\dot{C}HNR_2 + H^+ \text{(solvated)} \\ R\dot{C}HNR_2 + ClO_2 &\longrightarrow RCH{=}\overset{+}{\overset{..}{N}}R_2 + ClO_2^- \\ RCH{=}\overset{+}{\overset{..}{N}}R_2 + H_2O &\longrightarrow RCHO + HNR_2 \end{aligned} \quad (7.69)$$

Here the radical cation easily loses a proton because the resulting radical is stabilized by the powerful $-E$ substituent NR_2 (see Table 4.3, p. 179).

The main difference between the reactions of radical cations and of neutral radicals thus arises from the acidity of the former, due to their positive charge. Radical cations have a very strong tendency to lose protons or to add anions, forming neutral radicals which in turn are further oxidized to cations. One does not therefore as a rule see products arising from bimolecular reactions of radical cations, e.g., combination or disproportionation, since their lives are too short under the conditions where they are formed.

B. Electrochemical (Anodic) Oxidation

Oxidation at an anode follows the same principles as oxidation by chemical electron transfer, with one exception. In the case of anodic oxidation, the radical cations are produced in a very small volume near the cathode. Even if their lifetimes are short, they are therefore liable to undergo bimolecular reactions. This effect is seen in the Kolbe synthesis. Electrolysis of a salt of a carboxylic acid, RCOOH, usually leads to good yields of the hydrocarbon R_2 by the following process:

$$\begin{aligned} RCOO^- &\longrightarrow RCOO\cdot + e^- \\ RCOO\cdot &\longrightarrow R\cdot + CO_2 \\ 2R\cdot &\longrightarrow R_2 \end{aligned} \quad (7.70)$$

Such salts can also be oxidized chemically by electron acceptors (A) in solution,

$$\begin{aligned} RCOO^- + A &\longrightarrow RCOO\cdot + A^- \\ RCOO\cdot &\longrightarrow R\cdot + CO_2 \end{aligned} \quad (7.71)$$

Here, however, products other than the dimer R_2 are usually formed. This is because the radicals R· are being produced throughout the solution, so that their concentration is low and the chance of bimolecular reactions between them correspondingly small. In the Kolbe synthesis, the radicals are produced in a very small volume next to the anode, in which their local concentration is consequently high and the chance of bimolecular reactions between them is correspondingly enhanced.

An interesting example is the oxidation of dimethylaniline (39). Oxidation of (39) with manganic acetate in acetic acid gives[16b] as the sole isolable product 4,4'-bis(dimethylamino)diphenylethane (40). Apparently the first step in the reaction is dealkylation by the mechanism of equation (7.69),

$$\underset{(39)}{PhNMe_2} + Mn^{III} \longrightarrow (PhNMe_2)^{+\cdot} + Mn^{II}$$

$$(PhNMe_2)^{+\cdot} \longrightarrow PhNMeCH_2\cdot + H^+ \text{(solvated)} \quad (7.72)$$

$$PhNMeCH_2\cdot + Mn^{III} \longrightarrow Ph\overset{+}{N}Me{=}CH_2$$
$$\longrightarrow PhNHMe + CH_2O$$

The formaldehyde then undergoes acid-catalyzed condensation with dimethylaniline to form the product:

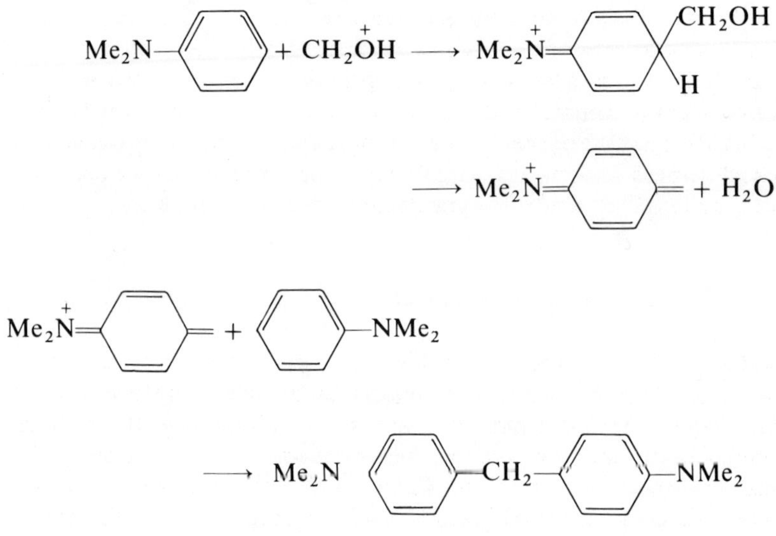

$$Me_2\overset{+}{N}-\!\!\!\left\langle\;\right\rangle + CH_2\overset{+}{O}H \longrightarrow Me_2\overset{+}{N}{=}\!\!\!\left\langle\;\right\rangle\!\!{<}^{CH_2OH}_{H}$$

$$\longrightarrow Me_2\overset{+}{N}{=}\!\!\!\left\langle\;\right\rangle{=} + H_2O \qquad (7.73)$$

$$Me_2\overset{+}{N}{=}\!\!\!\left\langle\;\right\rangle{=} + \left\langle\;\right\rangle\!-NMe_2$$

$$\longrightarrow \underset{(40)}{Me_2N-\!\!\!\left\langle\;\right\rangle\!-CH_2-\!\!\!\left\langle\;\right\rangle\!-NMe_2}$$

(The two condensations with dimethylaniline are of course typical electrophilic substitutions, the ring being activated by the $-E$ Me_2N group, see p. 316.)

Anodic oxidation of dimethylaniline can give different products, depending on the conditions.[16d] Thus under conditions where the initial radical cation readily loses a proton, the product arises from oxidation of a

Reactions of Transient Ions

methyl group [cf. equation (7.69)]; e.g.,

$$PhNMe_2 \longrightarrow (PhNMe_2)^{+\cdot} + e^-$$
$$\longrightarrow PhNMeCH_2\cdot + H^+ \text{(solvated)}$$
$$\longrightarrow PhNMeCH_2^+ + e^-$$
$$\xrightarrow{+ \text{MeOH}} PhNMeCH_2OMe + H^+ \text{(solvated)}$$

(7.74)

Under conditions where deprotonation takes place less readily, the initial radical cation can, on the other hand, dimerize to form tetramethylbenzidine (41),

$$PhNMe_2 \longrightarrow (PhNMe_2)^{+\cdot} + e^-$$

$2(PhNMe_2)^{+\cdot} \longrightarrow$ [structure of dihydro intermediate with $Me_2\overset{+}{N}=$ ... $=\overset{+}{N}Me_2$ and two H atoms]

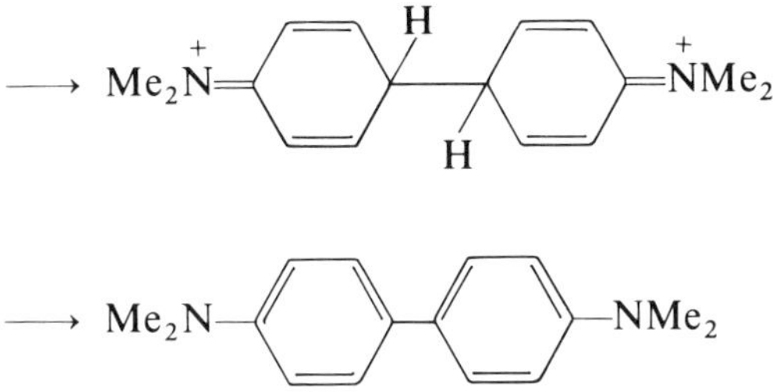

$\longrightarrow Me_2N-$[biphenyl]$-NMe_2$ (7.75)

(41)

An interesting additional point is that chemical oxidation of $PhNMe_2$ with lead dioxide (PbO_2) in benzene also gives tetramethylbenzidine. Lead dioxide is a powerful oxidizing agent, but is insoluble in all solvents. When it oxidizes $PhNMe_2$, presumably by electron abstraction, the reaction takes place rapidly on the surface of the oxide. The radical cations are therefore produced in a small volume near the surface, just as in anodic oxidation. Moreover, they are produced in a medium with little affinity for protons, lead dioxide being a weak base and the solvent being depleted of $PhNMe_2$ near the surface because of the rapid oxidation to $(PhNMe_2)^+$. The conditions are therefore ideal for dimerization of the radical cation [equation (7.70)].

C. Photochemical and Radiochemical Oxidation

Radical cations are produced photochemically by electron transfer transitions (p. 419) but only as ion pairs with the radical anion formed simultaneously. In most cases, the ion pair reverts to reactant by reverse electron transfer before any reaction can take place. Radical cations can, however, be prepared as stable entities by photolysis of compounds with low ionization potentials at low temperatures in glasses containing electron acceptors. Electrons expelled by photoionization wander through the glass until they are trapped by an acceptor. Since the acceptor is not in contact with the cation radical and since energy is required to extract the electron from the trap, the radical cation can survive as such for long periods. A classic example is the photoionization of tetramethyl-*p*-phenylenediamine (42) to Wurster's Blue (43),

$$Me_2N-\text{C}_6H_4-NMe_2 \xrightarrow[\text{glass}]{h\nu} [Me_2N\cdots\text{C}_6H_4\cdots NMe_2]^{+\cdot} + e^- \quad (7.76)$$

(42)   (43)

Reactions of this kind are analogous to those taking place in photographic emulsions. Here silver ions act as the traps, giving silver atoms that collect into tiny grains of silver. These grains act as nuclei for the reduction of silver halide by photographic developers.

A still more important process involving photochemical production of a radical cation is probably involved in the first step in photosynthesis. The chloroplasts at which photosynthesis takes place contain aggregates of molecules of chlorophyll stacked together so that their π MOs interact. J. J. Katz and his co-workers have demonstrated the existence of a water-bonded chlorophyll dimer that appears to be the active center in photosynthesis. Photoexitation causes electron transfer from one of the chlorophylls to the other (7.77):

$$\text{[chlorophyll dimer]} \xrightarrow{h\nu} \text{[chlorophyll dimer]}^{+\cdot\,-\cdot} \quad (7.77)$$

The oxidizing and reducing power thus generated is used to drive the photosynthetic unit. In the process H_2O is eventually converted to oxygen by a process whose details are obscure but which follows the overall equation

$$4H_2O^{+\cdot} \longrightarrow 2H_2O + 4H^+ + O_2 \quad (7.78)$$

The radical anion aggregate by another series of steps, involving ferredoxin among other intermediates, eventually reduces carbon dioxide to sugars.

While radical cations can in special cases be produced photochemically in solution by ultraviolet or visible light, the main province for such reactions lies in the field of radiation chemistry, where the electrons are ejected by X rays or γ rays. This has become a major field in recent years, partly because of the interest in the effects of radiation from nuclear bombs and partly because the operation of nuclear reactors yields as a by-product embarrassingly large amounts of radioisotopes which consequently provide a very cheap and almost unlimited source of γ rays. Processes of this kind necessarily produce radical cations together with stoichiometric amounts of electrons or radical anions. The charges tend to annihilate one another rapidly, the net effect being the production of radicals. The chemical reactions observed under these conditions are therefore normal radical reactions. Moreover, in cases where a solvent is used, most of the primary ionization will involve the solvent since ionization by high-energy radiation is quite unselective. In the case of aqueous solutions, the main primary step is the production of hydroxyl radicals and hydrogen atoms by radiolysis of water,

$$H_2O \rightsquigarrow H_2O^{+\cdot} + e^- \text{ (solvated)}$$
$$H_2O^{+\cdot} + H_2O \longrightarrow HO\cdot + H_3O^+ \quad (7.79)$$
$$H_2O + e^- \longrightarrow HO^- + H\cdot$$

The destructive effects of radiation on living organisms are due to consequent radical reactions, in particular oxidation by the hydroxyl radicals produced according to equation (7.79).

In this regard, it is interesting to note that the absorption spectra of both the radical anion and the radical cation of chlorophyll a have been recorded at 77°K in an organic glass.[18a] In this case the ions were produced by either X-ray or γ-radiolysis.

The reactions involved in radiochemistry involve no new principles and can be predicted from the known behavior of the primary radical ions and of the secondary radicals derived from them. For example, the radiolysis of single crystals of carboxylic acids at 77°K leads[18b] to the following reactions:

$$RCO_2H \rightsquigarrow RCO_2H^{+\cdot} + e^- \quad (7.80)$$

$$RCO_2H^{+\cdot} + RCO_2H \longrightarrow RCO_2\cdot + R\!-\!\overset{+}{C}\!\!\begin{array}{c}OH\\ \diagup\\ \diagdown\\ OH\end{array} \quad (7.81)$$

On warming, the radical $RCO_2\cdot$ loses carbon dioxide, and various products are formed from the resulting alkyl radicals.

7.7 Radical Anions in Solution

In the gas phase, radical cations are more stable and chemically more important than radical anions because the latter can too easily lose their extra weakly bound electrons. In solution, the reverse is the case. Here radical cations are difficult to obtain and so extremely reactive that they usually disappear almost as soon as they are formed. Only in a few cases are they stable enough to survive in solution, the exceptions involving species highly stabilized by mesomerism, such as Wurster's Blue (43) or the thianthrene radical cation (44).

The reason for this is that radical cations are usually formed by loss of an electron from a bonding MO or the AO of an electronegative heteroatom. Since such orbitals are low in energy, the resulting radical cation is extremely electrophilic and acts as an extremely powerful protic or Lewis acid. They therefore react very readily under almost all conditions, either by loss of a proton or by combining with some anion or other nucleophile.

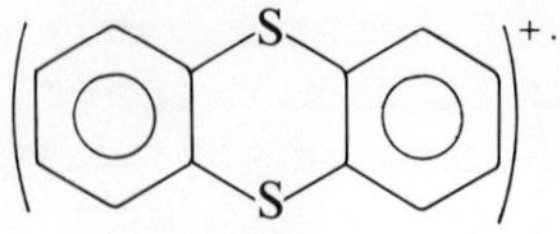

(44)

The situation is entirely different in the case of radical anions. These are stabilized in solution by solvation. Their formation by electron capture therefore takes place much more readily in solution than it does in the gas phase and once formed, they are more stable. They are also less reactive than radical cations because they cannot easily lose hydride ions. Spontaneous decomposition can usually occur only if the radical anion contains an atom or group that can easily be expelled as an anion. Thus electron capture by alkyl halides gives rise to radicals and halide ions; e.g.,

$$RBr + e^- \longrightarrow (RBr)^{-\cdot} \longrightarrow R\cdot + Br^- \qquad (7.82)$$

Otherwise, if acids and other electrophiles are excluded, radical anions can commonly exist in solution as stable species that can be prepared and studied at leisure.

Such solutions can often be prepared directly by reducing organic compounds RH with alkali metals M. The process

$$M\cdot + RH \longrightarrow M^+ + RH^{-\cdot} \qquad (7.83)$$

is in itself endothermic because the ionization potential of any metal, even cesium, is greater than the electron affinity of any organic molecule. In solution, however, the solvation energies of the ions are sufficient to make the process exothermic. Thus sodium reacts with aromatic hydrocarbons in ethers (e.g., tetrahydrofuran) to give the corresponding radical anions as stable solutions; e.g.,

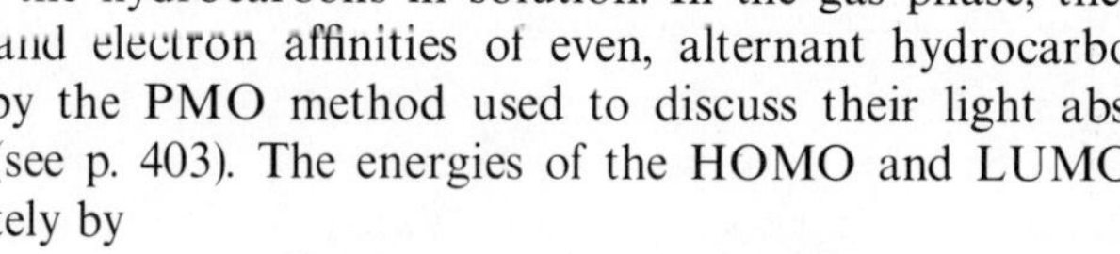

(45)   (7.84)

The ease with which such reactions occur should depend on the electron affinities of the hydrocarbons in solution. In the gas phase, the ionization potentials and electron affinities of even, alternant hydrocarbons can be estimated by the PMO method used to discuss their light absorption in Chapter 6 (see p. 403). The energies of the HOMO and LUMO are given approximately by

$$E(\text{HOMO}) = \alpha + \sum a_{0r}b_{0s}\beta \qquad (7.85)$$

$$E(\text{LUMO}) = \alpha - \sum a_{0r}b_{0s}\beta \qquad (7.86)$$

where a_{0r} and b_{0s} refer to the optimum dissection into a pair of odd AHs (see Table 6.1). From Koopmans' theorem, the ionization potential and electron affinity of a molecule should be equal to *minus* the orbital energies of the HOMO and LUMO. From equations (7.85) and (7.86) and the discussion of p. 403, it will be seen that there should be linear relations between the energy of the first $\pi \to \pi^*$ transition ($2\Sigma a_{0r}b_{0s}\beta$) for an even AH and its ionization potential and electron affinity. Such relations have been shown to hold for a number of such hydrocarbons.

The ionization potential and electron affinity of a molecule in solution are measured by the corresponding oxidation and reduction potentials (see Section 4.4). It has been shown that the oxidation potentials of even AHs run parallel to their gas-phase ionization potentials and the same seems to be true of the reduction potentials and electron affinities. The PMO method can therefore be used to estimate the relative reduction potentials of hydrocarbons in solution and hence the ease with which they react with alkali metals. One can also observe reversible electron transfer reactions between hydrocarbons and radical anions, e.g.,

$$R_1^{-\cdot} + R_2 \rightleftharpoons R_1 + R_2^{-\cdot} \qquad (7.87)$$

The equilibrium naturally favors formation of the radical anion corresponding to the hydrocarbon with the higher electron affinity. This can be correctly predicted by the PMO method indicated above.

Alkali metals can not only reduce organic compounds by electron transfer; they can also dissolve directly in certain solvents (notably liquid ammonia and ethylenediamine) to give colored solutions containing the metal cations together with solvated electrons,

$$Na + NH_3(\text{liquid}) \longrightarrow Na^+ + e^- (\text{solvated}) \qquad (7.88)$$

Such solutions are excellent reagents for the formation of radical anions by combination of the (solvated) electrons with added organic species. The same is true of the solutions of radical anions derived from hydrocarbons [cf. equation (7.84)]. These can reduce added compounds by electron transfer if their electron affinity is greater than that of the hydrocarbon used to form the radical anion. Some examples follow.

(a) Nitrogen can be reduced[19] to ammonia by sodium naphthalide [(45); equation (7.84)] in presence of titanium isopropoxide,

$$3\,(45)^{-\cdot} + \tfrac{1}{2}N_2 + Ti(OCHMe_2)_4 \longrightarrow 3\,(\text{naphthalene}) + (\text{Ti complex}) \xrightarrow{H^+} H_3N \qquad (7.89)$$

Under appropriate conditions, the reaction can be operated[19] as a continuous cycle, forming the closest analog yet discovered to the biological fixation of nitrogen.

(b) Sodium naphthalide or solvated electrons can be used to reduce cyclopropyl halides with retention of configuration,[20] e.g.,

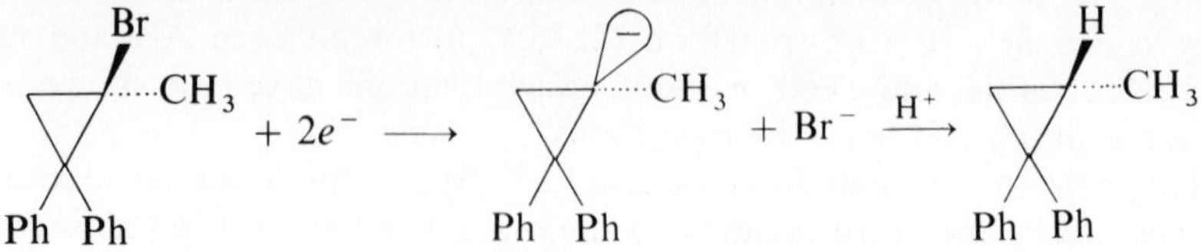

$$\text{(7.90)}$$

The retention of configuration could be explained in two ways. First, the reaction might take place in steps via a cyclopropyl radical (46),

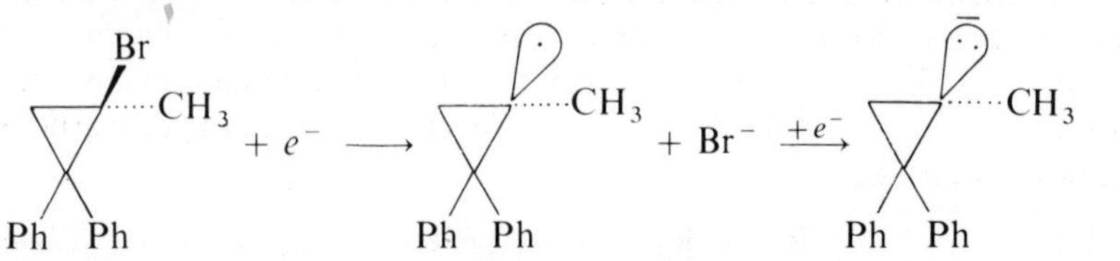

$$\text{(7.91)}$$

The cyclopropyl radical is known[21] to be pyramidal, so configuration could be retained if the second electron transfer were sufficiently fast. The barrier to inversion in the radical is, however, very low, so the second step would have to be *very* fast. The anion, once formed, is conformationally stable, the barrier to inversion being quite large. The second possibility would be that the primary radical anion is stable and undergoes loss of Br⁻ only on addition of the second electron. This could be the case if the first electron attaches itself to the benzene rings rather than the cyclopropyl ring. In this case, attack by the second electron would lead to the radical (46) with an extra electron attached to the CPh_2 group. This could indeed be transferred almost instantaneously to convert the pyramidal radical to the anion (47) before it had time to invert.

(c) Analogous reactions of sodium naphthalide with aryl halides leads to cleavage to aryl radicals which can undergo typical radical reactions, e.g.,[22]

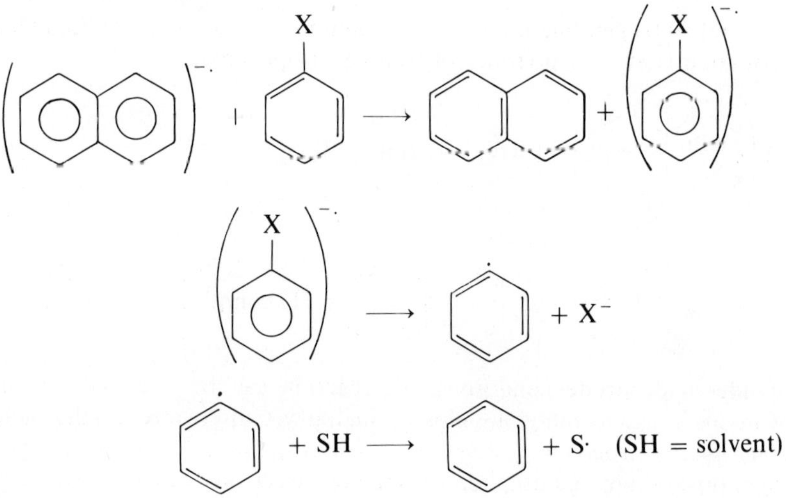

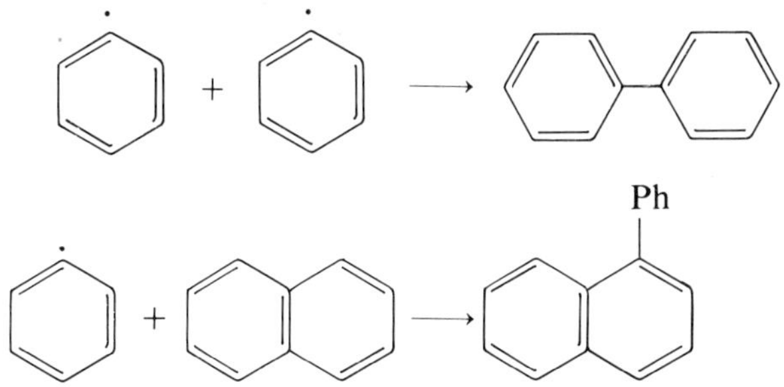

Reactions of Transient Ions

$$\text{Ph} \cdot + \text{Ph} \cdot \longrightarrow \text{Ph-Ph}$$

$$\text{Ph} \cdot + \text{naphthalene} \longrightarrow \text{1-phenylnaphthalene} \tag{7.92}$$

The fact that the intermediate phenyl radicals react before they are reduced seems to support the second explanation for the retention of configuration observed in the previous example. Since, however, phenyl radicals are much more reactive than alkyl radicals, it is just possible that the reduction of the intermediate radicals by electrons may be slow compared with the reactions of phenyl radicals but still fast compared with the rate of inversion of cyclopropyl. In this case the phenyl radicals would react before they could be reduced, while the cyclopropyl radicals would be reduced before they could invert.

(d) If the preceding reaction is carried out in the presence of amide ion, replacement of halogen by amino can take place[23] by a radical chain process involving electron transfer, e.g.,

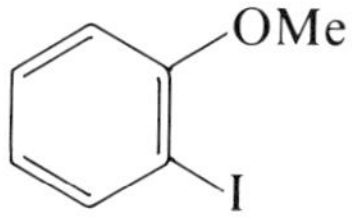

$$o\text{-MeO-C}_6\text{H}_4\text{-I} + e^- \longrightarrow (\text{ArI})^{-\cdot} \quad (\text{Ar} = o\text{-anisyl})$$

$$(\text{ArI})^{-\cdot} \longrightarrow \text{Ar} \cdot + \text{I}^- \tag{7.93}$$

$$\text{Ar} \cdot + \text{NH}_2^- \longrightarrow (\text{ArNH}_2)^{-\cdot}$$

$$(\text{ArNH}_2)^{-\cdot} + \text{ArI} \longrightarrow \text{ArNH}_2 + (\text{ArI})^{-\cdot}$$

The last step takes place in the direction indicated because the electron affinity of ArI is greater than that of ArNH_2, halogen having a greater $+I$ activity than nitrogen and also having some ability to stabilize adjacent anionic centers by $p\pi:d\pi$ bonding.

The reactions of equations (7.92) and (7.93) are special cases of the Wurtz and Wurtz–Fittig reactions that were once prominent in elementary texts. Such reactions involve the formation of radicals from halides through reductive fission by electron transfer,

$$\text{RX} + e^- \longrightarrow \text{R} \cdot + \text{X}^- \longrightarrow \text{R}_2 \tag{7.94}$$

Carbonyl compounds undergo similar reductions by electron transfer, the formation of pinacols (7.95) and the acyloin condensation (7.96) being typical examples:

$$R_2CO + e^- \longrightarrow (R_2CO)^{-\cdot}$$

$$2(R_2CO)^{-\cdot} \longrightarrow R_2\overset{O^-}{\underset{}{C}}-\overset{O^-}{\underset{}{C}}R_2 \xrightarrow{+2H^+} R_2\overset{HO}{\underset{}{C}}-\overset{OH}{\underset{}{C}}R_2 \qquad (7.95)$$

$$RCOOR + e^- \longrightarrow (RCOOR)^{-\cdot}$$

$$2(RCOOR)^{-\cdot} \longrightarrow R-\overset{O^-}{\underset{OR}{C}}-\overset{O^-}{\underset{OR}{C}}-R \longrightarrow R-\overset{O}{\underset{}{C}}-\overset{O}{\underset{}{C}}-R + 2RO^- \qquad (7.96)$$

$$R-\overset{O}{\underset{}{C}}-\overset{O}{\underset{}{C}}-R + 2e^- \longrightarrow R-\overset{O^-}{\underset{}{C}}=\overset{O^-}{\underset{}{C}}-R \xrightarrow{+2H^+} R-CHOH-CO-R$$

Radical anions can also act as bases in the presence of acids or electrophiles, combining with a proton or electrophile to form a neutral radical. The classic example of such a process is the Birch reduction of benzene derivatives to dihydrobenzenes by sodium in liquid ammonia, i.e., by solvated electrons, in the presence of an alcohol as proton donor. The first step in this reaction is a reduction of the benzene derivative to a radical anion,

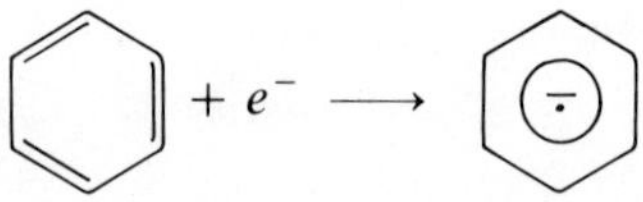

$$\bigcirc + e^- \longrightarrow \bigcirc^{\overline{\cdot}} \qquad (7.97)$$

The radical anion then reacts with the alcohol to give a cyclohexadienyl radical,

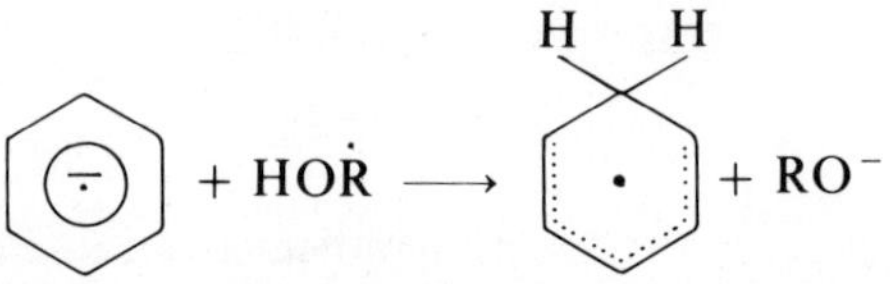

$$\bigcirc^{\overline{\cdot}} + HO\dot{R} \longrightarrow \bigcirc\!\!\overset{H\ H}{\underset{\cdot}{}} + RO^- \qquad (7.98)$$

Now the electron affinity of an odd AH radical is equal to *minus* the orbital energy of the NBMO, i.e., $-\alpha$. Comparison with equation (7.86) shows that this is greater than the electron affinity of any even AH. The radical formed in equation (7.98) is therefore rapidly reduced by more of the original radical anion, giving the cyclohexadienate anion (48):

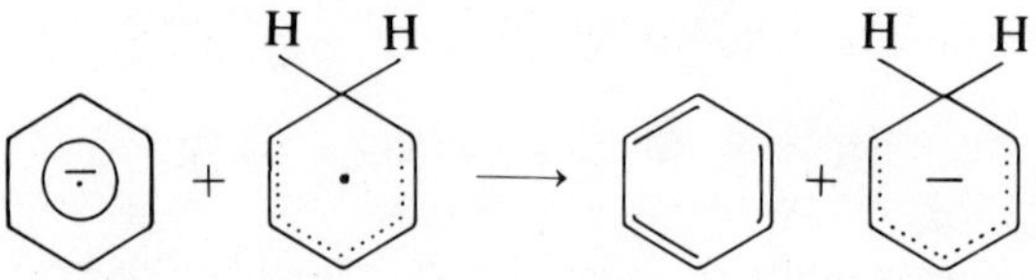

$$\bigcirc^{\overline{\cdot}} + \bigcirc\!\!\overset{H\ H}{\underset{\cdot}{}} \longrightarrow \bigcirc + \bigcirc\!\!\overset{H\ H}{\underset{-}{}} \qquad (7.99)$$
$$\text{(48)}$$

Protonation of (48) gives the final product, a dihydrobenzene.

The final product could be 1,3-cyclohexadiene (49) or 1,4-cyclohexadiene (50). Since the protonation is a typical reaction of an ambident ion and since it is very rapid, one would expect (see p. 251) that attack would take place at the point of maximum formal negative charge. Unfortunately, the PMO method does not help us much here since the first-order charges at the three positions in (48) are equal (51).

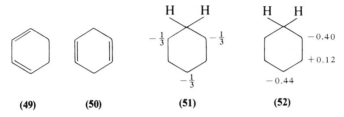

(49) (50) (51) (52)

This is indeed is a case where the PMO method proves inadequate because the distinction depends on factors (in particular electron repulsion) that are neglected in the first-order PMO treatment. Detailed calculations show that the charges at the various positions in (48) are not equal. The values given by the Pople method are shown in (52). It will be seen that they lead to the correct prediction that protonation of (48) should give (50) rather than (49).

Similar reactions are observed when other benzenoid hydrocarbons are reduced under the Birch conditions or when the corresponding radical anions are treated with proton donors such as alcohols or water. For example, treatment of the naphthalide ion with water gives a mixture of naphthalene and dihydronaphthalene by the following path[24]:

$$C_{10}H_8{}^{-\cdot} + H_2O \longrightarrow C_{10}H_9\cdot + HO^- \qquad (7.100)$$

$$C_{10}H_9\cdot + C_{10}H_8{}^{-\cdot} \longrightarrow C_{10}H_9{}^- + C_{10}H_8 \qquad (7.101)$$

$$C_{10}H_9{}^- + H_2O \longrightarrow C_{10}H_{10} + HO^- \qquad (7.102)$$

First we have to predict the point of primary proton attack [equation (7.100)]. This can be done very simply by the PMO method. The product of the reaction is a radical derived from the parent hydrocarbon (here naphthalene) by addition of a hydrogen atom. The ease of proton attack on $C_{10}H_8{}^{-\cdot}$ should therefore parallel that of hydrogen atom attack on naphthalene, i.e., that of a typical radical substitution (see p. 330). Attack should therefore take place at the point with the lowest reactivity number, i.e., at an α position,

$$\left[\bigcirc\!\bigcirc\right]^{-\cdot} + H^+ \longrightarrow \bigcirc\!\bigcirc\!\overset{H\ H}{\underset{\cdot}{}} \qquad (7.103)$$

Next we have to explain how so reactive a species as the naphthalide ion can survive long enough in the presence of an active proton donor to be able to reduce the radical formed in the first step. The reason for this is that the first step involves a change in hybridization of the carbon atom at the point of attack (from sp^2 to sp^3) and so should require activation; see p. 251.

The electron transfer of equation (7.101), on the other hand, should require no activation and should therefore be diffusion controlled. Even if $C_{10}H_8^-$ reacts much more exothermically with a proton than with the $C_{10}H_9$ radical, the latter reaction should be faster.

Finally, we have to predict the point of attack of the second proton. In some cases, the calculated first-order charge distribution in the intermediate anion enables an unambiguous prediction to be made. This is the case in the reduction of anthracene; see (53). In naphthalene, however, there is an ambiguity, the charges at the 2 and 4 positions being equal [see (54)]. Here second-order effects must be taken into account, as in benzene.

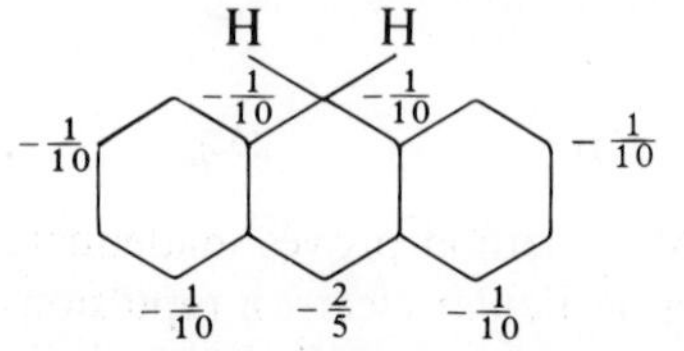

(53) 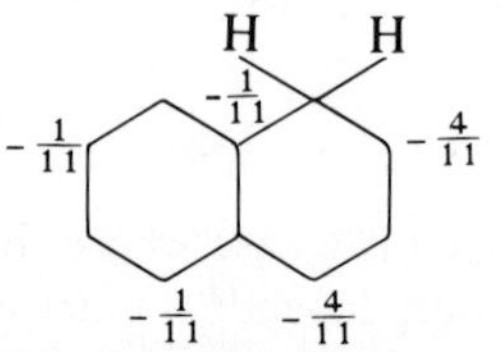 (54)

In such cases, the ambiguity can be resolved by using the following rule, derived from a second-order treatment.

Consider the atoms in the conjugated system separated from the position in question by an unstarred atom. The greater the sum of negative charges at those atoms, the more reactive is the position in question.

The application of this rule to benzene and naphthalene is shown in (55) and (56). In each case, it leads to the correct prediction that the product should be the 1,4-dihydro derivative.

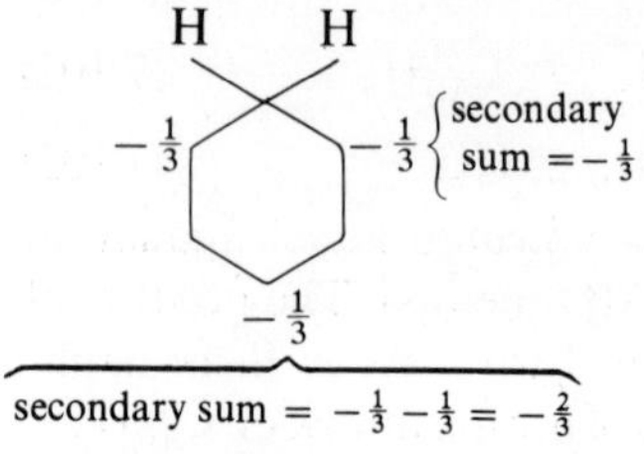

secondary sum = $-\frac{1}{3} - \frac{1}{3} = -\frac{2}{3}$

(55)

secondary sum = $-\frac{4}{11} - \frac{1}{11} - \frac{1}{11} = -\frac{6}{11}$

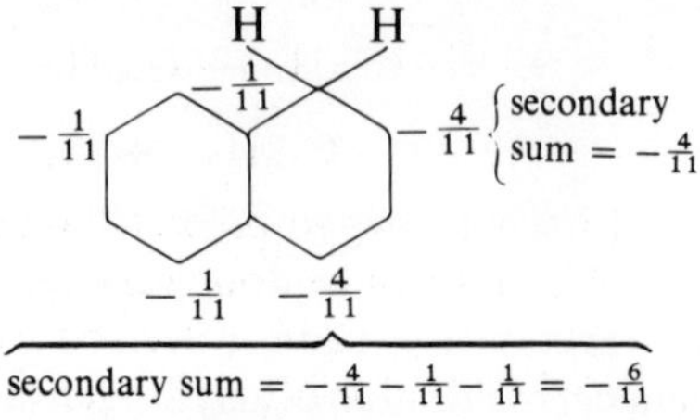

(56)

It should be added that detailed studies have completely confirmed[24] this mechanism for the protonation of naphthalide ion and that the orientation of reduction in other polycyclic aromatic hydrocarbons agrees with that predicted by this PMO approach.

Aromatic radical anions do not behave as radicals, because of the loss of resonance energy in forming nonaromatic products. Thus dimerization of the radical anion from benzene would give a nonaromatic dianion,

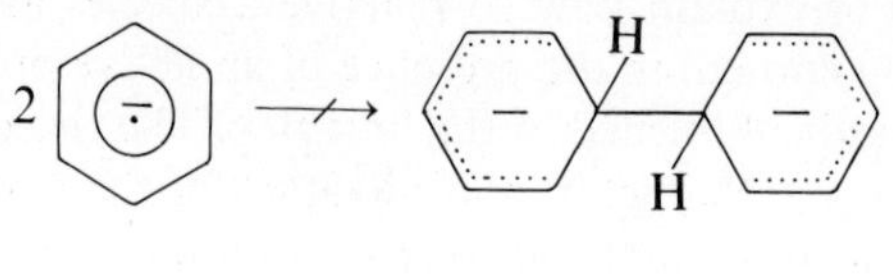

(7.104)

This diffiulty does not arise in the case of radical anions derived from olefins and so such radical anions usually dimerize very readily. The most important reactions of this kind are those where the radical anion is formed electrochemically by cathodic reduction, the reaction being somewhat analogous to a Wurtz synthesis, e.g.,

$$CH_2=CHCN + e^- \longrightarrow (CH_2CHCN)^{-\cdot}$$

$$2(CH_2CH_2CN)^{-\cdot} \longrightarrow \overset{\frown}{NC\cdots CH}-CH_2-CH_2-\overset{\frown}{CH\cdots CN}$$
$$\xrightarrow{+2H^+} NC-CH_2-CH_2-CH_2-CH_2-CN \qquad (7.105)$$

The resulting dinitrile can be hydrolyzed to adipic acid, $HOOC(CH_2)_4COOH$, or reduced to hexamethylenediamine, $H_2N(CH_2)_6NH_2$, the two components needed for the production of nylon.

For reactions of this kind to occur, it is necessary that the parent compound should be reducible at an accessible electrode potential, i.e., that it should have a sufficiently high electron affinity. Olefins cannot be reduced in this way and even 1,3-dienes are marginal. Since replacement of a carbon atom in a conjugated system lowers the energies of all the MOs (except those with nodes at the atom in question), hetero-analogs of dienes are, on the other hand, easily reduced (e.g., $CH_2=CH-CR=O$, $CH_2=CH-C\equiv N$).

In systems containing numerous stabilizing groups, the situation may be intermediate between those observed in aromatic and aliphatic radical anions. Here the radical anion may or may not dimerize. Since dimerization removes from the conjugated system the atoms through which combination takes place, the ease of the reaction should be greater, the lower are the localization energies at those positions.

Consider, for example, the radical anion from styrene, $PhCH=CH_2$. Dimerization of styrene to a biradical would remove the two terminal carbon atoms from conjugation. The resulting loss in π energy per styrene unit would then be $N_t\beta$, N_t being the corresponding reactivity number,* namely

$$\delta E_\pi = 2N_t\beta = 2(4/\sqrt{7})\beta \qquad (7.106)$$

The extra electron in the radical anion from styrene occupies the antibonding LUMO. Dimerization of the radical anion gives a dianion in which the two extra electrons occupy NBMOs. The π energy of dimerization of the radical

---

* The change in π energy of each molecule of styrene is clearly the same as in the combination of styrene with a neutral radical R to form $PhCHCH_2R$. The change in π energy is therefore given in each case by $N_t\beta$.

anion is therefore less than that of styrene by twice the difference in energy between the LUMO of styrene and an NBMO. From equation (7.85), this is equal to $2\Sigma a_{0r}b_{0s}\beta$, where the NBMO coefficients refer to the optimum dissection of styrene into odd AHs (see p. 404). This dissection is easily seen to be into benzyl and methyl radicals; hence

$$a_{0r} = 2/\sqrt{7}; \qquad b_{0s} = 1; \qquad a_{0r}b_{0s}\beta = 2/\sqrt{7} \qquad (7.107)$$

Thus the localization energy δE_π^- at the terminal carbon in the styrene radical anion is given [see equations (7.106) and (7.107)] by

$$\delta E_\pi^- = -\frac{4\beta}{\sqrt{7}} + \frac{2\beta}{\sqrt{7}} = -\frac{2\beta}{\sqrt{7}} = -0.756\beta \qquad (7.108)$$

Table 7.5 shows the localization energies calculated in this way for the

TABLE 7.5. Localization Energies δE_π^- for the Most Reactive Positions (○) in Phenyl Derivatives of Ethylene and the Behavior of the Corresponding Radical Anions in Solution

Compound	δE_π^-	Behavior[25]
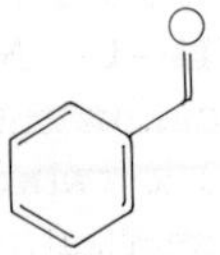	0.756	Monomer–dimer equilibrium
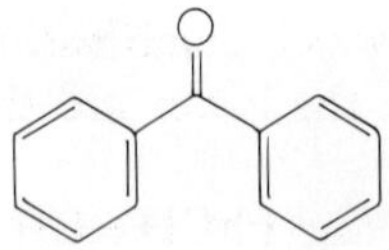	0.632	Dimer
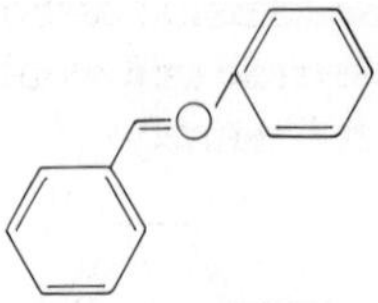	0.940	Monomer
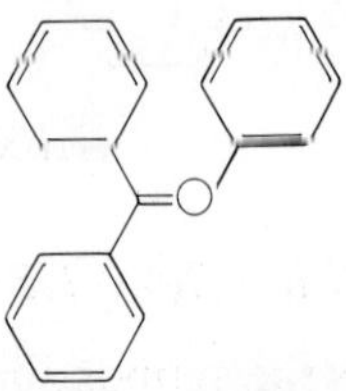	0.769	Monomer
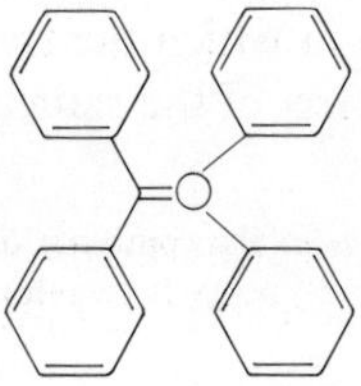	0.865	Monomer (in equilibrium with dianion)

radical anions derived from phenylated ethylenes. It will be seen that there is indeed a critical limit for the localization energy above which dimerization does not occur. Styrene is near this limit, the monomer and dimer being present in comparable amounts at equilibrium. In other cases, the equilibrium is almost entirely on one side or the other. An additional complication occurs in the case of tetraphenylethylene, where the radical anion tends to disproportionate,

$$2(Ph_2C=CPh_2)^{-\cdot} \longrightarrow Ph_2C=CPh_2 + Ph_2C-CPh_2 \quad (7.109)$$

Two other radical anions of interest are those derived from fulvene (57) and benzo[c,d,]pyrene-6-one (58). In the case of (57), the dimer (59) contains two aromatic rings. The radical anions derived from (57) and its derivatives therefore dimerize very easily. The radical anion from (58) is the only organic radical anion that has so far been observed to dimerize in the gas phase.[26] The dimerization of radical anions in the gas phase is usually inhibited by the very large Coulombic repulsion between the two negative charges in the dimer. Presumably (58) dimerizes because the negative charges are dispersed over two large conjugated systems.

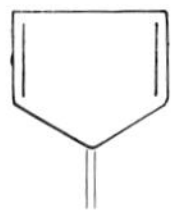

(57) 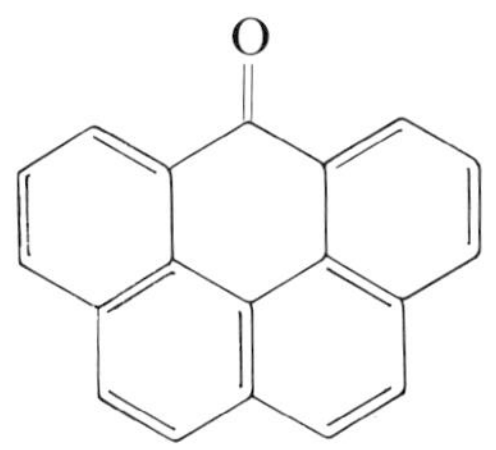 (58) 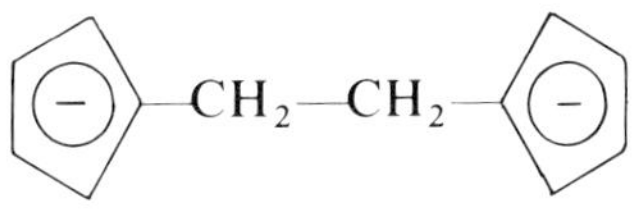 (59)

So far we have considered the formation of radical anions only under very strongly reducing conditions, the electrons being provided by alkali metals, by solutions of alkali metals in liquid ammonia, or by less stable radical anions. This, however, is not at all essential. All that is needed is an anion with weakly bound electrons to act as a reducing agent. A very nice example of such a process is provided by the replacement of the exocyclic group in (60) by thiophenol,

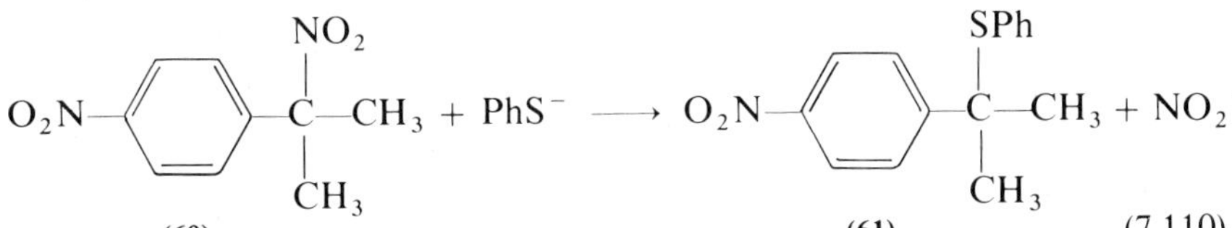

$$O_2N-C_6H_4-C(NO_2)(CH_3)_2 + PhS^- \longrightarrow O_2N-C_6H_4-C(SPh)(CH_3)_2 + NO_2^- \quad (7.110)$$

(60) (61)

The reaction does not involve a normal S_N1 or S_N2 substitution but takes place[27a-c] by the following chain mechanism involving intermediate radical anions:

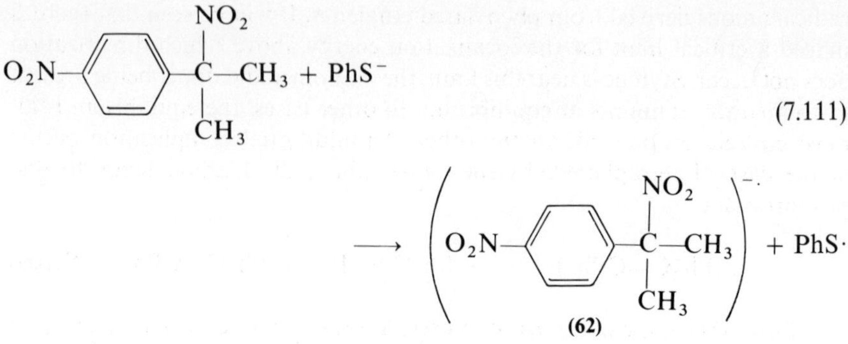

$$O_2N-\underset{}{\bigcirc}-\underset{\underset{CH_3}{|}}{\overset{\overset{NO_2}{|}}{C}}-CH_3 + PhS^- \quad (7.111)$$

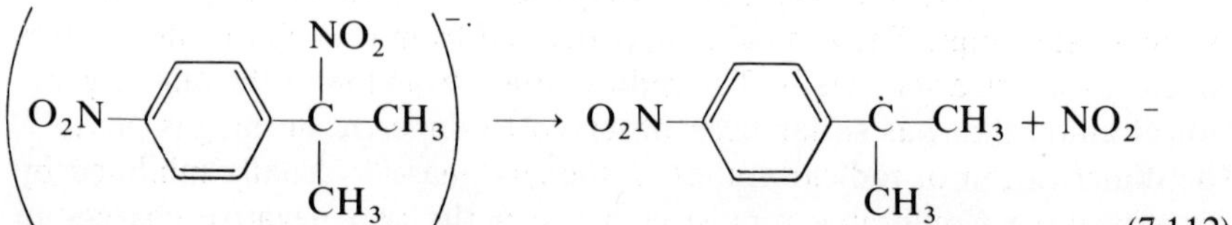

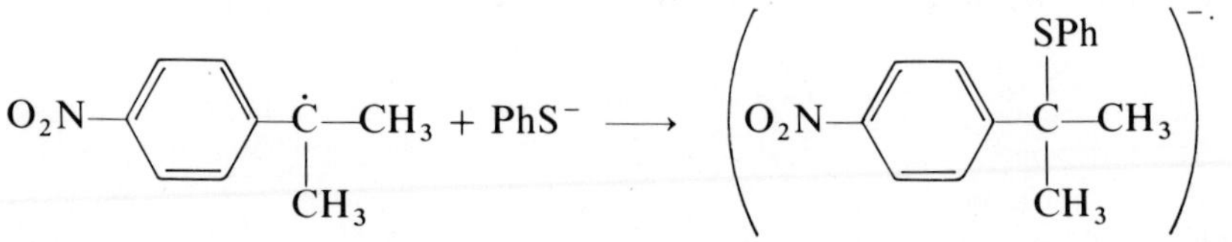

$$\longrightarrow \left(O_2N-\underset{}{\bigcirc}-\underset{\underset{CH_3}{|}}{\overset{\overset{NO_2}{|}}{C}}-CH_3 \right)^{-\cdot} + PhS\cdot$$

$$(62)$$

$$\left(O_2N-\underset{}{\bigcirc}-\underset{\underset{CH_3}{|}}{\overset{\overset{NO_2}{|}}{C}}-CH_3 \right)^{-\cdot} \longrightarrow O_2N-\underset{}{\bigcirc}-\underset{\underset{CH_3}{|}}{\overset{}{\dot{C}}}-CH_3 + NO_2^- \quad (7.112)$$

$$O_2N-\underset{}{\bigcirc}-\underset{\underset{CH_3}{|}}{\overset{}{\dot{C}}}-CH_3 + PhS^- \longrightarrow \left(O_2N-\underset{}{\bigcirc}-\underset{\underset{CH_3}{|}}{\overset{\overset{SPh}{|}}{C}}-CH_3 \right)^{-\cdot} \quad (7.113)$$

$$\left(O_2N-\underset{}{\bigcirc}-\underset{\underset{CH_3}{|}}{\overset{\overset{SPh}{|}}{C}}-CH_3 \right)^{-\cdot} + O_2N-\underset{}{\bigcirc}-\underset{\underset{CH_3}{|}}{\overset{\overset{NO_2}{|}}{C}}-CH_3$$

$$\longrightarrow O_2N-\underset{}{\bigcirc}-\underset{\underset{CH_3}{|}}{\overset{\overset{SPh}{|}}{C}}-CH_3 + (62) \quad (7.114)$$

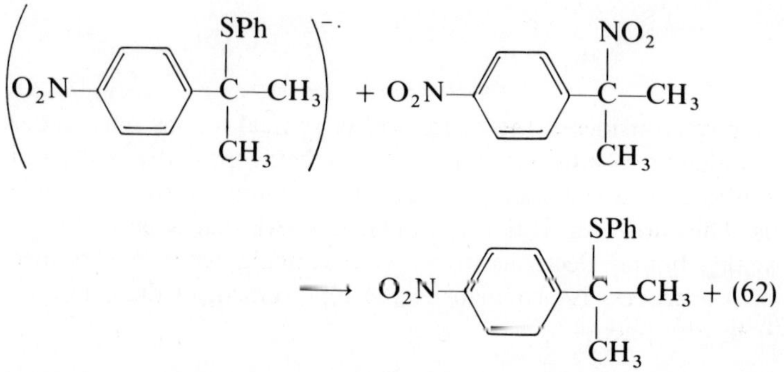

Equation (7.111) represents the initiation step and the three following equations the propagation steps of a typical chain process. The first radical intermediate can be trapped by oxygen and there is no evidence to suggest the intermediate formation of anions.

The essential feature of this mechanism is the presence of an aromatic moiety with a high affinity for electrons. Benzene itself is inadequate in this connection. The *p*-nitro group is necessary for the intermediate radical anion to be formed. An interesting illustration of this, and also of the difference between BEP and anti-BEP reactions, is provided by the reactions of the

2-nitro-2-propyl anion (63) with benzyl chloride (64) and with *p*-nitrobenzyl chloride (66):

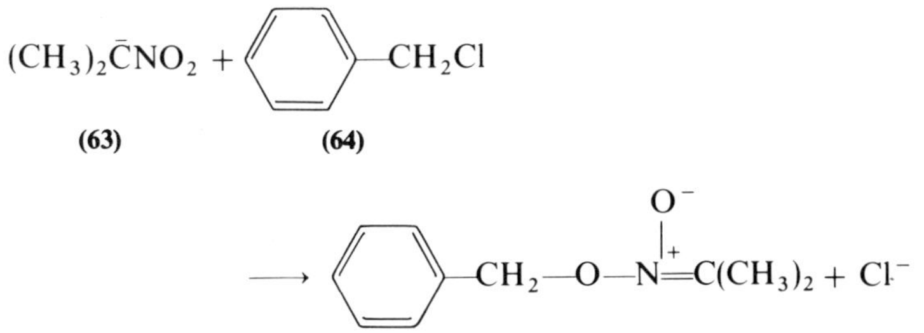

$$(CH_3)_2\bar{C}NO_2 + \text{C}_6\text{H}_5\text{-CH}_2\text{Cl}$$
(63)   (64)

$$\longrightarrow \text{C}_6\text{H}_5\text{-CH}_2\text{-O-}\overset{+}{N}\overset{O^-}{=}C(CH_3)_2 + Cl^- \quad (7.115)$$
(65)

$$(63) + O_2N\text{-C}_6\text{H}_4\text{-CH}_2Cl \longrightarrow O_2N\text{-C}_6\text{H}_4\text{-CH}_2\text{-}\overset{NO_2}{\underset{|}{C}}(CH_3)_2 + Cl^-$$
(66)   (67)   (7.116)

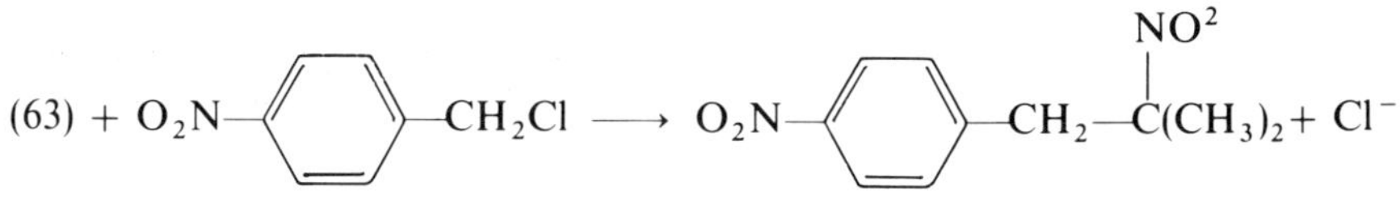

The first reaction is a straightforward S_N2 substitution, while the second takes place by a chain mechanism analogous to equations (7.111)–(7.114). Since (63) is an ambident anion, it can react via nitrogen or oxygen. Reaction through nitrogen leads to the more stable product since nitroparaffins are more stable than their *aci* isomers. The S_N2 reaction is therefore of anti-BEP type, leading to a less stable *aci* isomer (65), while the chain reaction leads to the more stable nitroparaffin (67).

The product-forming step in the chain reaction is combination of a *p*-nitrobenzyl radical with (63) [cf. equation (7.113)],

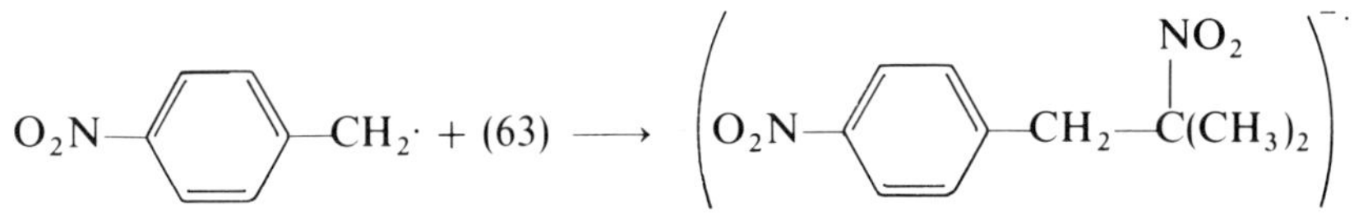

$$O_2N\text{-C}_6\text{H}_4\text{-CH}_2\cdot + (63) \longrightarrow \left(O_2N\text{-C}_6\text{H}_4\text{-CH}_2\text{-}\overset{NO_2}{\underset{|}{C}}(CH_3)_2\right)^{-\cdot}$$
(7.117)

Since this is a simple bond-forming process and presumably is exothermic, it should take place without activation in the absence of solvent. In solution, however, the oxygen atoms in (63) will be very strongly solvated since the negative charge is concentrated on them. Reaction at oxygen will therefore involve desolvation, effectively a bond-breaking process. The reaction therefore takes place preferentially at the carbon atom, which is presumably only weakly solvated. The fact that the resulting product is the more stable one is coincidential.

The S_N2 reaction leads to O-alkylation for the reasons discussed in the treatment of ambident ions (p. 251). The negative charge in (63) is concentrated on oxygen and its bond structure approximates to the *aci* configuration. Since benzyl chloride is a rather reactive halide, it reacts to give the *aci* isomer preferentially.

Problems

7.1. If *p*-nitrobenzyl chloride or benzyl chloride is added in small concentrations to methylene chloride in a high-pressure mass spectrometer, the major molecular ion relatives are

[Structures: p-NO₂-C₆H₄-CH₂⁻ ; (p-NO₂-C₆H₄-CH₂Cl)⁻· ; ĊHCl ; p-NO₂-C₆H₄-CH₂Cl₂⁻ ; C₆H₅-CH₂Cl₂⁻]

Use this observation and information in this chapter to devise a system for predicting when reactions will take a radical-ion course in solution.

7.2. When a reaction can proceed by an S_N2 or radical anion pathway, will increasing the anion solvating power of the solvent favor the S_N2 or the radical anion pathway? Explain.

7.3. List the requirements for a solvent that would preserve radical cations in solution. Give examples of the best known solvents with those properties.

7.4. How would you go about estimating or measuring the electronic spectrum of a molecular ion?

7.5. Estimate interelectronic repulsion in the cyclooctatetraene dianion due to the interaction of the two charges only (use the benzene bond length and a reasonable average for the interelectronic distance).

7.6. What would be the major impediments to the success of a solution mixing reaction of the type

$$M^{-\cdot}_{solv} + M'^{+\cdot}_{solv} \longrightarrow \text{products?}$$

If the reaction could be arranged, would it be likely to yield M—M' as a product?

7.7. The following β-ketoester anion gives both C- and O- (ketone) alkylation[28]:

[Structure: benzene ring fused with a 5-membered ring containing CO, C–COOEt, O]

When the alkylating agent is changed from benzyl chloride to β-nitrobenzyl chloride, the ratio of carbon to oxygen alkylation is roughly reversed. When the alkylating agent is benzyl iodide or *p*-nitrobenzyl iodide, the ratio of carbon to oxygen alkylation is roughly constant.
a. Predict which product is the dominant one for *p*-nitrobenzyl chloride alkylation of the anion.
b. Predict which product is the dominant one for benzyl iodide alkylation of the anion.
c. The initial reactions were run in DMF at 0°C; if the solvent shifted to THF, what would happen to the ratios of products in the two cases above?

7.8. Prove that $4n - 1$ Hückel and $4n + 1$ anti-Hückel odd cyclopolyenemethylenes are aromatic with reference to open-chain radical ions.

Selected Reading

Organic mass spectrometry:

D. H. WILLIAMS AND I. HOWE, *Principles of Organic Mass Spectrometry*, McGraw-Hill Book Co., London, 1972.

F. W. MCLAFFERTY, *Interpretation of Mass Spectra*, 2nd ed., W. A. Benjamin, New York, 1973.

R. W. KISER, *Introduction to Mass Spectrometry*, Prentice-Hall, New York, 1965.

C. E. MELTON, *Principles of Mass Spectrometry and Negative Ions*, Marcel Decker, New York, 1970.

J. H. BEYNON, R. A. SAUNDERS, AND A. E. WILLIAMS, *The Mass Spectra of Organic Molecules*, Elsevier, New York, 1968.

Reactions of ions in solution:

D. J. CRAM, *Fundamentals of Carbanion Chemistry*, Academic Press, New York, 1965.

L. KEVAN AND E. T. KAISER, *Radical Ions*, Wiley—Interscience, New York, 1968.

M. SZWARC, ed., *Ions and Ion Pairs in Organic Reactions*, Vol. I, Wiley—Interscience, New York, 1972.

Ion-molecule reactions:

P. AUSLOOS, ed., *Fundamental Processes in Radiation Physics*, Wiley—Interscience, New York, 1968.

E. W. MCDANIEL, V. VORMAK, A. DALGARNO, E. E. FERGUSON, AND L. FRIEDMAN, *Ion-Molecule Reactions*, Wiley—Interscience, New York, 1970.

J. L. FRANKLIN, ed., *Ion Molecule Reactions*, Vols. I, II, Plenum Press, New York, 1972.

References

1. (a) R. C. DOUGHERTY, *J. Am. Chem. Soc.* **90**, 5780, 5788 (1968).
   (b) R. C. DOUGHERTY, *Tetrahedron* **24**, 6755 (1968).
2. (a) E. P. SMITH AND E. R. THORTON, *J. Am. Chem. Soc.* **89**, 5079 (1967).
   (b) M. J. S. DEWAR AND S. D. WORLEY, *J. Chem. Phys.* **49**, 2454 (1968).
3. (a) A. KARPUTI, A. RAVE, J. DEUTSCH, AND A. MANDELBAUM, *J. Am. Chem. Soc.* **95**, 4244 (1973).
   (b) S. HAMMERUM AND C. DJERASSI, *J. Am. Chem. Soc.* **95**, 5806 (1973).
4. (a) A. G. LOUDON, A. MACCOLL, AND S. K. WONG, *J. Chem. Soc. B* 1727, 1733 (1970).
   (b) R. A. W. JOHNSTONE AND S. O. WARD, *J. Chem. Soc.* 1805, 2540 (1968).
5. R. C. DOUGHERTY, *Topics in Current Chemistry* **10**, 93 (1973).
6. R. C. DOUGHERTY AND C. R. WEISENBURGH, *J. Am. Chem. Soc.* **90**, 6570 (1968).
7. R. G. LAWLER AND C. T. TABIT, *J. Am. Chem. Soc.* **91**, 5671 (1969).
8. M. SZWARC, *Prog. Phys. Org. Chem.* **6**, 323 (1968).
9. G. A. OLAH, G. KLOPMAN, AND R. H. SCHLOSBERG, *J. Am. Chem. Soc.* **91**, 3261 (1969).
10. J. I. BROWMAN, J. M. RIVEROS, AND L. K. BLAIR, *J. Am. Chem. Soc.* **93**, 3914 (1971).
11. M. S. B. MUNSON, *J. Am. Chem. Soc.* **87**, 2332 (1965).

12. (a) F. H. Field, *J. Am. Chem. Soc.* **91**, 2827 (1969).
    (b) G. A. Olah and P. W. Westerman, *J. Am. Chem. Soc.* **95**, 7530 (1973).
13. (a) J. I. Browman and L. K. Blair, *J. Am. Chem. Soc.* **91**, 2126 (1969).
    (b) J. I. Browman and L. K. Blair, *J. Am. Chem. Soc.* **93**, 4315 (1971).
    (c) J. I. Browman and L. K. Blair, *J. Am. Chem. Soc.* **92**, 5986 (1970).
    (d) M. J. McAdams and L. I. Bone, *J. Phys. Chem.* **75**, 2226 (1971).
14. (a) R. Yamdagni and P. Kebarle, *J. Am. Chem. Soc.* **95**, 4050 (1973).
    (b) K. Hiraoka, R. Yamdagni, and P. Kebarle, *J. Am. Chem. Soc.* **95**, 6833 (1973).
15. (a) R. C. Dougherty, J. Dalton, and J. D. Roberts, *Org. Mass Spectrom.* **8**, 77 (1974).
    (b) R. C. Dougherty and J. D. Roberts, *Org. Mass. Spectrom.* **8**, 81 (1974).
    (c) R. C. Dougherty, *Org. Mass Spectrom.* **8**, 85 (1974).
16. (a) P. J. Andrulis, Jr., M. J. S. Dewar, R. Dietz, and R. L. Hunt, *J. Am. Chem. Soc.* **88**, 5473 (1966).
    (b) T. Aratani and M. J. S. Dewar, *J. Am. Chem. Soc.* **88**, 5479 (1966).
    (c) P. J. Andrulis, Jr. and M. J. S. Dewar, *J. Am. Chem. Soc.* **88**, 5483 (1966).
    (d) L. L. Miller, *J. Chem. Ed.* **48**, 168 (1971).
17. (a) J. K. Kochi, R. T. Tang, and T. Bernath, *J. Am. Chem. Soc.* **95**, 7114 (1973).
    (b) A. McKillop and E. C. Taylor, *Adv. Organometal. Chem.* **11**, 147 (1973).
    (c) I. H. Elson and J. K. Kochi, *J. Am. Chem. Soc.* **95**, 5060 (1973).
18. (a) H. Seki, S. Arai, T. Shida, and M. Imamara, *J. Am. Chem. Soc.* **95**, 3404 (1973).
    (b) K. Toriyama, M. Iwasaki, S. Noda, and B. Eda, *J. Am. Chem. Soc.* **93**, 6415 (1971).
19. E. E. van Tamelen, *Acct. Chem. Res.* **3**, 361 (1970).
20. H. M. Walborsky, F. P. Johnson, and J. B. Pierce, *J. Am. Chem. Soc.* **90**, 5222 (1968).
21. R. W. Fessenden and R. H. Schuler, *J. Chem. Phys.* **39**, 2147 (1963).
22. T. C. Cheng, L. Headly, and A. F. Hatasa, *J. Am. Chem. Soc.* **93**, 1502 (1971).
23. J. K. Kim and J. F. Burnett, *J. Am. Chem. Soc.* **92**, 7464 (1970).
24. S. Bank and B. Bokrath, *J. Am. Chem. Soc.* **93**, 430 (1971), and papers cited therein.
25. M. Szwarc, *Carbanions, Living Polymers and Electron Transfer Processes*, Wiley—Interscience, New York, 1968.
26. R. C. Dougherty, *J. Chem. Phys.* **50**, 1896 (1969).
27. (a) N. Kornblum, T. M. Davies, G. W. Earl, N. L. Holy, R. C. Kerber, M. T. Musser, and D. H. Snow, *J. Am. Chem. Soc.* **89**, 7251 (1967).
    (b) N. Kornblum, T. M. Davies, G. W. Earl, G. S. Greene, N. L. Holy, R. C. Kerber, J. W. Manthey, M. T. Mussev, and D. H. Snow, *J. Am. Chem. Soc.* **89**, 5714 (1967).
    (c) N. Kornblum, T. M. Davies, G. W. Earl, N. L. Holy, J. W. Mathey, M. T. Mussev, and R. T. Swiger, *J. Am. Chem. Soc.* **90**, 6219 (1968).
28. N. Kornblum, R. E. Michel, and R. C. Kerber, *J. Am. Chem. Soc.* **88**, 5660 (1966).
29. R. N. Compton and R. H. Hulkner, in *Advances in Radiation Chemistry*, Vol. 2, M. Burton and J. L. Magee, ed., John Wiley and Sons, New York (1970).
30. P. B. D. de la Mare, L. Fowden, E. D. Hughes, C. K. Ingold, and J. D. H. Mackie, *J. Chem. Soc.* **1955**, 3200.

Answers to Selected Problems

Chapter 1

1.2. 0.55 Å

1.3.

		$+ - y^-, z^-$
$+ + y^+, z^+$	$+\!\!\!+ y^+, z^+$	$+\!\!\!+ y^+, z^+$
$+\!\!\!+ x^+$	$+\!\!\!+ x^+$	$+\!\!\!+ x^+$
$+\!\!\!+ s^-$	$+\!\!\!+ s^-$	$+\!\!\!+ s^-$
$+\!\!\!+ s^+$	$+\!\!\!+ s^+$	$+\!\!\!+ s^+$
BeO triplet	CO singlet	NO doublet

1.5. "Ionic bonds" using $6p$ electrons are stronger than those using $6s$ electrons ($Pb^{2+}2Br^- \gg Pb^{4+}4Br^-$). "Covalent bonds" with $6s$ electrons are stronger than those with $6p$ electrons ($PbMe_4 > PbMe_2$). The energy gap between $6p$ and $6s$ orbitals is larger than that for $2p$ and $2s$ orbitals.

1.6.

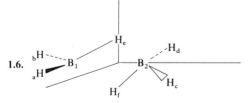

$SO_1 = C_1(a + b + c + d + e + f)$

$SO_2 = C_2(a + b - c - d + e + f)$

$SO_3 = C_3(a - b + c - d + e + f)$

$$SO_4 = C_4(a + b + c + d - e + f)$$
$$SO_5 = C_5(a + b - c - d - e + f)$$
$$SO_6 = C_6(a - b - c + d - e + f)$$

1.9. a.

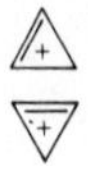

b.

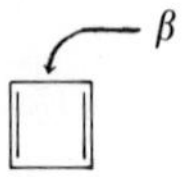

c.

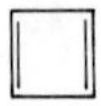

d.

1.10. There is no vibrational distortion of a homonuclear diatomic that can lower its symmetry.

Chapter 2

2.2. Charge density $q_i = 1$; π-bond order $p_{ii} = 1$.

2.3. $\Delta E = \Delta E_\gamma + \Delta E_\pi$; $\Delta E_\pi = 2(1 - \frac{1}{2}\sqrt{2})\beta$.

2.4. Second-order perturbations are not additive. In double union of even AHs, the second-order terms are large and cannot be ignored.

Chapter 3

3.1.

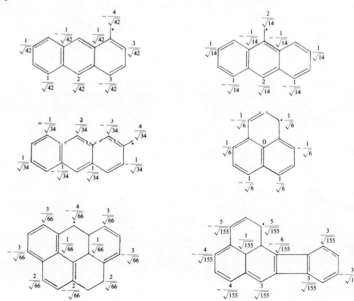

Answers to Selected Problems

3.2. a.

(i) 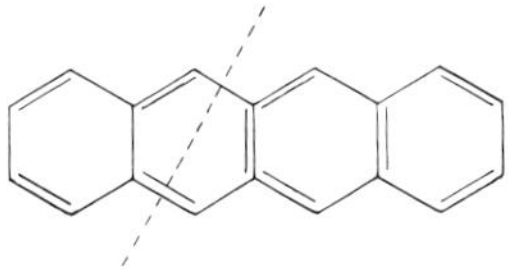 $\Delta E_{\pi_\beta} = 2\left(\dfrac{2}{\sqrt{17.7}} + \dfrac{3}{\sqrt{17.7}}\right)\beta = 0.92\beta$

(ii) 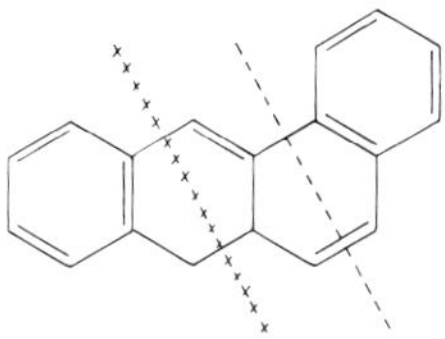 $\Delta E_{\pi_\beta} = 2\left(\dfrac{6}{\sqrt{17.7}} + \dfrac{1}{\sqrt{17.7}}\right)\beta = 1.28\beta$

$\Delta E_{\pi_\alpha} = 2\left(\dfrac{3}{\sqrt{20.7}} + \dfrac{2}{\sqrt{20.7}}\right)\beta = 0.84\beta$

(iii) 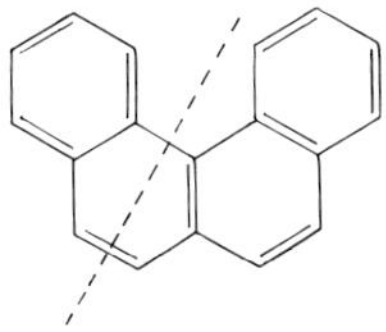 $\Delta E_{\pi_\beta} = 2\left(\dfrac{6}{\sqrt{17.7}} + \dfrac{2}{\sqrt{17.7}}\right)\beta = 1.47\beta$

(iv) 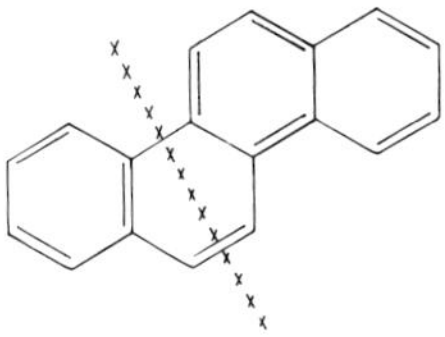 $\Delta E_{\pi_\alpha} = 2\left(\dfrac{6}{\sqrt{20.7}} + \dfrac{2}{\sqrt{20.7}}\right)\beta = 1.35\beta$

Thus (iii) > (ii) > (i); (iv) > (ii).

b.

(i)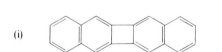

(ii)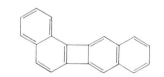

(iii)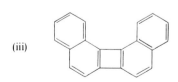

(iv)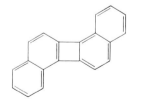

(i) > (ii) > (iv) ≲ (iii).

3.4. ΔE_π values: (i) $2(1 - 3/\sqrt{17})\beta = 0.54\beta$.

(ii) $2(1 - 3/\sqrt{20})\beta = 0.66\beta$.

(iii) $4/20\beta + 2(1 - 3/\sqrt{20})\beta = 0.86\beta$.

(iv) $4/17\beta + 2(1 - 3/\sqrt{17})\beta = 0.78\beta$.
(v) $4/20\beta + 2(1 - 3/\sqrt{20})\beta = 0.86\beta$.
(vi) $1/17\beta + 2(1 - 3/\sqrt{17})\beta = 0.60\beta$.

3.5.
a. $>$
b. $<$
c. $>$
d. $<$

3.6.

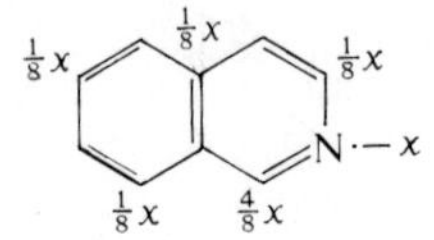

Isoquinoline with coefficients $\frac{1}{8}x$ at positions, $\frac{4}{8}x$ at N-adjacent, and $-x$ at N.
$$x = \frac{0.471}{\beta}\beta = 0.471$$

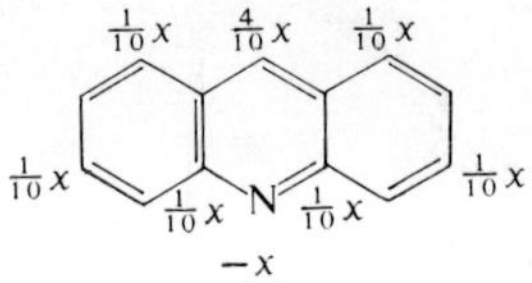

Acridine with coefficients $\frac{1}{10}x$ and $\frac{4}{10}x$, N bearing $-x$.
$x = 0.638$

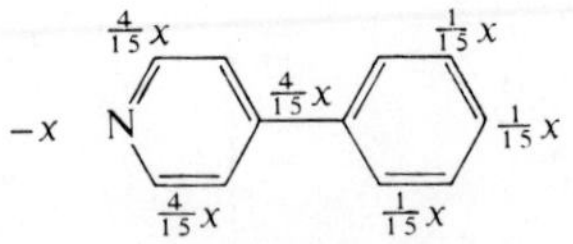

4-phenylpyridine with coefficients $\frac{4}{15}x$, $\frac{1}{15}x$, N bearing $-x$.
$x = 0.484$

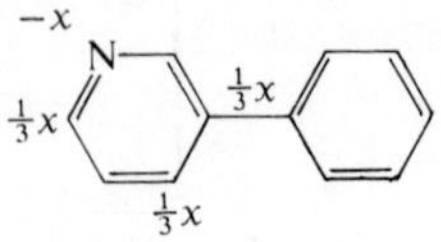

3-phenylpyridine with coefficients $\frac{1}{3}x$, N bearing $-x$.
$x = 0.433$

3.7. For example, ratio of numbers of classical structures: NBMO coefficient,

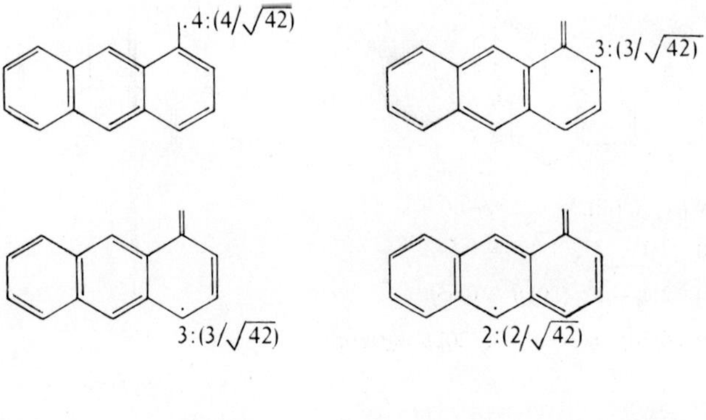

$4:(4/\sqrt{42})$

$3:(3/\sqrt{42})$

$3:(3/\sqrt{42})$

$2:(2/\sqrt{42})$

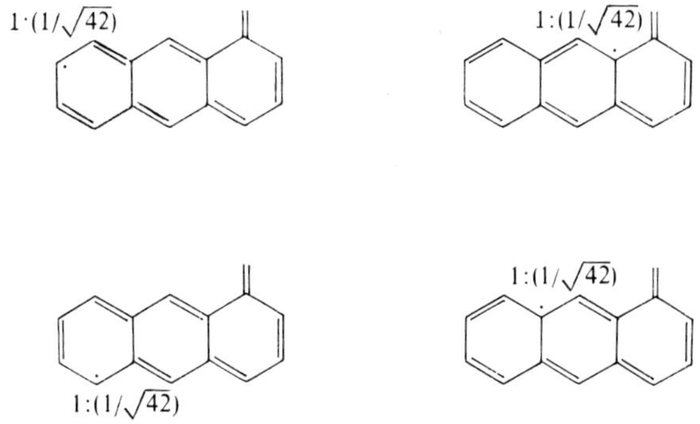

Answers to Selected Problems

1·(1/√42) [anthracene structure]

1:(1/√42) [anthracene structure]

1:(1/√42) [anthracene structure]

1:(1/√42) [anthracene structure]

However,

2:(0) [phenalene structure]

3:(1/√6) [phenalene structure]

3.8.

[stilbene structure]   >   [phenanthrene structure with −β]   >   [anthracene structure with −β]

$\Delta E_\pi = (8/7)\beta$ β 0

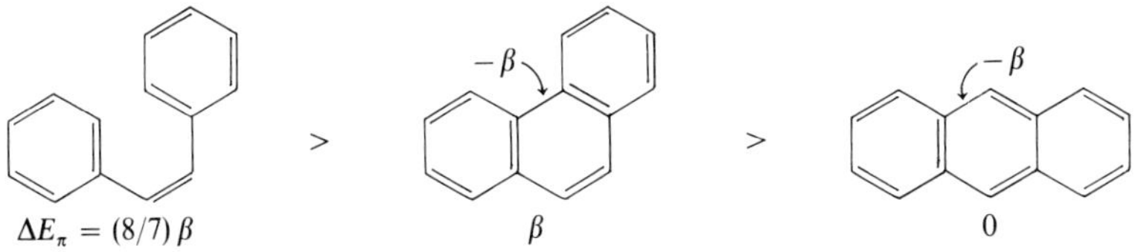

Chapter 4

4.2. They are stabilized by mutual conjugation.

4.3.

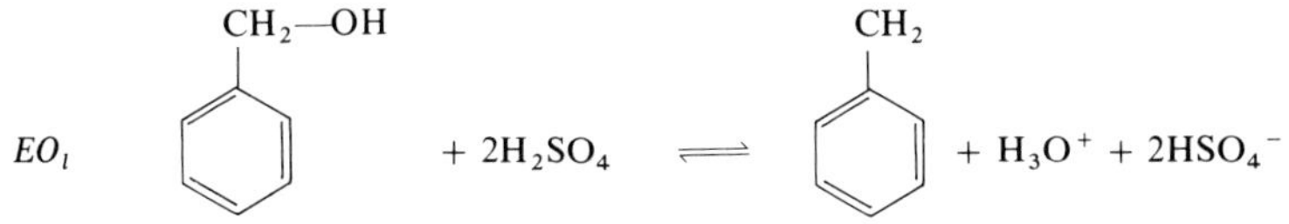

EO_l [PhCH$_2$OH] + 2H$_2$SO$_4$ $\rightleftharpoons$ [PhCH$_2^+$] + H$_3$O$^+$ + 2HSO$_4^-$

EO_s [benzene] + D$_2$SO$_4$ $\rightleftharpoons$ [protonated benzene with H, D] + DSO$_4^-$

EE 2 [cyclopentadiene] $\rightleftharpoons$ [dicyclopentadiene]

I [PhBr] + BuLi $\rightleftharpoons$ [PhLi$^+$] + BuBr

4.4. a.

ΔE_π(ionization)

(i) 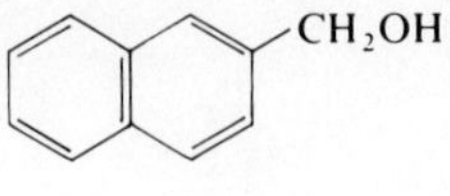 [naphthalene-2-CH$_2$OH] $2(1 - 3/\sqrt{17})\beta = 0.54\beta$

(ii) 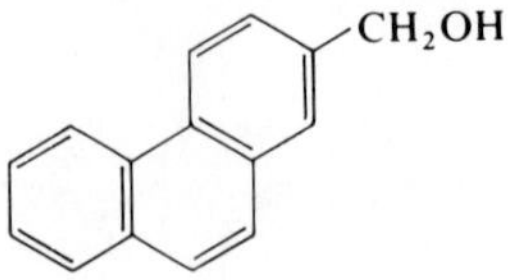 [phenanthrene-CH$_2$OH] $2(1 - 5/\sqrt{46})\beta = 0.53\beta$

(iii) 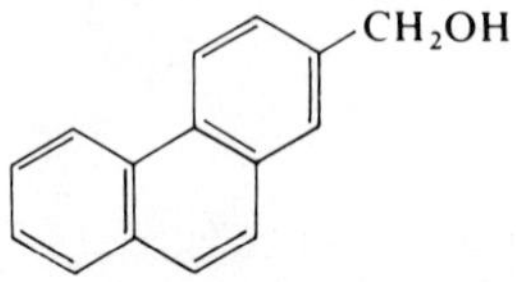 [phenanthrene isomer-CH$_2$OH] $2(1 - 5/\sqrt{51})\beta = 0.60\beta$

(iv) 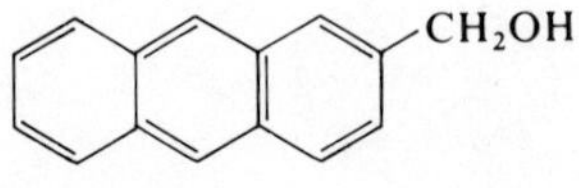 [anthracene-2-CH$_2$OH] $2(1 - 4/\sqrt{34})\beta = 0.63\beta$

(v) 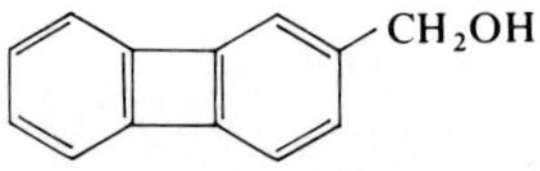 [biphenylene-CH$_2$OH] $2(1 - 3/\sqrt{21})\beta = 0.69\beta$

(vi) 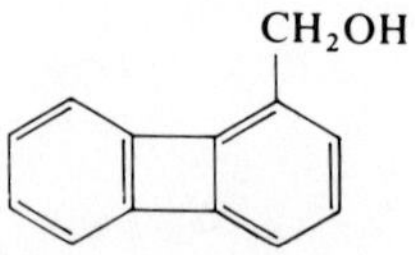 [biphenylene isomer-CH$_2$OH] $2(1 - 3/\sqrt{18})\beta = 0.59\beta$

Thus the order of equilibrium constants is

$$(v) > (iv) > (iii) \lesssim (vi) > (i) \lesssim (ii)$$

b. The anionic equilibrium constants should follow the same order because in the absence of heteroatoms, the change in π energy for anions, cations, and radicals should be the same.

4.5. a. 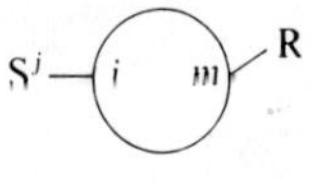 $S^j \!-\!\! \bigcirc\!\!\!\!-\!\! R$ (with i, m labeled)

$$\sigma_{Sim} = \frac{F_S}{r_{im}} + M_S q_{im} + MF_S \sum_{k \neq m} \frac{q_{ik}}{r_{km}}$$

(i) All bond lengths are in units of the standard bond length (1.40 Å).
(ii) The field effect F_S is obtained on the assumption that the substituent acts as a dipole with charge q at atom i and charge $-0.9q$ one standard bond length from i along the bond to the substituent. The magnitude of the charge is accounted for by the F_S parameter. Thus

$$\frac{F_S}{r_{im}} = F_S\left(\frac{1}{r_{in}} - \frac{0.9}{r_{jn}}\right)$$

for p-F,

$$F_F = 4.85, \qquad \frac{F_F}{r_{im}} = 4.85\left(\frac{1}{3} - \frac{0.9}{4}\right) = 0.525$$

(iii) $q_{i,m}$ is the formal negative charge at the reaction site when the substituent is replaced by CH_2^-; for the p-F this is 1/7.

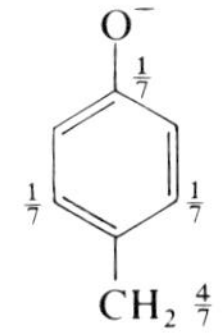

Thus,
$$M_F q_{im} = -2.10(1/7) = -0.3$$

(iv) The q's in the term for the mesomeric field effect are obtained in the same way.

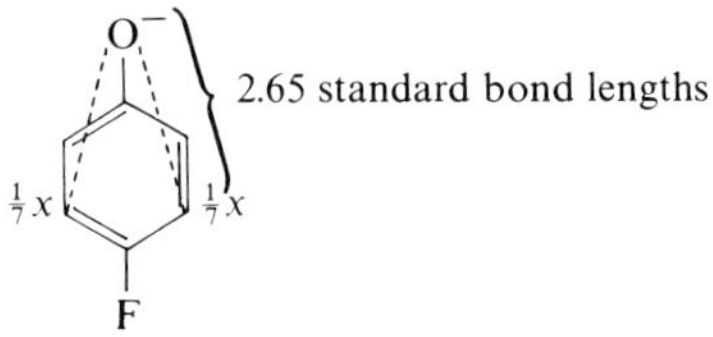

2.65 standard bond lengths

$$MF_F \sum_{k \neq m} \frac{q_{ik}}{r_{km}} = -0.577\left(\frac{1/7}{2.65} + \frac{1/7}{2.65}\right) = -0.062$$

$$\sigma_{F\,1,4} = 0.525 - 0.30 - 0.062 = 0.16$$

$\sigma_{1,4}:$ Br = 0.23, $CH_3 = -0.11$

$\sigma_{1,3}:$ Cl = 0.50, $CH_3O = 0.12$

b. $\log(K_F/K_H) = \rho\sigma$

$$(-pK_F) - (-pK_H) = 1.74(0.16) = 0.27$$

$$pK_F = 12.80 - 0.27$$

$$= 12.53 \text{ (calc)}$$

$$= 12.82 \text{ (obs)}.$$

4.6. These will follow the reactivity numbers, with appropriate statistical corrections: $N_t = 2(a_{0r} + a_{0s})$.

4.7. This is a long exercise. The structure of the dimer is not known.

4.8. Increasing σ_S should increase the half-wave potential in a linear way. The reactive center should be the center of the ring.

4.9. This is an OE_s reaction; the change in π energy is given by $\Delta E_\pi = 2(1 - a_{0r})\beta$.

Chapter 5

5.1. This reversal illustrates the conflict between the $-I$ effect of alkyl groups and polarization stabilization of charge. For carbanions, the two effects oppose each other in the gas phase; for cations, they are in the same direction. In solution, the polarizability effects on relative stability are much less important because the medium is highly polarizable, and access of the solvent to the charged center is more important than charge polarization within the molecule.

5.3. Product B would be favored by increasing the solvation power of the medium.

5.3. This question involves the effect of charge delocalization in the transition state on relative solvent effects. The $E1cb$ transition state has much more carbanionic character than that for the $E2$ reaction. Increasing solvation will favor $E1cb$ reactions.

5.4. Since the BEP principle holds for these reactions, this can be answered in the context of Problems 4.4 and 3.4:

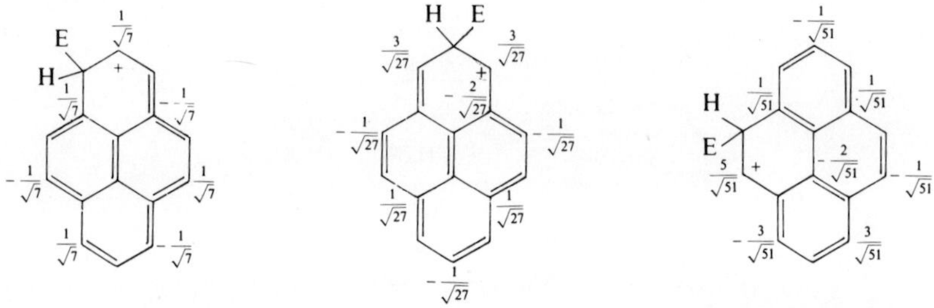

$N_t = 4/\sqrt{7} = 1.51$ $N_t = 12/\sqrt{27} = 2.31$ $N_t = 12/\sqrt{51} = 1.68$

Thus the relative rates are $1 > 4 > 2$; for nitration in acetic anhydride, the 1 isomer is formed fastest.

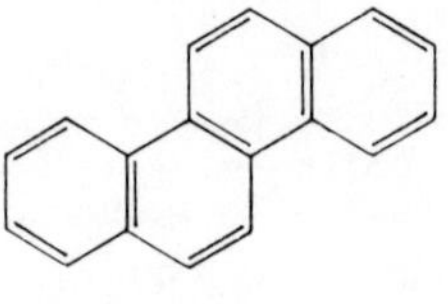

$N_t = 1.67$

Reaction is fastest at the 2 position.

5.6. **a.** See Problem 4.9; the same principle applies since this is a BEP process.

b. The least reactive molecule in the series will show the largest difference between S_N1 and S_N2 rates.

5.7. **a.**

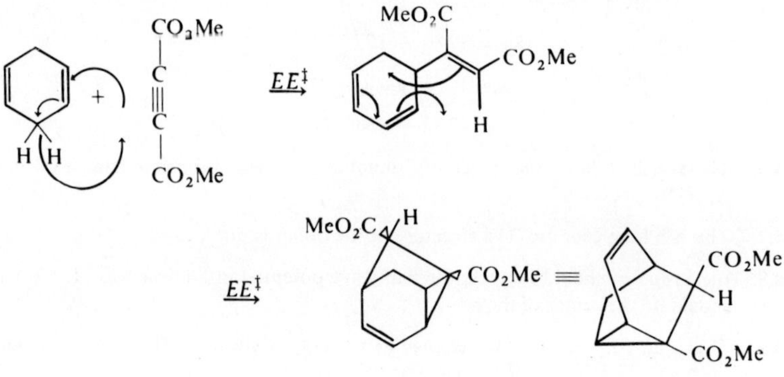

Answers to Selected Problems

549

b.

e., f., g., h.

Answers to Selected Problems

i. 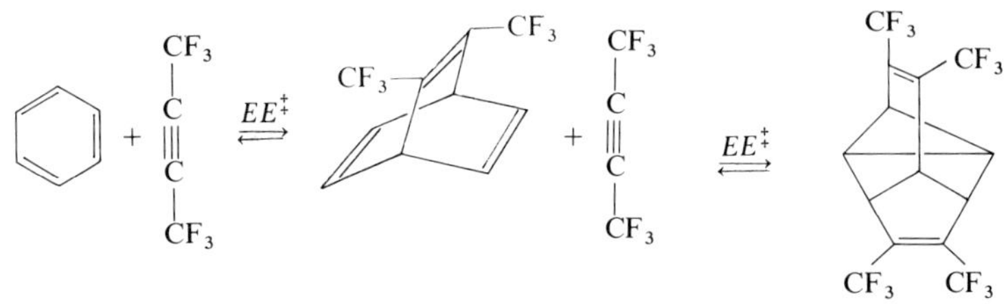 [reaction scheme: benzene + bis(CF$_3$)acetylene $\underset{EE^+}{\rightleftharpoons}$ bicyclic CF$_3$ adduct + bis(CF$_3$)acetylene $\underset{EE^+}{\rightleftharpoons}$ cage CF$_3$ adduct]

j. 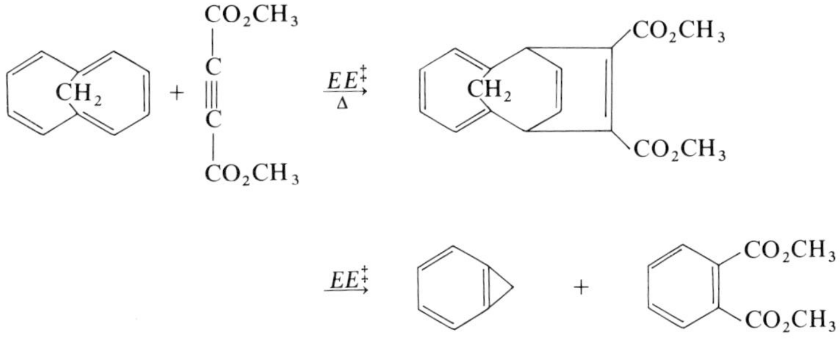 [reaction scheme: naphthalene-CH$_2$ + dimethyl acetylenedicarboxylate $\underset{\Delta}{\overset{EE^+}{\rightarrow}}$ cycloadduct with CO$_2$CH$_3$ groups]

[reaction: $\xrightarrow{EE^+}$ benzvalene + dimethyl phthalate]

5.8. a. 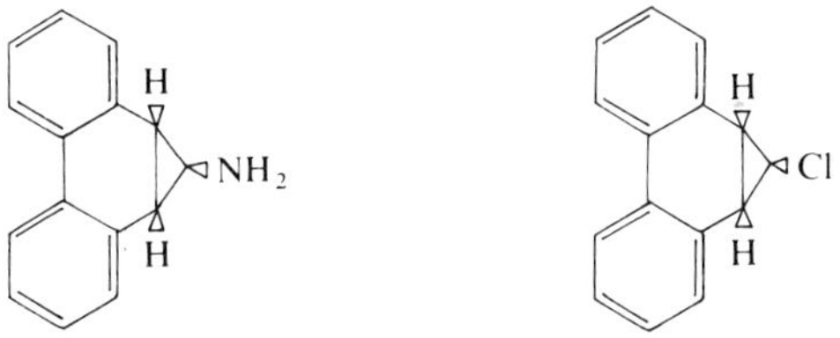 [phenanthrene-fused cyclopropane with NH$_2$; analogous with Cl]

b. 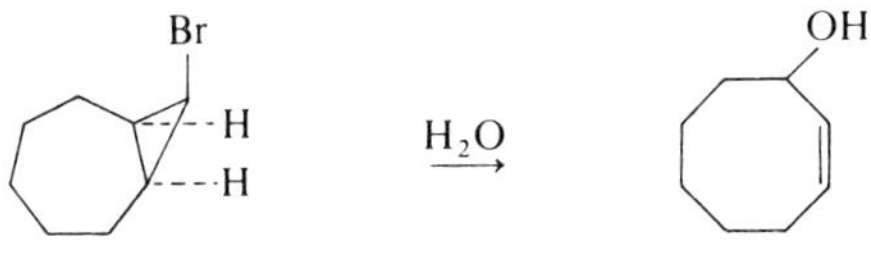 [bromo bicyclic compound with H, H $\xrightarrow{H_2O}$ cis cyclooctenol]

5.12. [phenanthrene with OH, Cl, NO$_2$ substituents]

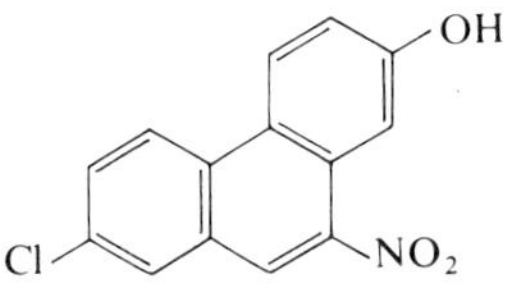

5.14. Consider the 1,4 localization energies in the starting material and the intermediate.

Chapter 6

6.1. The easiest way to examine the stabilities of open-chain and cyclic excited states using the PMO method is to compare the two by intramolecular union.

$$\Delta E_\pi = 2p_{rs}\beta_{rs}$$

552   Answers to Selected Problems

The problem is to determine the sign of p_{rs} in the first excited state. The pairing theorem (p. 75) provides a direct basis for doing this. The AO coefficients in open-chain polymers are determined by symmetry and the magnitude of the coefficients on the terminal atoms increases as one approaches α. Thus for butadiene in its first excited state, the orbital occupation and symmetry are

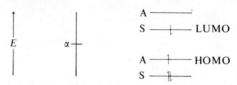

In this state, the 1,4 antibondingness of the singly occupied HOMO is exactly cancelled by the 1,4 bondingness of the singly occupied LUMO since the AO coefficients in these two orbitals are the same by the pairing theorem. The only remaining contribution to $p_{1,4}$ is a positive one from the doubly occupied symmetric orbital. This will be the case in all $4n$-polyene excited states because the magnitude of the terminal coefficients increases on approaching α, so the symmetric (positive) contributions to $p_{\alpha-\omega}$ will always be largest. Since β is negative, terminal intramolecular union will always decrease the energies of these systems, i.e., the cyclic systems are aromatic when compared to the excited states of their open-chain analog.

6.3. (See Ref. 22):

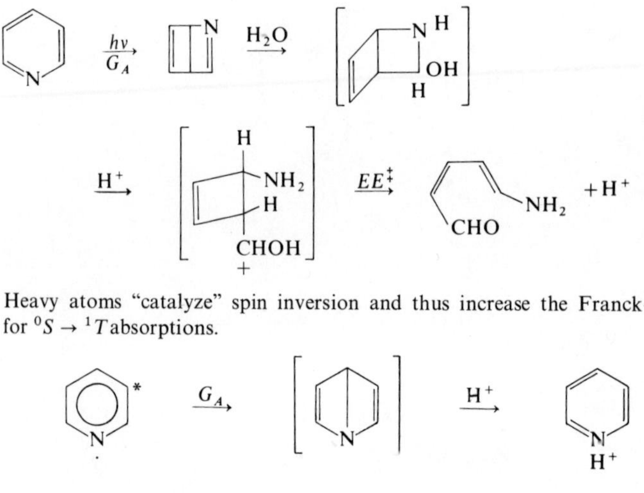

6.4. Heavy atoms "catalyze" spin inversion and thus increase the Franck–Condon factors for $^0S \to {}^1T$ absorptions.

6.5.

6.6. a.

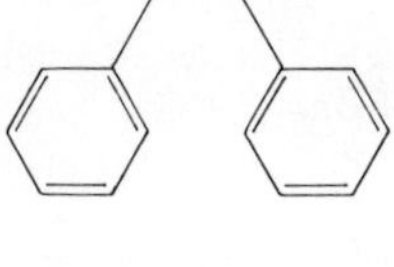

  b.

Answers to Selected Problems

6.8.

a. $CH_3-\overset{O}{\underset{\|}{C}}-\overset{O}{\underset{\|}{C}}-OC_2H_5 \xrightarrow{h\nu}$ [excited state with O^*, CH$_3$, HCHCH$_3$ and C=O, O groups] $\xrightarrow{G_R}$ [radical pair with OH, CH-CH$_3$, CH$_3$, C=O, O]

$\longrightarrow CH_3-\underset{\underset{OH}{|}}{C} + CH_3CHO + CO \longrightarrow CH_3\overset{O}{\underset{\|}{C}}H$

b. 2 [methyl-benzoquinone] $\xrightarrow{h\nu}$ [excited O^* form] $\xrightarrow{X}$ $\left(\text{[excited }O^*\text{ form]} \right)_2$

$\xrightarrow{G_A}$ [dimer cage structure] $\xrightarrow[2.\ G_A]{1.\ h\nu}$ product

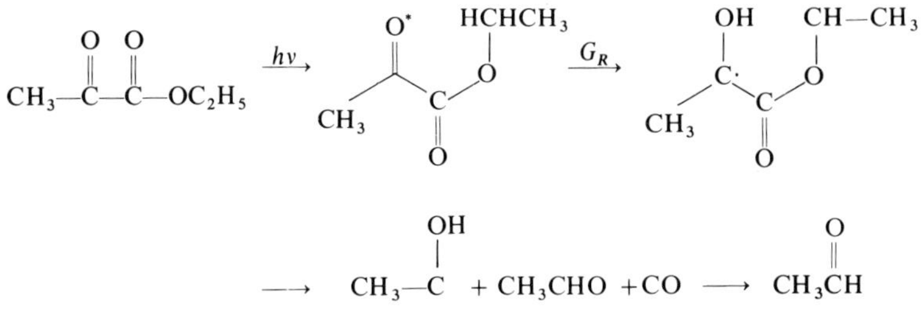

c. [cyclohexenone] $\xrightarrow{h\nu}$ ^1S [cyclohexenone] $\xrightarrow{G_A}$ [with isobutylene] [bicyclic product with radical]

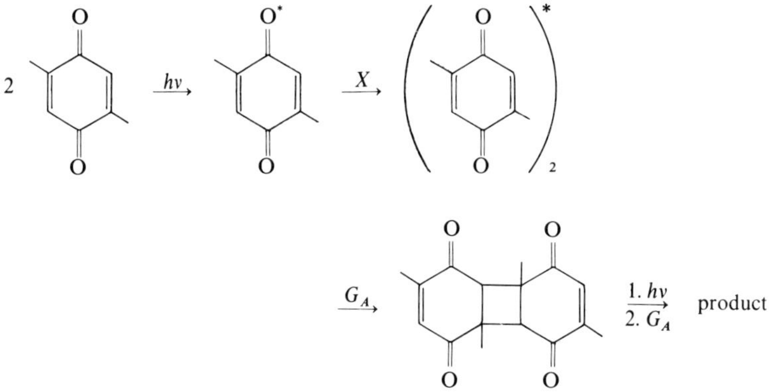

$\downarrow$

^1T [cyclohexenone] $\xrightarrow{G_R}$ [biradical O^*] $\longrightarrow$ [bicyclic product]

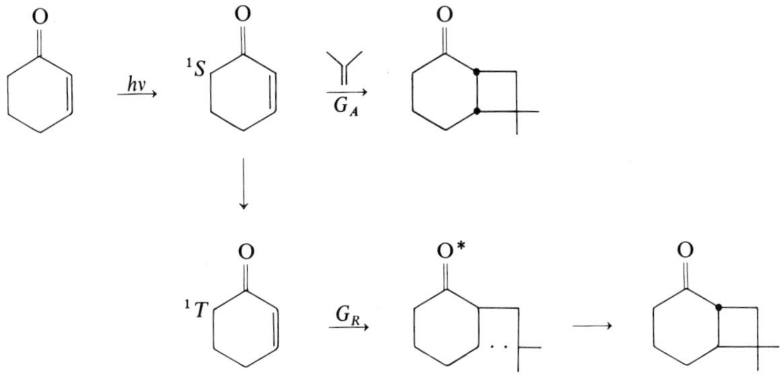

d. [fused tetracyclic structure] $\xrightarrow[2.\ G_A]{1.\ h\nu}$ [benzene] $+$ [cage compound]

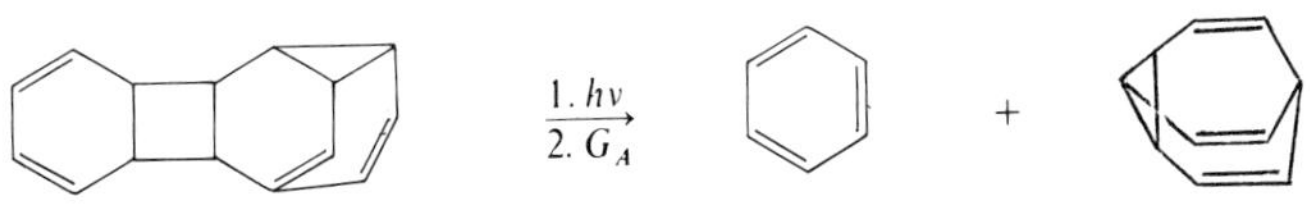

e. [PhCO-CH$_2$-CH$_2$-CH(CH$_3$)-] $\xrightarrow[2.\ G_R]{1.\ h\nu}$ [PhC(OH)(CH$_2$)(CH$_2$)(ĊH$_3$) biradical] $\longrightarrow$

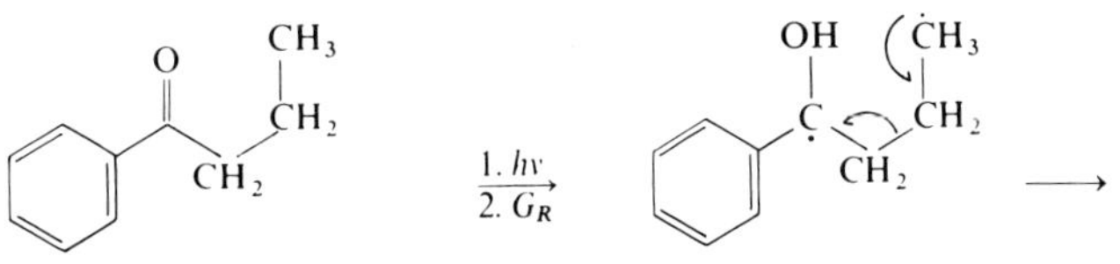

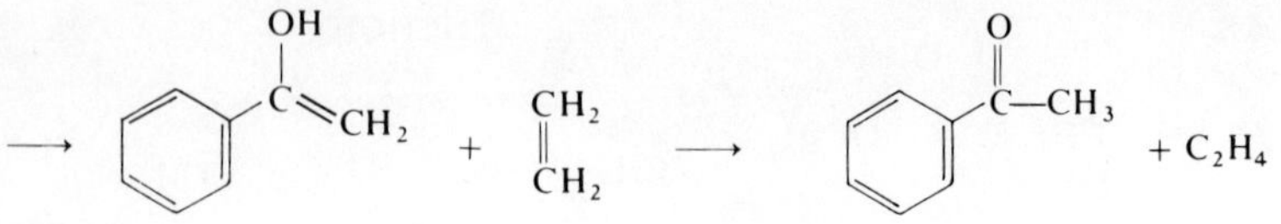

6.9.

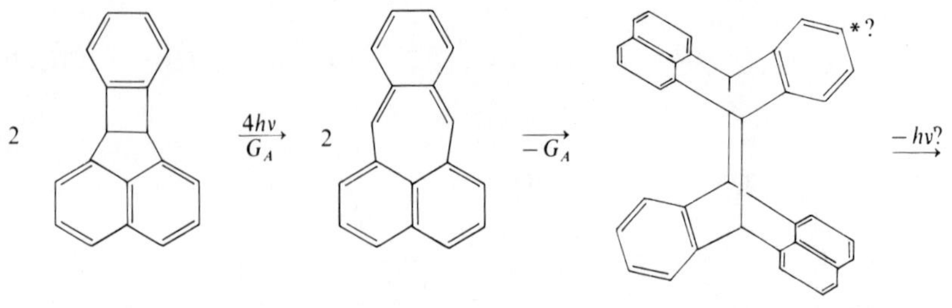

The energy available in the reaction (aromatization of two benzene rings) should be sufficient to put the product in one of the naphthalene triplet states.

6.10.

a.

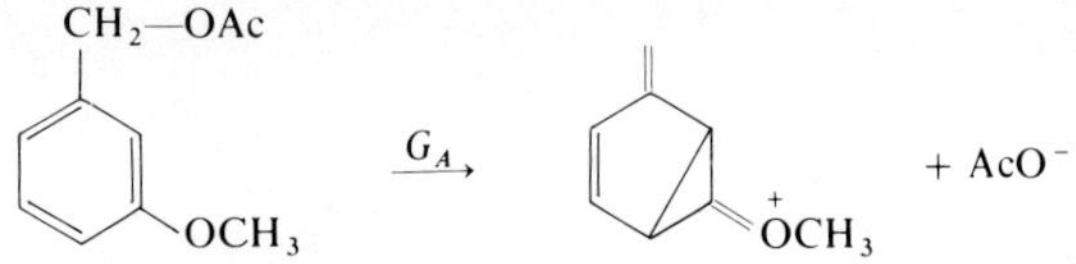

b.

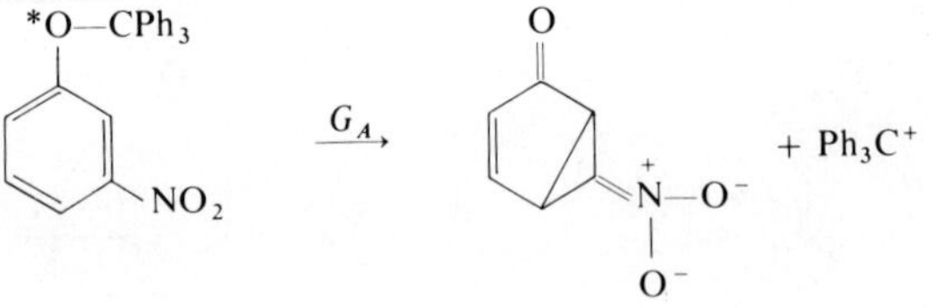

Chapter 7

7.2. The relative solvation energies of the S_N2 transition state and the radical anion are crucial here. Increases in anion solvating power should generally favor the S_N2 reaction.

7.3. High ionization potential; low Brønsted basicity. SbF_5, FSO_3H, $SOCl_2$.

7.5. 3.66 Å charge separation.   90.8 kcal/mole.

7.7.
a. O-alkylation.
b. C alkylation.
c. They would remain approximately constant unless the concentrations were high enough to make ion pairs important in THF. In this case, the radical ion pathway (O-alkylation) might be suppressed.

Index

Abrahamson, 368
Absorption
 electronic, 392
 and emission bands, relation between, 397
 spectra of AHs, calculated and observed, 405
 spectrum, 392
Acetaldehyde, 221
Acetic acid, 244
Acetone
 bromination of, 252
 deuteration of, 252
Acetophenone
 G_R photoprocess of, 433
 nitration of, 326
Acetylacetone, acidity of, 245
Acetylene, 45
 condensation reactions, 387
 gas-phase acidity, 512
 photoelectron spectrum, 46
Acetylenedicarboxylate dimethyl ester, 348
Aci isomers, 251
Acid–base properties of ground and excited states, 445
Acidities
 of alcohols in gas phase, 512
 of aryl methanes, 249
 of carboxylic acids in gas phase, 515
 of excited states, 443
 gas phase, 509, 511
 kinetic, 243
 of phenols, 167
Acridine
 electrophilic substitution in, 323
 excited-state basicity, 445

Acridine orange, 411
Acridinium ion, 162
Acrylic ester, 310
Acrylonitrile, 313
Activated complex, 200
Activation energy, 199, 203
 for aromatic substitution, 318, 324
 for *cis*–*trans* isomerization, 430
 differences in Cope rearrangements, 357
 and heat of reaction, 213, 218
 and heats of reaction for S_N2 reactions in the gas phase and in solution, 516
 for radical reactions, 276
 and S_N2 reactions, 258
Acylation of ambient anions, 269
Acyloin condensation, 529
Addition
 of aldehydes to double bonds, 312
 Br_2 to C_2H_4, 197
 electrophilic, 297
 of ketones to olefins, photochemical, 457
 radical, 310
 reactions of aromatics, 332
 vs. substitution, 331
Addition-substitution in aromatic systems, 317
Addition-substitution reactions, Claisen condensation, 309
Additivity of first-order perturbations, 69
Adipic Acid, 533
AH cations, 115
AH electron affinities, 506
Aldehyde, addition, 312
Alder rule, 350
Aldol condensation, 308

Aliphatic substitution
　electrophilic, 270
　nucleophilic, 253
　transition states, 205
Alkali metal reductions, 527
Alkyl amine basicity in gases and liquids, 510
Alkyl group stabilization of anions, 515
Alkyl halides, electron capture by, 526
Alkyl iodides, photochemical cleavage, 454
Allowed electronic transitions, 399
Allowed pericyclic reactions, 367, 375
　mistaken notions concerning, 368
Allyl anion, 111, 152
　allyl radical, anion, or cation, 105
1,3-Allylic migration, 352
Alternant heteroconjugated systems, 83
Alternant hydrocarbons, 74
　unique features of, 76
Alternant systems, 73
Ambident alkylation, factors controlling rates, 338
Ambient anion alkylation, 537
Ambient anions, 250
　as nucleophiles, 268
　reaction with acyl halides, 269
Amine basicity, 169
　gas phase, 510
Amine oxidation, 521
m-Aminobenzoic acid, 187
Aminophthalhydrazide, 475
Ammonia, 119
Anet, 93
Angular momentum, electron, 398
Aniline, 113, 121, 166
Anilinium ion, pK_A, 169
Anion radicals in solution, 525
Anion reactivity, second-order rule for, 532
Anionic polymerization, 309
Anions, alkyl group stabilization of, 515
Annulenes, 91
　[10], 93
　[12], 95
　[18], 92
Anodic oxidation, 518; electrochemical, 521
Antarafacial migration, 353
Anthracene, 95, 148
　absorption spectrum, 392
　addition reactions of, 332
　1,2-bislocalization energy, 151
　Diels–Alder reactions, 379
　diethylamine photochemistry, 443
　oxidation of, 520
　photodimerization, 469
　reactivity numbers, 140

Anthracene (cont'd)
　tertiary amine exiplexes, 442, 443
Antiaromatic, 97
　AH π complexes with metals, 371
　AHs, 371
　BO holes, 437
　pericyclic reactions, 462
　stabilization by π complex formation, 372
　transition states, 434
Antiaromaticity, 89
　in the anti-Hückel series, 109
　basic consequences of, 436
Anti-BEP reactions, 238, 536
Antibonding electrons, role of in photochemistry, 426
Antibonding MO, 15
Antibonding π MOs (π^*), 394
Anti-Hückel
　chelotropic transition states, 365
　cycloaddition reactions, 349
　naphthalene, 468
　phenanthrene, 469
　π-systems, 342
　sigmatropic rearrangement, 354
　six-electron systems, photochemical pericyclic reactions of, 464
　systems, 106
Anti-Markownikov addition of HBr, 313
7-$Anti$-norbornenyl chloride, solvolysis, 292
AO coefficients, calculation of, 78
Aprotic solvents, 517
Arenonium ions, 318
Aromatic ring, 97
　energies of individual rings, 148
　hydrocarbons, 146
Aromatic substition, 316
　nucleophilic radical chain mechanism, 529
Aromaticity, 89, 331
　of anti-Hückel systems, 109
　general rules for, 99
　heteroconjugated molecules, 110
　radical cation, 499
　relationship to ring size, 91
　of transition states, 345
Aromatics, polycyclic, 93
Arrhenius equation, 199
Aryl halides, radical anion reactions of, 528
Aryl methane acidity, 247
Arylmethyl chlorides
　rates of S_N2 reactions, 259
　S_N2 reactions in, 257
　solvolysis rates, 261
Arylmethyl radicals, NBMO coefficients in, 144

Index

Asymmetric reactions, 368
Atomic orbital coefficients, 73
 by symmetry, 23
Atomization, heat of, 58
Autoionization
 of anions, 504
 of luminferin, 475
Autooxidation, 277
Aziridine ring opening, 305
7-Azaindole phototautomerism, 446
Azimuthal quantum number, 399
Azulene, 81, 93, 100, 348
 electrophilic substitution in, 327

Back-coordination in π complexes, 302
Banana bonds, 300
Bands in electronic spectra, 392
Bartlett, 375
Base catalysis, 215
Basicity
 of amines, 169
 of aniline, substituent effects on, 178
 of azaaromatics, 138
 of excited states, 443
 gas phase, 509
 inversions between gas and liquid systems, 512
 of nitrogen heterocycles, 245
 and nucleophilicity, 265
Basis set, 28, 39
Bathochromic effect, 414
 shifts, 412
Bell, 215
Bell–Evans–Polanyi (BEP)
 principle, 212
 plots, 218
Bent bonds, 300
Benzaldehyde, 221
 uv spectrum, 410
Benzene, 47, 89, 92, 106
 1,2 bislocalization energy, 150
 photochemistry, 466
 photochemistry, with pyrole and naphthalene, 460
 photoisomerizations, 465
 and pyridine, 409
 radical anion, 507, 530
 reactivity number, 140
Benzenonium ion, 317
Benzo(a)anthracene, 1,2-bislocalization energy, 151
Benzo-bicyclo-(2.2.2)-octatriene photochemical reactions, 458
Benzocyclobutadiene, 375, 369

Benzo(a)perinaphthenyl, NMBO coefficients, 80
Benzo(a)pyrene, 1,2-bislocalization energy, 151
Benzo(c,d)pyrene-6-one radical anions, 535
Benzoquinone, 153
 ortho and *para,* 111
Benzotropylium, 99
Benzvalene, 465
Benzyl anion, 81, 82
Benzyl cations, 81, 511
Benzyl chloride, S_N2 reactions, 262
Benzyl radical, anion or cation, 105
BEP plot
 deprotonation reactions, 247
 for photochemical reactions, 436
BEP principle, 219
 deviations from, 239
 failures of, 235
 limitations of, 230
 selectivity effects, 320
BEP reactions, 138, 536
Berson, 354
Beryllium, 12
Biacetyl, intersystem crossing in, 402
Bianthrone photochromism, 462
3,3'-Bibenzyl, 103
Bicyclobutadiene, 361
Bicyclobutane, 368
 ring opening, 361
Bicyclo-[2.2.0]-hexatriene, 356, 465
Bicyclopentane electrocyclic, opening of, 375
Bimolecular reactions, 198
 substitution reactions, 253
Binschedler's green, 411
Biphenyl, 66, 125
Biphenylene, 100, 125, 372
1-Biphenylenylmethyl, 82
Biradical, 432, 447, 450
 intermediates, 373
 molecules, 103
Birch reduction: intermediate, 507
 mechanism, 530
 of naphthalene, 532
1,2-Bislocalization energies, 149, 150
 in hydrocarbons, 151
1,4-Bislocalization energy, 146, 380
1,5-Bismethylenecyclopentadienate anion, 97
Bismethylenecyclobutene reaction with TCNE, 375
BO
 approximation failure, 424
 hole, 431

BO (cont'd)
  hole in an excited potential surface, 425
Bond alternation, 114
  in annulenes, 91
  in antiaromatic rings, 434
  in polyenes, 88
Bond-breaking and bond-forming reactions, 217
Bond dissociation energy, 43, 276
Bond energies, 40
  for HX and CX bonds, 264
  relationship to classical structures, 124
  of second-row diatomics, 428
  X–H for second row elements, 428
Bond energy, 16
  essential double, 100
  essential single, 100
Bond length, 16
  acetylene, 46
  in aromatics and antiaromatics, 92
  ethylene, 42
  in paraffins, 41
  in polyenes, 89
Bond localization, definition, 86
Bond order, 63, 65, 97, 145, 390
  and localization energies, 149
  neutral AHs, 75
  odd AH anions, 77
  odd AH cations, 77
  π-bond order, 145, 149
Bond reorganization as a factor in activation energy, 338
Bond strengths, 174
Bonding
  MO, 15
  in paraffins, 39
π-Bonds, 33, 41
σ-Bonds, 33
Bone, 514
10,9-Borazaronaphthalene, 160
Born–Oppenheimer approximation, 16, 25, 53, 395
  role of in photochemistry, 422
Born–Oppenheimer breakdown, 17, 27
Boron, 12
Breslow, 296
Bromine addition to ethylene, 197
Bromoalkyl esters, solvolysis of, 335
Bromonium ions, 299, 335
Bromonaphthalene color, 401
Brønsted, 215
  relation, 216, 217, 220
  failure of relation, 250

Brown's selectivity rule, 220, 321
Butadiene, 48, 85, 313
  addition of SO_2, 362
  sensitized photodimerization of, 402

C_2, ground state, 36
Calculations of $A^\pm$, effect of $\pm E$ substituents, 170
Calomel electrode, 154
Carbanionoid substituents, 176
Carbene, generation by photolysis, 422
Carbene addition reactions, 366
Carbon, 12
Carbonium ion
  formation, 169
  rearrangements, 286
  stability, 167
  stability in gases and liquids, 511
  substituent effects on, 175
Carbonyl addition reactions, 308
Carboxylic acids
  acidities of in the gas phase, 515
  radiolysis of, 525
Catalysis, 369
  acid–base, 215
Catalytic constant, Brønsted, 216
Cation radical reactions, 492
Chain length in radical reactions, 274
Chain radical reactions, 311
Chain reactions, 537
Charge
  and dipole interaction, 224
  distribution in AHs, 75
Charge densities, 172, 116
  neutral AHs, 75
  odd AH anions and cations, 77
Charge-induced dipole interactions, 514
Charge-transfer transitions, 394, 420
Chelotropic reactions, 362
  exceptions, 366
Chemical conversion of electronic excitation, 403
Chemical ionization mass spectrometry, 488, 504
Chemiluminescence, dioxetanes, 474
Chemiluminescent
  oxidation of lophine, 474
  oxidation of luminol, 475
  reactions, 471
  reactions, G-type, 473
  reaction types, 472
Chichibabin amination reaction, 328
Chiral photochemical reactions, 469

Index

Chlorine
 addition to benzene, 333
 addition to stilbene, 300
 bromine addition to olefins, 311
Chlorine dioxide oxidations, 521
Chloroacetaphenone, 235
ω-Chloroalkylamines, cyclization of, 334
Chloroethyl ketones, elimination reactions of, 381
Chloroform, 224
 autooxidation, 278
Chlorohydrin formation, 297
1-Chloroisoquinoline, hydrolysis, 330
Chloromethyl methyl ether, 263
2-Chloro-*N*-methylpiperidine, 254
Chlorophyll, 524
 radical anion and radical cation, 525
 Katz special pair, 524
Chloroplasts, 524
4-Chloropyridine, hydrolysis, 330
Chlorquinoline, hydrolysis, 330
β-Chlorovinyl ketone nucleophilic addition/substitution, 309
Cinnamate ion, reaction with bromine, 316
Circular dichroism, 469
Cis–trans isomerization, 430, 452
Cis-stilbene, 469
 conversion to phenanthrene in a mass spectrometer, 503
Claisen condensation, 309
Claisen rearrangement, 357
Clar, 102
Classical norbornyl cation, 289
Classical structures and bond energies, 124
Classification of reactions, 235
α-Cleavage, photochemical, 455
β-Cleavage, 454
 in heteroaromatic mass spectra, PMO description of, 495
Cleavage reactions of gas-phase ions, 492
Closed shell system, 74
Closs, 362
Cobaltic acetate oxidations, 518
Collective properties of molecules, 38
Collision frequency, 198
Competitive method for finding relative rate constants, 319
π-Complexes, 287, 370, 371
 of aromatics, 332
 mechanism for carbene addition, 366
 mechanism for electrophilic addition, 298
 description of neighboring group assistance, 335

π-Complexes *(cont'd)*
 with metal ions, 294
 MO description, 371
 protonated epoxides, 306
 reactions, 285
 stability, 288
 vs. three-membered rings, 300
σ-Complexes in aromatic substitutions, 318
Concerted and nonconcerted paths for gas-phase pericyclic reactions, 499
Concerted and stepwise cyclization reaction continuum, 378
Condon, 395
Conformations of cyclohexanone, 356
Conjugated molecule, 148
Conjugation
 cross, 179, 181
 mutual, 180, 186, 188
Conrotatory, 360
 chelotropic reactions, 365
 ring opening or closing, 343
 transition state, electrocyclic opening of cyclobutene, 344
Conservation of orbital symmetry, 339, 357, 362, 365
 foundations, 368
Conservation of total spin angular momentum, 401
σ Constants for α-naphthoic acids calculated and experimental, 190
Constants of the motion, 2
Conversion, internal, 403
Coordinate dissection analysis of reactions, 213
Cope rearrangement, 229, 339, 352, 368
 of 2,5-diphenyl-2,5-hexadiene, 378
 transition state for, 355
Copolymerization, 315
Core, 20
Coronene cation, 501
Coulomb integrals, 63, 110
 empirical nature of, 73
Coulson, 73
Covalent bonds, 18
Craig, 107
Cross-conjugation, 103, 179, 181
Cross-propagation in copolymerization, 316
Crystal Violet, 178
Cuminyl chlorides, solvolysis rates, 190
Cyanine dyes, 414
Cyanoanilines, absorption spectra, 416
Cyanohydrin reaction, 308

Cycloaddition
  2 + 2, 350
  2 + 2 π photochemical, 464
  4 + 6, 348
  8 + 2, 348
  14 + 2, 350
  biradical mechanism, 341
  reactions, anti-Hückel, 349
Cyclobutadiene, 90, 361
  antiaromaticity, 341
  complexes with metals, 372
  dimer, 369
  MOs of, 434
Cyclobutane, strain energy, 376
Cyclobutanol, photochemical formation of, 455
Cyclobutene
  disrotatory opening, 369
  ring opening, 343
Cyclodecapentaene, 327, 348
Cyclodimerization of ethylene, 368
Cycloheptatrieneone, 348
1,4-Cyclohexadiene, photoaddition to, 467
1,3-Cyclohexadiene, ring opening, 361
Cyclohexadienones, rearrangement of 4,4-disubstituted, 446
Cyclohexane conformations, 356
Cyclohexenone rearrangements, 448
Cycloöctatetraene, 92, 93, 125, 471
Cyclopentadienate
  addition to maleic anhydride, 350
  anion, 98, 114
  radical, 97
Cyclopentadiene, 157, 229
2-Cyclopentenyl cation, ring opening, 361
Cyclopropane
  bonding in, 243
  edge-protonated, 295
Cyclopropenium ion stability, 295
Cyclopropyl anion, 152
Cyclopropyl halides reduction by sodium naphthalide, 528
Cyclotron resonance frequency, 488

Dacron, 518
Dative bonds, 18
Davisson, 4
Decarboxylation reactions, 272
Deexcitation, 401
  during photochemical reactions, 422
  of upper excited states, 403
Degeneracy, orbital, 434
Degenerate
  level, 27

Degenerate *(cont'd)*
  orbitals, 19
  states, 7
Delocalization energy
  of activation, 205, 236
  of activation for radical polymerization, 315
  effect on equilibria, 137
Delocalized bonds, 50
Deprotonation of aryl methanes, 248
Desolvation of nucleophiles, 266
Dewar, 121
Dewar benzene, 465
Dewar pyridine, 465
Diatomic molecules, 32
  C_2, 35
  F_2, 37
  HeH^+, 18
  HHe, 19
  Li_2, 35
  LiH, 20
  N_2, 36
  O_2, 36
Diazomethane photolysis, 422
1,3-Dibenzanthracene, 417
Diborane, 47
2,3-Dibromobutane, I^- catalyzed bromine elimination, 280
Dichlorodicyanoquinone, 517
Dichlorodifluoroethylene addition to trans-cycloöctene, 377
Dicobaltoctacarbonyl, 387
Dielectric constant, 184
Diels–Alder reactions, 144, 240, 339
  application of ±E substituents technique, 379
  of heteroatomic systems, 160
  rates of, 381
  rates correlation with 1,4-bislocalization energies, 382
  solvent effects on, 228
  special features of, 350
Dienophile, 147
Diethylamine photochemistry with anthracene, 443
Diffusion-controlled reactions, 241
9,10-Dihydronaphthalene photocyclization, 468
α-Diketones, use as sensitizers, 403
Dimerization
  of ethylene, orbital description, 342
  of olefins, photochemical, 462
Dimethylaniline, oxidation of, 522
3,4-Dimethylcyclobutene, ring opening, 344

Dimethylformanide, solvation of anions, 267
Dimethylsulfoxide, 242, 517
  solvation of anions, 267
Dioxetane chemiluminescence, 474
Diphenyl azulene, 348
Diphenylacetylene, 348
Diphenylmethane, pK_A, 169
Di-π-methane photoisomerization, 458
Di-π-methane photoreactions, 450
Dipole–dipole interactions, 224
Dipole-induced dipole interactions, 227
Dipole moment, 21
  of even AHs, 81
Dispersion forces, 227
Displacement reactions
  in mass spectra, 496
  radical anion, 508
Disproportionation of radical cations, 509
Disrotatory opening, 360
Disrotatory chelotropic reactions, 364
Disrotatory electrocyclization, 446, 447
Disrotatory ring opening or closing, 343
Disrotatory transition state, electrocyclic
    opening of cyclobutene, 344
Disrotatory–conrotatory ring opening of
    bicyclobutane, 362
Dissociation
  of ethanes, 143
  photochemical of a diatomic molecule, 431
Disulfides, photochemical cleavage, 454
Djerassi, 500
Doering, 356
Double bond, 43, 123
Double resonance, ion cyclotron, 489
Doublet states, 12, 398
Dye industry, 410
Dye spectra, 418
  heteroatom effects on, 411

$\pm E$ activity of a substituent, 171
$E1cB$ reactions, 281
Eberbach, 461
ECE reactions, 518
$EE_A{}^{\ddagger}$ reactions, pericyclic, 338
$EE_A{}^{\ddagger}$ reactions, sigmatropic rearrangements, 352
EE equilibria, 144
Electric dipole, 21
Electric vector, 398
Electrochemical (anodic) oxidation, 521
Electrochemistry, 518
Electrocyclic cyclization of radical cations, 503
Electrocyclic reaction, 360

Electrode potential, standard, 153
Electrolysis, Kolbe, 521
Electromeric substituent, 165, 167
  effects on electronic spectra, 414
Electron affinity, 490
  of alternant hydrocarbons, 526
  calculation of, 505
  of even AHs, calculated and observed, 506
  of H_2O and $OH^{\cdot}$, 265
Electron bombardment, 486
Electron capture, 490
  by alkyl halides, 526
Electron correlations, 10, 13
Electron-deficient molecules, 47
Electron density, 62
  in heteroatomic systems, 114
π-Electron density
  in heteroconjugated systems, 116
  q_i, 62
Electron distribution, 14
π-Electron energy, total, 63
Electron interactions through space, 459
Electron repulsion, 8
  approximations, 9
  in excited states, 427
  magnitude, 9
Electron spin, 10
  resonance, 12
Electron transfer
  chemical reaction/electron transfer reactions, 518
  chemiluminescence, 473
  oxidation, 517
  photosensitized, 419
  processes, 152
Electrons, solvated, 525
Electronegativity
  and acidity, 242
  effects on carbonium ions structure, 298
Electronic energy, effect on equilibria, 137
Electronic spectra
  approximate calculation of, 404
  of even heteroconjugated systems, 409
  of odd heteroconjugated systems, 410
  of polynuclear aromatic hydrocarbons, 405
  rules for heteroatom effects on, 411, 412, 414, 415, 417
Electronic transitions, 391
  in atoms, 399
  and orbital overlap, 398
Electronically excited states, 393
Electrophile, 270
Electrophilic addition, 297

Electrophilic aliphatic substitution, 270
    examples of reactions, 272
Electrophilic aromatic substitution, 316
Electrophilic radicals, 277
Electrophilic substitution, 206
    relative rates, 322
    in naphthalene excited states, 445
    in toluene, isomer distribution, 322
Electrophilic vs. radical addition of halogens to olefins, 311
Electrostatic analyzer, 486
Electrostatic energies in solvation, 223
Electrostatic energy, 184
Electrostatic interactions, 222
Elimination of HCN, retrocycloaddition, 348
Elimination reactions, 279
    mechanistic continuum, 282
    orientation in, 284
    regiospecificity, 283
Emission
    and absorption bands, relation between, 397
    of light, 396
Enamine reactions, 337
Endo–cis addition, 351
Energy
    of activation, 199, 204
    electrostatic, 184
    and entropy relationships, 134
    first-order effect, 64
    internal, 134
    internal, and temperature, 199
    level diagram for π MOs of ethylene, 427
    of localization, 139
    of reaction, 135
    relative, of excited states, 398
    solvation, 513
    solvation, relation to ion size, 225
    of union, 67
    of union of AHs and heteroatomic substituents, 177
    of union for even AHs, 85
    and visible light quanta, 393
π Energy
    of activation in cycloaddition reactions, 380
    changes in radical reactions, 275
    charge on protonation, 244
    of dimerization of radical anions, 533
    of reaction
        for aromatic substitution, 318
        formation of arenonium intermediates, 324
Enolization of ketones, 253

Entropy
    of activation, 217, 377
    effect on photochemical reactions, 456
    and energy relationships, 134
    neglect of, 134
    of reaction, 132, 133
    translational, 334
E_N1 reactions, 281
E_N2 elimination, *cis*, 281
E_N2 reactions, 279
$EO_A{}^\ddagger$ reactions, 256, 270
    elimination, 279
$EO_B{}^\ddagger$ reactions, 254
    addition, 297
    aromatic substitution, 316
    nucleophilic addition, 306
    prototropic type, 245
    radical addition, 310
    radical substitution, 274
    S_E1, 272
EO_s equilibria, 139
EO_t equilibria reactions, 141
$E\pi_A{}^\ddagger$ reactions, carbene addition, 267
$E\pi_B$ mechanism for solvolysis of phenethyl halides, 292
$E\pi_B{}^\ddagger$ reactions, 285, 291
    addition, 297
Equilibria
    heteroatom effects, 159
    heteroconjugate, 139
    isoconjugate, 138
    statistical factors, 141
    term symbols for, 139
Equilibrium
    chemical, 131
    EE, 145
    EO_s, 139
    EO_2, 142
    OO, 150
Equilibrium constants, 132
    for excited state reactions, 443
    prediction of, 137
Equivalent orbitals, 30, 43
$-E$ reaction, 472
Essential single bonds, 123
    and double bonds, 99
$+E$ substituent, 172
$-E$ substituent, 175
$\pm E$ substituent technique, 378
    application of, 379
Ethane dissociation, 168
Ether autooxidation, 278
Ethyl acetoacetate, 250
    anion, 268

Ethylene, 41, 85
  energy-level diagram for π MOs, 427
  excited state of, 429
  photoelectron spectrum, 45
Ethylene dimerization, 374
  to cyclobutane, 340
Ethylene oxide, MO description, 300
E-type photochemical reactions, 428
Evans, 213, 339
  principle for pericyclic reactions, 345, 499
Evans–Polanyi plot, 214
Even AH, 74
  union multiple, 86
Excimer, 460
  formation, 439
  PMO description of formation, 441
Excitation, 401
Excitation energy
  calculated and observed of even AHs, 405
  estimation of for AHs, 404
  of even AHs, 403
  and probability of transitions, 424
  of odd AHs, 408
Excitation transfer, 401, 419
Excited dimers, formation of, 439
Excited state, 391
  of ethylene, 429
  failure of the localized bond model, 447
  lifetime, 396
  lower, 393
  potential surface, 423
  potential surfaces — comparison to ground state surfaces, 430
  production of radicals from $n \to \pi^*$, 455
  proton transfer reactions in, 443
  reactions giving excited state products, 421
  reactions leading to products in the ground state, 421
Exiplexes, 442, 466
Exo–endo ratios in the Diels–Alder reaction, 350
Exonorbornyl ester, solvolysis of, 291
Extinction coefficient, 392, 398
  dependence on transition moment, 400
  effects on the course of photochemical reactions, 453

Faraday equivalent, 155
Favorskii rearrangement, 447
Fenton's reagent, 475
Ferricyanide oxidations, alkaline, 520
Ferrous ion, 420
Feynman, 228
Field, 511

Field effect, 182
  estimation of constant, 188
  positive and negative, 184
  substituent constant, 185
Finkelstein reaction, 235
  solvent effects on, 230
First-order perturbation, 19, 64
  additivity, 88
  of doubly occupied degenerate π MOs, 86
  on union of odd AHs, 68, 88
First-order substitution reactions, 253
Fission in isomeric ions $ArCH_3$, PMO description, 495
Fittig, 529
  flash photolysis, 445
Fluorenyl
  anion, 99, 252
  cation, 256
Fluorescence, 401, 402, 421, 438, 448, 451, 460
  excimer, 439
  from photoreaction products, 445
  quantum yield, 441
  quenching of, 439
  uv chemiluminescence, 477
Fluorinated ethylene, cycloaddition reactions of 375
FMMF method, 187, 546
Forbidden pericyclic reactions, 367, 375
  mistaken notions concerning, 368
Formal charge, 117
  in AHs, 81
  distribution in transition states for electrophilic substitution, 323
Formamide, 224
Four-electron systems, photochemical pericyclic reactions, 463
Fragmentation
  of ions in a mass spectrometer, 487
  patterns, 487
Franck, 395
Franck–Condon principle, 25, 395, 423
Franklin, 295
Free energy
  of activation, 204, 217
  of reaction, 131
  of solvation, 225
Free energy–entropy relationships, 134
Frequency of first absorption band for AHs, calculation of, 404
Frontier orbital method, 357, 362, 367
Frontier orbital theory, 339
Fukui, 339, 367
Fulvene, 61, 81, 97

G_A photochemical reactions, 434, 462
G_I photochemical reactions, 433, 459
G_N photochemical reactions, 461
G_R photochemical reactions, 431, 452
G_S photochemical reactions, 437, 461
G-type
  chemiluminescent reactions, 473
  photochemical reactions, 431
Gammexane, 331
Gas phase
  ion–molecule reactions in, 488, 509
  radical anions in the, 504
  S_N2 reactions in the, 516
Gas-phase acidities, 509, 511
Gas-phase basicity, 509
Gas-phase ions, 485
General acid–base catalysis, 216
Geometry differences between ground and excited states, 396
Germer, 4
Gleicher, 276
Goldstein, 357
Gomberg reaction, 329
Grignard addition to carbonyls and Schiff bases, 308

Half-cells, electrochemical, 153
Haller–Bauer reaction, 273
Halogenation rates, effect of halogen type on, 278
α-Haloketones, S_N2 reactivity of, 516
Hammerum, 500
Hammett equation, 185
Hammett relation, application to elimination mechanisms, 282
Hammett substituent constants, calculation by FMMF method, 190
Hammond postulate, 220
Hartree–Fock, 10
Haselbach, 461
Heat
  of atomization, 40, 58, 85, 100
  of π-complex formation, 288
  of hydrogenation, 59
Heat of reaction, 132, 213, 218
  and activation energies for S_N2 reactions in the gas phase and in solution, 516
  ethylene dimerization, 374
Heavy-atom effect in singlet–triplet transitions, 401
Heilbronner, 52, 107
Heisenberg, 1
Heitler–London dispersion forces, 227
Helium, 7

Hellmann–Feynman theorem, 222, 228
Heptafulvene, 159, 498
  addition, 349
Heptalene, 95
Heteroaromatic mass spectra, PMO description of β-Cleavage in, 495
Heteroaromatic systems, 109
  charge density, 116
Heteroatomic AH cations, 115
Heteroatom effects on electronic spectra, rules for, 411, 412, 414, 415, 417
Heteroatoms, effects on equilibria, 159
Heteroconjugate equilibria, 138
Heteroconjugated anions, 112
Heteroconjugated systems, electronic spectra of, 409
Hexachlorocyclohexane, 331
β-1,2,3,4,5,6-Hexachlorocyclohexane, elimination of HCl from, 279
Hexahelicene, 469
  rearrangement, 501
Hexamethylenediamine, 533
Hexamethyl-7-oxo-norbornadiene, 365
Hexamethylphosphoramide, 517
Hexaphenylethane, 168
Hexatriene, 85, 89
  photochemical cyclization, 464
Higher excited states, lifetimes of, 403
High-resolution mass spectrometers, 486
HMO calculations of absorption spectra, 405
Hofmann's rule for elimination reactions, 283
Holes in excited-state surfaces, 425
Homoallylic assistance in solvolysis reaction, 292
Houk, 455
Hückel's rule, 89, 91, 98
  for polycyclic systems, 95
Hückel systems, 106
  four-electron systems, 463
  π systems as pericyclic transition states, 342
Hughes, 253
Hund's rule, 12, 55, 398
Hybrid, resonance, 122
Hybrid AO, 28
  accuracy, 31
  ethane, 40
  ethylene, 42
  overlap, 29
  shape, 29
  sp, 46
  sp^2, 42

Index

Hybridization, 28
  and bond strength, 301
Hydration of olefins, photochemical, 437
Hydrocarbons, relative rates of nitration, 319
Hydrogen atom, 1
Hydrogen bonding, 224, 512
Hydrogen bromide addition to olefins, 312
Hydrogenlike orbitals, 8
Hydrogen molecule, 13
γ-Hydrogen rearrangement, 455
  in mass spectra, 496
Hydroquinone, 153, 182
Hyperconjugation, 276, 514
Hypsochromic shift, 414

Iminoether rearrangement to amides, 339
Inactive atoms, 103
Inactive segments in odd AHs, 103
Indenyl anion, 100
Inductive and polarizability effects, competition between, 515
Inductive effects, π electron, 164
π-Inductive effect, 186, 325
  on equilibria, 164
Inductive substituents, 165
Inductoelectromeric effect, 186
Ingold, 206, 253
Integral,
  coulomb, 63
  resonance, 63
Interference
  constructive, 15
  destructive, 15
  effect, 63
  wave mechanical, 4
Intermediates, 212
  in chemical reactions, 220
Intermolecular perturbations, 59, 61
  on union, 61, 65
  on union of even AHs, 83
Internal conversion, 242, 402, 421, 423, 448, 492
  in gas-phase ions, 491
  relationship to excitation energy, 424
Internal displacement reactions in mass spectra, 496
Internal energy, 134
  distribution, 199
Intersystem crossing, 401, 403, 458
Intramolecular perturbation, 59, 60
Intramolecular union, 60, 64, 96
Inversion barrier, 120
Iodoform reaction, 272
1-Iodonaphthalene, 478

Iodonaphthalene color, 401
Iodonium ions as intermediates in ICl addition, 299
Ion cyclotron resonance, 489
  cell, 489
  double, 489
  spectroscopy, 485, 504
Ion residence time in mass spectrometer sources, 488
Ion–dipole interactions, 224
Ionization in the gas phase, 486
Ionization potentials, 24, 155
  acetylene vs. ethylene, 47
  and ion intensities in mass spectra, 493
  relation with redox potentials, 156
Ion–molecule clustering, 514
Ion–molecule reactions in the gas phase, 488, 509
Ion-pair
  formation in photochemistry, 394
  photoprocesses, 433
Ions
  gas phase, 485
  radical, 491
  reactions of short-lived, 485
Isoconjugate equilibria, 138
Isoconjugate systems, 50
  correlation of monocentric perturbations, 83
Isoconjugate σ-type delocalized structures, 207
Isoelectronic, 60, 113
Isomerization
  photochemical of ethylenes, 427
  photochemical of olefins, 429
Isoprene, 229
Isoquinoline, 118
  electrophilic substitution in, 323
$ISO_B{}^{\ddagger}$ reaction types, 240
Isotope clusters in mass spectra, 496
Isotopic labeling in mass spectra, 497

Jahn–Teller distortion of cyclobutadiene, 435
Jahn–Teller effect, 28
Johnstone, 503

Kaplan, 466
Kasha, 478
Katz, 524
Kebarle, 515
Kekule structures, 99, 338
Keto–enol tautomerism, 251, 253
Ketone cleavage by sodamine, 273

Ketone photolysis, 422
Ketones
 photochemical rearrangement, 446
 use as sensitizers, 403
Ketonization, photochemical of methyl salicylate, 421
Kinetic acidity
 failures of, 245
 of hydrocarbons, 243
Kinetic activation of ion–molecule reactions, 489
Kinetic control, 220, 251
 vs. thermodynamic, 537
Kolbe synthesis, 521
Koopmans' theorem, 26, 156, 227, 505
Kornblum, 269, 508

Lapworth, 338
Lead dioxide oxidations, 523
Least motion, 367
Leaving group effects
 on S_N1 reactivity, 265
 on S_N2 reactions, 264
Lennard-Jones, 43
Leuco base, 459
Lifetimes of states, 401
Light absorption, 391
 of aromatic hydrocarbons, 405
Light energy conversion to chemical energy, 419
Linear combination of atomic orbitals (LCAO), 14, 28
Lithium, 10
Localization energy, 139, 171
 ortho-, 149
 para-, 145, 146
 of radical anions, 534
Localization of bonds in polyenes, 84
Localized bond, 28
 contributions to energies of reaction, 136
 definition, 86
 model failure in excited states, 447
Localized bond analog, 38
 breakdown, 47
 utility, 41
Lone pair, 109
Longuet-Higgins, 47, 83, 121, 368, 371
 rule, 78, 92
Lophine, chemiluminescent oxidation of, 474
Loudon, 500
Luminescence, 475
 vs. chemiluminescence, 477
Luminferin, autooxidation of, 475

Luminol, chemiluminescent oxidation of, 475

Maccoll, 500
Mackor, 141
Magnetic analyzer, 487
Magnetic moment, 12, 398
Malachite Green, 178, 459
Maleic anhydride, 229
Malonic acid, pK_A, 137, 183
Mandelbaum, 500
Manganic acetate oxidations, 518
Markownikov's rule, 304
Mass action, law of, 132
Mass defect, 486
Mass law, 132
Mass spectra
 paraffin, 492
 PMO description of β-cleavage in heteroatomic, 495
Mass spectral pericyclic reactions, stereochemistry of, 502
Mass spectral rearrangements, 496
Mass spectrometer, 485
 high resolution, 846
 substitution reaction parallels in a, 508
 pericyclic reactions in a, 497
 stereospecificity of retro-Diels–Alder reactions in, 500
Mass spectrometry
 chemical ionization, 488
 and photochemistry relationships, 502
 and thermochemistry relationships, 493
Matrix isolation of radical cations, 523
McAdams, 514
McLafferty, 496
Mechanism, 197
Mechanism probes using $\pm E$ substituents, 378
Mechanistic continuum in cycloaddition and pericyclic reactions, 378
Mechanistic probes, use of $\pm E$ substituents, 282
Meisenheimer complexes, 331
Meso-1,2-dibromo-1,2-diphenylethane, elimination of HBr, 279
Meso-3,4-dimethyl-1,5-hexadiene, Cope rearrangement of, 356
Mesomeric effect, 187
Mesomeric field effect, 187
 estimation of, 188
Metal complexes of cyclobutadiene, 372
Metastable ions, 488, 502
 and reaction mechanisms in a mass spectrometer, 500

Index

Methane, 21, 28
  deprotonation, 168
  pK_A, 169
  photoelectron spectrum, 26
  protonation of, 509
Methoxide cleavage of propylene oxide, 304
1-Methoxynaphthalene, nitration of, 326
2-Methoxynaphthalene, nitration of, 326
δ-Methoxyvaleryl chloride rearrangement, 334
Methyl acrylate, 313
4-Methyl-4,3-borazaroisoquinoline, mass spectrum, 496
10-Methyl-10,9-borazarophenanthrene, nitration of, 325, 367
Methyl methacrylate, 313
Methyl radical, 90
Methyl salicylate
  photochemical ketonization of, 421
  photoenolization, 451
Methyl substitution reactions of aromatic hydrocarbons, relative rates, 330
2-Methylallyl cation addition to cyclopentadiene, 346
Methylation of aromatics by CH_3, 329
Methylcyclohexylmercuric bromide S_E2 reactions, 271
Methylene blue, 420
Methyleneammonium ions, 338
1-Methylisoquinoline, 162
8-Methylisoquinoline, 162
1-Methylnaphthalene, 250
α-Methylnaphthalene, oxidation of, 520
Methylpyridines, acidity, 161
4-Methylquinoline, 162
5-Methylquinoline, 162
3-Methylquinoline, mass spectrum, 496
Methyltropylium, 159
Michael reaction, 307
Michler's hydrol blue, 411
Microscopic reversibility, 131, 402, 471
Migratory aptitude, alkyl groups, 290
Minimum gradient reaction path (MGRP), 210
Möbius orbitals, 51
Möbius π system, 108
Molecular orbitals, 13
  symmetry of, 22
Molecular weights, exact, 486
Molecularity of reactions, 200
Molecule ions, 490
Monocentric perturbations, 59, 61, 109
Monocyclic polyenes, 89
Monomers for vinyl polymerization, 313

Morse potential, 396
Mulliken, 6
Mulliken approximation, 16
Multiplicity, 12
Multiplicity of electronic states, 398
Muszkat, 462
Mutual conjugation, 180, 186, 188
  in copolymerization transition states, 316
  in Diels–Alder transition states, 352
  effect on rates of radical reactions, 276
  in S_N2 transition states, 262
  negative, 182

Naphthalene, 93
  Birch reduction of, 532
  1,2-bislocalization energy, 150
  Diels–Alder reaction, 147
  photochemistry, pyrole and benzene, 460
  photoinduced hydrogen–deuterium exchange, 445
  reactivity numbers, 140
  reduction through radical anions, 531
Naphthalocyclobutadiene, 372
α-Naphthyl radical, anion or cation, 106, 144
α-Naphthylamine, 119
Naphthyldiphenylmethyl radical, 172
α-Naphthylmethyl anion, 119, 161
α-Naphthylmethyl, NBMO coefficients, 79
NBMO, 77
  coefficients
    calculation of, 78
    parallel with resonance structures, 122
    sign of, 80
  of cyclobutadiene, 435
  molecules with two, 101
Negative ions in the gas phase, 490
Neighboring group participation, 290, 333
Newman projections, 279
Nitramide, 215
Nitration
  of aza-aromatics, 324
  of benzene, 317
  of benzenoid hydrocarbons, 319
Nitroaminodurene, 181
Nitrobenzene, nitration of, 326
Nitrogen fixation, 527
Nitromethane, 228
1-Nitronaphthalene, nitration of, 326
2-Nitronaphthalene, nitration of, 326
Nitrophenols, absorption spectra, 416
Nitrous oxide, 215
Node
  angular, 6
  radial, 6

No-mechanism reactions, 339
Nonalternant cyclic polyenes, 93
Nonalternant hydrocarbon polarity, 81
Nonalternant hydrocarbons, 96
Nonalternant systems, 73, 157
Nonaromatic ring, 97
  pericyclic reaction, 373
  transition state, 447
Nonbonding MO, 77
Nonclassical AHs, 102
Nonclassical carbonium ions, 295, 296
Nonclassical carbonium ions, π-complexes, 288
Nonclassical molecules, 104, 422, 461
  as photoproducts, 433
Non-least motion transition state, 367
7-Norbornadienyl cation, 294
Norbornyl cation, 288
Normalization of NBMOs, 79
Normalized orbitals, 32
Norrish type I photoprocesses, 455
Norrish type II photoprocesses, 455
Norrish type II reactions in a mass spectrometer, 496
Nonstereospecific *cis–trans* addition, 377
Nuclear charge, effective, 11, 12
Nucleophilic addition, 297, 306
  to carbonyls, 308
  to carbonyls, solvent effects on, 267
Nucleophilic aliphatic substitution, 253
Nucleophilic aromatic substitution, 328
  radical chain mechanism, 529
Nucleophilic displacement, photochemical, 438
Nucleophilic order for carbonyl addition reactions, 268
  $S_N 2$ reactions, 268
Nucleophilic substitution, 206
Nucleophilicity, 260
  in aprotic solvents, 267
  and basicity, 265
  effects on addition reactions, 307
  solvent effects on, 517
Nullpunktsenergie, 395
Nylon, 533

2,7-Octadiene, 356
Octatetraene, 498
Odd AH, 74
$OE^\ddagger$ reactions, reactions of enamines and enolates, 336
Olah, 509
Olefin
  addition to photoexcited benzene, 466

Olefin *(cont'd)*
  dimerization, pericyclic, 350
  isomerization, photochemical, 452
  MO description, 371
  photochemical dimerization of, 462
  photochemical hydration of, 437
Olefin–metal π-complexes, 294
Omegatron, 489
OO equilibria, 150
$OO_A{}^\ddagger$ reactions
  pericyclic, 338
  sigmatropic rearrangements, 352
Orbital, 6
  approximation, 7
  crossing, 435
  degeneracy, 11, 434
  mixing, 34
Orbital overlap
  π bond, 43
  σ bond, 43
  and electronic transitions, 398
Orbital phase, 51
Orbital space, 30
Orbital symmetry, 357, 362, 368
  conservation of, 365
Orbital topology, 33, 50, 107, 207, 209, 238
  in chelotropic reactions, 363
  in π-complexes, 287
  in pericyclic transition states, 340
  in $S_N 2$ reactions, 263
  trimethyl aluminum, 295
Orbitally degenerate systems, spin inversions in, 475
Order of reaction, 200
Organic chemistry, size of the subject, i
Orgel, 371
Orientation differences in acid- and base-catalyzed opening of epoxides, 305
Orientation in epoxide cleavage, 304
Ortho and *para* directing substituents, 327
Orthogonal orbitals, 31
Ortholocalization energy, 149
Osmium tetroxide, 148
Overlap, 15, 20
  in phase 31, 51
  out of phase, 31
Overlap integral S_{AB}, 16
Oxetanes, photochemical formation, 457
Oxidation
  of amines, 521
  of arylmethanes, 518
  of dimethylaniline, 522
  effects and dye color, 420
  electrochemical, 155, 521

Oxidation *(cont'd)*
  by electron transfer, 518
  of lophine, chemiluminescent, 474
  of luminol, chemiluminescent, 475
  photochemical–radiochemical, 523
  photosensitized, 421
Oxidation potentials, 527
Oxidation–reduction, photochemical, 456
Oxidative decarboxylation of cinnamate ions, 316
O_2 ground state, 36

Pairing theorem, 75
  odd AHs, 77
Paraffins, 38
  mass spectra, 492
Paralocalization energy, 145, 380
  for aromatic HCs, 146
Parent peak, 487
Parity of AH positions, 74
Partial rate factor, 319
Pauli principle, 10, 52, 397
Pauling, 121
Pentacene photodimerization, 442
Pentadienate
  ion, 157
  radical, 106
1,3-Pentadiene, 158
Peri hydrogen interactions, 250, 255
Pericyclic pseudo reactions, 478
Pericyclic reactions, 109
  definition of, 342
  in a mass spectrometer, 497
  metal catalyzed, 370
  nonaromatic, 373
  photochemical, 462
    four-electron Hückel systems, 463
    of ions, 463, 464
    involving eight or more delocalized electrons, 467
    stereochemistry of mass spectra, 502
    thermal, 338
    rules for, 345
Pericyclic transition states, ionic, 359
Perinaphthenyl, 82
Peroxides, 312
Perturbation, first-order, 64
Perturbations
  additivity of, 69
  between degenerate MOs, 30
  equilibria as, 58
  intermediate between first- and second-order, 104
  intermolecular, 59, 61

Perturbations *(cont'd)*
  intramolecular, 59
  monocentric, 59, 61
  reactions as, 58
  second-order, 66
Perturbation theory, 17
  usefulness, 57
Pettit, 370, 372
Phase discontinuity, 344
Phase dislocations, 51, 107
Phase topology, overlap, 53
Phases of the basis set, 106
Phenanthrene, 148
  addition of chlorine to, 331
  1,2-bislocalization energy, 151
Phenethyl cations, 290
Phenol, 244
Phenol-phenate ion equilibria in the excited state, 444
Phenonium ions, π-complex description, 290
Phenyl allyl ether rearrangements, 357
Phenyl anion, 138
Phenylated ethylenes, radical anions of, 535
Phenylation of aromatics, 329
Phenylbutadiene, 93
Phenyl butadienyl ether rearrangement, 358
3-Phenyl-1,5-hexadiene, 229
Phosphorescence, 401, 402, 421, 438, 448
  lifetimes for, 402
Photochemical bromination of paraffins, 274
Photochemical chlorination of benzene, 333
Photochemical cleavage
  of π bonds, 452
  of σ bonds, 454
Photochemical dimerization of ethylene, 422
Photochemical dissociation of a diatomic molecule, 431
Photochemical hydration of olefins, 437
Photochemical intermediates, generation by ground state reactions, 450
Photochemical isomerization
  of ethylenes, 427
  of olefins, 429
Photochemical ketonization of methyl salicylate, 421
Photochemical nonclassical molecules, 433
Photochemical nucleophilic displacement, 438
Photochemical oxidation, 523
Photochemical oxidation–reduction, 456
Photochemical processes, types of, 419
Photochemical racemization, 467
Photochemical reaction
  control by excited-state potential surfaces, 423

Photochemical reaction *(cont'd)*
  coordinate, 424
  examples, 438
  rates, 438
  rules, 426
Photochemical reactions
  BEP plot for, 436
  classes of, 419
  classification of, 428
  entropy effects on, 456
  G_A, 434, 462
  G_I, 433, 459
  G_N, 461
  G_R, 452
  G_S, 437, 461
  of ions, solvent effects on, 437
  pericyclic reactions, 462
    of four-electron Hückel systems, 463
    involving eight or more delocalized electrons, 467
    of ions, 463
    of six-electron anti-Hückel systems, 464
  PMO method and, 477
  prototropic reactions, 439
  X-type, 438
Photochemical solvolysis, 480
Photochemical transition states, 425, 452
Photochemistry, 391
  basic principles of, 419
  mass spectrometry relationships, 502
  with polarized light, 469
  of upper excited states, 470
Photochromism, bianthrone, 452
Photo-Cope rearrangement, 455
Photocyclization of aryl vinyl amines, 470
Photocyclization reactions of polycyclic aromatic systems, 469
Photodimerization, pentacene, 442
Photoelectron spectra, 461
  vibrational levels in, 37
Photoelectron spectrometer, 24
Photoelectron spectrum
  acetylene, 46
  ethylene, 45
  H_2, 26
  methane, 27
  N_2, 36
Photoelectron spectroscopy, 24, 487, 490, 499
Photoenolization, *o*-methylbenzophenone, 458
Photographic dye industry, 410
Photographic emulsions, reactions in, 524

Photoisomerizations
  benzene, 465
  di-π-methane, 458
Photolysis of fabric dyes, 421
Photolysis of halogen, 432
Photosensitized electron transfer, 419
Photosensitized oxidation, 421
Photosynthesis, 524
Phototautomerism, 446
Photothermal reactions, 422
Photovoltaic effect, 420
α-Picoline, 112
β-Picoline, 161, 164
γ-Picoline, 161, 164
P_{ij}, bond order, 63
Pinacol, photochemical formation, 456
Pinacols, formation, 529
pK's of amines in terms of NBMO coefficients, 163
pK_A's
  anilinium ion, 167
  for arylmethyl systems, 143
  of ground and excited state acids, 445
  values for hydrocarbons, 142, 243
Planck's constant, 4
Platinum electrode, 154
PMO method, 75
p-Nitroaniline, 181, 186
p-Nitrobenzyl chloride, S_N2 reactions, 262
p-Nitrocymyl halides, substitution reactions in, 508
Polanyi, 213
Polar bonds, 18
Polar solvent in S_N1 reactions, effect of, 237
Polarity of heteroaromatics, 118
Polarizability
  effects and inductive effects, competition between, 515
  of π electrons, 117
Polarization
  of electronic absorptions, 399
  of π electrons in heteroconjugated systems, 115
Polarized light, photochemistry, 469
Polycyclic AHs, aromaticity of individual rings, 95
Polycyclic aromatic systems, photocyclization reactions, 469
Polycyclic polyenes, 93
Polymer formation, 310
Polymer molecular weight, 315
Polymerization
  nucleophilic, 310

Polymerization *(cont'd)*
  radical, 310
  of styrene, uncatalyzed, 374
Pople, 531
Pople method, 531
Potassium ferricyanide oxidations, 521
Potential energy, molecular, 17
Potential surface, 203
  calculations, 475
  excited state, 423
  for eximers, 442
  for reactions, 211
Potentiometric cell, 154
Probability, energy, 3
Promotion of electrons, 13, 119
Propagation step in polymerizations, 314
Propene, 313
Propyl, 497
Propylene oxide, cleavage, 304
Proton affinity, 266
Proton transfer reactions in the excited state, 443
Protonated epoxides as π complexes, 306
Protonation
  of aromatic π-systems, 133
  of even AHs, 140
  of methane, 509
  of olefins, mechanism, 299
Prototropic reactions, 240, 245
  activation, 246
  change in hybridization, 252
  photochemical, 439
Pyrene, 101
  1,2-bislocalization energy, 151
  fluorescence, 439
Pyridine, 109, 138
  luminescence, 478
  nitration of, 322
  photoisomerization, 465
Pyrrole, 114, 121
  photochemistry, with benzene and naphthalene, 360
Pyran, 382

Quantum number
  azimuthal, 6
  principal, 6
Quantum theory, 5
Quantum yield of monomer fluorescence, 441
Quasiequilibrium in mass spectra, 493
Quenching
  of fluorescence, 439
  photochemical, 460
Quinoline, 118, 138
  electrophilic substitution in, 323
  excited-state basicity, 444
Quinone reactions with biradicals, 374

Racemization, photochemical, 467
Radiation chemistry, 518, 524
Radical addition, 297, 310
  requirements for chain propagation, 311
Radical anion
  chain reactions, 537
  dimerization, 533
  disproportionation, 535
  π energy of dimerization of, 533
  in the gas phase, 504
  of phenylated ethylenes, 535
  and radical cation of chlorophyll *a,* 525
  reactions of aryl halides, 528
  in solution, 525
  stability, solvent effects on, 507, 526
  substitution reactions, 508
Radical aromatic substitution, 328
  relative rates of, 330
Radical cations, 491, 497
  aromaticity, 499
  disproportionation of, 509
  electrocyclic cyclization of, 503
  matrix isolation of, 523
  reactions, 492
  in solution, 517
  stability in solution, 525
Radical chain
  autooxidation, 519
  mechanism, nucleophilic aromatic substitution, 529
  polymerization, 313
  reactions, 274
Radical inhibitors, 312
Radical ions, structure of, 490
Radical polymerization, 310
Radical rearrangments, 449
Radical substitution reactions, 274
  rates, 277
Radicals
  photochemical generation, 455
  in photochemical reactions, 432
Radiochemical photochemical oxidation, 523
Radiolysis, 514
  of carboxylic acids, 525
  of water, 525
Rate constant, 199
Rate-determining step, 202
Rate law, 201

Rate of reaction, 198, 213
Reaction centers, 183
Reaction constants, 185
Reaction coordinate, 202, 203, 210, 212
　for aromatic substitution, 318
　photochemical, 424
Reaction of AHs, 138
Reaction order, 200
Reaction paths, 210
Reaction rates, 198
Reactivity, chemical, 197
Reactivity numbers, 171, 318, 380, 533
　examples of, 140
　and paralocalization energy, 145
　phenanthrene, 325
　for polycyclic aromatics, 381
Rearrangements, mass spectral, 496
Redox potential, 152, 155
　change on photoexcitation, 420
Redox systems, 154
Reduction, electrochemical, 155
Regiospecificity
　of the Birch reduction, 531
　in Diels–Alder reactions, 351
　of the di-π-methane rearrangement, 459
　of elimination reactions, 283
　of ICl additions, 303
　in metal-catalyzed pericyclic reactions, 372
　of photoaddition of ketones to olefins, 457
Relative rates of substitution reactions, 319
Relativity, special theory of, 131
Relaxation methods for chemical kinetics, 241
Resonance, relation to PMO method, 82
Resonance active substituents, 165
Resonance integral, 63, 430
　between two sp^3 AOs of a carbon atom, 363
　empirical nature of, 73
　variation with bond length, 91
Resonance theory, 121
Resorcinol, 182
Retro-aldol reactions, 272
Retrocycloaddition, elimination of HCN, 348
Retro-Diels–Alder reactions, 499
　in a mass spectrometer, sterospecificity of, 500
　photochemical, 462
Retro (2 + 2)-π-cycloaddition, 474
Reversible redox reaction, photolysis effects, 420
Reversibility, microscopic, 131
Ring opening in epoxides, 303
Robinson, 338

Roth, 356
Rules
　for heteroatom effects on electronic spectra, 411, 412, 414, 415, 417
　for photochemical reactions, 426
Rushbrooke, 73

Saytzeff's rule for elimination reactions, 283
Schiff base formation, 308
Schmidt, 357
Schrödinger, 5
Screening, nuclear, 11
Second-order perturbations, 19, 66, 226
　nonadditivity, 70, 86
　simplification of, 126
　in union of even AHs, 84
　union of two even systems, 67
Second-order PMO rule for orientation in ionic reactions, 532,
Selectivity
　and aromatic substitution, 320
　and exothermicity, 218
Self-consistent field (SCF) approximation, 10
Self-polarizability, 117
Semicarbazones, 220
　reactions, rates, and equilibrium constants, 221
Sensitization, 401
Sensitized excitation, 448
Sensitizers, 403
S_E1 reactions, 206, 270
S_E2 reactions, 206, 270
S_E2 transition state, orbital description, 209
Sigmatropic reactions, 352
Sigmatropic rearrangements, 1, 3, 5, 354, 448
　of anions, 358
Singlet emission, 402
Singlet states, 12, 397
Singlet–triplet separation, 445
Singlet–triplet transitions, 400
Skell, 366
Smith, 499
S_N1 and S_N2
　reaction continuum, 261
　relative rates, 208
S_N1 reactions, 206, 253
　limiting type, 262
　solvent effects on, 229
S_N2 intermediate, 49
S_N2 reactions, 206, 256
　BEP principle and, 231
　charges and bond orders, 253
　gas phase, 208

Index 573

$S_N 2$ reactions *(cont'd)*
 limiting types, 262
 nucleophilicity effects, 265
 rates for arylmethyl chlorides, 259
 in solution and in the gas phase, 234, 516
$S_N 2$ transition state, orbital description of, 207, 257
Sodium β-naphthoxide reaction with benzyl chloride, 269
Sodium flame reactions, 213
Sodium naphthalide, 526
Solvated electrons, 525, 527
Solvation effects
 on acid–base properties, 242
 on amine basicity, 510
 on electronic spectra, 409
 on electronic spectra of ions, 413
 on gas-phase reactions, 234
 on photochemical reactions, 443, 445
 on photochemical reactions of ions, 437
 on radical anion stability, 507
 on radical reactions, 278
 steric effects on, 513, 516
Solvation energy, 513
 of activation, 516
 effect on equilibria, 136
 relation to ion size, 225
Solvent effects, 222
 on ambient anion reactivity, 269
 on carbonium ion stability, 511
 on competition between $S_N 2$ and $E_N 2$ reactions, 285
 on elimination reactions, 285
 on enamine reactions, 337
 on ionic reactivity, 485
 on nucleophilicity, 517
 on π-complex stability, 288
 on radical anion stabilities, 526
 on radical reactions, 310
 on rates of reaction, 230
 on redox potentials, 156
 on $S_N 1$ and $S_N 2$ reactions, 263
 on solvolysis mechanisms with neighboring group participation, 292
Solvent nucleophilicity, 260
Solvolysis
 of arylmethyl chlorides, rates of, 255
 of β-arylethyl tosylates, rates, 293
 of phenethyl halides, 292
 photochemical, 438, 480
Spectra
 of dyes, 418
 electronic, rules for heteroatom effects on, 411, 412, 414, 415, 417

Spectrometer, uv–visible, 392
Spin inversions in orbitally degenerate systems, 475
Spin momentum, 398
Spin orbitals, 11
$S_R 2$ reactions, 275
Standard electrode, 154
Standard states, 132
Starring of hydrocarbons, 101
Starring procedure for AHs, 74
State lifetimes, 401
 and extinction coefficients, 402
Stationary states in reaction kinetics, 201
Stereochemistry
 of addition reactions, 297
 of assisted $S_N 1$ reactions, 291
 of carbanions, 271, 273
 of chelotropic reactions, 365
 of chlorine addition, 300
 of the Claisen rearrangement, 357
 of the Cope rearrangement, 356
 of electrocyclic reactions, 360
 of elimination reactions, 279
 of ER_A cycloadditions, 377
 of halide reduction, 528
 of mass spectral pericyclic reactions, 502
 of nitrogen, 119
 in pericyclic reactions, 345
 of photoaddition of ketones to olefins, 458
 of ring opening of bicyclobutane, 362
 of $S_E 2$ reactions, 270
 in solvolysis of homoallylic systems, 293
 of substitution reactions, 210
 of Wagner–Meerwein rearrangements, 293
Stereospecificity of retro-Diels–Alder reactions in a mass spectrometer, 500
Steric effects
 in epoxide ring opening, 305
 perihydrogen, 250
 on $S_N 1$ reactions, 255
 on solvation, 513
Steric hindrance of solvation, 516
Steric repulsion, 120
Stern, 441
Stern–Volmer relationship, 441
Stevens rearrangement, 383
 intramolecular, 369
Stilbene, photochemical isomerization, 453
 photoisomerization of, 432
Strain energy
 in cyclobutane, 376
 and nonclassical carbonium ions, 289
Strecker amino acid synthesis, 309

Streitwieser, 143, 242
Styrene, 125, 313
  dimerization, 374
  radical anion, 533
Substituents
  activities of ±E groups, 170
  carbanionoid, 176
  classification, 165
  constants, 185
  constants in the FMMF method, 189
  +E, 172
  −E, 175
  electromeric, 165, 167
  inductive, 165
  σ^+ constants, 191
Substituent effects, 226
  additivity of, 181
  on aromatic substitution, 326
  calculation of, 546
  on carbanion stereochemistry, 273
  on carbonium ions, 175
  on cation stability in gases and liquids, 511
  on electrophilic aromatic substitution, 321
  on electronic spectra, MO description, 415
  FMMF method, 187
  on gas-phase acidity, 512
  on gas-phase basicity, 514
  general equation for, 189
  Hammett equation for, 185
  on homolytic cleavage, 373
  on light absorption, 413
  on mass spectra, 494
  as mechanistic probes, cycloaddition reactions, 380
  on nonclassical carbonium ion stability, 289
  on nucleophilic addition, 307
  on nucleophilic aromatic substitution, 330
  on oxidation reactions, 518
  on radical aromatic substitution, 329
  rules for, 163, 166, 168, 174, 177
  on S_E1 reactions, 270
  sign convention, 166
  on S_N1 reactions, 254
  on S_N2 reactions, 260
  on stability of cyclopropenium ions, 296
  summary of, 178
Substitution
  vs. addition, 331
  electrophilic aliphatic, 270
  nucleophilic aliphatic, 253
  nucleophilic aromatic radical chain mechanism, 529
Substitution reactions
  aliphatic, 205

Substitution reactions *(cont'd)*
  nucleophilic and electrophilic, 206
  parallels in a mass spectrometer, 508
  radical, 274
  radical anion, 508
Sulfur dioxide addition to butadiene, 362
Suprafacial migration, 353
  of hydrogen, 356
Symmetry, 14
  pericyclic reactions
    analysis of, 368
    unimportance to, 369
  restrictions on MOs, 32
Symmetry orbital, 21, 23
  ethylene, 43

Tautomerism, prototropic, 250
Tendering of fabrics by light, 421
Terephthalic acid, 518
Termination of radical chain reactions, 274
Terylene, 518
Tetracyanoethylene, cycloaddition reactions of, 375
Tetracyanoethylene addition, 349
Tetrahydronaphthalene derivatives, 501
Tetramethylbenzidine, 523
Tetramethyl-1,2-dioxetane chemiluminescence, 475
Tetramethyl-p-phenylenediamine photoionization, 524
Thallation of aromatic compounds, 521
Thallium tris-(trifluoroacetate), 521
Thermal pericyclic reactions, 338
Thermochemistry–mass spectrometry relationships, 493, 501
Thermodynamic control, 220, 251
Thermodynamic vs. kinetic control, 537
Thianthrene radical cation, 525
Thioepoxide ring opening, 305
Thiol addition to olefins, 312
Thornton, 499
Three-center bond, 17
Three-electron bond, 19
Three-membered rings and π complexes, 303
Titanium isopropoxide, 527
Tolan, 348
Toluene
  acidity, 161
  gas phase acidity, 512
  pK_A, 169
Topologically equivalent orbitals, 50
Topology of orbital overlap, 33, 108
Trans-butadiene ionization potential, 500
Transient ions, reactions of, 485

Index

Transition, lowest electronic, 391
Transition energy, charge-transfer transitions, 395
Transition metal π complexes, 370
$n \to \pi^*$ transitions, 394, 409, 446
  in odd, heteroconjugated systems, 410
  polarization of, 400
Transition moment integral, 399
Transition moments, 398
Transition state, 200
  for aliphatic substitution, 205
  for chelotropic reactions, 364
  for chemiluminescent reactions, 476
  for the Cope rearrangment, 355
  for the Diels–Alder reaction, 340, 351, 364
  for Diels–Alder reaction, orbital description of, 341
  energy of, 211
  for radical addition, 314
  for 1,3 sigmatropic rearrangements, 354
  structure and reaction exothermicity, 219
  theory of, 202
Transition states
  antiaromatic, 434
  antiaromatic in photochemical reactions, 437
  for pericyclic reactions, 342
  photochemical, 425, 434, 452
  for photochemical pericyclic processes, 464
Translational entropy effects on cyclization rates, 334
Transmission coefficient, 204
Triafulvene, 499
Triangulene, 101
Trifluoroethylene reaction with butadiene, 376
Trifluoromethyl substitution reactions of aromatic hydrocarbons, relative rates, 330
Trifluoromethylation of aromatics by CF_3, 329
Trimethylenemethane, 433, 447
Triphenylcarbinol, 167
Triphenylene, 1,2-bislocalization energy, 151
1,2,2-Triphenylethyl derivatives, solvolysis of, 296
Triphenylmethane, pK_A, 169
Triphenylmethyl, absorption spectrum of, 408
Triphenylmethyl cation, 167
  substituent effects on formation of, 177
Triphenylmethyl radicals, dimerization, 160
Triplet chemiluminescence, 475
Triplet states, 13, 397, 458

Trishomocyclopropenium ion, 295
Tropone cycloadditions, 348
Tropylium, 498
  cation, 97
  chloride, 256
Tropyliumlike transition states, 346

Ultraviolet spectra, 410
Uncertainty, energy, 2
Uncertainty principle, 1
Unimolecular reactions, 198
  and FO methods, 367
Union
  of an AH with a heteroatom, 114
  of even AHs, 125
  of odd AHs, 68, 86
  of an odd AH with an even AH, 104
  intermolecular, 64, 65
  multiple, 69
  of π systems, 60
  of the pentadienate radical and methyl to form benzene, 90
UV spectra, 409

Vacuum ultraviolet, 393
Valence, 121
Valence shell AOs, 32
Valence state, 119
Valence structures
  classical, 101
  failure in excited states, 447
van der Waals
  forces, 227
  radii, 226
Van Tamelen, 527
Vibration amplitude, 395
Vibrational levels in photoelectron spectra, 37
Vibrational relaxation, 396
Vibrational structure in absorption bands, 392
Vibrationally excited states, 396
Vibronic energy, 392
Vinyl acetate, 313
Vinyl chloride, 313
Vinyl cyclopropanes, photochemical formation, 458
Vinyl polymerization, 313
Vinylcyclohexane retro-Diels–Alder reactions, 499
Volmer, 441

Wagner–Meerwein rearrangement, 285, 336, 449

Walborsky, 528
Walden inversion, 210
Ward, 503
Water, radiolysis of, 525
Wave function, 5
  sign, 51
Wavelengths of first absorption bands for HCs, calculated and experimental, 405
Wavelike behavior, of particles, 3
Wheland intermediates, 379
  in aromatic substitution, 318
Wilzbach, 466
Wolfsberg–Helmholtz approximation, 16
Woodward, 343, 352, 360, 365, 366, 367
Woodward–Hoffman rules, 339, 345, 369

Wurster's blue, 524, 525
Wurtz, 529
Wurtz synthesis, 533
Wurtz–Fittig reaction, 529

X-type photochemical reactions, 428, 438, 443, 446

Ziese's salt, 294
Zero-point energy, 395
Zeroth-order reactions, 252
Zero–zero transition, 396
Zimmerman, 447
Zwitterions, 225, 448